TABLE C.1 Cumulative probabilities and percentiles of the standard normal distribution

(a) Cumulative probabilities

Entry is area a under the standard normal curve from $-\infty$ to $z(a)$.

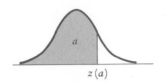

z	.00	.01	.02	.03	.04	.05	.06	.07	.08	.09
2.5	.9938	.9940	.9941	.9943	.9945	.9946	.9948	.9949	.9951	.9952
2.6	.9953	.9955	.9956	.9957	.9959	.9960	.9961	.9962	.9963	.9964
2.7	.9965	.9966	.9967	.9968	.9969	.9970	.9971	.9972	.9973	.9974
2.8	.9974	.9975	.9976	.9977	.9977	.9978	.9979	.9979	.9980	.9981
2.9	.9981	.9982	.9982	.9983	.9984	.9984	.9985	.9985	.9986	.9986
3.0	.9987	.9987	.9987	.9988	.9988	.9989	.9989	.9989	.9990	.9990
3.1	.9990	.9991	.9991	.9991	.9992	.9992	.9992	.9992	.9993	.9993
3.2	.9993	.9993	.9994	.9994	.9994	.9994	.9994	.9995	.9995	.9995
3.3	.9995	.9995	.9995	.9996	.9996	.9996	.9996	.9996	.9996	.9997
3.4	.9997	.9997	.9997	.9997	.9997	.9997	.9997	.9997	.9997	.9998

(b) Selected percentiles

Entry is $z(a)$ where $P[Z \leq z(a)] = a$.

a:	.10	.05	.025	.02	.01	.005	.001
$z(a)$:	−1.282	−1.645	−1.960	−2.054	−2.326	−2.576	−3.090
a:	.90	.95	.975	.98	.99	.995	.999
$z(a)$:	1.282	1.645	1.960	2.054	2.326	2.576	3.090

EXAMPLE: $P(Z \leq 1.96) = 0.9750$ so $z(0.9750) = 1.96$.
TEXT REFERENCE: Use of this table is discussed on pp. 214–220.

APPLIED STATISTICS

THIRD EDITION

JOHN NETER   University of Georgia
WILLIAM WASSERMAN   Syracuse University
G.A. WHITMORE   McGill University

ALLYN AND BACON, INC.
Boston   London   Sydney   Toronto

To Dottie, Cathy, and Lonnie

Editorial-production service: Technical Texts, Inc.
Text designer: Sylvia Dovner
Cover designer: Hemenway Design Associates
Cover administrator: Linda Dickinson
Production administrator: Lorraine Perrotta
Manufacturing buyer: Ellen Glisker
Cover art: "View (Gray)" by Tetsuro Sawada. Published and distributed exclusively by Buschlen/Mowatt Fine Arts Ltd., 1467 West 47th Avenue, Vancouver, Canada V6M 2L9.

Library of Congress Cataloging-in-Publication Data

Neter, John.
  Applied statistics.
  Includes index.
  1. Social sciences — Statistical methods.
2. Management — Statistical methods. 3. Economics —
Statistical methods. 4. Statistics. I. Wasserman,
William. II. Whitmore, G. A. III. Title.
HA29.N437 1988   519.5   86-14147
ISBN 0-205-10328-6

Printed in the United States of America.
  10 9 8 7 6 5 4 3 2 1   91 90 89 88 87

Contents

UNIT SIX: Bayesian Decision Making   749

UNIT SEVEN: Time Series Analysis and Index Numbers   811

Preface

Applied Statistics, Third Edition, is written for students taking basic statistics courses in management, economics, and other social sciences, as well as for people already engaged in these fields who desire an introduction to statistical methods and their application. Our aim has been to offer a balanced presentation of fundamental statistical concepts and methods, along with practical advice on their effective application to real-world problems. Conceptual rigor is not sacrificed, however. The conceptual foundation of each subject is developed carefully up to that level needed for prudent and beneficial use of statistical methods in practice.

The third edition differs from the second in a number of important ways. We have extensively revised the chapters on statistical testing to unify the treatment of statistical tests throughout the book. As a result, all statistical testing is now carried out by means of standardized test statistics. In addition, we have expanded the coverage of *P*-values and integrated their use into the discussion of statistical testing. We have also made greater use of computer outputs and plots and have further emphasized the uses of statistical computer packages for data management and analysis.

The updating in the third edition to reflect current needs and practices has led to the addition or expansion of several important topics. The treatment of statistical quality control has been expanded by considering several additional types of control charts and their uses. In regression analysis, we have enlarged the treatment of indicator variables, curvilinear models, regression models with interaction effects, and statistical tests, and we have added a section on the use of the all-possible-regressions method for selecting the regression model. In time series analysis, we now include exponential smoothing for time series with trend and with trend and seasonal components. We also discuss the Durbin-Watson test and have added an introduction to ARIMA models.

In addition, we have greatly expanded the number of problems at the ends of the chapters and have added three more data sets in the appendix. We have also made many revisions in the text to improve the clarity of the presentation.

As in the second edition of this text, explanation of important principles and concepts always precedes discussion of detailed statistical procedures. Each new technique is illustrated by one or more examples drawn from real life. In addition, many case applications are presented, so that statistical concepts and methods can be understood in the context of their actual use. These case applications demonstrate the usefulness of statistical methods and statistically rigorous thinking in management, economics, and other social sciences.

The topical coverage of the text is broad, but it is in no sense a miscellany of statistical tools. Topics have been selected on the basis of two criteria: (1) their importance in real applications of statistics, and (2) their contribution to the development and understanding of material presented subsequently in the book. Units, chapters, and sections have been organized and sequenced in a way that always keeps the main track of the subject clear to the reader. Technical notes and secondary observations are presented in *Comments* sections. Major topics of interest that are not essential to the main development of statistical ideas are presented in *Optional Topic* sections and may be included in the basic statistics course at the instructor's discretion to meet particular course objectives. These sections always appear at the ends of chapters and can be omitted without loss of continuity. Extensive use is made of figures and tables. Important definitions and formulas are set out in a distinctive manner to aid in learning and to facilitate ready reference.

Use of this book requires knowledge of college-entrance algebra but not calculus. We believe that fundamental statistical ideas can be introduced with little mathematics, and that this approach in an introductory text leads to a fuller appreciation of the uses of statistical methods than a more mathematical approach. Mathematical demonstrations are included where they make a significant contribution to the reader's understanding of the subject. All such demonstrations are self-contained and illustrated by numerical examples. Occasionally, a mathematical demonstration is presented that requires calculus. These sections are marked "calculus needed" and may be omitted without loss of continuity. Those readers who will continue their study of statistics and, therefore, must become more familiar with the mathematical basis of the subject will find that this book provides a strong foundation upon which they can build.

We assist the reader with the mathematical and computational aspects of the subject in a number of ways. Notation is used only where needed, and then a uniform and straightforward notational system is employed throughout the text. A mathematical review is included in Appendix A for readers desiring a brief summary of: (1) summation notation, (2) rules for exponents and logarithms, (3) set notation, operations, and rules, and (4) the basics of permutations and combinations. Because statistical computer packages play a major role in real applications of statistics, we have sought to familiarize readers with ways in which they assist in statistical analyses (such as regression) and in data handling (such as tabulations). Computer printout is presented throughout the text to illustrate its form and its usefulness in various practical contexts.

Large numbers of problems and questions are given at the ends of chapters. The *Problems* sections contain basic problems and drill questions. The *Exercises* sections present questions dealing with mathematical concepts and extensions of ideas developed in the chapters. Finally, the *Studies* sections contain major, comprehensive problems and case studies. The problems, exercises, and studies each follow the sequence of topics in the text. Numerical answers for selected problems (identified by asterisks) are given at the end of the book to facilitate immediate checking by the reader. The problems, exercises, and studies have been designed to assist the reader's understanding of concepts and to enable him or her to obtain experience in applying statistical techniques in practical situations and in interpreting results of statistical investigations. Many different types of applications are employed in order to expose the reader to the rich and varied settings in which statistics is applied in real life. A number of the problems and studies draw on the data sets in Appendix D.

Several teaching and learning aids, specifically keyed to this edition of *Applied Statistics,* are available. The aids include: (1) an Instructor's Manual containing fully worked numerical and discussion answers for all problems, exercises, and studies; (2) a self-learning Study Guide for students, prepared by Professor Kenneth C. Schneider; (3) a Test Manual for instructors containing a large collection of multiple-choice test questions, also prepared by Professor Schneider; and (4) a computer diskette containing the four data sets in Appendix D and other data sets of the text, to support computer-aided learning and instruction.

The book is divided into seven units, as shown in Figure I.

Unit One, Data (Chapters 1–3). However sophisticated a statistical procedure, its successful application depends on data. This unit is concerned with the acquisition, classification, and summarization of data. Appropriately, the unit emphasizes large data sets, computerized data handling, and exploratory data analysis. Since a large portion of day-to-day dealings with statistics for many persons is concerned with raw data and their examination by descriptive (as opposed to inferential) means, the chapters in this unit are designed to prepare the reader for these common encounters. The important role of computer data-processing packages, data banks, and retrieval systems is also recognized in this unit, and the groundwork is laid for data handling in subsequent chapters.

Unit Two, Probability (Chapters 4–7 and Appendix B). Probability theory is the foundation of statistical inference, so this unit comes next in the book. The chapters of this unit contain an integrated presentation of basic probability concepts, random variables, and common probability distributions utilized in applications. The last topic is covered in two chapters, the first presenting common discrete distributions and the second, common continuous distributions. Appendix B covers the χ^2, t, and F distributions and explains their relationships to one another and to the standard normal distribution. These three distributions are presented in the appendix as a handy reference so that they can be accessed readily from any place in the main text where they might be encountered for the first time in a course. Extensive probability tables, prepared in a consistent fashion, are provided in Appendix C to support this and later units.

Unit Three, Estimation and Testing — I (Chapters 8–14). This unit opens with two chapters discussing sampling and the sampling distribution of $\overline{X}$. In contrast to conventional theoretical explanations, an empirical demonstration of the sampling distribution of $\overline{X}$ and the central limit theorem is presented first, using simulated, repeated sampling from an actual population. Then the key theorems are stated and explained. This approach allows the reader to anticipate the theoretical results on the basis of empirical experience, and it should lead to greater understanding of these conceptually important topics. Next, estimation and testing for a population mean are presented, followed by a discussion of the sampling distribution of $\overline{p}$ and inference procedures for a population proportion. Inference procedures in comparative studies involving two population means or proportions and inferences for population variances are discussed in the next-to-last chapter of this unit. The final chapter considers statistical quality control procedures, as well as sampling of finite populations and various sampling procedures other than simple random sampling.

Unit Four, Estimation and Testing — II (Chapters 15–17). In the first chapter of this second unit on estimation and testing, nonparametric procedures that have found extensive use are discussed. The approach is not one of presenting a grab bag of statistical tools — rather, procedures are presented which match those discussed in the previous unit, but which do not depend on large samples or restrictive distributional assumptions. Goodness-of-fit procedures and techniques for multinomial populations are taken up in the next two chapters. Separation of these topics into two chapters emphasizes the separate classes of problems involved and is a departure from the usual textbook coverage.

Unit Five, Linear Statistical Models (Chapters 18–21). This unit meets a pressing need for a thorough treatment of regression and analysis of variance to prepare the reader for the extensive use of these techniques in management, economics, and social sciences in general. The topics of regression and analysis of variance are handled in a unified fashion in these chapters. Computer-assisted analysis is stressed, as is the facility to interpret standard printout from a computer regression package. Also stressed are practical aspects in using linear statistical models, such as study of the aptness of the model employed.

Unit Six, Bayesian Decision Making (Chapters 22 and 23). This unit presents a discussion of Bayesian decision analysis, showing how the decision-theoretic aspects of a statistical problem can be handled formally. The relation between statistical decision making and statistical testing is made clear by our initial emphasis on normal-form analysis and later presentation of extensive-form analysis.

Unit Seven, Time Series Analysis and Index Numbers (Chapters 24–26). A compact treatment of the classical time series model is presented in the first chapter of this unit. The second chapter is devoted to exponential smoothing, regression time series models, and an introduction to ARIMA models. The emphasis again is on concepts and procedures needed for effective application. In this same spirit a third chapter is devoted to price and quantity indexes, which play an important role today for many administrators, economists, and social scientists.

This book can be used for a wide variety of one- or two-quarter or one- or two-semester courses. Figure I shows the logical interdependencies of the units and their chapters, with the arrows indicating prerequisite chapters. As can be seen, various chapter sequences may be used, depending on the length of the course and the emphasis desired. Here are four examples:

1. A course covering descriptive statistics and the basics of statistical inference might include data (Unit One), probability (Unit Two), inferences for population mean and proportion (Chapters 8–12), simple linear regression (Chapters 18 and 19), and the descriptive portions of time series analysis and index numbers (Chapters 24 and 26).

2. A course on inferential statistics emphasizing linear statistical models might include data (Unit One), probability (Unit Two), inferences for population mean (Chapters 8–11), linear statistical models (Unit Five), and regression models for time series analysis (Chapter 25).

3. A course covering both parametric and nonparametric inferences might include data (Unit One), probability (Unit Two), parametric inferences (Unit Three), nonparametric inferences (Unit Four), and as much of linear statistical models (Unit Five) as time permits.

FIGURE I **Structure of *Applied Statistics*.** Chart shows chapter interdependencies—arrows point to prerequisite chapters.

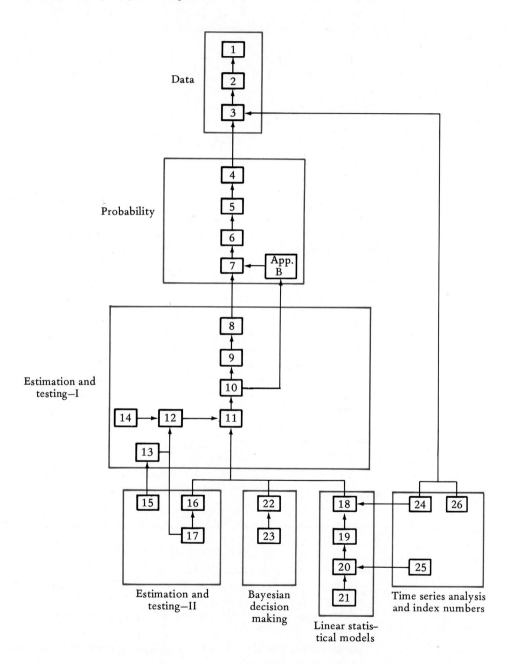

FIGURE I **Structure of *Applied Statistics* (continued)**

Unit One: Data
1. Data Acquisition
2. Data Analysis I: Classification and Distribution Patterns
3. Data Analysis II: Summary Measures

Unit Two: Probability
4. Basic Probability Concepts
5. Random Variables
6. Common Discrete Probability Distributions
7. Common Continuous Probability Distributions

Unit Three: Estimation and Testing — I
8. Statistical Sampling
9. Sampling Distribution of $\overline{X}$
10. Estimation of Population Mean
11. Tests for Population Mean
12. Inferences for Population Proportion
13. Comparisons of Two Populations and Other Inferences
14. Quality Control and Other Applications of Sampling

Unit Four: Estimation and Testing — II
15. Nonparametric Procedures
16. Goodness of Fit
17. Multinomial Populations

Unit Five: Linear Statistical Models
18. Simple Linear Regression
19. Inferences in Simple Linear Regression
20. Multiple Regression
21. Analysis of Variance

Unit Six: Bayesian Decision Making
22. Bayesian Decision Making I: No Sample Information
23. Bayesian Decision Making II: Sample Information

Unit Seven: Time Series Analysis and Index Numbers
24. Time Series and Forecasting I: Classical Methods
25. Time Series and Forecasting II: Exponential Smoothing and Regression Methods
26. Price and Quantity Indexes

Appendices
A. Mathematical Review
B. Chi-Square, t, and F Distributions
C. Tables
D. Data Sets

4. A course emphasizing decision theory might include data (Unit One), probability (Unit Two), the major elements of statistical inference (Chapters 8–13), Bayesian decision making (Unit Six), and other topics as time permits.

We are greatly indebted to many individuals and organizations who have helped us in the preparation of this book. Our sincere thanks go to all who have provided us with case materials and illustrations that demonstrate the usefulness of statistical methods in management, economics, and other social sciences. Many persons, including colleagues and reviewers, have made helpful suggestions on the manuscript and otherwise assisted us, for which we are most grateful. We particularly wish to thank for their help Robert F. Berner, R. V. Erickson, Edgar Hickman, Oswald Honkalehto, H. K. Hsieh, Allan Humphrey, Raj Jaganathan, William Meeker, Robert Norland, Jr., Thomas Pray, Thomas Rothrock, Barbara Rufle, J. Michael Ryan, Kenneth C. Schneider, Randolph Shen, Erland Sorensen, Stephen Vardeman, Dean Wichern, and Morty Yalovsky.

We are also indebted to our respective universities for providing an environment in which the undertaking could be brought to fruition. A number of people were most helpful in working with us in preparing and checking the manuscript. We wish to express our appreciation to Brenda Adams, Karen Robertson, and Johanne Thiffault for their valuable assistance. The staffs of Technical Texts, Inc. and Allyn and Bacon, Inc. helped us in many ways. Of course, we thank our families for their patience and encouragement while we completed this book.

Introduction

What is Statistics?

Almost everyone uses statistics and is affected by applications of statistics. The word *statistics* refers in common usage to numerical data. Vital statistics, for example, are numerical data on births, deaths, marriages, divorces, and communicable diseases; business and economic statistics are numerical data on employment, production, prices, and sales; social statistics are numerical data on housing, delinquency and crime, education, and social security and welfare.

Statistics has an additional meaning that is more specialized. In this second sense, *statistics* refers to the methodology for the collection, presentation, and analysis of data, and for the uses of such data. Unless data are accurate, properly presented, and correctly analyzed, they may be dangerously misleading. Since we all are "consumers" of statistics, it is important for all of us, not only professional statisticians, to acquire some knowledge of statistical methodology.

The word *statistician* also has several meanings. It can refer to: (1) a person who performs routine operations with statistical data; or (2) an analyst who is highly trained in statistical methodology and uses this methodology in the collection and interpretation of data; or finally, (3) an applied mathematician who utilizes advanced mathematics in the development of new statistical methods. Statisticians are needed in all these capacities in order to make statistical data most useful.

The Expanding Role of Statistics

Statistical data have been used for many centuries by governments as an aid in administration. In antiquity, statistics were compiled to ascertain the number of citizens liable for military service and taxation. After the Middle Ages, governments in Western Europe were interested in vital statistics because of the widespread fear of devastating epidemics and the belief that population size could affect political and military power. As a result, data were compiled from registrations of christenings, marriages, and burials. In the sixteenth through eighteenth centuries, when mercantilistic aspirations set nation-states in search of economic power for political purposes, data began to be collected on such economic subjects as foreign trade, manufacturing, and food supply.

Today, data are collected, classified, stored, and retrieved in diverse and comprehensive information systems that supply individuals and organizations with the statistical intelligence required to carry out their activities. The expansion in the collection, transmission, storage, and retrieval of statistical data, facilitated by computers, has been accompanied by the rapid development of statistical methodology and data analysis.

Statistical concepts have exercised a profound influence in almost every field of human activity and have been incorporated into the basic principles of such sciences as physics, genetics, meteorology, and economics. Statistical methods have been used to improve agricultural products, to design space equipment, to plan traffic control, to forecast epidemics, and to attain better management in business and in government. Students of the natural and social sciences study statistics to become better scientists; students of economics study statistics to become better economists; students of administration study statistics to become more effective administrators.

Uses of Statistics

Some knowledge of statistics is essential today for people pursuing careers in almost every area of industry, government, public service, or the professions. Not only are more comprehensive networks of data available to serve as a basis for drawing valid conclusions and making decisions, but the purpose in assembling the data has shifted from record keeping to evaluation and action, based on timely information. Until recently, statistics were collected primarily as a record of past events. Although such statistics were analyzed to gain insights into current problems, the emphasis was essentially upon the past. At the present time, the collection of numerical information for the record still takes place, but because of the needs for improved planning and control, data-collection systems and data repositories have been designed to provide data that are as up-to-date as possible. Statistical analysis, in turn, has become chiefly concerned with the present and the future, rather than the past.

The increasing uses of statistics are part of the trend toward basing evaluations and making decisions on the most objective and scientific foundations possible. Modern organizations are becoming more dependent on statistical data to obtain factual information about their internal operations and their social, economic, and ecological surroundings. Statistical data are concise, specific, capable of being analyzed objectively with powerful formal procedures, and well suited for making comparisons. Hence, they are especially useful in such key organizational functions as choosing among alternatives, setting goals, evaluating performance, measuring progress, and locating weaknesses.

Uses of statistics have been greatly facilitated by the extensive array of statistical computer software packages now available. These packages not only expedite the application of statistical methods and the handling of data at lower cost, but they also permit the use of more powerful and complex statistical techniques. Statistical procedures and models are now integral parts of many management computer systems, such as decision support systems and expert systems. This expanded computerization of statistical methods increases the need for users of statistics to have an adequate understanding of statistical concepts and methods, making the subject of statistics an even more essential component of the professional education of businesspersons, economists, and administrators.

UNIT ONE

Data

1 DATA ACQUISITION

Data are the raw material of statistics. Statistical analysis cannot proceed until the data of interest in an investigation are assembled and organized in a useful manner.

1.1 DATA SETS

Our dictionary defines data as follows.

(1.1)

> *Data* are facts or figures from which conclusions may be drawn (Ref. 1.1; *see reference at end of chapter*).

We shall refer to the data collected for a particular study as a *data set*. Data sets are found all around us. The financial section of our daily paper contains price data for securities and commodities; an economic report shows inflation rates for different countries; a newspaper article contains data on achievement test scores for all schools in a city.

EXAMPLE

Figure 1.1 shows a data set for a study of physical-fitness profiles of the six police officers in a township. Information about age, sex, and various physical and fitness characteristics of each officer was obtained in this study.

Characteristics of Data Sets

A data set represents a collection of facts and figures. We now consider the key characteristics of data sets.

Element. A data set provides data about a collection of elements and contains, for each element, information about one or more characteristics of interest. In Figure 1.1, an element of the data set is a particular police officer.

FIGURE 1.1   **Data set for a study of physical-fitness profiles of police officers in a township**

Variable

Officer	Age	Sex	Systolic blood pressure	Diastolic blood pressure	Triceps skinfold thickness[1] (cm)	Number of sit-ups	Fitness rank
Anders	32	M	120	80	1.50	100	1
Colm	28	F	118	75	1.96	35	3
Greene	46	M	138	90	1.79	45	4
Keene	23	F	121	75	2.30	29	5
Osman	36	M	141	95	3.05	18	6
Waldorn	22	M	123	75	1.91	75	2

Case

Element   Observation

[1]Index of body fat

Variable.   A variable is a characteristic of interest about an element. In Figure 1.1, one characteristic of interest is age of police officer. This characteristic takes on different values for different officers; hence, it is called a variable. Age is a quantitative variable because its outcomes are numerical in nature. On the other hand, the variable sex in Figure 1.1 is a qualitative variable because its outcomes are nonnumerical. Of the seven variables in the data set in Figure 1.1, six are quantitative and one (sex) is qualitative. The following are formal definitions.

(1.2)

A characteristic that can take on different possible outcomes is called a *variable*. When the outcomes are expressed numerically, the variable is said to be *quantitative*. When the outcomes refer to nonnumerical qualities or attributes, the variable is said to be *qualitative*.

Case.   The information on all variables for one element in the data set is called a *case* or a *record;* occasionally, it is called an *observation vector*. Thus, the information on the seven variables for Officer Colm constitutes a case. In Figure 1.1, each row of data represents a case, so there are six cases in the data set.

Observation.   The information for one element of the data set about a single variable is called an *observation*, a *reading*, or an *outcome*. Thus, age 32 of Officer Anders is an observation about the variable age for this officer.

Comments

1.   Sometimes, a quantitative variable is converted into a qualitative one by grouping the possible numerical outcomes into nonnumerical classes. This conversion is done to

facilitate reporting, interpreting, or analyzing the data. Thus, with eggs for retail sale, the quantitative variable weight (measured in grams) is converted into a qualitative variable by partitioning all possible weights into categories such as extra large and large. At other times, the reverse is done to convert a qualitative variable into a quantitative one by assigning a numerical value or code to each nonnumerical category. For instance, the variable sex in Figure 1.1 might be quantified by using the number 0 for male and the number 1 for female.

2. Data sets may be distinguished on the basis of the number of variables they contain. A *univariate* data set contains one variable; a *bivariate* data set, two variables; and a *multivariate* data set, three or more variables. Thus, the data set in Figure 1.1 is multivariate.

Types of Data

Statistical data are obtained in a variety of ways. For instance, the data on blood pressure in Figure 1.1 were obtained by employing a blood-pressure-measuring instrument, and these data are called *measurement data*. The data on number of sit-ups were obtained by counting and are called *count data*. The data on fitness were obtained by ranking the officers, with the most fit officer assigned rank 1 and the least fit officer, rank 6. These data are called *rank data*. Finally, the data on sex involve a classification process where classes or categories (male, female) are set up and each data set element is assigned to the appropriate category. These data are called *classification data*.

1.2 DATA SOURCES

Statistics is concerned not only with organizing and analyzing data once they are assembled but also with the sources of data and how data are collected for study. The first stage of any investigation involves a specification or definition of the problem to be studied. From this specification comes an identified need for particular types of data to clarify the problem. At this point, the question of where to obtain the necessary data is posed.

Internal and External Sources

Some data sets may be partially or totally available from an *internal source,* such as records of operating and accounting data, which most organizations maintain routinely. Indeed, many organizations store such data in records that are grouped into computerized data files for efficient entry and retrieval of information. These data files, together with the associated computer programs, constitute the organization's internal data base. Design and maintenance of the data base have become important managerial functions in many firms and other organizations.

Other data are obtainable from an *external source.* Such external data may be published in a reference book or a statistical periodical, or the data may be in a computerized form such as disks, tapes, or on-line access to an external data bank. The external-source organization may be a government agency, a trade association, or a private specialized service company.

To illustrate the use of these sources of data, consider the case of a market researcher who wishes to study the geographic distribution of customers who use a particular company product. The researcher might find much of the required data in the internal accounting records of the business. On the other hand, a store location analyst for a retail-outlet chain who is concerned with crime patterns in different cities might find the necessary data in external sources containing criminal and judicial statistics such as government publications and computerized police records.

Cautions in the Use of Data Sources

When using any data source, the user should be thoroughly acquainted with the nature and limitations of the data. Limitations may include imperfect or improper methods of data collection, recording, and classification, as well as errors of omission or commission when data are transferred from one record to another. The user also needs to determine whether the definitions employed in compiling the data are appropriate for the purpose. It is also important to check whether changes in concepts, definitions, and data collection methods have occurred over the time period of interest and, if so, to determine the effect of these changes on the data. The user should have access to pertinent information and explanatory comments about the manner in which the data were compiled, the data's accuracy, and how the data should be interpreted. The need for caution in the use of data applies to all data sources, whether they be internal or external, manual or computerized.

1.3   EXPERIMENTAL AND NONEXPERIMENTAL STUDIES

When the data needed for an investigation are not available in existing sources, one must consider some method for obtaining them directly. Two major methods of data collection are experimental studies and nonexperimental or observational studies. We now illustrate the distinction between these two types of studies.

EXAMPLE

An insurance company hired a large number of new programmers when it moved its headquarters to another city. The director of training for the company set up a voluntary training program that could be taken by the new programmers. Later, the director compared data on the job progress of the new programmers who took the training program and those who did not. He found that the programmers with the training, on the whole, had made more progress on the job than the others.

The director could not, however, assess the contribution of the training program to job progress since it is likely that the programmers who volunteered for the training were ones who were better motivated and more achievement-oriented.

Such individuals might have made more job progress than the others even without the training program.

The fact that the training program was voluntary so that no control was exercised over any factors that might affect progress on the job makes this a *nonexperimental* or *observational study*.

Some years later, the director of training had the opportunity to assess the effect of the training program more definitively. The company expanded its data-processing staff substantially and hired 50 new programmers. Twenty-five of these were selected at random and assigned to the training program, and the other 25 programmers were immediately assigned to operations. Follow-up studies showed that the job progress was about the same for both groups of programmers.

The second study is an *experimental study* because control was exercised over the factor under study (training program) and randomization was employed to balance out all other factors that might affect job progress. Randomization here tended to balance highly motivated programmers in both groups and similarly tended to balance programmers by age, experience, and any other factors that might affect job progress. Thus, if any difference in job progress had been observed between the two groups, it could have been attributed to the training program because of the experimental nature of the study.

We now consider experimental and nonexperimental studies in more detail.

Experimental Studies

The use of experiments for collecting data in the fields of business, economics, and the social sciences continues to expand.

(1.3)

> An *experimental study* is a study where the factors under consideration are controlled so as to obtain information about their influence on the variable of interest.

In statistically designed experiments, the control over the factors under study is accompanied by randomization to balance out the influence of all extraneous factors that might affect the variable of interest.

EXAMPLES

1. Thirty stores were randomly assigned one of three store displays to study the effect of the display on store sales of a product.

2. Four hundred consumers were randomly assigned one of four brands of all-purpose flour to study consumer reactions to each brand in routine home use over a period of one month.

Nonexperimental Studies

In spite of the increasing use of experimental studies, researchers and analysts in business, economics, and the social sciences must often rely on nonexperimental or observational studies.

(1.4)
> A *nonexperimental* or *observational study* is a study where no special controls are exercised in the collection of the data over any of the factors influencing the variable of interest.

EXAMPLES

1. A survey of 1000 residents of a metropolitan area was conducted to obtain information about frequency of attending concerts and plays (the variable of interest) and about possible explanatory factors, including income, age, and education of the resident.

2. A study was undertaken by a large corporation of five plants to obtain information about plant productivity (the variable of interest) and about possible explanatory factors, including type and age of machinery used in the plant, education and age characteristics of the production employees in the plant, and type of wage incentive program in the plant.

Choice Between Experimental and Nonexperimental Studies

Both experimental and nonexperimental studies can be extremely useful for studying the effects of one or more factors on the variable of interest. However, as noted earlier, experimental studies provide stronger evidence of these effects than nonexperimental studies. Experiments are especially advantageous in investigating cause-and-effect patterns, such as determining whether or not a high cholesterol level tends to cause atherosclerosis.

Despite the advantages of experimental studies, much of statistical analysis in business, economics, and the social sciences is based on nonexperimental studies. One reason is that most available data, such as internal data on company operations and external data on the economy and consumer behavior, are nonexperimental data. Another reason is that it is often not feasible or may not be desirable to exercise the experimental controls required in experimental studies. For example, an economist interested in the effect of family size on the proportion of income saved cannot select a group of newlyweds and tell each couple to have a family of certain size. Observational studies of existing families, on the other hand, can provide useful information on family size, income, and expenditures, from which the relation between size of family and proportion of income saved can be studied.

1.4 DATA ACQUISITION

A variety of procedures for acquiring data are employed in experimental and nonexperimental studies. Three commonly used ones are observation, interview, and self-enumeration.

Data Acquisition Procedures

Observation. Observation entails direct examination and recording of an ongoing activity.

EXAMPLES

1. In a study of family decision making, a researcher observed and recorded interactions between husband and wife as they decided on the make of home computer to buy.
2. In an engineering study, data about the internal temperature of a kiln were obtained by reading an instrument inserted in the kiln.
3. In a marketing study, an analyst monitored customer flow in a department store by means of closed-circuit television. □

The observation procedure has certain advantages:

1. The directness of the procedure avoids problems such as incomplete or distorted recall.
2. Data can be gathered more or less continuously over an extended time period.

Limitations of the method include the following:

1. The observer (or the instrument) must be free of bias and must accurately record the events of interest. Human observers usually require thorough training so that they will record precisely what they observe and so that different observers will record the same events in the same manner.
2. Individuals who are under observation and aware of this fact may alter their behavior, and as a result, observations of their behavior may be biased.

Interview. In an interview, an interviewer asks questions that are printed on a questionnaire and records the respondent's answers in designated spaces on the questionnaire form.

EXAMPLES

1. A household member was interviewed at home about purchases of toothpastes and mouthwashes and about family characteristics, such as income and family size.
2. A household member was interviewed over the telephone about television viewing, including viewing at the moment of call, the station viewed, and the number of persons viewing.

3.   A company's financial executive was interviewed at the office by a representative of a trade association about the firm's plans for capital expenditures in the coming year.   ☐

Both the advantages and the limitations of securing data through interviews arise from the direct contact between the respondent and the interviewer. Advantages include the following:

1.   Persons will tend to respond when they are approached directly; hence, the interview procedure usually yields a high proportion of usable returns from those persons who are contacted.
2.   The direct contact generally enables the interviewer to clear up misinterpretations of questions by the respondent, to observe the respondent's reactions to particular questions, and to collect relevant supplementary information.

There are several limitations of the interview method:

1.   The interviewer may not follow directions for selecting respondents. For instance, if a member of the family other than the one designated is interviewed, a bias may be introduced into the results.
2.   The interviewer may influence the respondent by the manner in which the questions are asked or by other actions. A slight inadvertent gesture of surprise at an answer, for example, can exert subtle, undetected pressures on the respondent.
3.   The interviewer may make errors in recording the respondent's answers.

Self-Enumeration.   With self-enumeration, the respondent is provided with a questionnaire to complete, which often also contains necessary instructions.

EXAMPLES

1.   A recent high school graduate received a self-enumeration questionnaire through the mail that requested information about educational activities since graduation. A page of the questionnaire is shown in Figure 1.2.
2.   A new magazine subscriber received a questionnaire through the mail to provide information about age, type of job held, income, and amounts of money spent last year on specified recreational activities.
3.   A person completed a certificate of registration for a motor vehicle, supplying information on make, model, and year of car.
4.   A purchaser of a toaster filled out the warranty card, giving information on family characteristics and on the primary method by which attention was directed to this appliance (for example, word of mouth, television commercial).   ☐

Both the advantages and the limitations of the self-enumeration procedure arise from the elimination of interviewers. The types of interviewer errors discussed earlier are thus avoided. On the other hand, the absence of interviewers creates two serious problems:

FIGURE 1.2 **Example of part of a self-enumeration questionnaire**

3. Since you left high school, have you taken any courses in a technical or trade school, hospital school, beauty school, business school, or other vocational school? (DO NOT include regular college courses.)	(115) 1 ☐ Yes — Continue with question 4 2 ☐ No — Go directly to question 8
4. When did you first begin classes at a vocational, technical, trade, or business school?	Month ┊ Year (116) _____ ┊ 19 _____
5. What was your field of study when you began taking these classes (for example, beautician, auto mechanics, accounting, etc.)?	_____ OR (117) 0 ☐ No specific field of study ☐☐ **CENSUS USE ONLY**
6. What is the full name and address of the school you attended? If you attended more than one school, enter the one you attended longest.	Name _____ Address (Number and street) _____ City State ZIP code (118) ☐☐☐☐☐ **CENSUS USE ONLY**
7a. At the school you entered in question 6, were you enrolled in a program leading to a degree or certificate?	(119) 1 ☐ Yes — Continue with question 7b 2 ☐ No — Go directly to question 8
7b. How many years of continuous full-time study does it usually take to complete the requirements for the degree or certificate for which you were enrolled? (Count only years after high school.)	(120) 1 ☐ Less than ½ year 2 ☐ ½ to 1 year 3 ☐ 2 years (13–24 months) 4 ☐ 3 years (25–36 months) 5 ☐ 4 years (37–48 months) 6 ☐ 5 years or more
7c. Did you complete that program?	☐ YES→When did you complete that program? (121) Month ┊ Year ┊ 19 } Go directly to question 8 ☐ NO → When did you last attend classes at the school you entered in question 6? (122) Month ┊ Year ┊ 19 } Continue with question 7d
7d. What are the reasons that you did not complete the program? (Mark (X) all that apply)	(123) 1 ☐ Financial reasons * 2 ☐ Health reasons 3 ☐ Marriage or pregnancy 4 ☐ Family or household responsibilities 5 ☐ Other personal or family reasons 6 ☐ Academic problems (124) 7 ☐ Took a job * 8 ☐ Entered military service 9 ☐ Wanted to leave school 0 ☐ Other — Specify_____
8. Since you left high school, have you been enrolled in a college or university?	(125) 1 ☐ Yes — Continue with question 9 on page 4 2 ☐ No — Go directly to question 17a on page 5

SOURCE: U.S. Bureau of the Census.

1. When a questionnaire is sent to a household or an organization, there is no control over which person answers the questions.
2. The absence of interviewers can lead to low response rates. A low response rate can be a source of serious bias in survey results, because the persons who *do* answer the questionnaires are often not representative of the entire group contacted. The user of data collected by a self-enumeration procedure should therefore know the rate of nonresponse as one factor affecting the magnitude of the potential bias.

Where there is a material rate of nonresponse, some follow-up of nonrespondents will be valuable. In most well-conducted mail surveys, some or all of the nonrespondents are contacted as a routine procedure. Nonrespondents may be contacted by means of "reminder" letters, telephone calls, or special personal interviews.

Questionnaire Design

Since questionnaires are a major vehicle in data collection, we now examine briefly some important considerations in designing and using a questionnaire so that accurate data are obtained.

Types of Questions.   The two basic types of questions employed in questionnaires are multiple-choice and free-answer or open-end questions.

The *multiple-choice question* presents the respondent with a choice from among two or more prespecified answers. In Figure 1.2, Questions 7a and 7b are of this type.

The alternatives in a multiple-choice question should be clear-cut; mutually exclusive; and when dealing with an issue, more or less evenly distributed on both sides of the issue. Ideally, the choices should cover all answers likely to arise in response to the question. If too many alternatives are presented, however, they may not be clear-cut enough for the respondent to make a meaningful selection. One disadvantage of multiple-choice questions is that they tend to suggest an answer to the respondent in terms of the alternatives stated when actually the question would have been answered differently if no choices had been suggested.

This danger is avoided by the other type of question, the *free-answer* or *open-end question,* which elicits answers in the respondent's own words. The following question is of the free-answer type:

> In your opinion, what things are good or bad about the Blank Company as a place to work?

In requesting answers in the respondent's own words, the free-answer question yields a multitude of replies that must be classified into a limited number of categories before statistical analysis can be undertaken. This classification becomes a difficult task when thousands of replies to a free-answer question have been received, since the wording of each reply must be interpreted carefully. Hence, free-answer questions are usually employed only in small-scale studies or in the exploratory stages of large-scale ones to determine the patterns into which the responses to a question tend to fall. The information obtained through the use of exploratory free-answer questions in the early stages of a large-scale study is then used to construct the multiple-choice questions utilized later for the bulk of the data collection.

Multiple-choice and free-answer formats sometimes are combined in constructing a question. An example is the use of a category titled "other" among the multiple choices, which allows the respondent to provide an answer other than the ones given. When the respondent chooses the "other" category, usually he or she is asked to provide additional details. Question 7d in Figure 1.2 is of this type.

Order of Questions. A questionnaire consists of a battery of questions arranged in a certain order. The first questions should establish rapport with the respondent, and all early questions should be simple. Where several topics are involved, it is usually best to complete one topic before going to another rather than to jump back and forth. The order of questions often affects the answers given by the respondent, because questions draw the respondent's attention to a particular complex of thoughts or feelings in the context of which later questions are answered. In market research surveys, for example, questions mentioning a specific product or firm tend to bias answers to questions that follow; consequently, such identifying questions should be placed toward the end whenever possible.

Directness of Approach. Many respondents are prone to rationalize or to exaggerate their replies when they are questioned directly about their motives, accomplishments, or other subjects involving their prestige or self-esteem. In order to avoid biased data, questionnaire designers often take an indirect approach in framing "prestige questions." For example, instead of asking, "Did you complete high school?" interviewers might be instructed to ask, "What grade did you complete when you left school?" In the latter question, an attempt is made to avoid implications adverse to a respondent who did not graduate from high school.

Need for Clarity in Questions. A question must have approximately the same meaning for all respondents if the data obtained from the replies are to be meaningful. Vaguely defined terms should be avoided in framing questions, since such terms will be interpreted differently by different persons. In addition, questions should be stated simply. If long and involved questions are asked, there will usually be some respondents who do not interpret them correctly.

Need to Avoid Bias in Question Framing. The replies of respondents are frequently influenced in a one-sided manner by the way in which questions are phrased. One type of question that tends to produce biased replies is a *leading question* — that is, one suggesting a particular answer. The question "Wouldn't you agree that the extra quality obtained in this brand is well worth the few extra pennies in cost?" so obviously suggests a particular answer that few persons will take the replies seriously. In many instances, though, the suggestive quality of a question is less easily recognized.

Value of Pretesting. Once a questionnaire has been designed, it is often pretested in a preliminary small-scale study. Pretesting, usually involving between a few dozen and several hundred respondents, brings out unforeseen difficulties, such as the layout of the questionnaire, arrangement and wording of the questions, and even the clarity of instructions to interviewers when interviewing is used. The difficulties can then be corrected before the full-scale study is begun.

1.5 ERRORS IN DATA

Data sets often contain errors. They may arise in the data acquisition stage or in other ways. It is important for users of data to be on guard against possible errors. In this section, we discuss errors in data and show examples of how errors arise.

Errors Arising in Data Acquisition

Errors in data can be introduced by defects in the acquisition procedure.

(1.5)   An *error in data acquisition* is any discrepancy between the actual result obtained and the correct result that would be provided by an ideal procedure.

EXAMPLES

1. Some Canadians completing a travel questionnaire overlooked trips to the United States in answering a question on the number of foreign trips taken in the past year. These errors led to incorrect counts of total number of foreign trips.

2. In answering a question relating to occupation, some respondents who are "retired" mistakenly recorded themselves as "unemployed," possibly because the definitions of the categories were misunderstood. These errors are classification errors, which led to incorrect counts.

3. Wind velocity and direction measured on an offshore drilling rig were distorted as a result of the air turbulence created by the rig itself, and consequently, incorrect measurements were recorded.

4. An industrial scale showed incorrect weights for loaded trucks because it was not calibrated properly, and thereby, it provided incorrect measurements.   □

As the examples show, errors in acquiring data may be caused by imperfect recall by respondents, inaccurate measurements by an instrument, misinterpretations of a question, misunderstandings of a definition—the list is long. In acquiring data, one must select a procedure that will keep errors at tolerable levels. Thus, a personal interview may be preferred over a mailed self-enumeration questionnaire in a study involving many technical definitions, since the interviewer can help respondents avoid misunderstandings. Of course, not all errors in data are of equal importance. The procedure for data acquisition must be chosen to control the important errors so that the data set obtained will be useful.

Other Errors in Data

Errors in data acquisition are not the only kind of errors found in data sets. Other errors may also be present, such as when a data set is incomplete or contains typographical errors.

EXAMPLES

1. In a survey of families residing in a community to obtain information about labor skills, families away on vacation at the time were not included. As a result, the data set is incomplete.

2. A data set on financial operating characteristics of small businesses in a region contained some incorrect codings of the survey results, some data entry errors in the survey computer file, and some typographical errors made when the data were published in the survey report. □

CITED REFERENCE

1.1   *Funk & Wagnalls Standard College Dictionary.* Canadian ed. Toronto: Fitzhenry and Whiteside, 1976.

PROBLEMS

1.1 The following data set, compiled by a firm's senior management committee, relates to candidates for promotion to product manager:

Name	Age	Sex	Number of Training Programs Attended	Committee's Order of Preference
Bixby, Milton	46	M	2	2
Morris, Wm.	31	M	0	3
Parker, Kim	42	F	1	1

a. What constitute the elements of this data set? How many elements are there?

b. Identify the variables in the data set. For each variable, indicate whether the data were obtained by measurement, counting, ranking, or classification and whether the variable is quantitative or qualitative.

c. What constitutes the case for Milton Bixby? What is the observation for Bixby on the age variable?

1.2 **Municipal Debt.**   The following data set was compiled by a financial analyst while examining the public debt burden of several municipalities:

Municipality	Population (000)	Per Capita Debt ($000)	Ratio of Debt to Property Value (percent)
Carullo	457	4.8	6.9
Edwards	1322	5.9	7.7
Girvan	181	3.4	8.3
Marois	665	5.5	8.0
Morency	690	5.9	7.3
Pratt	234	4.9	10.1

a. What constitute the elements of this data set? How many elements are there?

b. Identify the variables in the data set. For each variable, indicate whether it is quantitative or qualitative. Is the data set univariate or multivariate?

c. What constitute the observations on the per capita debt variable?

1.3 The following data set pertains to five houses that are listed for sale:

Listing Number	Subdivision	Asking Price ($000)	Number of Bedrooms	Selling Agency
80344	Maxwell	99	4	Peabody
15671	Pine	76	4	Peabody
14006	Maxwell	195	5	Realax
73317	Bayview	105	4	Realax
28180	Bedford	85	3	Peabody

a. Identify the variables in the data set. For each variable, indicate whether the data were obtained by measurement, counting, ranking, or classification.

b. What constitute the observations on the selling agency variable?

c. What constitutes the case for the house with listing number 73317?

1.4 For each of the following variables, indicate whether it is quantitative or qualitative and whether the observations on the variable are obtained by measurement, counting, ranking, or classification.

a. Position of a company among the top 500 companies, ordered by sales revenue.

b. Meat plant assignment of a government inspector.

c. Number of students enrolled in a university.

d. Dividend income reported by a taxpayer in the current year's tax return.

1.5 For each of the following variables, indicate whether it is quantitative or qualitative and whether the observations on the variable are obtained by measurement, counting, ranking, or classification.

a. Number of employees of a company who quit last month.

b. Country of registry of an oil tanker.

c. Store price of a brand of canned peaches.

d. League standing of a baseball team.

1.6 A public utility company has distributed its annual financial statement to each of its industrial customers. Is this statement an internal or external source of data for one of its industrial customers? For the public utility company itself? Explain.

1.7 A multinational corporation operates in 27 countries. Would the *Monthly Bulletin of Statistics* published by the United Nations be an internal or external source of data for this corporation? Explain.

1.8 A computer tape contains production data for the various divisions of a firm. Is this tape an internal or external source for the firm? Explain.

1.9 A company has been selling one of its popular beverage products in bottles in supermarkets and in cans in other outlets. An analyst is examining sales data for the beverage to see which type of container is preferred by customers.
   a. Are these data nonexperimental or experimental? Why?
   b. Why does the distinction in **a** matter to the analyst? Explain.

1.10 Thirty sales trainees were grouped into 15 pairs, each pair being similar in age, sales experience, and educational background. One trainee in each pair was selected at random to attend a sales training program based on role playing; the other trainee was assigned to a sales training program based on case studies. At the end of the programs, the trainees were tested for sales skills and knowledge. The test scores are now being evaluated to determine whether one type of program produces better scores than the other.
   a. Is this study experimental or nonexperimental? Why?
   b. An observer asks why the trainees in each pair could not simply have decided between themselves who would attend each program. Answer the observer's question.

1.11 Refer to the **Power Cells** data set (Appendix D.2). These data are from an experimental study.
   a. Which variables represent factors that were controlled in this experimental study?
   b. Which were the response variables in this study that were believed to be influenced by the factors identified in **a**?
   c. Would operating the same number of power cells until they fail without attempting to control any of the factors identified in **a** yield the same information as the experiment? Discuss.

1.12 Each purchaser of a new personal computer receives an information card that is to be completed by the purchaser and returned to the equipment manufacturer.
   a. Which method of data acquisition (observation, interview, or self-enumeration) is involved here?
   b. Would your answer change if the information card were completed by the salesperson on the basis of questions put to the purchaser?
   c. Is the response rate likely to be affected by whether or not the postage for the information card is prepaid by the manufacturer? Explain.

1.13 For each of the following, state which method of data acquisition (observation, interview, or self-enumeration) is most appropriate for obtaining the information, and justify your choice.
   a. Quarterly data from employers on the number of employees who worked at any time during the quarter.
   b. Data from technicians of a large research organization on hazardous aspects of their jobs.
   c. Data on sequence in which shoppers purchase items in supermarkets.

1.14 For each of the following, state which method of data acquisition (observation, interview, or self-enumeration) is most appropriate for obtaining the information, and justify your choice.
   a. Data from persons in a national register of scientists concerning field of specialization, nature of present employment, and academic training.
   b. Data on blood cholesterol levels of Canadians.
   c. Data from manufacturers on sales, inventory levels, and new orders.

1.15 In a mail survey of 2000 practicing physicians conducted by a pharmaceutical firm, a series of questions related to prescriptions for an antidepressant drug code-named ADV. The proportion of responding physicians who stated they had prescribed the drug at least once in the previous month was 0.40.

    a. Suppose that 95 percent of the questionnaires had been completed and returned. By how much might the proportion 0.40 be changed if every questionnaire had been completed and returned?

    b. Answer **a** on the supposition that the response rate was 20 percent.

    c. What are the implications of your answers to **a** and **b** for interpreting data from studies with low response rates?

    d. What factors might cause a physician to not respond in this type of mail survey? Might any of these factors be related to whether or not the physician prescribes this particular drug? Does it matter, for the purposes of the survey, whether or not such a relation exists? Discuss.

1.16 A questionaire on wine consumption asked the following question:

What is the typical quantity of wine that you personally drink with a restaurant meal?

_____ None; _____ one glass; _____ half bottle; _____ whole bottle; _____ more than a whole bottle.

    a. Is this question multiple choice or free answer?

    b. What difficulties, if any, might a respondent have in replying to this question?

1.17 Identify any major faults in each of the following questions, and reword the question to eliminate the faults. Indicate any assumptions you made when rewording the question.

    a. How many calls for telephone directory assistance did you make in the past 24 months?

    b. Is your family income low, average, or high compared with incomes of other families living on this block?

1.18 Identify any major faults in each of the following questions, and reword the question to eliminate the faults. Indicate any assumptions you made when rewording the question.

    a. Does Sony or some other brand come to mind when a videocassette recorder is mentioned?

    b. Have you suffered from an elevation of bodily temperature during the past week?

1.19 Consider the following alternative questions for ascertaining consumer purchasing plans: (1) Do you plan to buy a new small car within the next year? (2) Do you plan to buy a new small car within the next year, and how certain are you of this plan on a scale from 1 (highly uncertain) to 10 (highly certain)?

    a. Which question do you think will yield more useful information? Explain.

    b. Would the clarity of the question be improved if "within the next year" were replaced by "within the next 12 months"? Explain.

1.20 An instructor divided a class on opinion research at random into two groups. One group was asked the following question: "Should university officials take primary responsibility for reducing rowdyism at basketball games?" The second group was asked the same question, except that "student leaders" was substituted for "university officials." About 80 percent of the students in each group answered affirmatively.

    a. What fault in question design is illustrated by the instructor's experiment?

    b. Design a better question to obtain student opinions on who should take primary responsibility to reduce rowdyism at basketball games. Explain how your approach avoids the fault identified in **a**.

1.21 Explain whether each of the following entails an error arising from defects in the data acquisition procedure or some other type of error.

    a. A schoolchild, when asked to indicate the number of brothers or sisters at home who have not yet started school, overlooked a new baby brother and wrote "1" instead of "2" on the questionnaire.

    b. An incorrect entry of data to a hospital computer file caused a baby's weight to be shown as 8400 grams when the actual weight on the hospital birth form was 4800 grams.

    c. In filling out a questionnaire sent to owner-operated restaurants, an owner misinterpreted the definition of *employee* in answering a question on number of full-time employees and counted himself as one of the employees.

1.22 In a consumer mail survey, respondents were asked to name their regular brand of toothpaste. In follow-up personal interviews conducted with 60 of the initial respondents, each interviewee was again asked to name the regular brand of toothpaste. Twelve of the 60 respondents gave different answers to this question in the interview than they did in the mail questionnaire. Do the different answers represent errors in data acquisition, other types of errors, or something else? Comment.

STUDIES

1.23 For each of the following, ascertain a library source where data for the most recent period can be obtained.

    a. U.S. annual birthrate.

    b. U.S. monthly unemployment rate.

    c. U.S. monthly Consumer Price Index for All Urban Consumers.

1.24 Refer to Problem 1.23. Answer it for data pertaining to Canada.

1.25 For each published data source assigned from the following list, state who publishes the source, how often it is published, and give three specific items of information from the most recent issue of that source.

    —*Business Conditions Digest*   —*Monthly Labor Review*
    —*Federal Reserve Bulletin*   —*Moody's Industrial Manual*
    —*International Financial Statistics*   —*Survey of Current Business*

1.26 For each published data source assigned from the following list, state who publishes the source, how often it is published, and give three specific items of information from the most recent issue of that source.

    —*Bank of Canada Review*   —*Direction of Trade Statistics*
    —*Canadian Life Insurance Facts*   —*The Conference Board Statistical Bulletin*
    —*Current Economic Analysis*   —*Toronto Stock Exchange Review*

1.27 A study indicated that high school students who took a voluntary driver training course tended later to have better driving records than students who did not take the course. However, the study team noted that this difference could be due simply to differences in personal qualities of the students in the two groups. Among other things, relatively more women than men took the course, and young women tend to have better driving records than young men in any case. The team also noted that students taking the course might have been better motivated to be good drivers than students not taking the course.

    a. Why was this study not an experimental study?

    b. Develop a design for an experimental study in a high school to evaluate the impact of the driver training course. Your design should control the effects of sex and motivation of students.

1.28 A professional association composed of economists and statisticians surveyed its members to study how they liked the association's journal. A high proportion of the responses indicated an unfavorable opinion. Subsequently, it was learned that the response rate was much higher for economists than for statisticians. Construct a numerical example to show how the differential response rates could have biased the survey results.

1.29 It is frequently stated that errors in response are not too serious because they will tend to balance out. One researcher conducted a survey of 1643 library users in a community to study the nature of response errors. He compared responses for the number of books borrowed from the library in the preceding three months with the library's computerized records. The results were as follows:

Respondent Statement	Percent of Respondents	Average Number of Books Borrowed	
		Library Record	Respondent's Recollection
Accurate	23	6.1	6.1
Overstated	43	2.8	5.6
Understated	34	6.4	5.4

    a. By what percent did overstaters overstate their borrowings? Did understaters understate their borrowings to the same extent?

    b. What is the overall relative bias in the reported average number of books borrowed? Did the response biases largely balance out?

    c. What are some factors that might account for the biases in response here?

DATA ANALYSIS I: CLASSIFICATION AND DISTRIBUTION PATTERNS

Once the data needed for an investigation are in hand, their analysis can begin. The analysis usually starts with a study of the underlying pattern of the data. To this end, data sets (unless they are very small) must be organized and reduced to manageable proportions before any study of them can be made. The human mind is simply not capable of assimilating and interpreting a large number of facts and figures in raw form. In this chapter, we present a number of methods for studying data for purposes of analyzing, interpreting, and reporting their distribution pattern.

2.1 QUALITATIVE DATA

We begin with a discussion of methods for the analysis of distribution patterns of data for qualitative variables — that is, for variables having nonnumerical outcomes.

Qualitative Distributions

Observations on a qualitative variable can be classified into a qualitative distribution to facilitate the study of their distribution pattern.

(2.1) | A *qualitative distribution* is the classification of the elements of a data set by a nonnumerical characteristic.

EXAMPLE

An analyst for a retailer wished to study the occupational profile of the company's credit card customers. Her intention was to compare this profile with the occupational profile of the total labor force as reported in a recent government publication. The analyst expected that the comparison would provide insights useful for credit policy, advertising, and billing.

The analyst had available in a computer file occupational data obtained from a recent survey of the company's customers. Altogether, information was available for 131,845 customers who were in the labor force. The analyst concluded that a simple computer listing of the customers' occupations would be of little use since she could not digest such a large collection of raw data nor effectively compare these data with the occupational profile of the total labor force. Instead, she decided to classify all occupations into the seven categories utilized for the government labor force study. These seven categories are shown in Table 2.1.

She employed a routine in a statistical software package that read each customer's occupation from the file (carpenter, dietitian, civil engineer, and so on), assigned it to the appropriate occupational category, and counted the number of customers in each category. The results are shown in Table 2.1, column 1. The routine also computed the percentage of customers who are in each occupational category, shown in column 2 of Table 2.1. As an illustration of the latter calculations, $100(38,835/131,845) = 29.5$ percent of customers have managerial occupations.

The analyst added, in column 3 of Table 2.1, the corresponding percentages for the total labor force in order to facilitate a comparison of the occupational profiles of the company's customers and the total labor force. In making this comparison, the analyst discovered, among other things, that over half of the customers (53.2 percent) fall into the managerial and professional and technical categories, whereas these categories comprise only 19.1 percent of the total labor force. This one fact alone has significant implications for the company.   □

The analyst used classification and counting here to simplify the data set. She established a set of classes or categories and counted the number of elements (customers) that belong in each one. The data classification scheme shown in Table 2.1 is called a qualitative distribution because the classes refer to a characteristic that is non-numerical (occupation).

TABLE 2.1   Distribution of customers by major occupational categories — An example of a qualitative distribution

Occupational Category	(1) Number of Customers	(2) Percent of Customers	(3) Percent of Total Labor Force
Managerial	38,835	29.5	9.2
Professional and technical	31,262	23.7	9.9
Service and recreation	14,011	10.6	10.9
Clerical and sales	12,797	9.7	20.7
Crafts and production workers	11,090	8.4	24.2
Laborers and unskilled workers	1,577	1.2	5.0
Others	22,273	16.9	20.1
Total	131,845	100.0 (131,845)	100.0

Systems of Classification

Any system of classification used for a qualitative distribution must meet certain formal requirements.

(2.2) The classes in any system of classification must be *mutually exclusive* and *exhaustive*.

In other words, every element in the data set must fall into *one* and *only one* class of the system. The classification system used in Table 2.1 satisfies this requirement. Each customer can be assigned to *one* occupational category (so the categories are exhaustive), and each customer qualifies for assignment to *only one* of the categories (so the categories are mutually exclusive — that is, not overlapping).

Whenever data are classified, some information is lost. The amount lost depends on the system of classification employed. For example, the classification in Table 2.1 has resulted in loss of exact information about customers' occupation types. We know, for instance, that 31,262 customers have professional and technical occupations, but the qualitative distribution does not tell us how many of these are architects or economists.

The determination of the classes to be utilized in constructing a qualitative distribution therefore requires care. The objective is to select a system of classification that loses as little essential information as possible while still achieving an effective summarization of the data. For example, in the classification of inspection results for a production process, the class defective would be too coarse if it were important to distinguish between defects caused by faulty material and those caused by poor workmanship. Often, several systems of classification need to be tried before a final selection is made.

Graphic Presentation of Qualitative Distributions

Often, it is useful to study a qualitative distribution by displaying it pictorially. The display usually takes the form of a *bar chart,* such as the one shown in Figure 2.1a for the qualitative distribution of customers' occupations given in Table 2.1. Each bar corresponds to one class of the qualitative distribution. The length of each bar corresponds to the number (or percentage) of customers in the class.

Several aspects of the arrangement of the bars should be noted: (1) The bars differ only in length, not in width; (2) a space has been left between each bar to make it easier to identify each bar by its label; (3) the bars have been ranked by order of magnitude to facilitate analysis. The order may be a decreasing one, as here, or it may be an increasing one. If an "others" or "miscellaneous" category is present, it is usually shown as the lowest bar regardless of its magnitude, because this type of category most often is a collection of relatively unimportant classes.

When a pictorial comparison of two or more qualitative distributions is desired, it is often possible to combine their bar charts. Such a combination is shown in Figure 2.1b, where the occupational profiles of the retail-store customers and the total

FIGURE 2.1   Examples of bar charts of qualitative distributions

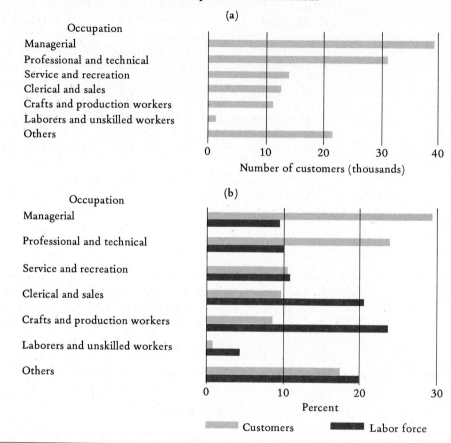

labor force are compared. Note that both distributions are expressed in percentage terms to provide a meaningful comparison.

Modifications can be made in the bar charts in Figure 2.1 to show additional detail or to highlight particular features of the data. For example, the bars in Figure 2.1a could be segmented to show the numbers of male and female customers in each occupational category.

Graphics subsystems of statistical software packages enable users to readily generate various types of bar charts for display and analysis of qualitative distributions.

2.2   QUANTITATIVE DATA

We now discuss methods for the analysis of distribution patterns of data for quantitative variables—that is, for variables having numerical outcomes.

Arrays

A useful first step in organizing data for a quantitative variable and for discerning a pattern in them, when the number of observations is not too large, is to list the observations in increasing or decreasing order of magnitude. Such an ordering is called an *array* and can greatly facilitate inspection of the data.

EXAMPLE

Semiconductor Failure. An industrial engineer examined failure data for the bond between a wire and a semiconductor wafer. An array of the breaking stresses (in milligrams) of 17 bonds, arranged in increasing order of magnitude, follows:

1   4   31   43   47   50   51   51   58   59   62   63   66   71   75   88   113

Observe how the smallest and largest observations in the data set are easily ascertained from the array.   □

Stem-and-Leaf Displays

A *stem-and-leaf display* provides information about the pattern of observations in a data array and assists in the discovery of concentrations or clusters of particular values in the data set. The name of the display comes from the way in which it is constructed. The following two examples illustrate the method of construction.

EXAMPLES

1. **Concert Subscribers.** An analyst, commissioned to study characteristics of concert subscribers in a large metropolitan area, pretested the questionnaire on a sample of 70 subscribers. An array of the responses to the question on subscriber's age is as follows (only part of the data are shown):

20   25   25   25   29   30 $\cdots$ 76   77

Figure 2.2a shows a stem-and-leaf display for the reported ages. The digits to the left of the vertical line are the "stems" and the digits to the right are the "leaves." The digit 2 constitutes the first stem. All age responses from 20 to 29 inclusive are recorded on this stem. Only the second digit in each age response is written because the first digit is given by the stem. We see that the leaves on the first stem consist of the digits 0, 5, 5, 5, 9, indicating that five respondents reported ages in the interval 20–29, with the reported ages being 20, 25, 25, 25, and 29.

The analyst, viewing the display, concluded that it fitted the usual age pattern for concert subscribers. In turning to the specific responses, however, he noted a disproportionately large number of 0s and 5s among the second digits (leaves) in the age responses. This result was unanticipated since he had expected each final digit from 0 to 9 to be roughly equally represented. We shall further consider this finding in a later section.

2. A stem-and-leaf display for the 17 breaking stresses in the semiconductor failure example appears in Figure 2.2b. The display is readily constructed from the data

FIGURE 2.2   **Stem-and-leaf displays for the concert subscribers and semiconductor failure examples**

(a) Subscribers' Ages (years)

```
2 | 05559
3 | 001233455678
4 | 000012234455677899
5 | 0000011233455567889
6 | 0023455679
7 | 015567
```

(b) Breaking Stresses (milligrams)

```
 0 | 14
 1 |
 2 |
 3 | 1
 4 | 37
 5 | 01189
 6 | 236
 7 | 15
 8 | 8
 9 |
10 |
11 | 3
```

array given earlier. The tens' and hundreds' digits of the stress observations form the stems of the display, and the units' digits form the leaves.

The engineer, upon examining the display, could see immediately that two of the semiconductor bonds had broken under negligible stress (1 and 4 mg), possibly because of defective manufacture. He also noted that the bulk of the observations were concentrated between 31 and 88. Finally, one bond had withstood considerable stress before breaking (113 mg). This bond was subsequently subjected to special study because of its superior performance.   □

Frequency Distributions

Observations on a quantitative variable can also be classified into a distribution to facilitate the study of the distribution pattern. The method is analogous to that for qualitative data. This type of distribution is called a frequency distribution.

(2.3)   A *frequency distribution* is the classification of the elements of a data set by a quantitative (that is, numerical) characteristic.

EXAMPLE

Table 2.2 shows a frequency distribution of the balances of a bank's 30,794 savings accounts. The classes in this case are defined in terms of the amount of the balance, and each class is expressed as an interval. Thus, the first class contains all accounts that have balances under $5000. We readily see from Table 2.2 that there are 10,196 such accounts.

The frequency distribution in Table 2.2 was prepared to assist management in revising the service charge schedule for savings accounts transactions. It shows, for example, that over 17 percent of the savings accounts had balances of $10,000

TABLE 2.2   Distribution of savings accounts by the amount of the balance — An example of a frequency distribution

Amount of Balance (dollars)	Number of Accounts	Percent of Accounts
0–under 5,000	10,196	33.1
5,000–under 10,000	15,335	49.8
10,000–under 15,000	1,812	5.9
15,000–under 20,000	1,798	5.8
20,000–under 25,000	1,653	5.4
Total	30,794	100.0 (30,794)

or more. Two years ago, when a similar study was conducted, the corresponding figure was only 6 percent. ☐

The number of elements in a given class of a frequency distribution is called the *frequency* of that class. When the number of elements in each class is expressed as a percentage of the total number of elements in the data set, the resulting figure for a class is called the *percent frequency* of that class, and the resulting distribution is called a *percent frequency distribution*. When a percent frequency distribution is given for a data set, we customarily show the total number of elements in the data set. Thus, in Table 2.2 the total number of accounts (30,794) has been shown in parentheses under the 100.0 percent total. Showing the total number of elements for a percent distribution is also customary for qualitative distributions, as illustrated in Table 2.1.

Construction of Frequency Distributions

The general principles of classification discussed for qualitative distributions apply equally to frequency distributions. Thus, the classes of a frequency distribution must be mutually exclusive and exhaustive. Also, the number of classes utilized should be small enough to provide an effective summary yet large enough to avoid losing essential information.

In general, the construction of a satisfactory frequency distribution requires experimentation with alternative classifications, since there are no rules available that fit all situations. Nevertheless, there exist some general considerations that are helpful for constructing a frequency distribution. We take these up now.

Number of Classes.   The larger the number of classes in a frequency distribution, the more detail is shown. If the number of classes is too large, though, the classification loses its effectiveness for summarizing the data. Too few classes, on the other hand, condense the information so much as to leave little insight into the pattern of the distribution. The best number of classes in a frequency distribution often needs to be determined by experimentation. Usually, an effective number of classes is somewhere between 4 and 20.

EXAMPLE

In a study of the capability of untrained subjects to discriminate between different food preparations, 40 subjects were each given two food samples and asked for a rating of each sample. Unknown to the subjects, both food samples were identical in every respect. Each subject tasted both food samples and gave each sample a rating between 0 and 10. The difference between each subject's ratings for the first and second samples was taken as a measure of preference for the first sample over the second. Tables 2.3a, 2.3b, and 2.3c show the subjects' rating differences classified into 3, 7, and 21 classes, respectively.

At one extreme, the frequency distribution in Table 2.3c has so many classes that it is difficult to discern a meaningful pattern in the rating differences. At the other extreme, the frequency distribution in Table 2.3a has so few classes that any pattern present is almost lost. Finally, in Table 2.3b we have a frequency distribu-

TABLE 2.3   Three frequency distributions of subjects' rating differences of two food samples

(a) 3 classes		(b) 7 classes		(c) 21 classes	
Rating Difference	Number of Subjects	Rating Difference	Number of Subjects	Rating Difference	Number of Subjects
−10 to −4	13	−10 to −8	1	−10	0
−3 to +3	19	−7 to −5	7	−9	1
+4 to +10	8	−4 to −2	12	−8	0
Total	40	−1 to +1	4	−7	2
		+2 to +4	11	−6	3
		+5 to +7	3	−5	2
		+8 to +10	2	−4	5
		Total	40	−3	3
				−2	4
				−1	2
				0	0
				+1	2
				+2	3
				+3	5
				+4	3
				+5	1
				+6	2
				+7	0
				+8	1
				+9	0
				+10	1
				Total	40

tion that reveals the pattern of variation in the rating differences in a meaningful way. We note first that the rating differences are approximately symmetrically distributed about zero, which is not unexpected because both food samples were identical. An even more interesting observation is that the frequency of the class −1 to +1 is much lower than the frequencies of the classes immediately above and below it. It would therefore appear that the subjects tended to give different ratings to the two food samples, even though the two food samples were identical. The subjects apparently believed the two samples must be different (for what other reason would the study be undertaken?) and consequently gave different ratings to them. These important characteristics of the rating differences, which are readily discernible from Table 2.3b, are not readily apparent from the other two frequency distributions.   □

The very clear statistical lesson provided by this example is that consideration of alternative numbers of classes is helpful for finding the frequency distribution that best yields insights into the data set.

Width of Classes.   The choice of the class width or class interval is related to the determination of the number of classes. It is generally best if all the classes have the same width. If the classes are not equally wide, one often cannot tell readily whether differences in class frequencies are due mainly to differences in the concentration of items or to differences in the class widths.

Sometimes, one must use unequal class intervals. Consider the construction of a frequency distribution of annual salaries of employees in a company that range from $10,000 to $180,000. Suppose that about six classes are to be used in the distribution. If equal class intervals were used, each would have a width of about $30,000, the first class perhaps being $0–under $30,000. This grouping provides equal class intervals but destroys the usefulness of the classification. Most of the employees earn less than $30,000 and hence would be included in the first class. Thus, no information would be provided about the distribution of salaries of the majority of the employees except that they earned less than $30,000. Subsequent classes would not reveal much information either, since relatively few employees earn over $30,000 in this company, yet five of the six classes would be devoted to a classification of their salaries.

In cases of this nature, unequal class intervals are generally used. For instance, equal class intervals of, say, $10,000 in width might be used for the range wherein most of the salaries fall, after which the interval might increase to, say, $30,000. In fact, when all but a few of the salaries have been classified, an open-end class might be used to account for the remainder. An *open-end class* is one that has only one limit, either upper or lower. Thus, if only three officials in the company earned over $100,000, the last class might be $100,000 and more. In this way, only one class is needed to account for these top three salaries. Open-end classes may be necessary at the upper end of the distribution, as just illustrated, sometimes at the lower end of the distribution, and occasionally at both ends.

Class Limits.   Still another problem is the choice of class limits. Calculations from a frequency distribution often use the midpoint of each class to represent all the items in the class. The *midpoint* of a class is the value halfway between the two class limits. It is usually suggested as good statistical practice that the class limits be chosen so that the midpoint of each class is approximately equal to the arithmetic average of the values falling in that class. Often, this objective can be accomplished without special concern about the class limits. If, however, the values tend to be bunched at certain periodic points throughout the range of the data — for example, rents are often multiples of $5 or $10 — one may have to experiment in deciding upon class limits so that each midpoint will approximately equal the arithmetic average of the values in that class.

In stating class limits, one should be careful to be unambiguous. For instance, the limits $300–$400, $400–$500 are not clear because one cannot be sure in which class $400 is included. Stating limits as $300–$399, $400–$499 is clear when the data are expressed in dollars. When they are, the midpoint of the first class is $(300 + 399)/2 = 349.50, or for most practical purposes, $350.

The limits $300–under $400, $400–under $500 are clear. However, without additional information, it may not be possible to determine the midpoints accurately. If no additional information is provided, we shall follow the convention of considering the midpoint of a class to be the arithmetic average of the two limits. Thus, the midpoint of the class $300–under $400 would be taken to be $(300 + 400)/2 = 350.

In some frequency distributions, each class corresponds to a single value, as shown in Table 2.3c earlier. As another example, in a frequency distribution of number of automobiles owned by a family, the classes might be the numbers 0, 1, 2, etc. In these cases, the individual values correspond to the class midpoints.

Graphic Presentation of Frequency Distributions

Graphic displays of frequency distributions often are useful for presentation and analysis. Two common methods of graphic display are the histogram and the frequency polygon.

(2.4)   A *histogram* is a bar graph of a frequency distribution. A *frequency polygon* is a line graph of a frequency distribution.

The method of constructing these graphic displays depends upon whether equal or unequal class intervals are utilized in the frequency distribution. We shall consider each of these two cases in turn.

Equal Class Intervals.   Figure 2.3 illustrates both graphic methods when the frequency distribution has equal class intervals. Figure 2.3a presents a percent frequency distribution of ages of 27,303 farm operators classified in equal, 10-year intervals. The age data were obtained from a recent survey of farm operators conducted by a farm equipment manufacturer. Figures 2.3b and 2.3c show a histogram and frequency

FIGURE 2.3   **Age distribution of farm operators — An example of the graphic presentation of a frequency distribution with equal class intervals**

(a) **Frequency Distribution**

Age (years)	Percent of farm operators
15–under 25	2
25–under 35	10
35–under 45	19
45–under 55	27
55–under 65	25
65–under 75	16
75–under 85	1
Total	100
	(27,303)

(b) **Histogram**

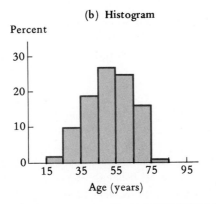

(c) **Frequency Polygon**

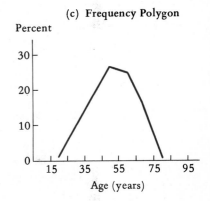

polygon, respectively, of this frequency distribution. In the histogram, adjoining bars are drawn with a width spanning the class interval and a height corresponding to the frequency (or percent frequency) of that class. In the frequency polygon, each class frequency (or percent frequency) is plotted corresponding to the midpoint of that class. The plotted points are then connected by a series of straight lines. Both the histogram and frequency polygon show readily that the ages of the farm operators tend to be concentrated in the range 35 to 75 years.

In both graphic presentations, the magnitudes of the class frequencies are conveyed visually by areas — in the case of the histogram, by the areas of the bars; and for the frequency polygon, by the area under the line graph. The reason is that when the class widths are equal, as they are in Figure 2.3, and the heights of the histogram bars or polygon points are equal to the class frequencies (or percent frequencies), the areas are automatically proportional to these frequencies.

| Comment | Stem-and-leaf displays, as illustrated earlier in Figure 2.2, are presentations of frequency distributions that combine the features of arrays and of histograms for equal class intervals. |

Unequal Class Intervals.   When the class intervals of the frequency distribution are not equal, the heights of the histogram bars or the polygon points must be adjusted to make the areas proportional to the frequencies (or percent frequencies) of the classes. In this procedure, we choose a class width as the unit width and express all frequencies (or percent frequencies) in terms of this unit width. The procedure is illustrated in Figure 2.4. Figure 2.4a contains the percent frequency distribution of incomes of taxpayers in a state who have incomes under $40,000. The distribution is of interest to a legislative committee reviewing income tax exemptions for lower- to middle-income taxpayers of the state. The interval of $2000 has been chosen as the unit width, and all percent frequencies have been expressed in terms of this unit width. Thus, the percent frequency for the class 6000–under 10,000 in terms of the unit width is $29.3/2 = 14.65$, because that class has a width of two $2000 units. Likewise, the percent frequency for the class 10,000–under 20,000 in terms of the unit width is $36.4/5 = 7.28$, because that class has a width of five $2000 units. The procedure then is to plot the percent frequencies per unit width in the usual fashion, which has been done in Figures 2.4b and 2.4c.

Comparison of Frequency Polygons.   Two or more frequency polygons can be compared readily if (1) they have the same class intervals, and (2) they have the same total frequency (or are both expressed in percentage form). Figure 2.5 provides an illustration. It shows the age distribution of farm operators (taken from Figure 2.3c) as well as that of the nonfarm civilian labor force (as reported in a recent economic study), each expressed in percent frequencies. It is easy to compare the two distributions from Figure 2.5. For example, we readily see that the farm operators are relatively much older than the nonfarm civilian labor force.

| Comment | In contrast to frequency polygons, it is quite difficult to obtain an effective comparison of two or more frequency distributions by superimposing their histograms. |

Cumulative Frequency Distributions

When one wishes to know what proportion of the elements of a data set have values above or below certain levels, the cumulative form of the frequency distribution is a very effective device for summarizing this information.

(2.5)   The cumulative form of a frequency distribution is called a *cumulative frequency distribution*. A graph of a cumulative frequency distribution is called an *ogive*.

FIGURE 2.4   Income distribution of taxpayers—An example of the graphic presentation of a frequency distribution with unequal class intervals

(a)

Total income (dollars)	Percent of taxpayers	Number of $2000 widths	Percent of taxpayers per $2000 width
0–under 2000	0.6	1	0.60
2000–under 4000	10.2	1	10.20
4000–under 6000	13.4	1	13.40
6000–under 10,000	29.3	2	14.65
10,000–under 20,000	36.4	5	7.28
20,000–under 40,000	10.1	10	1.01
Total	100.0 (1,049,916)		

(b)

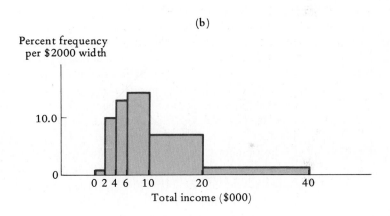

(c)

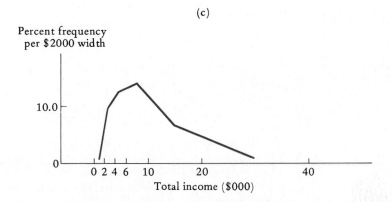

FIGURE 2.5   **Age distributions of farm operators and nonfarm civilian labor force**

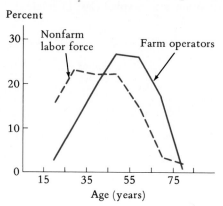

Figure 2.6a repeats the frequency distribution of the ages of farm operators from Figure 2.3a, Figure 2.6b shows the cumulative distribution, and Figure 2.6c contains the ogive. The cumulative distribution in Figure 2.6b is called a *less-than cumulative frequency distribution* because, for each value of the quantitative characteristic (age in this example), the number (or percentage) of elements having less than that value is recorded. For instance, the cumulative age distribution indicates that 2 percent of farm operators are less than 25 years of age and 12 percent are less than 35 years of age.

In a graph of the cumulative distribution, the cumulative number (or percentage) of elements is plotted against the corresponding value of the variable. Thus, the points plotted begin with (15, 0) and (25, 2) and end with (85, 100). The points are then connected by straight lines to permit interpolation. Thus, one can estimate from Figure 2.6c (see the dashed line) that 50 percent of farm operators are less than 52 years of age.

Comments   1.   When straight lines are used to connect the plotted points of the ogive, it is assumed that the observations within each class are uniformly distributed throughout the class. Since this condition is usually not met exactly, the resulting graph is only an approximation to the exact distribution of the values. The straight-line approximation is sufficiently good in many cases.

2.   A situation where it is *not* appropriate to use straight lines to connect the known points of an ogive occurs when the discrete nature of the data set must be recognized. In this case, the ogive should be drawn as a step function. For example, Figure 2.7 shows the cumulative frequency distribution and the ogive of family size for families belonging to a community recreation center. Since family size must be a whole number, the ogive has a "step" at each whole number. Note that the cumulative frequency distribution in this case shows the percentage of families with size equal to or less than the given number.

FIGURE 2.6 Less-than cumulative age distribution of farm operators

(a) Frequency Distribution

Age (years)	Percent of farm operators
15–under 25	2
25–under 35	10
35–under 45	19
45–under 55	27
55–under 65	25
65–under 75	16
75–under 85	1
Total	100

(b) Cumulative Frequency Distribution

Less than this age (years)	Cumulative percent of farm operators
15	0
25	2
35	12
45	31
55	58
65	83
75	99
85	100

(c) Ogive

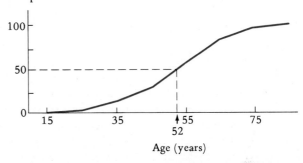

FIGURE 2.7 Cumulative distribution of family size — An example of an ogive for discrete data.

The points for the cumulative distribution are plotted and the ogive is drawn as a step function.

(a) Cumulative Frequency Distribution

This size or less	Cumulative percent of families
1	0
2	33
3	57
4	76
5	87
6	100

(b) Ogive

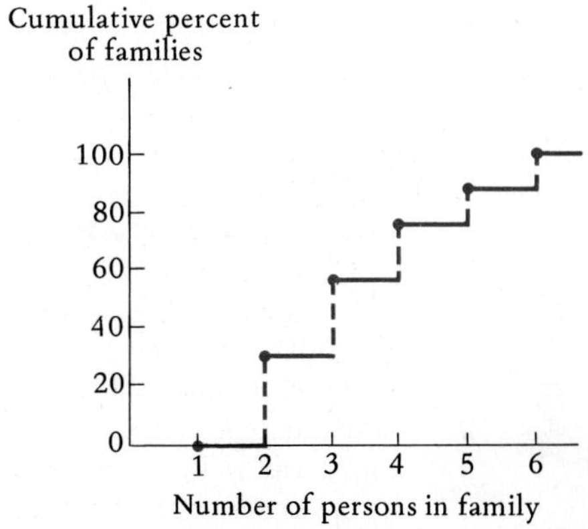

3. An ogive will not reach the 100 percent level if the frequency distribution has an open-end class at the upper end. The reason is that the maximum value of the elements is not known in this case. A similar situation arises with the 0 percent level when there is an open-end class at the lower end of the distribution.

4. A cumulative distribution and ogive can be constructed so that they indicate the percentage (or number) of items having magnitudes "more than" any specified value. The resulting distribution is called a *more-than cumulative frequency distribution*. The method of construction is analogous to that for the less-than cumulative distribution and ogive. For example, the more-than cumulative distribution of the ages of farm operators, based on the data in Figure 2.6a, is as follows:

This Age or More (years)	Cumulative Percent of Farm Operators
15	100
25	98
35	88
45	69
etc.	etc.

5. In the plot of an ogive, no special procedures are required when the class intervals are unequal. (Recall that unequal class intervals entail special procedures for constructing a histogram or frequency polygon.)

6. Ogives of different frequency distributions can be compared readily on the same graph if the total frequencies are the same or if percent frequencies are used. Class intervals of the respective distributions need not be the same.

 ## BIVARIATE DATA

The analysis of bivariate data also usually begins with a study of the underlying pattern of the data, one of the major objectives being to gain insights into the nature of any relationship between the two variables. As in the case of data sets with only a single variable, the initial analysis of bivariate data sets often employs classification and graphic presentation.

Bivariate Distributions

When the elements of a bivariate data set are classified simultaneously in terms of both variables, the classification process is called *cross-classification* or *cross-tabulation*. The principles discussed earlier for classification systems for univariate data sets apply equally to bivariate data sets. As we shall show, cross-classification can be very helpful in studying the nature of the statistical relationship, if any, between the two variables.

Bivariate data sets may be based on two qualitative variables, two quantitative variables, or one variable of each type. We shall illustrate cross-classification for each of these types of bivariate data sets by an example.

EXAMPLE 1

Two Qualitative Variables. In a recent year, 2504 oil and gas wells were drilled in the four districts of a region. Each well was one of three possible types: new-field wildcat well, other exploratory well, or development well. Table 2.4a shows the bivariate cross-classification of these well-drilling data, classified by type of well and district. Each of the 2504 wells is counted in the cell of the system of classification that corresponds to its type and district. The resulting cross-classified distribution is called a *bivariate qualitative distribution* because the variables, type and district, are both qualitative variables.

Observe in Table 2.4a that the marginal totals of the columns form the univariate qualitative distribution of the district variable. Similarly, the marginal totals of the rows form the univariate qualitative distribution of the type-of-well variable.

An examination of the cross-classified data in Table 2.4a shows, for example, that development wells were the most common type of well drilled in each district. Moreover, for each type of well, more drilling was done in district I than in all other districts combined.

TABLE 2.4 **Oil and gas wells drilled, cross-classified by type of well and district — An example of a bivariate qualitative distribution**

(a) Bivariate distribution

Type	I	II	III	IV	Total
New-field wildcat	338	60	113	10	521
Other exploratory	223	11	28	3	265
Development	1131	72	461	54	1718
Total	1692	143	602	67	2504

(b) Percent distributions for each district and all districts

Type	I	II	III	IV	Total
New-field wildcat	20.0	42.0	18.8	14.9	20.8
Other exploratory	13.2	7.7	4.7	4.5	10.6
Development	66.8	50.3	76.6	80.6	68.6
Total	100.	100.	100.	100.	100.
	(1692)	(143)	(602)	(67)	(2504)

NOTE: Percentages may not add to 100 because of rounding.

Differences in patterns can often be seen more readily by constructing percent distributions. For example, Table 2.4b presents the percent distributions of type of well for each district. It can readily be seen that district II differs from the other three districts, with relatively more new-field wildcat wells and relatively fewer development wells.   □

EXAMPLE 2

Two Quantitative Variables.   A European government tourist bureau compiled a list of 322 escorted tours of eight days or less, which are offered by various private agencies. There are two characteristics of interest for each tour in the data set: its duration in days and its cost in U.S. dollars. To summarize the data set with respect to these two characteristics, the bureau set up the classification system shown in Table 2.5. Each tour from the data set is counted in the cell of the table that corresponds to its duration and its cost. The cross-classified distribution in the table is called a *bivariate frequency distribution* because the two variables, duration and cost, are both quantitative variables. Note that the marginal totals of the columns and rows form the univariate frequency distributions of the duration and cost variables, respectively.

The bivariate frequency distribution in Table 2.5 has summarized the data set in a way that brings a number of interesting facts to light. One is that an escorted tour of three days, costing up to $300, is the most commonly offered type of tour. Another is that the cost of an escorted tour tends to vary directly with its duration—a very reasonable relationship. Note that this relationship between the two variables is here readily evident without requiring the construction of percent frequency distributions for tours of different durations.   □

EXAMPLE 3

One Qualitative and One Quantitative Variable.   In Figure 2.8a, a computer-generated bivariate distribution is shown for 214 patients suffering from Parkinson's disease, cross-classified by age at onset of the disease (ONSETAGE) and patient's sex (SEX). In this cross-classification, four values are given in each cell of the distribution. The topmost value represents the actual count or frequency for

TABLE 2.5   **Escorted tours cross-classified by cost and duration — An example of a bivariate frequency distribution**

Cost (U.S. $)	Duration (days)						Total
	3	4	5	6	7	8	
0– 300	53	43	9	–	–	–	105
301– 600	20	45	37	12	26	1	141
601– 900	4	9	11	6	17	12	59
901–1200	–	–	1	1	5	10	17
Total	77	97	58	19	48	23	322

Scatter Plots

A simple graphic display that is helpful in studying a relationship between two quantitative variables is a *scatter plot*. In a scatter plot, the observations for the two variables for each element in the data set are plotted on a two-dimensional graph. The pattern of the scatter of the points provides insights into the existence and nature of the relationship between the two variables.

EXAMPLE

Delivery Fleet. A company's accountant examined bivariate data on the annual cost of maintaining and operating the eight vehicles in the company's delivery fleet and the distance that each vehicle traveled during the year. The accountant plotted the bivariate data as shown in Figure 2.9. The horizontal and vertical coordinates of each point are the distance traveled (in thousand miles) and the annual cost (in thousand dollars), respectively, of the vehicle. The pattern of the points suggests that the annual cost is related to the distance traveled, the relation tending to be approximately linear.   □

2.4 DATA MODELS AND RESIDUAL ANALYSIS

We have seen that qualitative and frequency distributions are effective means of summarizing data so that the basic patterns in the data emerge. Often, we wish to compare these observed patterns with patterns that are anticipated on the basis of a theoretical model or from previous experience. Residual analysis is an effective technique for making such comparisons.

A *residual* is the difference between an *observed value* and the corresponding *anticipated value,* where the value may be a measurement, count, or rank. A graphical presentation of residuals, called a *residual plot,* is useful for highlighting major depar-

FIGURE 2.9   Scatter plot for the delivery fleet example.   The two variables are annual cost and distance traveled.

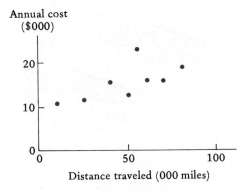

tures between the observed and anticipated patterns in a data set. We shall illustrate residual analysis and residual plots by three examples.

1. In the concert subscribers example, we mentioned earlier that the analyst noted that respondents frequently reported their ages with second digits of 0 or 5. Figure 2.10a repeats the stem-and-leaf display given initially in Figure 2.2a. Figure 2.10b shows a residual plot for the analyst's anticipation that the final digits in age would occur with equal frequency. Since the data set contains 70 readings and 10 digits are involved, each digit should appear about 7 times in the leaves under the expectation of equal representation. The digit 0 appears 15 times, so the residual for this digit is $15 - 7 = 8$. This residual is represented by the left-most dot in the residual plot. The digit 1 appears 5 times; hence, its residual is $5 - 7 = -2$. This residual is represented by the second dot from the left. The residuals for the other digits are calculated and plotted in the same manner. Note how readily the residual plot shows the excessive numbers of 0 and 5 final digits and the correspondingly lower-than-anticipated frequencies for the other digits.

    The analyst concluded that the most likely explanation for this behavior of the residuals is end-digit preferences by some respondents who tend to round their reported ages to multiples of 5 or 10. The analyst therefore decided to ask for year of birth on the questionnaire to be used in the large-scale study and then calculate a subscriber's age from the year of birth while editing the questionnaire.

2. **Motor Rod.**   (Adapted from Ref. 2.1, pp. 280–281.) A company that manufactures small motors learned that some of the motors produced recently were breaking down. One possible cause was a rod in the motor being too loose in the

---

FIGURE 2.10   **Stem-and-leaf display and residual plot for the concert subscribers example**

(a) Stem–and–Leaf Display   (b) Residual Plot

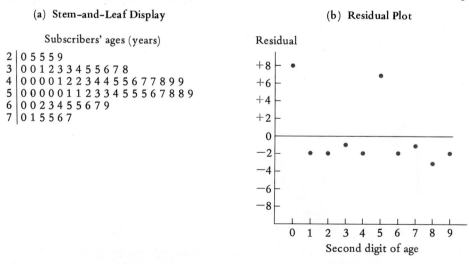

bearing within which it rotates. Other possible causes included defects in the bearing. A consultant who was asked to locate the cause of the breakdowns learned that the diameter of each rod is inspected by the manufacturing department and that any rod with a diameter less than the lower tolerance limit of 1.000 centimeter is to be scrapped. A histogram of the diameters of rods recently inspected, based on records kept by the inspectors, is shown in Figure 2.11a. Also shown there is a smooth curve that the consultant drew as the expected model for the data. This model reflects the pattern of variability found by the consultant in similar situations in other companies.

Note that the histogram and the model are in close accord except in the interval around the lower tolerance limit, which is marked by an arrow in the figure. The residuals for each class were obtained and are graphed in Figure 2.11b. The consultant was interested in the size of the large negative residual just below the tolerance limit and the large positive residual in the next-higher class. Some systematic factor appeared to be operating here that tended to create large residuals in opposite directions in the two adjacent classes. Upon further investigation, the consultant found that the inspectors were under pressure to keep the scrappage rate low and had been misclassifying borderline-defective rods by recording their diameters as borderline-acceptable. This misclassification led to too few rods recorded just below 1.000 centimeter and too many rods recorded just above the tolerance limit. Further investigation showed that the misclassified rods were in most cases the cause of the motors breaking down.

3. In the delivery fleet example, the accountant believed that the annual cost for a vehicle is the sum of a fixed-cost component and a variable-cost component that

FIGURE 2.11   Histogram and residual plot for the motor rod example

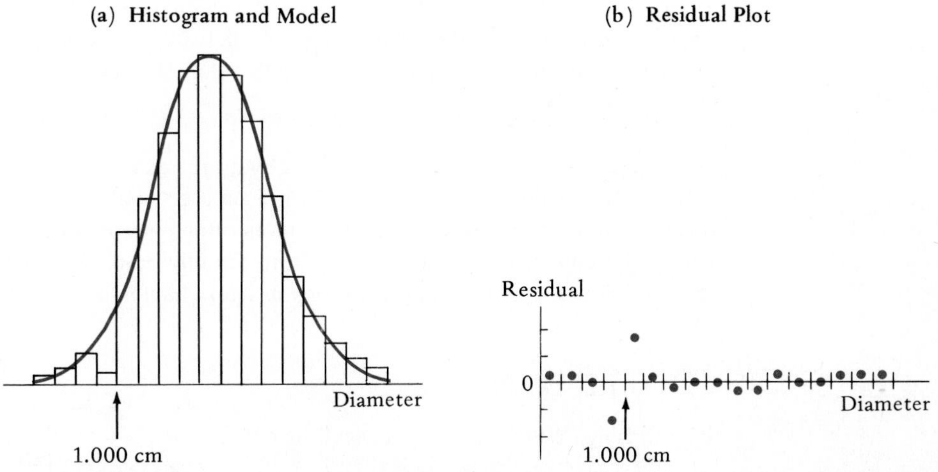

(a) Histogram and Model   (b) Residual Plot

FIGURE 2.12   **Fitting a model to bivariate data, and the associated residual plot for the delivery fleet example**

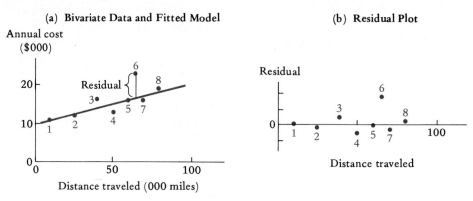

varies linearly with the distance traveled. Hence, the accountant anticipated that the data points would follow the straight-line model fitted in Figure 2.12a. This figure repeats the scatter plot in Figure 2.9, with each vehicle identified by number $(1, 2, \ldots, 8)$. A residual was calculated for each vehicle by computing the difference between the actual annual cost for the vehicle and its anticipated annual cost according to the straight-line model. The residual for vehicle 6 is marked in Figure 2.12a. The residuals for the eight vehicles are plotted in Figure 2.12b against the distance traveled. The accountant could see clearly that, compared with the other vehicles, vehicle 6 had a cost that was out of line in relation to its distance traveled.

In an attempt to understand this finding, the accountant ascertained the age of each vehicle and found that the age of vehicle 6 is a key factor in explaining why that vehicle's actual cost deviates so much from its anticipated cost based on the model considering only the distance traveled. The company subsequently replaced vehicle 6 by a new one.   □

These examples illustrate how effectively residual plots can point to major differences between the observed patterns in a data set and anticipated patterns based on theory or related experience. Residual plots often also provide useful clues for investigating the causes of the departures from anticipations. We shall discuss the use of residual plots for some other situations in later chapters.

CITED REFERENCE

2.1   Deming, W. Edwards. "Making Things Right." Pages 279–286 in *Statistics: A Guide to the Unknown,* 2nd ed. Edited by J. W. Tanur. San Francisco: Holden-Day, 1978.

PROBLEMS

2.1 Rider Complaints. A bus company in a northeastern city has categorized complaints from riders for last May and June as follows:

	Number of Complaints	
Type of Complaint	May	June
Buses too hot	14	31
Fare too high	11	12
Buses sometimes don't stop at bus stops	17	16
Bus stops have no shelters	10	21
Bus travel too expensive	12	13
Buses should be air-conditioned	14	41
Drivers are rude	13	11
Bus schedules make transfers difficult	14	14
Route maps not available	25	21
Total	130	180

a. For each month, classify the complaints data using the complaint categories comfort, service, cost.
b. Would other persons necessarily assign the complaints to the same complaint classes in **a** as you have? If not, what is the implication for interpreting qualitative distributions for free-answer data classified by others?
c. Construct a bar chart patterned after Figure 2.1b to compare the distributions for the two months. Briefly report your findings.

2.2 Refer to **Rider Complaints** Problem 2.1.
a. For each month, classify the complaints data using the complaint categories bus schedules, bus drivers, equipment, other.
b. Construct a bar chart patterned after Figure 2.1b to compare the distributions for the two months. Briefly report your findings.

*2.3 Sixty individuals with eye troubles were examined by an ophthalmologist. The examination outcomes are listed below, coded as astigmatism (A), cataract (C), glaucoma (G), injury (I), strabismus (S), and other (O):

A I G A A A A I A S A C A C C A A A A A A O I A A A A A O A C
O A A C A A I A A A S I A I A O A A A A A A O I A A S G A A A

a. Construct a qualitative distribution of individuals by eye examination outcome. Also give the corresponding percent distribution.
b. Construct a bar chart of the percent distribution in **a**.
c. What was the most common examination outcome for this group of individuals? What percentage of nonastigmatic outcomes was accounted for by eye injuries?

2.4 Refer to the **Financial Characteristics** data set (Appendix D.1). The qualitative distribution of firms classified by industry follows.

Industry		Number
Code	Name	of Firms
1	Crude oil producers	60
2	Textile products	52
3	Textile apparel manufacturers	75
4	Paper	24
5	Electronic computer equipment	37
6	Electronics	21
7	Electronic components	57
8	Auto parts and accessories	57
	Total	383

a. In which industry class is firm 261 counted?

b. Construct the percentage form of this qualitative distribution, and display it as a bar chart.

c. Which industry has the greatest representation among firms in the data set? If the three electronics industry classes were combined to form a single class, would your answer change? If so, what is the implication for the construction and interpretation of qualitative distributions?

2.5   A questionnaire asks how many automobiles a family owns, giving the following answer choices: none, 1 or fewer, 2 or fewer, more than 2.

a. Do these four choices constitute an exhaustive set? A mutually exclusive set?

b. Do the answer choices constitute a system of classification as defined in (2.2)? Explain.

2.6   An apple sorter is instructed to put sound red apples in one bin, sound green apples in another bin, and all others in a third bin. Does this sorting rule satisfy the formal requirements of a system of classification? Explain.

2.7   Refer to Figure 1.2.

a. Do the response classes for Question 7b satisfy the formal requirements of a system of classification as defined in (2.2)? Explain.

b. Do the choices in Question 7d constitute a system of classification as defined in (2.2)? Explain.

*2.8   Twenty-five merchants donated the following dollar amounts in a fund-raising campaign to beautify a nearby park:

70   90   25   50   30   100   75   50   125   100   150   15   80

100   50   25   50   65   60   100   50   100   75   90   80

a. Array the observations by increasing order of magnitude.

b. Present the observations in a stem-and-leaf display.

c. (1) What are the smallest and largest donations? (2) What percentage of the donations are for $100 or more? (3) What is the most frequently contributed dollar amount?

2.9   A communications engineer is studying error bursts in the transmission of digital data. The following are the number of erroneous data bits transmitted in each hour during the past 24 hours over one transmission link:

5   24   36   2   1   3   19   55   1   2   4   20

4   1   24   30   0   1   8   2   27   0   2   22

a. Array the observations by increasing order of magnitude.

b. Present the observations in a stem-and-leaf display.

c. Is there any evidence in the display in **b** that the numbers of erroneous bits tend to cluster in particular value ranges? Comment.

2.10 **Urban Wages.** Weekly wages (in dollars) of 60 wage earners in a plant during the week of January 5 were as follows:

409	455	386	395	410	391	427	375	375	407
401	382	425	398	402	421	379	446	409	435
392	383	410	389	427	403	401	387	431	386
404	410	398	421	400	397	410	372	431	437
369	382	408	405	399	405	378	418	453	409
425	389	400	450	376	365	415	445	415	385

a. Make a stem-and-leaf display for these data.

b. Is there any evidence in the display in **a** that the weekly wages tended to cluster at particular values?

c. What are the minimum and maximum wages?

2.11 **Hospital Costs.** The average daily costs (in dollars) of hospital care last year in 50 hospitals were as follows:

257	315	327	392	318	305	342	308	384	309
274	313	267	312	272	254	257	245	276	286
319	368	318	265	235	271	252	252	341	268
282	306	326	249	241	287	282	318	289	335
253	230	255	276	309	258	267	331	249	278

a. Make a stem-and-leaf display for these data. Use as stems 23, 24, and so on. Some of the stems contain no leaves. Did you include such stems in your display or did you omit them? Justify your procedure.

b. Do any cost observations appear to be unusually small or large?

2.12 For each of the following systems of classification, indicate whether or not it meets the formal requirements (2.2).

a. Classes $0–$5.00, $5.00–$10.00, $10.00 or more, for hourly wage rates.

b. Classes 3 or less, 4 or more, for number of children in family.

c. Classes 0°–under 90°, 100°–under 190°, 200° and over, for the temperature of ovens.

2.13 Refer to the response classes for Question 7b in Figure 1.2.

a. Do the classes have equal widths?

b. Is there an open-end class?

c. What is the midpoint of the second class? Of the third class? Explain any assumptions you made in your answers.

2.14 A frequency distribution of monthly rents in middle-income housing is to be constructed. The rents tend to be in multiples of $50 and range from $350 to $750. Class intervals of $50 are to be employed. What class limits would be most appropriate? Why?

*2.15 Refer to **Urban Wages** Problem 2.10.

a. Construct a frequency distribution for these data, using equal class intervals 360–under 380, and so on.

b. Plot the frequency distribution as a frequency polygon and as a histogram, in separate graphs.

2.16 Refer to **Hospital Costs** Problem 2.11.

    a. Construct a percent frequency distribution for these data, using equal class intervals 225–under 250, and so on.

    b. Plot the percent frequency distribution as a frequency polygon and as a histogram, in separate graphs.

2.17 Universal Health Care. Eighteen persons are selected from each of two socioeconomic groups. Each person is asked to rate his or her approval of universal health care on an eight-point scale ranging from 1 (strong disapproval) to 8 (strong approval). The ratings follow.

Group 1:	5	2	3	8	5	3	4	3	5	3	3	6	3	4	3	5	3	5
Group 2:	7	6	7	3	6	6	7	7	6	7	7	4	5	8	6	4	6	6

    a. Construct a frequency distribution of ratings for each group, using classes $1, 2, \ldots, 8$.

    b. What differences are evident in the rating distributions of the two groups?

    c. Plot the frequency distribution for group 1 as a histogram. Would it be appropriate to plot the distribution as a frequency polygon? Explain.

*2.18 Response Times. The distribution of response times for alarms answered last year by the volunteer fire company in a township follows.

Response Time (minutes)	Number of Alarms
2.5–under 7.5	5
7.5–under 12.5	18
12.5–under 17.5	15
17.5–under 22.5	9
22.5–under 27.5	3
Total	50

    a. (1) What is the width of the first class? (2) What is the midpoint of the first class? (3) Are any of the classes open-ended?

    b. Convert the frequency distribution into a percent distribution and plot it as a frequency polygon and as a histogram, in separate graphs.

    c. Which of the two graphs would be more effective for comparing the distribution of response times with that for a fire company in a nearby township? Why?

2.19 Horse Trials. The age distribution of horses competing in the Winston County Trials during a recent year follows.

Age (years at last birthday)	Number of Horses	Age (years at last birthday)	Number of Horses
5	26	10	11
6	60	11	8
7	51	12	3
8	44	13	0
9	23	14	1
		Total	227

    a. (1) What is the width of age class 5? (2) What is the midpoint of age class 5?

b. Convert this frequency distribution to a percent distribution, and graph it as a frequency polygon and as a histogram, in separate graphs.

c. Does it appear from the graphs that horses under 5 years of age were not permitted to compete? Horses over 14 years of age? Explain.

d. Suppose the frequencies for the age classes 5 and 6 had been reversed to give 60 horses aged 5 and 26 horses aged 6. Would this new pattern indicate that an unusual factor is affecting the age data? Explain.

2.20 **Product Sales.** An industrial chemical company sells 102 different products. The distribution of the annual sales amounts last year for these products follows.

Annual Sales ($000)	Number of Products
0–under 10	19
10–under 50	44
50–under 100	17
100–under 200	14
200–under 500	8
Total	102

a. An executive of the company, upon studying this distribution, states that products with annual sales between $10 and $50 thousand are the most common. Do you agree or disagree? Explain.

b. Construct a frequency polygon for the distribution, using $100 thousand as the unit width. Does your graph support your answer in **a**? Explain.

2.21 The age distribution (to nearest birthday) of employees who were absent more than 10 days last year follows.

Age Class	Percent of Employees	Age Class	Percent of Employees
15–19	7	40–49	7
20–24	18	50–64	4
25–29	30	65–79	1
30–39	33	*Total*	100

a. Would you agree that employees in the 30–39 age class had the worst absentee record of all the employees, and that the employees in the 25–29 age class were a close second? Explain.

b. Construct a histogram of the age distribution, using five years as the unit width. Does your graph support your answer in **a**? Explain.

*2.22 Refer to **Response Times** Problem 2.18.

a. Construct a less-than cumulative percent distribution of the response times. Use classes less than 2.5, less than 7.5, and so on.

b. Plot the ogive of the cumulative percent distribution in **a**. Reading from the ogive, what percentage of calls had a response time of less than 20 minutes?

2.23 Refer to **Product Sales** Problem 2.20.

a. Construct a less-than cumulative frequency distribution of the annual sales amounts. Use classes less than 0, less than 10, and so on.

b. Plot the ogive of the cumulative frequency distribution in **a**. Reading from the ogive, what number of products had annual sales of less than $80 thousand last year?

c. The plotted points of the ogive in **b** are connected by straight lines. What does this use of straight lines imply about the distribution of annual sales amounts within each class of the frequency distribution?

2.24 The distribution of a company's stock according to the number of shares held (including fractional shares) follows.

Number of Shares Held	Percent of Stockholders
0–under 25	41
25–under 50	25
50–under 100	19
100–under 500	12
500 and over	3
Total	100

a. Construct a less-than cumulative percent distribution. Use classes less than 0, less than 25, and so on.

b. Plot the ogive of the cumulative distribution in **a**. (1) What percentage of the stockholders held fewer than 75 shares each? (2) The 50 percent of the stockholders with the most shares each held what number of shares or more?

2.25 Refer to **Horse Trials** Problem 2.19.

a. Construct a more-than cumulative percent distribution. Use classes 5 or more, 6 or more, and so on.

b. Plot the ogive of the more-than cumulative percent distribution in **a**. Reading from the ogive, the oldest 40 percent of horses in the trials were what age or older?

2.26 Refer to **Universal Health Care** Problem 2.17.

a. Construct a less-than cumulative frequency distribution for the ratings of group 1. Use classes 1 or less, 2 or less, and so on.

b. Plot the step function ogive for your cumulative distribution in **a**, as in Figure 2.7b.

*2.27 Refer to Table 2.4a.

a. What is an element of this data set? How many elements are there?

b. Identify the variables in this data set. Present the univariate qualitative distribution for each variable.

c. If you had *only* the univariate qualitative distributions in **b**, could you construct Table 2.4a? Explain.

2.28 Refer to Table 2.5.

a. What is an element of this data set? What are the variables in the data set?

b. (1) How many tours cost $300 or less? (2) How many tours have a duration of three days? (3) How many tours cost $300 or less *and* have a duration of three days?

c. What is the second most commonly offered type of tour?

2.29 Refer to Figure 2.8a.

a. (1) What is the most common age interval for onset of the disease for male patients? (2) What percentage of patients who experienced onset of the disease at 70 years or more are female? (3) What percentage of all patients experienced onset of the disease before 60 years of age?

b. Present the univariate frequency distribution of all patients by age at onset.

*2.30 **Flight Simulation.**   The coded observations that follow pertain to errors made in handling "emergencies" in training sessions on a flight simulator. The observations are cross-classified by phase of flight—takeoff (T), cruising (C), or landing (L)—and by cause of error—misinterpretation of instrument readings (M), or other cause (O). For instance, the coded observation TM involves a takeoff error caused by misinterpretation.

TM	TO	LM	LO	CO	TM	CM	LO	TM	CO
LO	CM	LO	TM	LO	CM	TM	CO	LO	LM
TM	TO	TO	LM	TM	LO	LO	TM	TM	LO
LO	CO	LO	LM	TM	LM	TM	LM	TM	CM
TM	TM	TO	LO	TO					

  a. Construct the bivariate qualitative distribution for these observations.

  b. (1) What percentage of the M-type errors occurred in the landing phase? (2) What percentage of the landing phase errors were of the O type?

  c. Construct percent distributions of causes of error for each phase of flight. Describe the relationship between cause of error and phase of flight for these observations.

2.31 **Failure Times.**   An aluminum reduction cell has two modes of failure, metal contamination (C) or structural distortion (D). The failure times (in days) for 25 cells, arrayed by magnitude within each mode of failure, follow.

Mode C:	909	1293	1601	1616	2012	2016	2180	2201	2442
Mode D:	824	1082	1135	1308	1359	1372	1401	1412	
	1601	1638	1641	1674	1709	1805	1947	2208	

  a. Construct the bivariate distribution of modes of failure (C, D) and failure times (using equal class intervals 500–under 1000, and so on) for these data.

  b. Construct percent distributions of failure times for each mode of failure. Describe the relationship between mode of failure and failure time for these data.

2.32 The following table shows a bivariate frequency distribution of 125 unemployed construction workers, by age and duration of the current spell of unemployment.

Duration (days)	Age (years)	
	Under 35	35 or more
1– 7	36	21
8–30	30	5
More than 30	14	19

  a. What percentage of workers who have been unemployed 30 days or less are under 35 years of age?

  b. Obtain the univariate frequency distribution of the durations of unemployment for all workers.

  c. Construct percent distributions of unemployment durations for each age class. Describe the relationship between age and duration of unemployment for these construction workers.

2.33 Refer to **Flight Simulation** Problem 2.30. Construct a table of frequencies and percentages for these bivariate data patterned after Figure 2.8a, where the rows and columns of the table correspond to phase of flight and cause of error, respectively.

2.34 Refer to **Municipal Debt** Problem 1.2.

 a. Construct a scatter plot of per capita debt and population for the six municipalities. Plot population on the horizontal axis.

 b. Describe the relationship between the two variables indicated by the scatter plot in **a** for the six municipalities.

2.35 Refer to Figure 1.1.

 a. Construct a scatter plot of systolic blood pressure and age for the six officers. Plot age on the horizontal axis.

 b. Describe the relationship between the two variables indicated by the scatter plot in **a** for the six officers.

2.36 **Grain Production.** Data on sales revenue (in $000) and volume (in metric tons) for wheat and barley grown on a farm during the last five crop years follow.

Year:	1	2	3	4	5
Wheat Revenue:	12.7	20.7	11.8	9.8	8.9
Wheat Volume:	64	119	76	63	50
Barley Revenue:	13.6	5.7	20.8	15.6	16.3
Barley Volume:	103	51	138	109	136

 a. Display the bivariate data for the two grains on the same scatter plot. Use a different plotting symbol for each grain. Plot volume on the horizontal axis.

 b. Compare the relationships between revenue and volume for the two grains. Are they similar or do they differ? Explain.

*2.37 Eight packers in a mail-order department fill orders for shipment to customers. Packers 1–4 work full-time during the day, and packers 5–8 work full-time at night. Orders are assigned to packers at random, and all packers fill orders at about the same speed. Of the orders filled by the department during a recent four-month period, 60 proved to be incorrectly filled. The packers' identification numbers for these orders are as follows (number 1 refers to packer 1, and so on):

```
2   7   4   1   7   1   3   4   4   7
1   5   7   3   8   7   6   3   7   2
3   8   3   8   3   7   3   5   3   4
7   3   1   7   3   8   7   8   5   5
8   6   7   2   3   2   6   3   1   7
5   3   6   2   7   1   7   3   8   4
```

 a. What are the anticipated numbers of incorrectly filled orders for each packer if all packers are equally accurate and time of the shift does not affect the accuracy of filling orders? Explain your reasoning.

 b. Construct a residual plot to assess the comparative accuracy of the packers during this four-month period. What does your plot show?

2.38 Technicians in a health survey were instructed to take and record blood pressure readings for each subject to the nearest millimeter. The distribution of end digits in 500 blood pressure readings expressed to the nearest millimeter follows.

End Digit:	0	1	2	3	4	5	6	7	8	9
Number of Readings:	60	0	80	0	112	40	88	0	120	0

 a. What are the expected frequencies for the end digits if all were equally likely?

   b. Construct a residual plot to assess whether each end digit actually was equally likely. What does your plot show?

2.39 A new product was expected to show a 10 percent sales increase each quarter for the first three years, starting with a sales volume of 40 thousand units. Actual and expected sales during this period were as follows (in thousands):

Quarter:	1	2	3	4	5	6	7	8	9	10	11	12
Actual:	37.9	46.2	51.1	50.4	55.7	67.0	74.4	74.8	84.0	96.5	105.7	111.4
Expected:	40.0	44.0	48.4	53.2	58.6	64.4	70.9	77.9	85.7	94.3	103.7	114.1

   Obtain the residuals and plot them against time. What does your plot show?

2.40 The observed percent frequency distribution of the numbers of defects detected in 90 separate pieces of optical fiber and the expected percent frequency distribution if the process control standards are being adhered to are as follows:

	Percentage of Pieces	
Defects per Piece	Observed	Expected
0	78	90
1	6	9
2	1	1
3	4	0
4 or more	11	0
Total	100	100
	(90)	

   Obtain the residuals for the percent frequencies. Does it appear that the process control standards are being adhered to?

EXERCISES

2.41 Refer to **Urban Wages** Problem 2.10. Suppose one were to construct a stem-and-leaf display of weekly wages during the week of January 5 of the wage earners in 3 plants, including the one in Problem 2.10. How would you expect this display to compare with the one constructed in Problem 2.10? What generalization does this comparison suggest?

2.42 In a large training experiment for a new welding technology, trainees were classified by previous welding experience — no previous experience (A_1) or previous experience (A_2) — and half of each group was randomly assigned to a concentrated program (B_1) while the other half was assigned to an extended program (B_2). The following four tables show different possible outcome patterns for the percentages of trainees in each experience-training group who later used the new technology successfully:

Pattern 1	B_1	B_2		Pattern 2	B_1	B_2		Pattern 3	B_1	B_2		Pattern 4	B_1	B_2
A_1	60%	60%		A_1	60%	90%		A_1	60%	60%		A_1	60%	80%
A_2	60%	60%		A_2	60%	90%		A_2	90%	90%		A_2	70%	90%

Interpret each of the four possible outcome patterns as to the effects of previous experience and type of training on the likelihood of successful use of the new technology.

STUDIES

2.43 Refer to **Hospital Costs** Problem 2.11. An analyst wishes to investigate whether the frequency distribution of costs has one or two peaks and whether any of the hospitals have relatively extreme costs. To determine the most effective number of classes, construct frequency distributions with 3, 5, 7, 9, and 11 classes, each using equal class intervals. Plot the histograms for these frequency distributions on different graphs. Which frequency distribution appears to you to be most effective here? Explain.

2.44 The frequency distributions of numbers of days of bed care for recent burn cases in two treatment centers follow.

Number of Days	Center A	Center B
Under 5	174	52
5–under 10	225	120
10–under 15	372	210
15–under 25	159	36
25–under 50	180	80
Total	1110	498

a. Plot frequency polygons of these distributions on the same graph for ready comparison. Use 0 as the lower limit of the first class, and employ a unit width of five days.
b. Compare the frequency polygons plotted in **a** and state your findings.
c. Plot less-than ogives of the two distributions on the same graph for ready comparison. Explain how your findings in **b** are indicated by the ogives.
d. Reading from your ogives, (1) What percentage of the cases entailed a bed stay of less than eight days in center A? In center B? (2) The longest 25 percent of the bed stays in center A entailed a stay of how many days or more? What is the comparable figure for center B?

2.45 Refer to the **Financial Characteristics** data set.
a. Construct a bivariate frequency distribution of net sales in years 1 and 2 for textile apparel manufacturers. Consider such points as (1) number of classes to be used for each variable, (2) use of equal or unequal class intervals, and (3) possible use of open-end classes.
b. Construct a scatter plot of net sales in the two years for the textile apparel manufacturers. Describe the relationship between net sales in years 1 and 2 for these manufacturers.
c. Which provides more direct information about the nature of the relationship between sales in the two years, the bivariate frequency distribution in **a** or the scatter plot in **b**? Explain.

2.46 Refer to the **Power Cells** data set.
a. Construct a bivariate frequency distribution of the numbers of cycles before failure (CYCLES) and the ambient temperatures (TEMP). Make an appropriate choice of classes for the CYCLES variable.
b. Construct a scatter plot of the number of cycles before failure and the ambient temperature. Describe the relationship between the two variables.

2.47 A city police official has released the following data on the outcomes of charges for serious crimes by type of crime, for last year and five years ago. The data are annual data and exclude pending cases.

	Five Years Ago		Last Year	
Outcome of Charge	Violent Crimes	Property Crimes	Violent Crimes	Property Crimes
Adults — guilty	2499	10,402	3152	15,704
Adults — acquitted or dismissed	1427	3,754	4461	10,543
Referred to juvenile court	1159	14,970	1925	13,461
Total	5085	29,126	9538	39,708

Construct appropriate tables and graphs to answer the following questions, and discuss your findings.

a. For adults only, has the proportion of charges that led to acquittal or were dismissed changed between five years ago and last year, for each type of crime?

b. Combining both types of crimes, what changes have taken place in the relative occurrence of each outcome of charge between five years ago and last year?

c. Are your findings in **b** largely the result of changes in violent crimes, in property crimes, or in both types of crimes?

3 DATA ANALYSIS II: SUMMARY MEASURES

In the preceding chapter, we have examined a number of methods for studying data for purposes of analyzing, interpreting, and reporting their distribution pattern. In this chapter, measures that summarize important characteristics of univariate data sets are presented. These measures fall into three broad categories: measures of position, variability, and skewness. Summary measures are useful for exploratory analysis of a data set as well as for the reporting of final results of a study.

3.1 MEASURES OF POSITION

Measures of position for a data set describe where the values are concentrated. They can be particularly useful for comparing several data sets.

EXAMPLE

Oil Spills. In 1979, a government agency was established to monitor and control pollution of an inland waterway. Since then, the agency has recorded, among other things, occurrences of oil spills in the waterway. An agency analyst now wishes to compare the time intervals between oil spills after the agency was established with comparable data available for a limited period prior to the agency's establishment. The intention is to determine whether or not the agency's controls have lengthened the period between oil spills, which, of course, would imply that the frequency of oil spills has been reduced.

The data sets for the two periods are given in Table 3.1a; the values shown represent time intervals in days between oil spills. The data sets are classified into percent frequency distributions in Table 3.1b to facilitate comparison.

A comparison of the two frequency distributions indicates that the intervals between oil spills have tended to be longer since the creation of the agency. This conclusion is evident because (1) the proportion of cases where the time interval is under 15 days has decreased from 37.5 percent to 30.3 percent and (2) the proportion of cases where the time interval is long has increased. A simple summary measure of this lengthening of the intervals between oil spills is desired. ☐

TABLE 3.1   Time intervals between oil spills in an inland waterway

(a) Number of days between oil spills

Before Pollution Controls			After Pollution Controls							
17	16	4	37	38	87	6	20	11	22	23
3	1	7	3	11	4	39	17	23	37	21
5	37	15	21	10	12	20	15	2	20	15
31	2	22	3	18	45	36	9	15	51	24
20	36	6	31	23	49	33	5	36	161	12
64	16	45	1	2	4	33	31	31	5	163
31	11	1	13	22	23	1	5	35	32	2
2	24	38	0	6	18	32	16	25	131	31
14	41	18	1	27	25	33	0	27	34	20
48	19	1	17	19	29	31	21	17	74	1
21	73	20	19	32	32	17	38	1	21	135
28	19	5	12	27	56	24	34	120	12	18
3	30	8	25	15	39	7	22	64	27	46
33			12	21	7	36	2			

(b) Percent frequency distributions

Time Interval (days)	Percent of Cases	
	Before Controls	After Controls
0–under 15	37.5	30.3
15–under 30	32.5	35.8
30–under 45	20.0	22.0
45–under 60	5.0	4.6
60–under 75	5.0	1.8
75–under 90		0.9
90–under 105		
105–under 120		
120–under 135		1.8
135–under 150		0.9
150–under 165		1.8
Total	100.0	100
	(40)	(109)

NOTE: Percentages may not add to 100 because of rounding.

A summary measure of the position of a frequency distribution is a number that provides information about where the distribution is located. A comparison of such a measure for two frequency distributions can indicate in direct fashion a shift in the frequency distribution, such as the one noted in the oil spills example.

We now discuss several useful measures of position.

Mean

One measure of position is the arithmetic average, or mean.

(3.1)

The *mean* $\overline{X}$ (read "X bar") of a set of values $X_1, X_2, \ldots, X_n$ is the sum of the values divided by the number of items, that is:

$$\overline{X} = \frac{\sum_{i=1}^{n} X_i}{n}$$

(Note that Appendix A, Section A.1, contains a review of the summation symbol Σ.)

EXAMPLE

For the data set before pollution controls in Table 3.1a, there are 40 time intervals, whose sum is 835. Therefore, the mean time interval between oil spills in the period before the agency was created is $\overline{X} = 835/40 = 20.9$ days. By a similar calculation, it is found that the mean of the 109 intervals after pollution controls is 27.6 days. Thus, the two means summarize the lengthening of the time intervals between oil spills after pollution controls quickly and easily, much more so than a comparison of the two frequency distributions. ☐

Properties of Mean. The mean has a number of unique properties that should be considered when deciding on an appropriate measure of position.

1. The sum of the values of a set of items is equal to the mean multiplied by the number of items.

(3.2)

$$\sum_{i=1}^{n} X_i = n\overline{X}$$

This result is obtained by multiplying both sides of (3.1) by n.

EXAMPLE

A restaurant owner is informed that the mean expenditure per customer was \$9.800 for the past week when 650 customers visited the restaurant. Total receipts for the week therefore were $650(9.800) = \$6370$. ☐

2. The sum of the deviations of the X_i values from their mean $\overline{X}$ is zero.

(3.3)

$$\sum_{i=1}^{n} (X_i - \overline{X}) = 0$$

EXAMPLE

Consider the set of three values 2, 6, and 13. Their mean is $\overline{X} = 7$, and their deviations sum to zero:

$$(2 - 7) + (6 - 7) + (13 - 7) = -5 - 1 + 6 = 0$$

Figure 3.1 illustrates these deviations $X_i - \overline{X}$ and shows that the X_i values deviate from the mean $\overline{X}$ in a balanced fashion. Thus, $\overline{X}$ is in this sense centered among the X_i values. $\square$

To derive (3.3), we utilize (3.2) as follows:

$$\sum_{i=1}^{n} (X_i - \overline{X}) = \sum_{i=1}^{n} X_i - \sum_{i=1}^{n} \overline{X} = \sum_{i=1}^{n} X_i - n\overline{X} = 0$$

3. In the next section, we shall be concerned with the variability of values in a data set. One measure of variability we shall consider depends on an expression of the form:

$$\sum_{i=1}^{n} (X_i - A)^2 \quad \text{where } A \text{ is a fixed value}$$

Note that this expression represents the sum of the squared deviations of the X_i values from the value A. An important property of the mean is that the sum of these squared deviations is a minimum when $A = \overline{X}$.

(3.4)

$$\sum_{i=1}^{n} (X_i - A)^2 \text{ is a minimum when } A = \overline{X}.$$

EXAMPLE

Consider again the three values 2, 6, and 13. For $A = \overline{X} = 7$, we have:

$$\sum (X_i - 7)^2 = (2 - 7)^2 + (6 - 7)^2 + (13 - 7)^2 = 62$$

FIGURE 3.1   **Illustration of the balancing of the deviations around the mean**

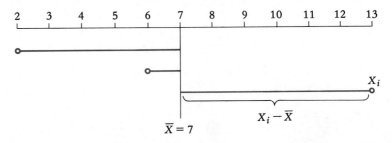

while for any other value of A this sum is larger. For $A = 0$, for instance:

$$\sum(X_i - 0)^2 = (2 - 0)^2 + (6 - 0)^2 + (13 - 0)^2 = 209$$ □

4. When a set of items is divided into two or more subsets, the overall mean is a weighted average of the means of the subsets.

(3.5)
$$\overline{X} = \frac{f_1\overline{X}_1 + f_2\overline{X}_2 + \cdots + f_k\overline{X}_k}{n} = \frac{\displaystyle\sum_{i=1}^{k} f_i\overline{X}_i}{n}$$

where:

k is the number of subsets

$\overline{X}_i$ is the mean of the ith subset

f_i is the number of items in the ith subset

$n = \displaystyle\sum_{i=1}^{k} f_i$ is the total number of items

EXAMPLE

A trade union has 1500 male members whose mean age is 41 years and 300 female members whose mean age is 25 years. The mean age of the entire trade union membership therefore is:

$$\overline{X} = \frac{1500(41) + 300(25)}{1500 + 300} = 38.3 \text{ years}$$ □

Calculation of Mean from Frequency Distribution.   When data are presented only in terms of a frequency distribution, an approximation of the mean $\overline{X}$ can be obtained as follows.

(3.6)
$$\overline{X} \simeq \frac{\displaystyle\sum_{i=1}^{k} f_i M_i}{n}$$

where:

k is the number of classes in the frequency distribution

M_i is the midpoint of the ith class

f_i is the frequency of the ith class

$n = \displaystyle\sum_{i=1}^{k} f_i$

(Note that the symbol $\simeq$ is used throughout the text to denote approximate equality.)

EXAMPLE

In Table 3.2, the mean of the frequency distribution of time intervals between oil spills before pollution controls is calculated. This distribution was presented in percentage form in Table 3.1b. Note that the calculated mean $\overline{X} = 23.6$ days is simply a weighted average of the midpoints M_i with the class frequencies f_i used as weights. ☐

TABLE 3.2 Calculation of the mean from the frequency distribution for the oil spills example

Time Interval (days)	Number of Cases f_i	Class Midpoint M_i	$f_i M_i$
0–under 15	15	7.5	112.5
15–under 30	13	22.5	292.5
30–under 45	8	37.5	300.0
45–under 60	2	52.5	105.0
60–under 75	2	67.5	135.0
Total	40		945.0

$$\overline{X} = \frac{945.0}{40} = 23.625$$

Comments

1. Formula (3.6) for computing $\overline{X}$ from a frequency distribution assumes that the midpoint of each class is approximately equal to the mean value of the items included in that class. This assumption is widely used and is often, but not always, appropriate. In Table 3.2, for instance, the approximation of $\overline{X}$ computed from the frequency distribution (23.6 days) differs substantially from the exact mean of the 40 time intervals, namely, 20.9 days.

   In general, the more detailed the frequency distribution (i.e., the smaller the class widths), the better is the approximation of $\overline{X}$ obtained from (3.6). If the midpoints of the classes in the frequency distribution coincide exactly with the means of the items in the respective classes, then (3.6) becomes identical to (3.5) and the weighted average of the midpoints gives the exact mean of the data set.

   Whenever there is a choice, the mean should be calculated from the original data set rather than from a frequency distribution to avoid approximation errors. Ready access to computers makes this calculation routine, even for large data sets.

2. Unequal class widths in a frequency distribution do not affect the computation of the mean by formula (3.6). On the other hand, if the distribution has an open-end class, it is not possible to calculate the mean by (3.6) without additional information because an open-end class has no defined midpoint.

3. The mean of a percent frequency distribution can be computed by rewriting formula (3.6) as follows.

(3.6a)

$$\overline{X} \simeq \sum_{i=1}^{k} \left(\frac{f_i}{n}\right) M_i$$

Here, f_i/n is the percent frequency of the ith class expressed in decimal form.

Trimmed Mean.   A major disadvantage of the arithmetic mean is that it is strongly influenced by extreme values in a data set. Consider the following array of 12 values:

1   1   2   4   5   7   8   9   12   12   13   130

for which the arithmetic mean is $\overline{X} = 17$. Note that 11 of the 12 values in the data set fall below the mean. The mean is so large here because of the influence of the extreme 12th value, 130.

To lessen the effect of extreme values on the mean, a modified mean called the trimmed mean is employed at times.

(3.7)

The *50 percent trimmed mean* is the mean of the central 50 percent of the values in an array of the data set.

Thus, to calculate a 50 percent trimmed mean, we simply eliminate 25 percent of the values at each end of the array and calculate the mean of the remaining middle items.

EXAMPLE

Obtain the 50 percent trimmed mean for the data set given earlier, repeated here for convenience:

1   1   2   4   5   7   8   9   12   12   13   130

Fifty percent of this data set represents six items, so we eliminate the first three and last three values in the array. The 50 percent trimmed mean is:

$$\frac{4 + 5 + 7 + 8 + 9 + 12}{6} = 7.5$$

Note that the 50 percent trimmed mean falls near the middle of the array.   □

Comment

A trimmed mean can be based on percentages other than the central 50 percent. For example, to calculate the 80 percent trimmed mean of a data set consisting of 50 values, we trim the five smallest and five largest values from the set, and the trimmed mean is the mean of the remaining central 80 percent of items.

Median

In frequency distributions that are not symmetrical, such as the oil spill distributions, the mean tends to be located somewhat away from the concentration of items. Thus, the mean time interval between oil spills for the period before pollution controls, namely, 20.9 days, is only exceeded in 16 of the 40 cases. Similarly, the mean of 27.6 days for the period after pollution controls is only exceeded in 38 of the 109 cases. Also, when the data set contains an outlying value, as in the trimmed-mean example, the mean can be unduly influenced by the outlying observation.

A measure of position that is more central in the distribution, in the sense that it divides the arrayed items into two equal parts, is the median.

(3.8)

> The *median* of a set of items is the value of the middle item in an array of the items. Expressed symbolically, the median Md is the value of the item with rank $(n + 1)/2$ when the n items in the data set are ranked from 1 to n by order of magnitude.

For large data sets that are stored in a computer file, the values of the data set can be easily sorted and arrayed by order of magnitude. The median of the data set can then be determined readily from the array.

EXAMPLE

Convenience Stores. Changes in annual sales between 1984 and 1985 for five convenience stores operated by a single owner were as follows (data in thousands of dollars): 14, 17, -13, 41, 12. By arranging the items in ascending order and giving them ranks, we obtain the following:

Value:	-13	12	14	17	41
Rank:	1	2	3	4	5

Rank $(n + 1)/2 = (5 + 1)/2 = 3$ corresponds to the value 14. Hence, $Md = 14$. Note that the median divides the array of items equally, with two items being to one side and two to the other of the median. □

When the data set contains an even number of items, rank $(n + 1)/2$ is not an integer, and the median is taken to be the average of the middle two values—that is, the average of the values with ranks $n/2$ and $(n/2) + 1$.

EXAMPLE

In the convenience stores example, suppose that the owner operated six stores, with the following changes in annual sales: 14, 17, -13, 41, 12, 66. By arranging the values in ascending order and giving them ranks, we obtain the following:

Value:	-13	12	14	17	41	66
Rank:	1	2	3	4	5	6

Rank $n/2 = 6/2 = 3$ and rank $(n/2) + 1 = 4$ correspond to values 14 and 17, respectively. Hence, $Md = (14 + 17)/2 = 15.5$. Again, we find that the median divides the array of items equally, three being to one side and three being to the other side of the median. □

Medians are useful for comparing several data sets.

EXAMPLE

In the oil spills example, the two data sets presented in Table 3.1 were arrayed by use of a descriptive statistics computer software package (see Figure 3.8 on p. 85 for the array of the data set before pollution controls). For the data set before pollution controls, for which $n = 40$ and $(n + 1)/2 = (40 + 1)/2 = 20.5$, its array showed that ranks 20 and 21 correspond to average value $(18 + 19)/2 = 18.5$. For the data set after pollution controls, for which $n = 109$, its array showed that rank $(109 + 1)/2 = 55$ corresponds to value 21. Thus, the medians of the two data sets are 18.5 and 21 days, respectively. Consequently, comparing the two medians shows, just as did the comparison of the means, that the time intervals between oil spills have tended to be longer after the pollution controls were instituted. □

Properties of Median

1. As definition (3.8) indicates, the median Md of a set of items is in the middle of the values ordered by magnitude. For large data sets where the values do not tend to repeat themselves extensively, the median can be interpreted to indicate that about one-half of the items are smaller than the median and about one-half are larger. Thus, if the median family income in a community is \$30,000, then about one-half the families have incomes under \$30,000 and about one-half have incomes over \$30,000.

   When the values in a data set repeat themselves frequently, this interpretation may not be proper. For example, suppose 20 percent of families in a community consist of two persons, 40 percent consist of three persons, and the other 40 percent consist of four persons. The data set appears as follows:

   $$2 \quad 2 \quad 2 \quad \cdots \quad 2 \quad 3 \quad 3 \quad 3 \quad \cdots \quad 3 \quad 4 \quad 4 \quad 4 \quad \cdots \quad 4$$

   The median value is $Md = 3$, but it does not indicate here that about one-half the family sizes are less than 3. In fact, only 20 percent of family sizes are smaller.

2. Consider the sum of absolute deviations around an arbitrary fixed value A for the values in a data set, that is, consider the expression:

   $$\sum_{i=1}^{n} |X_i - A|$$

   An important property of the median is that this sum is a minimum when A is set equal to the median, that is, when $A = Md$.

(3.9)

$$\sum_{i=1}^{n} |X_i - A| \text{ is a minimum when } A = Md.$$

EXAMPLE

Consider the values 14, 17, -13, 41, 12. For $A = Md = 14$, we have:

$$\sum |X_i - 14| = |14 - 14| + |17 - 14| + |-13 - 14| + |41 - 14| + |12 - 14| = 59$$

For any other value of A, this sum would be larger. For instance, if A were set equal to the mean of these values ($\overline{X} = 14.2$), the sum would equal:

$$\sum |X_i - 14.2| = |14 - 14.2| + |17 - 14.2| + |-13 - 14.2| + |41 - 14.2| + |12 - 14.2| = 59.2$$

3. The median is not affected by any extreme items in a data set. For this reason, it is often preferred to the mean as a measure of position for frequency distributions containing a few extreme values. Income distributions are a case in point, often containing some extremely large values. The median in this situation tends to be more indicative of typical income levels than the mean.

4. The median can be viewed as a special case of a trimmed mean, where all items other than the middle one or two are trimmed away or excluded. Thus, the 50 percent trimmed mean is a compromise between the mean, which is based on all items in the data set, and the median, which excludes all but the middle one or two items.

Calculation of Median from Frequency Distribution. When data are presented only as a frequency distribution, an approximation of the median Md can be obtained as follows.

(3.10)

$$Md \simeq L + \left(\frac{n_1}{n_2}\right)I$$

where:

L is the lower limit of the median class (class containing the middle item)
n_1 is the number of items that must be covered in the median class in order to reach the middle item
n_2 is the frequency of the median class
I is the width of the median class

EXAMPLE

The calculation of Md for the frequency distribution of time intervals between oil spills before pollution controls is illustrated in Table 3.3. Note that $n/2$ is used to locate the middle of the distribution [rather than using the rank $(n + 1)/2$ as we did for locating the middle item in the original data set]. In Table 3.3, the middle of the distribution corresponds to a cumulative frequency of $n/2 = 40/2 = 20$, which lies in the second class.

TABLE 3.3   Calculation of the median from the frequency distribution for the oil spills example

Time Interval (days)	Number of Cases	Cumulative Number of Cases
0–under 15	15	15
15–under 30	13	Class containing 20 → 28
30–under 45	8	36
45–under 60	2	38
60–under 75	2	40
Total	40	

$$Md = 15 + \frac{5}{13}(15) = 20.77$$

since:

$$\text{middle} = \frac{n}{2} = \frac{40}{2} = 20$$

$L = 15$

$n_1 = 5$

$n_2 = 13$

$I = 30 - 15 = 15$

Comment

Formula (3.10) for computing Md from a frequency distribution assumes that the items in the median class are spread evenly throughout that class. This assumption is adequate in many cases, but not always. For the oil spills example, the approximation is only fair, the median for the original data set being 18.5 while the approximate value calculated from the frequency distribution is 20.8. If there is a choice, the median should be calculated from the original data set to avoid approximation errors.

Mode

The mode is still another measure of position that may be used in investigating a data set. Consider, for instance, the age distribution of farm operators shown previously in Figure 2.3 (p. 31). Interest centers on the most common age of farm operators. We see from the frequency distribution that ages in the class 45–under 55 are the most common. The class 45–under 55 is called the modal age class.

(3.11)

The *modal class* in a frequency distribution with equal class intervals is the class with the largest frequency. If the frequency polygon of a distribution has only a single peak, it is said to be *unimodal*. If the frequency polygon has two peaks, it is said to be *bimodal*.

Consideration of whether a frequency polygon is unimodal or bimodal can sometimes provide insights into the nature of the underlying data set, as illustrated by the
following example.

Melting Points.   A metal alloy manufacturer was concerned with customer complaints about the lack of uniformity in the melting points of one of the firm's alloy
filaments. Sixty filaments were selected from the production process and their
melting points determined. The resulting frequency polygon is shown in Figure
3.2a. Note its bimodal nature. Closer examination of the data revealed that 26 of
the filaments were produced by the first shift and the other 34 by the second shift.
The two frequency polygons for the melting points of the filaments for each shift
separately are shown in Figure 3.2b. It is now clear that the bimodal nature of the
original frequency polygon is the result of a difference in the positions of the distributions for the two shifts. An investigation revealed that the second shift was
using the wrong alloy mixture specification. Correction of the specification resulted in production of filaments with more uniform melting points.   □

Percentiles

The median, as we saw, divides a set of items ordered by magnitude into two equal
parts and may therefore be considered the 50th percentile. Other percentiles, such as
the 25th and 75th percentiles, often are useful for indicating the positions of the
noncentral values in a data set. For small data sets, percentiles can be easily found
from a step function ogive, such as the one illustrated in Figure 2.7, or from a line
diagram.

FIGURE 3.2   Bimodal frequency polygon for the melting points example

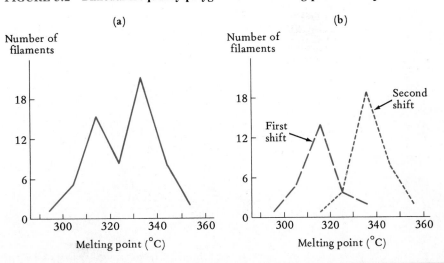

EXAMPLE

In the convenience stores example, the owner of the five stores wanted to compare the changes in annual sales for these stores with industry data, which showed that 25 percent of convenience stores had annual sales increases of $10 thousand or less, and 75 percent of stores had annual sales increases of $15 thousand or less. The array of changes in annual sales for the five stores is repeated in Figure 3.3a. A step function ogive has been plotted in Figure 3.3b. Reading from this graph, we see that the 25th percentile for the five stores is $12 thousand, and the 75th percentile is $17 thousand. Thus, the five convenience stores performed comparably

FIGURE 3.3   Determination of percentiles from the step function ogive and the line diagram for the convenience stores example

(a) Array

−13 12 14 17 41

(b) Step Function Ogive

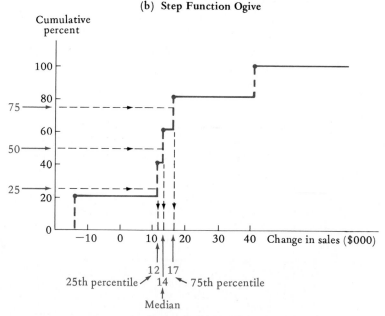

(c) Line Diagram

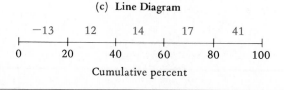

with the industry. Note also from Figure 3.3b that the median, or 50th percentile, is $14 thousand, as we determined previously.

The correspondences employed in Figure 3.3b can also be expressed by the *line diagram* of Figure 3.3c. Since there are five items in the data set, the cumulative percent scale ranging from 0 to 100 is divided into five equal intervals and the corresponding value from the array is associated with each interval. For example, the smallest value (-13) is placed in the percent interval from 0 to 20. The value associated with any percentile can then be read from the line diagram. Thus, the 25th percentile is $12 thousand because the value 12 is in the interval in which the cumulative percent 25 lies. Similarly, the 50th percentile is $14 thousand and the 75th percentile is $17 thousand.

Occasionally, the desired percentile falls on the boundary between two intervals in the line diagram in Figure 3.3c or, equivalently, falls at the top of a step in the ogive in Figure 3.3b. In that case, the percentile is not uniquely defined. We shall follow the convention employed for the median of taking the mean of the two adjacent values. For example, the 40th percentile of the data set in Figure 3.3a is $(12 + 14)/2 = 13 thousand.   ☐

Meaning of Percentiles. In large data sets where values do not tend to repeat themselves extensively, the 25th percentile indicates that about 25 percent of the items are less than this value and about 75 percent are more. Other percentiles are interpreted correspondingly. When the values in a data set do tend to repeat themselves, this interpretation is no longer appropriate, as was illustrated for the median.

Comments

1. A single value can correspond to more than one percentile. For instance, the value 12 in Figure 3.3 is not only the 25th percentile but is also every percentile between 20 and 40.

2. Many percentiles are known by other names. Percentiles that are multiples of 25 are called *quartiles*. Thus, the 25th percentile is also called the first quartile, the 50th percentile the second quartile (also the median), and the 75th percentile the third quartile. Similarly, percentiles that are multiples of 10 are called *deciles*. For instance, the 70th percentile is also called the 7th decile.

3. If data are only available in the form of a frequency distribution, an approximation of the pth percentile can be obtained in the same manner as for the median, using the following formula.

(3.12)

$$p\text{th percentile} \simeq L + \left(\frac{n_1}{n_2}\right)I$$

where:

L is the lower limit of the pth percentile class (class containing the pth percentile)
n_1 is the number of items that must be covered in the pth percentile class in order to reach the pth percentile
n_2 is the frequency of the pth percentile class
I is the width of the pth percentile class

As for the median, the formula assumes that the items are spread evenly throughout the pth percentile class.

For instance, the 80th percentile for the before-controls oil spill distribution in Table 3.3 corresponds to cumulative frequency 0.80 (40) = 32. To reach this cumulative frequency, 4 items must be covered in the interval 30–under 45. This interval contains 8 items altogether, so:

$$\text{80th percentile} \simeq 30 + \frac{4}{8}(15) = 37.5$$

Thus, the estimate of the 80th percentile based on the frequency distribution is 37.5 days.

4. Plotting two ogives together makes it very easy to compare the percentiles of two different distributions. For instance, Figure 3.4 presents the ogives for the two cumulative percent frequency distributions of the age at onset of Parkinson's disease for male and female patients presented in Figure 2.8b (p. 39). Note that the ogive for female patients lies entirely to the right of the ogive for male patients. It follows, therefore, that all percentiles for the age at onset for female patients are larger than the corresponding percentiles for male patients, over the range of cumulative percentages covered by the two ogives. A look at the positions of the two ogives reveals, in fact, that the ogive for female patients is shifted about four years to the right of that for male patients, suggesting that onset of the disease tended to come about four years later for the female patients than for the male patients.

 ## 3.2   MEASURES OF VARIABILITY

When data sets are summarized, the variability of the values is often an important feature of interest. We now consider several summary measures of variability, sometimes also called measures of dispersion or spread.

Range

A data set consists of the amount of active ingredient in each of 300 tablets in a shipment from a pharmaceutical manufacturer. One factor that is important in assessing the quality of these tablets is the maximum extent of variation in the amount of active ingredient per tablet.

(3.13)   The *range* is the difference between the largest and smallest values in a set of items.

EXAMPLES

1. The largest value in the data set for the 300 tablets is 1.38 grams, and the smallest value is 0.96 gram. The range then is 1.38 − 0.96 = 0.42 gram, indicating a relatively large extent of variability in the amounts of active ingredient.

FIGURE 3.4   **Comparison of percentiles of two cumulative percent frequency distributions for the Parkinson's disease example**

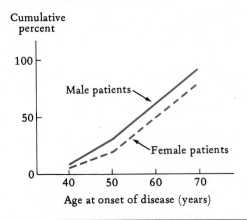

2. In the oil spills example, it can be ascertained from Table 3.1a that the range for the before-controls data set is $73 - 1 = 72$ days, while that for the after-controls data set is $163 - 0 = 163$ days. Thus, the after-controls data set has a substantially larger spread. ☐

Comments

1. A limitation of the range as a measure of the variability of a set of items is that it depends only on the extreme items and does not consider the others. Thus, a data set might contain items quite close to each other with the exception of one extreme value. Despite the concentration of almost all the items, the range would be large since it is based only on the extreme values. Another limitation of the range as a measure of variability is that it depends on the number of items in the set. The larger the number of items, the larger the range tends to be.

2. When data are classified in a frequency distribution, the range of the original items usually cannot be determined exactly. For instance, from the frequency distribution in Table 3.3, it can only be seen that the range of the 40 time intervals is not larger than $75 - 0 = 75$ days. Actually, from the original data in Table 3.1a, the range is $73 - 1 = 72$ days, as noted in Example 2.

3. Sometimes, the two extreme values in a set of items are presented instead of the difference between the two. Thus, stock market summaries in newspapers often provide the high and low prices for the day. This form of presentation has the advantage of not only providing information about the variability of a stock's price during the day, since the range can be easily calculated, but also providing information about the location of the stock's price distribution. Another advantage is that the daily highs and lows can be used to determine the price range for a longer period, such as a week.

Interquartile Range

Because the range depends only on the two extreme values in a data set, a modified range is sometimes used that reflects the variability of the middle 50 percent of items in the array of the data set. This modified range is called the interquartile range.

(3.14)   The *interquartile range* is the difference between the third and first quartiles of the data set.

EXAMPLES

1. An economist, studying the variation in family incomes in a community, found that the middle 50 percent of families have incomes which vary from $22,400 to $29,100, a range of $6700. Here, the first quartile is $22,400, and the third quartile is $29,100. The range of $6700 is the interquartile range, reflecting the extent of variation among the middle 50 percent of family incomes.

2. In the oil spills example, the first and third quartiles and the interquartile ranges for the two data sets are as follows:

	Before Controls	After Controls
Third Quartile:	31	33
First Quartile:	5.5	11.5
Interquartile Range:	25.5	21.5

The upward shift of the quartiles reflects the lengthening of the durations between oil spills after pollution controls were instituted. The decline in the interquartile range indicates that the middle 50 percent of items in the after-controls data set are somewhat more concentrated than those in the before-controls data set.   □

Comment   The interquartile range may be considered to be approximately the range for a trimmed data set in which the smallest 25 percent and the largest 25 percent of values have been removed. It can thus be viewed, approximately, as a 50 percent trimmed range.

Variance

The most commonly used measure of variability in statistical analysis is called the variance. It is a measure that takes into account all the values in a set of items.

EXAMPLE

In the melting points example, we considered the situation where customer complaints about increased variability in the melting points of alloy filaments led the manufacturer to investigate the cause of the problem. One of the customers had selected a sample of 15 filaments from a shipment received in April and another sample of 15 filaments from a shipment received the following October. The data on the melting points for these two samples are given in Figure 3.5.

To compare the variability of melting points for the two data sets, the customer considered the deviations of values in each data set around their respective means. These two sets of deviations are shown graphically in Figure 3.5. The means of the two data sets are 330.8 and 339.3, respectively. A comparison of the two sets of deviations in Figure 3.5 shows readily that there was greater variability around the center of the distribution (as measured by the mean) for the October filaments than for the April filaments.

The variance is a measure that provides quantified information about the variability in a data set. The first step in calculating the variance is to square the deviations around the mean. Then, to obtain an overall measure of variability per item, we take the average of the squared deviations. This mean squared deviation is called the variance. For reasons to be explained in a later chapter, we divide by $n - 1$ rather than by n in obtaining the mean of the squared deviations. The cal-

FIGURE 3.5 Deviations around the mean for two data sets — Melting points example

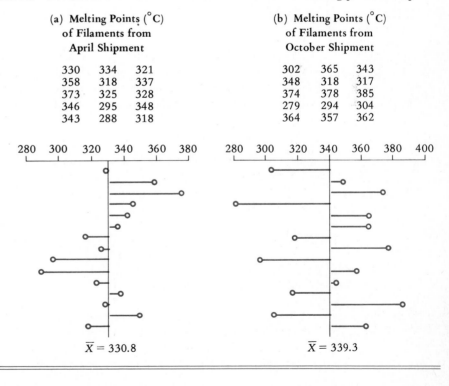

(a) Melting Points (°C) of Filaments from April Shipment

330	334	321
358	318	337
373	325	328
346	295	348
343	288	318

$\overline{X} = 330.8$

(b) Melting Points (°C) of Filaments from October Shipment

302	365	343
348	318	317
374	378	385
279	294	304
364	357	362

$\overline{X} = 339.3$

culation of the variance for each of the two data sets proceeds as follows:

April shipment: $\dfrac{(330 - 330.8)^2 + \cdots + (318 - 330.8)^2}{15 - 1} = 487.5$

October shipment: $\dfrac{(302 - 339.3)^2 + \cdots + (362 - 339.3)^2}{15 - 1} = 1161.1$

The variance for the October filaments is more than twice that for the April filaments. It was this observation that prompted the customer to complain about the increased variability of the melting points of the alloy filaments. □

We now define the variance formally.

(3.15)

The *variance* s^2 of a set of values $X_1, X_2, \ldots, X_n$ is defined:

$$s^2 = \frac{\sum\limits_{i=1}^{n} (X_i - \overline{X})^2}{n - 1}$$

where:
$\overline{X}$ is the mean of the X_i values

An equivalent formula for calculating the variance s^2 is the following.

(3.15a)

$$s^2 = \frac{\sum\limits_{i=1}^{n} X_i^2 - \dfrac{\left(\sum\limits_{i=1}^{n} X_i\right)^2}{n}}{n - 1}$$

EXAMPLE

Consider the data set 5, 17, 12, 10, whose mean is $\overline{X} = 11$. We can calculate s^2 by (3.15):

$$s^2 = \frac{(5 - 11)^2 + (17 - 11)^2 + (12 - 11)^2 + (10 - 11)^2}{4 - 1}$$

$$= \frac{(-6)^2 + 6^2 + 1^2 + (-1)^2}{3} = 24.67$$

or by (3.15a):

$$s^2 = \frac{5^2 + 17^2 + 12^2 + 10^2 - \dfrac{(5 + 17 + 12 + 10)^2}{4}}{4 - 1} = 24.67$$

Frequently, the latter formula is easier for manual calculations. □

Comments
1. The greater the variability of the values in a data set, the greater the variance is. If there is no variability of the values—that is, if all are equal and hence all are equal to the mean—then $s^2 = 0$.
2. Many calculators contain a preprogrammed function to produce the variance. Some calculator models employ a different definition of the variance than the one in (3.15), using n as the divisor instead of $n - 1$. Unless the data set is small, either n or $n - 1$ in the denominator gives essentially the same result. To determine which divisor is being used in a calculator, consult the user's manual or run a small sample problem. A variance with n in the denominator can be converted to one with $n - 1$ by multiplying by $n/(n - 1)$.

Calculation of Variance from Frequency Distribution. When the data are presented in a frequency distribution, only an approximation to the variance can be obtained.

(3.16)

$$s^2 \simeq \frac{\sum_{i=1}^{k} f_i (M_i - \overline{X})^2}{n - 1}$$

where:

k is the number of classes in the frequency distribution
f_i is the frequency of the ith class
M_i is the midpoint of the ith class
$\overline{X}$ is the mean of the frequency distribution as approximated by (3.6)

$$n = \sum_{i=1}^{k} f_i$$

TABLE 3.4 Calculation of the variance from the frequency distribution for the oil spills example

Time Interval (days)	Number of Cases f_i	Class Midpoint M_i	Deviation $M_i - \overline{X}$	Deviation Squared $(M_i - \overline{X})^2$	$f_i(M_i - \overline{X})^2$
0–under 15	15	7.5	−16.125	260.0156	3,900.234
15–under 30	13	22.5	−1.125	1.2656	16.453
30–under 45	8	37.5	13.875	192.5156	1,540.125
45–under 60	2	52.5	28.875	833.7656	1,667.531
60–under 75	2	67.5	43.875	1,925.0156	3,850.031
Total	40				10,974.374

$$\overline{X} \simeq 23.625 \qquad \text{(from Table 3.2)}$$

$$s^2 \simeq \frac{10,974.374}{40 - 1} = 281.39$$

Comments	1.	Formula (3.16) for computing s^2 from a frequency distribution assumes that each item in a class has a value equal to the midpoint of that class. This assumption may appear somewhat crude, but frequently, it yields results close to the variance that would have been obtained from the original data. If there is a choice, the variance should be calculated from the original data to avoid approximation errors.
	2.	Since the mean cannot be computed from an open-end frequency distribution without further information, neither can the variance. Also as for the mean, the calculation of the variance is not affected by unequal class intervals.
	3.	The following formula is algebraically equivalent to (3.16) and is often easier to use for manual calculation.

(3.16a)

$$s^2 \simeq \frac{\sum\limits_{i=1}^{k} f_i M_i^2 - \dfrac{\left(\sum\limits_{i=1}^{k} f_i M_i\right)^2}{n}}{n-1}$$

EXAMPLE

Table 3.4 contains the frequency distribution of the time intervals between oil spills before pollution controls from Table 3.2 and illustrates the calculation of the variance. Note that the approximate value of $\overline{X}$ as computed from the frequency distribution in Table 3.2 is used in the calculation of s^2. ☐

Standard Deviation

The variance s^2 is expressed in units that are the square of the units of measure of the characteristic under study. For instance, each variance in the melting points example is expressed in Celsius degrees squared. Often, it is desirable to return to the original units of measure. We obtain the original units by taking the positive square root of the variance. The resulting value, called the standard deviation, is also used as a measure of variability.

(3.17)

The positive square root of the variance is called the *standard deviation* and is denoted by s:

$$s = \sqrt{s^2}$$

As with the variance, the greater the variability of the values in a data set, the larger the standard deviation is. If there is no variability, that is, if all values are equal, the standard deviation is zero.

EXAMPLE

For the melting points example, the standard deviations are $s = \sqrt{487.5} = 22.1°C$ for the April filaments and $s = \sqrt{1161.1} = 34.1°C$ for the October filaments. The standard deviations again show that there was greater variability of the filament melting points in the October sample than in the April sample. □

Coefficient of Variation

The standard deviation is a measure of the *absolute variability* in a set of items. Frequently, however, the *relative variability* is a more significant measure. The most commonly used measure of relative variability is the coefficient of variation.

(3.18)

The *coefficient of variation,* denoted by C, is the ratio of the standard deviation to the mean expressed as a percentage:

$$C = 100\left(\frac{s}{\overline{X}}\right)$$

EXAMPLES

1. A sales analyst was studying the variability of a company's sales revenues over time periods of different lengths. She determined that weekly sales revenues during the past year had a mean value of $1.36 million and a standard deviation of $0.28 million, while sales revenues for four-week intervals during the past year had a mean value of $5.44 million and a standard deviation of $0.50 million. The analyst felt that a direct comparison of the two standard deviations would not be appropriate because the absolute variability in revenues is affected by the fact that sales revenues for four-week periods, on the average, are four times as large as weekly sales revenues. Instead, the analyst computed the coefficient of variation for each case as follows:

Weekly Sales Revenues	Four-Weekly Sales Revenues
$C = 100\left(\dfrac{0.28}{1.36}\right) = 20.6\%$	$C = 100\left(\dfrac{0.50}{5.44}\right) = 9.2\%$

From these results, the analyst concluded that sales revenues for four-week periods ($C = 9.2\%$) were relatively less variable than weekly sales revenues ($C = 20.6\%$) during the past year. In particular, she noted that measuring revenues over a period four times as long cut the relative variability of revenues approximately in half.

2. A traffic analyst for an air carrier needed to compare the variability in volume and weight of containerized air-cargo shipments. The analyst found that the shipments had a mean volume of 12.6 cubic feet and standard deviation of 2.1 cubic feet, while the corresponding measures for weight were 64.0 pounds and 18.3 pounds, respectively. The standard deviations cannot be compared here because volume is measured in cubic feet and weight in pounds. The coefficients of variation, in

contrast, are dimensionless, because the units of measurement cancel when the standard deviation is divided by the mean. The coefficients of variation here are as follows:

Volume	Weight

$$C = 100\left(\frac{2.1}{12.6}\right) = 16.7\% \qquad C = 100\left(\frac{18.3}{64.0}\right) = 28.6\%$$

Hence, the relative variability of the container volumes is smaller than that of the container weights.   ☐

Comment   The coefficient of variation is usually employed only when all values are positive, such as for sales, volume, and weight data. When the values can be both positive and negative, as for profit-and-loss data, the coefficient of variation is not an appropriate measure of relative variability.

3.3   MEASURES OF SKEWNESS

Measures of position are concerned with the location around which items are concentrated, whereas measures of variability consider the extent to which the items vary. Measures of skewness are still another kind of measure, summarizing the extent to which the items are symmetrically distributed.

(3.19)   A *skewed* frequency distribution is one that is not symmetrical.

Nature of Skewness

One way of studying the skewness of a frequency distribution is to compare the values of the mode, median, and mean. The mode is the position on the scale that has the greatest concentration of items. For the median, we know that generally about half the items will lie below it and about half above it. In contrast, the mean tends to be pulled in the direction of the extreme values because of the disproportionate influence of very large or very small values on it. The three frequency polygons in Figure 3.6 show the comparative positions of the mean, the median, and the mode in unimodal frequency distributions with differing degrees and directions of asymmetry.

Figure 3.6a is an example of a symmetrical unimodal frequency distribution. The values of the mean, median, and mode in any such distribution are identical.

Figures 3.6b and 3.6c show, respectively, an example of a unimodal frequency distribution skewed to the left (negatively) and one skewed to the right (positively).

FIGURE 3.6 **Examples of symmetrical and skewed unimodal frequency distributions.** In skewed unimodal distributions, the mean is typically furthest out toward the tail, and the median falls between the mean and the mode.

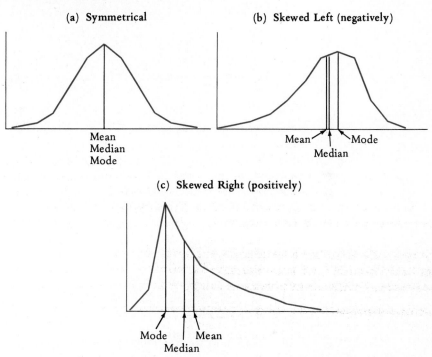

These figures indicate the relationship typically existing between the mean, the median, and the mode in unimodal frequency distributions that are moderately skewed: The mean is furthest out toward the tail of the distribution, and the median is between the mean and the mode.

Standardized Skewness Measure

We have seen that the variance s^2, as defined in (3.15), is the average of the squared deviations about the mean, i.e., the average of the squared deviations $(X_i - \overline{X})^2$. The variance is sometimes called the *second moment about the mean* and denoted by m_2. A corresponding measure of skewness for a data set is obtained by averaging the cubed deviations about the mean, i.e., averaging the values $(X_i - \overline{X})^3$. The resulting measure is called the third moment about the mean and is denoted by m_3. As with the variance s^2, the average is computed by using $n - 1$ in the denominator.

(3.20)   The *third moment about the mean* m_3 of a set of values $X_1, X_2, \ldots, X_n$ is defined:

$$m_3 = \frac{\sum_{i=1}^{n} (X_i - \overline{X})^3}{n - 1}$$

where:

$\overline{X}$ is the mean of the X_i values

Because of the cubing operation, large deviations $X_i - \overline{X}$ tend to dominate the sum in the numerator of m_3. If the large deviations are predominately positive, m_3 will be positive because $(X_i - \overline{X})^3$ has the same sign as $X_i - \overline{X}$. On the other hand, if the large deviations are predominately negative, m_3 will be negative. Since large deviations are associated with the long tail of a frequency distribution, we can see from Figure 3.6 that m_3 will be positive or negative according to whether the direction of skewness is positive (right) or negative (left). If the values in the data set are symmetrical about the mean, the third moment m_3 will be zero.

EXAMPLE   For the data set $X_1 = -3$, $X_2 = 5$, and $X_3 = 40$, we have $\overline{X} = 14$. Hence:

$$m_3 = \frac{(-3 - 14)^3 + (5 - 14)^3 + (40 - 14)^3}{3 - 1} = 5967$$

Since m_3 is positive, the third moment indicates that the data set is skewed to the right. ☐

The magnitude of the third moment about the mean for a data set reflects the unit of measure of the observations in the data set. A standardized skewness measure for a data set is generally more useful because its magnitude does not depend on the unit of measure. It is obtained by dividing m_3 by the cube of the standard deviation. The resulting quantity is called the standardized skewness measure.

(3.21)   The *standardized skewness measure* m_3' for a data set is defined:

$$m_3' = \frac{m_3}{s^3}$$

where:

m_3 is given by (3.20)
s is given by (3.17)

A data set is skewed positively or negatively according to whether m_3' is positive or negative. If the values in the data set are symmetric about the mean, $m_3' = 0$.

EXAMPLES

1. For the data set $X_1 = -3$, $X_2 = 5$, and $X_3 = 40$, we computed earlier $m_3 = 5967$. Also, $s = 22.87$ for this data set. Hence, $m_3' = 5967/(22.87)^3 = 0.499$, indicating some positive skewness in the data set.

2. For the data set before pollution controls in the oil spills example (Table 3.1a), $m_3 = 5397.9$ and $s = 17.352$ (calculations not shown). Hence, $m_3' = 5397.9/(17.352)^3 = 1.033$, so the data set is positively skewed. This skewness can be seen also from the percent frequency distribution in Table 3.1b. □

Comments

1. The standardized skewness measure m_3' can be unreliable when an unusual data set is encountered, such as one containing extreme values.
2. The fourth moment about the mean of a data set, denoted by m_4, is sometimes computed as a measure of the *peakedness* or *kurtosis* of the data set. A standardized form of this measure is often found in computer printouts.

(3.22)

The *standardized kurtosis measure* m_4' for a data set is defined:

$$m_4' = \frac{m_4}{s^4}$$

where:

$$m_4 = \frac{\displaystyle\sum_{i=1}^{n} (X_i - \bar{X})^4}{n - 1}$$

s is given by (3.17)

Interpreting m_4' is difficult with data sets that are not simple, and hence, this measure should be used with particular caution.

3.4 USE OF STANDARDIZED VALUES

It is often helpful to examine the pattern of variation of values in a data set in relation to their mean and standard deviation. This approach is also useful for judging whether some observations are outlying in relation to the other observations in the data set. An analytical procedure useful for both of these purposes is called *standardizing* the observations.

(3.23)

The *standardized values* $Y_1, \ldots, Y_n$ corresponding to a set of observations $X_1, \ldots, X_n$ are defined:

$$Y_i = \frac{X_i - \overline{X}}{s} \qquad i = 1, 2, \ldots, n$$

where:

$\overline{X}$ is the mean of the values in the data set as given by (3.1)
s is the standard deviation of the values in the data set as given by (3.17)

We see that the standardized value Y_i simply measures the distance that X_i lies from the mean $\overline{X}$ in units of the standard deviation s.

EXAMPLE

Figure 3.7a repeats the April shipment data for the melting points example from Figure 3.5a, as well as the mean $\overline{X} = 330.8$ and the standard deviation $s = 22.1$

FIGURE 3.7   Standardized values for the melting points example (April shipment data)

(a) Observations and Standardized Values
$(\overline{X} = 330.8, s = 22.1)$

i	Observation X_i	Standardized value Y_i	i	Observation X_i	Standardized value Y_i	i	Observation X_i	Standardized value Y_i
1	330	−0.04	6	334	0.14	11	321	−0.44
2	358	1.23	7	318	−0.58	12	337	0.28
3	373	1.91	8	325	−0.26	13	328	−0.13
4	346	0.69	9	295	−1.62	14	348	0.78
5	343	0.55	10	288	−1.94	15	318	−0.58

(b) Histogram of Data Set and Correspondence of
Original and Standardized Scales

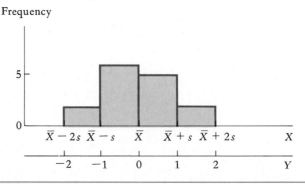

for the 15 observations in this data set. The standardized values for this set are also given in Figure 3.7a and were calculated by using (3.23). For example, the calculations of Y_1 and Y_2 were done as follows:

$$Y_1 = \frac{X_1 - \overline{X}}{s} = \frac{330 - 330.8}{22.1} = -0.04$$

$$Y_2 = \frac{X_2 - \overline{X}}{s} = \frac{358 - 330.8}{22.1} = 1.23$$

The standardized value $Y_1 = -0.04$ indicates that $X_1 = 330$ lies 0.04 standard deviation below $\overline{X} = 330.8$, while $Y_2 = 1.23$ indicates that $X_2 = 358$ lies 1.23 standard deviations above $\overline{X} = 330.8$.

Figure 3.7b shows a histogram of the April shipment data. Note that the horizontal scale of the histogram is given both in terms of the original units (X) and, equivalently, in terms of the standardized units (Y). The histogram shows that all but 4 of the 15 observations lie within one standard deviation of $\overline{X}$ ($-1 \leq Y_i \leq 1$), and none lies beyond two standard deviations from $\overline{X}$ ($Y_i < -2$ or $Y_i > 2$).   □

An observation whose standardized value is large in absolute value (say larger than 3 or 4) is often called an *outlier* of the data set.

EXAMPLE

In the melting points example in Figure 3.7, the data set for the April shipment does not contain any outliers because no absolute standardized value exceeds 2 there.   □

Comments

1.   The standardized value $Y_i = (X_i - \overline{X})/s$ is a unitless quantity because the numerator deviation $X_i - \overline{X}$ and the denominator standard deviation s are expressed in the same unit of measure (dollars, degrees Celsius, etc.).

2.   The mean of the standardized values of any data set is always zero, and the variance is always one.

Chebyshev Inequality

An important theorem relating to standardized values is called the Chebyshev inequality.

(3.24)

Chebyshev Inequality.   For any data set, the proportion of standardized values that are larger than k in absolute value cannot exceed $1/k^2$.

This inequality assures us that, no matter what the pattern of variation for a data set, the proportion of items falling beyond k standard deviations from the mean is at most $1/k^2$.

1. For the April shipment data in Figure 3.7a, the Chebyshev inequality tells us that the proportion of the 15 standardized values that are larger than, say, 1.5 in absolute value ($Y_i < -1.5$ or $Y_i > 1.5$) cannot exceed $1/(1.5)^2 = .44$. In fact, we see that exactly 3 of the 15 Y_i values (proportion $3/15 = 0.20$) are larger than 1.5 in absolute value—namely, $Y_3 = 1.91$, $Y_9 = -1.62$, and $Y_{10} = -1.94$.

2. A data set consists of the waiting times of customers at the checkout counter of a food store during the past week. The mean waiting time was 4.0 minutes, and the standard deviation was 0.9 minute. From the Chebyshev inequality we know that for, say, $k = 2$, at most $1/k^2 = 1/2^2 = 0.25$, or 25 percent, of the waiting times differed by more than two standard deviations from the mean. Hence, at most 25 percent of the customers waited less than $4.0 - 2(0.9) = 2.2$ minutes or more than $4.0 + 2(0.9) = 5.8$ minutes. The Chebyshev inequality does not, however, provide any information about how these customers are divided between very short and very long waits.   □

3.5   COMPREHENSIVE DATA SUMMARIES

In the preceding sections we have explained and illustrated the principal summary measures of position, variability, and skewness for a data set. In exploratory data analysis and in reporting, all or many of these measures are often calculated as a collection to provide an overview or description of the main characteristics of the data set. Such a collection of measures is generally referred to as the *descriptive statistics* for the data set.

We shall now discuss a typical computer output of descriptive statistics for a data set, and we shall also describe a graphic method for presenting key descriptive statistics that is particularly useful for comparing several data sets.

Descriptive Statistics Computer Output

Most statistical computer packages contain a routine for producing key descriptive statistics for a data set. Figure 3.8 contains a typical output from such a routine for the data set before pollution controls in the oil spills example, given initially in Table 3.1a. The output is self-explanatory; all of the summary measures reported were discussed in earlier sections. Note that the output also shows the original data set in array form to facilitate subsequent analysis.

Box Plots

A variety of graphical devices can be used to give an effective visual summary of the descriptive statistics for a data set. One of these is known as a *box plot* or a *box-and-whisker plot*. As will be seen, the names are quite descriptive of the plot.

FIGURE 3.8 Computer output of descriptive statistics for the oil spills example data set before pollution controls

```
            ---- DESCRIPTIVE STATISTICS ----


NUMBER OF CASES (N)    40            3RD MOMENT      5397.858
              MEAN    20.875    (DIVISOR N - 1)
    STD. DEVIATION    17.352        4TH MOMENT    343727.189
           MINIMUM     1        (DIVISOR N - 1)
           MAXIMUM    73        STD. 3RD MOMENT        1.033
                                STD. 4TH MOMENT        3.792




             ---- DATA ARRAY ----


    1     1     1     2     2     3     3     4     5     5
    6     7     8    11    14    15    16    16    17    18
   19    19    20    20    21    22    24    28    30    31
   31    33    36    37    38    41    45    48    64    73
```

Figure 3.9 shows the main elements of a box plot. The plot is displayed vertically in this figure, but it also may be displayed horizontally. The extremities of the plot (the ends of the whiskers) correspond to the smallest and largest observations in the data set. The ends of the box are positioned at the first and third quartiles of the data set.

FIGURE 3.9 Main elements of a box plot

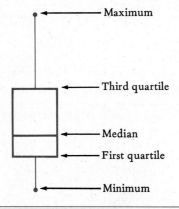

The partitioning line within the box is drawn at the median of the data set. The following example illustrates the use of this type of plot and shows how effective it is in giving a snapshot of the distribution pattern of a data set and in facilitating the comparison of several data sets.

| EXAMPLE |

Work Force.   A regional economic report presented the box plots shown in Figure 3.10a to summarize data on the number of employees in establishments in four

---

FIGURE 3.10   **Illustration of the use of box plots in a comparative study — Work force example**

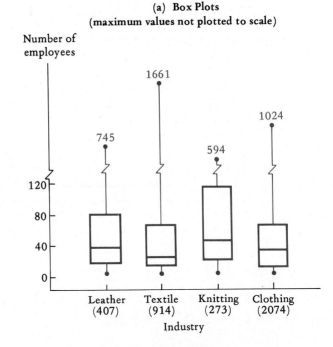

(a) **Box Plots**
(maximum values not plotted to scale)

(b) **Summary Measures**

Number of employees

Industry group	Minimum	1st quartile	Median	3rd quartile	Maximum	Number of establishments
Leather	1	9	30	74	745	407
Textile	1	7	18	59	1661	914
Knitting	2	16	39	107	594	273
Clothing	1	4	26	59	1024	2074

related industries: leather, textile, knitting, and clothing. The summary measures upon which the plots are based are given in Figure 3.10b.

To illustrate the construction of one of the box plots, consider the 407 leather industry establishments. The minimum and maximum numbers of employees in these 407 establishments are 1 and 745, respectively. These two values determine the lower and upper ends of the whiskers in the plot. Note that the upper whisker has been broken to facilitate the plotting of the maximum because of its large value. The first and third quartiles (9 and 74) define the length of the box. The location of the line inside the box corresponds to the median (30).

The box plot for the leather industry establishments gives a concise visual summary of their work force sizes. For instance, we see from the upper end of the box (the third quartile) that about 75 percent of establishments in this industry group employ fewer than 74 employees. We also see that the distribution of number of employees in this industry is heavily right-skewed (in the positive direction). This feature is evident from the length of the whisker in the positive direction and from the median being so much closer to the first quartile than to the third quartile. When the median lies closer to the third quartile than to the first, the distribution is left-skewed, and when the median is centered between the first and third quartiles, the distribution is symmetrical in the central range.

The box plots for the four industries provide an effective means of comparing the work force sizes in the four industries. We see that the median number of employees is largest for the knitting industry. However, the sizes of the establishments vary considerably in this industry, the interquartile range being the largest among the four industries (though the total range is the smallest). The textile and clothing industries have quite similar distributions with respect to position, variability, and skewness. Indeed, the distributions of all four industry groups are skewed to the right. These and many other comparative features of the distributions are grasped quickly from a scan of the box plots. ☐

3.6   OPTIONAL TOPIC — GEOMETRIC AND HARMONIC MEANS

Geometric Mean

When a data set contains ratios, a geometric mean may be a useful measure of position. Suppose that beginning-of-year prices of a product in four consecutive years were $30.00, $33.00, $33.66, and $41.74, respectively. Consider the ratio of the price at the beginning of one year to the price at the beginning of the previous year. The first price ratio is $33.00/30.00 = 1.10$, and the other two are 1.02 and 1.24, respectively. If a single measure is desired to describe the typical value of the three price ratios, the geometric mean may be appropriate.

(3.25)

The *geometric mean* G of a set of values $X_1, X_2, \ldots, X_n$ is the antilogarithm of the arithmetic mean of the logarithms of the values; that is:

$$\log G = \frac{\log X_1 + \log X_2 + \cdots + \log X_n}{n}$$

The logarithms in (3.25) are to the base 10, but they can be to any base. Many calculators are preprogrammed to produce logarithms. (Note that Appendix A, Section A.2, contains a review of logarithms.)

EXAMPLE

The geometric mean price ratio in the example where $X_1 = 1.10$, $X_2 = 1.02$, and $X_3 = 1.24$ is:

$$\log G = \frac{\log 1.10 + \log 1.02 + \log 1.24}{3} = \frac{0.04139 + 0.00860 + 0.09342}{3}$$

$$= 0.04780$$

$$G = \text{antilog } 0.04780 = 1.11635$$

The geometric mean here is an average ratio showing that the annual price increase, on the average, was 11.635 percent. It is a property of the geometric mean that the price in the final period can be obtained from the price in the initial period and the average year-to-year price ratio, as follows:

$$30.00(1.11635)(1.11635)(1.11635) = \$41.74 \qquad \square$$

Comments

1. By taking the antilogarithm of both sides of formula (3.25), one obtains the following equivalent formula for the geometric mean.

(3.25a)

$$G = (X_1 \cdot X_2 \cdot \cdots \cdot X_n)^{1/n}$$

2. The geometric mean is undefined for a data set that contains zero or negative values.

Harmonic Mean

When a data set contains values that represent rates of change, the harmonic mean may be a useful measure of position. For instance, in a study of traffic flow, it was noted that a vehicle traveled at 50 miles per hour over one 10-mile stretch of highway and at 30 miles per hour over a second 10-mile stretch. Suppose it is desired to obtain a measure of the average of the two speeds, reflecting the average speed traveled over the entire 20-mile stretch. In this case, the harmonic mean may be appropriate.

(3.26)

The *harmonic mean H* of a set of values $X_1, X_2, \ldots, X_n$ is the reciprocal of the arithmetic mean of the reciprocals of the values; that is:

$$\frac{1}{H} = \frac{\frac{1}{X_1} + \frac{1}{X_2} + \cdots + \frac{1}{X_n}}{n}$$

EXAMPLE

In the example where $X_1 = 50$ and $X_2 = 30$, the harmonic mean speed is:

$$\frac{1}{H} = \frac{\frac{1}{50} + \frac{1}{30}}{2} = 0.0267 \quad \text{so} \quad H = \frac{1}{0.0267} = 37.5 \text{ miles per hour}$$

The harmonic mean is an appropriate measure here because the vehicle took 0.533 hour to cover the entire 20 miles of highway, and this represents an average speed of $20/0.533 = 37.5$ miles per hour. $\square$

Comments

1. By taking reciprocals of both sides of the equation in (3.26), one obtains the following equivalent formula for the harmonic mean.

(3.26a)

$$H = \frac{n}{\frac{1}{X_1} + \frac{1}{X_2} + \cdots + \frac{1}{X_n}}$$

2. The harmonic mean is undefined for a data set that contains zero or negative values.

PROBLEMS

*3.1 Travel Expenditures. Last year's travel expenditures by the 12 members of a university's physics department were (in dollars):

0   0   173   378   441   733   759   857   958   985   1434   2063

Calculate the mean travel expenditures per member. Also, calculate the mean travel expenditures of the 10 members who did some traveling.

3.2 Emergency Admissions.   The daily number of admissions to the emergency ward of a hospital during the evening shift for the past three weeks were:

$$\begin{array}{cccccccccc}
6 & 0 & 3 & 1 & 5 & 7 & 4 & 2 & 1 & 0 \\
2 & 3 & 2 & 9 & 0 & 3 & 5 & 3 & 1 & 4 & 2
\end{array}$$

Calculate the mean number of admissions per evening shift. Is the mean necessarily a whole number?

3.3 Lot Defectives.   The numbers of defectives in eight recent production lots of 100 items each were:

$$\begin{array}{cccccccc}
3 & 1 & 0 & 0 & 2 & 21 & 4 & 1
\end{array}$$

   a. Calculate the mean number of defectives per lot.
   b. Verify that identity (3.3) holds for these data.
   c. The mean number of defectives per lot in a group of 16 lots from another plant was 3.5. What was the total number of defectives in these 16 lots?

3.4 Enrollment Changes.   The changes in enrollment between this year and last year for seven colleges are:

$$\begin{array}{ccccccc}
-614 & -103 & 41 & 248 & 313 & 387 & 490
\end{array}$$

   a. Calculate the mean enrollment change per college.
   b. Does the fact that the mean calculated in **a** is positive guarantee that the combined enrollment for all seven colleges is larger this year than last year? Comment.

3.5   In a shipment of 500 frozen turkeys, the mean weight is guaranteed to be at least 6 kilograms. What can be said about the total weight of these turkeys?

*3.6   A police department has 125 officers whose mean length of service is 8.5 years and 23 civilian employees whose mean length of service is 6.1 years. Calculate the mean length of service of all 148 members of the department.

3.7   The mean age of the 25 women employees in a department is 28.3 years, and the mean age of the 41 male employees is 39.2 years. Calculate the mean age of all employees in the department.

*3.8   Refer to **Response Times** Problem 2.18. Calculate the mean of the frequency distribution.

3.9   Refer to **Product Sales** Problem 2.20. Calculate the mean of the frequency distribution. Is the calculation complicated by the unequal class intervals? Explain.

3.10 Bonus Pay.   The frequency distribution of the amounts of bonus pay earned during the last year by 65 salespersons follows.

Bonus Pay ($)	Number of Salespersons
0–under 100	4
100–under 200	17
200–under 400	27
400–under 600	11
600–under 1000	5
1000–under 2000	1
Total	65

     a. Calculate the mean bonus pay per salesperson.

     b. Calculate the total of bonuses earned by all 65 salespersons.

     c. Why are the values in **a** and **b** both approximations?

3.11 Hourly Earnings. The percent frequency distribution of hourly earnings for the 2100 plant employees of the Calder Corporation follows.

Hourly Earnings ($)	Percent of Employees
10–under 12	6
12–under 14	14
14–under 16	44
16–under 18	31
18–under 20	5
Total	100
	(2100)

Calculate the mean hourly earnings per employee. What assumption did you make in your calculation?

*3.12 Refer to **Travel Expenditures** Problem 3.1.

     a. Calculate the 50 percent trimmed mean for the data set.

     b. Compare the values of the trimmed mean and the arithmetic mean. Is the difference between them relatively small here? Will it always be? Discuss.

3.13 Refer to **Lot Defectives** Problem 3.3.

     a. Array the observations in the data set. Calculate (1) the 50 percent trimmed mean and (2) the 75 percent trimmed mean.

     b. Why can one not calculate the total number of defectives in the eight lots by using either of the trimmed means in **a**?

3.14 Refer to the array of 40 values in Figure 3.8. Calculate the 50 percent trimmed mean of this data set. Is the trimmed mean substantially different from the untrimmed mean here? Explain.

*3.15 Refer to **Travel Expenditures** Problem 3.1.

     a. Obtain the median travel expenditures per department member.

     b. Did half of the department's members spend more than this median amount on travel? Does the median always have this property? Explain.

3.16 Refer to **Emergency Admissions** Problem 3.2.

     a. Array the observations in ascending order, and obtain the median number of admissions per evening shift.

     b. Is the median equal to an actual value in the data set here? Must it be for every data set? Explain.

3.17 Refer to **Lot Defectives** Problem 3.3. Obtain the median of this data set. Why does the median differ substantially from the mean here?

3.18 Refer to **Enrollment Changes** Problem 3.4.

     a. Obtain the median enrollment change for the seven colleges.

     b. If the median in **a** is multiplied by 7, does one obtain the total enrollment change for all seven colleges combined? Explain.

*3.19 Refer to **Response Times** Problem 2.18. Calculate the median of the frequency distribution. Explain how the median is interpreted here.

3.20 Refer to **Bonus Pay** Problem 3.10. Calculate the median bonus pay per salesperson. Why is the median less than the mean here?

3.21 Refer to **Hourly Earnings** Problem 3.11.
   a. To calculate the median of this frequency distribution, must one first convert the percent frequencies to actual frequencies? Explain.
   b. Calculate the median hourly earnings. What assumption did you make in calculating the median?

*3.22 Refer to **Response Times** Problem 2.18.
   a. What is the modal class of the frequency distribution? Is the distribution unimodal?
   b. If one wanted to use a class interval to predict the response time for an individual alarm answered by this fire company, would the modal class in **a** be a reasonable choice? Would the use of a combination of two classes be preferable? Explain.

3.23 Refer to **Hourly Earnings** Problem 3.11.
   a. What is the modal class of this frequency distribution? Is the distribution bimodal?
   b. Does the modal class contain the median of the distribution? Must this be the case for every frequency distribution?

3.24 A commentator, learning that the modal number of children in U.S. families is zero, stated: "The birthrate has fallen so low that the majority of U.S. families have no children at all!" Comment.

3.25 A service worker stated: "The Internal Revenue Service has calculated that my average tip is $2.50. I have to admit they are correct. But I'll tell you, most of my tips are below the average!" What type of average must the Internal Revenue Service have used here? Would some other average be more appropriate for their purposes? Discuss.

3.26 A promotional brochure reported: "The average family in our condominium development operates 1.79 automobiles." What type of average must have been used? Would some other type of average have been more meaningful here? Discuss.

3.27 An analyst stated: "When a data set has one or several extreme outcomes that might have arisen from errors in data acquisition, I use the median or a trimmed mean in preference to the arithmetic mean." Why might the analyst take this approach?

3.28 A chain of sound equipment stores sells four stereo systems. Data on sales during the past six months follow.

System	Price ($)	Number Sold
1	250	68
2	400	110
3	650	91
4	1000	56
Total		325

When one calculates the mean or median price for this frequency distribution, why will they be exactly the same as when these measures are calculated from the actual 325 observations?

*3.29 Refer to **Travel Expenditures** Problem 3.1. Obtain the 80th percentile for this data set by using a line diagram. Interpret the meaning of the 80th percentile here.

3.30 A newspaper reports that exchange rates for seven major currencies have changed by the following percentages during the past six weeks:

   −3.5   1.8   2.1   2.1   5.1   6.6   9.2

    a. Obtain the first quartile of this data set by using a line diagram.

    b. What range of percentiles corresponds to the observation 5.1?

3.31 Utility Stocks. Rates of return (in percent) earned on ten utility stocks during the past year were:

$$15.7 \quad 21.5 \quad 22.7 \quad 29.2 \quad 21.8 \quad 20.5 \quad 24.2 \quad 20.8 \quad 28.4 \quad 20.9$$

    a. Array these observations, and plot their step function ogive.

    b. Obtain the 25th and 75th percentiles from the ogive.

3.32 Refer to **Emergency Admissions** Problem 3.2.

    a. Plot the step function ogive for this data set.

    b. Obtain the 20th percentile and 6th decile from the ogive.

    c. Do the repeated values in the data set need to be taken into account in interpreting either of the measures obtained in **b**? Discuss.

*3.33 Refer to **Response Times** Problem 2.18.

    a. Calculate the first and third quartiles of the frequency distribution. Interpret their meanings.

    b. The shortest 90 percent of response times were how many minutes or less?

3.34 Refer to **Hourly Earnings** Problem 3.11. Obtain the 10th percentile hourly earnings. Interpret its meaning here.

3.35 The G. Stokes Company, as part of a profit improvement program, studied customer order sizes. The distribution of order sizes for the past fiscal year follows.

Order Size ($)	Number of Orders	Order Size ($)	Number of Orders
5–under 25	1852	100–under 250	354
25–under 50	1120	250–under 500	471
50–under 100	693	500–under 1000	530
		Total	5020

    a. Calculate the 40th and 80th percentiles of the distribution of order sizes. Interpret the meaning of these measures.

    b. What percentage of orders were for less than $250? What percentage of the total dollar volume do these orders represent?

*3.36 Refer to **Travel Expenditures** Problem 3.1.

    a. Obtain the range and the interquartile range for the data set.

    b. Must the interquartile range always be smaller than the range for a data set? Explain.

3.37 Refer to **Lot Defectives** Problem 3.3.

    a. Obtain the range and the interquartile range for the data set.

    b. If the observation 21 happened to be an incorrect count of defectives in that lot, would the range be affected by this error? Would the interquartile range be affected?

3.38 Refer to **Enrollment Changes** Problem 3.4.

    a. Obtain the range and the interquartile range for the data set.

    b. If the two negative changes in the data set had been positive, would your answers in **a** have changed? If so, how?

    c. If the enrollment change for an eighth college were added to the data set, might your answers in **a** be different? Explain.

3.39 Swimming team A has nine divers whose scores had a range of 2.3 in a recent competition. Swimming team B has five divers whose scores had a range of 1.6 in the same competition. A sports reporter commented that team B has more uniform diving talent because its scores had a smaller range. Do you agree with the reporter? Discuss.

*3.40 Refer to **Response Times** Problem 2.18. Calculate the interquartile range for the frequency distribution. In what units is this measure expressed?

3.41 Refer to **Bonus Pay** Problem 3.10.
   a. Calculate the interquartile range for the frequency distribution.
   b. Does the interquartile range in **a** measure the actual extent of spread in the middle 50 percent of arrayed bonus pay amounts? Explain.

3.42 Refer to **Hourly Earnings** Problem 3.11.
   a. Calculate the interquartile range for the frequency distribution.
   b. If every employee received a 10 percent increase in hourly earnings, what would be the interquartile range for the hourly earnings distribution after the increase? Explain.

*3.43 Refer to **Travel Expenditures** Problem 3.1. Calculate the variance and the standard deviation of the data set. In what units is each measure expressed?

3.44 Refer to **Lot Defectives** Problem 3.3.
   a. Calculate the variance and the standard deviation of the data set.
   b. Which observation in the data set makes the largest contribution to the magnitude of the variance through the sum of squared deviations? Which observation makes the smallest contribution? Explain.

3.45 Refer to **Enrollment Changes** Problem 3.4. Calculate the standard deviation of this data set. In what units is this measure expressed?

3.46 A scientist made four measurements of the ionization energy of helium (in electron volts), obtaining 24.547, 24.623, 24.590, and 24.571.
   a. Calculate the standard deviation of these measurements.
   b. Would the same standard deviation be obtained if 24 electron volts were subtracted from each measurement before making the calculation? Explain.

*3.47 Refer to **Response Times** Problem 2.18. Calculate the variance and the standard deviation of the frequency distribution.

3.48 Refer to **Bonus Pay** Problem 3.10.
   a. Calculate the standard deviation of the frequency distribution. What assumption did you make in calculating this measure?
   b. The mean and the standard deviation of the bonus pay distribution two years ago were $198.4 and $151.7, respectively. What changes in the bonus pay distribution appear to have taken place?

3.49 Refer to Table 2.5.
   a. Calculate the standard deviation of the costs of the 77 tours that have a duration of three days. What assumption did you make in calculating this measure?
   b. If the costs of all 77 tours were increased by $50 because of an added tax, would the standard deviation of the new cost figures be the same as in **a**? Explain.

3.50 A student entered the values -1 and 1 in a pocket calculator, pressed the key for the standard deviation, and obtained 1.414 in the display. Does this calculator use n or $n-1$ in the divisor of the formula for the standard deviation?

*3.51 Refer to **Travel Expenditures** Problem 3.1.
   a. Calculate the coefficient of variation of the data set.
   b. In the university's biology department, the mean travel expenditures per member last year were about the same as in the physics department, but the coefficient of variation was sub-

stantially smaller. What does this comparison tell about the travel expenditures patterns in the two departments?

3.52 Refer to **Grain Production** Problem 2.36. The mean and the standard deviation of the revenue and volume data for each grain are as follows:

	Revenue		Volume	
Grain	Mean	Standard Deviation	Mean	Standard Deviation
Wheat	12.78	4.68	74.4	26.58
Barley	14.40	5.53	107.4	35.20

a. For each grain, calculate the coefficients of variation for the revenue and volume data.
b. Which grain has the greater absolute variability in volume? Which has the greater relative variability in volume? Is the same result true for revenue?
c. Is revenue or volume relatively more variable for wheat? Is the same result true for barley?

3.53 A survey of college students engaged in summer work last year gave the following statistics for hours worked and total earnings:

Statistic	Hours Worked	Earnings ($)
Mean	487	2985
Standard deviation	53	261

a. Calculate the coefficients of variation for hours worked and for earnings. In what units is each coefficient expressed?
b. Which exhibits greater variability, hours worked or earnings? Be specific, and explain your choice of the measure of variability used in the comparison.

*3.54 Refer to **Travel Expenditures** Problem 3.1.
a. What does a comparison of the mean and the median indicate about the direction or absence of skewness in this data set?
b. Calculate the third moment about the mean and the standardized skewness measure for this data set. Are your observations about skewness in **a** consistent with the signs of these two measures? Explain.

3.55 Refer to **Lot Defectives** Problem 3.3.
a. What does a comparison of the mean and the median indicate about the direction or absence of skewness in this data set?
b. Calculate the third moment about the mean. Which of the eight observations makes the largest contribution to the magnitude of this measure?
c. Calculate the standardized skewness measure for this data set. In what units is the measure expressed?

3.56 Refer to **Utility Stocks** Problem 3.31.
a. Calculate the third moment about the mean and the standardized skewness measure for this data set. In what units are these measures expressed?
b. What direction of skewness is indicated by the sign of the standardized skewness measure in **a**? Are the relative positions of the mean and the median consistent with the sign of the standardized skewness measure? Explain.
c. Plot the ten observations of the data set on a line. Is the pattern of asymmetry seen in the plot consistent with your conclusion in **b**? Comment.

3.57 Refer to **Horse Trials** Problem 2.19. Identify the modal class of this frequency distribution. Identify the class that contains the median. What does a comparison of these two classes indicate about the symmetry or lack of symmetry of this frequency distribution?

*3.58 Refer to **Travel Expenditures** Problem 3.1. Which of the 12 observations lies farthest from the mean? What is its standardized value? Is this observation an outlier? Explain.

3.59 The standardized values of the grades of nine students in a graduate course were:

$$-2.01 \quad -0.62 \quad -0.27 \quad -0.15 \quad -0.15 \quad 0.19 \quad 0.54 \quad 1.12 \quad 1.36$$

    a. How many standard deviations apart are the lowest and highest grades in this course?
    b. The mean and the standard deviation of the nine grades are 73.3 and 8.6, respectively. What was the highest grade in the course?

3.60 Refer to **Utility Stocks** Problem 3.31.
    a. Calculate the standardized values of this data set.
    b. A financial analyst uses standardized values of stock returns as indicators of their relative performance. He calls a stock with a standardized value exceeding 2 in absolute value an "extreme performer." Are any of the ten utility stocks extreme performers?
    c. According to the Chebyshev inequality, what is the maximum proportion of observations in a data set that can have standardized values exceeding 2 in absolute value?

*3.61 Consider the responses of homeowners to a survey question about their total expenditures last year on air-conditioning and heating equipment. What does the Chebyshev inequality guarantee is the maximum proportion of these expenditures that fall beyond ±3 standard deviations from the mean? Does the Chebyshev inequality provide any information about the proportion of outlying expenditures below the mean?

3.62 A research chemist reported in an article that in 48 repeated measurements on the yield of a particular chemical reaction, the mean yield was 103.31 grams and the standard deviation was 0.25 gram. What is the minimum number of measurements that were within ±0.5 gram of the mean? Within ±1.0 gram?

3.63 A data set contains 20 observations as follows:

Observation:	−2	−1	0	1	2	3
Frequency:	2	1	13	2	1	1

The mean and the standard deviation of this data set are 0.1000 and 1.1192, respectively.
    a. According to the Chebyshev inequality, what is the minimum proportion of observations that can lie within ±1.5 standard deviations of the mean? Within ±2.5 standard deviations?
    b. What are the actual proportions of observations lying within ±1.5 and ±2.5 standard deviations of the mean in this data set? Are the actual proportions consistent with the Chebyshev inequality?

*3.64 Produce the descriptive statistics presented in the computer output of Figure 3.8 for the following data set:

$$21 \quad 16 \quad 8 \quad 53 \quad 108 \quad 73$$

3.65 Produce the descriptive statistics presented in the computer output of Figure 3.8 for the following data set:

$$10.8 \quad -4.6 \quad 0.8 \quad 21.8 \quad -17.2$$

*3.66 Refer to **Travel Expenditures** Problem 3.1.
    a. Construct a box plot for this data set.
    b. Does the box plot reveal whether the data set is skewed? Explain.

3.67 Refer to **Lot Defectives** Problem 3.3.
    a. Construct a box plot for this data set.
    b. In the box plot in **a**, the median lies closer to the first quartile than to the third quartile. What does this fact imply about the symmetry of the data set?

3.68 Refer to **Response Times** Problem 2.18. Construct a box plot for the data in this frequency distribution. Assume that 2.5 and 27.5 are the minimum and maximum observations in the data set, respectively.

*3.69 Refer to **Universal Health Care** Problem 2.17.
    a. Construct the box plots for the rating data of the two groups on the same graph.
    b. What does a comparison of the two box plots in **a** reveal about differences in the rating distributions for the two groups?

3.70 Refer to **Failure Times** Problem 2.31.
    a. Construct the box plots for the failure times data of each mode of failure on the same graph.
    b. What does a comparison of the two box plots in **a** reveal about differences in the failure times distributions for the two modes of failure?

*3.71 The charge per day for a semiprivate room in a hospital was as follows in four consecutive years:

Year:	1	2	3	4
Charge ($):	150	160	184	205

    a. Obtain the ratio of the charge in one year to that in the preceding year for years 2, 3, and 4.
    b. Obtain the geometric mean of the three ratios in **a**. Show how the charge in year 4 can be obtained from knowledge of the charge in year 1 and the geometric mean.

3.72 The monthly rates of return on a stock for the past six months follow. The rates of return are based solely on price changes for the stock and ignore dividend payouts.

Month:	1	2	3	4	5	6
Rate of Return (%):	1.8	−0.6	1.3	2.6	3.6	−1.1

    a. The rate of return for the first month, 1.8 percent, indicates that the ratio of the closing stock price for month 1 to the closing stock price for the preceding month (month 0) was 1.018. Record the corresponding price ratios for the other months (that is, the ratio of each month's closing price to the preceding month's closing price).
    b. Calculate the geometric mean of the six price ratios recorded in **a**. On the basis of the geometric mean, what has been the average monthly rate of return on this stock over the past six months?
    c. The closing stock price for month 0 was $50. Use this price and the geometric mean in **b** to calculate the closing stock price for month 6.

*3.73 In flying a distance of 1200 miles between two airports, a cargo aircraft encountered strong head winds and made the trip at a ground speed of 300 miles per hour. The 1200-mile return flight was made at a ground speed of 400 miles per hour. Calculate the harmonic mean of the ground speeds.

3.74 The accident record for a plant, showing number of man-hours worked per accident during a four-year period in which 4 million man-hours were worked in each year, follows.

Year:	1	2	3	4
Man-Hours per Accident:	12,535	10,810	11,691	14,735

   a. Calculate the harmonic mean number of man-hours per accident for the four-year period.
   b. Is the harmonic mean an appropriate measure of position here? (*Hint:* How many accidents occurred during the four-year period, in which 16 million man-hours were worked?)
   c. Would your answer in **b** be affected if the annual number of man-hours worked was not constant during the four-year period? Explain.

EXERCISES

3.75 Prove property (3.4). [*Hint:* Write the sum as $\Sigma(X_i - \bar{X} + \bar{X} - A)^2$.]

3.76 Use (3.9) to demonstrate that the mean distance (irrespective of direction) of observations in a data set from their median is smaller than the mean distance from any other value.

3.77 Rearrange (3.15a) to show that $\Sigma X_i^2 = n\bar{X}^2 + (n - 1)s^2$.

3.78 If $X_1, \ldots, X_n$ are n temperature readings expressed in degrees Celsius (°C) and s is their standard deviation, then what is the standard deviation of these same n readings expressed in degrees Fahrenheit (°F)? Recall that °F $= (9/5)$°C $+ 32$.

3.79 Show that the third moment of the standardized values of a data set equals the standardized skewness measure (3.21) for the data set.

3.80 Prove that the mean and the variance of the standardized values of a data set are 0 and 1, respectively.

3.81 Refer to Problem 3.73. Show that a weighted arithmetic mean of the two ground speeds, using the number of hours flown at each speed as weights, is equivalent to the harmonic mean.

STUDIES

3.82 The price movements of common stocks traded on an exchange such as the New York Stock Exchange are of importance to government, business, academicians, and the public. Effective summarization of the mass of price data generated each trading day requires careful selection of appropriate statistical measures.
   a. Describe the principal statistical measures used by newspapers to report the general trend of common stock prices on a daily basis and to report the price movements of any single stock on a daily, weekly, monthly, and yearly basis.
   b. How appropriate are the measures used? What are their advantages and disadvantages compared with other measures that might be used for summarizing the price data?

3.83 In a sample survey about family health insurance, the following data on family income, health insurance coverage, and age of family head were obtained:

Family Income ($)	Percent of All Families	Percent of Families with Health Insurance	Mean Age of Family Head
Under 15,000	18	24	28
15,000–under 25,000	43	55	39
25,000–under 40,000	27	83	51
40,000 and over	12	91	57

a. What percentage of all families had health insurance?

b. What is the mean age of the family head for all families?

c. How is it possible that the mean age can be calculated in **b**, but the mean income of all families cannot because of the open-end class?

d. How does the median income of families with health insurance compare with the median income of all families? Make specific reference to the data to support your answer.

3.84 A management-labor team initiated a training program to increase productivity and thereby raise the average earnings of piecework employees in a garment plant by $16.00 per week without any change in the piecework rate. Before and after data on hourly earnings follow for the 1668 employees involved in the program.

Hourly Earnings ($)	Number of Employees	
	Before	After
9–under 10	146	148
10–under 11	395	400
11–under 12	735	731
12–under 13	332	23
13–under 14	43	146
14–under 15	17	220
Total	1668	1668

a. If you used the mean hourly earnings as the measure to assess the impact of the training program, how would you evaluate the success of the program?

b. Would you reach the same conclusion if the median were used to measure the impact of the program? Discuss fully.

c. The management-labor team had two objectives in running the program: (1) to improve the output of employees and (2) to improve the economic welfare of employees. Evaluate the extent to which these goals were achieved by comparing the means, the medians, and the ogives of the two distributions. State your findings.

3.85 (Computer needed.) Refer to the **Financial Characteristics** data set. Consider the data on net sales of firms in the paper and electronics industries (industries 4 and 6) in both years.

a. Obtain the descriptive statistics presented in the computer output of Figure 3.8 for the sales data for each year for each of the two industries.

b. Construct box plots to compare the sales data for the two industries for each year. Also, construct box plots to compare the sales data for the two years for each industry.

c. Use the results in **a** and **b** to compare changes in sales between years 1 and 2 and differences in sales between the two industries. Briefly state noteworthy differences and similarities that are revealed by the comparisons.

UNIT TWO

Probability

BASIC PROBABILITY CONCEPTS

The collection, summarization, and exploratory analysis of data, discussed in the preceding chapters, are important first steps in using data for analysis and decision making. We now take up probability theory. This theory is important in its own right, being used by managers and analysts in assessing odds, in constructing probability models, and in selecting courses of action where risk must be taken into account. In addition, the theory of probability is a major building block in the logical foundation of statistical inference, which is the central topic in later units of this book.

We begin with basic probability concepts.

 ## RANDOM TRIALS, SAMPLE SPACES, AND EVENTS

Random Trial

Activities that have uncertain or chance outcomes are common.

EXAMPLES

1. An auditor selects a voucher and examines it. Whether or not the voucher contains an error is uncertain in advance of the examination. If it does contain an error, the type of error is uncertain in advance, since a number of different types of errors may occur.

2. A farmer plants a crop and eventually harvests it. The crop yield is uncertain in advance because of the chance influences of weather and other natural factors.

3. A laboratory centrifuge experiences operational breakdowns from time to time. The cause of the next breakdown is uncertain, there being several possible causes, including different kinds of mechanical and electrical failures. □

Each of the foregoing examples involves an activity referred to as a random trial.

(4.1)   A *random trial* is an activity having two or more different possible outcomes, with uncertainty in advance as to which outcome prevails.

In Example 1, for instance, the random trial is the auditor's examination of a voucher. The possible outcomes are no error, or one or more errors of particular types. In Example 2, the random trial is the planting and harvesting of a crop, and the possible outcomes are different crop yields. In each example, the outcome is uncertain until the trial has been conducted.

Sample Space

The outcomes of a random trial usually can be defined in different ways, depending on the purpose of the study.

EXAMPLE

Delivery Service.   A department store is studying its delivery service. The delivery of a customer's order is the random trial. If interest lies in whether or not the delivery is made on the day the customer's order is placed, the outcomes of interest are the two shown in Figure 4.1a. Here, day 1 signifies delivery on the same day, and delayed delivery signifies delivery after day 1. One of these two outcomes will result from the random trial.

If the purpose of the study is to investigate in detail the nature of delivery delays, interest might be in knowing whether delayed delivery occurred on the second day, on the third day, or after the third day. This more detailed array of outcomes is shown in Figure 4.1b. ☐

We refer to the given set of outcomes of interest for a random trial as the sample space.

FIGURE 4.1   Two univariate sample spaces for the random trial, delivery of customer's order — Delivery service example

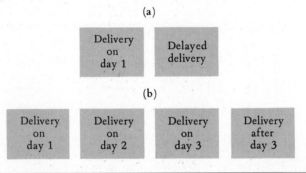

(4.2)

> The different possible outcomes of a random trial are called the *basic out-comes* of the trial. The set of all basic outcomes for a random trial is called the *sample space* of the trial.

Figures 4.1a and 4.1b contain two different sample spaces for the delivery service random trial. The first has two basic outcomes; the second is more detailed and has four basic outcomes.

Comments   1.   The basic outcomes of a sample space are mutually exclusive and exhaustive — that is, the outcome of a random trial will always be one and only one of the basic outcomes in its sample space. In Figure 4.1b, for example, the timing of delivery of a customer's order must be one and only one of the four basic outcomes shown.

2.   A sample space and its basic outcomes are equivalent in structure to a system of classification and the classes of the system, respectively. This correspondence carries over to other characteristics as well, such as the dimension of a sample space, discussed next.

Univariate, Bivariate, and Multivariate Sample Spaces.   The sample spaces in Figures 4.1a and 4.1b are called *univariate sample spaces* because the basic outcomes refer to a single characteristic — in this case, the timing of delivery. For many random trials, each basic outcome refers to two, or more than two, characteristics. The corresponding sample spaces are then called *bivariate* and *multivariate sample spaces,* respectively.

EXAMPLE

In the delivery service example, the timing of the delivery may not be the only characteristic of interest. There may also be concern about whether or not the correct order is delivered to the customer. The resulting bivariate sample space can be displayed in a tabular arrangement, such as in Figure 4.2a. Note that the sample space here contains eight basic outcomes and that they have been arrayed in the same manner as in a bivariate cross-classification system. The univariate sample spaces for the two characteristics considered separately, status of order and timing of delivery, are found at the margins of the bivariate sample space.   □

A *tree diagram* is a useful graphic device for visualizing bivariate or multivariate sample spaces. It is especially helpful when there are more than two characteristics, since multivariate sample spaces are difficult to portray in a table.

EXAMPLE

Figure 4.2b shows a tree diagram for the bivariate sample space in Figure 4.2a. Note that the first branching (on the left) shows the two possible outcomes for the status of the order. Each of these outcomes has a second branching corresponding to the four possible outcomes for the timing of delivery. Thus, the tree ends on the right with eight branches corresponding to the eight basic outcomes of the bivariate sample space. Of course, the tree diagram could have been drawn with the first

FIGURE 4.2 Bivariate sample space for the random trial, delivery of customer's order— Delivery service example. A tree diagram representation is shown in part b.

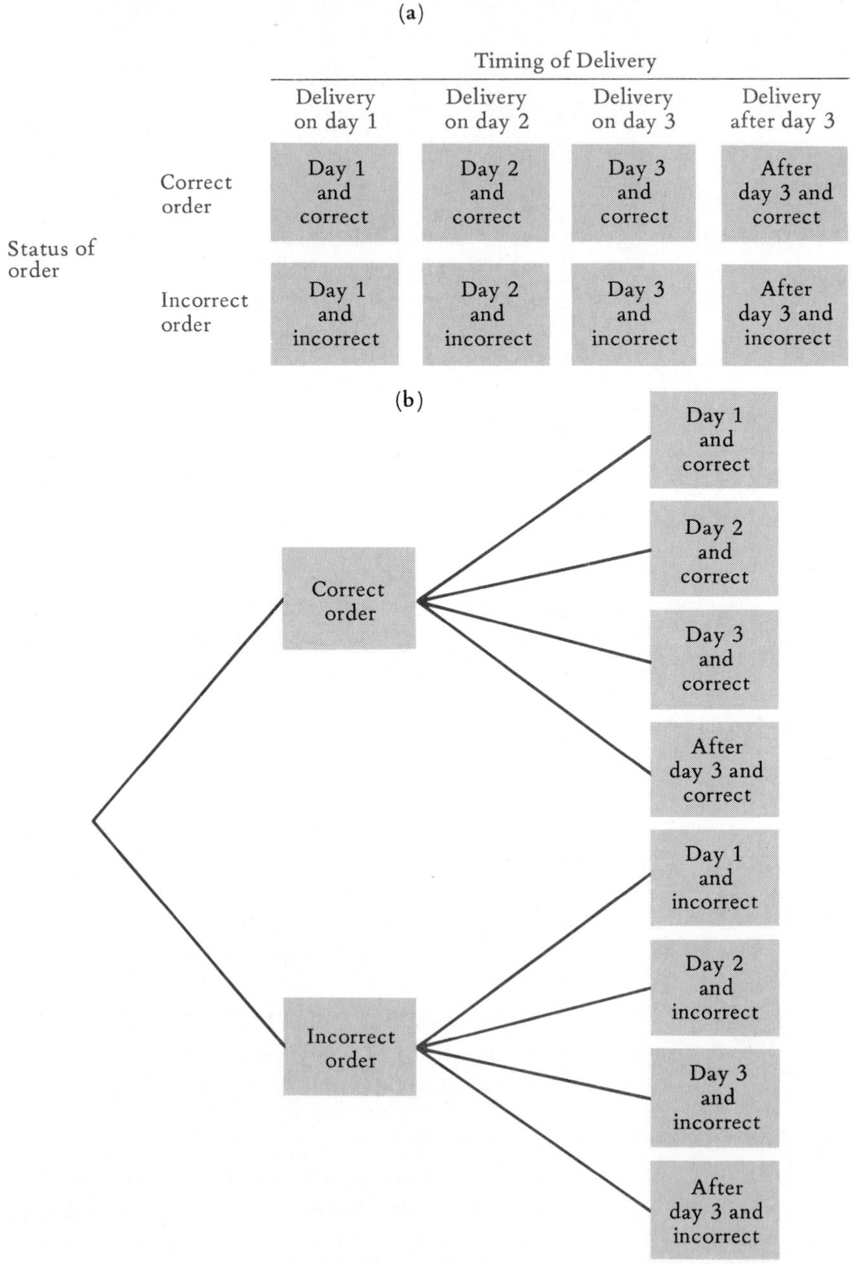
(a)
Timing of Delivery
Delivery on day 1
Delivery on day 2
Delivery on day 3
Delivery after day 3
Status of order
Correct order
Incorrect order
Day 1 and correct
Day 2 and correct
Day 3 and correct
After day 3 and correct
Day 1 and incorrect
Day 2 and incorrect
Day 3 and incorrect
After day 3 and incorrect
(b)
Correct order
Incorrect order
Day 1 and correct
Day 2 and correct
Day 3 and correct
After day 3 and correct
Day 1 and incorrect
Day 2 and incorrect
Day 3 and incorrect
After day 3 and incorrect

branching for the timing of delivery and the second for status of the order and it would still represent the same sample space.   ☐

Event

Interest frequently centers on subsets of basic outcomes of the sample space.

1. The sample space of a student's grade in a course consists of the basic outcomes corresponding to the letter grades A, B, C, D, and F. A subset of basic outcomes of interest consists of those corresponding to a pass in the course (grades A, B, C, and D).

2. In the delivery service example, the department store is concerned with poor delivery service. The store defines poor service as the subset of basic outcomes that involve either delivery of an incorrect order, or delivery after two days, or both.   ☐

Figure 4.3 shows the six basic outcomes constituting poor delivery service in Example 2. Such a subset of basic outcomes is called an event of the sample space. We say that the poor-service event occurs if any one of these six outcomes is realized in the random trial.

(4.3)

An *event* is a subset of basic outcomes of a sample space. An event is said to *occur* if one of its basic outcomes is realized in the random trial.

FIGURE 4.3   **Poor-service event for the random trial, delivery of customer's order —** **Delivery service example**

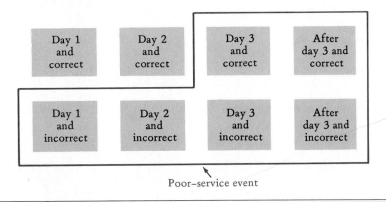

Poor–service event

Many events can be defined for the same sample space. In Example 1, another event of interest might be failure in the course. This event consists of a single basic outcome, grade F. In Example 2, another event of interest might be delivery of incorrect order (whatever the day of delivery).

Set Notation and Terminology

To facilitate subsequent discussion, we introduce some mathematical notation for representing basic outcomes and events of a sample space. (For a review of mathematical set notation and operations, refer to Appendix A, Section A.3.)

Sample Space and Basic Outcomes.   The sample space of a random trial will be denoted by S, and the basic outcomes contained in it will be denoted by $o_1, o_2, \ldots, o_k$, where o_i represents the ith basic outcome and k the number of basic outcomes in S. We view the sample space S as the set containing all the basic outcomes and represent it as $S = \{o_1, o_2, \ldots, o_k\}$.

EXAMPLE

Figure 4.4a shows the sample space S consisting of the eight basic outcomes for the delivery service example.   □

Events.   Symbols such as E and F, or E_1 and E_2, will be used to denote events.

EXAMPLE

The poor-service event described in Figure 4.3 will be denoted by E. Since this event E consists of the basic outcomes o_3, o_4, o_5, o_6, o_7, and o_8 in Figure 4.4a, we will write $E = \{o_3, o_4, o_5, o_6, o_7, o_8\}$. Event E is shown in Figure 4.4b.   □

Complementary Event.   Occasionally, interest centers on all the basic outcomes of a sample space that are *not* contained in some specific event.

(4.4)

The set of all basic outcomes not contained in an event E is called the *complementary event to E* and is denoted by E^*.

EXAMPLE

Figure 4.4c shows the basic outcomes in the complement to the poor-service event E defined in Figure 4.4b. In this case, complementary event E^* consists of all basic outcomes that constitute good delivery service; that is, $E^* = \{o_1, o_2\}$.   □

Event Intersection.   Often, we are interested in the basic outcomes that are common to two events.

(4.5)

The set of basic outcomes that belong to both of the events E_1 and E_2 of a sample space is called the *intersection of E_1 and E_2* and is denoted by:

$$E_1 \cap E_2 \qquad \text{or equivalently} \qquad E_2 \cap E_1$$

FIGURE 4.4 Illustrations of various types of events — Delivery service example

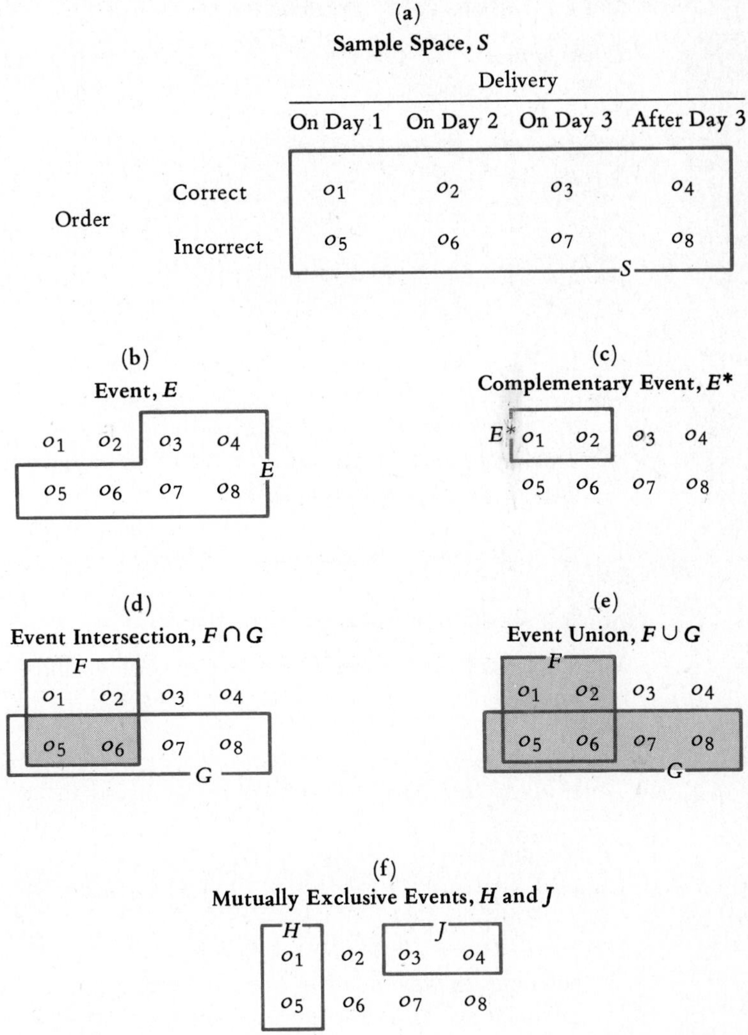

(a)

Sample Space, S

Delivery

	On Day 1	On Day 2	On Day 3	After Day 3
Correct	o_1	o_2	o_3	o_4
Incorrect	o_5	o_6	o_7	o_8

Order

S

(b)

Event, E

o_1 o_2 o_3 o_4

o_5 o_6 o_7 o_8

E

(c)

Complementary Event, E^*

E^* o_1 o_2 o_3 o_4

o_5 o_6 o_7 o_8

(d)

Event Intersection, $F \cap G$

F

o_1 o_2 o_3 o_4

o_5 o_6 o_7 o_8

G

(e)

Event Union, $F \cup G$

F

o_1 o_2 o_3 o_4

o_5 o_6 o_7 o_8

G

(f)

Mutually Exclusive Events, H and J

H J

o_1 o_2 o_3 o_4

o_5 o_6 o_7 o_8

EXAMPLE

In Figure 4.4d, two events are shown: $F = \{o_1, o_2, o_5, o_6\}$ and $G = \{o_5, o_6, o_7, o_8\}$. Here, F represents all basic outcomes in which delivery occurs within two days, and G represents all basic outcomes in which the order is incorrect. The intersection $F \cap G$ is shaded in Figure 4.4d. It is an event consisting of the two basic outcomes that F and G have in common — that is, $F \cap G = \{o_5, o_6\}$. Thus, this intersection represents basic outcomes with both delivery within two days *and* an incorrect order. □

If the intersection of several events, say $E_1, E_2, \ldots, E_n$, is of interest, it is written symbolically as $E_1 \cap E_2 \cap \cdots \cap E_n$ and it represents the set of all basic outcomes that the n events have in common.

Event Union.   At times, we are concerned with all of the basic outcomes that are present in two events.

(4.6)

The set of basic outcomes that belong to at least one of the events E_1 and E_2 of a sample space is called the *union of E_1 and E_2* and is denoted by:

$$E_1 \cup E_2 \quad \text{or equivalently} \quad E_2 \cup E_1$$

EXAMPLE

In Figure 4.4e, the union of events F and G defined previously is shown by the shaded region. It is an event consisting of the six basic outcomes found in one or the other of events F and G or in both — that is, $F \cup G = \{o_1, o_2, o_5, o_6, o_7, o_8\}$. Thus, this union represents the set of basic outcomes with delivery within two days *or* an incorrect order *or both*.

If the union of several events, say $E_1, E_2, \ldots, E_n$, is of interest, it is written symbolically as $E_1 \cup E_2 \cup \cdots \cup E_n$ and it represents the set of all basic outcomes each of which is found in at least one of the n events.

Mutually Exclusive Events.   Some events of a sample space have no basic outcomes in common.

(4.7)

If events E_1 and E_2 of a sample space have no basic outcomes in common, they are said to be *mutually exclusive events*. In this case, $E_1 \cap E_2 = \varnothing$, where $\varnothing$ denotes the empty set.

EXAMPLE

Events H and J in Figure 4.4f are mutually exclusive. Here, H represents all basic outcomes in which delivery occurs on day 1, and J represents all basic outcomes in which the correct order is delivered on day 3 or later.

Any event E and its complement E^* are mutually exclusive. Thus, in the delivery service example, the good- and poor-service events E and E^* are mutually exclusive. Similarly, in the course grade example, the events pass and fail are mutually exclusive.

Infinite Sample Space

The preceding definitions have assumed implicitly that a random trial has only a finite number of basic outcomes. This assumption is convenient for the presentation of basic concepts. There are, however, many settings in which an infinity of possible outcomes is conceivable. For example, the weight of the wheat crop in North America next year can be any of an infinity of values in an interval; thus, the sample space has an infinity of possible outcomes. With only a few qualifications, the concepts for finite sample spaces carry over directly to infinite sample spaces.

PROBABILITY AND ITS POSTULATES

Probability Measure

We have seen that the outcome of a random trial will be one and only one of the basic outcomes in its sample space. However, in advance of the random trial, it is not known with certainty which particular basic outcome will be realized. In probability theory, we assign to each basic outcome o_i a value called its *probability,* which is a measure of how likely it is that o_i will be the realized outcome of the random trial. We shall denote the probability of o_i by $P(o_i)$. In like manner, we shall let $P(E)$ denote the probability that event E will occur, that is, that one of the basic outcomes of event E will be the realized outcome of the random trial. We temporarily postpone discussion about the meaning of the probability measure and how probability values are obtained.

EXAMPLES

1. If the probability of outcome o_1 in Figure 4.4a is 0.57, we write $P(o_1) = 0.57$. Thus, the probability that the delivery is correct and on day 1 is 0.57.

2. If the probability of the event E in Figure 4.4b is 0.25, we write $P(E) = 0.25$. Thus, the probability that the outcome is either o_3, o_4, o_5, o_6, o_7, or o_8—one of the poor-service outcomes—is 0.25. □

Probability Postulates

Mathematicians have developed a formal structure whereby certain properties of probability are defined and from which numerous important consequences flow. One set of postulates on which this formal structure can be built follows.

(4.8)

> *Postulate 1.* $0 \leq P(o_i) \leq 1$ for any basic outcome o_i of the sample space S.
> *Postulate 2.* For any event E of a sample space S:
>
> $$P(E) = \sum_E P(o_i)$$
>
> where the summation is over the basic outcomes contained in E.
> *Postulate 3.* $P(S) = 1$ and $P(\varnothing) = 0$.

The first postulate states that the probability value of each basic outcome in a sample space is a number between 0 and 1. If the occurrence of basic outcome o_i is impossible, then $P(o_i) = 0$. If basic outcome o_i is certain to occur, then $P(o_i) = 1$. The closer the probability measure is to 0, the less likely it is that o_i will occur; and the closer $P(o_i)$ is to 1, the more likely it is that o_i will occur.

The second postulate states that the probability of any event equals the sum of the probability values of the basic outcomes that constitute the event. For instance, if $E = \{o_1, o_5, o_7\}$, then $P(E) = P(o_1) + P(o_5) + P(o_7)$.

The third postulate states that the probability value associated with the entire sample space S is 1 and the probability associated with the empty set $\varnothing$ is 0.

Some Consequences of the Postulates.   A number of important consequences follow directly from the probability postulates. Several of these are given now. Others are presented later as formal probability theorems.

1.   Since the sample space S contains all the basic outcomes, that is, $S = \{o_1, o_2, \ldots, o_k\}$, it follows from postulate 2 that $P(S) = P(o_1) + P(o_2) + \cdots + P(o_k)$. Furthermore, from postulate 3, we have $P(S) = 1$. Therefore, the sum of the probabilities of all the basic outcomes is 1.

(4.9)
$$\sum_{i=1}^{k} P(o_i) = 1$$

2.   If all basic outcomes contained in an event E_1 are also contained in an event E_2, so E_1 is a subset of E_2, it follows from postulate 2 and from the fact that probability values for basic outcomes are nonnegative (postulate 1) that $P(E_1)$ cannot exceed $P(E_2)$.

(4.10)
$$P(E_1) \leq P(E_2) \text{ when event } E_1 \text{ is a subset of event } E_2.$$

For instance, in the delivery service example, E denotes poor delivery service and G denotes an incorrect order. Event G is a subset of E, as can be seen from Figure 4.4. Hence, $P(G) \leq P(E)$.

3.   Since every event E is a subset of the sample space S on which it is defined, it follows from (4.10) that $P(E) \leq P(S)$. Furthermore, postulates 1 and 2 imply that the probability of an event E cannot be negative, and postulate 3 states that $P(S) = 1$. Hence, every event has a probability value between 0 and 1.

(4.11)
$$0 \leq P(E) \leq 1 \qquad \text{for any event } E$$

4. If E_1 and E_2 are any two mutually exclusive events of a sample space, then by definition, they have no basic outcomes in common; that is, $E_1 \cap E_2$ is the empty set $\varnothing$. Thus, by postulate 3, $P(E_1 \cap E_2) = 0$.

(4.12)

> If E_1 and E_2 are mutually exclusive events of a sample space:
>
> $$P(E_1 \cap E_2) = 0$$

Meaning of Probability

We have noted already that the probability for any basic outcome or event is a number between 0 and 1. For instance, suppose that event E denotes the occurrence "It rains in Seattle on August 17," and that the probability value assigned to this event is $P(E) = 0.3$. How is this probability value interpreted, and how does one determine such a value in practice? We now present two interpretations of probability, which will allow the reader to view its meaning from two useful points of view. We shall follow this presentation with a brief discussion of how probability values are assessed.

Objective Interpretation. In the *objective* interpretation of probability, the probability value of an event is equated with the relative frequency of occurrence of the event in the long run under constant causal conditions. Let us again consider the event E "It rains in Seattle on August 17." With the objective interpretation, the probability statement $P(E) = 0.3$ is understood to mean that in an indefinitely large number of August 17s in Seattle, with basic climatic conditions the same as at the present time, there will be rain on about 30 percent of them.

The objective interpretation of probability applies only to repeatable events and not to unique events. For instance, from this point of view, one would not talk about the probability that John William Jones, aged 40, will die during the next year; he either will or won't. John William Jones' 41st year is not a repeatable event. But it does make sense, according to this interpretation of probability, to say that the probability of a male of age 40 dying during the next year is 0.002. This probability is interpreted to mean that among many males of age 40, about 0.2 percent of them will die during their 41st year. Thus, probability is interpreted from the objective point of view with reference to a large number of random trials under the same causal conditions, not with reference to a specific random trial.

Subjective Interpretation. Since many events of interest are not repeatable under constant causal conditions, there are many situations where the objective interpretation of probability cannot be applied. The *subjective* interpretation of probability (also called the *personal* interpretation) relates probability to degree of personal belief. It is not restricted to repeatable events but applies also to unique events. Thus, under this approach, one would be willing to consider the probability that John William Jones, aged 40, will die this coming year, or the probability that the marketing of a new breakfast cereal will be successful. The subjective interpretation even allows one to assign a probability number to an event for a random trial that has occurred in the past. For instance, a police detective investigating the burglary of a store last week might assign a probability of $P(A) = 0.9$ to suspect A having committed this burglary. For all of these events, it would not be possible to interpret probabilities as relative frequencies in the long run since these events are inherently nonrepeatable.

Personal probability is intimately related to the person making the probability evaluation, including subjective feelings and the information available to him or her at the time. Thus, two insurance examiners who have identical information about John William Jones might still arrive at different personal probabilities that he will die during his 41st year. Indeed, each insurance examiner might revise his or her probability assessment upon more reflection. Even with reference to a repeatable event, two individuals might subjectively assess the probabilities differently, depending on the information available to each and how this information is evaluated. On the other hand, with the objective probability interpretation, the probabilities of repeatable events are considered to be determined by the causal conditions that are operating and therefore do not depend on personal factors.

Probability Expressed as Odds Ratio.   Often, probability values are expressed in terms of odds. We say that the *odds in favor* of an event E are a to b if:

$$P(E) = \frac{a}{a + b}$$

1.   If the odds in favor of candidate A winning an election are 3 to 2, then the probability of candidate A winning is $3/(3 + 2) = 0.6$.

2.   If the probability of dying in the next year is 0.002, then the odds of death can be expressed as 0.002 to 0.998, or 1 to 499.   ☐

Assessment of Probability Values

The probability postulates provide conditions that a probability measure for any random trial must satisfy, but by themselves, they do not provide probability numbers. We now consider one method for obtaining probability values consistent with the objective interpretation of probability, and then we take up two methods consistent with the subjective interpretation.

Observed Relative Frequency.   The objective interpretation of probability, as we have seen, equates the probability of an event to the relative frequency of occurrence of the event in a long series of trials conducted under the same conditions. Hence, the observed relative frequency of an event in a large number of trials under approximately constant causal conditions may be used as an estimate of the probability of occurrence. For instance, if a production process has produced 30 defective items in the last 5000 produced and conditions for this production run have been stable and are likely to remain unchanged, then $30/5000 = 0.006$ is an estimate of the probability that the next item produced will be defective.

The use of observed relative frequency is not without difficulties. It requires repeated experience with the random process and constant causal conditions for the repeated trials. It is also assumed that the observed relative frequency has not departed substantially from the "true" long-run relative frequency as the result of chance influences. Statistical procedures (covered in subsequent chapters) exist for determining the precision with which an observed relative frequency approximates the underlying probability value. Nevertheless, the application of any of these statistical procedures involves assumptions, so it is impossible to avoid judgmental factors completely even with the use of observed relative frequencies.

Direct Questioning.   If a subjective probability for an event is required, the most straightforward way to obtain it is to ask a direct question about it. There are various ways in which a probability assessment question can be phrased. We illustrate three questions for a product manager about the likelihood that the sales of a new product will exceed the sales target.

1.   "What is the probability that this new product will exceed its sales target?"
2.   "How many chances out of ten does this new product have of exceeding its sales target?"
3.   "What are the odds in favor of this new product exceeding its sales target?"

Like observed relative frequency, the direct-question approach to probability assessment has some inherent difficulties. The respondent may find it difficult to express probabilistic beliefs in quantitative terms, or the answer may vary depending on when the question is asked and how much thought is given to it. There is also the problem of who should be the source of the probability assessment. Should the company president ask the sales manager for a probability assessment about the sales of a new product, or should the marketing research director be asked? If the president asks both and different probability assessments are furnished, how are they to be combined?

Another difficulty is the tendency of some respondents to provide biased probability assessments. For instance, the director of a computer center assigned a probability of 0.3 each week to the possibility of a computer breakdown the following week. Yet there were computer breakdowns in only 10 percent of the weeks during the past two years, suggesting that the director tended to overestimate the probability of a computer breakdown.

A final difficulty is that the probability assessments provided by respondents may not satisfy the probability postulates. For instance, a precious-metals trader responded that the probability of gold having a higher price next Monday is 0.6 and yet also responded that the probability of *both* gold *and* silver having higher prices next Monday is 0.7. These responses are inconsistent with the inequality in (4.10).

Indirect Methods.   A host of methods exist for eliciting subjective probabilities by indirect means. We shall illustrate one method for ascertaining the sales manager's assessment that the sales of a new product will exceed the sales target. The first question is:

> Would you accept the gamble of receiving $5 if sales exceed the target and paying $5 if they do not?

If the answer is yes, one may impute that the subjective probability of sales exceeding the target is at least 0.5. A second question then might be:

> Would you accept the gamble of receiving $4 if sales exceed the target and paying $5 if they do not?

If the answer again is yes, one may impute that the manager views the odds as at least 5 to 4 in favor of sales exceeding the target. Hence, the subjective probability is at least $5/(5 + 4) = 5/9$ that sales will exceed the target.

By considering additional gambles with still other payoffs, one can narrow the sales manager's subjective probability of sales exceeding the target to a small interval.

4.3   PROBABILITY DISTRIBUTIONS

To employ probability theory, one must assign probability numbers to all possible outcomes of a random trial.

(4.13)

> An assignment of probabilities to each of the basic outcomes in a sample space is called a *probability distribution* for the sample space.

EXAMPLE

Consider the sample space in Figure 4.1b relating to the delivery service example. Probabilities have been assigned in Table 4.1a to each of the four basic outcomes in this sample space, on the basis of relative frequency in the recent past. Notice that none of the probability values is less than 0 and that the sum of all the probabilities equals 1, in accordance with postulates 1 and 3 in (4.8).   ☐

Univariate, Bivariate, and Multivariate Probability Distributions

The probability distribution in Table 4.1a is called a *univariate probability distribution* because it is based on a univariate sample space. Probability distributions for bivariate and multivariate sample spaces can be constructed as well. They are referred to as *bivariate* and *multivariate probability distributions,* respectively.

EXAMPLE

Table 4.1b contains a bivariate probability distribution for the bivariate sample space given earlier in Figure 4.2a. The two variables forming the bivariate sample space are status of order and timing of delivery.   ☐

TABLE 4.1 Univariate and bivariate probability distributions—Delivery service example

(a)

Basic Outcome	Delivery	Probability
B_1	On day 1	0.60
B_2	On day 2	0.20
B_3	On day 3	0.10
B_4	After day 3	0.10
	Total	1.00

(b)

Status of Order	Timing of Delivery				Total
	Delivery on Day 1 B_1	Delivery on Day 2 B_2	Delivery on Day 3 B_3	Delivery after Day 3 B_4	
A_1: Correct order	0.57	0.18	0.08	0.07	0.90
A_2: Incorrect order	0.03	0.02	0.02	0.03	0.10
Total	0.60	0.20	0.10	0.10	1.00

(c)

	B_1	B_2	B_3	B_4	Total
A_1:	$P(A_1 \cap B_1)$	$P(A_1 \cap B_2)$	$P(A_1 \cap B_3)$	$P(A_1 \cap B_4)$	$P(A_1)$
A_2:	$P(A_2 \cap B_1)$	$P(A_2 \cap B_2)$	$P(A_2 \cap B_3)$	$P(A_2 \cap B_4)$	$P(A_2)$
Total	$P(B_1)$	$P(B_2)$	$P(B_3)$	$P(B_4)$	1.00

Joint, Marginal, and Conditional Probability Distributions

Joint Probabilities.   As discussed earlier, in bivariate and multivariate sample spaces, each basic outcome refers to two or more characteristics. Thus, in Table 4.1b, the basic outcome of delivery of a correct order on day 1 refers to the two characteristics (1) correct order (A_1) and (2) delivery on day 1 (B_1). Each of these characteristics can, in our new terminology, be called an event of the bivariate sample space. Thus, the basic outcome of delivery of a correct order on day 1 can be considered a *joint outcome* of the two events A_1 and B_1.

1. The basic outcome of delivery of a correct order on day 1 is denoted symbolically by $A_1 \cap B_1$, since it is the joint outcome of events A_1 and B_1. The probability of this joint outcome is denoted by $P(A_1 \cap B_1)$ in Table 4.1c. From Table 4.1b, we see that this probability is $P(A_1 \cap B_1) = 0.57$.

2. The basic outcome of an incorrect order delivered on day 3 is the joint outcome of the event incorrect order (A_2) and the event delivery on day 3 (B_3). This joint outcome is denoted by $A_2 \cap B_3$, and the corresponding probability according to Table 4.1b is $P(A_2 \cap B_3) = 0.02$.   ☐

The probability of a joint outcome is called a *joint probability*. Thus, $P(A_1 \cap B_1) = 0.57$ is a joint probability.

A bivariate or multivariate probability distribution, such as that in Table 4.1b, is called a *joint probability distribution* when it is desired to emphasize the fact that the basic outcomes are determined jointly by two or more characteristics. Table 4.1c shows the joint probability distribution for the delivery service example in symbolic form.

Marginal Probabilities.   With a bivariate probability distribution, one is also frequently interested in the probability distributions of the individual variables considered separately. Thus, management may wish to know the probability that an order is incorrect, irrespective of the timing of delivery, or to know the probability that an order is delivered on day 1, irrespective of whether or not it is correct. The univariate probability distribution for each of the individual variables can be obtained by summing the joint probabilities across the columns or rows, as the case may be.

1. The univariate probability distribution in Table 4.1a corresponding to the variable timing of delivery may be obtained by summing each column in Table 4.1b. The resulting probabilities are shown in the bottom row of Table 4.1b and are boxed for emphasis. For example, we have $P(B_1) = 0.60$.

2. The univariate probability distribution for the variable status of order is obtained by summing each row in Table 4.1b. The resulting probabilities are shown in the rightmost column and are boxed for emphasis. For example, we have $P(A_2) = 0.10$.   ☐

The rationale for summing will be explained for $P(B_1)$, the probability of delivery on day 1. Since event B_1, if it occurs, must occur jointly with correct order (A_1) or incorrect order (A_2) but not both, it follows that the joint outcomes $A_1 \cap B_1$ and $A_2 \cap B_1$ are mutually exclusive and exhaustive of the possible outcomes leading to the occurrence of B_1. Therefore:

$$P(B_1) = P(A_1 \cap B_1) + P(A_2 \cap B_1) = 0.57 + 0.03 = 0.60$$

Since the probabilities obtained by summing across either one of the classifications are shown in the margins of Table 4.1b, they are often called *marginal probabilities,* and the resulting probability distributions are then called *marginal probability distributions.*

EXAMPLE

Table 4.1a contains the marginal probability distribution for the timing of delivery shown in the horizontal box of Table 4.1b. The marginal probabilities for the second variable, status of order, are shown in the vertical box in Table 4.1b. □

In general, the marginal probabilities are found from a bivariate probability distribution as follows.

(4.14)

$$P(A_i) = \sum_j P(A_i \cap B_j)$$

$$P(B_j) = \sum_i P(A_i \cap B_j)$$

where the summations are over all events B_j and A_i, respectively.

Table 4.1c illustrates the appropriate summations to be performed for obtaining the marginal probabilities.

Conditional Probabilities. Often, it is desired to know the probability of one event occurring, given that a second event occurs. For instance, referring to Table 4.1b, we may wish to know the probability that an order is incorrect (A_2), given that the order is delivered on the third day (B_3). We shall denote this probability by $P(A_2|B_3)$. The vertical rule in the notation is read as "given." In other words, $P(A_2|B_3)$ means "probability of A_2 occurring, given that B_3 occurs." This type of probability is called a conditional probability.

(4.15)

If E_1 and E_2 are any two events of a sample space and $P(E_2)$ is not equal to zero, the *conditional probability of E_1, given E_2,* is denoted by $P(E_1|E_2)$ and defined:

$$P(E_1|E_2) = \frac{P(E_1 \cap E_2)}{P(E_2)}$$

We illustrate the rationale of the definition of conditional probability by an example.

EXAMPLE

Job Placement. Table 4.2 shows, for job placement at a college, the joint probability distribution for the two variables (1) location of a graduate's first job (C_i) and (2) length of time the graduate remains in this job (D_j). We wish to obtain $P(D_1|C_1)$, the conditional probability that a graduate remains in the first job for less than two years, given that the job is in the private sector. In accordance with the relative frequency interpretation of the probability $P(C_1) = 0.60$ in Table 4.2, we know that, of every 100 placements in first jobs, on the average 60 are in

TABLE 4.2   Joint probability distribution — Job placement example

| Job Location C_i | Job Duration D_j | | Total |
	Less Than Two Years D_1	Two Years or Longer D_2	
C_1: Private sector	0.45	0.15	0.60
C_2: Public sector	0.35	0.05	0.40
Total	0.80	0.20	1.00

the private sector. Furthermore, we can state, on the basis of the probability $P(C_1 \cap D_1) = 0.45$, that of every 100 first-job placements, on the average 45 are in the private sector *and* have a duration of less than two years. Thus, of every 60 first-job placements in the private sector, on the average 45 entail a stay of less than two years. The proportion $45/60 = 0.75$ is the conditional probability that a graduate remains in the first job for less than two years, given that the job is in the private sector.

Formula (4.15) provides this same result formally, as follows:

$$P(D_1|C_1) = \frac{P(C_1 \cap D_1)}{P(C_1)} = \frac{0.45}{0.60} = 0.75$$

The direct computation of conditional probabilities by (4.15) is illustrated by two additional examples.

EXAMPLES

1. For the delivery service example in Table 4.1b, we wish to obtain $P(A_2|B_3)$, that is, the conditional probability of A_2 occurring, given that B_3 occurs. Since $P(B_3) = 0.10$ and $P(A_2 \cap B_3) = 0.02$, we obtain:

$$P(A_2|B_3) = \frac{P(A_2 \cap B_3)}{P(B_3)} = \frac{0.02}{0.10} = 0.20$$

2. For the delivery service example in Table 4.1b, we wish to find the conditional probability of delivery on day 1, given that the order is correct. We obtain:

$$P(B_1|A_1) = \frac{P(A_1 \cap B_1)}{P(A_1)} = \frac{0.57}{0.90} = 0.63$$

Conditional Probability Distributions.   A conditional probability distribution is the set of conditional probabilities conditioned on a given event.

EXAMPLES

1. For the job placement example, we wish to obtain the conditional probability distribution for duration of first job (D_j), given that the job is in the private sector (C_1). We require the set of conditional probabilities $P(D_j|C_1)$. From previous

TABLE 4.3 Two conditional probability distributions — Delivery service example

(a)

Status of Order	Conditional Probability, Given B_3
A_1: Correct order	$P(A_1 \mid B_3) = 0.08/0.10 = 0.80$
A_2: Incorrect order	$P(A_2 \mid B_3) = 0.02/0.10 = \underline{0.20}$
	Total 1.00

(b)

Timing of Delivery	Conditional Probability, Given A_1
B_1: On day 1	$P(B_1 \mid A_1) = 0.57/0.90 = 0.63$
B_2: On day 2	$P(B_2 \mid A_1) = 0.18/0.90 = 0.20$
B_3: On day 3	$P(B_3 \mid A_1) = 0.08/0.90 = 0.09$
B_4: After day 3	$P(B_4 \mid A_1) = 0.07/0.90 = \underline{0.08}$
	Total 1.00

work, we know that $P(D_1 \mid C_1) = 0.75$. We calculate, from Table 4.2, that $P(D_2 \mid C_1) = P(D_2 \cap C_1)/P(C_1) = 0.15/0.60 = 0.25$. Hence, the conditional probability distribution for duration of first job, given that the job is in the private sector, is:

Duration	Conditional Probability, Given C_1
D_1: Less than two years	0.75
D_2: Two years or longer	$\underline{0.25}$
Total	1.00

Note that the conditional probability distribution has the attributes of any probability distribution: It involves outcome probabilities between 0 and 1, and these probabilities sum to 1.

2. For the delivery service example in Table 4.1b, we wish to obtain two conditional probability distributions: (1) for status of order, given that delivery is on day 3, and (2) for timing of delivery, given that the order is correct. Table 4.3 shows these two conditional distributions and how they were obtained. ☐

 4.4 BASIC PROBABILITY THEOREMS

Several useful theorems follow from the probability postulates and concepts presented previously.

Addition Theorem

We often wish to obtain the probability that one of two events, or both, occur.

EXAMPLES

1. In the delivery service example in Table 4.1b, what is the probability that the delivery is correct, or the delivery is on day 1, or both; that is, what is $P(A_1 \cup B_1)$?

2. In the job placement example in Table 4.2, what is the probability that the first job is in the private sector, or the graduate remains in the job for less than two years, or both; that is, what is $P(C_1 \cup D_1)$? □

The addition theorem may be used to find the probability that one of two events, or both, occur.

(4.16)
> *Addition Theorem.* For any two events E_1 and E_2 of a sample space:
>
> $$P(E_1 \cup E_2) = P(E_1) + P(E_2) - P(E_1 \cap E_2)$$

EXAMPLES

1. Refer to Table 4.1b. We wish to find $P(A_1 \cup B_1)$. We have $P(A_1) = 0.90$, $P(B_1) = 0.60$, and $P(A_1 \cap B_1) = 0.57$. Hence, by (4.16):

   $$P(A_1 \cup B_1) = 0.90 + 0.60 - 0.57 = 0.93$$

2. Refer to Table 4.2. We wish to obtain $P(C_1 \cup D_1)$. We have $P(C_1) = 0.60$, $P(D_1) = 0.80$, and $P(C_1 \cap D_1) = 0.45$. Thus, by (4.16):

   $$P(C_1 \cup D_1) = 0.60 + 0.80 - 0.45 = 0.95$$

3. Refer to the sample space in Figure 4.5a. The six basic outcomes of the sample space are indicated by circles, and the probability of each outcome is shown in its circle. Two events are defined on this sample space. Note that E contains four basic outcomes, F contains three, and two outcomes are common to both events. For these two events, we have:

   $$P(E) = 0.34 + 0.15 + 0.10 + 0.07 \qquad = 0.66$$
   $$P(F) = \qquad\qquad 0.10 + 0.07 + 0.22 = 0.39$$
   $$P(E \cap F) = \qquad\qquad 0.10 + 0.07 \qquad = 0.17$$

   Substituting these values into addition theorem (4.16), we obtain:

   $$P(E \cup F) = 0.66 + 0.39 - 0.17 = 0.88$$

   By adding the probabilities for the basic outcomes in the event $E \cup F$ directly from Figure 4.5a, we obtain:

   $$P(E \cup F) = 0.34 + 0.15 + 0.10 + 0.07 + 0.22 = 0.88$$

   which is the result provided by the addition theorem. □

FIGURE 4.5   Illustrations of addition and complementation theorems

(a)   (b)

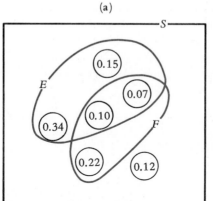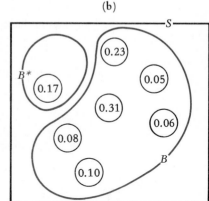

From the third example, it is clear why $P(E \cap F)$ must be subtracted from $P(E) + P(F)$ to obtain $P(E \cup F)$. Note that when $P(E)$ and $P(F)$ are added, the probabilities of the two basic outcomes these events have in common are included twice. Since $E \cap F$ is the set of basic outcomes the two events have in common, $P(E \cap F)$ is subtracted to correct for the double counting.

Mutually Exclusive Events.   The addition theorem takes on a special form when events E_1 and E_2 are mutually exclusive. From (4.12), we know that $P(E_1 \cap E_2) = 0$ in this case. Therefore, the addition theorem simplifies as follows.

(4.17)   For any two mutually exclusive events E_1 and E_2 of a sample space:

$$P(E_1 \cup E_2) = P(E_1) + P(E_2)$$

EXAMPLE   In Table 4.1b, the events delivery on day 1 (H) and delivery of correct order after day 2 (J) are mutually exclusive. Since $P(H) = 0.57 + 0.03 = 0.60$ and $P(J) = 0.08 + 0.07 = 0.15$, we have, by (4.17), that $P(H \cup J) = 0.60 + 0.15 = 0.75$. □

Formula (4.17) may be extended as follows.

(4.18)   For n mutually exclusive events $E_1, E_2, \ldots, E_n$ of a sample space:

$$P(E_1 \cup E_2 \cup \cdots \cup E_n) = P(E_1) + P(E_2) + \cdots + P(E_n)$$

Complementation Theorem

Sometimes, it is more difficult to find the probability of an event than to find the probability of its complement. One can then obtain the probability of the event by the complementation theorem.

(4.19)

Complementation Theorem.   For any event E of a sample space:

$$P(E) = 1 - P(E^*)$$

where:

E^* is the complementary event to E

EXAMPLES

1. For the delivery service example in Table 4.1b, we wish to find the probability that service is not perfect, that is, that the correct order is not delivered on day 1. This event E consists of all outcomes other than $A_1 \cap B_1$. Hence, $E^* = A_1 \cap B_1$. We have $P(E^*) = 0.57$, and by the complementation theorem, $P(E) = 1 - 0.57 = 0.43$.

2. In Figure 4.5b, $P(B) = 1 - P(B^*) = 1 - 0.17 = 0.83$.   □

Multiplication Theorem

Frequently, we wish to obtain the joint probability of two events.

EXAMPLES

1. Sally McQue has applied to two law schools. What is the probability that she will be accepted by both schools?

2. At a police road checkpoint, each vehicle is checked for having a current safety inspection sticker, and the driver is checked for having a current driver's license. What is the probability that a vehicle will have a current inspection sticker and the driver a current driver's license?   □

The joint probability of two events may be obtained by rearranging the definition of conditional probability (4.15). The result is the multiplication theorem.

(4.20)

Multiplication Theorem.   For any two events E_1 and E_2 of a sample space:

$$P(E_1 \cap E_2) = P(E_1)P(E_2 | E_1) = P(E_2)P(E_1 | E_2)$$

EXAMPLES

1. In the law school admissions example, let H and B denote acceptance by schools 1 and 2, respectively. The probability that Sally McQue is accepted by law school 1

is known to be $P(H) = 0.6$. Given acceptance by law school 1, the conditional probability of acceptance by law school 2 is $P(B|H) = 0.9$. We wish to find the probability of acceptance by both schools. Using (4.20), we obtain:

$$P(H \cap B) = P(H)P(B|H) = 0.6(0.9) = 0.54$$

2. For the traffic checkpoint example, let events S and D denote current safety inspection sticker and current driver's license, respectively. It is known that 90 percent of vehicles have current safety inspection stickers and that 95 percent of drivers of vehicles with current inspection stickers have current driver's licenses. We wish to find the probability that both driver and vehicle will be found to have current documents. We have $P(S) = 0.90$ and $P(D|S) = 0.95$. Hence, we obtain by (4.20):

$$P(S \cap D) = P(S)P(D|S) = 0.90(0.95) = 0.855 \qquad \square$$

When one is using the multiplication theorem to find several related joint probabilities, a *probability tree* is often helpful to portray the situation.

EXAMPLE

Computer Chip Inspection.   A computer manufacturer inspects memory chips 100 percent before they enter assembly operations. Let D denote that a chip is defective and D^* that it is acceptable (that is, not defective). Also, let A denote that a chip is approved for assembly by the inspector, and A^* that it is not approved. From past experience, it is known that $P(D) = 0.10$. Also, it is known that the probability of an inspector passing a chip, given it is defective, is $P(A|D) = 0.005$, while the corresponding probability, given the chip is acceptable, is $P(A|D^*) = 0.999$. The tree diagram in Figure 4.6 is called a probability tree and summarizes the situation.

We use multiplication theorem (4.20) to find the joint probability that a chip is defective and is approved for assembly:

$$P(D \cap A) = P(D)P(A|D) = 0.10(0.005) = 0.0005$$

Similarly, the joint probability that a chip is acceptable and is approved for assembly is:

$$P(D^* \cap A) = P(D^*)P(A|D^*) = 0.90(0.999) = 0.8991$$

The other joint probabilities in Figure 4.6 have been obtained in similar fashion.

Note that since a chip is either acceptable or defective, we can find the probability that a chip is approved for assembly as follows:

$$P(A) = P(D \cap A) + P(D^* \cap A) = 0.0005 + 0.8991 = 0.8996 \qquad \square$$

The multiplication theorem can be extended to more than two events. For three events, the multiplication theorem is as follows.

FIGURE 4.6   **Illustration of a probability tree — Computer chip inspection example**

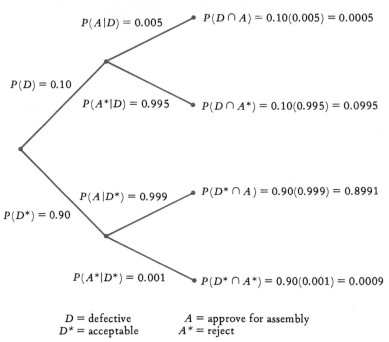

$P(A|D) = 0.005$ $P(D \cap A) = 0.10(0.005) = 0.0005$

$P(D) = 0.10$

$P(A^*|D) = 0.995$ $P(D \cap A^*) = 0.10(0.995) = 0.0995$

$P(A|D^*) = 0.999$ $P(D^* \cap A) = 0.90(0.999) = 0.8991$

$P(D^*) = 0.90$

$P(A^*|D^*) = 0.001$ $P(D^* \cap A^*) = 0.90(0.001) = 0.0009$

D = defective A = approve for assembly
D^* = acceptable A^* = reject

(4.21)   For any three events E_1, E_2, and E_3 of a sample space:

$$P(E_1 \cap E_2 \cap E_3) = P(E_1)P(E_2|E_1)P(E_3|E_1 \cap E_2)$$

The three events in (4.21) can be ordered in any fashion desired.

4.5   STATISTICAL INDEPENDENCE AND DEPENDENCE

The concept of statistical independence, and its contrast with statistical dependence, is the subject of the following examples.

EXAMPLES

1. Table 4.4 contains, for the computer chip inspection example, the probability distributions of inspector's action for a chip inspected, conditional on whether or not the chip is defective. Note that the probability of the chip being approved for assembly differs, depending on whether or not it is defective. This difference sug-

TABLE 4.4 **Conditional probability distributions for the inspector's action, given the quality of the chip — Computer chip inspection example.** The differences between the two conditional probability distributions indicate that quality of chip and inspector's action are not independent.

	Conditional On:	
Action	Chip Defective D	Chip Acceptable $D*$
Approve (A)	0.005	0.999
Reject ($A*$)	0.995	0.001
Total	1.000	1.000

gests that inspector's action is related to the quality of the chip. One would hope, of course, that such a relation exists so that it is both highly likely that acceptable chips are approved and that defective ones are rejected.

2. **Length of Hospital Stay.** A hospital administrator is studying the relationship between length of stay in hospital and whether or not the patient has hospital insurance, for patients with a certain illness. The joint probability distribution is shown in Table 4.5a. The conditional probabilities for length of stay, given insurance status, are shown in Table 4.5b. The conditional probabilities were obtained in the usual manner. For example, $P(A_1|B_1) = 0.42/0.70 = 0.60$. Note that for both the insured and uninsured, the conditional probabilities of any given length of stay are the same and are equal to the marginal probability of that class; that is, $P(A_i|B_j) = P(A_i)$ for all i and j. This result indicates that insurance status of the patient is unrelated to length of stay. A similar result is found if we look at conditional probabilities of insurance status, given the length of stay of the patient. □

Before considering statistical independence of two or more variables, we must first take up statistical independence of events.

Independent Events

We say that two events are unrelated statistically if the probability that one event occurs is unaffected by whether or not the other event occurs. When two events are unrelated in this sense, they are said to be statistically independent.

Symbolically, the statistical independence of two events E_1 and E_2 means that $P(E_2|E_1) = P(E_2)$ or, equivalently, that $P(E_1|E_2) = P(E_1)$. Substituting either of these relations into multiplication theorem (4.20) yields the formal definition of statistical independence.

TABLE 4.5   **Joint and conditional probability distributions for the length of hospital stay example.**   The identity of the conditional probability distributions indicates that length of stay and insurance status are independent variables.

(a) Joint probability distribution

Length of Stay (days)	Insurance Status		
	Insured B_1	Uninsured B_2	Total
A_1: Less than 5	0.42	0.18	0.60
A_2: 5–10	0.21	0.09	0.30
A_3: Over 10	0.07	0.03	0.10
Total	0.70	0.30	1.00

(b) Conditional probability of length of stay, given insurance status

Length of Stay (days)	Conditional On:		
	Insured B_1	Uninsured B_2	Total
A_1: Less than 5	0.60	0.60	0.60
A_2: 5–10	0.30	0.30	0.30
A_3: Over 10	0.10	0.10	0.10
Total	1.00	1.00	1.00

(4.22)

Two events E_1 and E_2 of a sample space are said to be *statistically independent* if:

$$P(E_1 \cap E_2) = P(E_1)P(E_2)$$

Thus, two events are statistically independent if their joint probability equals the product of their marginal probabilities.

EXAMPLE

For the length of hospital stay example in Table 4.5a, A_1 and B_1 are statistically independent events since $P(A_1 \cap B_1) = 0.42$ and $P(A_1)P(B_1) = 0.60(0.70) = 0.42$. In other words, a length of stay of less than five days is statistically independent of a patient being insured.   ☐

Formula (4.22) may be extended as follows.

(4.23)

> For n independent events $E_1, E_2, \ldots, E_n$ of a sample space:
>
> $$P(E_1 \cap E_2 \cap \cdots \cap E_n) = P(E_1)P(E_2) \cdots P(E_n)$$

EXAMPLE

In a new security system being tested at a military base, visitors wishing to enter a secure area must pass three security checks in succession — a voice pattern check, a fingerprint check, and a handwriting check. Studies indicate a probability of 0.02 that an unauthorized visitor will pass any one check. The three checks may be considered to be statistically independent. We wish to find the probability that an unauthorized visitor passes all three checks.

Let V, F, and H denote the events that an unauthorized person passes the voice, fingerprint, and handwriting checks, respectively. We have $P(V) = P(F) = P(H) = 0.02$. Hence, by (4.23):

$$P(V \cap F \cap H) = P(V)P(F)P(H) = 0.02(0.02)(0.02) = 0.000008$$

Of course, a materially different probability might apply if independence does not hold for the three checks.　□

Independent Variables

The concept of two independent events extends to the concept of two statistically independent variables or characteristics. We denote the classes of the two variables by A_1, A_2, and so on, and B_1, B_2, and so on, respectively.

(4.24)

> Two variables A and B are said to be *statistically independent* if:
>
> $$P(A_i \cap B_j) = P(A_i)P(B_j) \qquad \text{for all } A_i \text{ and } B_j$$

Note that *all* joint probabilities must equal the product of their corresponding marginal probabilities for the two variables to be statistically independent.

EXAMPLE

In Table 4.5a for the length of hospital stay example, length of stay and insurance status are statistically independent variables since the joint probability of any two events A_i and B_j is equal to the product of the respective marginal probabilities. For example, $P(A_1 \cap B_2) = 0.18$, and $P(A_1)P(B_2) = 0.60(0.30) = 0.18$.

Statistical independence of the two variables implies that the conditional distribution of length of stay for insured patients is the same as the conditional distribution of length of stay for uninsured patients, and that both conditional distributions are equal to the marginal distribution of length of stay, as we noted earlier in Table 4.5b.　□

Dependent Events and Variables

Dependent Events.   If two events are not statistically independent, they are related statistically and are said to be *statistically dependent events*.

EXAMPLE

For the computer chip inspection example in Figure 4.6, the events D and A are statistically dependent because $P(D \cap A) = 0.0005$ while $P(D)P(A) = 0.10(0.8996) = 0.08996$. [Recall that $P(A) = 0.8996$ was computed earlier.]   □

Comment

Mutually exclusive events are generally statistically dependent. For mutually exclusive events E_1 and E_2, we have, from (4.12), that $P(E_1 \cap E_2) = 0$. Hence, whenever both $P(E_1) > 0$ and $P(E_2) > 0$, condition (4.22) for statistical independence cannot be met, and E_1 and E_2 are statistically dependent. Intuitively, the statistical dependence is reasonable since if one event occurs, we know that the other has not occurred because they are mutually exclusive.

Dependent Variables.   If two variables are not statistically independent, they are related statistically and are said to be *statistically dependent variables*. Statistical dependence is present in many bivariate probability distributions.

EXAMPLE

In the delivery service example in Table 4.1b, status of order and timing of delivery are statistically dependent variables. This dependence can be seen quickly by multiplying corresponding pairs of marginal probabilities. For instance, $P(A_2)P(B_1) = 0.10(0.60) = 0.06$, which does not agree with $P(A_2 \cap B_1) = 0.03$. To be sure, some of the joint probabilities in Table 4.1b are equal to products of corresponding marginal probabilities. However, since not *all* are equal, there is statistical dependence between the two variables.   □

Nature of Statistical Dependence.   The nature of the statistical relationship between two dependent variables can be studied readily by examining the conditional probability distributions for one variable, conditional on each possible outcome of the other variable. This method of analysis is analogous to the one discussed in Section 2.3 for examining the nature of the relationship between two variables in a bivariate data set.

EXAMPLES

1. For the delivery service example, the conditional probability distribution for the timing of delivery, given that the order is correct, was derived in Table 4.3b and is repeated in Table 4.6. The corresponding conditional probability distribution for the timing of delivery, given that the order is incorrect, is also presented in Table 4.6. A comparison of these two conditional probability distributions facilitates the study of the statistical relationship between timing of delivery and status of the order. We see that incorrect orders are much less likely to be delivered on day 1 than correct orders and much more likely to be delivered on day 3 or thereafter.

2. For the computer chip inspection example in Table 4.4, we noted earlier that an inspector's action and the quality of a chip are statistically dependent variables.

TABLE 4.6 **Conditional probability distributions—Delivery service example.** The distributions show the nature of the statistical dependence between timing of delivery and status of order.

Timing of Delivery		Conditional on Status of Order	
		Correct A_1	Incorrect A_2
B_1:	On day 1	0.63	0.30
B_2:	On day 2	0.20	0.20
B_3:	On day 3	0.09	0.20
B_4:	After day 3	0.08	0.30
	Total	1.00	1.00

We can see from the conditional probability distributions in Table 4.4 that the probability of a chip being rejected or approved by the inspector is very close to 1, in accordance with whether the chip is defective or acceptable, respectively. This form of statistical relationship is, of course, exactly what should be found in an effective quality control system. ☐

4.6 OPTIONAL TOPIC—BAYES THEOREM

Bayes theorem represents a special application of conditional probabilities. The theorem describes how marginal probabilities for a variable, called *prior probabilities,* may be revised to reflect the acquisition of additional information about the variable. The revised probabilities are called *posterior probabilities.*

EXAMPLES

1. **Martin Wilson.** Martin Wilson has submitted an application for a trainee position. It is known that the company hires 4 percent of applicants. We shall denote the event that applicant is hired by A_1, so $P(A_1) = 0.04$. We shall denote the event that applicant is not hired by A_2, so $P(A_2) = 1 - 0.04 = 0.96$, because A_1 and A_2 are complementary events. Only some of the applicants are called for a formal interview. We shall denote the event called for interview by B_1 and the event not called for interview by B_2. It is known that among all applicants hired, 98 percent receive interviews (the other 2 percent being hired on the basis of written credentials alone). Hence, we know that $P(B_1|A_1) = 0.98$. Furthermore, it is known that among all applicants not hired, only 1 percent are interviewed; hence, $P(B_1|A_2) = 0.01$. Martin Wilson has been called for an interview and wishes to know the probability that he will be hired; that is, he wishes to determine $P(A_1|B_1)$.

The marginal probabilities $P(A_1)$ and $P(A_2)$ constitute initial information about the chances of being hired and are the prior probabilities. The desired conditional probability $P(A_1|B_1)$ is the revised or posterior probability because it is based on additional information, namely, that Martin Wilson has been called for an interview.

2. **Test Market.**   A new frozen food product can either turn out to be a failure in the market (A_1), a marginal success (A_2), or a major success (A_3). It is known from past experience with similar products that the prior probabilities are $P(A_1) = 0.80$, $P(A_2) = 0.15$, and $P(A_3) = 0.05$. The product has been introduced in a test market to see whether it would achieve a small market share (B_1) or a large market share (B_2) in the test market. It was found that the frozen food product obtained a large market share (B_2) in the test market. It is known from past experience that the conditional probabilities of a large share in the test market (B_2), given the ultimate success of the product (A_i), are as follows:

$$P(B_2|A_1) = 0.02 \qquad P(B_2|A_2) = 0.24 \qquad P(B_2|A_3) = 0.98$$

The marketing manager now wishes to know the probabilities of ultimate success of the product, given the test market results. In other words, the posterior probabilities $P(A_1|B_2)$, $P(A_2|B_2)$, and $P(A_3|B_2)$ are desired. ☐

We denote the outcomes of the variable of interest (job hiring, ultimate market success) by A_1, A_2, and so on. The outcomes of the additional information variable (call for interview, test market results) are denoted by B_1, B_2, and so on. The prior probabilities $P(A_1)$, $P(A_2)$, and so on, are known, as are the conditional probabilities $P(B_j|A_1)$, $P(B_j|A_2)$, and so on, for the known outcome B_j of the additional information variable. The conditional probability $P(A_i|B_j)$ of outcome A_i occurring, given that B_j occurs, is then given by *Bayes theorem,* as follows.

(4.25)
> *Bayes Theorem:*
>
> $$P(A_i|B_j) = \frac{P(A_i)P(B_j|A_i)}{\sum_i P(A_i)P(B_j|A_i)}$$

EXAMPLES

1. In the Martin Wilson example, we know that $P(A_1) = 0.04$, $P(A_2) = 0.96$, $P(B_1|A_1) = 0.98$, and $P(B_1|A_2) = 0.01$. Substituting into Bayes theorem (4.25), we obtain:

$$P(A_1|B_1) = \frac{P(A_1)P(B_1|A_1)}{P(A_1)P(B_1|A_1) + P(A_2)P(B_1|A_2)}$$

$$= \frac{0.04(0.98)}{0.04(0.98) + 0.96(0.01)} = \frac{0.0392}{0.0488} = 0.803$$

Thus, the posterior probability of Martin Wilson being hired, given that he has been called for an interview, is $P(A_1|B_1) = 0.803$. Note how much greater this

Figure this
one out !!!

TABLE 4.7   Calculation of conditional probabilities using Bayes theorem — Test market example

(1) Ultimate Market Success A_i	(2) Prior Probability $P(A_i)$	(3) $P(B_2|A_i)$	(4) Joint Probability $P(A_i \cap B_2)$ (2) × (3)	(5) Posterior Probability $P(A_i|B_2)$ (4) ÷ 0.1010
A_1: Failure	0.80	0.02	0.0160	0.158
A_2: Marginal success	0.15	0.24	0.0360	0.356
A_3: Major success	0.05	0.98	0.0490	0.485
Total	1.00		0.1010	1.00

posterior probability of being hired is than the prior probability, $P(A_1) = 0.04$, as a result of Wilson having been called for an interview.

Since A_2 is the complementary event of A_1, the posterior probability that Martin will not be hired, given that he has been called for an interview, is:

$$P(A_2|B_1) = 1 - P(A_1|B_1) = 1 - 0.803 = 0.197$$

2. In the test market example, the desired posterior probabilities can best be obtained by means of systematic calculations, such as those shown in Table 4.7. Column 2 contains the prior probabilities $P(A_i)$ of ultimate success for the new product. Column 3 contains the known conditional probabilities $P(B_2|A_i)$ of large test market share, given the ultimate success of the product. Column 4 contains the products of columns 2 and 3, representing the joint probabilities $P(A_i \cap B_2)$. The sum of this column is the denominator of Bayes theorem (4.25). The ratio of each entry in column 4 to the sum of column 4 is the posterior probability of each ultimate success outcome, given large test market share. Accordingly, the probability of major ultimate success, given large test market share, is $P(A_3|B_2) = 0.0490/0.1010 = 0.485$.

    The probabilities in column 5 constitute the posterior probability distribution of marketing success, given large test market share. Note how this posterior probability distribution differs from the prior probability distribution in column 2. As a result of the test market findings, it is much more likely now that the new product will be either a marginal success or a major success. □

Comment   Bayes theorem (4.25) is a compact formula for finding a posterior probability, and it is the equivalent of the definition of a conditional probability in (4.15).

(4.26)
$$P(A_i|B_j) = \frac{P(A_i \cap B_j)}{P(B_j)}$$

In the numerator, the equivalence follows from multiplication theorem (4.20).

(4.27) $$P(A_i \cap B_j) = P(A_i)P(B_j \mid A_i)$$

In the denominator, the equivalence follows from the definition of a marginal probability in (4.14) and the result in (4.27).

(4.28) $$P(B_j) = \sum_i P(A_i \cap B_j) = \sum_i P(A_i)P(B_j \mid A_i)$$

EXAMPLE

We illustrate by a numerical example how the basic probability theorems are combined in Bayes theorem. We return to the Martin Wilson example, where we wished to find $P(A_1 \mid B_1)$. We start with the following information about the joint probability distribution of job hiring (A_i) and call for interview (B_j):

	B_1	B_2	Total
A_1			0.04
A_2			0.96
		Total	1.00

$$P(B_1 \mid A_1) = 0.98 \qquad P(B_1 \mid A_2) = 0.01$$

We use this information to complete the joint probability distribution. Multiplication theorem (4.20) enables us to obtain:

$$P(A_1 \cap B_1) = P(A_1)P(B_1 \mid A_1) = 0.04(0.98) = 0.0392$$

$$P(A_2 \cap B_1) = P(A_2)P(B_1 \mid A_2) = 0.96(0.01) = 0.0096$$

The remaining joint probabilities can be obtained by subtraction:

$$P(A_1 \cap B_2) = 0.04 - 0.0392 = 0.0008$$

$$P(A_2 \cap B_2) = 0.96 - 0.0096 = 0.9504$$

We have therefore obtained the following joint probability distribution:

	B_1	B_2	Total
A_1	0.0392	0.0008	0.04
A_2	0.0096	0.9504	0.96
Total	0.0488	0.9512	1.00

The numerator of the Bayes theorem formula for $P(A_1|B_1)$ is equivalent to $P(A_1 \cap B_1) = 0.0392$, and the denominator is equivalent to $P(B_1) = 0.0488$. Hence, we obtain:

$$P(A_1|B_1) = \frac{P(A_1 \cap B_1)}{P(B_1)} = \frac{0.0392}{0.0488} = 0.803$$

This result, of course, is the same result obtained by using formula (4.25). □

PROBLEMS

4.1 Explain why each of the following situations involves a random trial, and list two possible outcomes of the trial.

 a. A cake is baked at 200°C and taken from the oven. The cake's internal temperature 10 minutes later is of interest.

 b. A pilot is about to start the three engines of the aircraft. The number of engines that start without trouble is of interest.

 c. A woman shopper is approached in a store and asked which of four perfume scents she finds most pleasant. Her response is of interest.

4.2 Explain why each of the following situations involves a random trial, and list two possible outcomes of the trial.

 a. A police squad is about to begin its night patrol. The number of incidents the squad will investigate that night is of interest.

 b. A student turns on her calculator prior to an examination to check it. Its working order is of interest.

 c. A person mails a letter. How many days later the letter reaches its destination is of interest.

4.3 **Food Inspection.** A food product is inspected at the processing plant by an inspector and given a quality grade of A, B, C, or D. Major food stores sell only the grade A product. Grade B and C products are sold only through discount outlets. Grade D products are not fit for human consumption and are sold to processors of animal foods.

 a. Develop sample spaces that describe the following: (1) different *quality* outcomes for inspected food product, (2) different channels of *market distribution,* and (3) whether or not the food product is *fit for human consumption.*

 b. Are the sample spaces in **a** univariate or bivariate? Explain.

*4.4 Refer to **Food Inspection** Problem 4.3. Let grade A be denoted by o_1, grade B by o_2, and so on.

 a. Let E_1 denote the event that the product is sold through discount outlets. What basic outcomes constitute E_1? What basic outcomes constitute E_1^*?

 b. Let E_2 denote the event that the product is fit for human consumption. What basic outcomes constitute E_2? Are E_1 and E_2 mutually exclusive events? Complementary events? Explain.

4.5 **Manuscript Review.** Two reviewers for a publishing house independently screen manuscripts that arrive unsolicited in the mail. Each reviewer gives a grade of good, fair, or poor to a manuscript.

 a. Develop a sample space to describe the possible outcomes of the joint review of a manuscript.

 b. Is the sample space in **a** univariate or bivariate? Explain.

4.6 Refer to **Manuscript Review** Problem 4.5. Label the basic outcomes in the sample space o_1, o_2, and so on, following the pattern in Figure 4.4a.

    a. Let E_1 denote the event that at least one reviewer assigns a grade of good. What basic outcomes constitute E_1? What basic outcomes constitute E_1^*?

    b. Let E_2 denote the event that the reviewers give different grades. What basic outcomes constitute E_2?

    c. Are E_1 and E_2 mutually exclusive events? Complementary events? Explain.

4.7 **Canal Lock.** A canal has a single lock. Ships may be waiting upstream or downstream. Assume that the number waiting on a given side never exceeds three.

    a. Develop a sample space to describe both the number of ships waiting upstream and the number waiting downstream.

    b. How many basic outcomes does the sample space in **a** contain? Is the sample space univariate or bivariate? Explain.

    c. Develop a sample space to describe the total number of ships waiting on both sides of the lock. Is this sample space univariate or bivariate?

4.8 Refer to **Canal Lock** Problem 4.7. Label the basic outcomes of the sample space in **a** as o_1, o_2, and so on, following the pattern in Figure 4.4a.

    a. Let E_1 denote the event that one or more ships wait upstream and one or more ships wait downstream. What basic outcomes constitute E_1? What basic outcomes constitute the complementary event to E_1?

    b. Let E_2 denote the event that no ships are waiting upstream and/or no ships are waiting downstream. What basic outcomes constitute E_2?

    c. Are E_1 and E_2 mutually exclusive events? Complementary events? Explain.

    d. Let E_3 denote the event that a total of three ships are waiting at the lock. What basic outcomes constitute E_3?

4.9 **Muscle Deterioration.** Two astronauts are given medical examinations after a long space flight. The degree of muscle deterioration from weightlessness is assessed for each astronaut as negligible, moderate, or severe.

    a. Develop the univariate sample space that describes the degree of muscle deterioration for one of the astronauts.

    b. Develop the bivariate sample space that describes the degrees of muscle deterioration for each of the two astronauts. How many basic outcomes does this sample space contain?

4.10 Refer to **Muscle Deterioration** Problem 4.9. Label the basic outcomes of the bivariate sample space in **b** as o_1, o_2, and so on, following the pattern in Figure 4.4a.

    a. Let E_1 denote the event that neither astronaut has a severe degree of muscle deterioration. What basic outcomes constitute E_1?

    b. Let E_2 denote the event that at least one of the astronauts has moderate or severe muscle deterioration. What basic outcomes constitute E_2?

    c. Are E_1 and E_2 mutually exclusive events? Complementary events? Explain.

4.11 Refer to Figure 4.2b.

    a. Construct the corresponding tree diagram, where the first branching is for timing of delivery.

    b. Is the validity of the tree diagram affected by which variable is involved in the first branching? Explain.

4.12 Refer to **Canal Lock** Problem 4.7a. Construct a tree diagram representing the sample space. Associate the first branching with the number of ships waiting upstream.

4.13 Refer to **Muscle Deterioration** Problem 4.9b. Construct a tree diagram representing the sample space. Associate the first branching with the first astronaut.

*4.14 Refer to Figure 4.4. List the basic outcomes for each of the following sets: (1) H^*, (2) $E \cap H$, (3) $H \cup J$, (4) $J \cap G$.

4.15 Refer to **Manuscript Review** Problems 4.5 and 4.6. List the basic outcomes for each of the following sets: (1) $E_1 \cap E_2$, (2) $E_1 \cup E_2$, (3) $E_1^* \cap E_2$, (4) $E_1^* \cup E_2^*$.

4.16 Refer to **Canal Lock** Problems 4.7 and 4.8. List the basic outcomes for each of the following sets: (1) E_2^*, (2) $E_1 \cap E_2$, (3) $E_1 \cup E_2$, (4) $E_1 \cup E_2^*$.

4.17 Refer to **Muscle Deterioration** Problems 4.9 and 4.10. Use set notation involving E_1 and/or E_2 to represent the following: (1) the event that at least one of the astronauts has a moderate degree of muscle deterioration but neither astronaut has a severe degree of muscle deterioration; (2) the event that one or both astronauts has a severe degree of muscle deterioration.

4.18 For each of the following probabilities, indicate whether it is amenable to an objective interpretation, a subjective interpretation, or both. Also, comment on how one might make the probability assessment for each.
   a. Probability that a certain type of automobile fuel pump will not fail during the first 1000 kilometers of operation.
   b. Probability that the common stock price of Sabine, Inc., will fall more than $5 per share during the next five trading days.

4.19 For each of the following probabilities, indicate whether it is amenable to an objective interpretation, a subjective interpretation, or both. Also, comment on how one might make the probability assessment for each.
   a. Probability that leukemia contracted by John Doe was caused by exposure to hazardous chemicals during a period of employment overseas ten years ago.
   b. Probability that a shipment of tires from a regular supplier will contain no defectives.

*4.20 An economist states that the odds are 6 to 1 in favor of the prime interest rate being higher a year from now than it is at present.
   a. What is the probability that the economist is assigning to the event of a higher prime interest rate a year from now?
   b. Is the economist's probability in **a** an objective or subjective probability assessment? Explain.

4.21 An engineer in a construction company states: "I'm willing to bet $3 against $1 that we'll get the contract." A cost analyst replies: "I'll take you up on that!" What does this exchange imply about the engineer's probability assessment of getting the contract? About the cost analyst's assessment?

4.22 A university president stated: "The probability that a graduating science student remains at our university for graduate study is 0.15."
   a. Express the president's probability statement as an odds ratio.
   b. Is the president's probability statement consistent with the statement that fewer than 2 out of 15 graduating science students remain at this university for graduate study? Explain.
   c. Can you tell whether the president's probability statement is based on an objective or subjective interpretation of probability? Explain.

4.23 For each of the following statements, give your probability assessment that it is correct, and explain whether the probability is objective or subjective. Do not consult any outside sources in arriving at your assessments. (1) Ottawa, Canada, lies farther east than Washington, D.C. (2) Cleopatra's first child was male. (3) A thumbtack will land point up if tossed to the floor.

4.24 A young accounting professor has sent an article to a professional journal and states: "There is 1 chance in 5 that this article will be accepted for publication."
  a. Define the sample space here, and give the professor's probability for each basic outcome.
  b. Since the submission of this article to the journal is not a repeatable event, what interpretation can be given to the probabilities in a?
  c. A colleague states: "I'll bet $5 against your $5 that the article will be accepted." Does there appear to be a discrepancy between the probability assessments of the two individuals? Explain.
  d. The professor's statement is based on the fact that this journal accepts 20 percent of the articles submitted. Would you now conclude that the professor's probability assessment is more credible than the colleague's? Discuss.

*4.25 An engineering report gives the following probability distribution for the number of errors that occur in the transmission of a given quantity of data by satellite:

Errors:	0	1	2	3 or more
Probability:	0.75	0.15	0.05	0.05

  a. Is this probability distribution univariate or multivariate?
  b. What is the probability that two or more errors occur in a transmission?
  c. What is the probability that three or more errors occur in a transmission, given that one or more errors occur?

4.26 **In-Flight Service.** A questionnaire on in-flight service asks passengers to rate the quality of the meal served. The following probability distribution applies to passengers' ratings:

Rating:	Poor	Fair	Good	Excellent
Probability:	0.1	0.2	0.5	0.2

  a. Verify that this probability distribution satisfies condition (4.9).
  b. What is the probability that a passenger rates the meal as either good or excellent?
  c. What is the probability that a passenger rates the meal as poor, given that the passenger has not rated the meal as excellent?

*4.27 **Internal Welds.** A device for checking internal welds in metal kegs is designed to signal when the weld is defective. The probability distribution for the status of the weld and the response of the device follows.

Status of Weld	Response of Device		Total
	Signals B_1	Does Not Signal B_2	
A_1: Defective	0.2	0.0	0.2
A_2: Not defective	0.1	0.7	0.8
Total	0.3	0.7	1.0

  a. Give the symbolic notation for the probability of each of the following events: (1) The weld is defective; (2) the weld is defective and the device signals; (3) the device signals, given that the weld is defective; (4) the weld is defective, given that the device signals.
  b. Explain what each of the following probabilities means here: (1) $P(A_1 \cap B_2)$, (2) $P(B_1|A_2)$, (3) $P(B_2)$.
  c. Determine each of the probabilities in a and b.

*4.28 Refer to **Internal Welds** Problem 4.27.
   a. Obtain the marginal probability distribution for the response of the device.
   b. Obtain the conditional probability distribution for the response of the device, given that the weld is not defective.

4.29 **Traffic Movements.**   The probability distribution for directions of traffic movements through an intersection in late afternoon follows.

| | | Vehicle Going To: | | | | |
Vehicle Coming From:		North B_1	South B_2	East B_3	West B_4	Total
A_1:	North	0.00	0.28	0.06	0.07	0.41
A_2:	South	0.12	0.00	0.03	0.03	0.18
A_3:	East	0.08	0.04	0.00	0.05	0.17
A_4:	West	0.07	0.06	0.11	0.00	0.24
	Total	0.27	0.38	0.20	0.15	1.00

   a. Give the symbolic notation for the probability of each of the following events: (1) The vehicle goes north; (2) the vehicle comes from the west and goes south; (3) the vehicle goes north, given that it comes from the west.
   b. Explain what each of the following probabilities means here: (1) $P(A_3)$, (2) $P(A_3 \cap B_4)$, (3) $P(A_3 \cup B_4)$.
   c. Determine each of the probabilities in **a** and **b**.

4.30 Refer to **Traffic Movements** Problem 4.29.
   a. Obtain the marginal probability distribution for the direction from which a vehicle enters the intersection.
   b. Obtain the conditional probability distribution for the direction from which a vehicle enters the intersection, given that it leaves to the south.

4.31 **Stock Movements.**   The joint probability distribution for the directions of price change for two common stocks on a given trading day follows.

| | | Stock B | | | |
Stock A		Increase B_1	No Change B_2	Decrease B_3	Total
A_1:	Increase	0.20	0.10	0.05	0.35
A_2:	No change	0.10	0.25	0.10	0.45
A_3:	Decrease	0.05	0.05	0.10	0.20
	Total	0.35	0.40	0.25	1.00

   a. Give the symbolic notation for the probability of each of the following events: (1) Both stock prices increase; (2) the price of stock A increases; (3) the price of stock B decreases, given that the price of stock A is unchanged.
   b. Explain what each of the following probabilities means here: (1) $P(A_2 \cap B_2)$, (2) $P(A_3 \cup B_1^*)$, (3) $P(B_3 | A_1 \cup A_2)$.
   c. Determine each of the probabilities in **a** and **b**.

4.32 Refer to **Stock Movements** Problem 4.31.
   a. Obtain the marginal probability distribution for the direction of price change of stock A. What is the most probable outcome for this distribution?

b. Obtain the conditional probability distribution for the direction of price change of stock A, given that the price of stock B increases. What is the most probable outcome for this distribution?

4.33 Given:

A_1: family owns home B_1: family income is under \$20,000
A_2: family rents home B_2: family income is \$20,000–under \$40,000
                          B_3: family income is \$40,000 or more

and that:

$$P(A_2) = 0.52 \qquad P(A_1 \cap B_1) = 0.10$$

$$P(B_1) = 0.50 \qquad P(A_1 \cap B_3) = 0.08$$

$$P(B_3) = 0.10$$

a. Develop the joint probability distribution of variables A and B.
b. Determine the probability that (1) a family owns its home; (2) a family owns its home, given that its income is under \$20,000; (3) a family's income is under \$20,000, given that the family owns its home.
c. Obtain the conditional probability distribution for family income, given that the family rents its home.

*4.34 Refer to **Internal Welds** Problem 4.27. Find the following probabilities by the multiplication, complementation, or addition theorems, and verify your answers by obtaining the probabilities directly from the joint probability distribution: (1) $P(A_1 \cup B_1)$, (2) $P(A_2 \cap B_1)$, (3) $P(A_2^*)$.

4.35 Refer to **Traffic Movements** Problem 4.29. Find the following probabilities by the multiplication, complementation, or addition theorems, and verify your answers by obtaining the probabilities directly from the joint probability distribution: (1) $P(A_2 \cup B_1)$, (2) $P(A_2 \cap B_4)$, (3) $P(A_1^*)$.

4.36 Events A, B, and C are mutually exclusive outcomes for a corporation implementing a new marketing strategy. If $P(A) = 0.4$ and $P(B) = 0.2$, obtain the following: (1) $P(A^*)$, (2) $P(A \mid C)$, (3) $P(A \cup B)$, (4) $P(A \mid B^*)$.

*4.37 An auditor is checking the payroll records of a company. The probability that a randomly chosen record has an error is 0.05. An error in a record is either large or small. Two out of ten errors, on average, are large.
a. What is the probability that a randomly chosen record has a large error?
b. Given that a randomly chosen record does not have a large error, what is the probability that it does not have an error?

4.38 An electronic system contains two parallel circuits (A, B). The system will operate if at least one of the two circuits is operational. Each circuit has a probability of 0.85 of operating at least 1000 hours. If one circuit operates at least 1000 hours, the probability is 0.90 that the other circuit will also operate at least 1000 hours. What is the probability of the system operating at least 1000 hours?

4.39 Several leading law schools have developed an information exchange under which applicants may apply to, at most, two of the schools for admission. Applicant Blanque wishes to study at one of three schools (A, B, C), all of which are participants in the information exchange program. On the basis of experiences of similar graduates from his college, Blanque makes the following probability assessments, where $P(A)$ denotes the probability of his being accepted by law school A and the other probabilities are interpreted analogously: $P(A) = 0.60$, $P(B) = 0.80$, $P(C) = 0.50$, $P(B \mid A) = 0.90$, $P(C \mid A) = 0.20$, $P(C \mid B) = 0.50$.

a. To which two of the three schools should Blanque apply to have the greatest probability of being accepted by one or the other (or both)? Show your calculations.

b. Do schools A and B appear to have similar or dissimilar criteria for admission? Schools A and C? Discuss.

4.40 A speculator specializing in commodities A and B has made the following subjective probability assessments about the prices of these commodities during the next two days: (1) The probability is 0.90 that the price of A will rise; (2) the probability is 0.80 that the price of B will rise; (3) if the price of A rises, the conditional probability is 0.65 that the price of B will rise.

a. Are these subjective probability assessments mutually consistent? Explain.

b. Would the assessments be consistent if the probability in (2) were 0.50 instead of 0.80?

4.41 Refer to **Internal Welds** Problem 4.27. Display the joint probability distribution in the form of a probability tree, as in Figure 4.6. Place status of weld in the first branching.

4.42 Refer to **Stock Movements** Problem 4.31. Display the joint probability distribution in the form of a probability tree, as in Figure 4.6. Place stock A price change in the first branching.

4.43 Refer to Figure 4.6. Reconstruct the probability tree, placing inspector's action in the first branching.

4.44 An early version of an intercontinental ballistic missile was deemed to have a probability of 0.10 of destroying an enemy missile silo when it was fired at the silo. The probability of destroying the silo if two such missiles were fired was stated to be 0.19. Were the outcomes of the two firings taken to be statistically independent? Explain.

*4.45 Refer to **In-Flight Service** Problem 4.26. Assume that passengers' ratings of the meal are statistically independent.

a. What is the probability that two passengers will both rate the meal as excellent? That three passengers will all rate it as excellent?

b. For the ratings by two passengers, what is the probability that one of the two will rate the meal as excellent and the other will rate it as good?

4.46 A salesperson is evaluated according to whether the week's sales exceed the sales quota (A), meet the sales quota (B), or fall below the quota (C). The probabilities for an experienced salesperson are $P(A) = 0.3$, $P(B) = 0.6$, $P(C) = 0.1$. Consider the performances of an experienced salesperson in two successive weeks, and assume these performances are statistically independent.

a. Obtain the joint probability distribution. What is the probability that the salesperson will exceed the quota in both weeks?

b. If the performances in two successive weeks were not independent, could you obtain the joint probability distribution with no additional information? Explain.

4.47 In handling a customer's order, the order department will fill it incorrectly with probability 0.04. Also, the order will be delivered to the wrong address with probability 0.02. Assume that the two variables filling of order and delivery of order are statistically independent.

a. Obtain the bivariate probability distribution.

b. What is the probability that a customer's order is filled incorrectly and delivered to the wrong address? That it is filled correctly and delivered to the right address?

c. If the two variables were statistically dependent, then what would be the maximum probability that a customer's order is filled incorrectly and delivered to the wrong address?

*4.48 Refer to **Internal Welds** Problem 4.27.

a. Show that the two variables are statistically dependent.

b. Obtain the conditional probability distributions for the response of the device, given that the weld is defective and given that it is not defective. Describe the nature of the statistical dependence of the variables.

4.49 Refer to **Stock Movements** Problem 4.31.
   a. Show that the two variables are statistically dependent.
   b. Obtain three conditional probability distributions for the stock A price change, given that the price of stock B increases, does not change, and decreases. Describe the nature of the statistical dependence of the two variables.
   c. Why might price changes of two common stocks exhibit the type of statistical dependence noted in **b**?

4.50 The conditional probability distributions for the numbers of days absent in a month for employees from each of the two divisions of a firm follow. Also presented is the marginal probability distribution for this same variable for both divisions as a whole.

| | Conditional On: | | |
Days Absent	Division A	Division B	Both Divisions
0	0.50	0.40	0.44
1	0.35	0.30	0.32
2	0.10	0.15	0.13
3	0.05	0.10	0.08
4	—	0.05	0.03
Total	1.00	1.00	1.00

   a. Are the variables division of employee and days absent statistically dependent or independent? Explain.
   b. Develop the joint probability distribution for the two variables.

*4.51 Refer to the text security check example (p. 129). Suppose (1) the probability an unauthorized visitor will pass the voice pattern check is 0.02; (2) the probability an unauthorized visitor will pass the fingerprint check if he or she passes the voice pattern check is 0.50; and (3) the probability an unauthorized visitor will pass the handwriting check if he or she passes both the other two checks is 0.90.
   a. Use (4.21) to obtain the probability that an unauthorized visitor will pass all three checks.
   b. How does the probability in **a** compare with the corresponding probability in the text example? What are the implications of this result for the security system?

4.52 The contents of a computer file will be lost only if all three of the following events occur: The file is erased from current memory (A), no backup electronic copy of the file exists (B), no printed copy of the file exists (C).
   a. If the probability of each of these events for a file is 0.05 and they are statistically independent events, what is the probability that a file will be lost?
   b. In contrast to **a**, suppose (1) the probability is 0.05 that a file is erased from current memory; (2) the probability is 0.15 that no electronic copy exists for a file erased from memory; (3) the probability is 0.60 that no printed copy exists for a file that is erased from memory and for which no electronic copy exists. Using (4.21), calculate the probability that a file will be lost.
   c. For the situation in **b**, what is the probability that a file will be erased from current memory and no electronic copy exists but a printed copy does exist?

*4.53 Refer to **Internal Welds** Problem 4.27.
   a. Obtain the following probabilities: $P(A_1)$, $P(A_2)$, $P(B_1|A_1)$, $P(B_1|A_2)$. Using these values, calculate $P(A_1|B_1)$ and $P(A_2|B_1)$ by means of Bayes theorem (4.25).

b. Which probabilities in **a** are the prior probabilities? Which are the posterior probabilities? Interpret the prior and posterior probabilities for this application.

c. Verify the probability value $P(A_1 | B_1)$ obtained in **a** by using (4.15) directly.

4.54 Sixty percent of the graduates of a driver-training school pass the official driver's test on the first attempt, and the other 40 percent fail. The school gives a pretest to graduates before they take the official test. Of the graduates who pass the official test on the first attempt, 80 percent passed the pretest. Of the graduates who fail the official test on the first attempt, 10 percent passed the pretest. Let A_1 denote that a graduate passes the official test on the first attempt, A_2 that the graduate fails, and B_1 that the graduate passed the pretest.

a. Give the values of the following probabilities: $P(A_1)$, $P(A_2)$, $P(B_1 | A_1)$, $P(B_1 | A_2)$.

b. A graduate has passed the pretest. Use Bayes theorem to obtain the posterior probabilities $P(A_1 | B_1)$ and $P(A_2 | B_1)$, and interpret them. Has the information provided by the pretest led to a substantial modification of the prior probabilities? Discuss.

4.55 The probabilities that an offshore tract has no gas (A_1), a minor gas deposit (A_2), or a major gas deposit (A_3) are 0.70, 0.25, and 0.05, respectively. A test well drilled in the tract will yield no gas (B_1) if none is present and will yield gas (B_2) with probability 0.3 if a minor deposit is present and with probability 0.9 if a major deposit is present. Using a tabular layout such as that in Table 4.7, calculate the posterior probability distribution for the size of the gas deposit if a test well drilled in the tract yields gas.

4.56 An electronic instrument can fail in one of two ways (T_1, T_2) with equal probability. The two types of failures occur, however, in different locations (L_1, L_2, L_3) with different probabilities. The following are the conditional probability distributions for failure location, conditional on the type of failure:

Location	Conditional On:	
	T_1	T_2
L_1	0.5	0.6
L_2	0.1	0.3
L_3	0.4	0.1
Total	1.0	1.0

a. Using a tabular layout such as that in Table 4.7, calculate the posterior probability distribution for the type of failure if the failure is known to have occurred in location (1) L_1, (2) L_2, (3) L_3.

b. From the results in **a**, does it appear that knowledge of failure location provides substantial information about the type of failure that has occurred? Comment.

EXERCISES

4.57 For each of the following probability statements, state whether it is always true, always false, or neither, for all pairs of events E and F of a sample space. Justify each answer.

a. $P(E) < P(E \cap F)$

b. $P(E) > P(E \cup F)$

c. $P(E \cup F) \leq P(E) + P(F)$

4.58 For each of the following probability statements, state whether it is always true, always false, or neither, for all pairs of events E and F of a sample space. Justify each answer.

    a. Both $P(E \mid F) < P(E)$ and $P(E \mid F^*) < P(E)$

    b. $P(E^* \mid F) = 1 - P(E \mid F)$

    c. $P(E^* \cap F^*) = 1 - P(E \cup F)$

4.59 Prove extension (4.21) of the multiplication theorem (4.20).

4.60 Extended Addition Theorem. Addition theorem (4.16) extends to more than two events. For any three events E_1, E_2, and E_3 of a sample space, use (4.16) to show that:

$$P(E_1 \cup E_2 \cup E_3) = P(E_1) + P(E_2) + P(E_3) - P(E_1 \cap E_2) - P(E_1 \cap E_3)$$
$$- P(E_2 \cap E_3) + P(E_1 \cap E_2 \cap E_3)$$

[*Hint:* Denote $E_2 \cup E_3$ by F initially, and expand $P(E_1 \cup F)$.]

4.61 The probability that a household has a color television is 0.55, that it has one automobile is 0.56, that it has two or more automobiles is 0.22, that it has a color television and one automobile is 0.28, and that it has a color television and two or more automobiles is 0.21. Use the extended addition theorem in Exercise 4.60 to find the probability that a household has either a color television or at least one automobile.

4.62 Two students will be selected at random from the top three students in a political science class to attend a mock political convention. The first student selected will be the voting delegate, and the second one selected will be the alternate. Before the selection begins, one of the three students states that she has one chance in three of being the voting delegate and one chance in two of being the alternate. Do you agree? (*Hint:* Before the selection begins, is the probability of being chosen second a marginal probability, a joint probability, or a conditional probability?)

4.63 A mathematical statistics book states that if E and F are two events of a random trial with nonzero probabilities, then E and F cannot be both mutually exclusive and statistically independent. Prove this statement.

4.64 Refer to Bayes theorem (4.25). If additional information B_j yields a posterior probability for outcome A_i that is identical to the prior probability for A_i, what is the nature of the statistical relationship between events B_j and A_i?

STUDIES

4.65 A supervisor of a large construction firm is responsible for preparing project bids. In an internal report on each bid, he states his subjective probability that the bid will be successful. A tabulation of his record on the last 205 bids is shown at the end of this problem.

    a. For bid probability level 0.10, what is the anticipated proportion of successful bids if the supervisor's probability assessments are accurate? What, in fact, was the observed proportion of successful bids at this bid probability level? What is the residual here?

    b. Obtain the residuals for all bid probability levels, and prepare a residual plot. Is there any range of bid probability levels where the supervisor's assessment performance is particularly poor?

    c. Overall, how does the total number of bids that were anticipated to be successful compare with the observed number (87)?

d. Since the firm has won 87 of the last 205 bids, it has been suggested that probability $87/205 = 0.42$ be used as the probability of winning each bid rather than relying on the supervisor's probability assessment. Evaluate this proposal.

Supervisor's Subjective Probability of Successful Bid	Number of Bids	Number of Successful Bids
0.10	3	0
0.20	10	2
0.30	29	9
0.40	48	18
0.50	37	18
0.60	35	22
0.70	21	10
0.80	11	5
0.90	7	2
1.00	4	1
Total	205	87

4.66 Consider the event: "The next U.S. president will hold office for less than four years."
  a. Use the indirect method described in the text to elicit the subjective probability for this event from ten persons not in your class. Record your results.
  b. Are the ten probability assessments highly variable? Calculate a suitable measure, and interpret it.

4.67 The probability distribution of the week of settlement of strikes in a particular industry follows. (Week 1 is the week when the strike begins, and so on.)

Week of Settlement:	1	2	3	4	5 or later
Probability:	0.63	0.23	0.09	0.03	0.02

  a. A strike has just begun. What is the probability that it will be settled in week 1? In week 5 or later?
  b. If a strike is just entering week 2, what is the probability that it will be settled this week?
  c. Calculate the probability that a strike *is not* settled in week 1 and *is* settled in week 2.
  d. For each of weeks 2, 3, and 4, obtain the probability that a strike will be settled in that week, given that it has not been settled before that week. Does it appear to become more or less difficult to settle strikes as the duration of the strike increases?

4.68 On the basis of a physical examination and symptoms, a physician assesses the probabilities that the patient has no tumor, a benign tumor, or a malignant tumor as 0.70, 0.20, and 0.10, respectively. A thermographic test is subsequently given to the patient. This test gives a negative result with probability 0.90 if there is no tumor, with probability 0.80 if there is a benign tumor, and with probability 0.20 if there is a malignant tumor.
  a. What is the probability that a thermographic test will give a negative result for this patient?
  b. Obtain the posterior probability distribution for this patient when the test result is negative. Interpret this probability distribution. What is the most likely state for the patient?
  c. Obtain the posterior probability distribution for this patient when the test result is positive. How does this probability distribution differ from the one in **b**? Discuss.

5 RANDOM VARIABLES

In Chapter 4, we considered probability distributions in general. In this chapter, we are concerned with probability distributions for sample spaces relating to quantitative characteristics. Special interest exists in these distributions because of the importance and pervasiveness of quantitative characteristics in practical problems and the relative ease with which they lend themselves to statistical analysis.

5.1 BASIC CONCEPTS

Definition of Random Variable

When the sample space of a random trial relates to a quantitative characteristic, a number or value can be associated with each basic outcome.

EXAMPLE

Tire Molding. Consider a random trial involving automobile tires that are molded in pairs. The quantitative characteristic of interest is the number of defective tires in a pair. Thus, the basic outcomes in the sample space are o_1: zero defective tire, o_2: one defective tire, and o_3: two defective tires. If X is used to denote the number of defective tires in the pair, clearly X in a random trial can assume one and only one of the values 0, 1, and 2, corresponding to basic outcomes o_1, o_2, and o_3, respectively. Which value X will assume, however, is uncertain. For this reason, X is called a random variable. ☐

(5.1)

A *random variable* is a variable whose numerical value is determined by the outcome of a random trial.

Initially, it is useful to distinguish notationally between a random variable and the possible values it can assume. We shall use an uppercase letter, such as X, to denote the random variable and the corresponding lowercase letter, x in this case, to denote a particular value assumed by the random variable.

EXAMPLES

1. In the tire-molding example, random variable X denotes the number of defective tires in a pair, a number that is uncertain before the random trial occurs. The nota-

tion $X = x$ designates that the actual trial outcome is x, where x is either the value 0, 1, or 2 here.

2. Let random variable X denote the number of games a football team will win in a nine-game regular season. The notation $X = x$ designates that the team actually wins x games, where x is either 0, 1, . . . , or 9 here.   □

Discrete Random Variable

A random variable that only takes on distinct values is called a *discrete* random variable. For example, the number of defective tires in a pair is a discrete random variable because it may assume one of three distinct values: 0, 1, 2. Similarly, the number of games the football team wins is a discrete random variable that may assume one of ten distinct values: 0, 1, . . . , 9.

Some discrete random variables are treated as if they can assume one of an infinity of possible distinct values. For instance, one may treat the number of traffic violations committed in a large city during a one-month period as having possible outcomes 0, 1, 2, . . . , ad infinitum. While actually there is a realistic upper limit to the number of violations, it is often useful to model the number of possible outcomes as potentially infinite.

Continuous Random Variable

A random variable that may take on any value on a continuum is called a *continuous* random variable. For example, the temperature in a warehouse may be any value on the temperature continuum between, say, −40°C and 45°C. Of course, any measurement can be made only to a finite number of significant digits, and so, strictly speaking, measured temperature is a discrete random variable. However, it is often conceptually advantageous to view measured temperature as being inherently continuous. Additional examples of measured variables that are often treated as continuous are family income, IQ, and height of person.

We consider discrete random variables first, and then we take up continuous random variables in Section 5.7.

5.2   DISCRETE RANDOM VARIABLES

Probability Distribution

Let X be a discrete random variable and let x_i, $i = 1, 2, . . . , k$, denote the k distinct values that X may assume. Since each x_i corresponds to a basic outcome of the sample space of the random trial, a probability distribution for the sample space will associate a probability value with each x_i. The probability that random variable X will assume value x_i will be denoted by $P(X = x_i)$. For instance, $P(X = 0)$ denotes the probability that X will assume the value 0.

(5.2)

The *probability distribution* for a discrete random variable X associates with each of the distinct outcomes x_i $(i = 1, 2, \ldots, k)$ a probability $P(X = x_i)$.

EXAMPLE

In the tire-molding example, the probabilities assigned to the basic outcomes are $P(o_1) = 0.75$, $P(o_2) = 0.10$, and $P(o_3) = 0.15$. The probability distribution for X therefore is:

x:	0	1	2
$P(X = x)$:	0.75	0.10	0.15

Figure 5.1a presents this probability distribution in graphic form, and Figure 5.1b shows an alternative tabular form. □

Comments

1. Any probability distribution for a discrete random variable X has the usual properties of a probability distribution:

    a. $0 \leq P(X = x_i) \leq 1$ $i = 1, 2, \ldots, k$

    b. $\sum_{i=1}^{k} P(X = x_i) = 1$

2. The probability distribution for a discrete random variable is also called *probability mass function* or simply *probability function*.

Notation.   For convenience, we will often abbreviate the notation $P(X = x_i)$ to $P(x_i)$. Then $P(0)$ denotes $P(X = 0)$, the probability that the value of random variable X is 0; $P(1)$ denotes $P(X = 1)$, the probability that the value of X is 1; and so on.

FIGURE 5.1   Graphic and tabular representations of a probability distribution for a discrete random variable — Tire-molding example

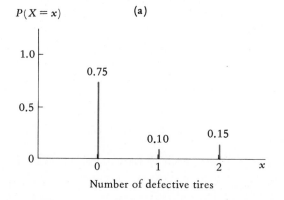

x	$P(X = x)$
0	0.75
1	0.10
2	0.15
Total	1.00

Cumulative Probability Distribution

In some applications, we are interested in the probability that a random variable X takes a value less than or equal to some specified value x. Such a probability is called a *cumulative probability* and is denoted by $P(X \leq x)$.

EXAMPLE

In the tire-molding example, a production trial is profitable if the number of defective tires X is one or less. Thus, a trial is profitable if the event $X \leq 1$ occurs. This event happens if either 0 or 1 defective tire occurs in the trial. Using the abbreviated notation $P(0)$ and $P(1)$ to denote the probabilities $P(X = 0)$ and $P(X = 1)$, respectively, we see from Figure 5.1b that:

$$P(X \leq 1) = P(0) + P(1) = 0.75 + 0.10 = 0.85$$

Thus, the probability is 0.85 that a trial will be profitable.   ☐

Just as a probability distribution gives probabilities $P(X = x_i)$ for all outcomes x_i, a cumulative probability distribution gives cumulative probabilities $P(X \leq x)$ for various values of x.

(5.3)

> The *cumulative probability distribution* for a discrete random variable X provides the cumulative probabilities $P(X \leq x)$ for all values x.

EXAMPLE

Clinic Visits. A health systems analyst obtained the probability distribution tabulated in Figure 5.2a for the annual number of visits by families to a clinic. The cumulative probability distribution for this random variable is developed in Figure 5.2b. The cumulative probabilities are $P(X \leq 0) = P(0) = 0.37$, $P(X \leq 1) = P(0) + P(1) = 0.37 + 0.40 = 0.77$, and so on. We thus see that the probability is 0.97 that a family makes three or fewer visits during a year.

A graph of the cumulative probability distribution for clinic visits is shown in Figure 5.2c. The graph is a step function because the random variable is discrete. The step function is constructed in exactly the same way as a step function ogive for a cumulative frequency distribution. Each step of the graph occurs at an outcome x_i and has a size equal to $P(X = x_i)$. For instance, the step size at $x = 2$ is $P(X = 2) = 0.15$. Figure 5.2c shows how one can graphically determine any cumulative probability by reading the height of the step function. We see that $P(X \leq 1) = 0.77$.   ☐

Comments   1. The definition of a cumulative probability distribution in (5.3) associates cumulative probability values with all values of x, not just those constituting the basic outcomes of the random variable. For the clinic visits example, we note in Figure 5.2c that $P(X \leq 2.6) = 0.92$, even though 2.6 visits cannot occur. Thus $P(X \leq 2.6)$ denotes the probability that 2.6 or fewer visits take place, and this happens when 0, 1, or 2 visits occur.

2. Two important properties of any cumulative probability distribution are as follows:
    a. $P(X \leq x)$ is always a value between 0 and 1.
    b. The cumulative distribution never decreases as x increases.
    These two properties are illustrated in Figure 5.2c, where we see that the cumulative distribution starts at 0 on the bottom left and ends at 1 on the top right.

3. The cumulative probability distribution is also called *cumulative probability function*.

FIGURE 5.2   **Cumulative probability distribution — Clinic visits example**

(a) **Probability Distribution**

Number of visits x	$P(X = x)$
0	0.37
1	0.40
2	0.15
3	0.05
4	0.03
Total	1.00

(b) **Cumulative Probability Distribution**

Number of visits x	$P(X \leq x)$
0	0.37
1	0.77
2	0.92
3	0.97
4	1.00

(c) **Graph of Cumulative Probability Distribution**

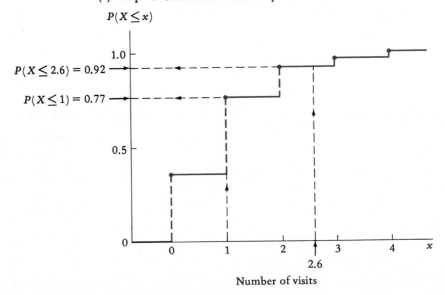

Number of visits

Joint, Marginal, and Conditional Probability Distributions

In Chapter 4, joint, marginal, and conditional probability distributions were defined for bivariate and multivariate sample spaces. When the sample spaces relate to discrete random variables, the joint, marginal, and conditional probability distributions play the same roles as before. We review the main ideas by means of an example.

EXAMPLE

A *bivariate probability distribution* for two discrete random variables is illustrated in Table 5.1. The two variables are the number of trucks from a delivery fleet of three trucks that are in a repair depot on two consecutive nights. Variable X denotes the number of trucks in the repair depot during the first night, and Y denotes the number during the second night. Thus, the possible outcomes for X are 0, 1, 2, 3, and the same holds for Y.

The notation for joint probabilities for two discrete random variables follows that used in Chapter 4 for events. The joint probability that x_i trucks are in the depot the first night and that y_j are in the depot the second night is written $P(X = x_i \cap Y = y_j)$. We see from Table 5.1 that $P(X = 0 \cap Y = 0) = 0.58$ and $P(X = 1 \cap Y = 2) = 0.03$.

The two *marginal probability distributions* are found in the margins of Table 5.1—that for X in the rightmost column and that for Y in the bottom row. The marginal probabilities are obtained as usual, in accordance with (4.14). Thus, we have:

$$P(X = 0) = P(X = 0 \cap Y = 0) + P(X = 0 \cap Y = 1)$$
$$+ P(X = 0 \cap Y = 2) + P(X = 0 \cap Y = 3)$$
$$= 0.58 + 0.06 + 0.01 + 0.00 = 0.65$$

TABLE 5.1 **Example of a bivariate probability distribution for discrete random variables**

Number of Trucks in Depot During First Night	Number of Trucks in Depot During Second Night y				
x	0	1	2	3	Total
0	0.58	0.06	0.01	0.00	0.65
1	0.06	0.10	0.03	0.01	0.20
2	0.01	0.03	0.05	0.01	0.10
3	0.00	0.01	0.01	0.03	0.05
Total	0.65	0.20	0.10	0.05	1.00

Similarly:

$$P(Y = 2) = P(X = 0 \cap Y = 2) + P(X = 1 \cap Y = 2)$$
$$+ P(X = 2 \cap Y = 2) + P(X = 3 \cap Y = 2)$$
$$= 0.01 + 0.03 + 0.05 + 0.01 = 0.10$$

Thus, the probability of 0 trucks in the depot the first night is 0.65 and that of 2 trucks in the depot the second night is 0.10.

Conditional probability distributions are obtained analogously to (4.15). Thus, the conditional probability distribution for the number of trucks in the depot during the second night, given that one truck is in the depot during the first night, is as follows:

Number of Trucks in Depot During Second Night y	Conditional Probability $P(Y = y \mid X = 1)$
0	0.06/0.20 = 0.30
1	0.10/0.20 = 0.50
2	0.03/0.20 = 0.15
3	0.01/0.20 = 0.05
Total	1.00

Note, for instance, that $P(Y = 0 \mid X = 1) = P(X = 1 \cap Y = 0)/P(X = 1)$ according to definition (4.15). Therefore, we obtain $P(Y = 0 \mid X = 1) = 0.06/0.20 = 0.30$.   □

Statistical Independence

We can apply the definition of statistical independence in (4.24) to random variables.

(5.4)

Two random variables X and Y are said to be statistically independent if:

$$P(X = x_i \cap Y = y_j) = P(X = x_i)P(Y = y_j) \qquad \text{for all } x_i \text{ and } y_j$$

EXAMPLE

In Table 5.1, the two random variables are not independent. For example, $P(X = 0 \cap Y = 1) = 0.06$, while $P(X = 0)P(Y = 1) = 0.65(0.20) = 0.13$. An examination of Table 5.1 indicates the nature of the relation, namely, that the number of trucks in the depot on two successive nights is likely to be the same.   □

5.3 EXPECTED VALUE AND VARIANCE

Expected Value of Random Variable

Often, we are interested in the mean value of a random variable in many trials. This value is known as the expected value of a random variable.

(5.5)

> The *expected value* of a discrete random variable X is denoted by $E\{X\}$ and defined:
>
> $$E\{X\} = \sum_{i=1}^{k} x_i P(x_i)$$
>
> *where:*
>
> $P(x_i)$ denotes $P(X = x_i)$
>
> The notation $E\{\ \}$ (read "expectation of") is called the *expectation operator*.

EXAMPLES

1. In the clinic visits example, the probability distribution of X, the annual number of family visits, is given in Figure 5.2a and is as follows:

x:	0	1	2	3	4
$P(x)$:	0.37	0.40	0.15	0.05	0.03

The expected value of X is obtained by using (5.5):

$$E\{X\} = 0(0.37) + 1(0.40) + 2(0.15) + 3(0.05) + 4(0.03) = 0.97 \text{ visit}$$

2. A piece of jewelry worth \$10,000 has probability 0.003 of being stolen or lost during a year. Let X denote the amount of loss during a year. The probability distribution of X is as follows:

Loss x:	0	10,000
$P(x)$:	0.997	0.003

The expected loss during a year is, using (5.5):

$$E\{X\} = 0(0.997) + 10,000(0.003) = \$30$$

Meaning of $E\{X\}$. The expected value of a random variable may be likened to a measure of position for the probability distribution. Indeed, $E\{X\}$ is simply a weighted mean of the possible outcomes, with the probability values used as weights. For this reason, $E\{X\}$ is often called the *mean* of the probability distribution of X, or simply the mean of X.

There is another way of understanding the meaning of $E\{X\}$. If, for instance, the random trial associated with the annual number of family clinic visits is repeated independently many times, the relative frequency interpretation of probability suggests that about 37 percent of families will have no visit in a year, 40 percent will have one visit, and so on. The mean outcome over many independent trials would be about $E\{X\} = 0.97$ visit per family.

Observe that $E\{X\}$, like any mean, may be a number that does not correspond to any of the possible outcomes. In the clinic visits example, we found $E\{X\} = 0.97$ visit, yet all the outcomes of X are whole numbers.

Variance of Random Variable

Since the outcomes of a random variable vary, it is useful to have a measure of their variability. A key measure is the variance of a random variable.

(5.6)

The *variance* of a discrete random variable X is denoted by $\sigma^2\{X\}$ and defined:

$$\sigma^2\{X\} = \sum_{i=1}^{k} (x_i - E\{X\})^2 P(x_i)$$

where:

$P(x_i)$ denotes $P(X = x_i)$

The notation $\sigma^2\{\ \}$ (read "variance of") is called the *variance operator*.

EXAMPLES

1. In the clinic visits example, the probability distribution of X, the annual number of family visits, is as follows:

x:	0	1	2	3	4
$P(x)$:	0.37	0.40	0.15	0.05	0.03

    We found earlier that $E\{X\} = 0.97$. Hence, the variance of X according to (5.6) is:

    $$\sigma^2\{X\} = (0 - 0.97)^2(0.37) + (1 - 0.97)^2(0.40) + (2 - 0.97)^2(0.15)$$
    $$+ (3 - 0.97)^2(0.05) + (4 - 0.97)^2(0.03)$$
    $$= 0.9891$$

2. In the jewelry loss example, the probability distribution of the dollar loss is as follows:

Loss x:	0	10,000
$P(x)$:	0.997	0.003

We found previously that $E\{X\} = 30$. Hence:

$$\sigma^2\{X\} = (0 - 30)^2(0.997) + (10{,}000 - 30)^2(0.003) = 299{,}100$$

Meaning of $\sigma^2\{X\}$. The variance is a weighted average of squared deviations, the deviations being the differences between the outcomes of X and their expected value or mean $E\{X\}$, and the weights being the respective probabilities of occurrence. Therefore, $\sigma^2\{X\}$ measures the extent to which the outcomes of X depart from their expected value in the same way that the variance of a data set measures the variability of values in the set about their mean.

Comment

Since the variance of a random variable X is a weighted average of the squared deviations, $(X - E\{X\})^2$, it may be defined equivalently as an expected value.

(5.7)

$$\sigma^2\{X\} = E\{(X - E\{X\})^2\}$$

An algebraically identical expression is given next.

(5.7a)

$$\sigma^2\{X\} = E\{X^2\} - (E\{X\})^2$$

Standard Deviation of Random Variable

Note that $\sigma^2\{X\}$ is expressed in the squared units of X. When we take its positive square root, we return to the original units of measure and obtain the standard deviation of X.

(5.8)

The positive square root of the variance of X is called the *standard deviation* of X and is denoted by $\sigma\{X\}$:

$$\sigma\{X\} = \sqrt{\sigma^2\{X\}}$$

The notation $\sigma\{\ \}$ (read "standard deviation of") is called the *standard deviation operator*.

EXAMPLES

1. In the clinic visits example, we found that the variance is $\sigma^2\{X\} = 0.9891$. Hence, the standard deviation of X is $\sigma\{X\} = \sqrt{0.9891} = 0.9945$ clinic visit per family.

2. In the jewelry loss example, we found the variance to be $\sigma^2\{X\} = 299{,}100$. Hence, $\sigma\{X\} = \sqrt{299{,}100} = \$546.90$. We see that the standard deviation is much larger than the mean, $E\{X\} = \$30$. □

5.4   STANDARDIZED RANDOM VARIABLES

In Chapter 3, we introduced the concept of the standardized value of an observation in a data set. The standardized value expresses the distance between the observation and the mean of the data set in units of the standard deviation of the data set. A corresponding standardized expression for random variables is frequently useful.

(5.9)

If X is a random variable with expected value $E\{X\}$ and standard deviation $\sigma\{X\}$, then:

$$Y = \frac{X - E\{X\}}{\sigma\{X\}}$$

is known as the *standardized* form of random variable X.

The standardized form of a random variable is called a *standardized random variable*. Each standardized random variable has expected value 0 and variance 1. Thus, for Y defined in (5.9), we have $E\{Y\} = 0$ and $\sigma^2\{Y\} = 1$.

EXAMPLE

For the clinic visits example, we wish to obtain the probability distribution for the standardized form of the random variable X from the probability distribution for X given in Figure 5.2a. We determined earlier that $E\{X\} = 0.97$ visit and $\sigma\{X\} = 0.9945$ visit for this example. Hence, the standardized value y corresponding to the outcome $x = 0$ visit equals:

$$y = \frac{0 - 0.97}{0.9945} = -0.98$$

Likewise, the standardized value for the outcome $x = 1$ visit equals:

$$y = \frac{1 - 0.97}{0.9945} = 0.03$$

Table 5.2 shows the probability distribution of the standardized random variable Y for this example. Note that each outcome x and its associated standardized value y necessarily have the same probability, because they both pertain to the same basic outcome. For instance, $P(X = 0) = P(Y = -0.98) = 0.37$. One can verify by direct calculation, using (5.5) and (5.6), that $E\{Y\} = 0$ and $\sigma^2\{Y\} = 1$. □

TABLE 5.2 **Illustration of a standardized random variable — Clinic visits example**

Number of Visits x	Standardized Value $y = \dfrac{x - E\{X\}}{\sigma\{X\}}$	Probability $P(X = x) = P(Y = y)$
0	-0.98	0.37
1	0.03	0.40
2	1.04	0.15
3	2.04	0.05
4	3.05	0.03
	Total	1.00

where:

$$E\{X\} = 0.97 \quad \text{and} \quad \sigma\{X\} = 0.9945$$

Chebyshev Inequality

The Chebyshev inequality for a data set was presented in (3.24). A corresponding inequality applies to random variables. The Chebyshev inequality for random variables provides information about the probability that any standardized random variable will be large in absolute value.

(5.10)

> *Chebyshev Inequality.* For the standardized random variable Y corresponding to any random variable X, the probability that Y is larger than k in absolute value cannot exceed $1/k^2$.

This inequality assures us that, no matter what the pattern of the probability distribution of a random variable X, the probability that X will fall beyond k standard deviations from its mean is at most $1/k^2$.

EXAMPLES

1. For the clinic visits example, the Chebyshev inequality for $k = 2$ states that the probability that the standardized random variable Y will have an absolute value larger than $k = 2$ cannot exceed $1/k^2 = 1/2^2 = 0.25$. We see from Table 5.2 that two outcomes of Y are larger than $k = 2$ in absolute value; $y = 2.04$ corresponding to $x = 3$ visits, and $y = 3.05$ corresponding to $x = 4$ visits. The total probability associated with these two outcomes is $0.05 + 0.03 = 0.08$, which, as expected, is smaller than the upper bound of 0.25 provided by the Chebyshev inequality.

2. Consider the number of trucks arriving at a warehouse in a day. For this random variable X, it is known that $E\{X\} = 18$ and $\sigma\{X\} = 2$. The Chebyshev inequality for $k = 3$ then tells us that the probability is at most $1/3^2 = 1/9$ that on any one day fewer than $18 - 3(2) = 12$ trucks or more than $18 + 3(2) = 24$ trucks will arrive.   ☐

5.5   LINEAR FUNCTION OF RANDOM VARIABLE

Often, interest centers on a random variable W that is a linear function of a random variable X, that is, $W = a + bX$, where a and b are constants.

EXAMPLES

1. Let X denote the number of local calls made from a pay telephone in a day. Let W denote the daily revenue from the local calls. Each call costs $0.25, so we have $W = 0.25X$. Here, $a = \$0$ and $b = \$0.25$. If $x = 3$, for instance, $w = \$0.75$.

2. Let X denote the quantity of output of a plant in a day, and let W denote the total cost of this output. It is known that $W = 200 + 4X$, where $a = \$200$ is the fixed setup cost and $b = \$4$ is the cost per unit produced. If $x = 40$, for instance, $w = \$360$.   ☐

Expected Value of Linear Function of Random Variable

To find the expected value of W, where $W = a + bX$, one can obtain the probability distribution of W from that of X and then use (5.5) to find $E\{W\}$. In Example 1, for instance, the probability distribution of the number of local calls X is known to be as follows:

x:	0	1	2	3
$P(x)$:	0.2	0.4	0.3	0.1

Since each call costs $0.25, the probability distribution of the total revenue from these calls (W) must correspond to that of the number of calls (X), as follows:

w:	0	0.25	0.50	0.75
$P(w)$:	0.2	0.4	0.3	0.1

Using (5.5) to obtain $E\{W\}$, we find:

$$E\{W\} = 0(0.2) + 0.25(0.4) + 0.50(0.3) + 0.75(0.1) = \$0.325$$

When $E\{X\}$ is already known, there is no need to find the probability distribution of W first. The following theorem can be used to find $E\{W\}$ from $E\{X\}$ directly when $W = a + bX$.

(5.11)
$$E\{a + bX\} = a + bE\{X\}$$

This theorem says that $E\{W\}$ is the same linear function of $E\{X\}$ as W is of X. There are three important special cases of (5.11).

(5.11a) $E\{a\} = a$

(5.11b) $E\{bX\} = bE\{X\}$

(5.11c) $E\{a + X\} = a + E\{X\}$

EXAMPLES

1. In the telephone calls example, $E\{X\} = 1.3$ (calculations not shown). Hence, using theorem (5.11b), we obtain for $W = 0.25X$:

   $$E\{W\} = 0.25(1.3) = \$0.325$$

   This result, of course, is the same result as that obtained when $E\{W\}$ was calculated from the probability distribution of W.

2. In the plant output example, $E\{X\} = 50$. Hence, using theorem (5.11), we obtain for $W = 200 + 4X$:

   $$E\{W\} = 200 + 4(50) = \$400$$

Variance of Linear Function of Random Variable

To find the variance of W, where $W = a + bX$, one can construct the probability distribution of W from that of X and then use (5.6) to compute $\sigma^2\{W\}$. Alternatively, when the variance of X is known, one can obtain the variance of W directly from the following theorem.

(5.12) $\sigma^2\{a + bX\} = b^2\sigma^2\{X\}$

Two important special cases of (5.12) are given next.

(5.12a) $\sigma^2\{a + X\} = \sigma^2\{X\}$

(5.12b) $\sigma^2\{bX\} = b^2\sigma^2\{X\}$

EXAMPLES

1. In the telephone calls example, the variance of X is $\sigma^2\{X\} = 0.81$ (calculations not shown). Hence, using theorem (5.12b), we find for $W = 0.25X$:

   $$\sigma^2\{W\} = (0.25)^2(0.81) = 0.0506$$

   The standard deviation of W is $\sigma\{W\} = \sqrt{0.0506} = \$0.225$.

2.   the plant output example, $\sigma^2\{X\} = 400$. Hence, using (5.12), we obtain for $= 200 + 4X$:

$$\sigma^2\{W\} = (4)^2(400) = 6400$$

so $\sigma\{W\} = \sqrt{6400} = \$80$. Note that the constant setup cost $a = \$200$ had no effect on the variance of W.   □

Comments

1.   Example 2 illustrates that the constant a does not affect the variance of $a + bX$. This result is intuitively reasonable, since adding a constant to a variable shifts the position of the distribution but does not affect the variability of the distribution.

2.   The standard deviation of $W = a + bX$ is related to $\sigma\{X\}$ as follows.

(5.12c)
$$\sigma\{W\} = |b|\sigma\{X\}$$

In Example 1, for instance, $\sigma\{X\} = \sqrt{0.81} = 0.90$ and $W = 0.25X$. Hence:

$$\sigma\{W\} = |0.25| (0.90) = \$0.225$$

3.   A standardized random variable, defined in (5.9), is a linear function of X:

$$Y = \frac{X - E\{X\}}{\sigma\{X\}} = -\frac{E\{X\}}{\sigma\{X\}} + \frac{1}{\sigma\{X\}}X$$

where $a = -E\{X\}/\sigma\{X\}$ and $b = 1/\sigma\{X\}$. To show that a standardized random variable Y has variance 1, we have, by (5.12):

$$\sigma^2\{Y\} = b^2\sigma^2\{X\} = \left(\frac{1}{\sigma\{X\}}\right)^2 \sigma^2\{X\} = 1$$

5.6   SUMS AND DIFFERENCES OF INDEPENDENT RANDOM VARIABLES

Sum and Difference of Two Independent Random Variables

The sum or difference of two independent random variables is frequently encountered.

EXAMPLES

1.   Let X denote the bonus amount received by salesperson A and Y the bonus amount received by salesperson B. Then $T = X + Y$ represents the total bonus amount received by the two salespeople.

2.   Let X denote the number of responses to a classified advertisement in city 1 and Y the number of responses to the advertisement in city 2. Then $R = X + Y$ represents the total number of responses to the advertisement in the two cities.

3. Let X denote quarterly sales revenues and Y denote quarterly direct costs. Then $P = X - Y$ represents quarterly gross profit. ☐

Probability Distribution.   The probability distribution of the sum or difference of two independent random variables can be readily derived from the probability distributions of the individual random variables.

EXAMPLE

Sales Bonus.   The probability distribution of X, the bonus amount received by salesperson A, is as follows:

Bonus x:	0	500
$P(x)$:	0.6	0.4

Variable Y, the bonus amount received by salesperson B, follows the same probability distribution as X. Further, X and Y are independent random variables. It then follows that the joint probability distribution is:

	y	
x	0	500
0	0.36	0.24
500	0.24	0.16

These joint probabilities are obtained from the independence definition (5.4). For instance, we have $P(X = 0 \cap Y = 0) = P(X = 0)P(Y = 0) = 0.6(0.6) = 0.36$.

The probability distribution of $T = X + Y$, the total bonus amount, is obtained by recognizing that $T = 0$ if neither salesperson receives a bonus. Hence, $P(T = 0) = 0.36$. Similarly, $P(T = 1000) = 0.16$, since both salespeople must receive a bonus if the total amount is to be $1000. Finally, $P(T = 500) = 0.24 + 0.24 = 0.48$, since the total bonus amount is $500 when either salesperson, but not the other, receives the bonus. Thus, the probability distribution of T is as follows:

Total Bonus t:	0	500	1000
$P(t)$:	0.36	0.48	0.16

☐

Expected Values and Variances.   If the probability distribution of the sum or difference of two random variables is available, the expected value and the variance of the sum or difference can be found by direct calculation in the usual fashion.

EXAMPLE

In the sales bonus example, we find the expected total bonus amount as follows:

$$E\{T\} = 0(0.36) + 500(0.48) + 1000(0.16) = 400$$

We can calculate similarly that the variance is $\sigma^2\{T\} = 120,000$. ☐

When the expected values and the variances of both random variables are already known, one need not obtain the probability distribution of the sum since the expected value and the variance of the sum can be obtained directly from the following theorem.

(5.13)   If X and Y are two independent random variables, then the expected value and
              the variance of $X + Y$ are as follows:

(5.13a)   *Expected Value:*

$$E\{X + Y\} = E\{X\} + E\{Y\}$$

(5.13b)   *Variance:*

$$\sigma^2\{X + Y\} = \sigma^2\{X\} + \sigma^2\{Y\}$$

Note that the expected value of the sum of two independent random variables is simply
the sum of the expected values for each of the two random variables, and similarly for
the variance.

The expected value and the variance of a difference are presented next.

(5.14)   If X and Y are two independent random variables, then the expected value and
              the variance of $X - Y$ are as follows:

(5.14a)   *Expected Value:*

$$E\{X - Y\} = E\{X\} - E\{Y\}$$

(5.14b)   *Variance:*

$$\sigma^2\{X - Y\} = \sigma^2\{X\} + \sigma^2\{Y\}$$

Note that the expected value of the difference of two independent random variables is
simply the difference of the expected values of the two random variables. Note also
that the variance of the difference is the same as the variance of the sum when the two
random variables are independent.

EXAMPLES

1. In the sales bonus example, $E\{X\} = E\{Y\} = 200$ and $\sigma^2\{X\} = \sigma^2\{Y\} = 60,000$
    (calculations not shown). Hence, for $T = X + Y$, we obtain using (5.13):

    $$E\{T\} = 200 + 200 = 400 \qquad \sigma^2\{T\} = 60,000 + 60,000 = 120,000$$

    These are the same results, of course, as those obtained from the probability distri-
    bution of T. The standard deviation of T is $\sigma\{T\} = \sqrt{120,000} = 346.4$.

2. In the advertisement response example, $E\{X\} = 40$, $E\{Y\} = 70$, $\sigma^2\{X\} = 15$,
    $\sigma^2\{Y\} = 10$, and X and Y are independent. For $R = X + Y$, the total number of
    responses, we obtain using (5.13):

    $$E\{R\} = 40 + 70 = 110 \qquad \sigma^2\{R\} = 15 + 10 = 25$$

    The standard deviation of R is $\sigma\{R\} = \sqrt{25} = 5$.

3. In the gross profit example, $E\{X\} = 100{,}000$, $E\{Y\} = 70{,}000$, $\sigma^2\{X\} = 8{,}000{,}000$, $\sigma^2\{Y\} = 4{,}000{,}000$, and X and Y are independent. For $P = X - Y$, the gross profit, we obtain using (5.14):

$$E\{P\} = 100{,}000 - 70{,}000 = 30{,}000$$

$$\sigma^2\{P\} = 8{,}000{,}000 + 4{,}000{,}000 = 12{,}000{,}000$$

The standard deviation of P is $\sigma\{P\} = \sqrt{12{,}000{,}000} = 3464$. □

Sum of More Than Two Independent Random Variables

In statistical analysis, we are frequently concerned with the sum of more than two random variables. For example, suppose $X_1, X_2, \ldots, X_s$ represent the amounts an insurance company must pay out for fire losses during the year on s policies. Then $T = X_1 + X_2 + \cdots + X_s$ represents the total disbursement for fire losses during the year on these policies.

The expected value and the variance of a sum of independent random variables are given in the next theorem.

(5.15)   If $T = X_1 + X_2 + \cdots + X_s$ is the sum of s independent random variables, then the expected value and the variance of T are as follows:

(5.15a)   *Expected Value:*

$$E\{T\} = \sum_{i=1}^{s} E\{X_i\}$$

(5.15b)   *Variance:*

$$\sigma^2\{T\} = \sum_{i=1}^{s} \sigma^2\{X_i\}$$

EXAMPLE

A carton contains 100 cans of peas. The weight X_i of the peas in the ith can $(i = 1, 2, \ldots, 100)$ has expected value $E\{X_i\} = 8$ ounces and variance $\sigma^2\{X_i\} = 0.017$. In other words, all cans have the same mean and variance. In addition, the weights X_i are independent random variables. Let $T = X_1 + X_2 + \cdots + X_{100}$ denote the total weight of the peas in the carton. We obtain by (5.15):

$$E\{T\} = 100(8) = 800 \text{ ounces} \qquad \sigma^2\{T\} = 100(0.017) = 1.7$$

The standard deviation of T is $\sigma\{T\} = \sqrt{1.7} = 1.3$ ounces. □

Effect of Statistical Dependence

The influence of statistical dependence on the variance formulas for sums and differences of random variables is taken up in detail in Section 5.8. We point out here, however, that formulas (5.13a), (5.14a), and (5.15a) for expected values of sums and differences are valid even if the random variables are statistically dependent. In contrast, formulas (5.13b), (5.14b), and (5.15b) for the corresponding variances generally will not be valid if the random variables are not independent.

5.7   CONTINUOUS RANDOM VARIABLES

Probability Density Function

As noted at the beginning of the chapter, a continuous random variable assumes values on a continuum. It is not meaningful, however, to speak of a probability value being associated with each possible point on the continuum. Instead, we associate probability values with intervals on the continuum. To illustrate this point, consider a continuous random variable X that represents the yield of a crop in tons per acre. Suppose the yield can be any value between 0 and 1 ton. Figure 5.3 shows a mathematical function that has the property that, for any interval, the area between the axis and the curve of the function corresponds to the probability that X will have a value in this interval. For example, the area for the interval 0.5–0.7 has been shaded and equals 0.229. Therefore, the probability that the yield will lie somewhere in this interval is 0.229. Symbolically, we state $P(0.5 \leq X \leq 0.7) = 0.229$. The probability corresponding to any other interval is found in a similar manner, namely, by determining the area under the curve corresponding to that interval.

The mathematical function of a curve of the type shown in Figure 5.3 is called a probability density function.

(5.16)   The *probability density function* of a continuous random variable X is a mathematical function for which the area under the curve corresponding to any interval is equal to the probability that X will take on a value in the interval. The probability density function is denoted by $f(x)$. The value $f(x)$ is called the *probability density* at x.

EXAMPLE   The probability density function for the curve shown in Figure 5.3 is as follows.

(5.17)
$$f(x) = \begin{cases} 12x(1-x)^2 & 0 \leq x \leq 1 \\ 0 & \text{elsewhere} \end{cases}$$

For instance, the probability density at $x = 0.5$ (and therefore the height of the curve at $x = 0.5$) is:

$$f(0.5) = 12(0.5)(1 - 0.5)^2 = 1.5$$

FIGURE 5.3 **Example of a probability density function:** $f(x) = 12x(1 - x)^2$ **for** $0 \leq x \leq 1$. The area under the curve represents probability.

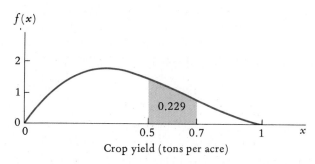

Crop yield (tons per acre)

Comments

1. A probability density function can never lie below the x axis because the probability that X will lie in any interval cannot be negative. Thus, $f(x)$ is always nonnegative. Further, since X must assume some value on the axis under the curve, the total area under a probability density function must be 1.

2. The probability that a continuous random variable will take on any particular value on the continuum is zero. This fact follows because a single point corresponds to an interval of zero width, and hence there is zero area under the probability density function. As a result of this fact, in any probability statement of the type $P(a \leq X \leq b)$ for a *continuous* random variable, it is immaterial whether the endpoints of the interval are included; that is:

$$P(a \leq X \leq b) = P(a < X \leq b) = P(a \leq X < b) = P(a < X < b)$$

3. The probability density function of a random variable is often simply called *probability function, density function,* or *probability distribution.*

Cumulative Probability Function

Sometimes, we wish to work with cumulative probabilities for continuous random variables.

(5.18) The *cumulative probability function* of a continuous random variable is denoted by $F(x)$ and is defined:

$$F(x) = P(X \leq x)$$

where:

$-\infty < x < +\infty$

In other words, the cumulative probability function $F(x)$ indicates the probability that the outcome of X in a random trial will be less than or equal to any specified value x. Thus, $F(x)$ corresponds to the area under the probability density function to the left of x.

Two probability density functions and their cumulative probability functions are shown in Figure 5.4. Note that $F(x)$ is a smooth continuous curve when the random variable is continuous, rather than a step function as for discrete random variables. However, the properties of cumulative probability functions for discrete random variables apply also for continuous random variables. Thus, $F(x)$ always has values between 0 and 1 and never decreases as x increases.

Calculational Procedures with Continuous Probability Functions (Calculus Needed)

Determining Probabilities.   Integral calculus provides a convenient means for determining probabilities for continuous random variables. To find an area under a probability function $f(x)$ corresponding to the interval from a to b, we determine the value of the definite integral.

FIGURE 5.4   **Examples of probability density functions and their cumulative probability functions**

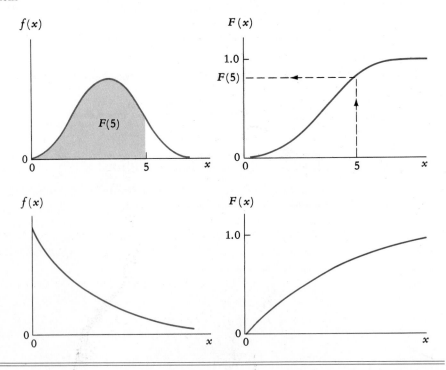

$$(5.19) \qquad P(a \leq X \leq b) = \int_a^b f(x)\, dx$$

EXAMPLE

For the probability function in Figure 5.3, we wish to find $P(0.5 \leq X \leq 0.7)$, the area illustrated in the figure. We obtain:

$$P(0.5 \leq X \leq 0.7) = \int_{0.5}^{0.7} 12x(1 - x)^2\, dx = 0.229$$

Cumulative Probability Function.   In calculus notation, the cumulative probability function $F(x)$ for a continuous random variable is defined as follows.

$$(5.20) \qquad F(x) = \int_{-\infty}^{x} f(u)\, du$$

where:

u denotes the variable of integration

EXAMPLE

The cumulative probability function for probability function (5.17) is:

$$F(x) = \int_0^x 12u(1 - u)^2\, du = 3x^4 - 8x^3 + 6x^2 \qquad 0 \leq x \leq 1$$

We see, for instance, that $F(0.5) = 3(0.5)^4 - 8(0.5)^3 + 6(0.5)^2 = 0.688$.

Expected Value.   The expected value of a continuous random variable is defined as follows.

$$(5.21) \qquad E\{X\} = \int_{-\infty}^{+\infty} xf(x)\, dx$$

Note that (5.21) is analogous to (5.5); integration has simply replaced summation in the definition.

EXAMPLE

The expected value for probability function (5.17) is:

$$E\{X\} = \int_0^1 x[12x(1 - x)^2]\, dx = 0.4$$

Variance.   The variance of a continuous random variable is defined as follows.

(5.22)
$$\sigma^2\{X\} = \int_{-\infty}^{+\infty} (x - E\{X\})^2 f(x)\,dx$$

Note that (5.22) is analogous to (5.6); integration has simply replaced summation.

EXAMPLE

The variance for probability function (5.17) is (recall that $E\{X\} = 0.4$):

$$\sigma^2\{X\} = \int_0^1 (x - 0.4)^2 [12x(1 - x)^2]\,dx = 0.04$$

5.8 OPTIONAL TOPIC — COVARIANCE AND CORRELATION

In later chapters, we shall be concerned with the extent to which two random variables are linearly associated with each other. We now present two measures of linear association for a pair of random variables, the *covariance* and the *coefficient of correlation*.

Covariance

We begin with the concept of concurrent variation or covariation.

EXAMPLES

1. In a midwestern city, the expected midday temperature on July 4 is 25°C and the expected relative humidity is 45 percent. Last July 4, the midday temperature was 30°C and the relative humidity was 70 percent. The covariation of temperature and humidity on that day is the product of the deviations from the expected values, that is, $(30 - 25)(70 - 45) = 125$. Covariation can be positive, as here, negative, or zero. The covariation is positive here because both temperature and humidity exceeded their respective expected values.

2. The expected annual yield of stock A is 5 percent and that of stock B is 10 percent. The actual yields last year were 8 percent each. Hence, the covariation is $(8 - 5)(8 - 10) = -6$. Here, the covariation is negative, indicating that one stock performed above its expected value and the other below its expected value.

As has been seen in the preceding examples, *covariation* for the outcomes x_i and y_j is defined as follows.

(5.23)
$$\text{Covariation} = (x_i - E\{X\})(y_j - E\{Y\})$$

The expected value of the covariation of two random variables, that is, their mean covariation over repeated trials, is called their covariance.

(5.24) The *covariance* of two discrete random variables X and Y is denoted by $\sigma\{X, Y\}$ and defined:

$$\sigma\{X, Y\} = \sum_i \sum_j (x_i - E\{X\})(y_j - E\{Y\})P(x_i, y_j)$$

where:

$P(x_i, y_j)$ denotes $P(X = x_i \cap Y = y_j)$

The notation $\sigma\{\ ,\ \}$ (read "covariance of") is called the *covariance operator*.

(Readers who are unfamiliar with double summation notation should refer to Appendix A, Section A.1.)

EXAMPLES

1. Random variables X and Y have the following joint probability distribution:

	y	
x	5	10
10	0.30	0.20
30	0.10	0.40

Note that the two most likely outcomes are $x = 10$, $y = 5$ and $x = 30$, $y = 10$, indicating that random variables X and Y tend to vary in the same direction, or positively. We find that $E\{X\} = 20$ and $E\{Y\} = 8$ from the marginal distributions of the random variables (calculations not shown).

The calculation of the covariance of X and Y is best carried out in a systematic fashion, as illustrated in Table 5.3a. Columns 1 and 2 contain the joint probability distribution of X and Y in a columnar form, where $P(x, y)$ in column 2 denotes $P(X = x \cap Y = y)$. Columns 3 and 4 contain the deviations $x - E\{X\}$ and $y - E\{Y\}$, respectively. The covariations $(x - E\{X\})(y - E\{Y\})$ appear in column 5. Finally, column 6 contains the covariations weighted by their probabilities $P(x, y)$. Thus, for the outcome $x = 10$, $y = 5$, the deviations are $(10 - 20) = -10$ and $(5 - 8) = -3$, and the covariation is $(-10)(-3) = 30$. This covariation is weighted by the probability $P(X = 10 \cap Y = 5) = P(10, 5) = 0.30$ to yield the weighted covariation $30(0.30) = 9.0$, as shown in column 6.

The sum of column 6 is the covariance. We see that $\sigma\{X, Y\} = +10.0$. Note that the covariance is positive because the positive covariations in column 5 are the most likely ones to occur.

TABLE 5.3   Calculation of covariance for two examples

(a) Example 1: E{X} = 20, E{Y} = 8

(1)		(2)	(3)	(4)	(5)	(6)
						Weighted
			Deviation	Deviation	Covariation	Covariation
x	y	$P(x, y)$	$x - E\{X\}$	$y - E\{Y\}$	(3) $\times$ (4)	(5) $\times$ (2)
10	5	0.30	$10 - 20 = -10$	$5 - 8 = -3$	$(-10)(-3) = 30$	9.0
10	10	0.20	$10 - 20 = -10$	$10 - 8 = 2$	$(-10)(2) = -20$	-4.0
30	5	0.10	$30 - 20 = 10$	$5 - 8 = -3$	$(10)(-3) = -30$	-3.0
30	10	0.40	$30 - 20 = 10$	$10 - 8 = 2$	$(10)(2) = 20$	8.0
	Total	1.00				$\sigma\{X, Y\} = +10.0$

(b) Example 2: E{W} = 10.5, E{Z} = 3

(1)		(2)	(3)	(4)	(5)	(6)
						Weighted
			Deviation	Deviation	Covariation	Covariation
w	z	$P(w, z)$	$w - E\{W\}$	$z - E\{Z\}$	(3) $\times$ (4)	(5) $\times$ (2)
5	2	0.10	-5.5	-1	5.5	0.55
5	4	0.20	-5.5	1	-5.5	-1.10
10	2	0.10	-0.5	-1	0.5	0.05
10	4	0.20	-0.5	1	-0.5	-0.10
15	2	0.30	4.5	-1	-4.5	-1.35
15	4	0.10	4.5	1	4.5	0.45
	Total	1.00				$\sigma\{W, Z\} = -1.50$

2. The joint probability distribution of W and Z is as follows:

	z	
w	2	4
5	0.10	0.20
10	0.10	0.20
15	0.30	0.10

We find $E\{W\} = 10.5$ and $E\{Z\} = 3$ from the marginal distributions of the random variables (calculations not shown). Table 5.3b contains the calculation of the covariance. We see that $\sigma\{W, Z\} = -1.50$. The covariance is negative because the negative covariations in column 5 are the most likely ones to occur.   ☐

Interpretation of Covariance.   The magnitude of the covariance measure $\sigma\{X, Y\}$ depends on the units of X and Y and generally will change when the units of X and Y are changed. Hence, the main information provided by the covariance measure about the association between X and Y is whether $\sigma\{X, Y\}$ is positive, negative, or zero.

In Table 5.4, we present three bivariate probability distributions exhibiting different amounts and direction of covariance. In all three distributions, $E\{X\} = 1$ and $E\{Y\} = 3$. Since covariance is expected covariation, the interpretation of covariance parallels that of covariation for an individual joint outcome. When $\sigma\{X, Y\}$ is positive, negative, or zero, we say that X and Y exhibit *positive, negative,* or *zero covariance,* respectively. Thus, Table 5.4a contains a bivariate probability distribution exhibiting negative covariance. Note, in this case, that X and Y tend to vary inversely. If X has a value greater than its mean 1, Y is likely to have a value less than its mean 3, and if X has a value less than 1, Y is likely to have a value greater than 3. The bivariate probability distribution in Table 5.4b exhibits positive covariance. Here, X and Y tend to vary in the same direction. Finally, Table 5.4c shows a bivariate probability distribution exhibiting zero covariance. Here, the covariations between X and Y tend to balance out.

TABLE 5.4 Three bivariate probability distributions with varying amounts of covariance and correlation

(a)

	$y = 1$	$y = 3$	$y = 5$	Total
$x = 0$	0.00	0.05	0.25	0.30
$x = 1$	0.05	0.30	0.05	0.40
$x = 2$	0.25	0.05	0.00	0.30
Total	0.30	0.40	0.30	1.00

$$\sigma\{X, Y\} = -1.0 \qquad \rho\{X, Y\} = -0.833$$

(b)

	$y = 1$	$y = 3$	$y = 5$	Total
$x = 0$	0.30	0.00	0.00	0.30
$x = 1$	0.00	0.40	0.00	0.40
$x = 2$	0.00	0.00	0.30	0.30
Total	0.30	0.40	0.30	1.00

$$\sigma\{X, Y\} = +1.2 \qquad \rho\{X, Y\} = +1.0$$

(c)

	$y = 1$	$y = 3$	$y = 5$	Total
$x = 0$	0.10	0.10	0.10	0.30
$x = 1$	0.10	0.20	0.10	0.40
$x = 2$	0.10	0.10	0.10	0.30
Total	0.30	0.40	0.30	1.00

$$\sigma\{X, Y\} = 0 \qquad \rho\{X, Y\} = 0$$

An important property of the covariance measure follows.

(5.25)   When X and Y are independent, $\sigma\{X, Y\} = 0$

Thus, if X and Y are statistically independent random variables, theorem (5.25) assures us that the covariance between them must be zero.

Comments   1.   Although statistical independence implies zero covariance, the converse is *not* always true. It is possible that $\sigma\{X, Y\} = 0$, yet X and Y are statistically dependent. This possibility is illustrated by Table 5.4c where $\sigma\{X, Y\} = 0$, yet X and Y are statistically dependent, Note, for example, that $P(X = 2 \cap Y = 1) = 0.10$, while the product of the marginal probabilities is $0.30(0.30) = 0.09$. Thus, two random variables may be associated in a fashion that the covariance measure does not reflect.

2.   The covariance can be defined formally as expected covariation.

(5.26) $$\sigma\{X, Y\} = E\{(X - E\{X\})(Y - E\{Y\})\}$$

An algebraically identical expression is as follows.

(5.26a) $$\sigma\{X, Y\} = E\{XY\} - E\{X\}E\{Y\}$$

Coefficient of Correlation

As we noted earlier, the main information provided by the covariance measure about the linear association between two random variables X and Y is whether the covariance is positive, negative, or zero. The magnitude of $\sigma\{X, Y\}$ depends on the units of X and Y and thus usually cannot be directly compared for different pairs of variables.

The coefficient of correlation is an adjustment of the covariance so that the resulting measure is unit-free.

(5.27)   The *coefficient of correlation* of two random variables X and Y is denoted by $\rho\{X, Y\}$ (Greek rho) and defined:

$$\rho\{X, Y\} = \frac{\sigma\{X, Y\}}{\sigma\{X\}\sigma\{Y\}}$$

where:
$\sigma\{X\}$ is the standard deviation of X
$\sigma\{Y\}$ is the standard deviation of Y
$\sigma\{X, Y\}$ is the covariance of X and Y

EXAMPLES	1. For the example in Table 5.3a, we found that $\sigma\{X, Y\} = +10.0$. Further, $\sigma\{X\} = 10$ and $\sigma\{Y\} = 2.449$ (calculations not shown). Hence, by (5.27):

$$\rho\{X, Y\} = \frac{+10.0}{10(2.449)} = +0.41$$

Note that $\rho\{X, Y\}$ is positive, just like $\sigma\{X, Y\}$, which must be the case because the standard deviations in the denominator of $\rho\{X, Y\}$ in (5.27) are always positive.

2. In Table 5.4, we have shown the coefficients of correlation for each of the three bivariate distributions. The coefficient of correlation in Table 5.4a is negative, as is the covariance, and the coefficient of correlation in Table 5.4b is positive, as is the covariance. ☐

Interpretation of Coefficient of Correlation. It can be shown that $\rho\{X, Y\}$ takes on values between -1 and $+1$.

(5.28)
$$-1 \leq \rho\{X, Y\} \leq +1$$

The coefficient -1 is attained only when Y is a linear function of X, that is, $Y = a + bX$, and b is negative. Likewise, the coefficient $+1$ is attained only when $Y = a + bX$ and b is positive. The latter case is illustrated in Table 5.4b, where all possible pairs of X and Y values (that is, pairs with positive probability) satisfy the linear relationship $Y = 1 + 2X$. Thus, for instance, the possible pair $(X, Y) = (2, 5)$ satisfies the relation $Y = 1 + 2(2) = 5$. Because the coefficients -1 and $+1$ only arise when X and Y are exactly linearly related, these coefficients are said to indicate perfect negative and positive linear association, respectively.

A coefficient of correlation of 0 indicates that there is no linear association between X and Y, as is illustrated in Table 5.4c. We say that X and Y are uncorrelated then. The closer $\rho\{X, Y\}$ is to -1 or to $+1$, the stronger is the degree of linear association between the two random variables.

Comment	It can be shown that the coefficient of correlation $\rho\{X, Y\}$ in (5.27) is equivalent to the covariance between the standardized forms of the random variables X and Y. Thus, $\rho\{X, Y\}$ may also be interpreted as the expected covariation between the standardized forms of the random variables X and Y.

Variances of Sum and Difference of Two Dependent Random Variables

One use of the covariance is found in the variances of the sum and the difference of two dependent random variables. We noted earlier that the expected values for the sum and the difference of two random variables in (5.13a) and (5.14a) apply whether the random variables are independent or dependent. The variances given earlier, however, do not apply to dependent variables in general.

If two random variables are dependent, the variances of the sum and the difference include the covariance term.

(5.29) $$\sigma^2\{X + Y\} = \sigma^2\{X\} + \sigma^2\{Y\} + 2\sigma\{X, Y\}$$

(5.30) $$\sigma^2\{X - Y\} = \sigma^2\{X\} + \sigma^2\{Y\} - 2\sigma\{X, Y\}$$

EXAMPLES

1. Let X denote output by machine 1 and Y output by machine 2. Suppose $E\{X\} = 200$, $E\{Y\} = 100$, $\sigma^2\{X\} = 400$, $\sigma^2\{Y\} = 80$, and $\sigma\{X, Y\} = -10$. Let $T = X + Y$ denote the total output. We then have:

$$E\{T\} = 200 + 100 = 300 \qquad \text{by (5.13a)}$$

$$\sigma^2\{T\} = 400 + 80 + 2(-10) = 460 \qquad \text{by (5.29)}$$

The standard deviation of T is $\sigma\{T\} = \sqrt{460} = 21.4$. Note that if the covariance had been positive, $\sigma^2\{T\}$ would have been larger. For instance, if $\sigma\{X, Y\} = +10$, $\sigma^2\{T\} = 400 + 80 + 2(10) = 500$.

2. In Example 1, suppose interest is in the difference D in output, where $D = X - Y$. We then have:

$$E\{D\} = 200 - 100 = 100 \qquad \text{by (5.14a)}$$

$$\sigma^2\{D\} = 400 + 80 - 2(-10) = 500 \qquad \text{by (5.30)}$$

Comment

When random variables X and Y are statistically independent, we know from (5.25) that $\sigma\{X, Y\} = 0$. Hence, (5.29) and (5.30) in that case reduce to (5.13b) and (5.14b) for independent variables.

PROBLEMS

5.1 In each of the following situations, indicate whether the random variable is discrete or continuous, and describe its sample space (that is, its set of possible outcomes).
   a. The number of passengers on a scheduled flight having a capacity of 220 passengers.
   b. The number of employees of a firm with 300 employees who are absent because of sickness on a given day.
   c. The length of time a machine is idle during an 8-hour working day.

5.2 In each of the following situations, indicate whether the random variable is discrete or continuous, and describe its sample space (that is, its set of possible outcomes).
   a. The height of a plant that never grows taller than 1 meter.

b. The current volume of mineral oil in a 5000-liter storage tank.

c. The number of pills left from a 20-pill prescription.

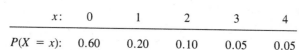

*5.3 Job Applicant. The probability distribution of X, the number of positions held previously by a job applicant, follows.

x:	0	1	2	3	4
$P(X = x)$:	0.60	0.20	0.10	0.05	0.05

a. Interpret each of the following expressions: (1) $P(0)$, (2) $P(X \le 1)$, (3) $P(1 \le X \le 3)$.

b. Obtain each of the probabilities in **a**.

5.4 Helicopter Shuttle. The probability distribution of X, the number of passengers on a daily helicopter shuttle run from airport A to airport B, follows.

x:	1	2	3	4
$P(X = x)$:	0.1	0.3	0.2	0.4

a. Interpret each of the following expressions: (1) $P(3)$, (2) $P(X \le 2)$, (3) $P(2 \le X \le 4)$.

b. Obtain each of the probabilities in **a**.

5.5 Telephone Call. The probability distribution of X, the length of a long-distance telephone call in minutes, follows.

x:	5	10	15	20
$P(x)$:	0.3	0.5	0.1	0.1

a. Interpret each of the following expressions: (1) $P(10)$, (2) $P(X \ge 10)$, (3) $P(5 \le X \le 10)$.

b. Obtain each of the probabilities in **a**.

5.6 Ceramic Filters. The probability distribution of X, the number of damaged ceramic filters in a shipment of three, follows.

x:	0	1	2	3
$P(x)$:	0.70	0.05	0.10	0.15

a. Give the notational representation for the probability of each of these events: (1) All filters in the shipment are damaged; (2) some but not all filters in the shipment are damaged; (3) fewer than two filters in the shipment are damaged.

b. Obtain the probability of each of the events in **a**.

c. The two most probable outcomes for X are zero and three damaged filters. What practical consideration in shipping ceramic filters might explain this fact?

*5.7 Refer to **Job Applicant** Problem 5.3.

a. Construct a graph of the probability distribution.

b. Construct a graph of the cumulative probability distribution. From your graph, find $P(X \le 3)$. Explain how this probability is interpreted.

5.8 Refer to **Helicopter Shuttle** Problem 5.4.

a. Construct a graph of the probability distribution.

b. Construct a graph of the cumulative probability distribution. From your graph, find (1) $P(X \le 2)$, (2) $P(X \le 3.5)$. Explain how the latter probability is interpreted.

5.9　Refer to **Telephone Call** Problem 5.5.
　　　a. Construct a graph of the probability distribution.
　　　b. Construct a graph of the cumulative probability distribution. From your graph, find
　　　　　(1) $P(X \leq 5)$, (2) $P(X \leq 25)$. Explain how the latter probability is interpreted.

5.10　Refer to **Ceramic Filters** Problem 5.6.
　　　a. Construct a graph of the probability distribution.
　　　b. Construct a graph of the cumulative probability distribution.
　　　c. From your graph in **b**, find (1) the probability that two or fewer filters in the shipment are
　　　　　damaged, (2) $P(X \leq 1.4)$.

*5.11　Refer to **Job Applicant** Problem 5.3. Obtain (1) $P(X = 0 | X \leq 3)$, (2) $P(X \geq 2 | X \geq 1)$.

5.12　Refer to **Ceramic Filters** Problem 5.6. Obtain (1) $P(X = 3 | X > 0)$, (2) $P(1 \leq X < 3 | X \leq 2)$.

*5.13　**Computer Operations.**　The bivariate probability distribution for number of computer stop-
　　　pages in a day (X) and number of computer operators sick for the day (Y) follows.

Number of Stoppages x	Number of Sick Operators y		
	0	1	2
0	0.40	0.15	0.02
1	0.30	0.05	0.01
2	0.04	0.03	0

　　　From inspection of the bivariate probability distribution, find and interpret each of the follow-
　　　ing probabilities: (1) $P(X = 1 \cap Y = 0)$, (2) $P(X = 2)$, (3) $P(X \leq 1)$, (4) $P(Y \geq 2)$,
　　　(5) $P(X = 1 | Y = 0)$.

5.14　**Bottle Flaws.**　The bivariate probability distribution of the number of bubble flaws (X) and
　　　solid-particle flaws (Y) in hand-blown decorative bottles follows.

Bubble Flaws x	Particle Flaws y	
	0	1
0	0.2	0.1
1	0.1	0.2
2	0.1	0.3

　　　From inspection of the bivariate probability distribution, find and interpret each of the follow-
　　　ing probabilities: (1) $P(X = 0 \cap Y = 1)$, (2) $P(Y = 0)$, (3) $P(X \leq 1)$, (4) $P(Y = 1 | X = 2)$,
　　　(5) $P(X \leq 1 | Y = 1)$.

5.15　Refer to **Helicopter Shuttle** Problem 5.4. Let X_1 be the number of passengers on day 1 and X_2
　　　the number of passengers on day 2. Assume that X_1 and X_2 are statistically independent, each
　　　with the probability distribution given in Problem 5.4.
　　　a. Construct the bivariate probability distribution for X_1 and X_2.
　　　b. From inspection of the bivariate probability distribution, find and interpret each of the fol-
　　　　　lowing probabilities: (1) $P(X_1 = 2 \cap X_2 = 1)$, (2) $P(X_2 = 4 | X_1 = 2)$, (3) $P(X_2 = 4)$,
　　　　　(4) $P(X_1 \leq 1 \cap X_2 \leq 1)$.

5.16 Refer to **Ceramic Filters** Problem 5.6. Let X_1 and X_2 denote the number of damaged ceramic filters in shipments 1 and 2, respectively. Assume that X_1 and X_2 are statistically independent, each with the probability distribution given in Problem 5.6.

    a. Construct the bivariate probability distribution for X_1 and X_2.

    b. From the bivariate probability distribution, give the notational representation and numerical value for the probability of each of these events: (1) One filter is damaged in shipment 1 and none is damaged in shipment 2; (2) two filters are damaged in shipment 2, given none is damaged in shipment 1; (3) three filters are damaged in shipment 1; (4) at least one filter is damaged in each shipment.

*5.17 Refer to **Computer Operations** Problem 5.13.

    a. Obtain the marginal probability distribution of the number of computer stoppages.

    b. Obtain the conditional probability distribution of the number of computer stoppages when no operator is sick.

    c. How does the conditional distribution in **b** compare with the marginal distribution in **a**? Does this result imply that X and Y are dependent? Explain.

5.18 Refer to **Bottle Flaws** Problem 5.14.

    a. Obtain the marginal probability distribution of the number of bubble flaws.

    b. Obtain the conditional probability distribution of the number of bubble flaws when no particle flaw is present. Also, obtain the conditional distribution when one particle flaw is present.

    c. Does it appear from the conditional distributions in **b** that the number of bubble flaws and the number of particle flaws in a bottle are statistically related? Explain.

5.19 Refer to **Helicopter Shuttle** Problems 5.4 and 5.15.

    a. From the joint probability distribution, obtain the conditional probability distribution of the number of passengers on day 2, given that there are two passengers on day 1.

    b. Why must the conditional distribution in **a** be the same as the marginal distribution of the number of passengers on day 2? Explain.

*5.20 Refer to **Job Applicant** Problem 5.3. Calculate $E\{X\}$. Interpret its meaning here in relative frequency terms.

5.21 Refer to **Helicopter Shuttle** Problem 5.4. Calculate $E\{X\}$, and interpret this measure. Should one expect the measure to be a whole number here?

5.22 Refer to **Ceramic Filters** Problem 5.6.

    a. Calculate the mean of the probability distribution. What notation is used in the text to denote this quantity?

    b. If the random trial represented by the probability distribution here were repeated independently for a very large number of shipments, in about what percentage of the shipments would no filters be damaged? What would be the approximate mean number of damaged filters per shipment?

5.23 Underwriting Syndicate. An underwriting syndicate will insure an offshore gas production platform for one year. The syndicate's potential loss X (in \$ million) has the following probability distribution:

x:	0	20	150
$P(x)$:	0.990	0.008	0.002

    a. What is the syndicate's expected loss? What is the probability that the syndicate's actual loss will be smaller than the expected loss?

b. The risk manager of the company that owns the platform has suggested that \$400,000 would be a fair premium for the syndicate to charge for assuming the potential loss under the insurance contract. Do you agree? Discuss.

*5.24 Refer to **Job Applicant** Problem 5.3. Calculate the variance and the standard deviation of the probability distribution. In what units is the standard deviation expressed?

5.25 Refer to **Helicopter Shuttle** Problem 5.4. Calculate $\sigma^2\{X\}$ and $\sigma\{X\}$. In what units is $\sigma\{X\}$ expressed?

5.26 Refer to **Ceramic Filters** Problem 5.6.
   a. Calculate $\sigma\{X\}$.
   b. Is the standard deviation here large relative to the mean? If an improvement in the quality of packaging changed the probability distribution of X so that $E\{X\}$ became smaller, would you expect $\sigma\{X\}$ to become smaller also? Discuss.

5.27 Refer to **Underwriting Syndicate** Problem 5.23.
   a. Calculate the variance and the standard deviation of the probability distribution. In what units is the standard deviation expressed?
   b. Which of the three possible outcomes of X contributes the most to the magnitude of the variance in **a**?

*5.28 Refer to **Job Applicant** Problem 5.3.
   a. Obtain the probability distribution of the standardized form of random variable X.
   b. Verify by direct calculation that the probability distribution in **a** has mean 0 and standard deviation 1.

5.29 Refer to **Underwriting Syndicate** Problem 5.23.
   a. Obtain the probability distribution of the standardized form of random variable X.
   b. How many standard deviations from the mean is the outcome of X that lies farthest from the mean of the probability distribution?

5.30 The probability distribution of the standardized form Y of a random variable X follows.

y:	-0.375	0.562	8.991
$P(y)$:	0.69	0.30	0.01

   a. How many standard deviations apart are the smallest and largest possible outcomes of X?
   b. The mean and the variance of X are 0.40 and 1.14, respectively. What is the value of the largest possible outcome of X?
   c. Verify by direct calculation that $E\{Y\} = 0$ and $\sigma\{Y\} = 1$ here, except for rounding differences.
   d. According to the Chebyshev inequality, what is the maximum probability that a standardized random variable will exceed 8 in absolute value? What is the actual probability here? Is this probability consistent with the Chebyshev upper bound?

*5.31 A report on rural water resources states that the nitrate level of wells in a certain groundwater system has a probability distribution whose mean and standard deviation are 5.2 and 2.1 parts per million (ppm), respectively. From the Chebyshev inequality, determine the minimum probability that a well in this system will have a nitrate level between 0.4 and 10.0 ppm.

5.32 A bank's study of midmonth balances of personal checking accounts shows that they follow a probability distribution with mean \$300 and standard deviation \$40.
   a. From the Chebyshev inequality, determine the minimum probability that the midmonth balance of an account is between \$180 and \$420.

b. The probability distribution of midmonth balances is skewed to the right, and large negative balances do not occur. Explain if either of these two facts affects the validity of the probability bound computed in **a**.

c. Consider the probability distribution of the end-of-month balances for these same accounts. Would the probability bound in **a** apply to the end-of-month balances? Explain.

*5.33 The probability distribution of X, the number of persons in a restaurant party, follows.

x:	1	2	3	4	5	6
$P(x)$:	0.05	0.15	0.25	0.40	0.10	0.05

The restaurant serves a buffet dinner costing $9.50 per person. Let Y denote the total cost of the buffet dinner for a party.

a. Express Y in terms of X.

b. Calculate $E\{X\}$ and $\sigma\{X\}$. Use these results to obtain $E\{Y\}$ and $\sigma\{Y\}$ by (5.11) and (5.12).

c. Verify your results in **b** by obtaining the probability distribution of Y and calculating $E\{Y\}$ and $\sigma\{Y\}$ directly from this probability distribution.

5.34 Refer to **Helicopter Shuttle** Problem 5.4. The fare for the trip is $48. Let Y denote the total fare revenue for the run.

a. Express Y in terms of X.

b. Calculate $E\{Y\}$ and $\sigma\{Y\}$ from $E\{X\}$ and $\sigma\{X\}$, using (5.11) and (5.12).

c. Verify your results in **b** by obtaining the probability distribution of Y and calculating $E\{Y\}$ and $\sigma\{Y\}$ directly from this probability distribution.

5.35 Refer to **Telephone Call** Problem 5.5. Let Y denote the cost of a long-distance telephone call in dollars. The initial connection charge is $1.60, and the cost per minute is $1.20.

a. Express Y in terms of X.

b. Calculate $E\{Y\}$ and $\sigma^2\{Y\}$ from $E\{X\}$ and $\sigma^2\{X\}$, using (5.11) and (5.12).

c. Which would increase the expected cost of a telephone call more, an increase in the initial connection charge to $2.50 or an increase in the cost per minute to $1.40?

d. Is the standard deviation of the cost of a telephone call affected by the initial connection charge? By the cost per minute? Explain.

5.36 A contractor has successfully bid for a construction job that will pay him $800 thousand. The total cost of completing the job (in $ thousand) is a random variable X, with $E\{X\} = 600$ and $\sigma\{X\} = 150$. The contractor's gross income from the job is therefore $Y = 800 - X$.

a. Obtain $E\{Y\}$ and $\sigma\{Y\}$. In what units is each of these measures expressed?

b. Compare the coefficients of variation (that is, the ratio of standard deviation to mean, expressed as a percentage) of X and Y. Which is relatively more variable, the total cost of the job or the gross income from the job?

c. If the total cost of the job were 10 percent higher because of poor cost control, thereby changing the contractor's gross income from the job to $Y' = 800 - 1.10X$, what would be the resulting values of the mean and the standard deviation of the gross income?

*5.37 **Sports Match.** The probability distributions of the number of goals scored by the home team (X) and the visiting team (Y) in a sports match follow.

x:	0	1	2		y:	0	1	2
$P(x)$:	0.4	0.3	0.3		$P(y)$:	0.5	0.3	0.2

Here, X and Y are independent random variables. Let $T = X + Y$ denote the total number of goals scored in the match.

      a. Calculate $E\{T\}$ and $\sigma\{T\}$, using (5.13). Interpret each measure in terms of relative frequency.

      b. Obtain the probability distribution of T. Verify your results in **a** by direct calculations from this probability distribution.

*5.38 Refer to **Sports Match** Problem 5.37.

      a. What does the difference $D = X - Y$ represent here? Calculate $E\{D\}$ and $\sigma\{D\}$, using (5.14).

      b. Obtain the probability distribution of D. What is the probability of a tie game? Of the home team winning?

5.39 Refer to **Helicopter Shuttle** Problems 5.4 and 5.15. Let $T = X_1 + X_2$ denote the total number of passengers on the two days' runs.

      a. Calculate $E\{T\}$ and $\sigma^2\{T\}$, using (5.13). Does either of these results depend on the fact that X_1 and X_2 are statistically independent?

      b. Obtain the probability distribution of T. Verify your results in **a** by direct calculations from this probability distribution.

5.40 Refer to **Helicopter Shuttle** Problems 5.4 and 5.15.

      a. What does the difference $D = X_2 - X_1$ represent here?

      b. Find $E\{D\}$ and $\sigma\{D\}$, using (5.14). Interpret each of these measures in terms of relative frequency.

5.41 The percentage grades in a history course that two students will obtain this semester are denoted by X_1 and X_2. Assume X_1 and X_2 are statistically independent, with the following means and standard deviations:

$$E\{X_1\} = 86 \qquad \sigma\{X_1\} = 5$$
$$E\{X_2\} = 75 \qquad \sigma\{X_2\} = 8$$

      a. Obtain the mean and the standard deviation of $D = X_1 - X_2$. Does either of these results depend on the fact that X_1 and X_2 are statistically independent?

      b. Obtain the mean and the standard deviation of $T = X_1 + X_2$.

      c. Use the results in **b** and (5.11) and (5.12) to obtain the mean and the standard deviation of the average grade of the two students, that is, of $T/2$.

*5.42 The daily output X in a bottling plant is a random variable, with $E\{X\} = 12,000$ bottles and $\sigma\{X\} = 500$ bottles. Assume that the outputs on successive days are statistically independent.

      a. What is the expected total output for 10 successive days? For 30 successive days?

      b. Obtain the standard deviation of total output for 10 days. Also, obtain the standard deviation of total output for 30 days. How do these two standard deviations compare relative to their respective means?

5.43 Refer to **Helicopter Shuttle** Problem 5.4. Let T denote the total number of passengers on five statistically independent runs.

      a. Obtain $E\{T\}$ and $\sigma^2\{T\}$, using (5.15).

      b. Is the standard deviation of T five times as large as the standard deviation of X? Discuss.

5.44 Let X_1, X_2, and X_3 denote the service lives (in operating hours) of three drive belts installed in a machine. The total service provided by all three belts is $T = X_1 + X_2 + X_3$. Assume X_1, X_2, and X_3 are statistically independent, with the following means and standard deviations:

	X_1	X_2	X_3
Mean:	410	530	590
Standard Deviation:	60	75	80

a. Obtain $E\{T\}$ and $\sigma^2\{T\}$, using (5.15).

b. If the first belt has an actual service life of $X_1 = 473$ hours, then what are the mean and the standard deviation of $X_2 + X_3$, the total service provided by the second and third belts?

*5.45 Refer to Figure 5.3. For this density function, $P(X \leq 0.5) = 0.688$.

a. Find (1) $P(0 \leq X \leq 0.7)$; (2) $P(X \geq 0.7)$. Interpret the probability in (2).

b. If you were asked to find $P(X > 0.7)$ rather than $P(X \geq 0.7)$, why would the answer not change? Explain.

5.46 **Suspension Cable.** When a 200-meter suspension cable breaks, it is equally likely that the break occurs at any point along its length. Let X denote the distance from one end of the cable to the break; then X has the probability density function:

$$f(x) = \begin{cases} \dfrac{1}{200} & 0 \leq x \leq 200 \\ 0 & \text{elsewhere} \end{cases}$$

a. Graph the probability density function for X. Does the area under your density function equal 1?

b. Use the geometric properties of the probability density function to find the following probabilities: (1) $P(X \leq 100)$, (2) $P(X > 50)$, (3) $P(40 \leq X \leq 80)$.

c. Obtain the cumulative probability function for X and graph it.

d. From your graph of the cumulative probability function in **c**, find the values of (1) $F(80)$, (2) $F(150)$. Interpret the meaning of the latter value.

5.47 **Shelf Life.** A melon's shelf life (X, in days) is treated as a continuous random variable. The density function of X is the right triangular probability function:

$$f(x) = \begin{cases} 0.005(20 - x) & 0 \leq x \leq 20 \\ 0 & \text{elsewhere} \end{cases}$$

a. Graph this probability function. Is the area under the density function equal to 1?

b. Use the geometric properties of $f(x)$ to find (1) $P(X \leq 10)$, (2) $F(5)$, (3) $P(5 < X \leq 10)$.

5.48 **Wind Velocity.** The cumulative probability function for the velocity of the wind (X, in kilometers per hour) at a given place is:

$$F(x) = \frac{5x}{100 + 4x} \qquad 0 \leq x \leq 100$$

a. Find (1) $F(30)$, (2) $P(10 \leq X \leq 80)$. Interpret the first probability.

b. Graph $F(x)$ on the interval $0 \leq x \leq 100$. Does $F(x)$ exhibit the properties of a cumulative probability function on this interval? Explain.

*5.49 **Product Demand.** The bivariate probability distribution of the price received (P) and quantity sold (Q) of a product follows.

	q (cartons)	
p ($ per carton)	30	40
6	0.2	0.4
8	0.3	0.1

a. Calculate $E\{P\}$ and $E\{Q\}$.

b. Obtain the covariation of $p = 6$ and $q = 40$. What information is provided by the sign of the measure of covariation?

c. Calculate $\sigma\{P, Q\}$. Does its value indicate that P and Q are linearly associated? Explain.

5.50 Refer to **Computer Operations** Problem 5.13.

a. Calculate $E\{X\}$ and $E\{Y\}$.

b. Obtain the covariation of $x = 2$ and $y = 0$. In what units is the covariation expressed?

c. Calculate $\sigma\{X, Y\}$. What does its sign tell us about the nature of the linear association between X and Y?

5.51 Refer to **Bottle Flaws** Problem 5.14.

a. Calculate $E\{X\}$, $E\{Y\}$, and $\sigma\{X, Y\}$.

b. Are the two random variables statistically independent? Discuss.

*5.52 Refer to **Product Demand** Problem 5.49.

a. Calculate the coefficient of correlation between price received and quantity sold. What range of values can this coefficient take on?

b. Using your result in **a**, describe the direction and the strength of the linear association between price and quantity.

5.53 Refer to **Computer Operations** Problem 5.13.

a. Calculate $\rho\{X, Y\}$. In what units is this measure expressed?

b. Using your result in **a**, describe the direction and the strength of the linear association between X and Y.

5.54 Refer to **Bottle Flaws** Problem 5.14.

a. Calculate $\rho\{X, Y\}$. Does $\rho\{Y, X\}$ necessarily have the same sign and magnitude as $\rho\{X, Y\}$? Explain.

b. Does the positive value of $\rho\{X, Y\}$ here guarantee that a bottle with more particle flaws than another must also have more bubble flaws? Discuss.

*5.55 Refer to Table 5.1. Here, $E\{X\} = E\{Y\} = 0.55$, $\sigma^2\{X\} = \sigma^2\{Y\} = 0.7475$, and $\sigma\{X, Y\} = 0.5675$.

a. Calculate $E\{X + Y\}$, $E\{X - Y\}$, $\sigma\{X + Y\}$, and $\sigma\{X - Y\}$. Interpret each of these numbers.

b. What would be the values calculated in **a** if X and Y had the marginal probability distributions shown in Table 5.1 but were statistically independent?

5.56 Refer to **Bottle Flaws** Problem 5.14. Let T denote the total number of flaws in a bottle. Find $E\{T\}$ and $\sigma\{T\}$. Interpret each of these measures.

5.57 Let X_1 and X_2 denote the earnings (in $ million) of a firm in two consecutive years; X_1 and X_2 have the same mean 14.8 and the same standard deviation 3.2.

a. Calculate the mean and the standard deviation of the total earnings $X_1 + X_2$ in the two years, assuming (1) $\rho\{X_1, X_2\} = -0.5$, (2) $\rho\{X_1, X_2\} = +0.5$. For which of these two correlations will the firm's total earnings in the two years be less variable?

b. Calculate the mean and the standard deviation of the difference in earnings $X_1 - X_2$, assuming (1) $\rho\{X_1, X_2\} = -0.5$, (2) $\rho\{X_1, X_2\} = +0.5$. For which of these two correlations will the difference in earnings for the two years be less variable?

EXERCISES

5.58 Refer to **Computer Operations** Problem 5.13, and find the following probabilities: (1) $P(X = 2 \mid Y \leq 1)$, (2) $P(X \leq 1 \mid Y \geq 1)$, (3) $P(X = Y \mid Y \leq 1)$.

5.59 Verify the identity (5.7a).

5.60 Verify the identities (5.11) and (5.12).

5.61 (Calculus needed.) Refer to **Suspension Cable** Problem 5.46. Obtain (1) $E\{X\}$, (2) $\sigma^2\{X\}$, (3) $F(x)$.

5.62 (Calculus needed.) Refer to **Shelf Life** Problem 5.47. Obtain (1) $E\{X\}$, (2) $\sigma^2\{X\}$, (3) $F(x)$.

5.63 (Calculus needed.) Refer to **Wind Velocity** Problem 5.48.
   a. Obtain the probability density function and graph it.
   b. Obtain $E\{X\}$. Interpret its meaning in this setting.
   c. Obtain the median wind speed. [*Hint:* $F(Md) = 0.5$.]

5.64 Verify property (5.25).

5.65 Verify the identity (5.26a).

5.66 Let Y_1 and Y_2 denote the standardized forms of random variables X_1 and X_2, respectively. Show that $\rho\{X_1, X_2\} = \sigma\{Y_1, Y_2\}$. [*Hint:* Use (5.26a).]

STUDIES

5.67 A potato farmer has a contract to sell the forthcoming harvest to a food cannery. The entire harvest will receive a grade of A, B, or C. If the harvest is graded A, the farmer will receive $6.20 per bushel. For grades B and C, the prices per bushel will be $5.80 and $5.20, respectively. The probabilities for the grade of the harvest, if the harvesting is done at the normal time, are $P(A) = 0.35$, $P(B) = 0.45$, and $P(C) = 0.20$. The yield of the harvest at the normal time is expected to be 60,000 bushels. If the farmer harvests earlier, the probabilities would be changed to $P(A) = 0.40$, $P(B) = 0.60$, and $P(C) = 0$, but the yield would be reduced to 57,000 bushels.
   a. Obtain the probability distribution of total revenue under each of the following alternatives: (1) harvest earlier, (2) harvest at the normal time.
   b. Which alternative in **a** leads to the larger expected total revenue? The larger standard deviation of total revenue?

5.68 A national chain operates 400 similar restaurants. The probability distribution of loss by fire in any one of these restaurants in a year (X, in dollars) follows.

x:	0	150,000
$P(x)$:	0.998	0.002

Assume that the losses in the different restaurants $(X_1, X_2, \ldots, X_{400})$ are statistically independent. Let $T = \Sigma\, X_i$ denote the total loss by fire for the whole chain in a year.

   a. Calculate $E\{X_i\}$ and $\sigma\{X_i\}$ for any one restaurant. Then, find $E\{T\}$ and $\sigma\{T\}$. Is random variable T relatively more predictable than the random variable X_i for any one restaurant? Comment.

   b. Use the Chebyshev inequality to obtain an upper bound on the probability that total loss by fire in a year for the chain will differ from the expected amount by more than \$380,000. Is this a bound on the probability $P(T \le 500,000)$ here? Explain.

   c. The chain self-insures its 400 restaurants against fire loss because the expected loss by fire for the chain is substantially less than the cost of fire insurance for a year. If a middle-income person were to own and operate a single one of these restaurants, why might he or she not be willing to self-insure against fire loss?

5.69 A truck driver's emergency kit includes four highway flares. The burning times of the flares are independent random variables, with mean 15.0 minutes and standard deviation 1.50 minutes for each flare. Let T denote the total burning time of the four flares when they are burned consecutively. From the Chebyshev inequality, determine the minimum probability that the flares will last between 54.0 and 66.0 minutes when burned consecutively.

5.70 A small investor is considering a \$200 investment in growth stocks for one year. Let X denote the gain in buying and holding for one year one \$100 share of stock 1, and let Y be defined similarly for one \$100 share of stock 2. Assume that $E\{X\} = 50$, $E\{Y\} = 50$, $\sigma^2\{X\} = 64$, $\sigma^2\{Y\} = 81$, and $\sigma\{X, Y\} = -60$. Consider the following options: (1) buy two shares of stock 1; (2) buy two shares of stock 2; (3) buy one share each of stocks 1 and 2.

   a. Is any one of these options better than the others in terms of expected gain?

   b. Calculate $\sigma^2\{2X\}$, $\sigma^2\{2Y\}$, and $\sigma^2\{X + Y\}$, and explain what each quantity represents. The investor states that option 3 "is less risky" than options 1 and 2. How is she interpreting risk here?

   c. Repeat the calculations in **b** for $\sigma\{X, Y\} = +60$. Which option now has the smallest variance? What does this finding suggest about the covariance conditions that will "spread the risk" when one is diversifying with a number of different securities?

5.71 Refer to **Product Demand** Problem 5.49.

   a. Suppose the dollar cost of producing quantity Q is given by $C = 100 + 3Q$. Obtain $E\{C\}$ and $\sigma^2\{C\}$.

   b. Denote the revenue from the sale of the product by R, so $R = PQ$. Find $E\{R\}$ and interpret its meaning. [*Hint:* Use (5.26a).]

   c. The profit from the sale of the product is $R - C$. Find $E\{R - C\}$.

COMMON DISCRETE PROBABILITY DISTRIBUTIONS

There exists an infinite variety of probability distributions. However, a limited number of types, or *families,* of probability distributions are used in a wide range of applications. In this chapter, we study several important families of distributions for discrete random variables. In the next chapter, we extend our study to distributions for continuous random variables.

6.1 BINOMIAL PROBABILITY DISTRIBUTIONS

Bernoulli Random Trial

A *Bernoulli random trial* is a random trial that has two basic outcomes of a qualitative nature. To quantify these outcomes, we assign one outcome the value 0 and the other the value 1. Which outcome is assigned the value 0 and which the value 1 is arbitrary. The random variable X associated with a Bernoulli random trial is called a *Bernoulli random variable*.

EXAMPLES

1. The calculation of a payroll check may be correct or incorrect. We define the Bernoulli random variable for this trial so that $X = 0$ corresponds to a correctly calculated check and $X = 1$ to an incorrectly calculated one.

2. A consumer either recalls a television commercial ($X = 1$) or does not recall ($X = 0$). Here, X is a Bernoulli random variable.

3. Screening of a credit card application gives either the rating $X = 1$ (accept) or $X = 0$ (reject). Here, X is a Bernoulli random variable. □

Bernoulli Process

Sequence of Bernoulli Trials. In statistical analysis, one is seldom interested in a single Bernoulli trial. More commonly, a sequence of Bernoulli trials is under consideration.

EXAMPLES

1. **Spoon Manufacturing.** In a process for manufacturing spoons, each spoon may either be defective or not. This process may be viewed as a sequence of Bernoulli

185

trials in which the ith spoon has an associated random variable X_i, with $X_i = 1$ if the ith spoon is defective and $X_i = 0$ if it is not defective. Interest may center on how many defective spoons are produced in the most recent production run of 1000 spoons.

2.   **Plant Specimens.**   Thirty plant specimens are treated for a particular fungus. At the end of the test period, each specimen is examined to see if it is fungus-free. This process may be viewed as a sequence of Bernoulli trials in which the ith specimen has an associated random variable X_i, with $X_i = 1$ if the ith specimen is found to be fungus-free and $X_i = 0$ if fungus is present. Although no time sequence is involved in this case, one may still view the 30 specimen outcomes as a sequence of 30 Bernoulli trials.   □

Bernoulli Process Postulates.   Suppose $X_1, X_2, \ldots, X_n$ are n Bernoulli random variables associated with a sequence of n trials. If these n trials meet the following conditions, we shall call the sequence of trials a *Bernoulli process*.

> *Postulate 1.*   The Bernoulli random trials in the sequence are statistically independent; that is, the Bernoulli random variables $X_1, X_2, \ldots, X_n$ are statistically independent.
>
> *Postulate 2.*   The probabilities that $X_i = 1$ and $X_i = 0$ are the same for all Bernoulli trials in the sequence; that is, $P(X_i = 1) = p$ and $P(X_i = 0) = 1 - p$ for $i = 1, 2, \ldots, n$, where p is the common probability that the trial outcome is 1.

EXAMPLE

For the spoon manufacturing example, the first postulate requires that whether or not one spoon is defective is statistically independent of the outcomes for the other spoons in the production run. The second postulate requires that the probability of a spoon being defective is the same for all spoons produced in the run. Of course, if the probability of a defective spoon is p, since there are only two possible outcomes, the probability of a spoon not being defective is $1 - p$.   □

The two postulates for a Bernoulli process describe two conditions that are frequently assumed for a sequence of random variables. The conditions are those of statistical independence and stationarity.

(6.1)

Let $X_1, X_2, \ldots, X_n$ be a sequence of random variables associated with a random process. The process will be said to be *independent and stationary* if the random variables are statistically independent of one another and if each has the same probability distribution. Under these conditions, the random variables will be said to be *independent and identically distributed*.

Binomial Random Variable

In a sequence of Bernoulli trials, interest often focuses on how many trials result in the outcome that has been assigned the value 1. In the plant specimens example, the number of specimens among the 30 that are fungus-free at the end of the test period may be the characteristic of interest. Similarly, for the spoon manufacturing example, interest may be in the number of defective spoons in the production run of 1000 spoons.

In the latter example, $X_i = 1$ if the ith spoon is defective and $X_i = 0$ otherwise. Hence, the sum $X_1 + X_2 + \cdots + X_{1000}$ contains as many 1s as there are defective spoons among the 1000, and 0s otherwise. Thus, $X_1 + X_2 + \cdots + X_{1000}$ equals the number of defective spoons in the run of 1000 spoons. Similarly, for the plant specimens example, $X_1 + X_2 + \cdots + X_{30}$ equals the number of fungus-free specimens among the 30 in the study, since $X_i = 1$ if the ith specimen is free of fungus and $X_i = 0$ otherwise.

(6.2)

Let the sum of n independent and identically distributed Bernoulli random variables be denoted by X:

$$X = X_1 + X_2 + \cdots + X_n$$

X is called a *binomial random variable*.

Binomial Probability Function

As we have seen in Chapter 5, the probability distribution for a discrete random variable X associates a probability $P(X = x)$ with each basic outcome x. Often, the probability function $P(X = x) = P(x)$ can be expressed by a mathematical formula so that the probability $P(x)$ can be calculated for each outcome x from the formula. The probability function for a binomial random variable is of this type.

(6.3)

The *binomial probability function* is:

$$P(x) = \binom{n}{x} p^x (1 - p)^{n-x}$$

where:

$P(x) = P(X = x)$
$x = 0, 1, \ldots, n$
$0 < p < 1$

Here, $\binom{n}{x}$ represents a *binomial coefficient*.

(6.4)

The binomial coefficient $\binom{n}{x}$ is defined:

$$\binom{n}{x} = \frac{n!}{x!\,(n-x)!}$$

where:

$$a! = a(a-1)\cdots(2)\,(1) \text{ and } 0! = 1$$

For example, we have:

$$\binom{5}{2} = \frac{5!}{2!\,3!} = \frac{5(4)\,(3)\,(2)\,(1)}{2(1)\,(3)\,(2)\,(1)} = 10$$

(A review of permutations and combinations can be found in Appendix A, Section A.4.)

The binomial probability distribution is a discrete probability distribution since X can take only one of the $n+1$ values $0, 1, \ldots, n$. Once values are assigned to p and n, we have identified one binomial probability distribution from the family of all such probability distributions; p and n are said to be the *parameters* of the binomial probability distribution.

EXAMPLE

In the plant specimens example, suppose that only three plant specimens are under study, that is, $n = 3$. Assume that the postulates for a Bernoulli process are satisfied by the test situation and that $p = 0.2$ is the probability of any individual specimen being free of fungus at the end of the test period. We compute the binomial probability distribution for the number X of fungus-free specimens in the group of $n = 3$, using (6.3):

x	$P(x)$
0	$\binom{3}{0}(0.2)^0(0.8)^3 = 0.5120$
1	$\binom{3}{1}(0.2)^1(0.8)^2 = 0.3840$
2	$\binom{3}{2}(0.2)^2(0.8)^1 = 0.0960$
3	$\binom{3}{3}(0.2)^3(0.8)^0 = \underline{0.0080}$
	Total 1.0000

We see, for instance, that the probability of all three specimens being fungus-free is $P(3) = 0.0080$. We also note that the probability of one or fewer specimens being fungus-free is $P(X \leq 1) = P(0) + P(1) = 0.5120 + 0.3840 = 0.8960$. □

Derivation.   To derive the binomial probability function (6.3), consider again the plant specimens example in which $n = 3$ and $p = 0.2$. The probability tree diagram in Figure 6.1 is helpful in this connection. It shows all possible outcome sequences for the three trials. Suppose we wish to find $P(X = 1) = P(1)$. Figure 6.1 shows that

===

FIGURE 6.1 Illustration of the derivation of the binomial probability function for $n = 3$ and $p = 0.2$

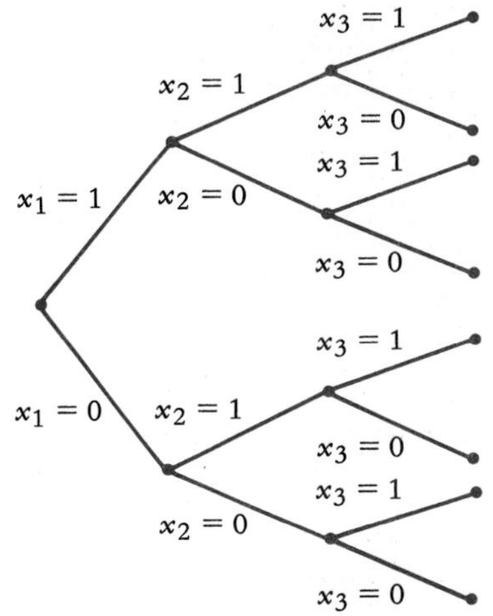

	Sequence	$x = x_1 + x_2 + x_3$	Probability of sequence
$x_3 = 1$	S_1	3	$(0.2)^3(0.8)^0$
$x_3 = 0$	S_2	2	$(0.2)^2(0.8)^1$
$x_3 = 1$	S_3	2	$(0.2)^2(0.8)^1$
$x_3 = 0$	S_4	1	$(0.2)^1(0.8)^2$
$x_3 = 1$	S_5	2	$(0.2)^2(0.8)^1$
$x_3 = 0$	S_6	1	$(0.2)^1(0.8)^2$
$x_3 = 1$	S_7	1	$(0.2)^1(0.8)^2$
$x_3 = 0$	S_8	0	$(0.2)^0(0.8)^3$

===

there are three sequences yielding exactly one fungus-free specimen, namely, S_4, S_6, and S_7.

Let us obtain the probability that sequence S_4 occurs. Since the random variables X_1, X_2, and X_3 are independent, and since $P(X_i = 1) = 0.2$ and $P(X_i = 0) = 0.8$ for each, it follows from a generalization of definition (5.4) for independent random variables that:

$$P(S_4) = P(X_1 = 1)P(X_2 = 0)P(X_3 = 0) = 0.2(0.8)(0.8) = (0.2)^1(0.8)^2$$

Similarly, we find:

$$P(S_6) = (0.2)^1(0.8)^2$$
$$P(S_7) = (0.2)^1(0.8)^2$$

Finally, since S_4, S_6, and S_7 are mutually exclusive events, we have, by (4.18):

$$P(X = 1) = P(1) = P(S_4) + P(S_6) + P(S_7) = 3(0.2)^1(0.8)^2 = 0.3840$$

This probability is the same as that obtained earlier.

Figure 6.1 shows the probabilities for all eight possible sample sequences. When they are combined appropriately, the binomial probabilities obtained directly from the probability function (6.3) are produced. For instance:

$$P(2) = P(S_2) + P(S_3) + P(S_5) = 3(0.2)^2(0.8)^1 = 0.0960$$

We generalize the derivation now.

1. When there are n independent Bernoulli trials, for each of which $P(X_i = 1) = p$, the probability of obtaining x 1s and $n - x$ 0s in a *specific sequence* or *permutation* is:

    $$p^x(1 - p)^{n-x}$$

2. The number of distinct sequences or permutations of x 1s and $n - x$ 0s is given by the binomial coefficient $\binom{n}{x}$.
3. Combining these results, we see that $P(x)$ is simply the sum of $\binom{n}{x}$ sequence probabilities, each of which is $p^x(1 - p)^{n-x}$:

    $$P(x) = \binom{n}{x}p^x(1 - p)^{n-x}$$

Probability Table

Binomial probabilities can be readily computed with many types of pocket calculators and statistical computer packages. In addition, they can be obtained from tables of binomial probabilities. Table C.5 (see Appendix C) gives binomial probabilities for selected values of p and selected values of n up to 20. For large n, approximation methods may be used. We shall take these up in Chapter 12.

EXAMPLES

1. In a psychological experiment, $n = 20$ children are each asked independently to solve a puzzle in a fixed amount of time. Interest is in the number of subjects who succeed in completing the task. Suppose the experimental conditions make it reasonable to assume that the number who succeed is a binomial random variable with $p = 0.4$. We wish to find the probability of exactly 6 successes. We see from Table C.5 that it is $P(6) = 0.1244$. This value is located in the column marked $p = 0.40$ at the top and in the row corresponding to $n = 20$ and $x = 6$ as labeled on the left-hand side of the table.

    If $p = 0.7$, we would then find $P(6) = 0.0002$ in the probability table. This time the value is located in the column marked $p = 0.70$ at the bottom and in the row corresponding to $n = 20$ and $x = 6$ as labeled on the right-hand side of the table. (Note, therefore, that the (n, x) values on the left-hand side correspond to p values of 0.5 or less, and those on the right-hand side correspond to p values of 0.5 or more.)

    To obtain cumulative binomial probabilities from Table C.5, we simply sum the appropriate individual probabilities. For example, if $P(X \le 2)$ is desired for the case in which $p = 0.40$ and $n = 20$, we find that $P(X \le 2) = P(0) + P(1) + P(2) = 0.0000 + 0.0005 + 0.0031 = 0.0036$.

2. In an insurance company, $n = 10$ claims adjusters are asked, as part of a continuing training program, to consider independently a hypothetical claim. Interest is in
the number of adjusters who handle the claim correctly. Suppose it is reasonable
to assume that the number who handle the claim correctly is a binomial random
variable with $p = 0.9$. We wish to find $P(X \leq 8)$. We use the complementation
theorem (4.19) to recognize that $P(X \leq 8) = 1 - P(X \geq 9)$. From Table C.5,
we find $P(X \geq 9) = P(9) + P(10) = 0.3874 + 0.3487 = 0.7361$. Hence,
$P(X \leq 8) = 1 - 0.7361 = 0.2639$. □

Characteristics of Binomial Probability Distributions

Mean and Variance. Two important characteristics of a binomial probability distribution are its mean and variance.

> The mean and the variance of a binomial probability distribution are:
>
> (6.5) $E\{X\} = np$
>
> (6.6) $\sigma^2\{X\} = np(1 - p)$

EXAMPLE

The number of substandard bindings produced in a production run of $n = 3000$ books is a binomial random variable X with probability $p = 0.02$ of a substandard binding. The expected number of substandard bindings in the production
run is then $E\{X\} = 3000(0.02) = 60$. The variance of the number of substandard
bindings in the production run is $\sigma^2\{X\} = 3000(0.02)(0.98) = 58.8$. Thus,
$\sigma\{X\} = \sqrt{58.8} = 7.67$ bindings is the standard deviation of the number of substandard bindings in the run. □

Comment

The mean and the variance of the binomial random variable can be readily derived. To show
that $E\{X\} = np$, for instance, we first find $E\{X_i\}$ for each Bernoulli random variable. Using
(5.5), we obtain $E\{X_i\} = 1(p) + 0(1 - p) = p$. Since $X = X_1 + X_2 + \cdots + X_n$, we have,
by (5.15a), that $E\{X\} = np$.

Distribution Shape. Binomial probability distributions can take on a variety of
shapes. Figure 6.2 shows three binomial probability distributions. The distribution is
skewed right if $p < 0.5$, is skewed left if $p > 0.5$, and is symmetrical if $p = 0.5$.
The closer p is to 0 or 1, for a given n, the more pronounced is the skewness.

Sum of Binomial Random Variables

In applications where interest is in the sum of two or more independent binomial random variables, the following theorem can be helpful.

FIGURE 6.2 **Three binomial probability distributions.** The binomial distribution is
skewed right when $p < 0.5$, skewed left when $p > 0.5$, and symmetrical when $p = 0.5$.

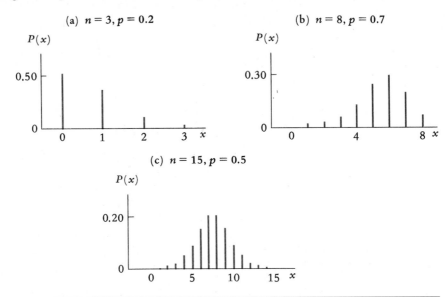

(a) $n = 3, p = 0.2$

(b) $n = 8, p = 0.7$

(c) $n = 15, p = 0.5$

(6.7)   If V and W are two independent binomial random variables with common
probability parameter p and based on n_1 and n_2 trials, respectively, the sum
$V + W$ is also a binomial random variable with parameters p and n, where
$n = n_1 + n_2$, the combined number of trials.

EXAMPLE

Two chess-playing computer programs will compete against one another in two
tournaments, consisting of $n_1 = 5$ and $n_2 = 7$ games, respectively. The probability that program A wins in any given game is 0.6 (the probability that program A
draws or loses is 0.4). The numbers of games to be won by program A in the two
tournaments are independent binomial random variables. What is the probability
that program A wins seven or more games in both tournaments?

By (6.7), the total number of wins by program A in both tournaments is a
binomial random variable with parameters $p = 0.6$ and $n = 5 + 7 = 12$. In consulting Table C.5 for $n = 12$ and $p = 0.6$, we find that $P(X \geq 7) = P(7)$
$+ P(8) + \cdots + P(12) = 0.2270 + 0.2128 + \cdots + 0.0022 = 0.6652.$ □

6.2 POISSON PROBABILITY DISTRIBUTIONS

The Poisson probability distribution applies to many random phenomena occurring in a period of time.

EXAMPLES

1. The number of machines in a plant that break down during a day (X) is a Poisson random variable. Management wishes to know how likely it is that there are no breakdowns in a day.

2. The number of units of an item sold from stock during a week (X) is a Poisson random variable. The inventory controller wishes to know the probability that more than five units will be sold in a week.

3. The number of persons arriving at a bank teller per quarter hour (X) is a Poisson random variable. For designing a staffing plan, a management consultant needs to know the probability that more than 10 persons will arrive in a 15-minute period.

The Poisson probability distribution also applies to other types of random phenomena such as the number of typographical errors on a page and the number of winning tickets, in a state lottery, held by persons residing in a certain county of the state.

Poisson Probability Function

A Poisson random variable is a discrete variable that can take on any integer value from 0 on up indefinitely.

(6.8)

The *Poisson probability function* is:

$$P(x) = \frac{\lambda^x \exp(-\lambda)}{x!}$$

where:
$P(x) = P(X = x)$
$x = 0, 1, \ldots, \infty$
$0 < \lambda < \infty$

Here and elsewhere in the text, $\exp(a)$ represents e^a, where $e = 2.71828\ldots$ is the base of natural logarithms. Thus, $\exp(-\lambda)$ represents $e^{-\lambda}$. (A review of exponentiation and natural logarithms can be found in Appendix A, Section A.2.)

Even though the Poisson random variable may assume any indefinitely large integer value while real-world phenomena are bounded, the Poisson probability distribution is often a reasonable model, as we shall see.

The Poisson probability distribution has only one parameter, λ (Greek lambda). Each different value of λ corresponds to a different member of the family of Poisson probability distributions. Observe in (6.8) that λ may be any positive number.

| EXAMPLE |

The number of crimes occurring in a city district in the 1-hour period between 1 A.M. and 2 A.M. is a Poisson random variable with $\lambda = 0.2$. We obtain a portion of the Poisson probability distribution using (6.8):

x	$P(x)$
0	$\dfrac{(0.2)^0 \exp(-0.2)}{0!} = 0.8187$ [Note that $\exp(-0.2) = 0.8187$]
1	$\dfrac{(0.2)^1 \exp(-0.2)}{1!} = 0.1637$
2	$\dfrac{(0.2)^2 \exp(-0.2)}{2!} = 0.0164$
3	$\dfrac{(0.2)^3 \exp(-0.2)}{3!} = 0.0011$
4	$\dfrac{(0.2)^4 \exp(-0.2)}{4!} = 0.0001$

We see, for instance, that the probability of no crime occurring during this hour is $P(0) = 0.8187$. Furthermore, the probability of two or fewer crimes during the hour is $P(X \leq 2) = P(0) + P(1) + P(2) = 0.8187 + 0.1637 + 0.0164 = 0.9988$.

For outcomes of five crimes or more, the probabilities of occurrence are so small that we have not shown calculations. Here is the reason why the Poisson probability distribution is often a useful model for bounded phenomena, namely, because the probabilities of large values are negligible. □

Probability Table

Poisson probabilities can be computed readily from many pocket calculators and statistical computer packages. Extensive Poisson probability tables also exist. Table C.6 contains Poisson probabilities for selected values of λ up to 20.

| EXAMPLES |

1. Daily demand for a certain replacement part of a videocassette recorder model is Poisson distributed with $\lambda = 0.9$. We see from Table C.6 that the probability of no demand for the part on a particular day is $P(0) = 0.4066$. This value is found in the column headed $\lambda = 0.9$ and the row marked $x = 0$. For cumulative probabilities, we need only add the appropriate individual probabilities. For example, we see the probability that 4 or fewer replacement parts are demanded is $P(X \leq 4) = P(0) + P(1) + \cdots + P(4) = 0.4066 + 0.3659 + \cdots + 0.0111 = 0.9977$.

2. The number of accidents in an office building during a four-week period is a Poisson random variable with $\lambda = 2$. Interest is in the probability that there is one or no accident in a four-week period. We find from the table that $P(X \leq 1) = P(0) + P(1) = 0.1353 + 0.2707 = 0.4060$. □

Characteristics of Poisson Probability Distributions

Mean and Variance. The mean and the variance of a Poisson probability distribution depend on the parameter λ in a simple way.

(6.9)

(6.10)

> The mean and the variance of a Poisson probability distribution are:
>
> $$E\{X\} = \lambda$$
>
> $$\sigma^2\{X\} = \lambda$$

Thus, the mean and the variance of a Poisson distribution are equal.

EXAMPLE

For the crime example, $\lambda = 0.2$, so $E\{X\} = 0.2$ crime and $\sigma^2\{X\} = 0.2$. The standard deviation therefore is $\sigma\{X\} = \sqrt{\lambda} = \sqrt{0.2} = 0.447$ crime. □

Distribution Shape. Figure 6.3 contains three Poisson probability distributions with parameter values $\lambda = 0.3$, 2.0, and 5.0. All Poisson probability distributions are skewed to the right, although they become more symmetrical as λ becomes larger.

Poisson Process

To obtain an understanding of the type of random phenomena described by the Poisson distribution, we examine the formal conditions or postulates that lead to this distribution.

A Poisson distribution is concerned with the number of occurrences of some event in a fixed length of time, fixed amount of space, and so on. For the sake of simplifying the exposition in setting out the postulates that follow, we shall consider phenomena or processes generating occurrences over time, such as number of phone calls received at a telephone switchboard in a 5-minute period. With minor alterations in the discussion, the postulates would apply to occurrences distributed in space, such as blemishes occurring in a roll of vinyl wallpaper or forest fires occurring in a wilderness region.

Poisson Process Postulates. Consider occurrences that happen randomly on the time continuum. Figure 6.4 illustrates three such occurrences. Imagine this continuum to be divided into many small nonoverlapping intervals of size Δt, as shown in Figure 6.4. The time between t and $t + \Delta t$ denotes a typical interval.

> *Postulate 1.* The numbers of occurrences in nonoverlapping time intervals are statistically independent.

FIGURE 6.3   **Three Poisson probability distributions.**   Right-skewness decreases as λ increases.

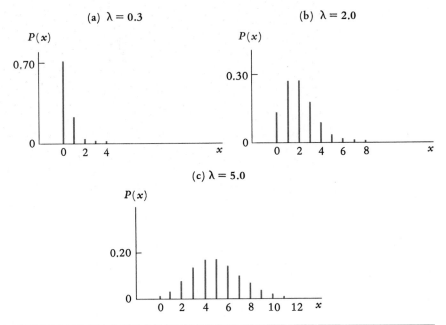

(a) λ = 0.3

(b) λ = 2.0

(c) λ = 5.0

FIGURE 6.4   **Illustration of a Poisson process**

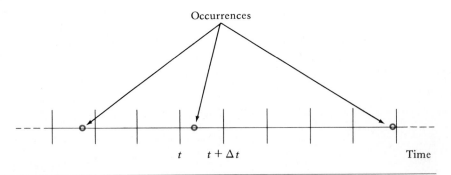

Postulate 2.   The number of occurrences in a time interval has the same probability distribution for all time intervals.

Postulate 3.   The probability of one occurrence in any time interval $(t, t + \Delta t)$ is approximately proportional to the size of the interval Δt. Specifically, if λ is the constant of proportionality (by postulate 2, it must be the same for all time intervals), the probability of one occurrence is approximately $\lambda \Delta t$.

Postulate 4.   The probability of two or more occurrences in any time interval $(t, t + \Delta t)$ is negligibly small, relative to the probability of one occurrence in the interval.

When these four postulates hold, the number of occurrences in a unit time interval follows a Poisson probability distribution with parameter λ. Any process that generates occurrences according to these four postulates is called a *Poisson process*.

The conditions specified in postulates 1 and 2, incidentally, are those of statistical independence and stationarity. We encountered these same conditions in our discussion of a Bernoulli process.

EXAMPLE

Consider the number of phone calls received at a switchboard in a 5-minute period. Divide this time period into a large number of nonoverlapping intervals of size Δt — for instance, into 3000 intervals of size $\Delta t = 0.1$ second. Postulate 1 requires that the number of phone calls received during each of these intervals be statistically independent of one another. Postulate 2 requires that the number of phone calls have the same probability distribution in each interval. This postulate would be violated, for instance, if some of the intervals correspond to busy periods and others to periods of slack telephone activity. Postulate 3 requires that the probability of receiving one phone call in any interval of size Δt is approximately proportional to Δt; for instance, the probability for $\Delta t = 0.1$ second must be about twice as large as the probability for $\Delta t = 0.05$ second. Finally, postulate 4 requires that the probability of receiving two or more phone calls in any interval is very small. If all four of these postulates are satisfied by the telephone-calling process, then the number of calls received in a time period of fixed length will be a Poisson random variable.   —□

Sum of Poisson Random Variables

In applications where interest is in the sum of two or more independent Poisson random variables, the following theorem is helpful.

(6.11)

If V and W are two independent Poisson random variables with parameters λ_1 and λ_2, respectively, the sum $V + W$ is also a Poisson random variable, with parameter $\lambda = \lambda_1 + \lambda_2$.

| EXAMPLE |

Broadcasting interruptions at a television station arise from equipment breakdowns and human errors. The numbers of interruptions per 1000 broadcasting hours from these two sources are independent Poisson random variables with parameters $\lambda_1 = 0.4$ and $\lambda_2 = 1.1$, respectively. By theorem (6.11), we know that the total number of interruptions from both sources is a Poisson random variable with parameter $\lambda = 0.4 + 1.1 = 1.5$. Thus, the probability of no interruptions from either source in the next 1000 hours of broadcasting is $P(0) = 0.2231$ (refer to Table C.6, column $\lambda = 1.5$, row $x = 0$).   □

 ## **6.3   DISCRETE UNIFORM PROBABILITY DISTRIBUTIONS**

A random variable that can take on integer values within a given interval with equal probabilities is useful for statistical sampling. This random variable is called a *discrete uniform random variable*.

| EXAMPLE |

The final digit in today's volume of stock transactions, X, can assume any of the 10 digits from 0 to 9, each with probability 0.10; that is:

x:	0	1	2	3	4	5	6	7	8	9
$P(x)$:	0.10	0.10	0.10	0.10	0.10	0.10	0.10	0.10	0.10	0.10

□

Discrete Uniform Probability Function

The discrete uniform probability distribution has two parameters, a and s. Parameter a denotes the smallest outcome, and s denotes the number of distinct outcomes.

(6.12)

The *discrete uniform probability function* is:

$$P(x) = \frac{1}{s}$$

where:

$P(x) = P(X = x)$

$x = a, a + 1, \ldots, a + (s - 1)$; a and s are integers, with $s > 0$

| EXAMPLE |

When the possible outcomes are the integers from 0 to 9, $a = 0$ and $s = 10$. A graph of this probability distribution is given in Figure 6.5.   □

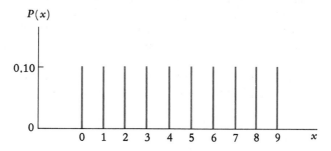

FIGURE 6.5   **Example of a discrete uniform probability distribution: $a = 0$, $s = 10$**

Characteristics of Discrete Uniform Probability Distributions

The mean and the variance of a discrete uniform probability distribution depend on the parameters a and s.

> The mean and the variance of a discrete uniform probability distribution are:
>
> (6.13) $$E\{X\} = a + \frac{s-1}{2}$$
>
> (6.14) $$\sigma^2\{X\} = \frac{s^2-1}{12}$$

EXAMPLE

For the discrete uniform probability distribution presented in Figure 6.5, $a = 0$ and $s = 10$, so:

$$E\{X\} = 0 + \frac{10-1}{2} = 4.5 \qquad \sigma^2\{X\} = \frac{10^2-1}{12} = 8.25$$

Thus, the standard deviation of the distribution is $\sigma\{X\} = \sqrt{8.25} = 2.87$.   □

OPTIONAL TOPIC—HYPERGEOMETRIC PROBABILITY DISTRIBUTIONS

The hypergeometric probability distribution is a discrete distribution that finds applications in statistical sampling. Unlike the binomial probability distribution, which is applicable when independent Bernoulli trials are made, the hypergeometric distribution applies when the trials are dependent.

EXAMPLES

1. **Senate Committee.**   A state senate consists of $N = 50$ senators, of whom $C = 10$ are women. What is the probability that a committee of $n = 5$ senators formed at random contains $X = 0$ women?

    The binomial probability distribution is not applicable here since the trials are not independent. The probability that the first senator assigned to the committee is female is 10/50. However, the probability that the second committee member selected at random is female depends on the first selection; it is 10/49 if the first committee assignment is a male senator and 9/49 if the first assignment is a female senator. Thus, the probabilities of selecting a female (or male) senator vary throughout the selection process, depending on the previous selections.

2. **Utility Generators.**   Of the $N = 15$ generators owned by a utility company, $C = 2$ have a defective rotor shaft. If $n = 4$ different generators are selected at random for inspection, what is the probability that exactly $X = 1$ is found to have a defective shaft?   ☐

Hypergeometric Probability Function

A hypergeometric random variable has as possible outcomes all the integers between two limits.

(6.15)

The *hypergeometric probability function* is:

$$P(x) = \frac{\binom{C}{x}\binom{N-C}{n-x}}{\binom{N}{n}}$$

where:
lower limit of x is *larger* of 0 and $n - N + C$
upper limit of x is *smaller* of n and C

The symbol $\binom{a}{b}$ is defined in (6.4). (A review of permutations and combinations can be found in Appendix A, Section A.4.)

EXAMPLES

1. In the senate committee example, we have $N = 50$, $C = 10$, and $n = 5$. The possible outcomes of X, the number of women on the committee, are $0, 1, \ldots, 5$. We wish to find $P(X = 0)$. Substituting $x = 0$ into (6.15), we obtain:

$$P(0) = \frac{\binom{10}{0}\binom{40}{5}}{\binom{50}{5}} = \frac{\dfrac{10!}{0!\,10!}\dfrac{40!}{5!\,35!}}{\dfrac{50!}{5!\,45!}} = 0.311$$

Thus, the probability is $P(0) = 0.311$ that a committee of five senators formed at random will not contain any women.

2. In the utility generators example, we have $N = 15$, $C = 2$, and $n = 4$. The possible outcomes of X, the number of inspected generators with a defective shaft, are 0, 1, or 2. We wish to find $P(X = 1)$. Using (6.15), we obtain, for $x = 1$:

$$P(1) = \frac{\binom{2}{1}\binom{13}{3}}{\binom{15}{4}} = \frac{\dfrac{2!}{1!\,1!}\dfrac{13!}{3!\,10!}}{\dfrac{15!}{4!\,11!}} = 0.419$$

Thus, the probability is $P(1) = 0.419$ that, among four generators selected at random, exactly one will be found to have a defective shaft.

3. If $N = 10$, $C = 8$, and $n = 3$, then the possible outcomes of X are 1, 2, and 3. To find $P(X = 3)$, we use (6.15) and obtain $P(3) = 0.467$ (calculations not shown). $\square$

Characteristics of Hypergeometric Probability Distributions

We shall denote the proportion C/N by p. In the senate committee example, $p = 10/50 = 0.20$ is the proportion of senators who are women. In the utility generators example, $p = 2/15 = 0.133$ is the proportion of generators with a defective shaft. Using this notation, we have the following results for the hypergeometric distribution.

The mean and the variance of a hypergeometric probability distribution are:

(6.16) $$E\{X\} = np$$

(6.17) $$\sigma^2\{X\} = \left(\frac{N - n}{N - 1}\right)np(1 - p)$$

where:

$$p = \frac{C}{N}$$

Note the similarity to the mean and the variance of a binomial random variable in (6.5) and (6.6), except for the term $(N - n)/(N - 1)$ in the variance expression in (6.17).

EXAMPLES

1. In the senate committee example, $p = 0.20$, $n = 5$, and $N = 50$. Hence:

$$E\{X\} = 5(0.20) = 1.0 \qquad \sigma^2\{X\} = \left(\frac{50-5}{50-1}\right)(5)(0.20)(0.80) = 0.735$$

Thus, the hypergeometric distribution here has mean 1.0 and standard deviation $\sigma\{X\} = \sqrt{0.735} = 0.86$.

2. The mean and the variance of the hypergeometric distribution for the utility generators example are:

$$E\{X\} = 4(0.133) = 0.53 \qquad \sigma^2\{X\} = \left(\frac{15-4}{15-1}\right)(4)(0.133)(0.867)$$

$$= 0.363 \qquad \square$$

Comment   When N is large relative to n, the hypergeometric probability function (6.15) can be approximated well by the binomial probability function (6.3) with the same values of n and p. In the senate committee example, for instance, $n = 5$ and $p = 0.20$. Hence, the approximating binomial probability for $P(X = 0)$ is found from Table C.5 to be 0.3277. The exact hypergeometric probability was calculated earlier as 0.311.

PROBLEMS

6.1   Which of the following random trials is a Bernoulli trial? (1) The inspection classification (accept, rework, or scrap) for a component produced on an assembly line. (2) The sex of a chick in a brood bred for egg production. (3) The weight of a package of ground beef in a supermarket.

6.2   Which of the following random trials is a Bernoulli trial? (1) The absence or presence of an employee at work on a particular day. (2) The number of patients who visit a hospital eye clinic during a week. (3) The status of an indicator light (on, off) in the control panel of an oil refinery.

6.3   There are 19 participants in a 10-mile run for charity. Let $X_i = 1$ if the ith participant finishes the race and $X_i = 0$ otherwise.
   a. Under what conditions does the sequence $X_1, \ldots, X_{19}$ constitute a Bernoulli process?
   b. What is represented by the random variable $X = X_1 + \cdots + X_{19}$? Under what conditions is X a binomial random variable?

6.4   A family has three children. Let $X_i = 1$ if the ith child in order of birth is a girl and $X_i = 0$ otherwise.
   a. Under what conditions does the sequence X_1, X_2, X_3 constitute a Bernoulli process?
   b. What is represented by the random variable $X = X_1 + X_2 + X_3$? Under what conditions is X a binomial random variable?

*6.5   **Inaccurate Gauges.**   The number of inaccurate gauges in a group of four is a binomial random variable X with $n = 4$ and $p = 0.25$.

    a. Describe the underlying Bernoulli trial. What is the sample space of X (that is, what are the different values that X can take on)?

    b. Using (6.3), obtain (1) $P(0)$, (2) $P(2)$, (3) $P(X \le 2)$. Confirm your answers by using Table C.5.

6.6 Transmission Work. The number of trucks in the fleet of a hauling firm that will require major transmission work next year is a binomial random variable X with $n = 8$ and $p = 0.2$.

    a. Describe the underlying Bernoulli trial. What different values can X take on (that is, what is its sample space)?

    b. Using (6.3), obtain (1) $P(0)$, (2) $P(1)$, (3) $P(X \le 3)$. Confirm your answers by using Table C.5.

6.7 The number of members of a university department who will incur less than \$1000 of research computing charges next term is a binomial random variable X with $n = 6$ and $p = 0.65$.

    a. Describe the underlying Bernoulli trial. What does the outcome $X = 5$ represent?

    b. Using (6.3), obtain (1) $P(4)$, (2) $P(X \le 2)$, (3) $P(X > 4)$. Confirm your answers by using Table C.5.

*6.8 Use Table C.5 to obtain the following binomial probabilities: (1) $P(1)$ for $p = 0.09$, $n = 10$; (2) $P(3)$ for $p = 0.35$, $n = 6$; (3) $P(8)$ for $p = 0.95$, $n = 9$; (4) $P(X \ge 4)$ for $p = 0.7$, $n = 6$.

6.9 Use Table C.5 to obtain the following binomial probabilities: (1) $P(2)$ for $p = 0.15$, $n = 12$; (2) $P(4)$ for $p = 0.25$, $n = 8$; (3) $P(3)$ for $p = 0.90$, $n = 7$; (4) $P(X \le 2)$ for $p = 0.6$, $n = 10$.

6.10 Use Table C.5 to obtain the following binomial probabilities: (1) $P(3)$ for $p = 0.20$, $n = 8$; (2) $P(2)$ for $p = 0.05$, $n = 10$; (3) $P(7)$ for $p = 0.70$, $n = 15$; (4) $P(X \ge 9)$ for $p = 0.90$, $n = 12$.

*6.11 Refer to **Inaccurate Gauges** Problem 6.5.

    a. Graph the probability distribution of X. Is the distribution symmetrical or skewed?

    b. Find $E\{X\}$, $\sigma^2\{X\}$, and $\sigma\{X\}$. In what units is $\sigma\{X\}$ expressed?

6.12 Refer to **Transmission Work** Problem 6.6.

    a. Graph the probability distribution of X. Is the distribution skewed? If so, in what direction?

    b. Find the mean and the standard deviation of the probability distribution of X. In what units are they expressed?

6.13 The number of persons in a university class who will be employed one year after graduation is a binomial random variable X with $n = 240$ and $p = 0.98$.

    a. What are the mean and the standard deviation of the probability distribution of X?

    b. Is the number of persons in the class who will *not be employed* one year after graduation a binomial random variable? What are the mean and the standard deviation of the number not employed?

6.14 The probability is 0.01 that a watch will require repair if it is dropped. Seven watches have just been dropped from a tray by a jewelry store clerk. Assume that the number of watches that require repair is a binomial random variable X.

    a. What is the probability that none of the seven watches requires repair? That two or fewer watches require repair?

    b. Obtain the probability distribution of X and graph it. Is the distribution highly skewed?

    c. What are the mean and the variance of X? Are these two measures almost equal? Is approximate equality necessarily the case for a binomial random variable when p is small? Explain.

6.15 The probability is 0.06 that a patient will cancel a dental appointment. Consider a group of 12 patients scheduled for appointments this morning, and let X denote the number of cancellations in this group. Assume that X is a binomial random variable.
   a. What is the underlying Bernoulli trial here?
   b. What is the probability that exactly 2 out of the 12 appointments will be cancelled? That 2 or fewer will?
   c. What is the expected number of cancellations? What is the standard deviation of the number of cancellations?

*6.16 Separate surveys will be undertaken of 8 manufacturers in one state and 12 in another state. Let V and W denote the counts for the number of manufacturers in the two surveys who are using a last-in, first-out (LIFO) accounting procedure for inventory. Assume V and W are independent binomial random variables, each with $p = 0.4$.
   a. Is $T = V + W$ a binomial random variable? What is the sample space of T?
   b. Find (1) $P(T = 6)$, (2) the most probable outcome of T, (3) $E\{T\}$.
   c. Would T be a binomial random variable if the values of p were 0.1 for V and 0.7 for W? Explain.

6.17 A firm has just hired four management trainees for its Houston office and five for its Dallas office. Let V and W denote the numbers of trainees who remain in their original offices for at least three years. Assume that V and W are independent binomial random variables, each with $p = 0.7$.
   a. What does $T = V + W$ represent here? What are the parameters of the binomial distribution for T?
   b. Find $E\{T\}$ and $\sigma\{T\}$.
   c. What is the probability that all nine trainees will remain in their original offices for at least three years?
   d. Would the probability distribution of T differ if six trainees had been hired by the Houston office and three by the Dallas office?

6.18 Which of the following random trials is possibly a Poisson trial? (1) The thickness of a coat of paint on a metal sheet. (2) The number of trout in a lake. (3) The number of blue pairs of socks among the next five pairs sold in a store.

6.19 Which of the following random trials is possibly a Poisson trial? (1) The number of women among six scholarship winners. (2) The length of a crack in a timber. (3) The pollen count in a cubic centimeter of air.

*6.20 **Police Calls.**   The number of calls to a police dispatcher between 8:00 P.M. and 8:30 P.M. on Fridays is a Poisson random variable X with $\lambda = 3.5$.
   a. Using (6.8), find the probability (1) of no calls during this period, (2) of three calls. Confirm your answers by using Table C.6.
   b. What is the expected number of calls received by the dispatcher during this period? What is the variance of the number of calls?
   c. Graph the probability distribution of X, using Table C.6 to obtain the needed probabilities. Is the distribution skewed? If so, in which direction?

6.21 The number of irregularities in 1 kilometer of optical fiber is a Poisson random variable X with $\lambda = 0.9$ irregularity per kilometer.
   a. Using (6.8), find (1) $P(0)$, (2) $P(1)$. Confirm your answers using Table C.6.
   b. Obtain $E\{X\}$ and $\sigma\{X\}$. In what units are they expressed?
   c. Graph the probability distribution of X, using Table C.6 to obtain the needed probabilities. Is the distribution skewed? If so, in which direction?

6.22 Word Processor Warranty. A new word-processing unit is fully warranted during its first year. The number of warranty service calls for a unit during its first year is a Poisson random variable X with $\lambda = 4.0$.

    a. Find the probability (1) of exactly four warranty service calls for a unit during the first year, (2) of four or more warranty service calls.

    b. Graph the probability distribution of X. Is the distribution skewed? If so, in which direction?

    c. What are the mean and the standard deviation of the probability distribution of X? In what units are these measures expressed?

6.23 Wallpaper Blemishes. The number of coating blemishes in 10-square-meter rolls of customized wallpaper is a Poisson random variable X with $\lambda = 0.3$.

    a. Find (1) $P(X = 2)$, (2) $P(X \leq 1)$, (3) $P(1 \leq X \leq 3)$.

    b. Find $E\{X\}$ and $\sigma\{X\}$. In what units are these measures expressed?

    c. Plot the probability distribution of X on a graph. Is the distribution skewed? If so, in which direction?

*6.24 Use Table C.6 to find the following: (1) $P(6)$ for $\lambda = 3.5$, (2) $P(X \leq 2)$ for $\lambda = 1.5$, (3) the value of λ for which $P(0) = 0.0041$.

6.25 Use Table C.6 to find the following: (1) $P(4)$ for $\lambda = 2.0$, (2) $P(X > 4)$ for $\lambda = 0.8$, (3) the most probable outcome for $\lambda = 7.5$.

6.26 Refer to **Police Calls** Problem 6.20. An observer noted that some of the calls to the dispatcher come in bunches because they relate to the same event (for example, separate calls from several witnesses of a road accident). Which postulates for a Poisson process are violated by this situation?

6.27 Refer to **Word Processor Warranty** Problem 6.22. For each of the following situations, explain which postulates of a Poisson process are violated: (1) Typically, the warranty calls for a unit are bunched toward the end of the warranty year. (2) When the production process has a serious undetected quality failure, at least 100 units are produced and sold that will require extensive warranty work before corrective action on the process is taken. (3) About half of the units have no warranty calls during the warranty year, while the other half have six or more calls.

6.28 The occurrences of demand for a replacement heating element of a water heater follows a Poisson process with $\lambda = 3$ elements per year. Let Δt denote a single day in a year having 365 days. State the approximate probability according to the postulates for a Poisson process that (1) exactly one element will be demanded on a given day, (2) no element will be demanded on a given day, (3) two or more elements will be demanded on a given day.

*6.29 The numbers of prescriptions for a certain medication written daily by two physicians are denoted by V and W. Assume V and W are independent Poisson random variables with $\lambda_1 = 2.0$ and $\lambda_2 = 1.5$, respectively. Let $T = V + W$.

    a. Obtain (1) $E\{T\}$, (2) $\sigma\{T\}$, (3) $P(T = 2)$.

    b. Is $P(T = 2)$ the same as the joint probability $P(V = 1 \cap W = 1)$? Explain.

6.30 Refer to **Wallpaper Blemishes** Problem 6.23. The number of printing blemishes in 10-square-meter rolls of customized wallpaper, Y, is also Poisson distributed but with $\lambda = 0.1$. Assume that Y is independent of X, the number of coating blemishes.

    a. What does the random variable $T = X + Y$ represent here?

    b. What are the mean and standard deviation of the probability distribution of T?

    c. What is the most probable total number of blemishes in a roll? If rolls with a total of more than one blemish are scrapped, what is the probability a roll will be scrapped?

*6.31 The number of students who participate on any day in discussion in Professor Bird's class is a discrete uniform random variable X with $a = 0$ and $s = 14$.
   a. Obtain each of the following probabilities: (1) $P(0)$, (2) $P(X > 10)$, (3) $P(1 \leq X \leq 5)$.
   b. What is the expected number of students who participate in discussion on any day? What is the standard deviation of X?
   c. Graph the probability distribution of X.

6.32 A computer random number generator produces one of the numbers $000, 001, \ldots, 999$ with equal probability.
   a. What are the parameter values of the discrete uniform probability function for this random variable?
   b. What is the probability that number 123 will be generated? That a number starting with the digit 1 will be generated?
   c. What are the mean and the standard deviation of the probability distribution?

6.33 **Insurance Calculation.**   The calculation of a home insurance premium involves seven consecutive computational steps. An auditor is reviewing a large number of premium calculations that contain an error. Let X denote the step at which the error is made. Assume X has a discrete uniform probability distribution.
   a. Graph the probability distribution of X.
   b. On the average, how many steps must the auditor check in each premium calculation to find the error?
   c. Obtain $\sigma\{X\}$. In what units is this measure expressed?

6.34 Refer to **Insurance Calculation** Problem 6.33.
   a. If locations of the errors for different premium calculations are statistically independent, what is the probability that two premium calculations have errors that occur at step 4?
   b. Would you expect an error to occur with equal probability at each step in actuality? Explain.

*6.35 An accounting department has six tenured and four untenured faculty members. A committee of $n = 3$ members is to be selected at random. Let X denote the number of committee members who are tenured.
   a. What is the probability that (1) exactly one of the committee members is tenured; (2) the number of tenured members on the committee exceeds the number of untenured members?
   b. Obtain the expected number of committee members who are tenured. Also, obtain the standard deviation of X.

6.36 Two of the 12 generators in a power plant have worn rotor shafts. Three of the 12 generators are selected at random for inspection. Let X denote the number of generators selected for inspection that have worn rotor shafts.
   a. What is the sample space of X?
   b. Obtain the probability distribution of X.
   c. Calculate the mean and the variance of X from the probability distribution in **b**. Verify that (6.16) and (6.17) give the mean and variance directly.

6.37 **Fluid Trading.**   An investment analyst is studying a group of 15 stocks for a trading condition she has termed "fluid trading." Of the 15 stocks, 8 exhibit fluid trading. The analyst will select 5 stocks at random from the group.
   a. What is the probability that (1) none of the fluid trading stocks will be selected, (2) exactly 4 of the fluid trading stocks will be selected?
   b. Find the mean and the standard deviation of X, the number of stocks selected that exhibit fluid trading.

6.38 Refer to **Fluid Trading** Problem 6.37. Approximate $P(4)$ in part **a** by a binomial probability. Is N large enough here for the binomial distribution to be a good approximation? Comment.

6.39 Consider the hypergeometric probability function with $N = 100$, $C = 10$, and $n = 3$.
   a. Calculate $P(0)$, using (6.15).
   b. Approximate $P(0)$ by a binomial probability. Is the approximation reasonably close?

EXERCISES

6.40 In a sports match, two teams (A and B) play each other in consecutive games until one team has won two games more than the other. Ties are impossible. Let $X_i = 1$ if team A wins game i and $X_i = 0$ otherwise. Assume that the X_i are statistically independent and that $P(X_i = 1) = 0.6$ for all i. Consider a match that ends in exactly four games.
   a. Is $X = X_1 + \cdots + X_4$ a binomial random variable *for this match*? Explain.
   b. Obtain the probability distribution of X in **a**.

6.41 Let X_i ($i = 1, 2, \ldots, n$) be independent and identically distributed Bernoulli random variables with $P(X_i = 1) = p$.
   a. Prove that $\sigma^2\{X_i\} = p(1 - p)$.
   b. Use the result in **a** to verify (6.6).

6.42 Each use of an electric switch is an independent Bernoulli trial in which the switch either functions ($X = 0$) or fails ($X = 1$). Consider the associated Bernoulli process, and let Y be a random variable representing the number of trials up to and including the trial in which the switch fails for the first time. Show that the probability function of Y is:

$$P(Y = y) = p(1 - p)^{y-1} \qquad y = 1, 2, \ldots$$

6.43 Prove (6.9). [*Hint:* $\exp(\lambda) = \sum_{x=0}^{\infty} \lambda^x / x!$.]

STUDIES

6.44 Let X be a binomial random variable with $p = 0.3$. Plot the probability distributions of X for $n = 5$ and $n = 20$ on two graphs, one below the other. What effect does the increase in n have on the shape of the distribution?

6.45 For binomial random variable X with $n = 20$, obtain $\sigma\{X\}$ for $p = 0, 0.1, \ldots, 0.9, 1.0$. Plot these $\sigma\{X\}$ values on a graph as a function of p. What effect does the magnitude of p have on the variability of the probability distribution?

6.46 A glass company has received an order for two large lenses that must be specially cast. The probability that a given lens will prove to be acceptable (that is, free from flaws) when the glass has cooled is 0.6. The number of acceptable lenses (X) in a casting run of n lenses is a binomial random variable. Management wishes to know how many lenses to cast in the run to have a probability of at least 0.95 that a minimum of two lenses are acceptable. Find the smallest size of the casting run that will provide this assurance.

6.47 Ten persons in a taste-test experiment will be served a portion of sausage prepared according to the current recipe (A) and another portion prepared from a recipe designed to give the same taste as the current recipe but having a longer shelf life (B). The order of the two servings will be randomized. Each person must independently identify which of the two servings is the preferred one. Let X denote the number of persons of the 10 who prefer serving B.

   a. Describe the underlying Bernoulli trial here. Does X necessarily follow a binomial distribution here?

   b. Suppose sausage B has exactly the same taste as sausage A, so a person will prefer A or B with equal probability. What is the probability, then, that the proportion X/n of persons in the study who prefer B is not less than 0.4 or greater than 0.6?

   c. How much larger would be the probability in **b** if 20 persons were included in the testing group? What are the implications of this result?

6.48 The number of typographical errors on a page is a Poisson random variable X with $\lambda = 0.8$.

   a. Obtain the expected number of errors on a page and the coefficient of variation of the probability distribution of X.

   b. If the numbers of errors on different pages are statistically independent, what is the expected total number of errors on 400 pages? What is the coefficient of variation of the probability distribution for the total number of errors on 400 pages? [*Hint:* Theorem (6.11) generalizes directly to more than two independent Poisson random variables.]

   c. From your results in **a** and **b**, what is the effect on the relative variability of the probability distribution of the total number of errors as the number of pages increases? Discuss.

6.49 The daily number of disabling breakdowns in cabs of a taxi fleet is a Poisson random variable X with $\lambda = 0.7$. The loss of revenue to the company on the day of the breakdown of a taxi is \$100 if a standby cab is not available and \$0 if a standby cab is available. Disabled cabs can be repaired by the next day. Standby cabs (without drivers) can be leased for \$35 per day. Assume that the probability of a breakdown in a standby cab is negligibly small.

   a. Let C denote the daily revenue loss from breakdowns plus the cost of leasing, and let m denote the number of standby cabs leased for the day. Express C as a function of X and m.

   b. Evaluate $E\{C\}$ for $m = 0, 1, 2,$ and 3.

   c. How many standby cabs should the company have on hand at the beginning of a day to minimize the expected value of C?

7 COMMON CONTINUOUS PROBABILITY DISTRIBUTIONS

In this chapter, we take up several important families of continuous probability distributions. Recall from Chapter 5 that the probability that a continuous random variable X takes on a value in a specified interval is found by determining the corresponding area under its probability density function $f(x)$.

7.1 CONTINUOUS UNIFORM (RECTANGULAR) PROBABILITY DISTRIBUTIONS

The discrete uniform probability distribution discussed in the previous chapter has a continuous analog known as the *continuous uniform* or *rectangular probability distribution*. The rectangular probability distribution is useful in a variety of situations, including as an approximation of the discrete uniform probability distribution.

Density Function

A *continuous uniform random variable* has uniform probability density over an interval.

(7.1)   The *continuous uniform probability density function* is:

$$f(x) = \frac{1}{b - a}$$

where:

$a \le x \le b$

EXAMPLE

An analyst has concluded that a continuous uniform probability distribution is a good approximation for the distribution of incomes of taxpayers in the $5000–$7500 income bracket. The density function here is:

$$f(x) = \frac{1}{2500} \qquad 5000 \le x \le 7500$$

In this example, $b = 7500$ and $a = 5000$. A graph of this probability distribution is shown in Figure 7.1. ☐

Note that the continuous uniform probability distribution has two parameters, a and b. The smallest value the random variable can assume is a and the largest is b. As the graph in Figure 7.1 illustrates, the use of the term *rectangular* in the name of this distribution is quite descriptive.

Characteristics of Continuous Uniform Probability Distributions

The mean and the variance of a continuous uniform probability distribution depend on the parameters a and b.

> The mean and the variance of a continuous uniform probability distribution are:
>
> (7.2)
> $$E\{X\} = \frac{b + a}{2}$$
>
> (7.3)
> $$\sigma^2\{X\} = \frac{(b - a)^2}{12}$$

EXAMPLE

The continuous uniform probability distribution in Figure 7.1 has mean and variance:

$$E\{X\} = \frac{7500 + 5000}{2} = 6250 \qquad \sigma^2\{X\} = \frac{(7500 - 5000)^2}{12} = 520{,}833$$

The standard deviation is $\sigma\{X\} = \sqrt{520{,}833} = 721.7$. ☐

FIGURE 7.1   **Example of a continuous uniform probability distribution: $a = 5000$, $b = 7500$**

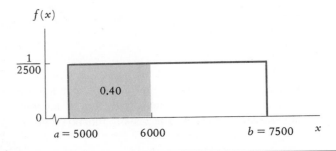

Determining Probabilities and Percentiles

Any desired probability or percentile for a continuous uniform probability distribution may be found by using its cumulative probability function.

(7.4)

> The *cumulative probability function* for a continuous uniform random variable is:
>
> $$F(x) = P(X \leq x) = \frac{x - a}{b - a}$$
>
> *where:*
> $a \leq x \leq b$

EXAMPLE

For the taxpayers' incomes example, the probability that a taxpayer in the $5000–$7500 income bracket has an income of $6000 or less is:

$$F(6000) = P(X \leq 6000) = \frac{6000 - 5000}{7500 - 5000} = 0.40$$

This probability is shown as the shaded area in Figure 7.1. Thus, the 40th percentile of incomes of taxpayers in this bracket is $6000. ☐

7.2 NORMAL PROBABILITY DISTRIBUTIONS

The family of normal probability distributions is one of the most important in statistics. Many types of random trials that arise in applications involve a *normal random variable*.

EXAMPLES

1. The temperature X at noon on August 15 in a southeastern city is a normal random variable. A weather forecaster wishes to know the probability that the noon temperature will exceed 35°C.

2. The weight X of a metal ingot produced in a smelter is a normal random variable. An industrial engineer needs to know the probability that the weight of an ingot will be between 500 and 540 pounds.

3. The height X of women who are 20–29 years old is a normal random variable. A clothing designer wishes to know the proportion of women in this age group who are less than 5 feet tall. ☐

In addition to the uses of the normal distribution to model many phenomena, we will see in later chapters that the normal distribution is also frequently utilized in drawing inferences from data.

Density Function

The normal random variable is a continuous random variable that may take on any value between $-\infty$ and $+\infty$. While real-world phenomena are bounded in magnitude, nevertheless the normal probability distribution, which is not bounded, is often a good model for bounded phenomena — as we shall explain.

(7.5)

The *normal probability density function* is:

$$f(x) = \frac{1}{\sqrt{2\pi}\,\sigma} \exp\left[-\frac{1}{2}\left(\frac{x - \mu}{\sigma}\right)^2 \right]$$

where:

$-\infty < x < \infty$

$-\infty < \mu < \infty$

$\sigma > 0$

$\pi = 3.14159\ldots$

The notation $\exp(a)$ represents e^a, where $e = 2.71828\ldots$ is the base of natural logarithms. (A review of exponentiation and natural logarithms can be found in Appendix A, Section A.2.)

Characteristics of Normal Probability Distributions

Distribution Shape.   The normal probability distribution has two parameters, μ (Greek mu) and σ (Greek sigma), with σ positive. Each different pair of (μ, σ) values corresponds to a different probability distribution of the family of normal distributions. Every normal distribution is bell-shaped and symmetrical, as illustrated in Figure 7.2. Each normal probability distribution is centered at μ, which is the mean of the distribution and determines its position on the x axis. The parameter σ is the standard deviation of the probability distribution and determines the spread of the distribution — the larger the value of σ, the more spread out is the distribution. The examples in Figure 7.2 illustrate these facts. The distributions in Figures 7.2a and 7.2b have the same standard deviation but different means, while the distributions in Figures 7.2b and 7.2c have the same mean but different standard deviations.

Figure 7.2 also shows that almost all of the probability in a normal probability distribution is located in a limited range about its mean, so the probability of an observation falling outside this range is practically zero. For this reason, the normal distribution can be used to model bounded phenomena, such as heights of persons and tensile strengths of iron bars.

Mean and Variance.   As the foregoing discussion has indicated, the mean and the variance of a normal distribution are related to its parameters as follows.

FIGURE 7.2 **Three normal probability distributions.** The mean μ determines the position and the standard deviation σ the variability of the distribution.

(a) $\mu = 56.0$, $\sigma = 2.7$

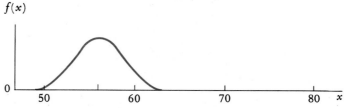

(b) $\mu = 66.5$, $\sigma = 2.7$

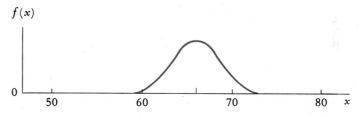

(c) $\mu = 66.5$, $\sigma = 4.1$

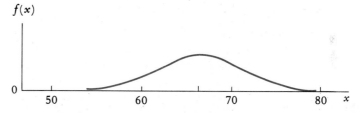

The mean and the variance of a normal probability distribution are:

(7.6) $E\{X\} = \mu$

(7.7) $\sigma^2\{X\} = \sigma^2$

Note that we have used the same Greek symbol for the variance operator, $\sigma^2\{X\}$, as for the variance of a normal probability distribution, σ^2. The context and the notation will make it clear whether the symbol refers to the variance operator or the variance parameter.

$N(\mu, \sigma^2)$ **Notation.**   The normal distribution is used frequently throughout the text, so we shall adopt the compact notation $N(\mu, \sigma^2)$ to denote a normal distribution with mean μ and variance σ^2. Thus, when we state that X is distributed as $N(100, 20)$, we mean that X is a normal random variable with mean $\mu = 100$ and variance $\sigma^2 = 20$.

Standard Normal Probability Distribution

The *standard normal probability distribution* is a particular member of the family of normal distributions, namely, that normal distribution with a mean of 0 and a standard deviation of 1, that is, $\mu = 0$ and $\sigma = 1$. The normal random variable corresponding to the standard normal distribution is called the *standard normal variable* and is denoted by Z. Thus, in terms of our compact notation, Z is distributed as $N(0, 1)$.

The significance of the standard normal distribution results from an important theorem.

(7.8)   Any linear function of a normal random variable is also a normal random variable.

The foregoing theorem permits any normal random variable to be transformed into the standard normal variable, and hence the probability table for the standard normal distribution can be used for all normal distributions. The transformation utilized is the standardized form of a random variable given in (5.9).

(7.9)   The standardized form of a normal random variable X with mean μ and standard deviation σ is:

$$Z = \frac{X - \mu}{\sigma}$$

We know from Chapter 5 that, for any standardized variable Z, we have $E\{Z\} = 0$ and $\sigma^2\{Z\} = 1$. Theorem (7.8) tells us that Z is normally distributed when X is a normal random variable. Thus, Z here is a standard normal random variable.

Determining Probabilities and Percentiles for Standard Normal Distribution

Standard Normal Probability Table.   As we mentioned earlier, probabilities for any normal random variable can be obtained from a table for the standard normal variable Z. We now explain Table C.1, the probability table for the standard normal variable, and then we consider how this table is used to obtain probabilities for any normal random variable.

In accordance with our earlier notation, we employ the lowercase letter z to denote a particular outcome of the standard normal random variable Z. The row and column labels in Table C.1 specify different z outcomes, the row label providing the first decimal place value, and the column label the second decimal place value. Each cell entry for a given value z is the cumulative probability $P(Z \leq z)$. This cumulative probability is represented by the shaded area labeled a in the figure at the top of Table C.1. The percentile corresponding to cumulative area a is denoted by $z(a)$.

We can use Table C.1 in two ways: (1) to determine the area a that corresponds to a specified value of z, and (2) to find the percentile $z(a)$ that corresponds to a specified cumulative probability a. We begin by explaining the first use.

Probabilities.   Figure 7.3 contains several examples of finding probabilities for the standard normal variable Z.

EXAMPLES

1. We wish to find $P(Z \leq 0.45)$; see Figure 7.3a. In entering Table C.1 at the row labeled 0.4 and the column labeled 0.05, we find, for $z = 0.45$, the cumulative area $a = 0.6736$. Hence, $P(Z \leq 0.45) = 0.6736$.

    Note, therefore, that $z = 0.45$ is the 67.36th percentile of the standard normal distribution.

2. We wish to find $P(Z \geq 1.00)$; see Figure 7.3b. Table C.1 shows areas to the left of any z value. To find areas to the right, we use the complementation theorem (4.19):

    $$P(Z \geq 1.00) = 1 - P(Z < 1.00)$$

    Remember that $P(Z < 1.00) = P(Z \leq 1.00)$ because Z is a continuous random variable. From Table C.1, we find that $P(Z < 1.00) = 0.8413$. Therefore, $P(Z \geq 1.00) = 1 - 0.8413 = 0.1587$.

3. We wish to find $P(Z \leq -1.00)$; see Figure 7.3c. Table C.1 does not show probabilities for negative values of z. The reason is the symmetry of the standard normal distribution about its mean 0. From this symmetry, we have:

    $$P(Z \leq -1.00) = P(Z \geq 1.00)$$

    Using the result from the previous example, we conclude that $P(Z \leq -1.00) = 0.1587$.

4. We wish to find $P(-1.00 \leq Z \leq 1.00)$; see Figure 7.3d. We know from Example 2 that $P(Z \leq 1.00) = 0.8413$. Furthermore, we know from Example 3 that $P(Z < -1.00) = 0.1587$. Hence:

    $$P(-1.00 \leq Z \leq 1.00) = P(Z \leq 1.00) - P(Z < -1.00)$$

    $$= 0.8413 - 0.1587 = 0.6826$$

Percentiles.   We stated earlier that $z(a)$ denotes the value on the z scale to the left of which the cumulative probability is a, as illustrated in Figure 7.4a. Thus, $z(a)$ is the $100a$ percentile of the standard normal distribution.

FIGURE 7.3   Determining probabilities for the standard normal distribution

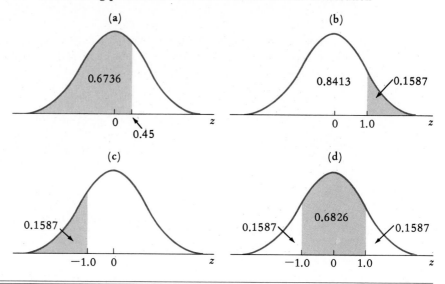

FIGURE 7.4   Determining percentiles for the standard normal distribution

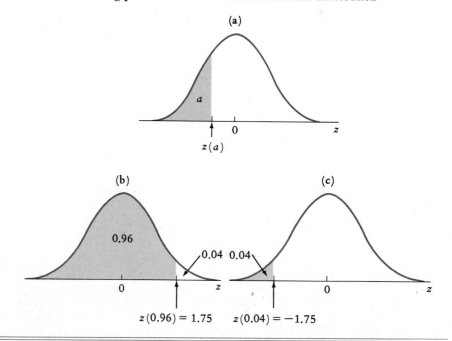

(7.10)

> The 100a percentile of the standard normal distribution, denoted by $z(a)$, is defined:
>
> $$P[Z \leq z(a)] = a$$

EXAMPLES

1. We wish to find the 67.36th percentile, that is, $z(0.6736)$. We found earlier in Figure 7.3a that for $z = 0.45$, the cumulative area to the left is 0.6736. Hence, $z(0.6736) = 0.45$, so 0.45 is the 67.36th percentile.

2. We wish to find $z(0.96)$, the 96th percentile. The cell entry nearest 0.96 in Table C.1 is $a = 0.9599$. The corresponding z value is 1.75 because entry 0.9599 is found in the row labeled 1.7 and the column labeled 0.05. Thus, $z(0.96) = 1.75$. This percentile is illustrated in Figure 7.4b.

3. We wish to find $z(0.95)$, the 95th percentile. The cell entries nearest 0.95 in Table C.1 are $a = 0.9495$ and $a = 0.9505$. The corresponding z values are 1.64 and 1.65, respectively. In interpolating linearly, we obtain 1.645 for the 95th percentile; that is, $z(0.95) = 1.645$. □

In the preceding examples we found percentiles above the 50th. Because of the symmetry of the standard normal distribution about its mean 0, percentiles below 50 are related to those above 50 in the following fashion.

(7.11)

> $$z(a) = -z(1 - a)$$

EXAMPLE

We wish to find the 4th percentile, that is, $z(0.04)$. By theorem (7.11), we know that $z(0.04) = -z(0.96)$. In Example 2, we found $z(0.96) = 1.75$; hence, $z(0.04) = -1.75$, as illustrated in Figure 7.4c. □

Some percentiles of the standard normal distribution are used so frequently that they are shown separately in part b of Table C.1. These percentiles may be obtained from the cumulative probabilities in part a of Table C.1 in the manner just illustrated, although not to the accuracy shown in part b of Table C.1.

EXAMPLES

1. We wish to find $z(0.98)$. From the table of selected percentiles, we obtain $z(0.98) = 2.054$.

2. We wish to find $z(0.001)$. From the table of selected percentiles, we obtain $z(0.001) = -3.090$. □

Determining Probabilities and Percentiles for Any Normal Distribution

The standard normal probability table may be used to calculate probabilities for any normal random variable X by transforming X into the standard normal variable Z, using the standardizing transformation (7.9). We illustrate the procedure by considering the normal distribution in Figure 7.5, which refers to the weight X of a metal ingot produced in a smelter. The mean and the standard deviation of X are $\mu = 520$ pounds and $\sigma = 11$ pounds, respectively.

EXAMPLES

1. We wish to find $P(X \leq 525)$. Figure 7.5 illustrates the relation between the area to the left of 525 and the corresponding area in the standard normal distribution. The value z for the standard normal distribution corresponding to $x = 525$ is obtained by the standardizing transformation (7.9):

$$z = \frac{x - \mu}{\sigma} = \frac{525 - 520}{11} = 0.45$$

    This value indicates that 525 is 0.45 standard deviation above the mean.
        Consequently, $P(X \leq 525) = P(Z \leq 0.45)$, as shown in Figure 7.5. We can find this probability directly from Table C.1. Actually, we found this probability earlier in Figure 7.3a, where it was seen to equal 0.6736. Hence, the probability that an ingot will weigh 525 pounds or less is $P(X \leq 525) = 0.6736$.

2. We wish to find $P(509 \leq X \leq 531)$. We calculate the z values corresponding to the specified values of x as follows:

$$z = \frac{x - \mu}{\sigma} = \frac{531 - 520}{11} = +1.00 \qquad z = \frac{x - \mu}{\sigma} = \frac{509 - 520}{11} = -1.00$$

    Hence, $P(509 \leq X \leq 531) = P(-1.00 \leq Z \leq +1.00)$. Figure 7.6a illustrates the correspondence between the probability distributions of X and Z by the alignment of the two scales. The desired probability equals 0.6826 and was found ear-

---

FIGURE 7.5   Relationship between the normal distribution with $\mu = 520$ and $\sigma = 11$ and the standard normal distribution

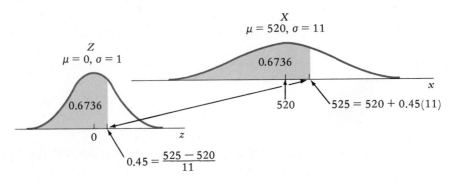

FIGURE 7.6 **Determining probabilities and percentiles for a normal probability distribution with $\mu = 520$ and $\sigma = 11$**

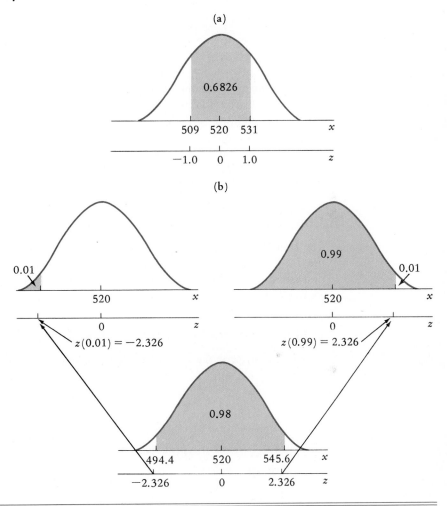

lier in Figure 7.3d. Hence, the probability that an ingot weighs between 509 and 531 pounds is $P(509 \le X \le 531) = 0.6826$.

3. We wish to find the central 98 percent probability limits of ingot weight. We see from Figure 7.6b that we must find the 1st and 99th percentiles of ingot weight. From part b of Table C.1, we find that the 99th percentile of the standard normal distribution is $z(0.99) = 2.326$ and that the 1st percentile is $z(0.01) = -2.326$.

The weight limits are therefore located 2.326 standard deviations on either side of the mean, that is, at the weights:

$$520 - 2.326(11) = 494.4 \text{ pounds}$$

$$520 + 2.326(11) = 545.6 \text{ pounds}$$

In other words, $P(494.4 \leq X \leq 545.6) = 0.98$.   □

Three Important Normal Probabilities.   Three sets of central probability limits for the normal distribution are used so frequently that it is worthwhile recording them. They are as follows:

1. $\mu \pm 1\sigma$ contains about 68.3 percent of the area under a normal distribution.
2. $\mu \pm 2\sigma$ contains about 95.4 percent of the area under a normal distribution.
3. $\mu \pm 3\sigma$ contains about 99.7 percent of the area under a normal distribution.

Cumulative Normal Probability Function

The cumulative probability function of a normal random variable is illustrated in Figure 7.7. Note the S-shaped appearance of the cumulative probability curve. The function is drawn by obtaining a series of cumulative probabilities of the form $P(X \leq x)$ from Table C.1 and then plotting them.

Sum of Independent Normal Random Variables

When we work with the sum of two independent normal random variables, the following theorem is helpful.

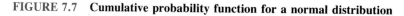

FIGURE 7.7   **Cumulative probability function for a normal distribution**

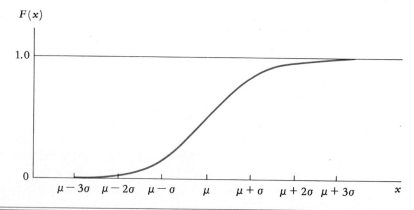

(7.12)

> If V and W are two independent normal random variables with means μ_1 and μ_2 and variances σ_1^2 and σ_2^2, respectively, the sum $V + W$ is also a normal random variable, with mean $\mu = \mu_1 + \mu_2$ and variance $\sigma^2 = \sigma_1^2 + \sigma_2^2$.

Theorem (7.12) extends directly to the sum of more than two independent normal random variables.

EXAMPLE

The cost of material for a construction project to be undertaken is a normal random variable V with mean μ_1 = \$60 million and standard deviation σ_1 = \$4 million. The cost of labor for this same project is an independent normal random variable W with mean μ_2 = \$20 million and standard deviation σ_2 = \$3 million. What is the probability that the total cost of material and labor for this project will not exceed \$85 million?

By (7.12), the total cost is a normal random variable with mean μ = $60 + 20$ = \$80 million and variance $\sigma^2 = (4)^2 + (3)^2 = 25$; hence, the standard deviation is $\sigma = \sqrt{25}$ = \$5 million. We employ the standardizing transformation (7.9) and obtain $z = (85 - 80)/5 = 1.00$. From Table C.1, we see that $P(Z \le 1.00) = 0.8413$. Hence, the probability is 0.8413 that the total cost of material and labor will not exceed \$85 million.

7.3 EXPONENTIAL PROBABILITY DISTRIBUTIONS

Exponential probability distributions are an important family of probability distributions useful in describing duration phenomena.

EXAMPLES

1. A police department has been criticized about the length of time required by police cars to answer calls. The mayor wishes to know what percentage of calls are answered within 15 minutes. Here, X, the number of minutes required to answer a call, has an exponential distribution. The desired probability is $P(X \le 15)$.

2. The length of a long-distance telephone call (X) has an exponential distribution. An analyst with a telephone company would like to know the proportion of calls completed within 3 minutes for the purpose of rate setting. She therefore requires $P(X \le 3)$.

Density Function

The *exponential random variable* is a continuous random variable that may take on any positive value.

(7.13)

> The *exponential probability density function* is:
>
> $$f(x) = \lambda \exp(-\lambda x)$$

continues

where:

$0 < x < \infty$

$\lambda > 0$

The notation $\exp(a)$ represents e^a, where $e = 2.71828\ldots$ is the base of natural logarithms. Thus, $\exp(-\lambda x)$ represents $e^{-\lambda x}$. (A review of exponentiation and natural logarithms can be found in Appendix A, Section A.2.)

Note that λ is the only parameter of the exponential probability distribution. It is always positive. Each different value of λ yields a different probability distribution of the exponential family.

EXAMPLE

The number of minutes (X) required by a police patrol car to respond to a call is an exponential random variable with $\lambda = 0.2$. The corresponding exponential density function is:

$$f(x) = 0.2 \exp(-0.2x)$$

Characteristics of Exponential Probability Distributions

The mean and the variance of an exponential probability distribution depend on the parameter λ in a simple way.

The mean and the variance of an exponential probability distribution are:

(7.14) $$E\{X\} = \frac{1}{\lambda}$$

(7.15) $$\sigma^2\{X\} = \frac{1}{\lambda^2}$$

Note that the standard deviation for an exponential distribution is $\sigma\{X\} = 1/\lambda$ and is equal to the mean.

EXAMPLE

In the police car example, where $\lambda = 0.2$, the mean time required by a patrol car to respond to a call is $E\{X\} = 1/0.2 = 5$ minutes. The standard deviation of the time required is the same, that is, $\sigma\{X\} = 5$ minutes, and hence the variance is $\sigma^2\{X\} = 5^2 = 25$.

Figure 7.8 contains two exponential probability distributions corresponding to parameter values $\lambda = 0.2$ and $\lambda = 0.4$. Note that the exponential distribution is markedly right-skewed, and that the probability density decreases steadily as x increases. Finally, note that the larger the value of λ, the less spread out is the distribution and the closer is the mean to the origin.

FIGURE 7.8 **Two exponential probability distributions.** The larger λ is, the smaller are the mean and the variance of the distribution.

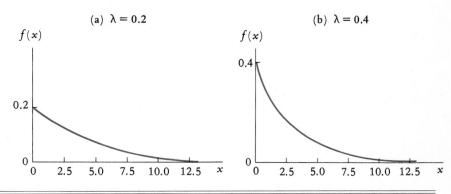

Determining Probabilities and Percentiles

Any desired probability or percentile for an exponential random variable may be calculated readily from its cumulative probability function.

(7.16)

The exponential cumulative probability function is:

$$F(x) = P(X \leq x) = 1 - \exp(-\lambda x)$$

where:

$0 < x < \infty$

The exponential cumulative probability function for λ = 0.2 is shown in Figure 7.9.

EXAMPLE

For the police car example, we wish to find the probability that a car's response time is within 15 minutes. Since λ = 0.2, this probability equals:

$$P(X \leq 15) = 1 - \exp[-0.2(15)] = 0.9502$$

In words, the probability is 0.95 that a police car's response time is within 15 minutes. □

Exponential cumulative probabilities are also tabulated. Table C.7 presents cumulative exponential probabilities for different values of λx.

EXAMPLES

1. For the police car example, we shall obtain the probability that a patrol car's response time will be less than or equal to 15 minutes. Since $x = 15$ and λ = 0.2

FIGURE 7.9   Cumulative probability function for the exponential distribution with $\lambda = 0.2$

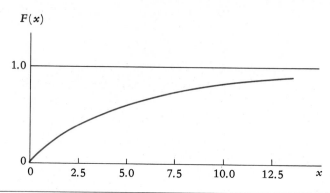

in this case, we enter Table C.7 at $\lambda x = 0.2(15) = 3$. The cumulative probability is found to be 0.9502, so $P(X \leq 15) = 0.9502$.

2. The waiting time (in minutes) at a ticket counter is an exponential variable with $\lambda = 0.5$. We wish to find the probability that a customer has to wait more than 5 minutes. Here, $\lambda x = 0.5(5) = 2.5$. We obtain:

$$P(X > 5) = 1 - P(X \leq 5) = 1 - 0.9179 = 0.0821$$

3. In Example 2, we also wish to find the 95th percentile of the exponential distribution with $\lambda = 0.5$. Referring to Table C.7, we find that the cell entry 0.9502 corresponds to $\lambda x = 3.00$. Since $\lambda = 0.5$, we have $0.5x = 3.00$, or $x = 6$ minutes. Hence, the 95th percentile waiting time is 6 minutes. $\square$

Comment   (Calculus needed.) The exponential cumulative probability function is readily derived from its probability function (7.13) as follows.

(7.17) $$F(x) = \int_0^x \lambda \exp(-\lambda u)\, du = 1 - \exp(-\lambda x)$$

OPTIONAL TOPIC — RELATION BETWEEN POISSON AND EXPONENTIAL DISTRIBUTIONS

Poisson and exponential probability distributions both pertain to occurrences generated by a Poisson process and are closely related. Figure 7.10 shows a time scale in which

FIGURE 7.10   Relation between Poisson and exponential probability distributions. The numbers of occurrences in the unit intervals are independent Poisson variables with parameter λ. The times between successive occurrences are independent exponential variables with parameter λ.

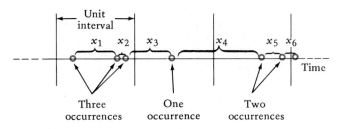

occurrences are indicated by small circles. The time scale is marked off in unit intervals. In the first interval, there are three occurrences; in the second, there is one; and so on. The lengths of time between consecutive occurrences are denoted by x_1, x_2, and so forth, and are shown by braces. Statistical theory gives us the following relationship between the number of occurrences and the times between successive occurrences in a Poisson process.

(7.18)

> If occurrences are generated by a Poisson process (that is, the numbers of occurrences in the different unit intervals are independent and have the same Poisson distribution), then the lengths of time between successive occurrences are statistically independent random variables having an exponential distribution. Moreover, the Poisson process and exponential distribution will have the same parameter value λ.

The last sentence of theorem (7.18) implies that if the mean number of occurrences per time interval is λ (the Poisson parameter), the mean length of time between successive occurrences is $1/\lambda$.

EXAMPLE

The numbers of accidents in a plant each month are statistically independent outcomes from a Poisson distribution with mean $E\{X\} = \lambda = 1.5$. The lengths of time between successive accidents are then statistically independent outcomes from an exponential distribution with mean $E\{X\} = 1/\lambda = 1/1.5 = 0.67$ month. □

PROBLEMS

*7.1 A team preparing a bid on an excavation project assesses that the lowest competitive bid is a continuous uniform random variable X with $a = \$250$ thousand and $b = \$300$ thousand.

a. State the density function for X and plot it on a graph.

b. Find the mean and the variance of the probability distribution of X.

c. Find the probability that the lowest competitive bid is (1) between \$250 thousand and \$270 thousand, (2) between \$275 thousand and \$300 thousand.

7.2 A service station is located at one end of a 4-kilometer tunnel. Any vehicle breakdown in the tunnel that requires servicing is equally likely to be at any point in the tunnel. Let X denote the position of a disabled vehicle (measured in kilometers from the service station).

a. Graph the probability distribution of X.

b. Find the probability that a service call to the tunnel will involve a vehicle located (1) in the first 2 kilometers, (2) between the first and third kilometers.

c. Obtain $E\{X\}$ and $\sigma\{X\}$. In what units are these measures expressed?

7.3 Refer to **Suspension Cable** Problem 5.46.

a. What are the values of the parameters a and b in this setting?

b. Obtain $E\{X\}$ and $\sigma\{X\}$. In what units are these measures expressed?

c. Obtain (1) $F(40)$, (2) $P(40 < X \leq 160)$.

*7.4 Consider a normal probability distribution with $\mu = 6$ and $\sigma = 3$.

a. Obtain the probability density at $x = 0, 3, 6, 9, 12$.

b. Graph the probability density function.

7.5 Consider a normal probability distribution with $\mu = 2.0$ and $\sigma = 0.7$.

a. Obtain the probability density at $x = 0.6, 1.3, 2.0, 2.7, 3.4$.

b. Graph the probability density function.

*7.6 a. Identify the probability distributions denoted by (1) $N(200, 100)$, (2) $N(0, 100)$, (3) $N(5, 1)$.

b. For each of the normal random variables in **a**, obtain the expression (7.9) for the standardized normal variable Z.

7.7 a. Identify the probability distributions denoted by (1) $N(40, 400)$, (2) $N(0, 400)$, (3) $N(-3, 1)$.

b. For each of the normal random variables in **a**, obtain the expression (7.9) for the standardized normal variable Z.

*7.8 Consider the standard normal variable Z and Table C.1.

a. Find the areas a that correspond to the z values (1) 0, (2) 0.42, (3) 1.06, (4) 2.50.

b. Find the following probabilities: (1) $P(Z \geq 0)$, (2) $P(Z \geq 1.50)$, (3) $P(Z < -1.25)$, (4) $P(-1.25 \leq Z \leq 1.50)$.

c. Find the following percentiles from Table C.1a: (1) $z(0.5000)$, (2) $z(0.9066)$, (3) $z(0.0934)$.

d. Find the following percentiles from Table C.1b: (1) $z(0.99)$, (2) $z(0.05)$, (3) $z(0.975)$.

7.9 Consider the standard normal variable Z and Table C.1.

a. Find the areas a that correspond to the z values (1) 0.57, (2) 1.73, (3) 2.46.

b. Find the following probabilities: (1) $P(Z \leq 0)$, (2) $P(Z \geq 1.41)$, (3) $P(Z \leq -2.40)$, (4) $P(-2.40 \leq Z \leq 1.41)$.

c. Find the following percentiles from Table C.1a: (1) $z(0.6179)$, (2) $z(0.8159)$, (3) $z(0.1841)$.

d. Find the following percentiles from Table C.1b: (1) $z(0.95)$, (2) $z(0.025)$, (3) $z(0.995)$.

7.10 Consider the standard normal variable Z and Table C.1.

a. Find the areas a that correspond to the z values (1) 0.22, (2) 1.46, (3) 3.10.

b. Find the following probabilities: (1) $P(Z \leq 1.00)$, (2) $P(Z \geq 0.53)$, (3) $P(Z > -1.49)$, (4) $P(-1.49 \leq Z \leq -0.53)$.

c. Find the following percentiles from Table C.1a: (1) $z(0.8980)$, (2) $z(0.5199)$, (3) $z(0.0571)$.

d. Find the following percentiles from Table C.1b: (1) $z(0.005)$, (2) $z(0.90)$, (3) $z(0.02)$.

*7.11 The weekly total long-distance telephone charge for a travel agency is a normal random variable X with $\mu = \$2800$ and $\sigma = \$150$.
   a. For a randomly selected week, what is the probability the total charge will be (1) less than $2950, (2) more than $3000, (3) between $2560 and $3040?
   b. Find the 75th percentile of the probability distribution, and interpret its meaning here.
   c. Within what interval centered about μ will the weekly total charge fall with probability 0.50?

7.12 The duration of a flight between two cities is a normal random variable X with mean $\mu = 3.6$ hours and standard deviation $\sigma = 0.15$ hour.
   a. Find the following probabilities: (1) $P(X \le 4.0)$, (2) $P(3.2 \le X \le 4.0)$, (3) $P(X > 3.7)$.
   b. Find the 80th percentile of the probability distribution. Interpret its meaning here.
   c. During 90 percent of the flights, within what interval centered about μ will the flight duration fall?

7.13 The temperature at noon on August 15 in a southeastern city is a normal random variable X with $\mu = 30.1°C$ and $\sigma = 2.3°C$.
   a. What is the probability the noon temperature will exceed $35.0°C$?
   b. What is the 95th percentile of the probability distribution?

7.14 The aptitude test scores of applicants to a university graduate program are normally distributed with mean $\mu = 400$ and standard deviation $\sigma = 100$.
   a. An applicant needs a test score higher than 525 to be admitted into the graduate program. What proportion of applicants qualify?
   b. The university wishes to set the cutoff score for graduate admission so that only the top 10 percent of applicants qualify for admission. What is the required cutoff score?
   c. What percentage of applicants have test scores within 2 standard deviations of the mean?

*7.15 In restoring the exterior of a historic mansion, the required number of carpentry manhours (V) is assessed to be $N(265, 144)$. The required number of painting manhours (W) is assessed to be $N(208, 25)$. Here, V and W are assumed to be independent random variables.
   a. What does $T = V + W$ represent here? Describe the probability distribution of T.
   b. The labor cost of either type of work is $15 per hour. A sum of $7000 has been budgeted for the total labor cost of carpentry and painting. What is the probability that the total labor cost will not exceed the budget?

7.16 A firm has two major products. Let V and W denote the annual sales (in $ million) of the two products. Here, V and W are independent random variables with distributions $N(38, 25)$ and $N(65, 39)$, respectively. Let $T = V + W$.
   a. Describe the probability distribution of T.
   b. Find the probability that total sales of these two products (1) will exceed $100 million, (2) will lie between $90 million and $110 million.
   c. Obtain the 10th percentile of the probability distribution of T. Interpret this quantity.

*7.17 The time required to check out a customer at a supermarket is an exponential random variable X with mean 200 seconds.
   a. What is the value of the parameter λ here?
   b. Obtain each of the following probabilities: (1) $P(X \le 100)$, (2) $P(X \le 250)$, (3) $P(X > 400)$.
   c. Obtain $\sigma\{X\}$.
   d. Obtain $F(x)$ for $x = 0, 60, 150, 400, 1000$. Sketch the cumulative probability function. How can you tell from it that the probability distribution is skewed?

7.18 The waiting time (in minutes) to place a call on an office long-distance telephone system with limited capacity is an exponential random variable X with $\lambda = 0.8$.

    a. Find the probability that a call is placed (1) with a wait under 1.0 minute, (2) with a wait between 0.6 and 2.0 minutes.

    b. Obtain $E\{X\}$ and $\sigma\{X\}$.

    c. Find the 25th and 75th percentiles of the probability distribution. What is the interquartile range, and how is this range interpreted here?

    d. Obtain $F(x)$ for $x = 0, 1, 2, 3, 5$. Sketch the cumulative probability function. How can you tell from your graph that the probability distribution is skewed?

7.19 The failure time (in years) of an electronic digital display is an exponential random variable X with $\lambda = 0.2$.

    a. Obtain $\sigma\{X\}$. In what units is this measure expressed?

    b. Obtain each of these probabilities: (1) $P(X \leq 4)$, (2) $P(X > 8)$, (3) $P(1 \leq X < 4)$.

    c. Obtain the following characteristics of the probability distribution of X: (1) the 90th percentile, (2) the median. Interpret the latter value.

    d. Obtain the probability density at $x = 0, 1.5, 3.0, 7.0, 15.0$. Sketch the probability density function.

7.20 The particle life (in days) of a radioactive isotope used in medical research is an exponential random variable X with mean 3.8 days.

    a. Find the probability that a particle of this isotope will survive (1) beyond 7 days, (2) beyond 14 days.

    b. What is the half-life (that is, median life) of this isotope?

*7.21 Calls to an emergency ambulance service in a large city are generated by a Poisson process with $\lambda = 3$ calls per hour.

    a. What is the nature of the probability distribution for the lengths of time between successive calls to the service?

    b. Find the mean length of time between successive calls to the service.

    c. What is the probability that less than 0.25 hour will elapse between successive calls?

7.22 In a canal serving oceangoing ships, the time durations between successive collisions ($X_1, X_2, \ldots$) are independent exponential random variables, each with parameter $\lambda = 0.5$ collision per year.

    a. Let Y denote the number of collisions in a year. What is the apparent nature of the probability distribution of Y?

    b. Obtain $P(X > 1)$ and $P(Y = 0)$. Why must these probabilities be the same?

EXERCISES

7.23 (Calculus needed.) Derive (7.2) and (7.3).

7.24 Let $X_1, X_2, \ldots, X_{12}$ be 12 independent continuous uniform random variables, each with $a = 0$ and $b = 1$. It can be shown theoretically that W is approximately a standard normal variable:

$$W = X_1 + X_2 + \cdots + X_{12} - 6$$

Verify that $E\{W\} = 0$ and $\sigma^2\{W\} = 1$.

7.25 Variable X is $N(10, 4)$. Find the following conditional probabilities: (1) $P(X \geq 14 \mid X \geq 12)$, (2) $P(9 \leq X \leq 13 \mid 8 \leq X \leq 14)$.

7.26 The 20th and 60th percentiles of a normal random variable are -6.3 and 15.8, respectively. Find the mean and the standard deviation of the random variable.

7.27 (Calculus needed.) Derive (7.14). (*Hint:* Use integration by parts.)

7.28 (Calculus needed.) Derive (7.16) by carrying out the integration shown in (7.17).

7.29 Variable X is an exponential random variable with parameter $\lambda = 0.004$.
  a. Find the following probabilities: (1) $P(X > 100)$, (2) $P(X > 200 \mid X > 100)$, (3) $P(X > 400 \mid X > 300)$.
  b. Find the following probabilities: (1) $P(X > 400)$, (2) $P(X > 600 \mid X > 200)$, (3) $P(X > 900 \mid X > 500)$.
  c. What property of an exponential probability distribution is illustrated by your results in **a** and **b**? Explain.

7.30 Show that, for any exponential random variable X, the following relation holds: $P(X > x + x_0 \mid X > x_0) = P(X > x)$ for any positive x and x_0.

STUDIES

7.31 Weights of canned hams are normally distributed with mean 4.15 kilograms and standard deviation 0.12 kilogram. The label weight is stated as 4.00 kilograms. Which one of the following two modifications in the canning process would lead to a greater reduction in the proportion of hams below the label weight? (1) Increase the process mean weight to 4.20 kilograms while keeping the standard deviation unchanged. (2) Decrease the process standard deviation to 0.10 kilogram while keeping the mean unchanged.

7.32 A safety light with two batteries is designed so that the second battery begins to operate when the first battery fails. The safety light fails when the second battery has failed. The lifetime of each battery is normally distributed with mean 200 hours and variance 16. Moreover, the lifetimes of the batteries are independent. Let T denote the lifetime of the safety light.
  a. Specify the nature of the probability distribution of T.
  b. Show that this light satisfies the current company specification that the probability is at least 0.95 of the light operating a minimum of 375 hours before failing.
  c. A new version of the light with three batteries is being designed to have a probability of at least 0.95 of operating for a minimum of 585 hours before failing. Will three batteries with independent lifetimes each distributed as $N(200, 16)$ satisfy this specification?

7.33 A machine fills 100-pound bags with white sand. The actual weight of the sand when the machine operates at its standard speed of 100 bags per hour follows a normal distribution with standard deviation 1.5 pounds. The mean of the distribution depends on the setting of the machine.
  a. At what mean weight should the machine be set so that only 5 percent of the bags are underweight, that is, contain less than 100 pounds of sand?
  b. If the machine operates at low speed, the standard deviation will be reduced to 1.0 pound, but only 50 bags will be filled in an hour. The sand costs the owner of the machine $0.16 per pound. An employee who operates the machine is paid $10.00 per hour. At what speed—standard or low—should the machine be operated so that the expected cost to the owner of filling 100 bags will be minimized? Assume that, with either setting, only 5 percent of the bags are to be underweight.

7.34 A large display will contain many colored light bulbs. Because the labor cost of replacing the bulbs individually when they burn out will be expensive, the operators plan to replace all the bulbs simultaneously. Two replacement schedules are under consideration: (1) Replace when 15 percent of the bulbs have burned out. (2) Replace when 30 percent of the bulbs have burned out.

   a. For bulbs of make A, the lifetimes in hours are independent and exponentially distributed with $\lambda = 0.0004$. Approximately how many operating hours will elapse before the bulbs are replaced under schedule 1? Under schedule 2?

   b. For bulbs of make B, the lifetimes are independent and normally distributed with $\mu = 1100$ hours and $\sigma = 250$ hours. Approximately how many operating hours will elapse before the bulbs are replaced under schedule 1? Under schedule 2?

   c. Which make of bulb do you recommend that the operators use under schedule 1? Under schedule 2? Why?

7.35 Refer to the frequency distribution in Table 3.1b of the number of days between oil spills after pollution controls were instituted. An analyst believes that these time intervals follow an exponential distribution. The mean of the 109 observations is 27.6 days. Using this mean value and ignoring the rounded nature of the data, find the exponential probabilities for the frequency classes in Table 3.1b; that is, find $P(0 \leq X < 15)$, $P(15 \leq X < 30)$, and so on. Express these probabilities in percentage form, and compare them with the actual percent frequencies. Obtain the residuals and plot them. Does the analyst's belief appear to be justified? Discuss.

UNIT THREE

Estimation and Testing — I

8 STATISTICAL SAMPLING

We are now ready to begin applying the principles of probability to the analysis of statistical data. Data sets are utilized in all statistical investigations. We now need to distinguish between those data sets that are considered to be populations and others that are considered to be samples.

In any statistical investigation, there is a collection of elements of interest, called the population. For instance, a U.S. congressional staff assistant was studying Medicaid abuses during the previous year to help in drafting new legislation. The data set consisting of all Medicaid claims in the previous year was the population of interest in this study.

Often, it is not possible to consider all elements in the population, and a subset, or sample, is selected to provide information about the population. For instance, the congressional staff assistant could not examine all Medicaid claims in the year because the number was much too large. Instead, she selected a probability sample of 1000 claims. This data set was then utilized to make inferences about the characteristics of Medicaid abuses in all claims for the year.

The use of probability principles to draw sound inferences from a sample about the entire population is the subject of *statistical inference*. In this chapter, we consider some basic concepts of sampling. We define a population and a sample, discuss why sampling is so widely used, and explain what a simple random sample is and how such a sample can be selected.

8.1 POPULATIONS

A formal definition of a statistical population or universe follows.

(8.1)
> A *population* or *universe* is the total set of elements of interest for a given problem. -

EXAMPLES

1. The manager of an automobile-leasing agency is interested in the fuel economy of the passenger cars in the company's fleet. Here, the population consists of all passenger cars in the fleet. The elements of the population are the individual cars.

2.   A candidate for the city council is concerned about the attitudes of voters toward a metropolitan transit agency. The population here consists of all eligible voters in the city. The elements of the population are the individual voters.

3.   A quality assurance manager wishes information about the current quality level of the firm's process for manufacturing computer memory chips. Here, the population consists of all the chips that could be produced by the current production process operating under the same conditions. The elements of the population are the individual chips.   ☐

Examples 1 and 2 are illustrations of finite populations, while Example 3 illustrates an infinite population. We consider each type of population in turn.

Finite Population

A *finite population* consists of a finite number of elements, such as the set of passenger cars owned by a leasing agency or the set of eligible voters in a city. Many of the populations of interest in business, economics, and the social sciences are finite, although they are often large. The set of all persons living in the United States is an example of a large finite population.

Mean and Variance.   When the variable of interest for the finite population is quantitative, we are often interested in the mean and the variance or standard deviation of the variable. In Example 1, for instance, we may be interested in the mean fuel economy per car in the company's fleet, as well as the variability of fuel economies between cars.

The mean, the variance, and the standard deviation for a data set representing a finite population are defined identically (with one small exception) as for a data set representing a sample. However, we shall use some new notation to distinguish the summary measures for population data sets from those for sample data sets, which were considered in Chapter 3.

The number of elements in a finite population will be denoted by N to distinguish it from the number of elements in a sample, denoted by n. Greek letters will be used to represent summary measures for populations. We shall use μ (Greek mu) to denote the population mean, σ (Greek sigma) for the population standard deviation, and σ^2 (read "sigma squared") for the population variance. The corresponding measures for a sample, $\overline{X}$, s, and s^2, were defined in (3.1), (3.17), and (3.15), respectively.

The definitions of a population mean, variance, and standard deviation follow.

(8.2)   Let $X_1, X_2, \ldots, X_N$ represent the values of the N population elements.

(8.2a)   *Population Mean:*

$$\mu = \frac{\sum_{i=1}^{N} X_i}{N}$$

(8.2b)   *Population Variance:*

$$\sigma^2 = \frac{\sum\limits_{i=1}^{N}(X_i - \mu)^2}{N}$$

(8.2c)   *Population Standard Deviation:*

$$\sigma = \sqrt{\sigma^2}$$

Note that the definition of a population mean μ is the same as that for a mean $\overline{X}$ of a sample; see the definition of $\overline{X}$ in (3.1). Also, a population variance σ^2 is defined almost the same as a sample variance s^2 in (3.15) except that the denominator is N and not $N - 1$. When N is large, the difference in the two definitions is negligible. We shall explain the reason for this definitional difference subsequently. For the moment, remember to use divisor N for the variance when the data set refers to a population and $n - 1$ when it refers to a sample.

EXAMPLE

A population consists of the $N = 5$ Ph.D. students currently enrolled in an economics department. Their grade point scores are:

$$X_1 = 3.74 \qquad X_2 = 3.89 \qquad X_3 = 4.00 \qquad X_4 = 3.68 \qquad X_5 = 3.69$$

The population mean is:

$$\mu = \frac{3.74 + 3.89 + 4.00 + 3.68 + 3.69}{5} = 3.80$$

The population variance is:

$$\sigma^2 = \frac{(3.74 - 3.80)^2 + (3.89 - 3.80)^2 + \cdots + (3.69 - 3.80)^2}{5} = 0.01564$$

Thus, the population standard deviation is $\sigma = \sqrt{0.01564} = 0.125$.   □

Comment

A finite population may be related to a probability distribution in the following way. Consider selecting one household from a population of N households such that each household in the population has equal probability of being chosen. The probability of any household being chosen must then be $1/N$. Let X be the random variable denoting the household income observed for the household selected at random. The random variable X can then take on the values $X_1, X_2, \ldots, X_N$ (the household incomes in the population) with probability $1/N$ each. Hence, by the definition of the expected value of a discrete random variable in (5.5), we have:

$$E\{X\} = \sum_{i=1}^{N} X_i \left(\frac{1}{N}\right) = \frac{\sum\limits_{i=1}^{N} X_i}{N} = \mu$$

Similarly, by the definition of the variance of a discrete random variable in (5.6), we have:

$$\sigma^2\{X\} = \sum_{i=1}^{N} (X_i - \mu)^2 \left(\frac{1}{N}\right) = \sigma^2$$

Thus, the population mean μ and variance σ^2 for a finite population correspond, respectively, to the expected value and variance of a random variable — namely, the random variable associated with the equal-probability selection of one population element.

Infinite Population

An *infinite population* consists of an indefinitely large number of elements. In general, an infinite population refers to a *process*, and its elements consist of all the outcomes of the process if it were to operate indefinitely under the same conditions. Thus, the process of manufacturing memory chips is represented by an infinite population whose elements are the chips that would be produced by the process if it were to operate indefinitely under the same conditions.

The characteristic of interest in an infinite population is described by a probability distribution. Consider the infinite population pertaining to the process of molding tires in pairs. Let X, the number of defective tires in a pair, be the characteristic of interest. This characteristic of the infinite population might then be described by the following probability distribution:

x:	0	1	2
$P(x)$:	0.75	0.10	0.15

The probability $P(0) = P(X = 0) = 0.75$ indicates that 75 percent of the trials in the infinite population lead to zero defective tires in the pair. Similarly, $P(1) = 0.10$ indicates that 10 percent of the trials in the infinite population lead to one defective tire in the pair.

Mean and Variance.   The population mean and variance for an infinite population correspond to the expected value and variance of the associated random variable.

(8.3)   Let X represent the random variable associated with the infinite population of interest.

(8.3a)   *Population Mean:*

$$\mu = E\{X\}$$

(8.3b)   *Population Variance:*

$$\sigma^2 = \sigma^2\{X\}$$

(8.3c)   *Population Standard Deviation:*

$$\sigma = \sqrt{\sigma^2} = \sigma\{X\}$$

| EXAMPLES | 1. | Consider the infinite population pertaining to the process of molding tires in pairs. The random variable X is the number of defective tires in the pair. The probability distribution for X is that previously given. The population mean then is, using definition (5.5) for the expected value of a discrete random variable: |

$$\mu = E\{X\} = 0(0.75) + 1(0.10) + 2(0.15) = 0.40$$

The population variance is found by using definition (5.6) for the variance of a discrete random variable:

$$\sigma^2 = \sigma^2\{X\} = (0 - 0.40)^2(0.75) + (1 - 0.40)^2(0.10) + (2 - 0.40)^2(0.15)$$

$$= 0.54$$

Hence, the population standard deviation is $\sigma = \sigma\{X\} = \sqrt{0.54} = 0.73$ tire.

2. The infinite population of interest pertains to the weight (X, in pounds) of metal ingots produced in a smelter. Suppose the probability distribution of X is $N(520, 121)$. Then, the population of interest is a normal distribution, and the population mean and variance are, respectively, $\mu = 520$ pounds and $\sigma^2 = 121$.

3. The infinite population of interest pertains to the response time (X, in minutes) of a police patrol car to a call. Suppose the probability distribution of X is exponential, with $\lambda = 0.2$. Then, the population of interest is an exponential distribution with population mean $\mu = 1/0.2 = 5$ minutes by (7.14) and population variance $\sigma^2 = 1/(0.2)^2 = 25$ by (7.15). □

8.2   CENSUSES AND SAMPLES

Information about a finite population can be obtained either by a census or by a sample.

Census

A census can only be taken of a finite population.

(8.4)

> A *census* of a finite population is a study that includes every element of the population.

In some cases, it is feasible to conduct a census of the population.

| EXAMPLES | 1. | An organization with 50 employees is interested in employees' preferences for a new pension plan. The population here is small, and it is easy to reach every employee; hence, a census is appropriate. |

2. A state employment office urgently needs information about the total number of hours devoted last week to processing new unemployment compensation claims by all of the offices in the state. Data are needed for the entire state as well as for the

five planning regions into which the state is divided. There are 23 such offices, and since they can be contacted readily, a census is practicable. A census is also necessary for obtaining reliable data for each of the five planning regions. □

Comment   An attempted census need not necessarily be successful and include every population element. Thus, the U.S. and Canadian population censuses are known to miss significant numbers of persons even though great efforts are made to include everyone.

Sample

When finite populations are large, a census is often not feasible because it would be too costly and time-consuming. Consequently, many studies of large finite populations utilize sampling.

(8.5)   A *sample* is a part of the population under study selected so that inferences can be drawn from it about the population.

EXAMPLES   1. An economist, as part of a study of Canadian fiscal policy, considers data on savings plans by households based on a sample of 1000 households from the population of Canada.

2. A state analyst uses data on the unemployment rate for the state and on the incidence of unemployment for different groups in the population in setting state priorities for public works. The data are based on a sample of households from all households in the state.

3. An inspector in an automobile assembly plant uses data from a sample of automobile tires selected from a shipment of tires to decide whether or not the shipment should be accepted. □

Reasons for Sampling

There are a number of reasons why sampling is utilized so often for finite populations:

1. *Cost.* Typically, a sample can provide reliable and useful information at much lower cost than a census. For example, the cost of a census of the population of a city to obtain information for an educational television station about the viewing habits of inhabitants would be extremely high. A sample of the population can provide data with sufficient reliability at a fraction of this cost.

2. *Timeliness.* A sample usually provides more timely information than a census, because fewer data have to be collected and processed. This feature is particularly

important when information is needed quickly, such as information about how much of a proposed tax reduction might be saved by families.

3. *Accuracy.* A sample often will provide information as accurate as, or even more accurate than, a census, because data errors typically can be controlled more effectively in a small undertaking than in a large one. Thus, use of sampling often permits better training and supervision of a smaller work force than would be possible with a census.

4. *Detailed information.* Frequently, more time can be spent in probing each respondent's attitudes and motivations and in getting more detailed information with a sample than with a census. This point is related to cost, because the cost per respondent of detailed probing or of getting more detailed information can be high.

5. *Destructive testing.* When a test involves the destruction of an item, sampling must be used. Thus, a manufacturer who wishes to study the life of batteries in a production lot by subjecting batteries to a life test must use a sample if there are to be any batteries left to sell.

When a population is infinite, information about it can be obtained only from a sample. Thus, observations for obtaining information about a process always provide sample information, no matter how large the number of observations.

Probability and Judgment Samples

We distinguish between two types of samples, probability samples and judgment samples, based on the manner in which population elements are selected for the sample. The following examples illustrate the distinction.

EXAMPLES

1. A sample of 36 members was selected from the 900 members of a professional association by giving each member probability 1/900 of being chosen as the first sample element, probability 1/899 of being chosen as the second sample element (if not already chosen before), and so on. This sample is a probability sample because the probabilities of selection are known for all population elements.

2. The director of a television show that was previewed in an auditorium selected 40 persons she judged to be representative of the audience of 435 for detailed interviews about their impressions of the show. The selection of persons here was not carried out by means of a probability mechanism. The sample is a judgment sample because the director judged it to be a "representative" sample.

Probability Sample. A formal definition of a probability sample follows.

(8.6)

> A *probability sample* is one where the selection of elements from the population is made according to known probabilities.

The selection of population elements by known probabilities allows no discretion about which particular elements in the population enter the sample. Probability selection has two major advantages:

1. The sample data can be evaluated by statistical methods to provide information about the margin of error in the results due to sampling.
2. Biases are avoided that could enter if judgment were used to select the population elements for the sample.

Throughout much of the remainder of this text, we shall be concerned with probability samples and their statistical evaluation.

Judgment Sample.   A formal definition of a judgment sample follows.

(8.7)   A *judgment sample* is one where judgment is used to select "representative" elements from the population or to infer that a sample is "representative" of the population.

Although the use of probability sampling has expanded greatly in recent years, judgment samples still are widely used. In contrast to probability sampling, where no discretion is allowed about which population elements enter the sample, in judgment sampling, expert opinion is used to select "representative" elements for the sample or to determine whether the sample is "representative" of the population.

Judgment samples may provide useful results. Statistical methods cannot, however, be utilized to evaluate the sample results for purposes of assessing the margin of error due to sampling. Also, some judgment samples may be very poor because the judgment is not sound and the sample turns out to be highly unrepresentative.

A common type of judgment sample used in surveys is a *quota sample*. For instance, interviewers in a household survey are given "quotas" to provide a cross section of the population under study with respect to certain characteristics such as age, sex, income, and residence. The actual selection of persons is left to the interviewers. While interviewers are supposed to use good judgment in selecting persons who meet the quotas, they often choose persons readily available such as homemakers at home during the day. Also, an interviewer can simply substitute another person from the same quota for a person who is not immediately available for interview. In these and other ways, biases can enter the survey. These biases do not enter a probability sample that is properly carried out.

Comments   1. Some samples designed as probability samples turn out to be judgment samples upon completion. For example, a probability sample of new-car purchasers was carefully selected. Unfortunately, only 50 percent of the purchasers in the sample replied to the questionnaire about the performance of their car. The market researcher decided, however, that the purchasers whose responses were obtained are typical of all new-car purchasers. Hence, the sample actually obtained should be viewed as a judgment sample because the market researcher made a key judgment about the "representativeness" of the responses obtained.

2.   Some studies are based on data that fulfill neither the definition of a probability sample nor that of a judgment sample. This situation occurs when a study is based on population elements that happen to be conveniently at hand, such as the students enrolled in a marketing course who were used in a study of the effectiveness of different advertisements. These students were not selected as representative from the relevant population, nor were they selected according to known probabilities. Such a study group is called a *chunk* or a *convenience sample*. A chunk can be useful for limited purposes, but it cannot be relied on for credible inferences about the population.

SIMPLE RANDOM SAMPLING FROM A FINITE POPULATION

There are many types of probability samples. The most basic is a simple random sample, which we discuss now. Other types of probability samples are discussed in Chapter 14.

(8.8)

> A *simple random sample* from a finite population is a sample selected such that each possible sample combination has equal probability of being chosen.

EXAMPLE

Consider the earlier population consisting of five Ph.D. students, and let us denote these by A, B, C, D, and E. If a sample of two students is to be selected from this population, there are altogether 10 possible sample combinations, as follows:

A, B	A, D	B, C	B, E	C, E
A, C	A, E	B, D	C, D	D, E

If one is to obtain a simple random sample of two students from the population of five, each of the 10 sample combinations must have probability 0.10 of being chosen as the actual sample.   □

Comments

1.   For brevity, a simple random sample is often called a *random sample*.

2.   Definition (8.8) refers to simple random sampling *without replacement,* where a population element can enter the sample only once. Simple random sampling *with replacement,* where the same population element may enter the sample more than once, is rarely used with finite populations.

3.   Definition (8.8) implies that each population element has an equal probability of being selected. However, simple random sampling requires more than this. Equal-selection probability for each population element is a necessary but not a sufficient condition for a sample to be a simple random one. For example, suppose that a random sample of 50 oranges is to be selected from a crate of 100 oranges. We could divide the

100 oranges into two groups consisting of the 50 largest and the 50 smallest oranges, respectively, and then select one of the two groups with equal probability. This procedure gives each orange in the population an equal (and known) probability of 0.5 of entering the sample, and hence it qualifies as probability sampling. Yet it does not qualify as simple random sampling because every combination of 50 oranges does not have equal probability of being selected. For example, a mixture of small and large oranges has no chance of being selected.

Selection of Simple Random Sample

A practical method of meeting the requirement that each possible sample combination have an equal probability of being chosen involves selecting sample elements one at a time.

(8.9)

The requirements of selecting a simple random sample of n elements without replacement from a finite population of N elements are met by the following procedure:

1. Select the first sample element by giving each of the N population elements equal probability of being chosen—that is, probability $1/N$.
2. Select the second sample element by giving each of the remaining $N - 1$ population elements equal probability of being chosen—that is, probability $1/(N - 1)$.
3. Repeat this process until all n sample elements have been selected.

Frame.   Selecting a simple random sample from a finite population requires a frame.

(8.10)

A *frame* is a listing of all the elements of the finite population.

Thus, a computer listing of the students at a university is a frame for sampling the population of students.

Sometimes, the frame available for sampling is not a perfect one for the population of interest.

EXAMPLE

County Lawyers.   A random sample of lawyers is to be selected from the population of all lawyers in a county. The frame for sampling is a listing of all lawyers who are members of the county bar association as of last month. This listing does not include lawyers who joined the association since last month nor any lawyers in the county who are not members of the county bar association.   □

In cases like the county lawyers example, we distinguish between the *target population* and the *sampled population*. The target population is the population of interest (all lawyers in the county), while the sampled population consists of the elements in the frame (all lawyers in the county who belong to the county bar association as of last month). When the frame available for sampling differs materially from the target population, the sample results may have only limited relevance. Hence, great efforts are made in practice to obtain a frame that coincides closely with the target population.

Use of Table of Random Digits. A table of random digits, such as the one in Table C.8, can readily be used to select a simple random sample by the procedure in (8.9). A table of random digits contains outcomes of independent random trials from the discrete uniform probability distribution in Figure 6.5, which has possible outcomes $0, 1, \ldots, 9$ with equal probability. Thus, for each position in the table, every digit from 0 to 9 has equal probability of appearing in that position, and the outcomes for the various positions in the table are independent.

Tables of random digits are generated by computer and are usually tested carefully to ensure close adherence to the required properties of equal probability and independence. These properties permit the user to form random numbers from 00 to 99, or from 000 to 999, or still larger numbers. Consider forming pairs of random digits. Since every digit has equal probability of appearing in a position and all positions are filled independently, pairs of positions are filled with numbers from 00 to 99 with equal probability and independently. It is this capability of forming random numbers of any size, which are equally likely and independent, that enables us to use a table of random digits for selecting a simple random sample.

EXAMPLE

In the county lawyers example, there are 950 lawyers in the frame. We assign numbers to the lawyers, for convenience from 001 to 950. Since the frame contains 950 elements, we shall select three-digit numbers. Table 8.1 contains an

---

TABLE 8.1   **Use of the table of random digits — County lawyers example**

Row	Columns 1–5	Lawyer
1	13284 ⟶	132
2	21224 ⟶	212
3	99052 ⟶	Disregarded
4	00199 ⟶	001
5	60578 ⟶	605
6	91240 ⟶	912
7	97458 ⟶	Disregarded
8	35249 ⟶	352
9	38980 ⟶	389
10	10750 ⟶	107

extract of Table C.8, consisting of the first 10 rows of columns 1–5. Suppose we use the first three digits in each row as the three-digit number and read downward. The procedure, then, is as follows:

1. The first number is 132. Lawyer 132 is therefore the first element for the sample. A number over 950 would have been disregarded. Thus, each of the 950 lawyers has equal probability of being the first sample element.

2. The second number is 212. Lawyer 212 is therefore the second element for the sample. A number over 950 would have been disregarded. Also, number 132 would have been disregarded since that lawyer is already in the sample. Thus, each of the remaining 949 lawyers has equal probability of being the second sample element.

3. This procedure is repeated until the required number of lawyers for the sample has been selected.

Table 8.1 shows the disposition for the first 10 random numbers.   ☐

Comments   1.   Numbers may be chosen from a table of random digits in any manner as long as the procedure is systematic and determined in advance. Thus, three-digit numbers in the county lawyers example might have been obtained by reading across in a row, or by taking the first digit in each of three columns, or by reading upward in a column.

2.   Sometimes, the elements in a frame are prenumbered, such as invoices that have serial numbers or students who have been assigned ID numbers. These numbers can be used for sampling identification provided that there are no duplicate assignments of the same number to several population elements. Often, these number assignments have gaps, such as when an invoice is voided or when a student has graduated. Such gaps cause no problem; when a selected random number refers to a blank, it is simply disregarded and the next random number is selected.

3.   At times, numbers need not be assigned explicitly to the elements in the frame. Suppose a sample of employees is to be selected. The frame consists of 8500 employee file folders organized in 85 files of 100 folders each. The employee designated by a random number can then be identified by counting. For example, random number 017 refers to the employee whose folder is the 17th one in the first file. Similarly, random number 317 refers to the employee whose folder is the 17th one in the fourth file.

Use of Computer Routine.   Random numbers often are produced directly by a computer routine using a random number generator. Using a computer is easier than using a table of random digits when the sample is large, especially since many computer packages also order the random numbers for the user. Typically, the user specifies the smallest and the largest values of the random numbers to be generated and the sample size. Thus, in the county lawyers example, the user would specify 1 and 950 as the smallest and the largest possible values and also would specify the sample size. Some calculators also generate random numbers.

In some computer applications, the random numbers generated by the computer are used internally to identify those elements of a computerized data base that are selected for the sample, and only the descriptions of the sample elements are printed out.

8.4 SIMPLE RANDOM SAMPLING FROM AN INFINITE POPULATION

When the population is infinite, the only available method for obtaining information about it involves sampling. However, one cannot use a probability mechanism for selecting the sample. The process associated with the infinite population simply furnishes sample observations. Thus, the process of molding tires in pairs provides sample observations on the number of defective tires in a pair day after day, without any probability selection mechanism.

Observations generated by a process are random variables prior to the random trial, and they constitute a simple random sample from an infinite population if two conditions are met.

(8.11)

> The n random variables $X_1, X_2, \ldots, X_n$ generated by a process constitute a *simple random sample from an infinite population* under the following conditions:
>
> 1. They come from the same probability distribution.
> ❋2. They are statistically independent.

Thus, the n random variables must be independent and identically distributed. Note that these are the conditions of an independent and stationary process as described in (6.1).

Statistical tests are available for examining whether a set of observations generated by a process meets the requirements of simple random sampling. One such test is discussed in Chapter 15.

Comment

A sample from an infinite population may, for some other purpose, be considered a finite population. Thus, the memory chips produced by a stable production process during a week represent a sample from the infinite population associated with the production process. When, however, these chips are sent to a computer manufacturer who wishes to sample them to determine whether this shipment is of acceptable quality, the chips in the shipment constitute a finite population for this purpose.

8.5   SAMPLE STATISTICS

Once a sample has been selected and observations on the sample elements have been made, the observations constitute a data set, and summary measures may be computed in the usual way, as explained in Chapter 3. For convenience, we repeat here the definitions of the mean, the variance, and the standard deviation for a sample.

(8.12)   Let $X_1, X_2, \ldots, X_n$ denote the n sample observations.

(8.12a)   *Sample Mean:*

$$\overline{X} = \frac{\displaystyle\sum_{i=1}^{n} X_i}{n}$$

(8.12b)   *Sample Variance:*

$$s^2 = \frac{\displaystyle\sum_{i=1}^{n} (X_i - \overline{X})^2}{n - 1}$$

(8.12c)   *Sample Standard Deviation:*

$$s = \sqrt{s^2}$$

Sample measures are usually called *sample statistics,* or *statistics* for short. Thus, $\overline{X}$ and s are examples of statistics. In contrast, population measures are called *population parameters,* or *parameters* for short. For example, μ and σ denote population parameters.

Intuition tells us that the sample statistic $\overline{X}$ provides information about the population mean μ and that the sample statistic s provides information about the population standard deviation σ. In succeeding chapters, we examine the nature of the information provided by these and other sample statistics about the population measures of interest.

PROBLEMS

8.1   In each of the following cases, identify the sampled population, and explain whether it is finite or infinite.
   a. Tax returns for last year are sampled to estimate the mean deduction for interest expenses.
   b. A company sends a large number of letters by mail to estimate the mean number of days required for delivery.
   c. Outgoing orders for books are sampled prior to mailing to estimate the proportion of orders filled incorrectly by the current process.

8.2   In each of the following cases, identify the sampled population, and explain whether it is finite or infinite.

a. A poultry specialist feeds a hen flock an experimental high-mineral diet to measure its impact on the strength of eggshells.

b. The children of Westview Elementary School are sampled to estimate the proportion who have had a dental checkup in the past year.

c. An auditor studies the accuracy of the total amount of fixed assets recorded by a large company by examining a sample of fixed assets.

*8.3   A population consists of $N = 7$ employees. Their ages are:

27   34   47   44   37   29   41

Find the population mean and standard deviation. Are these measures parameters or statistics? Explain.

8.4   A population consists of $N = 5$ assembly plants. The floor spaces in these plants (in thousand square meters) are:

10.1   6.8   12.4   7.9   8.0

Find the population mean and standard deviation. In what units are these measures expressed?

*8.5   A population consists of four resistors. Their resistances (in ohms) measured under standard conditions are:

108   103   104   112

One of these four resistors is chosen in a manner that gives each an equal probability of being selected. Let random variable X denote the resistance of the chosen resistor.

a. Describe the probability distribution of X.

b. Obtain $E\{X\}$ and $\sigma^2\{X\}$. Are these two measures the same as the population mean and variance, respectively? Explain.

8.6   The numbers of children in five families are:

3   1   1   2   2

One family is chosen in a manner that gives each an equal probability of being selected. Let random variable X denote the number of children in the chosen family.

a. Describe the probability distribution of X.

b. Obtain the mean and the variance of the probability distribution of X. Are these measures the same as the population mean and variance, respectively? Explain.

*8.7   The number of photocopying machines in a large office that are out of order at any one time is denoted by X. The probability distribution of X follows.

x:	0	1	2	3	4
$P(x)$:	0.80	0.11	0.05	0.03	0.01

a. Describe the infinite population associated with this probability distribution.

b. Find the population mean and variance.

8.8 An office building contains three elevators. The probability distribution of the number of elevators waiting on the ground floor at any time during business hours (X) follows.

x:	0	1	2	3
$P(x)$:	0.40	0.30	0.20	0.10

  a. Describe the infinite population associated with this probability distribution.
  b. Find the population mean and standard deviation. In what units are these measures
     expressed?

8.9 The number of telephone calls handled by a commodity trader each trading day is a Poisson random variable X with parameter value $\lambda = 32$.
  a. Describe the infinite population associated with this probability distribution.
  b. What are the population mean and standard deviation here?

8.10 For each of the following, discuss whether a sample or a census would be preferable. Indicate
any assumptions you make.
  a. A survey of Canadian households to obtain information on the proportion that used a
     microwave oven at least once during the preceding week.
  b. An examination of three turbines to be installed in a dam to obtain information about pos-
     sible damage in shipment.
  c. An examination of a large shipment of light bulbs to obtain information on the proportion
     that burn out prematurely in normal use.

8.11 For each of the following, discuss whether a sample or a census would be preferable. Indicate
any assumptions you make.
  a. A survey of the 25 universities in a regional association to obtain current enrollment data.
  b. A survey of the 3000 employees of a company to study their attitudes toward the company.
  c. An examination of dwellings in a large city to obtain information about the extent of
     present home insulation and the cost of providing additional insulation to bring substandard
     dwellings up to minimum standards.

8.12 An appliance manufacturer wishes to study consumers' color preferences for refrigerators.
  a. Identify the population of interest in this study.
  b. Would a sample or a census be preferable to obtain the desired information? Discuss.

8.13 Respond to each of the following arguments.
  a. "Samples are to be preferred to censuses because samples cost less."
  b. "Censuses are to be preferred to samples because censuses are more accurate."

8.14 An information specialist in O.K. Financial Services, Inc., selected six branch offices in which
to test a new computerized decision support system, because these offices represent the variety
of informational requirements found in the company's branch offices throughout the country.
  a. Do these six branch offices represent a probability sample, a judgment sample, or a conve-
     nience sample? Explain.
  b. Would your answer in **a** change if the offices were selected because they were within easy
     driving distance of the firm's home office? Discuss.

8.15 The names of 200 students were selected from a computer file that lists the names of all students
currently attending a university. The selection was done in a manner that gave each student in
the file an equal chance to be selected. Each selected student is now to be interviewed in depth
about the problems of coping with examination stress.

    a. Do the 200 selected students constitute a probability sample or a judgment sample? Explain.

    b. If some of the selected students cannot be contacted for an interview or refuse to be interviewed, will the completed study still be based on the kind of sample indicated in your answer to **a**? Explain.

8.16 An association of professional personnel managers has 7200 members of whom 830 attended the association's annual conference in Chicago. At the conference, each attending member was given a questionnaire concerning preferences for the location, format, and dates of the next conference. A total of 410 of these questionnaires were completed and returned.

    a. Do the respondents constitute a probability sample, a judgment sample, or a convenience sample of the association's membership? Explain.

    b. Do the respondents constitute a probability sample, a judgment sample, or a convenience sample of the members attending the conference in Chicago? Explain.

8.17 **Computer Users.** The computer users of a company have identification codes numbered consecutively from 001 to 330. Ten percent of the users are to be surveyed about their potential use of a new programming package. One of the 10 digits from 0 to 9 will be selected randomly, and every user with an identification code ending with this selected digit will be surveyed.

    a. State the probability that (1) digit 8 will be the one selected, (2) the user with identification code 228 will be selected.

    b. Does the set of users selected for the survey in this manner constitute a probability sample? A simple random sample? Explain.

*8.18 A simple random sample of two stores is to be selected without replacement from a population of four stores (A, B, C, and D).

    a. List the different possible sample combinations. How many are there?

    b. What probability of selection must each sample combination in **a** have?

    c. Must you list all the possible sample combinations in order to select the simple random sample? Explain.

8.19 A simple random sample of three cars is to be selected without replacement from a population of six cars (A, B, C, D, E, and F).

    a. List the different possible sample combinations. How many are there?

    b. What probability of selection must each sample combination in **a** have?

    c. Must you list all the possible sample combinations in order to select the simple random sample? Explain.

8.20 **Retail Establishments.** The retail establishments in a county in existence in March are to be sampled, using a directory published in October of the preceding year.

    a. What is the frame here?

    b. Describe several ways in which the sampled population might differ from the target population.

8.21 The adult population of a city is to be sampled next week. Some persons will be out of town during that time. What is the target population here? What is the sampled population?

8.22 Bulk-food stores that carry a certain line of exotic spices are to be sampled, using a list of all bulk-food stores. Each store selected from the list will be contacted initially to learn if it carries the line of spices. If it does, the store will be kept in the sample. Otherwise, the store will be discarded from the sample. What is the population from which the final sample is drawn here? Is it the same as the target population? Comment.

8.23 Refer to **Computer Users** Problem 8.17.

   a. Explain how you would use a table of random digits to select a simple random sample of 33 computer users without replacement.

   b. If the users' codes were divided into the three groups, 001–110, 111–220, 221–330, and a simple random sample of 11 codes were drawn without replacement and independently from each group, would the combined sample of 33 users constitute a simple random sample of the population of 330 users? Explain.

8.24 Refer to **Retail Establishments** Problem 8.20. The retail establishments directory contains 100 pages. Fifty establishments are listed to a page, except on the last page, where 31 establishments are listed.

   a. Explain how you would use a table of random digits to select a simple random sample of 100 establishments without replacement from the directory.

   b. If a table of random digits were used to select one establishment at random from each page of the directory, would the resultant sample be a probability sample? A simple random sample? Explain.

8.25 Explain how you would select a simple random sample without replacement by means of a table of random digits in each of the following cases.

   a. Fifty school buses from the 4000 school buses registered in a state.

   b. One hundred floor tiles from a warehouse floor laid with 150 rows of such tiles, 70 tiles to the row.

   c. One hundred points in time (designated in minutes) during working hours next week (9:00 A.M.–5:00 P.M. Monday through Friday).

8.26 Explain how you would select a simple random sample without replacement by means of a table of random digits in each of the following cases.

   a. Eighty members of the 760,000 members of an automobile association.

   b. Forty-five time points (designated in minutes) during the 5 hours of a production run.

   c. Five 1-cubic-centimeter cubes from a block of wood measuring (in centimeters) $15 \times 15 \times 120$.

8.27 A research assistant, explaining how he constructed a table of random digits on his own, stated that he never placed the same digit in adjoining positions in the table because such an arrangement would not be random. Comment.

8.28 Book orders leaving a publisher constitute a Bernoulli process with $p = 0.95$ with respect to whether each is correctly filled ($X = 1$) or not ($X = 0$).

   a. Describe the infinite population associated with this process. What are the mean and the standard deviation of this population?

   b. Does a sequence of five consecutive orders constitute a simple random sample from this infinite population? Explain.

8.29 Breakdowns of a computer occur according to a Poisson process with $\lambda = 2.5$ breakdowns per week. The time intervals between consecutive breakdowns are of interest.

   a. Describe the infinite population of interest. What are the mean and the standard deviation of this population?

   b. For 11 consecutive breakdowns, do the 10 time intervals between these breakdowns constitute a simple random sample from this infinite population? Explain.

8.30 The following numbers of wings were found in the five packages labeled "one cut-up frying chicken" in a supermarket on Tuesday morning: 1, 2, 0, 3, 1. Under what circumstances might this data set be viewed (1) as a finite population, (2) as a sample from an infinite population?

EXERCISES

8.31 Show for a population of $N = 4$ distinct elements that a random sample of $n = 2$ elements selected according to (8.9) yields equal probabilities for each possible sample combination. Draw a probability tree representing selection process (8.9) here, and identify the outcomes constituting a given sample combination.

8.32 A simple random sample of $n = 4$ elements is to be drawn without replacement from a population having $N = 11$ distinct elements.
  a. How many different sample combinations of size 4 can be drawn from the population?
  b. What is the probability that a particular one of the sample combinations will be chosen?
  c. What is the probability that a particular population element will appear in the selected random sample?

8.33 A two-digit number is to be selected from a table of random digits. What is the probability that (1) it will be 12, (2) it will be even, (3) the two digits will be the same?

8.34 In a table of random digits, what is the probability that (1) three successive digits are each 2, (2) three successive digits are the same?

8.35 A simple random sample of 200 vouchers is to be selected from a population of 3000 vouchers serially numbered from 0001 to 3000. The analyst plans to select four-digit random numbers. If the first digit is 0, 1, or 2, it will be treated as 0; if it is 3, 4, or 5, it will be treated as 1; if it is 6, 7, or 8, it will be treated as 2; and if it is 9, the four-digit number will be discarded. The other three digits in the number will remain unchanged. Thus, 1245 becomes 0245; 7391 becomes 2391; and 9218 is discarded. The transformed number 0000 will correspond to voucher 3000. Is this a valid way of obtaining a simple random sample in this case? Explain.

8.36 A city planning commission, using an up-to-date listing of all addresses in the city, will select a simple random sample of 300 addresses for a survey of households. The number of households that currently live at each address may be zero, one, two, or more than two.
  a. Suppose that all households living at the 300 addresses will be included in the survey. Does each household in the city have an equal chance of being included? Will the households at the 300 addresses constitute a simple random sample of households in the city? Explain.
  b. Alternatively, suppose that when a selected address contains more than one household, one of those households is chosen at random for inclusion in the survey and the other households at that address are not included. Will the households included in the survey constitute a simple random sample of all households in the city? Explain.

8.37 Explain why simple random sampling with replacement from a finite population is equivalent to simple random sampling from an infinite population, referring to definition (8.11).

STUDIES

8.38 The following 10 hospitals have agreed to participate in a joint research program. The research procedures are to be tested initially in a simple random sample of 3 of these 10 hospitals.

Carmel	Maple Hill	St. Francis
Central Valley	Mercy	Winchester General
Community	Northmount	
Franklin General	Southridge	

a. Using random numbers, select a simple random sample of 3 hospitals without replacement. Explain how you used the random numbers.

b. To study how a different random sample of $n = 3$ hospitals might be composed, select another simple random sample of 3 hospitals without replacement from the 10 hospitals in the population, using random numbers independent of those used in **a**. How much overlap is there between the hospitals in the samples in **a** and **b**? What is the probability of no overlap in the second sample, given the results for your first sample? What is the probability of no overlap in advance of selecting either of the two samples?

c. Continue selecting independent random samples of $n = 3$ hospitals from the population of 10 hospitals until you have selected 50 random samples of $n = 3$. For each hospital in the population, determine the number of samples in which it was selected. What is the expected number for each hospital? Are the actual frequencies close to the expected frequencies? Discuss.

8.39 Refer to the **Financial Characteristics** data set. The population of textile apparel manufacturing firms is to be sampled to obtain information about net sales in year 2. You are asked to take a simple random sample of 10 firms without replacement.

a. How many different possible sample combinations of 10 firms can be selected from this population?

b. Using random numbers, select a random sample of 10 firms. Explain how you used the random numbers.

c. Calculate the sample mean and the sample standard deviation.

d. The population mean is $\mu = \$153.43$ million, and the population standard deviation is $\sigma = \$256.00$ million. Given the large variability in the population [the coefficient of variation is $(256.00/153.43)100 = 167$ percent], do you expect the sample mean and standard deviation based on a sample of 10 firms to be close to the population parameters? How close are your sample statistics in **c** to the corresponding population parameters?

SAMPLING DISTRIBUTION OF $\overline{X}$

We noted in the previous chapter that samples are selected to provide information about population characteristics. Often, the characteristic of interest is the population mean μ.

<table>
<tr><td>EXAMPLES</td><td>

1. The unemployment rate in a western state is high. The governor needs to know whether the unemployed tend to be unemployed for a short time while shifting to another job, or for a long time unable to find a job. The mean duration of unemployment (μ) in the state during the week of January 12 is to be estimated from a random sample of unemployed persons.

2. An environmental control agency requires information about the mean pollution emission per car (μ) for a particular make of passenger car. A random sample of cars from the production line for this make of automobile will be selected and tested.

3. An auditor needs to estimate the population mean audit amount per account (μ) in a population of N accounts from a random sample of accounts. The total audit amount for all accounts in the population is given by $N\mu$, which can be estimated once the estimate of μ is in hand. □
</td></tr>
</table>

These examples illustrate the variety of problems where interest exists in the population mean μ. The sample mean $\overline{X}$ provides information about the population mean μ. Intuition tells us, however, that if we were to take a number of simple random samples of given size from a population and calculate the sample mean $\overline{X}$ for each, we would obtain different values of $\overline{X}$ for the several samples. The reason is that different samples usually will include different population elements. While, in practice, we select only a single sample, knowledge about the behavior of the sample statistic $\overline{X}$ in repeated samples from a population will enable us to assess and control the margin of error due to sampling for any given sample.

In this chapter, we study, by means of an experiment, the behavior of the sample mean $\overline{X}$ as it varies from one sample to another. Then, we present relevant results from statistical theory and show how the experiment illustrates the theoretical results.

9.1 **STUDY OF BEHAVIOR OF $\overline{X}$ BY EXPERIMENTATION**

Population

An auditor conducted an experiment involving the 8042 freight accounts receivable of a freight company. He wished to demonstrate the use of sampling to estimate the mean audit account balance in the population—that is, the mean audit amount receivable per account for the 8042 accounts. The audit amount may differ from the book amount for various reasons, including use of improper tariff, incorrect calculations, and dispute over damages. For purposes of this experiment, each of the 8042 accounts receivable was initially audited. Ordinarily, the auditor would only audit a sample of these accounts and base the conclusions on the evaluation of the sample results.

Table 9.1 contains a frequency distribution of the audit amounts of the 8042 accounts in the population. It also presents the population mean and standard deviation calculated from the actual 8042 observations. Figure 9.1 shows the frequency polygon of the distribution. The distribution is substantially skewed to the right, which is typical of many populations in business and economics.

Simple Random Sampling with $n = 3$

We begin the study of the behavior of the sample mean $\overline{X}$ by starting with a very small sample size, $n = 3$. A simple random sample of three accounts was selected from the population, and the audit amounts shown in the first row of Table 9.2 were obtained.

TABLE 9.1   Frequency distribution of the audit amounts of 8042 accounts receivable

Audit Amount ($)	Number of Accounts		
0–under 10	1008		
10–under 20	2856		
20–under 30	1926		
30–under 40	780		
40–under 50	326		
50–under 60	258	Population mean	$\mu = 30.303$
60–under 70	160	Population standard deviation	$\sigma = 30.334$
70–under 80	148		
80–under 90	99		
90–under 100	116		
100–under 150	244		
150–under 200	121		
Total	8042		

FIGURE 9.1   **Frequency polygon of the audit amounts of 8042 accounts receivable.**   The population is highly skewed to the right.

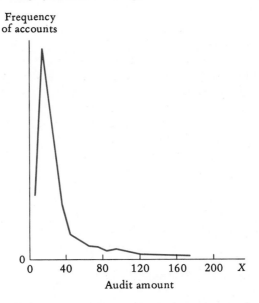

TABLE 9.2   **Results for the first five random samples of** $n = 3$

Sample	Observation 1	Observation 2	Observation 3	$\overline{X}$
1	30.96	38.20	22.45	30.537
2	18.91	6.75	15.45	13.703
3	10.60	14.08	9.15	11.277
4	51.82	20.76	50.79	41.123
5	23.05	31.20	25.15	26.467

The sample mean for this sample is:

$$\overline{X} = \frac{\Sigma X_i}{n} = \frac{30.96 + 38.20 + 22.45}{3} = 30.537$$

Additional random samples of size $n = 3$ were selected, always from the full population of 8042 accounts. Table 9.2 contains the results for five of these samples.

It is clear from Table 9.2 that all five sample means differ from one another and that none of them is equal to the population mean $\mu = 30.303$. In fact, some of the sample means differ substantially from the population mean. The departure of the sample mean from the population mean is called the sampling error.

(9.1)

> *Sampling error* is the difference between the result obtained from a sample and the result that would be obtained from a census of the population elements conducted under the same procedures as the sample.

Sampling error is distinguished from errors in data (discussed in Chapter 1) in that sampling error is due to sampling. In contrast, data errors — such as those due to faulty recall by a respondent or calculational errors by an auditor — can be present whether a census or a sample is conducted. Such errors often are called *nonsampling errors*.

Altogether, 600 samples of size $n = 3$ were selected for this experiment. Table 9.3 contains a frequency distribution of the 600 sample means, as well as the mean and standard deviation of the 600 $\overline{X}$ values. Figure 9.2a shows the frequency polygon of the 600 $\overline{X}$ values. Three important results emerge from Table 9.3 and Figure 9.2a.

1. While the 600 sample means are widely divergent, their mean of 30.68 is close to the population mean $\mu = 30.303$.
2. The standard deviation of the 600 $\overline{X}$ values, which is 17.60, shows that the variability of the sample means is substantially smaller than the variability of the audit account balances in the population ($\sigma = 30.334$). In fact, the standard deviation

TABLE 9.3   Frequency distribution of 600 $\overline{X}$ values; $n = 3$

Sample Mean	Number of Samples		
Under 5.80	1		
5.80–under 12.80	18		
12.80–under 19.80	188		
19.80–under 26.80	132		
26.80–under 33.80	77		
33.80–under 40.80	52	Mean of 600 $\overline{X}$ values	30.68
40.80–under 47.80	30	Standard deviation of 600 $\overline{X}$ values	17.60
47.80–under 54.80	29		
54.80–under 61.80	20		
61.80–under 68.80	26		
68.80–under 75.80	12		
75.80–under 82.80	9		
82.80 and more	6		
Total	600		

FIGURE 9.2 Frequency polygons of 600 $\overline{X}$ values for samples of $n = 3$ and $n = 10$ accounts receivable. As the sample size was increased, the sampling distribution of $\overline{X}$ has remained centered at μ and has become less variable and less skewed.

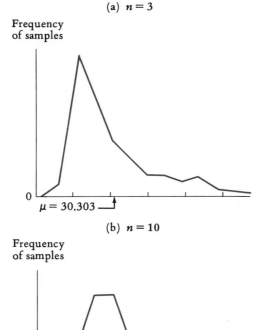

of the 600 $\overline{X}$ values is only about six-tenths as great as the population standard deviation.

3. The distribution of the 600 sample means is skewed to the right, like the population, but not as much.

Simple Random Sampling with $n = 10$

Next in the experiment, 600 random samples of size $n = 10$ were selected from the population of 8042 accounts receivable. Table 9.4 presents the frequency distribution

TABLE 9.4   Frequency distribution of 600 $\overline{X}$ values; $n = 10$

Sample Mean	Number of Samples
12.80–under 19.80	71
19.80–under 26.80	170
26.80–under 33.80	173
33.80–under 40.80	108
40.80–under 47.80	51
47.80–under 54.80	21
54.80–under 61.80	5
61.80–under 68.80	1
Total	600

Mean of 600 $\overline{X}$ values   30.23
Standard deviation of 600 $\overline{X}$ values   9.13

of the 600 $\overline{X}$ values, as well as the mean and standard deviation of the 600 sample means. Figure 9.2b shows the frequency polygon of the 600 $\overline{X}$ values. Four important results are evident.

1. The 600 sample means for sample size $n = 10$ have a mean of 30.23, which is close to the population mean $\mu = 30.303$. This was also true for sample size $n = 3$.
2. The standard deviation of the 600 $\overline{X}$ values, which is 9.13, is substantially smaller than the population standard deviation $\sigma = 30.334$. In fact, it is only about three-tenths as large.
3. The 600 $\overline{X}$ values for sample size $n = 10$ are less variable than the 600 $\overline{X}$ values for sample size $n = 3$. The respective standard deviations are 9.13 and 17.60.
4. The distribution of the 600 $\overline{X}$ values for sample size $n = 10$ is only moderately skewed to the right, in contrast to the marked positive skewness for the population.

Simple Random Sampling with $n = 100$

The final part of the experiment was to select 600 random samples of size $n = 100$. Table 9.5 and Figure 9.3 contain the relevant results. The following conclusions are of interest.

1. The mean of the 600 sample means, 30.31, again is close to the population mean $\mu = 30.303$.
2. The standard deviation of the 600 $\overline{X}$ values is smaller than that for both sample sizes $n = 3$ and $n = 10$ and is only about one-tenth as great as the population standard deviation.
3. The distribution of the 600 $\overline{X}$ values is quite symmetrical and appears, when smoothed, similar to a normal distribution.

TABLE 9.5   Frequency distribution of 600 $\overline{X}$ values; $n = 100$

Sample Mean	Number of Samples
22.80–under 25.80	39
25.80–under 28.80	157
28.80–under 31.80	213
31.80–under 34.80	148
34.80–under 37.80	37
37.80–under 40.80	5
40.80–under 43.80	1
Total	600

Mean of 600 $\overline{X}$ values   30.31
Standard deviation of 600 $\overline{X}$ values   3.05

FIGURE 9.3   Frequency polygon of 600 $\overline{X}$ values for samples of $n = 100$ accounts receivable.   As the sample size has become large, the sampling distribution of $\overline{X}$ has approached a normal distribution.

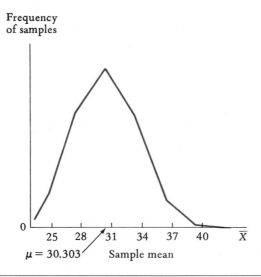

Empirical Conclusions

On the basis of this experiment, the following conjectures appear to be warranted:

1. The distribution of $\overline{X}$ values for simple random sampling is centered around the population mean, regardless of sample size.

2. The standard deviation of the $\overline{X}$ values decreases with increasing sample size — that is, the distribution of $\overline{X}$ values becomes more concentrated around the population mean as the sample size gets larger.
3. The distribution of the $\overline{X}$ values becomes more symmetrical as the sample size gets larger and is approximately normal for large sample sizes.

9.2   THEORY ABOUT BEHAVIOR OF $\overline{X}$

We now turn to theoretical results about the behavior of the sample mean $\overline{X}$ illustrated by the previous experiment. These results are applicable for infinite populations and for finite populations whenever the sample size n is a small portion of the population size N. This latter condition applies to the experimental study where $N = 8042$ and n varied from 3 to 100.

Sampling Distribution of $\overline{X}$

When a simple random sample of given size is selected from a population, the actual sample obtained is the result of a random trial, that is, the random selection process. Hence, the observed sample mean $\overline{X}$ is also the outcome of a random trial. In advance of sampling, therefore, the sample mean $\overline{X}$ is a random variable and has an associated probability distribution.

(9.2)   The probability distribution associated with the random variable $\overline{X}$ is called the *sampling distribution of $\overline{X}$* or the *sampling distribution of the mean.*

In the experiment, the 600 $\overline{X}$ values for a given sample size n represent 600 observations from the sampling distribution of $\overline{X}$ for that sample size.

There is a different sampling distribution of $\overline{X}$ for each population and random sample size. We now consider the characteristics of sampling distributions of $\overline{X}$ and how they relate to the population sampled and the sample size.

Expected Value of $\overline{X}$

The expected value of $\overline{X}$, or, equivalently, the mean of the sampling distribution of $\overline{X}$, is always equal to the population mean μ with simple random sampling.

(9.3)   For all populations and sample sizes, when simple random sampling is employed:

$$E\{\overline{X}\} = \mu$$

The results of the experiment certainly are consistent with theorem (9.3).

Sample Size	Mean of 600 $\overline{X}$ Values
3	30.68
10	30.23
100	30.31

Population mean $\mu = 30.303$

Of course, we would not expect the mean of 600 trials to equal μ exactly, since there is always some random variation present in a limited number of trials.

Comment   Theorem (9.3) can be proven readily when a simple random sample is selected from an infinite population. In that case, the expected value of any sample observation is $E\{X_i\} = \mu$, and we have:

$$E\{\overline{X}\} = E\left\{\frac{X_1 + X_2 + \cdots + X_n}{n}\right\} = \frac{1}{n}E\{X_1 + X_2 + \cdots + X_n\} \qquad \text{by (5.11b)}$$

$$= \frac{1}{n}(E\{X_1\} + E\{X_2\} + \cdots + E\{X_n\}) = \frac{n\mu}{n} = \mu \qquad \text{by (5.15a)}$$

Variance of $\overline{X}$

The variance of $\overline{X}$, or, equivalently, the variance of the sampling distribution of $\overline{X}$, is denoted by $\sigma^2\{\overline{X}\}$. It is often also called the *variance of the mean*. Similarly, the standard deviation of $\overline{X}$ is denoted by $\sigma\{\overline{X}\}$ and is often called the *standard deviation of the mean* or the *standard error of the mean*.

The variance $\sigma^2\{\overline{X}\}$ and the standard deviation $\sigma\{\overline{X}\}$ measure the variability of the possible sample means $\overline{X}$ that can be obtained with simple random sampling from a population. It turns out that these measures depend on the population variance σ^2 and sample size n in a simple way.

(9.4)   When simple random sampling is employed:

$$\sigma^2\{\overline{X}\} = \frac{\sigma^2}{n} \qquad \sigma\{\overline{X}\} = \frac{\sigma}{\sqrt{n}}$$

The results of the experiment certainly are consistent with theorem (9.4).

Sample Size n	Standard Deviation of 600 $\overline{X}$ Values	$\dfrac{\sigma}{\sqrt{n}}$
3	17.6	17.5
10	9.1	9.6
100	3.0	3.0

Population standard deviation $\sigma = 30.334$

Again, we must anticipate some variation from theoretical expectations in the results of a limited number of trials.

Effect of Sample Size.   Theorem (9.4) indicates that the standard deviation of $\overline{X}$ decreases proportionately with the square root of the sample size. Thus, the sampling distribution of $\overline{X}$ will be more concentrated the larger the sample size. As we shall see, this observation is the counterpart to our intuition that larger samples lead to more precise results.

However, since the standard deviation of $\overline{X}$ decreases proportionately with the *square root* of the sample size, it becomes increasingly difficult to reduce $\sigma\{\overline{X}\}$ by increasing n. Thus, if the standard deviation of $\overline{X}$ based on sample size $n = 100$ is to be cut in half, the sample size must be quadrupled to $n = 400$. In turn, if that standard deviation is to be cut in half, the sample size must be quadrupled to $n = 1600$.

Effect of Population Variability.   Theorem (9.4) indicates that, for any given sample size, the greater the population variability, the larger is the variability of the sampling distribution of $\overline{X}$. Thus, for any given n, $\overline{X}$ will tend to vary more about the population mean μ when the population is highly variable than when it is concentrated.

Comment   Theorem (9.4) can be readily proven for infinite populations. Simple random sampling from infinite populations requires independent and identically distributed observations, the variance of each being $\sigma^2\{X_i\} = \sigma^2$. Hence, we obtain:

$$\sigma^2\{\overline{X}\} = \sigma^2\left\{\frac{X_1 + X_2 + \cdots + X_n}{n}\right\} = \frac{1}{n^2}\sigma^2\{X_1 + X_2 + \cdots + X_n\} \qquad \text{by (5.12b)}$$

$$= \frac{1}{n^2}(\sigma^2\{X_1\} + \sigma^2\{X_2\} + \cdots + \sigma^2\{X_n\}) = \frac{n\sigma^2}{n^2} = \frac{\sigma^2}{n} \qquad \text{by (5.15b)}$$

Central Limit Theorem

One of the most important theorems in statistics is the central limit theorem. This theorem is important because from it we know the nature of the sampling distribution of $\overline{X}$ (approximately normal) when the random sample size is reasonably large, for almost any population.

(9.5)   *Central Limit Theorem.*   For almost all populations, the sampling distribution of $\overline{X}$ is approximately normal when the simple random sample size is sufficiently large.

The experimental results are consistent with this theorem. While the distribution of the 600 $\overline{X}$ values is skewed for $n = 3$, it is less skewed for $n = 10$ and is almost symmetrical for $n = 100$. Furthermore, the appearance of the distribution of the $\overline{X}$

values in Figure 9.3 for $n = 100$ suggests normality. To study this aspect more formally, recall that the population mean and standard deviation are $\mu = 30.303$ and $\sigma = 30.334$. Hence, we know that the sampling distribution of $\overline{X}$ for $n = 100$ has the following characteristics, according to (9.3) and (9.4):

$$E\{\overline{X}\} = \mu = 30.303 \qquad \sigma\{\overline{X}\} = \frac{\sigma}{\sqrt{n}} = \frac{30.334}{\sqrt{100}} = 3.033$$

If the sampling distribution of $\overline{X}$ for $n = 100$ is approximately normal, we can determine from Table C.1 the proportion of $\overline{X}$ values expected to fall in any interval around the population mean. Here are some results for the experiment with $n = 100$.

	Experimental Results		Theoretical Expectations
Interval	Number of $\overline{X}$ Values in Interval	Proportion of $\overline{X}$ Values in Interval	Normal Probability Based on Central Limit Theorem
30.30 ± 3.50	451	0.752	0.750
30.30 ± 5.50	559	0.932	0.930
30.30 ± 7.50	594	0.990	0.986
30.30 ± 9.50	599	0.998	0.998

The close accord between the experimental results and the theoretical expectations based on the central limit theorem strongly supports the applicability of the central limit theorem here.

Comments

1. What is a sufficiently large sample size in order for the central limit theorem to apply depends on the nature of the population and on the degree of approximation to the normal distribution required. In general, for skewed populations such as the one in the experiment, a larger random sample size is required for the sampling distribution of $\overline{X}$ to be approximately normal than is required for a population that is fairly symmetrical.

2. The central limit theorem applies whenever the population standard deviation σ is defined, which is the case in almost all problems encountered in practice.

3. The theoretical expectations for the experiment were calculated by using the standard deviation of the sampling distribution of $\overline{X}$ for $n = 100$, namely, $\sigma\{\overline{X}\} = 3.033$. For instance, in the first interval, the deviation 3.50 represents $3.50/3.033 = 1.15$ standard deviations. We find from Table C.1 that $P(-1.15 \le Z \le 1.15) = 0.750$.

Sampling Finite Populations

As we noted earlier, theorems (9.3) and (9.4) for the mean and the variance of the sampling distribution of $\overline{X}$ and central limit theorem (9.5) apply to the sampling of infinite and finite populations, provided in the latter case that the sample size n is not large relative to the finite population size N. The ratio n/N is referred to as the *sampling fraction*. In practice, the sampling fraction is usually small when finite populations are sampled. Thus, in the audit sampling experiment, the sampling fraction for

$n = 100$ was only $n/N = 100/8042 = 0.012$. Similarly, when a sample of $n = 1500$ households is selected from the population of a state containing $N = 1,000,000$ households, the sampling fraction is only $n/N = 1500/1,000,000 = 0.0015$. Even when a sample of $n = 75$ retail outlets is selected from the $N = 1800$ retail outlets of a chain, the sampling fraction is only 0.042.

Throughout this text, unless otherwise noted, we shall assume that finite population sizes are large relative to the sample sizes. In Chapter 14, we shall take up the special case where this condition does not hold.

9.3 PROBABILITY STATEMENTS ABOUT $\overline{X}$ USING NORMAL APPROXIMATION

Central limit theorem (9.5) can be utilized to make probability statements about $\overline{X}$ for reasonably large sample sizes. Use of the standard normal distribution in Table C.1 in this case requires the following standardized variable.

(9.6)
$$Z = \frac{\overline{X} - E\{\overline{X}\}}{\sigma\{\overline{X}\}} = \frac{\overline{X} - \mu}{\sigma/\sqrt{n}}$$

This variable corresponds to the earlier definition (7.9) of a standard normal variable, except that the variable standardized now is $\overline{X}$ instead of X.

We illustrate use of the central limit theorem to make probability statements about $\overline{X}$ with three examples.

EXAMPLES

1. Suppose that the auditor in the earlier illustration was actually planning to use a simple random sample of 250 accounts receivable. Invoking the central limit theorem, he would expect that the sampling distribution of $\overline{X}$ is approximately normal for that sample size. Theorem (9.3) indicates that the mean of the sampling distribution of $\overline{X}$ is equal to the population mean $\mu = 30.303$, and theorem (9.4) states that the standard deviation of the sampling distribution of $\overline{X}$ is:

$$\sigma\{\overline{X}\} = \frac{\sigma}{\sqrt{n}} = \frac{30.334}{\sqrt{250}} = 1.92$$

The auditor would like to know the probability that the sample mean will be within \$4 of the population mean audit amount, that is, between 26.30 and 34.30. Figure 9.4a shows the sampling distribution of $\overline{X}$ as approximated by the normal distribution. The shaded area represents the desired probability. We calculate the z values for the specified $\overline{X}$ values by (9.6) in the usual fashion:

$$z = \frac{26.30 - 30.30}{1.92} = -2.08 \qquad z = \frac{34.30 - 30.30}{1.92} = 2.08$$

We find that $P(26.30 \le \overline{X} \le 34.30) = P(-2.08 \le Z \le 2.08) = 0.96$.

FIGURE 9.4   **Sampling distributions of $\overline{X}$ for $n = 250$ and $n = 100$**

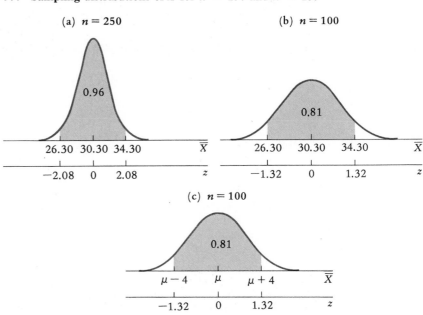

Thus, almost always the auditor will obtain a sample mean that is within $4 of the population mean. In ordinary practice, of course, the auditor takes only one random sample of 250 accounts, on the basis of which he estimates the population mean audit amount. Since the probability is so high that the sample mean is within $4 of the population mean for a sample of 250 accounts, the auditor has, by this knowledge, a measure of the margin of error due to sampling for the particular sample value of $\overline{X}$ that he will obtain.

In this way, the sampling distribution of $\overline{X}$ provides information about the margin of error due to sampling for the one sample result that is actually obtained—namely, by indicating the probability that the sample mean $\overline{X}$ will be within a given proximity of the population mean μ.

2. To study the effect of sample size, let us find the same probability as in Example 1 for a random sample of 100 accounts. Again, the sampling distribution of $\overline{X}$ would be approximately normal according to the central limit theorem. The mean of the sampling distribution of $\overline{X}$ would still be $\mu = 30.303$, because it is unaffected by sample size. The standard deviation of the sampling distribution of $\overline{X}$ would, however, be larger:

$$\sigma\{\overline{X}\} = \frac{30.334}{\sqrt{100}} = 3.03$$

Figure 9.4b shows the effect of the smaller sample size on the sampling distribution of $\overline{X}$. It is more variable than the distribution for the larger sample size shown in Figure 9.4a. Hence, the probability of $\overline{X}$ falling in a given interval around the population mean μ is smaller for $n = 100$ than for $n = 250$. Specifically, let us find the probability that $\overline{X}$ falls within \$4 of μ:

$$z = \frac{26.30 - 30.30}{3.03} = -1.32 \qquad z = \frac{34.30 - 30.30}{3.03} = 1.32$$

Hence, $P(26.30 \leq \overline{X} \leq 34.30) = P(-1.32 \leq Z \leq 1.32) = 0.81$. Recall that the corresponding probability for $n = 250$ is 0.96. The larger probability for the larger sample size corresponds to our intuition that the sample mean is more precise for larger samples.

3. Suppose that the auditor does not know the population mean μ but wishes to know the probability that the sample mean $\overline{X}$ based on $n = 100$ will fall within \$4 of μ. From Example 2, we know that $\sigma\{\overline{X}\} = 3.03$ for this sample size. Figure 9.4c portrays the situation. The z values are the same as for Example 2:

$$z = \frac{(\mu - 4) - \mu}{3.03} = -1.32 \qquad z = \frac{(\mu + 4) - \mu}{3.03} = 1.32$$

Hence, the probability that $\overline{X}$ based on a random sample of size $n = 100$ will fall within \$4 of the population mean μ is 0.81.   □

Comment   In actual practice, the population mean and standard deviation, which are required for obtaining the mean and the standard deviation of the sampling distribution of $\overline{X}$, are usually unknown. We shall explain in the next chapter how the sample mean $\overline{X}$ and the sample standard deviation s can be used to estimate the population mean μ and the population standard deviation σ, respectively, and how these estimates in turn enable us to estimate the mean and the standard deviation of the sampling distribution of $\overline{X}$.

 EXACT SAMPLING DISTRIBUTION OF $\overline{X}$

The central limit theorem tells us the approximate nature of the sampling distribution of $\overline{X}$ when the sample size is reasonably large. In certain cases, the exact sampling distribution of $\overline{X}$ can be ascertained.

Normal Population

When the population sampled is normal, the following theorem provides us with the exact sampling distribution of $\overline{X}$.

(9.7)

> When the population sampled is a normal probability distribution, the sampling distribution of $\overline{X}$ is exactly normal for *any* random sample size.

We shall utilize this theorem later when we consider sampling of normal populations.

Other populations for which statistical theory permits us to derive the exact sampling distribution of $\overline{X}$ include exponential and Poisson populations.

Construction by Enumeration

For simple discrete populations, the exact sampling distribution of $\overline{X}$ can be derived by enumeration of all possible samples. This procedure is tedious, however, and is rarely used. We present a very simple example to illustrate the basic principles.

EXAMPLE

The number of microcomputers sold in week 1 is denoted by X_1 and the number sold in week 2 by X_2. The random variables X_1 and X_2 are independent and have the same probability distribution.

x:	2	4
$P(x)$:	0.9	0.1

For this probability distribution, $E\{X\} = 2.2$ and $\sigma\{X\} = 0.6$ (calculations not shown). We wish to obtain the exact sampling distribution of $\overline{X}$, the mean number of microcomputers sold in the two weeks.

We list all possible sample outcomes and proceed as follows:

Sample Outcome X_1:	2	2	4	4
X_2:	2	4	2	4
Sample Mean $\overline{X}$:	2	3	3	4
Probability:	0.81	0.09	0.09	0.01

The probabilities are obtained by theorem (5.4) because X_1 and X_2 are independent random variables. Thus, $P(X_1 = 2 \cap X_2 = 2) = P(X_1 = 2)P(X_2 = 2) = 0.9(0.9) = 0.81$.

We can now obtain the exact sampling distribution of $\overline{X}$ for $n = 2$.

$\overline{X}$:	2	3	4
$P(\overline{X})$:	0.81	0.18	0.01

Note that $P(\overline{X} = 3) = 0.09 + 0.09 = 0.18$ because two sample outcomes yield this value of $\overline{X}$.

According to theorems (9.3) and (9.4), respectively, this sampling distribution of $\overline{X}$ has mean $E\{\overline{X}\} = \mu = 2.2$ and standard deviation $\sigma\{\overline{X}\} = \sigma/\sqrt{n} = 0.6/\sqrt{2} = 0.42$. We confirm these results by direct calculation of the mean and the standard deviation from the preceding sampling distribution, using (5.5) and (5.6):

$$E\{\overline{X}\} = 2(0.81) + 3(0.18) + 4(0.01) = 2.2$$

$$\sigma^2\{\overline{X}\} = (2 - 2.2)^2(0.81) + (3 - 2.2)^2(0.18) + (4 - 2.2)^2(0.01) = 0.18$$

Hence, $\sigma\{\overline{X}\} = \sqrt{0.18} = 0.42$, as anticipated. $\square$

PROBLEMS

*9.1 Refer to Table 9.2 for the accounts receivable example. The population mean audit amount is $\mu = 30.303$ for the $N = 8042$ accounts. What is the magnitude of the sampling error in $\overline{X}$ for the first sample?

9.2 A sample of 300 families was selected from the population of all families living in a city. The reported sample mean expenditures on children's footwear per family during the past 12 months were $157. Suppose that a census of this population, using the same survey methodology, would have yielded mean expenditures on children's footwear per family of $175.
  a. What is the magnitude of the sampling error in the sample mean?
  b. The sample responses in this survey were based on the unaided recalls of expenditures by respondents and in some instances were in error. If, in fact, the actual mean expenditures per family for the 300 sample families, based on accurate expenditures records, were $168, what is the magnitude of the nonsampling error in the reported sample mean?

9.3 The mean net weight of a random sample of five cartons of milk from a large shipment, weighed on an uncalibrated scale, is 1.26 kilograms. The same cartons have a mean weight of 1.20 kilograms when weighed on an accurate scale. Is the difference between 1.26 and 1.20 a sampling error or a nonsampling error? Explain.

9.4 A student, on studying sampling, was puzzled about how there can be a sampling distribution of $\overline{X}$ when, in practice, only a single sample is selected. Explain to this student the concept of a sampling distribution and its significance.

9.5 A statistics teacher used a computer routine to generate 1000 independent simple random samples, each of size 30, from an infinite theoretical population. The routine also computed the sample mean for each of the 1000 samples. The teacher then constructed a histogram of the 1000 sample means and stated that the histogram is "a graphical representation of the sampling distribution of the sample mean with a sample size of 30 for this population." Is the teacher's statement correct? Explain.

*9.6 **Brick Production.**   Bricks are produced with mean weight $\mu = 1.74$ kilograms and standard deviation $\sigma = 0.03$ kilogram. What are the mean and the standard deviation of the sampling distribution of $\overline{X}$ for a random sample of size (1) $n = 10$, (2) $n = 50$?

9.7 Scholastic aptitude scores of college-bound high school seniors have a mean of $\mu = 450$ and a standard deviation of $\sigma = 75$. What are the mean and the standard deviation of the sampling distribution of $\overline{X}$ for a random sample of size (1) $n = 100$, (2) $n = 40$?

9.8 **Bean Growth.**   The mean time from germination to harvest under laboratory conditions for a certain variety of bean is $\mu = 61$ days, and the standard deviation is $\sigma = 3$ days. What are the mean and the standard deviation of the sampling distribution of $\overline{X}$ for a random sample of size (1) $n = 8$, (2) $n = 20$?

9.9 Weekly soft drink purchases by teenagers have a mean of $\mu = \$7.50$ and a standard deviation of $\sigma = \$4.80$. What are the mean and the standard deviation of the sampling distribution of $\overline{X}$ for a random sample of size (1) $n = 121$, (2) $n = 49$?

9.10 Refer to **Brick Production** Problem 9.6.

    a. For which sample size ($n = 10$ or $n = 50$) is it more likely that the sample mean is within ± 0.01 kilogram of μ? Why?

    b. If a random sample of 10 bricks were selected from the process, and independently another random sample of 50 bricks were selected, would the sample mean based on the larger sample necessarily be closer to $\mu = 1.74$ kilograms? Discuss.

9.11 Refer to **Bean Growth** Problem 9.8.

    a. How many seedlings would be required in the sample if the standard error of the mean is to be $\sigma\{\overline{X}\} = 0.3$?

    b. What would be the answer in **a** if the population standard deviation were $\sigma = 6$ days instead of $\sigma = 3$ days? What is the effect of larger population variability?

9.12 Refer to Figure 9.2. The frequency polygons for $n = 3$ and $n = 10$ are located in almost identical positions on the $\overline{X}$ scale, but they differ in variability. What theorems are illustrated by this comparison?

9.13 An analyst remarked: "When the sample size is large, the sample mean is as likely to lie above the population mean as below it." Do you agree? Explain.

9.14 Refer to Table 9.1. A random sample of 200 accounts is to be selected from this population of accounts receivable.

    a. Obtain the mean and the standard deviation of the sampling distribution of $\overline{X}$.

    b. What is the approximate functional form of the sampling distribution?

9.15 On hearing an explanation of the central limit theorem, a listener states: "The theorem guarantees that the n values in a random sample will be approximately normally distributed if n is sufficiently large." Does the listener understand the central limit theorem? Explain.

*9.16 A sample of $n = 3000$ persons is to be taken from a population of 1,000,000 persons in the labor force. What is the sampling fraction here? Is it large or small?

9.17 A sample of $n = 200$ parts is to be taken from a lot of 50,000 parts. What is the sampling fraction here? Is it large or small?

*9.18 Consumers are to be asked to rate the quality of products of a major corporation on a four-point scale from 1 (very low quality) to 4 (very high quality). The population mean and standard deviation are $\mu = 2.30$ points and $\sigma = 0.75$ point.

    a. For a random sample of 150 consumers, find the probability that the sample mean will (1) exceed 2.15, (2) lie within ± 0.15 of μ.

    b. For a random sample of 150 consumers, within what interval centered around μ will $\overline{X}$ fall with probability 0.90? What is the corresponding interval when $n = 300$? Is the latter interval half as wide as the former? Comment.

9.19 A large population of music stores is to be sampled to estimate the mean loss per store due to theft during the past month. The population mean and standard deviation are $\mu = \$380$ and $\sigma = \$210$.

    a. For a random sample of 250 stores, what is the probability that the sample mean will be within $\pm\$15$ of the population mean?

    b. Would the probability in **a** be increased by 20 percent if the sample size were increased by 20 percent, to $n = 300$? Be specific.

   c. Would the probability in **a** be affected substantially if the population standard deviation were $\sigma = \$300$, for $n = 250$?

9.20 An automatic machine cuts ribbons of hot steel into bars of specified length. The variability in the lengths of the bars is $\sigma = 0.07$ meter. A random sample of 60 bars is to be selected. Assume that the central limit theorem applies.
   a. What is the probability that the sample mean $\overline{X}$ will not differ from the process mean μ by more than ± 0.01 meter?
   b. What is the probability that the sample mean will exceed the population mean?

9.21 Refer to **Bean Growth** Problem 9.8. A random sample of $n = 100$ seedlings is planted in the laboratory.
   a. What is the probability that the sample mean time from germination to harvest will lie within ± 0.5 day of the population mean?
   b. What would be the answer in **a** if the population standard deviation were $\sigma = 6$ days instead of $\sigma = 3$ days? What is the effect of larger population variability?

9.22 For each of the following cases, indicate whether the functional form of the sampling distribution of $\overline{X}$ is exactly normal, approximately normal, or not approximately normal, for sample sizes (1) $n = 3$, (2) $n = 300$.
   a. The mean duration of a random sample of satellite transmissions. The duration of satellite transmissions follows an exponential distribution.
   b. The mean weight of a random sample of rough gems. The population of gem weights is normally distributed.

9.23 For each of the following cases, indicate whether the functional form of the sampling distribution of $\overline{X}$ is exactly normal, approximately normal, or not approximately normal, for sample sizes (1) $n = 5$, (2) $n = 200$.
   a. The mean useful life of a random sample of drive chains. The useful life of drive chains is normally distributed.
   b. The mean number of interruptions per hour in an assembly line in a random sample of 1-hour intervals. Interruptions per hour in an assembly line follow a Poisson process with a mean of $\lambda = 0.2$ interruption per hour.

*9.24 Refer to **Brick Production** Problem 9.6. The weight of a brick (X) is normally distributed.
   a. For a random sample of 16 bricks, what is the probability that $\overline{X}$ will be within ± 0.01 kilogram of the population mean?
   b. For a random sample of 16 bricks, what is the interval centered around μ within which $\overline{X}$ will fall 95 percent of the time?
   c. If the population standard deviation were $\sigma = 0.07$ kilogram, what would be the interval in **b**? What is the effect of increased population variability?

9.25 The weight of solvent in a steel drum (X) is normally distributed, with mean $\mu = 69$ kilograms and standard deviation $\sigma = 0.9$ kilogram.
   a. For a random sample of six drums, what is the functional form of the sampling distribution of $\overline{X}$? What is the probability that the sample mean will not exceed 70 kilograms?
   b. For a random sample of six drums, what is the interval centered around μ within which $\overline{X}$ will fall 99 percent of the time?
   c. If the sample size were nine drums, what would be the interval in **b**? What is the effect of increased sample size?

9.26 The scores of subjects on an aptitude test (X) are normally distributed, with mean $\mu = 200$ points and standard deviation $\sigma = 36$ points.

a. For a random sample of nine subjects, what are the mean, the standard deviation, and the functional form of the sampling distribution of $\overline{X}$?

b. For a random sample of nine subjects, what is the probability that $\overline{X}$ will be within ± 25 points of μ? What is the probability that the score of a single subject will be within ± 25 points of μ?

c. Explain why a sample mean is more likely to fall within a given interval about the population mean than a single observation.

*9.27 The probability distribution of X, the number of computer breakdowns per week, follows.

x:	0	1	2
$P(x)$:	0.70	0.20	0.10

a. Obtain the mean and the standard deviation of this infinite population.

b. Obtain the exact sampling distribution of $\overline{X}$ for a random sample of $n = 2$ weeks.

c. Calculate the mean and the standard deviation of the sampling distribution of $\overline{X}$ in **b**, and verify that they agree with theorems (9.3) and (9.4).

9.28 The probability distribution of the number of rework cycles (X) before a machine component passes quality specifications or is scrapped has the following probability distribution.

x:	0	1	2
$P(x)$:	0.95	0.04	0.01

a. Obtain $E\{X\}$ and $\sigma\{X\}$.

b. Obtain the exact sampling distribution of $\overline{X}$ for a random sample of $n = 2$ components.

c. Calculate $E\{\overline{X}\}$ and $\sigma\{\overline{X}\}$ for the sampling distribution in **b**, and verify that they agree with theorems (9.3) and (9.4).

EXERCISES

9.29 Let τ denote a finite population total; that is:

$$\tau = \sum_{i=1}^{N} X_i = N\mu$$

Consider the sampling distribution of random variable $T = N\overline{X}$ when the sampling fraction n/N is small. Derive $E\{T\}$ and $\sigma\{T\}$, using (9.3) and (9.4).

9.30 The standard deviation of a population is $\sigma = 10$.

a. Calculate $\sigma\{\overline{X}\}$ for $n = 2, 5, 10, 20, 50, 100$.

b. Plot the $\sigma\{\overline{X}\}$ values obtained in **a** against n on a graph. Describe the effect of increasing the sample size on the variability of the sampling distribution of $\overline{X}$.

9.31 A population is normal with $\mu = 200$ and $\sigma = 20$.

a. Calculate $P(195 \leq \overline{X} \leq 205)$ for $n = 4, 10, 20, 50, 100$.

b. Plot the probabilities obtained in **a** against n on a graph. Describe the effect of increasing the sample size on the probability that $\overline{X}$ will be within ± 5 of the population mean μ.

   c. Calculate $P(195 \leq \overline{X} \leq 205)$ for $n = 20$ if $\sigma = 2, 4, 12, 16, 20, 30$. Plot the probabilities against σ on a graph, and describe the effect of a larger population standard deviation on the probability that $\overline{X}$ will be within ± 5 of the population mean μ.

9.32 Let X_1, X_2, X_3 be the outcomes of $n = 3$ independent Bernoulli trials, each with the same parameter p. Obtain the sampling distribution of $\overline{X}$.

≣ STUDIES

9.33 Consider the discrete uniform probability distribution (6.12) with possible outcomes $X = 0, 1, 2, 3,$ and 4.

   a. Obtain the mean and the standard deviation of this infinite population.

   b. Draw 50 independent random samples, each of size $n = 2$, from this infinite population, and obtain $\overline{X}$ for each sample.

   c. Construct a frequency distribution of the 50 sample means, and plot it as a frequency polygon. Does the theoretical shape of the sampling distribution of $\overline{X}$ appear to be symmetrical? Normal?

   d. Obtain the exact sampling distribution of $\overline{X}$ by enumeration. Calculate $E\{\overline{X}\}$ and $\sigma\{\overline{X}\}$ from the exact sampling distribution, and verify that they agree with theorems (9.3) and (9.4).

9.34 (Computer needed.)

   a. Use a uniform random number generator to generate 100 random samples of size $n = 2$ each from the continuous uniform probability distribution (7.1) with $a = 0$ and $b = 100$. Make a frequency distribution of the 100 $\overline{X}$ values, and plot it as a frequency polygon.

   b. Repeat **a** for $n = 5$ and $n = 10$.

   c. For which sample sizes does the sampling distribution of $\overline{X}$ appear to be approximately normal?

9.35 (Computer needed.) Refer to the **Financial Characteristics** data set. Consider the population of all firms.

   a. Determine the population mean and standard deviation for year 2 net income of all firms.

   b. Select 300 independent random samples, each of $n = 15$ firms. Use sampling with replacement — that is, when a firm is selected, do not remove it from the population. Calculate the sample mean $\overline{X}$ for each sample.

   c. Select 300 independent random samples, each of $n = 50$ firms. Again, use sampling with replacement. Calculate the sample mean $\overline{X}$ for each sample.

   d. Use your simulation results to demonstrate that the sampling distribution of $\overline{X}$ approaches the normal distribution as the sample size increases, and that its mean and standard deviation are given by (9.3) and (9.4), respectively.

10 ESTIMATION OF POPULATION MEAN

$\mathbb{S}$ tatistical estimation of population characteristics from sample data is needed in a wide variety of circumstances.

EXAMPLES

1. A market researcher wishes to estimate the mean shift in attitude induced by a television advertisement, based on a sample survey of television viewers.

2. In the development of inventory policy for a wholesaler, an operations analyst needs to estimate the mean levels of demand for stocked items, based on sample time periods.

3. A government energy agency requires an estimate of the total amount of heating oil burned in residential dwellings in a region during the last winter, based on a sample of dwellings using heating oil. □

In this chapter, we build on earlier theoretical foundations to explain how a population mean (or population total) is estimated from simple random sample data. This is a common type of statistical estimation problem. In later chapters, we shall consider the estimation of other population characteristics.

10.1 POINT ESTIMATION

We examine the main features of point estimation by means of an example and then summarize these features more formally.

EXAMPLE

Length of Service. A company requires an estimate of the mean length of service (μ) for the population of 3580 employees, for current collective-bargaining negotiations. Let us consider how to estimate the unknown parameter μ from a simple random sample of employees. One obvious way to estimate μ is to use the sample mean $\overline{X}$. Thus, the personnel office might select a simple random sample of 50 employees, determine the length of service for each, and calculate the sample mean $\overline{X}$ of the 50 lengths of service. If the sample mean is $\overline{X} = 6.3$ years, then 6.3 years would be the estimate of μ. □

The estimate 6.3 years in the length of service example is called a point estimate.

(10.1)

> When a population characteristic is estimated by a single number, the number is called a *point estimate* of the characteristic.

Features of Point Estimation

The length of service example has shown the main features of point estimation with simple random sampling. These features are as follows:

1. *Parameter.* An unknown population characteristic or parameter is to be estimated. We shall use the symbol θ (Greek theta) to represent this parameter in the general case.
2. *Estimate.* A simple random sample of n observations, $X_1, X_2, \ldots, X_n$, is selected from the population. Some statistic, which is a function of these n sample values, is used as a point estimate of the parameter θ. We shall use the symbol S to represent this statistic in the general case.
3. *Sampling distribution.* Prior to the selection of the actual sample, the sample observations $X_1, X_2, \ldots, X_n$ are random variables, and hence, the statistic S to be calculated is also a random variable. The probability distribution of S is called the sampling distribution of S.

EXAMPLE

With reference to the length of service example, the parameter θ corresponds to the population mean μ. The sample statistic S is $\overline{X}$, and from the definition of $\overline{X}$, we know it is the following function of $X_1, X_2, \ldots, X_n$:

$$\overline{X} = \frac{X_1 + X_2 + \cdots + X_n}{n}$$

Prior to the selection of the sample, $\overline{X}$ is a random variable whose probability distribution is called the sampling distribution of $\overline{X}$. We discussed this distribution in Chapter 9. □

Since a sample statistic prior to sample selection is a random variable but after sample selection is simply a number, statisticians use the terms *estimator* and *estimate* to distinguish between these two situations.

(10.2)

> An *estimator* is a random variable used to estimate a population characteristic. An actual numerical value obtained for an estimator is called an *estimate*.

EXAMPLE

In the length of service example, $\overline{X}$ is used as a point estimator of μ. For the particular sample selected in the study, $\overline{X} = 6.3$ years. The number 6.3 is an estimate of μ.

□

Alternative Point Estimators

One can estimate a population parameter by many point estimators. For example, one possibility for estimating the population mean μ is $\overline{X}$. Another possibility is the sample *midrange*.

(10.3)
$$\frac{X_S + X_L}{2}$$

The midrange is the mean of the smallest (X_S) and the largest (X_L) observations in the sample.

Another possibility is the sample median Md, and there are still many other estimators of μ. How does one choose between the different possible point estimators? Basically, one wishes to choose a point estimator with desirable properties. We now consider some important properties of good point estimators.

Properties of Good Point Estimators

The statistical properties of good point estimators that we now take up are related to the general criterion that a good estimator should provide reasonable assurance that the estimate obtained will be "close" to the population parameter being estimated.

Unbiasedness. The sampling distribution of an estimator should be located near the parameter to be estimated, rather than far from it. This is the motivation of the property of unbiasedness.

(10.4)
An estimator S is *unbiased* if the mean of its sampling distribution is equal to the population characteristic θ to be estimated; that is, it is unbiased if:

$$E\{S\} = \theta$$

If estimator S is biased, the amount of its *bias* is:

$$\text{Bias} = E\{S\} - \theta$$

Figure 10.1 shows two estimators, S_1 being unbiased and S_2 having substantial bias. Clearly, S_2 will tend to give estimates far from θ, while estimates obtained from S_1 will tend to be nearer θ.

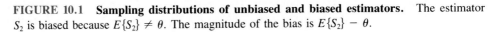

FIGURE 10.1   **Sampling distributions of unbiased and biased estimators.**   The estimator S_2 is biased because $E\{S_2\} \neq \theta$. The magnitude of the bias is $E\{S_2\} - \theta$.

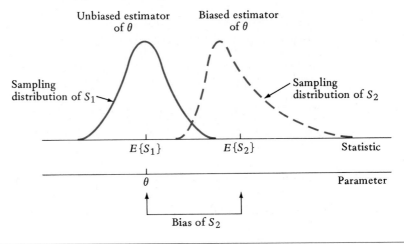

EXAMPLES

1. In Chapter 9, we showed that $E\{\overline{X}\} = \mu$. Hence, $\overline{X}$ is an unbiased estimator of μ.

2. The sample median Md is a biased estimator of μ when the population being sampled is skewed; that is, $E\{Md\} \neq \mu$ then. For example, median income in a random sample of families will, on the average, understate mean family income in the population when the income distribution is right-skewed.

3. The sample variance s^2, defined in (3.15), is an unbiased estimator of the population variance σ^2 for infinite populations; that is, $E\{s^2\} = \sigma^2$. Indeed, the reason for the denominator $n - 1$ in the sample variance is to obtain an unbiased estimator of σ^2.

4. The sample standard deviation s, defined in (3.17), is a biased estimator of the population standard deviation σ; that is, $E\{s\} \neq \sigma$. However, the bias is small for reasonably large samples.   □

Comments

1. Unbiasedness in point estimators refers to the tendency of sampling errors to balance over all possible samples. For any one sample, of course, the sample estimate will usually differ from the population parameter.

2. A biased estimator may still be a desirable estimator when the bias is not large, if the estimator has other desirable properties.

Efficiency.   If two estimators have no bias, we prefer the estimator that has the smaller variability (that is, the tighter sampling distribution) for the given sample size because its outcomes will tend to lie closer to the population characteristic. This observation leads to the property of relative efficiency.

(10.5)

The *efficiency* of an unbiased estimator is measured by the variance of its sampling distribution. If two estimators based on the same sample size are both unbiased, the one with the smaller variance is said to have greater *relative efficiency* than the other. Thus, S_1 is relatively more efficient than S_2 in estimating θ if:

$$\sigma^2\{S_1\} < \sigma^2\{S_2\} \quad \text{and} \quad E\{S_1\} = E\{S_2\} = \theta$$

EXAMPLE

The mean shelf life of a new breakfast cereal is to be estimated. The distribution of shelf lives of packages is normal, and the question is whether to use the mean life or median life of a sample of packages for estimating μ. For random sampling from a normal population, both $\overline{X}$ and Md are unbiased estimators of μ. We know from (9.4) that $\sigma^2\{\overline{X}\} = \sigma^2/n$. From statistical theory, it can be shown that $\sigma^2\{Md\} \simeq 1.57(\sigma^2/n)$ for random sampling from a normal population when n is large. Thus, $\sigma^2\{\overline{X}\} < \sigma^2\{Md\}$ for given sample size n, and consequently $\overline{X}$ is relatively more efficient than Md as an estimator of μ here. Figure 10.2 illustrates the situation. Note that $\overline{X}$ has a greater probability than Md of falling within any specified interval about μ. □

FIGURE 10.2 **Two unbiased estimators of μ that differ in relative efficiency when sampling a normal population.** Both estimators are unbiased, but the sampling distribution of $\overline{X}$ is tighter than that of Md.

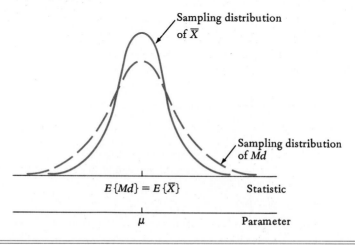

FIGURE 10.3 $\overline{X}$ **as a consistent estimator of** μ **when sampling a normal population.**   The sampling distributions of $\overline{X}$ for increasing sample sizes become more concentrated around μ.

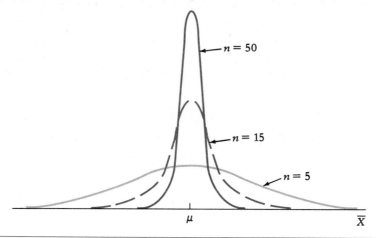

Consistency.   Also desirable is that an estimate tend to lie nearer the population characteristic as the sample size becomes larger. This is the basis of the property of consistency.

(10.6)

An estimator is a *consistent* estimator of a population characteristic θ if the larger the sample size, the more likely it is that the estimate will be close to θ.

EXAMPLE

The sample mean $\overline{X}$ is a consistent estimator of μ. Figure 10.3 illustrates, for sampling from a normal population, how the sampling distribution of $\overline{X}$ tightens around μ as the sample size increases.

Comments

1.   A consistent estimator may be biased. Consistency ensures, however, that the bias becomes smaller as the sample size becomes larger.
2.   A formal definition of consistency follows.

(10.6a)

S is a *consistent* estimator of population characteristic θ if for any small positive-value ε (Greek epsilon):

$$\lim_{n \to \infty} P(|S - \theta| < \varepsilon) = 1$$

The definition states that the probability is nearly 1 that a consistent estimator will have an outcome close to the parameter value (that is, within $\pm\varepsilon$) when n is sufficiently large. The only difficulty with this criterion is that the sample size may have to be very large before the assurance of closeness of the estimate takes effect.

10.2 INTERVAL ESTIMATION OF POPULATION MEAN

Need for Interval Estimate

Any point estimate has the limitation that it does not provide information about the *precision* of the estimate — that is, about the magnitude of the error due to sampling. Often, such information is essential for proper interpretation of the sample result. For instance, a television audience–rating agency reported that 24 percent of the sample households watched a particular show during the past four weeks. The sponsor of the program is assessing whether or not to continue sponsorship. If the margin of sampling error of the estimate were ±20 percent points, so that the percentage of households in the population watching the program could be as high as 44 percent or as low as 4 percent, the estimate would not be useful to the sponsor because of the wide margin of error due to sampling. On the other hand, if the margin of sampling error were ±0.5 percent point, the estimate would be quite useful since the sponsor would know that the proportion of households in the population viewing the program is somewhere between 23.5 and 24.5 percent. A point estimate by itself does not supply this information about its precision.

In this section, we consider statistical procedures for estimating the population mean in terms of an interval, where the width of the interval indicates the precision of the estimate.

(10.7)

An *interval estimate* of the population mean μ consists of two bounds within which μ is estimated to lie:

$$L \leq \mu \leq U$$

where L is the *lower bound* and U is the *upper bound*.

EXAMPLE

Accounts Receivable. In Chapter 9, we described the case of an auditor sampling the 8042 accounts receivable of a freight company to estimate the mean audit amount of the receivables in the population. The auditor has selected a simple random sample of 100 accounts to be used in estimating the population mean μ. A fragment of the sample data is presented in Table 10.1. The sample mean is $\overline{X} = 33.19$, and we know that this estimator has some desirable properties such as unbiasedness and consistency. Still, we also know that almost surely

TABLE 10.1   Sample results for a random sample of 100 accounts receivable — Accounts receivable example

i	X_i	$X_i - \overline{X}$	$(X_i - \overline{X})^2$
1	80.29	47.10	2,218.41
2	6.97	−26.22	687.49
3	4.55	−28.64	820.25
⋮	⋮	⋮	⋮
99	51.51	18.32	335.62
100	10.30	−22.89	523.95
Total	3318.73	0	117,674.67

$$n = 100 \qquad \overline{X} = \frac{3318.73}{100} = 33.19$$

$$s^2 = \frac{117,674.67}{100 - 1} = 1188.63 \qquad s = 34.48$$

$\overline{X} = 33.19$ is not equal to μ. To develop an interval estimate that reflects the margin of error due to sampling, we need to estimate the variability of the sampling distribution of $\overline{X}$. □

Estimated Standard Deviation of $\overline{X}$

Recall from Chapter 9 that the variability of the sampling distribution of $\overline{X}$ indicates how likely it is that the sample mean $\overline{X}$ is close to the population mean μ. The smaller the variability, the greater is the probability that $\overline{X}$ will fall within any specified interval around μ.

Even though we take only one sample from the population, we can estimate the variability of the sampling distribution of $\overline{X}$. The reason is that $\sigma^2\{\overline{X}\}$, the variance of $\overline{X}$, is a simple function of the population variance σ^2 as shown by (9.4) — namely, $\sigma^2\{\overline{X}\} = \sigma^2/n$.

An estimate of $\sigma^2\{\overline{X}\}$ is obtained by the following reasoning. We noted earlier that the sample variance s^2 is an unbiased estimator of the population variance when the population is infinite. Hence, an unbiased estimator of $\sigma^2\{\overline{X}\} = \sigma^2/n$ is s^2/n. We denote this estimator by $s^2\{\overline{X}\}$ and the estimated standard deviation of $\overline{X}$ by $s\{\overline{X}\}$. Thus, $s^2\{\overline{X}\} = s^2/n$ and $s\{\overline{X}\} = s/\sqrt{n}$. This latter standard deviation is often called the *estimated standard error of the mean*.

(10.8)   Estimators of the variance and the standard deviation of the sampling distribution of $\overline{X}$ are, respectively:

$$s^2\{\overline{X}\} = \frac{s^2}{n} \qquad s\{\overline{X}\} = \frac{s}{\sqrt{n}}$$

These estimators are also applicable when a finite population is sampled as long as the sample size n is not large relative to the population size N, that is, as long as the sampling fraction n/N is small.

EXAMPLE

In the accounts receivable example, we wish to estimate the variability of the sampling distribution of $\overline{X}$ from the sample results in Table 10.1; we have $s^2 = 1188.63$ and $n = 100$. The sampling fraction is small here so we can use (10.8) and obtain:

$$s^2\{\overline{X}\} = \frac{1188.63}{100} = 11.886 \qquad s\{\overline{X}\} = \sqrt{11.886} = 3.448$$

Since the sample size $n = 100$ is large, the sampling distribution of $\overline{X}$ is approximately normal by central limit theorem (9.5). Consequently, most of the time the sample mean $\overline{X}$ should be within, say, two standard deviations of the population mean. Thus, $s\{\overline{X}\} = 3.45$ implies that most of the time the sample mean $\overline{X}$ should be within $2(3.45) = 6.90$ of μ.

Incidentally, we know that $\sigma = 30.334$ here and $\sigma\{\overline{X}\} = 3.03$ (see Chapter 9), so our sample estimate $s\{\overline{X}\} = 3.45$ is reasonably close.   □

Interval Estimate

We just found for the accounts receivable example that $s\{\overline{X}\} = 3.45$ and concluded that most of the time the sample mean $\overline{X}$ should be within $2(3.45) = \$6.90$ of the population mean. If that is so, we might reason that starting with $\overline{X}$ and adding and subtracting $\$6.90$ should give us limits that are likely to include μ. Here, the limits would be:

$$L = \overline{X} - 2s\{\overline{X}\} = 33.19 - 2(3.45) = 26.29$$
$$U = \overline{X} + 2s\{\overline{X}\} = 33.19 + 2(3.45) = 40.09$$

and we would therefore state:

$$26.29 \leq \mu \leq 40.09$$

Thus, we would estimate that the mean audit amount per account in the population is somewhere between $\$26.29$ and $\$40.09$. Since we happen to know $\mu = \$30.303$ in this case, we see that this particular interval estimate is correct.

Note that the approach used involves the following steps:

Step 1. Select a random sample of size n.
Step 2. Obtain $\overline{X}$ and s for that sample.
Step 3. Estimate $\sigma\{\overline{X}\}$ by $s\{\overline{X}\} = s/\sqrt{n}$.
Step 4. Calculate the following interval.

(10.9)

$$\overline{X} - 2s\{\overline{X}\} \leq \mu \leq \overline{X} + 2s\{\overline{X}\}$$

The interval in formula (10.9) represents an interval estimate of the population mean μ, and the term $2s\{\overline{X}\}$ represents the calculated margin of error due to sampling.

Behavior of Interval Estimate

We shall study the behavior of interval estimate (10.9) by means of an experiment. We selected 600 independent samples of 100 accounts from the population of 8042 accounts receivable. The auditor's sample in Table 10.1 is the first of the 600. In the second sample, we obtained $\overline{X} = 31.89$ and $s = 34.94$. If the auditor had selected this second sample, we would have calculated $s\{\overline{X}\} = 34.94/\sqrt{100} = 3.494$ and $2(3.494) = 6.988$, giving limits of 31.89 ± 6.988. Hence, we would have obtained the interval $24.90 \leq \mu \leq 38.88$. Again, this interval contains $\mu = 30.303$. Figure 10.4 shows the interval estimates for a few of the 600 samples, including the two already mentioned. Note that all but one of the interval estimates displayed contain $\mu = 30.303$. The exception is sample 39, which provides an interval estimate lying entirely above μ. In all, 559 of the 600 intervals, or 93.2 percent, contain μ.

This percentage of correct intervals is a measure of the confidence we can have in the interval estimation procedure. Here, it is quite likely that the interval estimate obtained for the one sample actually selected will contain μ, because 93.2 percent of the 600 interval estimates were correct. Clearly, the percentage of correct intervals is a function of the multiple of $s\{\overline{X}\}$ used in calculating the margin of error due to sampling. In this study, we used a multiple of 2. In general, as we shall now see, we can specify the multiple to yield a desired probability of a correct interval.

Confidence Interval for μ — Large Sample

We now consider the construction of an interval estimate $L \leq \mu \leq U$ for any specified probability of a correct interval when the sample size is large.

(10.10)

> The probability that a correct interval estimate is obtained is called the *confidence coefficient* and is denoted by $1 - \alpha$. The interval:
>
> $$L \leq \mu \leq U$$
>
> is called a *confidence interval*. The limits L and U are called the *lower* and *upper confidence limits*, respectively.

The numerical confidence coefficient is often expressed as a percent. For example, coefficient 0.95 is frequently expressed as 95 percent. A confidence interval that has an associated confidence coefficient of $1 - \alpha$ is frequently called a $100(1 - \alpha)$ percent confidence interval. For instance, a 95 percent confidence interval is a confidence interval with a confidence coefficient of 0.95.

FIGURE 10.4   Interval estimates of the form $\overline{X} \pm 2s\{\overline{X}\}$ for several of the 600 accounts receivable samples.   The intervals differ in location because $\overline{X}$ varies from sample to sample, and they differ in width because s varies from sample to sample.

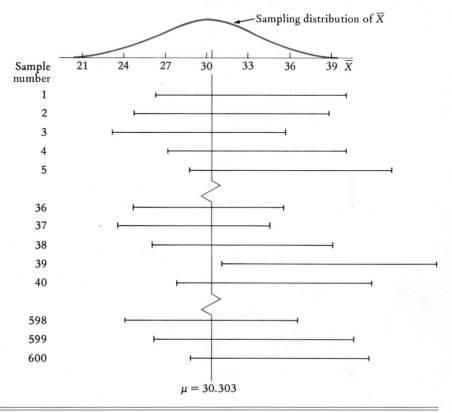

$\mu = 30.303$

(10.11)

The confidence limits L and U for the population mean μ with approximate confidence coefficient $1 - \alpha$, when the random sample size is reasonably large, are:

$$\overline{X} \pm zs\{\overline{X}\} \qquad \overline{X} \pm 2s\bar{x}$$

where:

$z = z(1 - \alpha/2)$
$s\{\overline{X}\}$ is given by (10.8)

The $100(1 - \alpha)$ percent confidence interval for μ is:

$$\overline{X} - zs\{\overline{X}\} \leq \mu \leq \overline{X} + zs\{\overline{X}\}$$

Recall from Chapter 7 that $z(1 - \alpha/2)$ denotes the $100(1 - \alpha/2)$ percentile of the standard normal distribution.

1. Refer to Table 10.1 for the accounts receivable example. We wish to construct a confidence interval for μ with confidence coefficient $1 - \alpha = 0.954$. Hence, $\alpha = 0.046$ and $1 - \alpha/2 = 0.977$. We find from Table C.1 that $z(0.977) = 2.0$. We previously obtained:

$$\overline{X} = 33.19 \qquad s\{\overline{X}\} = 3.448$$

Hence:

$$L = 33.19 - 2(3.448) = 26.29 \qquad U = 33.19 + 2(3.448) = 40.09$$

and:

$$26.29 \leq \mu \leq 40.09$$

We thus conclude, with 95.4 percent confidence, that the population mean audit amount per account receivable is between \$26.29 and \$40.09.

   We actually obtained this confidence interval before. Now, we know that it has an approximate confidence coefficient of 95.4 percent. Recall that the percentage of correct intervals in the 600 trials was close to 95.4 percent, namely, 93.2 percent.

2. Refer to the accounts receivable example in Table 10.1 again. Suppose we wish to use a confidence coefficient of $1 - \alpha = 0.90$. Hence, we require $z(0.95) = 1.645$. The confidence limits are then $33.19 \pm 1.645(3.448)$, and the confidence interval is:

$$27.52 \leq \mu \leq 38.86$$

3. In a random sample of $n = 40$ bricks from a production process, the mean weight was $\overline{X} = 3.724$ pounds and the standard deviation was $s = 0.071$ pound. We wish to estimate the process mean μ with a 99 percent confidence interval. We require:

$$s\{\overline{X}\} = \frac{0.071}{\sqrt{40}} = 0.011 \qquad z(0.995) = 2.576$$

and we obtain the confidence limits $3.724 \pm 2.576(0.011)$. The confidence interval therefore is:

$$3.70 \leq \mu \leq 3.75$$

We thus conclude, with 99 percent confidence, that the process mean weight of bricks is between 3.70 and 3.75 pounds.   □

Development of Confidence Interval

We now show the formal basis of confidence interval (10.11). First, we need a formal definition of a confidence interval.

(10.12)

> The interval $L \leq \mu \leq U$ is a $1 - \alpha$ confidence interval for the population mean μ if prior to sampling with a random sample of size n:
>
> $$P(L \leq \mu \leq U) = 1 - \alpha$$

This definition states that, in advance of selecting the sample, the probability is $1 - \alpha$ that the limits L and U of a $1 - \alpha$ confidence interval based on a random sample of size n will include μ. In other words, in many random samples of size n from a population, $100(1 - \alpha)$ percent of the confidence interval estimates will include μ and will therefore be correct.

The validity of confidence interval (10.11) depends on the following extension of central limit theorem (9.5).

(10.13)

> For almost any population, $(\overline{X} - \mu)/s\{\overline{X}\}$ follows approximately a standard normal distribution when the random sample size is sufficiently large; that is:
>
> $$\frac{\overline{X} - \mu}{s\{\overline{X}\}} \simeq Z$$

For populations that are not highly skewed, sample sizes need not be too large for theorem (10.13) to apply. For highly skewed populations, however, the sample size may need to be quite large.

Assuming n is large enough that $(\overline{X} - \mu)/s\{\overline{X}\}$ is distributed approximately as Z, we can state:

$$P\left[z(\alpha/2) \leq \frac{\overline{X} - \mu}{s\{\overline{X}\}} \leq z(1 - \alpha/2) \right] \simeq 1 - \alpha$$

Figure 10.5 illustrates why this probability is $1 - \alpha$. In each tail, the area is $\alpha/2$, so the central area is $1 - \alpha$. But by (7.11), $z(\alpha/2) = -z(1 - \alpha/2)$. Hence:

$$P\left[-z(1 - \alpha/2) \leq \frac{\overline{X} - \mu}{s\{\overline{X}\}} \leq z(1 - \alpha/2) \right] \simeq 1 - \alpha$$

Rearranging the inequalities, we obtain:

$$P[\overline{X} - z(1 - \alpha/2)s\{\overline{X}\} \leq \mu \leq \overline{X} + z(1 - \alpha/2)s\{\overline{X}\}] \simeq 1 - \alpha$$

FIGURE 10.5   **Two-sided confidence interval for μ — Large sample.**   Before the sample is taken, the probability is $1 - \alpha$ that the quantity $(\overline{X} - \mu)/s\{\overline{X}\}$ will fall in the shaded interval. The interval estimate $\overline{X} \pm z(1 - \alpha/2)\, s\{\overline{X}\}$ will be correct (that is, will contain μ) if $(\overline{X} - \mu)/s\{\overline{X}\}$ falls in the shaded interval.

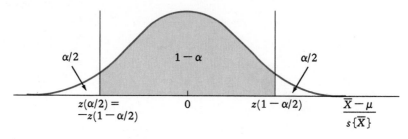

But this expression fits the definition of a confidence interval in (10.12), where the lower and upper confidence limits are:

$$L = \overline{X} - z(1 - \alpha/2)s\{\overline{X}\} \qquad U = \overline{X} + z(1 - \alpha/2)s\{\overline{X}\}$$

These are, of course, the limits given in (10.11).

Figure 10.5 summarizes the reasoning that underlies the confidence interval for μ in (10.11). In effect, the risk α of an incorrect confidence interval is divided equally in the two tails of the standard normal distribution.

Comment	We can now see why confidence interval (10.11) has a confidence coefficient of *approximately* $1 - \alpha$. The reason is that $(\overline{X} - \mu)/s\{\overline{X}\}$, in general, is only approximately normally distributed for large n.

Confidence Coefficient

Meaning of Confidence Coefficient.   From the definition of a confidence interval in (10.12), we know that, *in advance* of selecting the sample, the probability is $1 - \alpha$ that the confidence interval we obtain will contain the population mean μ. The particular confidence interval obtained, however, may or may not contain μ. Thus, any one confidence interval result will be either correct or incorrect, and we do not know for certain which is the case. We act as if the result is correct when $1 - \alpha$ is sufficiently large because the procedure gives us assurance that in $100(1 - \alpha)$ percent of the cases, the confidence interval result will be correct.

Selecting the Confidence Coefficient.   In the best of all worlds, we would like the confidence interval to be very precise (that is, very narrow) and would like to be very

FIGURE 10.6   **Relation between the confidence coefficient and the confidence interval width**

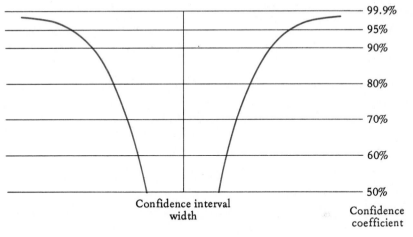

confident that it contains μ. Unfortunately, for any fixed sample size, the confidence coefficient can only be increased by increasing the width of the confidence interval. Figure 10.6 illustrates the relationship graphically. It shows how the width of confidence interval (10.11) varies with the confidence coefficient for any given sample result. Note how rapidly the confidence interval widens as the confidence coefficient gets near 100 percent.

The choice of $1 - \alpha$ will vary from case to case, depending on how much risk of obtaining an incorrect interval can be taken. Confidence coefficients of 90, 95, and 99 percent are often used in practice.

Comment	Confidence intervals provide an evaluation of the magnitude of the sampling error only. They do not take into account biases in the data, such as systematic understatements of income by respondents.

 ## 10.3   CONFIDENCE INTERVAL FOR μ — SMALL SAMPLE

When the sample size is small, $(\overline{X} - \mu)/s\{\overline{X}\}$ no longer follows the standard normal distribution and the construction of a confidence interval for μ depends on the nature of the underlying population.

Normal Population

When the population is normal, the following theorem helps us to construct confidence intervals for μ.

(10.14)

> For a simple random sample of size n from a normal population:
>
> $$\frac{\overline{X} - \mu}{s\{\overline{X}\}} = t(n - 1)$$

In other words, $(\overline{X} - \mu)/s\{\overline{X}\}$ has a t distribution with $n - 1$ degrees of freedom when the sampled population is normal. (If a review of the t distribution is needed, the reader should turn to Appendix B, Section B.2, before proceeding in this section.)

In view of theorem (10.14), the confidence limits for μ when n is small and the population is normal will be of exactly the same form as those for large samples, except that an appropriate percentile of the t distribution is used instead of a percentile from the standard normal distribution.

(10.15)

> The confidence limits for μ with confidence coefficient $1 - \alpha$, when the population is normal or the departure is not too marked, are:
>
> $$\overline{X} \pm ts\{\overline{X}\}$$
>
> *where:*
>
> $t = t(1 - \alpha/2; n - 1)$
> $s\{\overline{X}\}$ is given by (10.8)

EXAMPLE

A sample of five cans of tomatoes in puree was taken at random from a production line immediately after filling. The solid content of each can was weighed to serve as a basis for estimating the process mean drained weight per can (μ). Past experience indicates that the distribution of drained weights is normal. It is desired to construct a 99 percent confidence interval for the process mean μ.

The following are the sample results (in ounces).

i:	1	2	3	4	5
X_i:	23.0	23.5	23.5	25.0	24.5

$n = 5 \quad \overline{X} = 23.9 \quad s = 0.822$

Hence, $s\{\overline{X}\} = s/\sqrt{n} = 0.822/\sqrt{5} = 0.37$. The desired confidence coefficient is $1 - \alpha = 0.99$, so $\alpha = 0.01$ and $1 - \alpha/2 = 0.995$. The degrees of freedom for the t distribution are $n - 1 = 5 - 1 = 4$. From Table C.3, we find that

$t(0.995; 4) = 4.604$. Using (10.15), we obtain the confidence limits $23.9 \pm 4.604(0.37)$, which gives the confidence interval:

$$22.20 \le \mu \le 25.60$$

We can now state, with 99 percent confidence, that the mean drained weight for the current process is somewhere between 22.20 and 25.60 ounces. □

Nonnormal Populations

Confidence limits (10.15) are applicable even when the population is not exactly normal. Statistical theory shows that as long as the population does not depart too markedly from normality (for example, it is not highly skewed) and n is not exceedingly small, $(\bar{X} - \mu)/s\{\bar{X}\}$ is distributed approximately as $t(n - 1)$. Hence, confidence limits (10.15) may still be used in these cases, although the confidence coefficient now is only approximately $1 - \alpha$. This property of the t distribution, that it applies approximately for many other populations besides normal ones, is called *robustness*.

Comments

1. As the degrees of freedom $n - 1$ increase, we know from (B.6) in Appendix B that the t distribution becomes more concentrated. Hence, the t multiple in (10.15) becomes smaller for larger n. In fact, for large n, we use the standard normal distribution to approximate the t distribution.

2. The confidence limits (10.15) for sampling a normal population are obtained by using theorem (10.14). It follows from this theorem that:

$$P\left[-t(1 - \alpha/2; n - 1) \le \frac{\bar{X} - \mu}{s\{\bar{X}\}} \le t(1 - \alpha/2; n - 1)\right] = 1 - \alpha$$

when a normal population is sampled. Recall that $t(1 - \alpha/2; n - 1)$ is the $100(1 - \alpha/2)$ percentile of the t distribution with $n - 1$ degrees of freedom. Recall, also, that because of the symmetry of the t distribution about its mean 0, $t(\alpha/2; n - 1) = -t(1 - \alpha/2; n - 1)$. The preceding probability is analogous to the one illustrated in Figure 10.5 for the case of large n.

Rearranging the expression inside the brackets, as we did in the large-sample case, gives:

$$P[\bar{X} - t(1 - \alpha/2; n - 1)s\{\bar{X}\} \le \mu \le \bar{X} + t(1 - \alpha/2; n - 1)s\{\bar{X}\}] = 1 - \alpha$$

Thus, we have obtained $1 - \alpha$ confidence limits for μ.

3. Occasionally, the population standard deviation σ is known and need not be estimated by the sample standard deviation s. In this case, $\sigma\{\bar{X}\}$ is also known exactly. If the population is normal, then $(\bar{X} - \mu)/\sigma\{\bar{X}\}$ follows a standard normal distribution by (9.7) and (7.9). Consequently, the appropriate confidence limits in this case are $\bar{X} \pm z\sigma\{\bar{X}\}$, where $z = z(1 - \alpha/2)$, irrespective of sample size.

10.4 CONFIDENCE INTERVAL FOR POPULATION TOTAL

A confidence interval for the population total of a finite population is often desired. For example, in the study of the population of 8042 accounts receivable, the auditor was actually interested in the total audit amount of these accounts, and his concern with the mean audit amount (μ) was merely a step toward obtaining a confidence interval for the population total.

Let τ (Greek tau) denote the population total.

(10.16)
$$\tau = \sum_{i=1}^{N} X_i$$

In the accounts receivable example, τ denotes the total of all audit amounts for the $N = 8042$ accounts in the population.

Since, by definition (8.2a) of the population mean, we have $N\mu = \sum X_i$, it follows that:

(10.17)
$$\tau = N\mu$$

Consequently, confidence limits for τ can be obtained by multiplying the confidence limits for μ by N.

EXAMPLE

Refer to Table 10.1 for the accounts receivable example. We wish to obtain a 95.4 percent confidence interval for the total audit amount of the accounts in the population. Earlier, we found the 95.4 percent confidence interval for μ:

$$26.29 \le \mu \le 40.09$$

Since $N = 8042$, the 95.4 percent confidence limits for the total audit amount of the company's accounts receivable are:

$$26.29(8042) = 211{,}424 \qquad 40.09(8042) = 322{,}404$$

So the 95.4 percent confidence interval is:

$$211{,}424 \le \tau \le 322{,}404$$

We conclude, with 95.4 percent confidence, that the total audit amount of the 8042 accounts is between \$211.4 thousand and \$322.4 thousand. ☐

10.5 PLANNING OF SAMPLE SIZE

Need for Planning

The user of an interval estimate often would like to specify in advance the precision required in an estimate and the confidence level it should have.

EXAMPLES

1. A marketing manager specifies that she must know the mean income of households living in the vicinity of a shopping center within ±$500 with a 95 percent confidence coefficient. The number of households to be included in the sample needs to be determined.

2. In a merger negotiation, the mean number of days of sick leave accrued by the employees of one of the corporations must be estimated. Lawyers on both sides agree that it will be sufficient to estimate this mean within ±2 days with a confidence coefficient of 99 percent. The number of employee files to be included in the sample needs to be determined. ☐

From such specifications, appropriate sample sizes can often be determined that will approximately yield the desired precision at the specified confidence level.

This planning approach is often more reasonable than using an arbitrary sample size. With an arbitrary sample size, the width of the confidence interval may be tighter than needed, indicating that the sample size was too large, or it may be too wide, indicating that the sample size was inadequate. If too large a sample is taken, money will have been wasted in getting greater precision than necessary. If too small a sample is taken, it may be costly or even impossible to enlarge the sample subsequently.

Assumptions

The planning procedure we shall explain is based on the following assumptions:

1. The random sample size ultimately determined is reasonably large.
2. The population is infinite or, if finite, is large relative to the resulting sample size.

Planning Procedure

The planning procedure requires specifications of the desired margin of sampling error and the desired confidence coefficient. Furthermore, since we are planning the sample size, we do not yet have the sample standard deviation s available to compute the estimated standard deviation of the mean, $s\{\overline{X}\}$. Inasmuch as the population standard deviation σ is generally unknown, we shall also need to specify a planning value for σ to find the needed sample size.

Acreage Plans.   A farming region contains over 40,000 farms. In the early spring, it is planned to survey a simple random sample of farm operators to determine the acreage each operator intends to sow in summer wheat. The survey forms part of a larger study designed to estimate total agricultural production in the next 12 months. It is desired to have the wheat acreage survey produce an interval estimate of the mean wheat acreage per farm within ±5 acres with a confidence level of 95 percent. Ample recent historical data about farm wheat acreage in the region are available, and consultation with agricultural experts indicates that the variability of wheat acreages among farms in the coming year is not likely to differ sharply from that exhibited in the recent past. From the historical data, a planning value of $\sigma = 85$ acres seems reasonable.   ☐

As shown by the acreage plans example, the three specifications to be made for determining the needed sample size are as follows:

1. *The desired margin of sampling error.*   The margin is denoted by $\pm h$. This specification indicates how close the confidence limits must be around the point estimate $\overline{X}$. In the acreage plans example, $h = 5$ acres. Thus, the confidence limits $\overline{X} \pm h$ should be $\overline{X} \pm 5$. The symbol h denotes the desired *half-width* of the confidence interval, since the total width of the confidence interval is desired to be $2h$.
2. *The desired confidence coefficient* $1 - \alpha$.   In the acreage plans example, it is specified that $1 - \alpha = 0.95$.
3. *The planning value for the population standard deviation.*   Information about σ is often available from past experience with the same or a similar problem or can be obtained from a pilot study. Approximate information about the order of magnitude of σ is usually adequate for planning purposes. In the acreage plans example, the planning value specified is $\sigma = 85$ acres.

From (10.11), we know that the half-width h of the confidence interval is:

$$h = z\sigma\{\overline{X}\} \quad \text{where} \quad z = z(1 - \alpha/2)$$

Note that here we use the notation for the true standard deviation of the mean, $\sigma\{\overline{X}\}$, rather than that for the estimated standard deviation, $s\{\overline{X}\}$, because we are in the planning phase.

We have, by (9.4):

$$\sigma\{\overline{X}\} = \frac{\sigma}{\sqrt{n}}$$

Hence:

$$h = z\sigma\{\overline{X}\} = z\frac{\sigma}{\sqrt{n}}$$

Squaring both sides and solving for n, we obtain:

$$n = \frac{z^2 \sigma^2}{h^2}$$

(10.18)

> The necessary random sample size to achieve the desired half-width h for the specified confidence coefficient $1 - \alpha$, for a given planning value of σ, is:
>
> $$n = \frac{z^2 \sigma^2}{h^2}$$
>
> *where:*
> $z = z(1 - \alpha/2)$
> σ is the planning value for the population standard deviation

---

EXAMPLE

In the acreage plans example, $h = 5$ acres, the planning value for the population standard deviation is $\sigma = 85$ acres, and $1 - \alpha = 0.95$. Hence, $\alpha = 0.05$ and $1 - \alpha/2 = 0.975$. Thus, $z(0.975)$ is required. We find from Table C.1 that $z(0.975) = 1.960$. Substituting into (10.18) gives the following needed sample size for the survey:

$$n = \frac{(1.960)^2 (85)^2}{5^2} = 1110$$

If the planning value of σ is at all reliable, this sample size should produce a 95 percent confidence interval with a half-width of about 5 acres. Note that formula (10.18) is appropriate here because the resultant n is large while n/N is small. □

Once the sample size has been determined and the actual sample selected, the construction of a confidence interval proceeds in the usual manner. We use, of course, the sample standard deviation s and not the planning value for σ in constructing the interval.

EXAMPLE

In the acreage plans example, a random sample of 1110 farms was selected and yielded $\overline{X} = 75.6$ acres and $s = 80.1$ acres. We calculate $s\{\overline{X}\} = 80.1/\sqrt{1110} = 2.404$ by (10.8). Hence, the 95 percent confidence limits for the mean wheat acreage per farm are, by (10.11), $75.6 \pm 1.960(2.404)$. The confidence interval therefore is:

$$70.9 \leq \mu \leq 80.3$$

Observe that the attained half-width is 4.7 acres. This value is slightly smaller than the 5-acre value specified in the planning stage. The extra precision resulted because the sample standard deviation of 80.1 acres is a little smaller than the planning value of 85 acres.   ☐

Comment   Formula (10.18) must be modified when the needed sample size is large relative to the population size. We consider this case in Chapter 14.

10.6   OPTIONAL TOPIC—ONE-SIDED CONFIDENCE INTERVAL FOR μ

Thus far, our confidence intervals have had both upper and lower confidence limits. Sometimes, one-sided confidence intervals are useful. Examples are (1) a bank officer contemplating making a loan to a firm on its inventory is interested in an upper limit for the total value of obsolete inventory items, and (2) a purchasing agent is interested in a lower limit for the mean life of electronic components in a shipment just received.

One-sided confidence intervals are constructed in a similar fashion to two-sided intervals. The only difference is that the risk α is placed all in one side. This feature may be seen by examining Figure 10.4 again. If we were only interested in an upper confidence limit here, note that all upper limits shown are correct, including that for sample 39, where the lower limit was too high. Thus, when constructing a one-sided confidence interval, we incur risks of incorrect limits in only one side of Figure 10.5. Hence, for one-sided confidence intervals with confidence coefficient $1 - \alpha$, the risk α of an incorrect limit is placed entirely in one side of the distribution, and $100(1 - \alpha)$ percentiles are used in constructing the confidence limits.

We present the one-sided confidence limits for μ for large sample sizes.

(10.19)   One-sided confidence limits for μ with approximate confidence coefficient $1 - \alpha$, when the random sample size is reasonably large, are as follows:

Lower Confidence Limit:

$$L = \bar{X} - z(1 - \alpha)s\{\bar{X}\}$$

Upper Confidence Limit:

$$U = \bar{X} + z(1 - \alpha)s\{\bar{X}\}$$

EXAMPLE   Refer to the accounts receivable example in Table 10.1. We wish to obtain a lower 95 percent confidence interval for the mean audit amount per account receivable in the population. We know from before that $\bar{X} = 33.19$ and $s\{\bar{X}\} = 3.448$. For $1 - \alpha = 0.95$, we require $z(0.95) = 1.645$. Hence, the lower confidence limit is $33.19 - 1.645(3.448) = 27.52$. Thus, with 95 percent confidence, we can assert that $\mu \geq 27.52$. In terms of the total audit amount of the company's 8042 accounts receivable, the auditor can conclude, with 95 percent confidence, that $\tau \geq 8042(27.52) = 221,316$, or \$221.3 thousand.   ☐

Comments   1.   Note that the two-sided $1 - \alpha$ confidence limits for μ in (10.11) can be viewed as the respective one-sided lower and upper confidence limits in (10.19), each with confidence coefficient $1 - \alpha/2$.

2.   One-sided confidence limits for small sample sizes when the population is normal or the departure is not too marked are the same as in (10.19), except that $z(1 - \alpha)$ is replaced by $t(1 - \alpha; n - 1)$.

OPTIONAL TOPIC—PREDICTION INTERVAL FOR NEW OBSERVATION

Occasionally, one desires to use sample data to construct an interval estimate for a new observation. Such an interval estimate is called a *prediction interval*. One use of prediction intervals is to judge whether or not a process is in control.

EXAMPLE   **Chemical Process.**   A chemical process in each operating cycle converts a batch of raw material into final product. The yields of final product in a random sample of 20 operating cycles when the process was in control are to be used to construct a prediction interval for the yield in the next operating cycle. A comparison of this new yield with the prediction interval will permit the process operator to judge whether or not the process is in control in the next cycle.   ☐

Another use of prediction intervals is to help in identifying outliers.

EXAMPLE   A government meat inspector has measured the percentage of fat on 200 beef carcasses. He wishes to obtain a prediction interval for the percentage of fat on a new carcass so he can compare the fattiness of the new carcass with the prediction interval to determine whether the new carcass is an outlier with respect to its fattiness.   ☐

Assumptions

We shall let X_{new} denote the new observation for which a prediction interval is required. The procedure for constructing a prediction interval to be described is based on the following assumptions:

1.   The new observation X_{new} is selected independently of the random sample $X_1, X_2, \ldots, X_n$ from the same population.
2.   The population is normal.

These assumptions describe conditions that are encountered frequently in practice.

Population Parameters Known

To explain the construction of prediction intervals, we shall first consider the case where the parameters μ and σ of the normal population are known. The more common case where the parameters μ and σ are unknown will be taken up shortly.

When X_{new} is randomly drawn from a normal population with mean μ and standard deviation σ, we know from the normal probability calculations discussed in Figure 7.6b that the probability is $1 - \alpha$ that the following interval is correct:

$$\mu - z\sigma \leq X_{new} \leq \mu + z\sigma \quad \text{where} \quad z = z(1 - \alpha/2)$$

Hence, this interval is a $1 - \alpha$ prediction interval for X_{new}.

EXAMPLE

A new observation X_{new} is to be randomly drawn from a normal population with $\mu = 90$ and $\sigma = 15$. For a 95 percent prediction interval for X_{new}, we require $z(0.975) = 1.960$. Hence, the 95 percent prediction interval is:

$$90 - 1.960(15) \leq X_{new} \leq 90 + 1.960(15)$$

or

$$60.6 \leq X_{new} \leq 119.4$$

We can thus state, with 95 percent confidence, that the new observation will be between 60.6 and 119.4.   □

Population Parameters Unknown

In the more common case where the parameters μ and σ of the normal population are unknown, we use a random sample of size n from the normal population to provide the sample statistics $\overline{X}$ and s as estimators of the parameters μ and σ, respectively. The prediction interval uses the estimated variance $s^2\{X_{new}\}$.

(10.20)
$$s^2\{X_{new}\} = s^2 + \frac{s^2}{n} = s^2\left(1 + \frac{1}{n}\right)$$

Note that this estimated variance has two components related to two sources of variation. The first component, s^2, measures the variability of individual observations in the population. The second component, s^2/n, reflects the sampling error in using $\overline{X}$ as an estimator of μ.

It can be shown that prediction limits based on the t distribution are appropriate.

(10.2

The prediction limits for a new observation X_{new} with confidence coefficient $1 - \alpha$, when the sample and X_{new} are independently drawn from the same normal population, are:

$$\overline{X} \pm ts\{X_{new}\}$$

where:

$$t = t(1 - \alpha/2; n - 1)$$
$$s\{X_{new}\} \text{ is given by (10.20)}$$

EXAMPLE

In the chemical process example, the yields (in kilograms) of final product for $n = 20$ operating cycles were obtained. The sample statistics were $\overline{X} = 238.0$ and $s = 4.3$. Hence, from (10.20), we calculate:

$$s^2\{X_{new}\} = (4.3)^2\left(1 + \frac{1}{20}\right) = 19.41 \qquad s\{X_{new}\} = \sqrt{19.41} = 4.406$$

A 99 percent prediction interval for a new observation is desired. We require $t(1 - \alpha/2; n - 1) = t(0.995; 19) = 2.861$. Following (10.21), the prediction limits are $238.0 \pm 2.861(4.406)$, and, hence, the desired prediction interval is:

$$225 \leq X_{new} \leq 251$$

Thus, the process operator may predict, with 99 percent confidence, that the yield of the next operating cycle if the process is still in control will be between 225 and 251 kilograms. Should the yield of the next cycle be 213 kilograms, then the operator should conclude that the process was not in control during that cycle.   ☐

Comment

When predicting a new observation X on the basis of an earlier sample, the relevant variance is that of the deviation of the new observation from the sample mean, that is, the deviation $X - \overline{X}$. Because of the independence of the new observation and the earlier sample, we have:

$$\sigma^2\{X - \overline{X}\} = \sigma^2\{X\} + \sigma^2\{\overline{X}\}$$

Any observation from the population has variance σ^2; hence, $\sigma^2\{X\} = \sigma^2$. We further know from (9.4) that $\sigma^2\{\overline{X}\} = \sigma^2/n$. Hence, we have:

$$\sigma^2\{X - \overline{X}\} = \sigma^2 + \frac{\sigma^2}{n} = \sigma^2\left(1 + \frac{1}{n}\right)$$

Replacing σ^2 by the sample estimator s^2 yields (10.20), where the estimated variance for brevity is denoted by $s^2\{X_{new}\}$.

OPTIONAL TOPIC — MAXIMUM LIKELIHOOD METHOD OF ESTIMATION

Good point estimators for particular circumstances are not always intuitively obvious. We now discuss a general method of finding point estimators with desirable properties, namely, the *method of maximum likelihood*. We shall explain the rationale of this method by an example.

<table>
<tr><td>EXAMPLE</td></tr>
</table>

Calculator Demand. The weekly demand for a programmable calculator at a retail outlet follows a Poisson distribution with unknown parameter λ. Three weeks' observations are available; they are $X_1 = 2$, $X_2 = 6$, and $X_3 = 1$. We assume that these three observations are independent observations from the same Poisson distribution and hence constitute a random sample.

In choosing an estimate of the parameter λ, we clearly would like this estimate to be consistent with the observed sample results. Hence, we shall calculate the probability of obtaining the given sample results for different values of λ to find that value of λ that is most consistent with the sample results.

Recall that the Poisson probability function, given by (6.8), is:

$$P(x) = \frac{\lambda^x \exp(-\lambda)}{x!}$$

Because the three sample observations are independent, the joint probability of obtaining the sample equals the product $P(2)P(6)P(1)$. The individual probabilities in this product can be computed from the preceding formula for the probability function or can be obtained from Table C.6.

We shall proceed on a systematic search to see what value of λ yields the largest joint probability for the sample. First, we consider $\lambda = 4.5$. From Table C.6, we obtain $P(2) = 0.1125$, $P(6) = 0.1281$, and $P(1) = 0.0500$. Hence, the joint probability is:

$$P(2)P(6)P(1) = 0.1125(0.1281)(0.0500) = 0.000721$$

Next, we consider $\lambda = 3.0$. We now have:

$$P(2)P(6)P(1) = 0.2240(0.0504)(0.1494) = 0.001687$$

Since the joint probability of the observed sample outcomes varies as a function of λ, we shall denote this function by $L(\lambda)$. We have just computed two values of this function — namely, $L(4.5)$ and $L(3.0)$. These and other values of $L(\lambda)$ are plotted in Figure 10.7. It appears from this figure that $L(\lambda)$ is largest at $\lambda = 3.0$. Thus, $\lambda = 3.0$ is the estimate of λ that maximizes the joint probability of the observed sample outcomes and hence is the estimate that is most consistent with the sample results. It is no coincidence that $\overline{X}$ for this sample happens to equal 3.0 also, as we shall explain shortly. ☐

FIGURE 10.7   **Poisson likelihood function for the calculator demand example.**   The maximum likelihood estimate is that value of λ that maximizes the likelihood function.

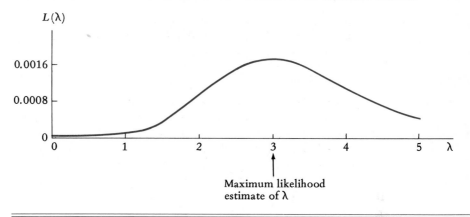

The function $L(\lambda)$ in the preceding example is called the *likelihood function*. The values of $L(\lambda)$ for different λ are called *likelihood values*—for instance, 0.000721 is the likelihood value for $\lambda = 4.5$. The value of the parameter λ that maximizes the likelihood function ($\lambda = 3.0$ in the example) is the desired estimate and is called the *maximum likelihood estimate* of λ.

EXAMPLE

We return to the calculator demand example to present the method of maximum likelihood more formally. To indicate explicitly that each sample outcome X depends on the value of the parameter λ, we shall now denote the Poisson probability function by $P(X; \lambda)$. The joint probability of obtaining the observations X_1, X_2, X_3, when viewed as a function of λ, is the likelihood function $L(\lambda)$. Here, $L(\lambda)$ equals:

$$L(\lambda) = P(X_1; \lambda)P(X_2; \lambda)P(X_3; \lambda) = \frac{\lambda^{X_1}\exp(-\lambda)}{X_1!} \cdot \frac{\lambda^{X_2}\exp(-\lambda)}{X_2!} \cdot \frac{\lambda^{X_3}\exp(-\lambda)}{X_3!}$$

$$= \frac{\lambda^{X_1+X_2+X_3}\exp(-3\lambda)}{X_1!X_2!X_3!}$$

For the specific sample outcomes, we have:

$$L(\lambda) = \frac{\lambda^{2+6+1}\exp(-3\lambda)}{2!\,6!\,1!} = \frac{\lambda^9\exp(-3\lambda)}{1440}$$

It is this function that is plotted in Figure 10.7.

General Approach

We now summarize the essence of the method of maximum likelihood when the random variable is discrete.

1. We start with the probability function of a sample observation X_i, that is, $P(X_i; \theta)$, where θ is the unknown parameter.
2. The joint probability function for the sample, because of the independence of the sample observations $X_1, X_2, \ldots, X_n$, is:

$$P(X_1; \theta)P(X_2; \theta) \cdots P(X_n; \theta)$$

This product, when viewed as a function of θ for given $X_1, X_2, \ldots, X_n$, is the *likelihood function* $L(\theta)$.

3. We maximize the likelihood function with respect to θ. This value of θ is a function of the sample observations and is the *maximum likelihood estimator* of θ.

(10.22)

> The method of maximum likelihood for estimating a parameter θ selects as a point estimator that value of θ that maximizes the likelihood function:
>
> $$L(\theta) = P(X_1; \theta)P(X_2; \theta) \cdots P(X_n; \theta)$$

The value of θ that maximizes $L(\theta)$ is usually obtained either by analytical methods or, when this approach is not feasible, by efficient, computerized numerical-search procedures. The next example illustrates the analytical approach and shows that $\overline{X}$ is the maximum likelihood estimator of λ for a random sample from a Poisson distribution.

EXAMPLE

(Calculus needed.) When a random sample of size n is selected from a Poisson distribution, $P(X; \lambda)$ is given by (6.8). Substituting into the likelihood function in (10.22), we obtain:

$$L(\lambda) = \left[\frac{\lambda^{X_1} \exp(-\lambda)}{X_1!} \right] \left[\frac{\lambda^{X_2} \exp(-\lambda)}{X_2!} \right] \cdots \left[\frac{\lambda^{X_n} \exp(-\lambda)}{X_n!} \right]$$

Algebraic simplification, using the fact that $\Sigma X_i = n\overline{X}$, leads to the likelihood function:

$$L(\lambda) = \frac{\lambda^{n\overline{X}} \exp(-n\lambda)}{X_1! X_2! \cdots X_n!}$$

To find the value of λ that maximizes $L(\lambda)$, we take the first derivative of $L(\lambda)$ with respect to λ and set it equal to zero, as follows:

$$\frac{dL}{d\lambda} = n\left(\frac{\overline{X}}{\lambda} - 1 \right) \frac{\lambda^{n\overline{X}} \exp(-n\lambda)}{X_1! X_2! \cdots X_n!} = 0$$

Setting λ equal to $\overline{X}$ solves this equation provided $\overline{X} > 0$. It can be shown that this solution corresponds to an absolute maximum of the likelihood function. If $\overline{X} = 0$ (because $X_1, X_2, \ldots, X_n$ are all zero), then setting λ equal to $\overline{X} = 0$ maximizes $L(\lambda)$. Hence, $\overline{X}$ is the maximum likelihood estimator of λ for a random sample from a Poisson distribution. $\qquad\qquad\qquad\qquad\qquad\qquad\square$

Properties of Maximum Likelihood Estimators

Under quite general conditions, maximum likelihood estimators are consistent estimators. Also, for reasonably large samples, they are usually approximately normally distributed and more efficient than other estimators. This last property makes the maximum likelihood estimators particularly useful in statistical analysis.

Comment   Another method that is widely used for finding point estimators with desirable properties is the method of *least squares*. We shall discuss this method in Chapter 18.

PROBLEMS

10.1   The mean claim in a random sample of 50 claims for railroad freight damages during the past year was $350. Distinguish between the estimator and the estimate here.

10.2   Cores are to be bored and the annual growth rings measured in a random sample of 25 trees in a large tract to estimate the mean growth per tree in the tract during the previous decade. Prior to the sample cores being obtained and measured, is the sample mean growth an estimator or an estimate? Explain.

10.3   A sample of four tires of a new type is subjected to forced-life testing. The first tire in the sample to wear out has X miles on it. Is X an unbiased estimator of the mean life of this type of tire? Explain.

10.4   A travel agents' association commissioned a survey of the winter travel plans and preferences of a random sample of retired persons. One item of particular interest was the mean anticipated expenditures per retired person for travel during the coming winter. The survey was conducted during the third week in September, and the response rate was 70 percent. Interviewers, when commenting on reasons for nonresponses, often noted that the nonrespondent was away on a trip. Explain what is the population mean μ here that $\overline{X}$, the mean anticipated expenditures per respondent in the sample survey, is estimating. Is μ here the quantity the travel agents' association wishes to estimate? Might a larger sample size have helped? Comment.

10.5   As noted in the text, $\overline{X}$ is relatively more efficient than Md as an estimator of μ in a normal population. Does this mean (1) that $\overline{X}$ will lie closer to μ than Md will in every random sample from a normal population? (2) That a smaller sample size is needed to estimate μ if $\overline{X}$ is used as the estimator than if Md is used? Explain.

10.6   The mean μ of a normal population is to be estimated by either the sample midrange (10.3) or the sample mean. The random sample size will be greater than 2. The sampling distribution of the sample midrange is symmetrical about μ in this case but has a larger variance than the sampling distribution of the sample mean.

a. Explain why the sample midrange is an unbiased estimator of μ here.

b. Which estimator — the sample midrange or the sample mean — (1) is relatively more efficient in this situation? (2) Would require the larger sample size to estimate μ with a given precision?

10.7   If a point estimator is biased but consistent, does this imply that the bias must approach zero as the sample size gets larger? Explain.

10.8   If two different statistics computed from the same sample are both consistent estimators of the same population parameter, does this imply that there is a high probability that the two statistics will be close in value when the sample size is large? Explain.

*10.9   **Parcel Postage.**   Amounts of postage were recorded for a random sample of $n = 400$ parcels handled by the postal system on a particular day. The sample mean and standard deviation were $\overline{X} = \$5.21$ and $s = \$3.70$, where X denotes the amount of postage on a parcel (in dollars). It is desired to obtain an interval estimate of μ, the mean amount of postage on all parcels handled by the postal system that day.

a. Estimate $\sigma\{\overline{X}\}$, the standard deviation of the sampling distribution of $\overline{X}$. What notation is used for this estimate?

b. Calculate the confidence limits $\overline{X} \pm 2s\{\overline{X}\}$. What confidence coefficient is associated with these limits?

c. Obtain a 90 percent confidence interval for μ.

d. If 100 independent random samples of 400 parcels each had been selected and a 90 percent confidence interval for μ calculated for each of the 100 samples, what proportion of the 100 confidence intervals would be expected to contain μ?

10.10   As part of a profitability analysis, a refrigeration company wished to estimate μ, the mean number of man-hours expended by company personnel per service call last year. A random sample of 144 service calls yielded $\overline{X} = 1.34$ man-hours and $s = 1.32$ man-hours, where X denotes the man-hours expended in a service call.

a. Calculate the confidence limits $\overline{X} \pm 1.0s\{\overline{X}\}$. How confident can one be that these limits cover μ?

b. A person, commenting on the analysis, said: "Since the confidence limits always include the mean $\overline{X}$, the confidence interval corresponding to $\overline{X} \pm s\{\overline{X}\}$ should always be correct." Do you agree? Comment.

c. Obtain a 90 percent confidence interval for the population mean, and interpret it.

10.11   **Trade Association.**   An apparel and shoe trade association has a large number of member firms. As part of a study of the impact of new minimum-wage legislation on association members, a random sample of 225 member firms was selected to estimate μ, the mean number of hourly paid employees in member firms. A computer analysis of the sample results showed that $\overline{X} = 8.31$ and $s = 4.80$, where X denotes the number of hourly paid employees in a firm.

a. Construct a 99 percent confidence interval for μ. Interpret your confidence interval.

b. Explain why the confidence coefficient of your interval estimate is only approximately 99 percent.

c. If you had constructed a 95 percent confidence interval, would it have been wider or narrower than the one in **a**? Which interval would involve the greater risk of an incorrect estimate?

10.12   **Airline Reservations.**   An airline researcher studied reservation records for a random sample of 100 days in order to estimate μ, the mean number of persons who fail to keep their reservations (no-shows) on the daily 4 P.M. commuter flight to New York City. The records revealed the following:

Number of No-Shows (X):	0	1	2	3	4	5	6
Number of Days:	20	37	23	15	4	0	1

a. Verify that $\overline{X} = 1.500$ and $s = 1.185$. Calculate $s\{\overline{X}\}$. What does $s\{\overline{X}\}$ estimate here?

b. Construct a 95 percent confidence interval for μ. Interpret the confidence interval.

c. Would a different random sample of 100 days have provided the same interval estimate as the one in b? Explain.

10.13 **Data Transmissions.** Data transmission errors tend to occur in clusters of $1, 2, 3, \ldots$ characters. A communications engineer selected a random sample of 200 clusters to estimate μ, the mean cluster size. The sample results showed $\overline{X} = 2.31$ characters and $s = 1.64$ characters, where X denotes the number of characters in the error cluster.

a. Construct a 90 percent confidence interval for μ. Interpret the confidence interval.

b. Explain why the confidence coefficient of the interval estimate is only approximately 90 percent.

10.14 A research report stated that the mean return on invested capital earned by furniture manufacturers is between 6.1 and 10.8 percent per year, with a confidence coefficient of 95 percent. One person interpreted this result as meaning that 95 percent of furniture manufacturers had investment returns between 6.1 and 10.8 percent. Another person interpreted it in the sense that if many random samples were taken, 95 percent of them would have sample means between 6.1 and 10.8 percent. Why are these interpretations incorrect?

10.15 Refer to Figure 10.6.

a. In the construction of a confidence interval for a population mean based on a large random sample, what is the ratio of the width of a 90 percent confidence interval to that of an 80 percent confidence interval? What is the ratio of the width of a 99 percent confidence interval to that of a 90 percent confidence interval? What generalization is suggested by these results?

b. Can a 100 percent confidence interval for a population mean be constructed? Comment.

*10.16 **Nutrition Study.** In a study of the effects of nutrition on work efficiency, a random sample of 16 workers were asked to follow a rigid dietary regimen. In one part of this study, the blood sugar level (X) of each sample worker was measured two hours after breakfast. The results (in milligrams of sugar per 100 cubic centimeters of blood) were $\overline{X} = 112.8$ and $s = 9.6$. Assume that X is normally distributed. Construct a 95 percent confidence interval for the mean blood sugar level under this regimen, and interpret your interval estimate.

10.17 **Vehicle Fleet.** An examination of the records for a random sample of nine motor vehicles in a large fleet shows the following operating costs (in cents per mile):

$$28.3 \quad 26.4 \quad 27.0 \quad 22.5 \quad 23.5 \quad 29.1 \quad 26.8 \quad 26.7 \quad 30.9$$

Let μ denote the mean operating cost in cents per mile for vehicles in the fleet. Assume that operating costs are nearly normally distributed. Construct a 90 percent confidence interval for μ. Interpret your interval estimate.

10.18 A scientist is studying the effect of communication modes on the problem-solving capability of teams. For one mode, a random sample of 16 teams completed a specific task in an average of 25.9 minutes, with a standard deviation of 3.6 minutes. Assume that completion times follow a distribution that does not depart from normality too markedly. Construct a 95 percent confidence interval for the mean time required to complete the task with this communication mode. Interpret your interval estimate.

10.19 A computer industry publication wished to estimate the mean percentage return on equity last year for all computer software firms. A random sample of $n = 14$ firms was drawn from the many firms in the software sector. The results were as follows:

8.37	31.38	−6.52	24.87	−8.47	24.83	17.78
−3.15	7.20	15.45	0.35	11.28	21.01	26.68

Assume that returns on equity follow a nonnormal distribution that does not depart from normality too markedly. Construct a 90 percent confidence interval for the population mean percentage return on equity, and interpret your interval estimate.

10.20 Refer to **Nutrition Study** Problem 10.16. A scientist doubts that the blood sugar levels are exactly normally distributed, though the departure should not be marked. If the scientist is correct, what does the robustness of the t distribution imply about the appropriateness of the confidence interval constructed in this problem? Comment.

10.21 A work study is undertaken to estimate, with a 95 percent confidence interval, the mean time required to weld a metal seam. Assume that the welding times are normally distributed, and consider a random sample of n trial welds.
   a. Obtain the t percentile for the confidence limits (10.15) if n is (1) 3, (2) 30, (3) 300.
   b. Does the t percentile in **a** shrink by much when n increases from 3 to 30? From 30 to 300? What is the implication of this finding for the effect of the t multiple on the width of a confidence interval with large sample sizes?

*10.22 Refer to **Parcel Postage** Problem 10.9. Two million parcels were handled by the postal system on the particular day. Obtain a 90 percent confidence interval for the total amount of postage on the parcels handled that day. Interpret your confidence interval.

10.23 Refer to **Trade Association** Problem 10.11. The association has 9100 member firms. Construct a 99 percent confidence interval for the total number of hourly paid employees of all member firms. Interpret your interval estimate.

10.24 Refer to **Vehicle Fleet** Problem 10.17. All of the vehicles in the fleet travel about the same number of miles during a month. Construct a 90 percent confidence interval for the expected total cost of operating the fleet in a month when a total of 5 million vehicle-miles is accumulated. Interpret your confidence interval.

*10.25 A sample survey of kindergarten children in a state is being planned to estimate, among other things, the mean number of older siblings of such children. It is desired to estimate this mean within ± 0.08, with a 90 percent confidence coefficient. A reasonable planning value for σ is 0.6.
   a. What sample size is needed to estimate the mean number of older siblings?
   b. If the desired precision were tightened to ± 0.06, would the required sample size be increased substantially?

10.26 A nationwide survey of practicing physicians is to be undertaken to estimate μ, the mean number of prescriptions written per day. The desired margin of sampling error is ± 0.75, with a 99 percent confidence coefficient. A pilot study revealed that a reasonable planning value for the population standard deviation is 5.
   a. How many physicians should be contacted in the survey to estimate μ?
   b. If the desired confidence coefficient were lowered to 95 percent, would the required sample size be reduced substantially?

10.27 Refer to **Airline Reservations** Problem 10.12. Sometime after this study, the airline instituted a new reservations system that discourages customers from holding simultaneous reservations on different flights to the same destination. It is now desired to reestimate the mean number of no-shows for the daily 4 P.M. flight to New York City. A 95 percent confidence interval, with a half-width of 0.25, is wanted. The standard deviation of the original sample ($s = 1.185$) is to be taken as the planning value for σ.

    a. How many days should be included in the sample to prepare the estimate?

    b. If the desired confidence coefficient were increased to 99 percent (other factors being unchanged), how many days would have to be included in the sample? Is the resultant change in sample size substantial relative to the increase in the confidence coefficient?

    c. What is the likely effect on the confidence interval if the planning value for σ is too large? Why can you not discuss the effect on the confidence interval for any one sample?

10.28 Refer to **Data Transmissions** Problem 10.13. Suppose that the sample size has not yet been specified. The communications engineer wishes to estimate the mean cluster size within ±0.1 character, with a 90 percent confidence coefficient. A reasonable planning value for σ is 1.6 characters. Find the number of clusters that must be observed to provide the desired interval estimate.

*10.29 Refer to **Parcel Postage** Problem 10.9.

    a. Construct a lower 95 percent confidence interval for μ. Explain why the limit of this lower 95 percent confidence interval is the same as the lower limit of the two-sided 90 percent confidence interval calculated in Problem 10.9c.

    b. Can the postal authorities confidently claim that the total amount of postage for the 2 million parcels handled that day exceeded $9 million? Explain.

10.30 Refer to **Trade Association** Problem 10.11.

    a. Obtain upper 99 percent confidence intervals for (1) the mean number of hourly paid employees per firm, (2) the total number of hourly paid employees in the 9100 member firms.

    b. The trade association is contemplating developing a mailing list containing all hourly paid employees and needs an indication of the minimum number of hourly paid employees that will be contained in the list. Obtain the appropriate 99 percent confidence limit.

10.31 Refer to **Nutrition Study** Problem 10.16.

    a. Construct a lower 95 percent confidence interval for the mean blood sugar level under this regimen.

    b. Why can one not conclude that 95 percent of workers' blood sugar levels 2 hours after eating the regimen breakfast will exceed the confidence limit in **a**?

10.32 When the soldering process for wire connections in an electric appliance meets quality assurance standards, the breaking strength of a connection has a normal distribution with $\mu = 1400$ milligrams and $\sigma = 350$ milligrams. An inspector will choose one connection at random for testing.

    a. Construct a 99 percent prediction interval for the breaking strength of the connection if the process standards are being met.

    b. If the breaking strength of the new connection is 450 milligrams, what does this imply?

*10.33 The monthly rates of return on a common stock over the past five months were (in percent):

$$-0.2 \quad 1.3 \quad 1.0 \quad -0.6 \quad 2.1$$

Assume that monthly rates of return for the stock constitute independent random observations from a normal population.

    a. Construct a 95 percent prediction interval for the rate of return on the stock for next month. Interpret the prediction interval.

    b. Would a 95 percent confidence interval for the mean monthly rate of return for this stock be wider or narrower than the prediction interval in **a**? Why?

10.34 The demand elasticity of a price promotion of a product is the ratio of the proportionate change in sales volume induced by the price promotion to the proportionate change in the price of the product. In three recent price promotions of a product, the demand elasticities were -1.80, -1.93, and -1.65. Assume that these three elasticities constitute a random sample from a normal population.

   a. Construct a 95 percent prediction interval for the elasticity of the next price promotion, assuming that this elasticity is an independent observation from the same population as the previous three observations.

   b. If the elasticity for the next promotion happens to be -0.95, would this outcome suggest that the assumption in **a** might not be valid? Discuss.

10.35 The mean and the standard deviation of the annual snowfalls in a northern city for the past 20 years are 2.03 meters and 0.45 meter, respectively. Assume that annual snowfalls for this city are random observations from a normal population. Construct a 90 percent prediction interval for the snowfall next year. Interpret the prediction interval.

10.36 An oil company executive was furnished a 95 percent prediction interval of $300 \leq X_{new} \leq 1550$ barrels per day for the flow of a new test well being drilled in a field, based on experience with four other test wells in the same field. He interpreted the interval as stating that the new well will have daily flows varying between 300 and 1550 barrels per day. Why is the executive's interpretation not accurate?

*10.37 The number of annual visits to a dentist by a child is a Poisson random variable X with unknown parameter value λ. In a random sample of two children, the numbers of visits made to the dentist last year were $X_1 = 0$ and $X_2 = 3$.

   a. Obtain the values of the likelihood function $L(\lambda)$ for this sample outcome at $\lambda = 0, 1, 1.5, 2, 3$.

   b. Sketch a graph of $L(\lambda)$, as in Figure 10.7. Does the sample mean $\overline{X}$ appear to be the maximum likelihood estimator of λ, as the theory in the text states? Explain.

10.38 The numbers of medal winners from a high school in three successive state gymnastic competitions were $X_1 = 1$, $X_2 = 0$, $X_3 = 2$. Assume that these three observations constitute a random sample from a Poisson distribution with unknown parameter value λ.

   a. Obtain the values of the likelihood function $L(\lambda)$ for this sample outcome at $\lambda = 0, 0.5, 1.0, 1.5, 2$.

   b. Sketch a graph of $L(\lambda)$, as in Figure 10.7. Which value of λ appears to have the greatest likelihood? Is this finding consistent with the theoretical result presented in the text? Explain.

EXERCISES

10.39 Consider the following probability distribution for an infinite population.

x:	0	2
$P(x)$:	0.5	0.5

   a. Verify that $\sigma = 1$ for this population.

   b. Construct the sampling distributions of s^2 and s for a simple random sample of $n = 2$ from this population.

   c. As noted in the text, s^2 is an unbiased estimator of σ^2 for an infinite population, but s is a biased estimator of σ. Verify this fact for the sampling distributions of s^2 and s constructed in **b**. What is the amount of bias in s here?

10.40 Given that $E\{s^2\} = \sigma^2$, prove that $s^2\{\bar{X}\}$ is an unbiased estimator of $\sigma^2\{\bar{X}\}$ for a random sample from an infinite population.

10.41 With simple random sampling, what distribution does each of the following statistics follow if (1) the population is normal and n is small? (2) The population is normal and n is large? (3) The population is not normal and n is small? (4) The population is not normal and n is large?

a. $\bar{X}$

b. $\dfrac{\bar{X} - \mu}{\sigma\{\bar{X}\}}$

c. $\dfrac{\bar{X} - \mu}{s\{\bar{X}\}}$

10.42 It is desired to estimate the mean μ of a Poisson population with a relative margin of sampling error of 10 percent (that is, within $\pm 0.1\mu$), using a 95 percent confidence coefficient. A reasonable planning value for μ is 1.5. What is the required sample size? (*Hint:* $\mu = \sigma^2$ for a Poisson population.)

10.43 Show that the interval estimate $\mu \geq L$, where $L = \bar{X} - t(1 - \alpha; n - 1)s\{\bar{X}\}$, is a lower $1 - \alpha$ confidence interval for μ based on a random sample of size n from a normal population. [*Hint:* Show that $P(L \leq \mu) = 1 - \alpha$ prior to sampling.]

10.44 The time until relapse for a patient suffering from a certain chronic disease has an exponential probability distribution. In a random sample of two patients, the numbers of days until relapse were $X_1 = 60$ and $X_2 = 100$.

a. State the likelihood function $L(\lambda)$ for this sample outcome, and evaluate it at $\lambda = 0.010$, 0.0125, 0.015, 0.020.

b. Sketch a graph of the likelihood function of λ for this sample outcome. Confirm by inspection that $1/80$ is the maximum likelihood estimate of λ.

c. Suppose that the relapse times of a random sample of n patients were observed. Verify that the likelihood function of λ in this case is $L(\lambda) = \lambda^n \exp(-\lambda n\bar{X})$, where $\bar{X}$ denotes the sample mean relapse time.

10.45 (Calculus needed.) Refer to Problem 10.44c. Show that $1/\bar{X}$ is the maximum likelihood estimator of λ in this case.

STUDIES

10.46 The 10 consecutive sets of five numbers in column 1 of Table C.9 are equivalent to 10 independent random samples of size $n = 5$ from $N(0, 1)$. The sample means and standard deviations for these 10 samples follow.

Sample	$\bar{X}$	s	Sample	$\bar{X}$	s
1	0.8852	0.6117	6	0.3080	1.2288
2	0.2532	1.2749	7	0.3896	0.5858
3	-0.2960	1.0780	8	-0.3112	1.0929
4	-0.3822	1.1650	9	-0.1750	0.7936
5	-0.8342	0.9706	10	0.2522	0.6594

a. For each sample, calculate an 80 percent confidence interval for μ of the form $\bar{X} \pm z\sigma\{\bar{X}\}$. (*Hint:* See Comment 3, p. 289.) How many of the 10 intervals are correct; that is, how many contain $\mu = 0$?

b. Redo **a**, using intervals of the form $\overline{X} \pm ts\{\overline{X}\}$.

c. Explain why, in both **a** and **b**, 8 out of 10 intervals are expected to contain $\mu = 0$. How do the actual numbers compare with the expected numbers in the two cases?

10.47 An aluminum company is experimenting with a new design for electrolytic cells in smelter pot-rooms. A major design objective is to maximize a cell's expected service life. Thirty cells of the new design were started and operated under similar conditions and failed at the following ages (in days):

634	976	1184	1355	1585	1820
751	1022	1252	1468	1608	1943
812	1120	1295	1477	1683	1980
855	1158	1310	1502	1698	1992
947	1168	1341	1535	1711	2192

Two items of concern to management are (1) the mean service life of this design, and (2) the comparative performance of this design with the standard industry design, which is known to have a mean service life of 1300 days. Management does not want to conclude that the new design is superior to the standard one unless the evidence is fairly strong.

a. Assuming that the distribution of service lives of the cells is normal, calculate an appropriate 95 percent confidence interval for the mean service life of cells of the new design. Justify your choice of a one- or a two-sided confidence interval. Should management conclude that the new design is superior to the standard one with respect to mean service life? Comment.

b. It has been suggested that the logarithms of service lives are more normally distributed than the original observations. Take logarithms (to base 10) of the service life data, and calculate the same type of confidence interval as in **a** for the mean log-service-life of cells of the new design. Cells of the standard design are known to have a mean log-service-life of 3.095 (to base 10). Can management claim with confidence that the mean log-service-life is greater for cells of the new design than for cells of the standard design?

c. Graph histograms of the original data and the log data. Does the distribution of log-service-lives appear to be more normal than the distribution of service lives, as suggested? Comment.

d. A management objective is to obtain a large total service life for the cells. Is the mean service life or the mean log-service-life the more relevant measure here? Explain.

10.48 A state employment office will survey business establishments in the state about their hiring plans for college students next summer. A questionnaire is to be mailed to a random sample of the 31,800 establishments in the state to obtain information for each sample establishment about the number of summer job positions it plans to create and the total number of student-weeks of employment for the summer positions.

a. The state employment office desires to estimate the total number of summer positions to be created and the total number of student-weeks for all establishments by means of 95 percent confidence intervals with half-widths of at most 3000 positions and 50,000 student-weeks, respectively. Planning values for the population standard deviations, based on similar surveys in previous years, are 1.1 positions and 19.3 student-weeks, respectively. Assume a 100 percent response rate for purposes of planning sample size. What is the smallest sample size that will give both interval estimates with the required precision at the desired confidence level?

b. It was finally decided to select a random sample of 650 establishments, and responses were obtained from each. The results were as follows:

Variable	Sample Mean	Sample Standard Deviation
Positions	1.040	1.212
Student-weeks	14.23	17.40

Calculate the desired 95 percent confidence intervals.

c. If, in fact, the survey questionnaire had been mailed to 1000 establishments selected at random but only the 650 establishments referred to in **b** actually replied, what interpretation could be given to the interval estimates in **b** under these circumstances? Would your answer be affected if the responding establishments were larger (as measured by the size of their work force), on average, than for the population as a whole? Discuss.

TESTS FOR POPULATION MEAN

Often, one desires to use sample data to draw a conclusion about whether a population parameter, such as the population mean, differs from a specified standard or has changed from its previous level. The statistical procedure used to draw an appropriate conclusion from sample data about a population parameter is called a *statistical test*. We now give three illustrations of test situations involving the population mean.

EXAMPLES

1. A manufacturing concern has received a sample of 36 modified machine components and measured their service lives. It wishes to test whether or not the mean service life of the modified component exceeds the mean service life of the original component.

2. A financial analyst has measured the current return offered by each mutual fund in a sample of 15 funds. He wishes to test whether or not mutual funds now offer a mean return that is lower than the mean return offered in earlier decades.

3. A psychologist who is concerned with job performance wishes to test whether or not the mean attention span for a task when an employee works alone is the same as the current mean attention span when several employees work in the same room. ☐

In this chapter, we discuss statistical tests for the population mean μ that involve two alternative conclusions and are based on simple random samples. This type of test is commonly encountered in practice and provides a basis for understanding other statistical tests to be discussed subsequently.

11.1 STATISTICAL TESTS

Specifying the Alternative Conclusions

The first step in conducting a statistical test is to specify the two alternative conclusions, one of which is to be chosen on the basis of the sample data. To illustrate how the alternatives are specified, let us look more closely at Example 1.

EXAMPLE

Machine Component. An ultrasonic-equipment manufacturer employs a component in one of its machines that must withstand considerable stress from vibration. An all-metal component has been available for some years from a supplier. Extensive past experience has shown that this component has a mean service life of 1100 hours. The research division of the supplier has just developed a modified component constructed from bonded metal and plastic. The manufacturer wishes to know whether or not the mean service life of the modified component (μ) exceeds the mean service life of 1100 hours of the original component. Thus, the manufacturer wishes to choose between the following two alternative conclusions:

1. The mean service life of the modified component is 1100 hours or less.
2. The mean service life of the modified component is more than 1100 hours.

We state these alternatives symbolically as follows:

H_0: $\mu \leq 1100$

H_1: $\mu > 1100$

Here, H_0 and H_1 denote the alternative conclusions. To conclude H_0 in this case is to conclude that the mean service life of the modified component (μ) is 1100 hours or less. To conclude H_1 is to conclude the opposite of H_0, namely, that μ is more than 1100 hours. ☐

Note three important aspects of this decision problem. It has been formulated as a choice between two alternatives that are mutually exclusive with respect to μ. Thus, either H_0 is true because $\mu \leq 1100$ or H_1 is true because $\mu > 1100$, but not both. Furthermore, which of the two alternatives is true is not known. Finally, the alternatives are defined by a special value (1100 in this case) that is the standard against which the population mean is to be compared. We denote this standard by μ_0. In this illustration, $\mu_0 = 1100$.

We next illustrate two additional types of alternatives encountered in tests of μ.

EXAMPLES

1. An engineer has observed the hourly output of a machine for 20 hourly periods selected at random. She is concerned with whether or not the mean hourly output is less than $\mu_0 = 40$ units. The alternatives of interest may be stated symbolically:

    H_0: $\mu \geq 40$

    H_1: $\mu < 40$

    To conclude H_0 here is to conclude that the mean hourly output (μ) is 40 units or more. To conclude H_1 is to conclude the opposite, namely, that μ is less than 40 units.

2. **Lake Acidity.** A scientist made a pH measurement of acidity on each of 18 water samples drawn from a lake. He wishes to know whether the mean pH

level for the lake differs from $\mu_0 = 7.0$ (the neutral pH level). The alternatives of interest may be stated symbolically:

$$H_0: \mu = 7.0$$

$$H_1: \mu \neq 7.0$$

To conclude H_0 here is to conclude that the mean pH level for the lake is 7.0 and, hence, that its water is neutral. To conclude H_1 is to conclude the opposite, namely, that the mean pH level of the lake is not 7.0 and, hence, that its water is not neutral.   □

In general, the three forms of alternatives we shall consider in this chapter are as follows.

(11.1a)   *One-Sided Upper-Tail Alternatives:*

$$H_0: \mu \leq \mu_0$$

$$H_1: \mu > \mu_0$$

(11.1b)   *One-Sided Lower-Tail Alternatives:*

$$H_0: \mu \geq \mu_0$$

$$H_1: \mu < \mu_0$$

(11.1c)   *Two-Sided Alternatives:*

$$H_0: \mu = \mu_0$$

$$H_1: \mu \neq \mu_0$$

Notice that the three forms of alternatives in (11.1) have been given names. Each name describes the range of values for μ given in alternative H_1. Thus, in (11.1c), H_1 includes values of μ on both sides of μ_0; hence, this form is called a *two-sided alternative*. Similarly, in (11.1a), H_1 includes values of μ that are larger than μ_0; hence, this form is called a *one-sided upper-tail alternative*. The reason for the term *upper-tail* will become apparent shortly.

Also notice in the three forms in (11.1) that we follow the convention of labeling the alternative containing the equality sign as H_0.

Comment   In statistical testing, the alternative conclusions H_0 and H_1 are often called *hypotheses* since each conclusion may be thought of as a hypothesis about the true value of the parameter. For instance, $H_0: \mu \leq 1100$ represents the hypothesis that μ is 1100 or less. In this terminology, H_0 is referred to as the *null hypothesis* of the test.

Statistical Decision Rule

As noted in the introduction, sample data are used to choose between the alternative conclusions H_0 and H_1 in a statistical test. For instance, in a test involving the alternatives H_0: $\mu \leq \mu_0$ and H_1: $\mu > \mu_0$, the sample mean $\overline{X}$ usually forms the basis for making the choice. If $\overline{X}$ is much smaller than μ_0, it is sensible to conclude that the population mean μ is smaller than μ_0 and, hence, that H_0 is true. On the other hand, if $\overline{X}$ is much larger than μ_0, one will conclude that μ is larger than μ_0 and, hence, that H_1 is true. We illustrate this reasoning with two examples.

EXAMPLES

1. In the machine component example, it is desired to choose between H_0 ($\mu \leq 1100$) and H_1 ($\mu > 1100$) on the basis of the sample mean life $\overline{X}$ of 36 modified machine components. If $\overline{X}$ is much smaller than 1100, it will be reasonable to conclude H_0, that is, that the modified component is not superior to the original component with respect to mean service life. If $\overline{X}$ is much larger than 1100, the opposite conclusion (H_1) will be drawn. A specific rule, called a *statistical decision rule,* will be employed for deciding on the basis of $\overline{X}$ which conclusion should be chosen. The statistical decision rule might be as follows:

    If $\overline{X} \leq 1150$, conclude H_0 ($\mu \leq 1100$).
    If $\overline{X} > 1150$, conclude H_1 ($\mu > 1100$).

    With this rule, H_0 is concluded if $\overline{X}$ is 1150 or less, and H_1 is concluded if $\overline{X}$ is greater than 1150.

2. For the lake acidity example, the alternative conclusions are H_0: $\mu = 7.0$ and H_1: $\mu \neq 7.0$. If $\overline{X}$, the mean pH of the 18 water samples, is close to 7.0, the scientist should conclude that the lake water is neutral, that is, conclude H_0. If $\overline{X}$ differs substantially from 7.0, then he should conclude that the lake water is not neutral, that is, conclude H_1. The statistical decision rule here might be as follows:

    If $6.8 \leq \overline{X} \leq 7.2$, conclude H_0 ($\mu = 7.0$).
    If $\overline{X} < 6.8$ or $\overline{X} > 7.2$, conclude H_1 ($\mu \neq 7.0$).

    With this rule, the scientist concludes H_0 if $\overline{X}$ is in the interval from 6.8 to 7.2 inclusive and concludes H_1 if $\overline{X}$ is either smaller than 6.8 or larger than 7.2. □

Before discussing how statistical decision rules are constructed, we must first consider the kinds of incorrect decisions that can be made in performing a test using a statistical decision rule because of the presence of sampling errors.

Errors of Inference

We know from our study of the sampling distribution of $\overline{X}$ in Chapter 9 that $\overline{X}$ is subject to sampling variability and hence generally differs from the population mean μ.

Consequently, when $\overline{X}$ is used in a statistical decision rule to choose between the two alternative conclusions about μ, there exist risks of drawing incorrect conclusions.

EXAMPLE

In the lake acidity example, the lake water may be neutral (that is, $\mu = 7.0$), but because of sampling error, the sample mean may be $\overline{X} = 7.4$. Using the earlier decision rule, the scientist would then mistakenly conclude H_1 ($\mu \neq 7.0$) because $\overline{X} = 7.4$ is larger than 7.2. Another possibility is that the lake is somewhat acidic with $\mu = 6.7$, yet because of sampling error, $\overline{X} = 7.1$. Using the same decision rule, the scientist would then mistakenly conclude H_0 ($\mu = 7.0$) because $\overline{X} = 7.1$ lies between 6.8 and 7.2. ☐

Drawing the incorrect conclusion in a statistical test is called an *error of inference* because it entails drawing an incorrect inference or conclusion from the sample about the population parameter. There are two possible types of errors when choosing between two alternatives of which one must be true. Table 11.1 shows the general situation.

Type I Errors.   When H_0 is the true alternative, we make a correct inference if we draw conclusion H_0 from the sample and an incorrect inference if we draw conclusion H_1. This type of incorrect inference is called a *Type I error*.

EXAMPLE

In the machine component example, the two alternatives are:

H_0: $\mu \leq 1100$

H_1: $\mu > 1100$

A Type I error corresponds to concluding that the modified component has a longer mean service life (H_1) when, in fact, it is not longer. ☐

As shall be demonstrated shortly, the probability or risk of making a Type I error can be ascertained mathematically for a given test situation. Risks of making Type I errors are called α *risks* (Greek alpha).

Type II Errors.   Referring again to Table 11.1, we see that when H_1 is the true alternative, we make a correct inference if we draw conclusion H_1 from the sample and an incorrect inference if we draw conclusion H_0. This type of incorrect inference is called a *Type II error*.

TABLE 11.1   **Two types of errors of inference**

Conclusion Drawn from Sample	True Alternatives	
	H_0	H_1
H_0	Correct conclusion	Type II error
H_1	Type I error	Correct conclusion

| EXAMPLE | In the machine component example, a Type II error corresponds to concluding that the modified component does not have a longer mean service life (H_0) when, in fact, it is longer. $\square$ |

As with Type I errors, the probability or risk of making a Type II error can be ascertained mathematically for a given test situation. Risks of making Type II errors are called β *risks* (Greek beta).

Definitions.   We now present formal definitions of the two types of inferential errors and their associated risks.

(11.2)

> When H_0 is true, the incorrect decision that might be made is to conclude H_1. This kind of incorrect decision is called a *Type I error,* and the probability of making that incorrect decision is called an α *risk.*
>
>   When H_1 is true, the incorrect decision that might be made is to conclude H_0. This kind of incorrect decision is called a *Type II error,* and the probability of making that incorrect decision is called a β *risk.*

Control of Error Risks.   In developing the statistical decision rule for a test, we would like to control both the α and β risks at tolerable levels. Some risks must be accepted, of course, whenever conclusions are based on sample data, but the magnitudes of the risks should be controlled. Control over both kinds of risks can be achieved when the sample size is not predetermined; we shall consider this case later. In the common case where the sample size is predetermined, only one of the two types of risks can be controlled at a specified level. In that case, it is reasonable to control at a low level the more serious type of risk, that is, the risk of error that would result in the more costly or undesirable outcome.

| EXAMPLE | A chemical processor is testing the mean toxicity level in a batch of product. The more costly of the two types of errors here is to conclude that the toxicity of the batch is within safety limits when, in fact, it is not. Hence, this risk must be kept low. On the other hand, concluding that the batch has excess toxicity when, in fact, it does not is a less costly error here. Hence, this risk need not be as low. $\square$ |

The seriousness of a Type I error or a Type II error generally will vary with the value of the parameter μ.

| EXAMPLE | In the machine component example, a conclusion that H_0 is true ($\mu \le 1100$) when, in fact, μ is 1500 hours represents a serious Type II error because of the substantial superiority of the modified component over the original one. On the other hand, concluding that $\mu \le 1100$ when, in fact, μ equals 1101 is not serious, even though technically a Type II error is committed. $\square$ |

We shall adopt the convention of formulating the alternative conclusions for a test so that the Type I error is the more serious type of error and consequently the α risk is to be controlled at a low level. The following examples illustrate how the alternatives are set up to make the Type I error the one to be controlled.

EXAMPLES

1. In the machine component example, the more serious error is to conclude that the modified component has a longer mean service life than 1100 hours (the mean for the original component) when, in fact, the mean life is not longer. This error is particularly costly because it would lead to an expensive conversion of the production process from the original to the modified component, and then a switch back to the original component when the error is confirmed by subsequent experience. In order to make the Type I error the more serious type of error here, we set up the alternatives in terms of $\mu_0 = 1100$ as follows:

    H_0: $\mu \leq 1100$

    H_1: $\mu > 1100$

    Note that a Type I error with these alternatives entails concluding $\mu > 1100$ when in reality $\mu \leq 1100$.

2. In the machine component example, suppose that the cost of shifting to the modified component and, if need be, back to the original component were comparatively small. Furthermore, management has determined that there is a substantial monetary gain in switching to the modified component when the mean service life of the modified component is 1200 hours or more. In this case, the alternatives should be formulated as follows in terms of $\mu_0 = 1200$ to make the Type I error the more serious one:

    H_0: $\mu \geq 1200$

    H_1: $\mu < 1200$

    The α risk here is the risk of concluding that the modified component has a mean service life of less than 1200 hours when, indeed, its mean service life is 1200 hours or more.

3. It is widely believed that eating normal amounts of a certain food additive does not increase systolic blood pressure. A scientist is setting out to test the validity of this belief. The mean blood pressure of subjects who do not consume any of the additive is known to be 120. Let μ denote the mean blood pressure of subjects who consume normal amounts of the additive. The scientist feels that the more serious error is to claim that consumption of normal amounts of the additive increases blood pressure when, in fact, it does not. She should therefore set up the test alternatives as follows in terms of $\mu_0 = 120$ to make the α risk the more serious one:

    H_0: $\mu \leq 120$

    H_1: $\mu > 120$

A Type I error is made here if the scientist concludes that consumption of normal amounts of the additive leads to a higher mean blood pressure (that is, concluding H_1) when, in fact, it does not (that is, when H_0 is actually true). ☐

11.2   NATURE OF STATISTICAL DECISION RULES

We illustrated in the preceding section how a statistical decision rule is used to choose between the two test alternatives, H_0 and H_1, on the basis of sample data. We now shall discuss the nature of statistical decision rules in more detail. We begin by again considering the machine component example.

EXAMPLE

In the machine component example, the alternatives are H_0: $\mu \leq 1100$ and H_1: $\mu > 1100$. The choice between H_0 and H_1 is to be based on a random sample of 36 modified components that have been prepared as a pilot batch by the supplier. Each of these components has been subjected to a forced-life test to determine its service life in hours. The sample mean $\overline{X}$ will be the relevant statistic for choosing between H_0 and H_1. If $\overline{X}$ is much smaller than 1100, it will be reasonable to conclude $\mu \leq 1100$, that is, to conclude H_0, because $\overline{X}$ is expected to lie somewhere in the neighborhood of the population mean life μ. If, on the other hand, $\overline{X}$ is much larger than 1100, it will be reasonable to conclude that $\mu > 1100$, that is, to conclude H_1. If, however, $\overline{X}$ lies near 1100, then the appropriate conclusion may not be readily apparent. The appropriate conclusion becomes apparent when we recall that the alternatives were formulated so that the α risk is the more serious of the two types of risks. Hence, one should avoid concluding H_1 unless the sample evidence clearly suggests that H_1 is true. The appropriate conclusion when $\overline{X}$ is near 1100 is therefore conclusion H_0, because this choice will reduce the risk of a Type I error (that is, reduce the probability of concluding H_1 when H_0 is true).

This commonsense argument tells us that a reasonable form for the statistical decision rule in this case is as follows:

If $\overline{X} \leq A$, conclude H_0 ($\mu \leq 1100$).
If $\overline{X} > A$, conclude H_1 ($\mu > 1100$).

Here, A is some number that is larger than 1100. We subsequently refer to A as the action limit of the decision rule and to $\overline{X}$ as the test statistic of the decision rule. ☐

Definition

A statistical decision rule involves a test statistic and an action limit.

(11.3)

> A *statistical decision rule* specifies for each possible outcome of the sample *test statistic* which alternative, H_0 or H_1, should be selected. The value A in the decision rule is called its *action limit*.

Acceptance and Rejection Regions

For one-sided upper-tail alternatives (the type encountered in the machine component example), the appropriate form of the decision rule is illustrated in Figure 11.1a. Note that the rule provides that $\overline{X}$ values in the neighborhood of μ_0 or lower lead one to conclude, or to accept, H_0. For this reason, this range of $\overline{X}$ values is called the acceptance region. Correspondingly, large values of $\overline{X}$ lead one to conclude H_1, or stated equivalently, to reject H_0. Hence, this range of $\overline{X}$ values is called the rejection region. The action limit A is the value that separates the acceptance and rejection regions.

FIGURE 11.1   General form of the statistical decision rule for three types of alternatives concerning population mean μ

(a) One–sided upper–tail alternatives

$H_0: \mu \le \mu_0$
$H_1: \mu > \mu_0$

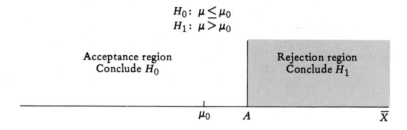

(b) One–sided lower–tail alternatives

$H_0: \mu \ge \mu_0$
$H_1: \mu < \mu_0$

(c) Two–sided alternatives

$H_0: \mu = \mu_0$
$H_1: \mu \ne \mu_0$

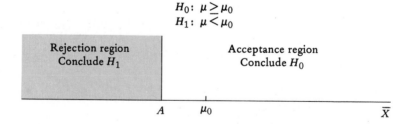

(11.4)

In a statistical decision rule, the range of values of the test statistic for which alternative H_0 is concluded is called the *acceptance region*. The range of values for which alternative H_1 is concluded is called the *rejection region*.

Similar reasoning leads us to suitable types of statistical decision rules for one-sided lower-tail alternatives and two-sided alternatives. The general forms of these rules are illustrated in Figures 11.1b and 11.1c, respectively. For one-sided lower-tail alternatives, the acceptance region includes $\overline{X}$ values in the neighborhood of μ_0 and above, while the rejection region has $\overline{X}$ values at the lower end of the scale. For two-sided alternatives, the acceptance region encompasses $\overline{X}$ values in the neighborhood of μ_0, while the rejection region includes all $\overline{X}$ values that lie either above or below this neighborhood of μ_0. In the two-sided case, the lower and upper ends of the acceptance region are defined by two action limits, which we denote by A_1 and A_2, respectively.

Comments

1. The rejection region is sometimes called the *critical region* of the statistical decision rule. Correspondingly, the action limit is sometimes called the *critical value*.
2. *Concluding H_0* is often referred to as *accepting H_0*. Both of these terms stand for acting as if H_0 is correct. Similarly, *concluding H_1* is often called *rejecting H_0* or *accepting H_1*, all of which stand for acting as if H_1 is correct.

11.3 CONSTRUCTION AND APPLICATION OF STATISTICAL DECISION RULES

Figure 11.1 shows the general form of the decision rule for each of the three types of alternatives. To actually construct the decision rule in a specific case, one must determine the position of the action limit(s) or, equivalently, precisely how far $\overline{X}$ must be from μ_0 in the direction of H_1 before H_1 is concluded. We now consider how to construct a statistical decision rule.

Assumptions

Our method of construction is based on two assumptions:

1. The sample data constitute a large simple random sample.
2. The risk of a Type I error, that is, the α risk, is to be controlled at $\mu = \mu_0$.

As a result of assumption 1, the sampling distribution of $\overline{X}$ is approximately normal. Furthermore, by the extension of the central limit theorem in (10.13), the quantity $(\overline{X} - \mu)/s\{\overline{X}\}$ has an approximate standard normal distribution. Assumption 2 states that the decision rule will be chosen to control the α risk for the case where the population mean (μ) equals the standard value μ_0.

Steps in Construction and Application of Decision Rule

Construction and application of any statistical decision rule involve the following four steps:

Step 1. Set up the test alternatives H_0 and H_1.
Step 2. Specify the desired α risk for the test.
Step 3. Set up the statistical decision rule.
Step 4. Calculate the test statistic from the sample data and draw the appropriate conclusion from the decision rule.

One-Sided Upper-Tail Alternatives

We return to the machine component example to illustrate the steps in the construction and the application of the decision rule for one-sided upper-tail alternatives.

EXAMPLE

Step 1. In the machine component example, the test alternatives are defined in terms of $\mu_0 = 1100$:

$$H_0: \mu \leq 1100$$

$$H_1: \mu > 1100$$

Thus, the α risk is to be controlled at $\mu_0 = 1100$.

Step 2. Management wishes to control the α risk at 0.01 when $\mu = \mu_0 = 1100$. This specification of the α risk implies that management wants only 1 chance in 100 of having $\overline{X}$ fall in the rejection region of the decision rule when the population mean service life of the modified component is 1100 hours.

Figure 11.2 illustrates the situation pictorially. The figure shows the sampling distribution of $\overline{X}$ centered at $\mu_0 = 1100$ because the probability that $\overline{X}$ falls in the rejection region (the α risk) is to be controlled at $\mu_0 = 1100$. The tail area in the rejection region, corresponding to the α risk, must be equal to 0.01. Note that the sampling distribution of $\overline{X}$ is approximately normal because the sample size is reasonably large.

Step 3. We shall measure how far $\overline{X}$ is from the standard μ_0 in units of $s\{\overline{X}\}$, the estimated standard deviation of $\overline{X}$. Thus, we shall use the standardized variable z^*, defined as follows:

$$z^* = \frac{\overline{X} - \mu_0}{s\{\overline{X}\}}$$

Since n is large, z^* is approximately a standard normal variable when $\mu = \mu_0$. We shall refer to z^* as the *standardized test statistic*. The z^* scale is also shown in Figure 11.2.

Because the tail area of the sampling distribution of $\overline{X}$ when $\mu = 1100$ must be equal to $\alpha = 0.01$, it follows that the action limit A must correspond to the 99th percentile of the standard normal distribution, that is, $z(0.99) = 2.326$.

FIGURE 11.2 **Illustration of the statistical decision rule for the machine component example — One-sided upper-tail test for μ**

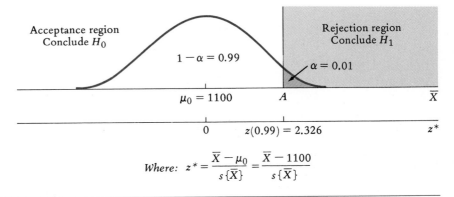

Where: $z^* = \dfrac{\overline{X} - \mu_0}{s\{\overline{X}\}} = \dfrac{\overline{X} - 1100}{s\{\overline{X}\}}$

Thus, the position 2.326 on the z^* scale corresponds to the position of the action limit A on the $\overline{X}$ scale and forms the boundary between the acceptance and rejection regions. The appropriate decision rule for the test is therefore as follows:

If $z^* \leq 2.326$, conclude H_0.
If $z^* > 2.326$, conclude H_1.

Step 4. The sample results for the service lives of the random sample of $n = 36$ modified components were $\overline{X} = 1121$ hours and $s = 222$ hours (sample data not shown). We now compute the estimated standard deviation of $\overline{X}$:

$$s\{\overline{X}\} = \frac{s}{\sqrt{n}} = \frac{222}{\sqrt{36}} = 37.0$$

We next compute the standardized test statistic z^*:

$$z^* = \frac{\overline{X} - \mu_0}{s\{\overline{X}\}} = \frac{1121 - 1100}{37.0} = 0.57$$

Referring to the decision rule, we see that $z^* = 0.57 \leq 2.326$, and hence we conclude H_0. Equivalently, we can see in Figure 11.2 that $z^* = 0.57$ lies in the acceptance region. We therefore conclude on the basis of the sample results that the modified machine component has a mean service life that is not longer than that of the original component. □

Test Statistic and Decision Rule. The test statistic developed here is appropriate for all tests concerning the population mean μ when the sample size is reasonably large.

(11.5)
Tests concerning the population mean μ, when the random sample size is reasonably large, are based on the standardized test statistic:

$$z^* = \frac{\overline{X} - \mu_0}{s\{\overline{X}\}}$$

where:

$$s\{\overline{X}\} = \frac{s}{\sqrt{n}}$$

α risk is controlled at $\mu = \mu_0$

When $\mu = \mu_0$, z^* follows approximately the standard normal distribution.

The decision rule for a test involving one-sided upper-tail alternatives about the population mean μ when n is reasonably large can be summarized as follows.

(11.6)
When the alternatives are:

$$H_0: \mu \leq \mu_0$$

$$H_1: \mu > \mu_0$$

and the random sample size is reasonably large, the appropriate decision rule to control the α risk at μ_0 is as follows:

If $z^* \leq z(1 - \alpha)$, conclude H_0.
If $z^* > z(1 - \alpha)$, conclude H_1.

where:

z^* is given by (11.5)

Comments

1. The reason why the test for the alternatives in (11.6) is called an upper-tail test is apparent from Figure 11.2. Conclusion H_1 is drawn when the observed value of $\overline{X}$ falls in the upper tail of the sampling distribution.
2. In Figure 11.2, it can be seen that action limit A lies $z(1 - \alpha)$ standard deviations to the right of μ_0, where the standard deviation is $s\{\overline{X}\}$. Thus, the value of A for a one-sided upper-tail test may be calculated as follows.

(11.7)
For a one-sided upper-tail test:

$$A = \mu_0 + z(1 - \alpha)s\{\overline{X}\}$$

For the machine component example, $A = 1100 + 2.326(37.0) = 1186$ hours.
3. The specified level of the α risk for a test is often called the *level of significance* of the test. Thus, the level of significance in the machine component test is $\alpha = 0.01$.
4. Because the sampling distribution of $z^* = (\overline{X} - \mu_0)/s\{\overline{X}\}$ is only *approximately* standard normal for large n when $\mu = \mu_0$, the actual α risk is not exactly at the specified level.

One-Sided Lower-Tail Alternatives

Figure 11.1b shows the general form of the decision rule for one-sided lower-tail alternatives. We construct the decision rule in this case similarly to the procedure shown in Figure 11.2 for an upper-tail test, using the same test statistic (11.5), except that the rejection region is in the lower tail of the sampling distribution. An example will make the method clear.

EXAMPLE

Bank Service. A bank recently installed an upgraded computer system for its exterior automatic tellers that extends the line of banking services offered on a 24-hour basis. After several months of operations of the new system, an operations analyst of the bank wishes to test whether or not the mean number of customers being served inside the main bank during regular hours (μ) has decreased from its level prior to the upgrading of the outside system. This previous level is known to have been $\mu_0 = 123$ customers per hour. The test is to be based on customer counts for a random sample of 64 hourly intervals during regular banking hours.

Step 1. The test alternatives are:

H_0: $\mu \geq 123$

H_1: $\mu < 123$

Step 2. The analyst wishes to control the α risk at 0.05 when $\mu = \mu_0 = 123$. Figure 11.3 illustrates the situation. Note that the sampling distribution of $\overline{X}$ is centered at $\mu_0 = 123$ and that the tail area in the rejection region, corresponding to the α risk, must be equal to 0.05.

Step 3. Since the sample size is large, the sampling distribution of $\overline{X}$ and that of its standardized equivalent, z^*, are approximately normal. As the lower-tail area of the sampling distribution in Figure 11.3 is to be 0.05, the action limit A must correspond to the 5th percentile of the standard normal distribution, that is, $z(0.05) = -1.645$ on the z^* scale. The appropriate decision rule for the test is therefore as follows:

If $z^* \geq -1.645$, conclude H_0.
If $z^* < -1.645$, conclude H_1.

**FIGURE 11.3   Illustration of the statistical decision rule for the bank service example —
One-sided lower-tail test for μ**

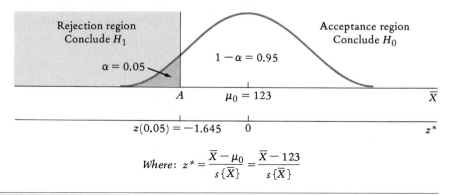

Step 4.   The customer counts for the random sample of 64 hourly intervals were obtained, and it was found that $\overline{X} = 115.8$ and $s = 24.4$ customers per hour. Hence, $s\{\overline{X}\} = 24.4/\sqrt{64} = 3.05$. From (11.5), we have:

$$z^* = \frac{\overline{X} - \mu_0}{s\{\overline{X}\}} = \frac{115.8 - 123}{3.05} = -2.36$$

Referring to the decision rule, we see that $z^* = -2.36 < -1.645$, and hence the analyst should conclude H_1. Equivalently, we can see in Figure 11.3 that $z^* = -2.36$ lies in the rejection region. The analyst can therefore conclude that the upgraded exterior teller system has reduced the mean number of customers per hour served inside the bank.   ☐

Decision Rule.   The decision rule for a test involving one-sided lower-tail alternatives about the population mean μ can be summarized as follows.

(11.8)

When the alternatives are:

  $H_0: \mu \geq \mu_0$

  $H_1: \mu < \mu_0$

and the random sample size is reasonably large, the appropriate decision rule to control the α risk at μ_0 is as follows:

  If $z^* \geq z(\alpha)$, conclude H_0.
  If $z^* < z(\alpha)$, conclude H_1.

where:

z^* is given by (11.5)

Comment

The position of the action limit A in a one-sided lower-tail test is given next.

(11.9)

For a one-sided lower-tail test:

$$A = \mu_0 + z(\alpha)s\{\overline{X}\}$$

For the bank service example, $A = 123 - 1.645(3.05) = 118.0$.

Two-Sided Alternatives

In a two-sided test, we wish to determine whether the population mean μ differs from the standard μ_0 in either direction. Figure 11.1c shows the general form of the decision rule for two-sided alternatives. The decision rule is based on the same test statistic (11.5) used for one-sided tests. The following example demonstrates how the decision rule is constructed in this case.

EXAMPLE

Art Valuation. The research department of an art-publishing house is interested in the ability of nonexpert subjects to judge the market value of art objects. In one test, 100 randomly selected subjects were asked to assess the market value of a thirteenth-century Andalusian vase that (unknown to the subjects) recently sold for $550. It is desired to test whether or not the mean dollar assessment of nonexperts (μ) differs from the actual market value of $\mu_0 = \$550$. If it were to differ, then nonexperts would be biased assessors of the market value of the vase.

Step 1. The test alternatives are:

H_0: $\mu = 550$

H_1: $\mu \neq 550$

Step 2. The research department wishes to control the α risk at 0.05 when $\mu = \mu_0 = 550$. Figure 11.4 illustrates the situation. The sampling distribution of $\overline{X}$ is centered at $\mu_0 = 550$ because the α risk is to be controlled there. Two action limits on the $\overline{X}$ axis (A_1 and A_2) define the acceptance region. These limits are placed equidistant from μ_0 in such a way that the probability of the observed $\overline{X}$ falling in either the upper or the lower rejection region when $\mu = \mu_0$ is $\alpha/2$. The total α risk is thus split equally in the two tails of the sampling distribution of $\overline{X}$.

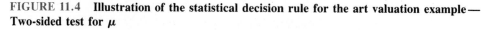

FIGURE 11.4   **Illustration of the statistical decision rule for the art valuation example —**
Two-sided test for μ

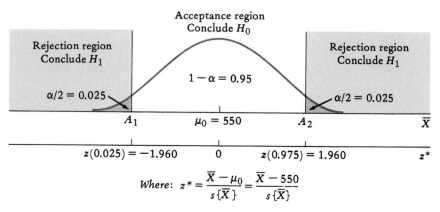

Step 3.   Since the sample size is large, the sampling distribution of $\overline{X}$ and
that of its standardized equivalent, z^*, are approximately normal. The lower-tail
area of the sampling distribution in Figure 11.4 is to be $\alpha/2 = 0.025$; hence, the
lower action limit A_1 must correspond to the 2.5th percentile of the standard
normal distribution, that is, $z(0.025) = -1.960$. Likewise, by symmetry, the
upper action limit A_2 must correspond to $z(0.975) = 1.960$. The acceptance region
includes all values of $\overline{X}$ between A_1 and A_2 or, equivalently, all values of z^*
between $z(0.025) = -1.960$ and $z(0.975) = 1.960$. The appropriate decision rule
for the test is therefore as follows:

If $-1.960 \leq z^* \leq 1.960$, conclude H_0.
If $z^* < -1.960$ or $z^* > 1.960$, conclude H_1.

The inequality $-1.960 \leq z^* \leq 1.960$ is mathematically equivalent to $|z^*| \leq$
1.960, where $|z^*|$ denotes the absolute value of z^*. Thus, the decision rule may
also be written as follows:

If $|z^*| \leq 1.960$, conclude H_0.
If $|z^*| > 1.960$, conclude H_1.

Step 4.   The judgments of market value of the vase from the random sample
of 100 nonexpert subjects yielded $\overline{X} = \$733$ and $s = \$787$. Hence, $s\{\overline{X}\} =$
$787/\sqrt{100} = 78.7$. From (11.5), we have:

$$z^* = \frac{\overline{X} - \mu_0}{s\{\overline{X}\}} = \frac{733 - 550}{78.7} = 2.33$$

Referring to the decision rule, we see that $|z^*| = 2.33 > 1.960$, and hence we conclude H_1. Equivalently, we can see in Figure 11.4 that $z^* = 2.33$ lies in the upper portion of the rejection region. It can be concluded, therefore, that nonexperts are biased assessors of the market value of this Andalusian vase. $\square$

Decision Rule. The decision rule for a test involving two-sided alternatives about the population mean μ can be summarized as follows.

(11.10)

When the alternatives are:

H_0: $\mu = \mu_0$

H_1: $\mu \neq \mu_0$

and the random sample size is reasonably large, the appropriate decision rule to control the α risk at μ_0 is as follows:

If $|z^*| \leq z(1 - \alpha/2)$, conclude H_0.
If $|z^*| > z(1 - \alpha/2)$, conclude H_1.

where:
z^* is given by (11.5)

Comment The positions of the action limits in a two-sided test are given next.

(11.11)

For a two-sided test:

$A_1 = \mu_0 + z(\alpha/2)s\{\overline{X}\}$

$A_2 = \mu_0 + z(1 - \alpha/2)s\{\overline{X}\}$

For the art valuation example, $A_1 = 550 - 1.960(78.7) = 396$, and $A_2 = 550 + 1.960(78.7) = 704$.

Summary of Decision Rules

Figure 11.5 summarizes the construction of the decision rules for the three types of tests about the population mean μ based on a large random sample.

FIGURE 11.5   Summary of decision rules for tests of μ — Large-sample case

Alternatives	Decision Rule

(a) **One-sided**
    Upper-tail

$H_0: \mu \leq \mu_0$
$H_1: \mu > \mu_0$

If $z^* \leq z(1-\alpha)$, conclude H_0
If $z^* > z(1-\alpha)$, conclude H_1

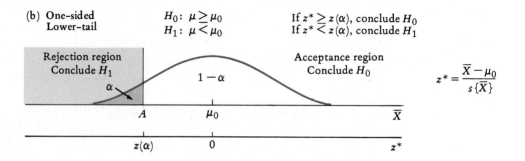

$$z^* = \frac{\overline{X} - \mu_0}{s\{\overline{X}\}}$$

(b) **One-sided**
    Lower-tail

$H_0: \mu \geq \mu_0$
$H_1: \mu < \mu_0$

If $z^* \geq z(\alpha)$, conclude H_0
If $z^* < z(\alpha)$, conclude H_1

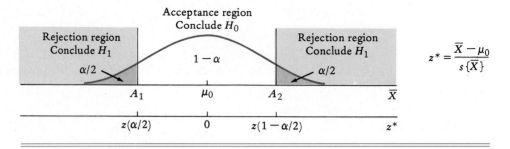

$$z^* = \frac{\overline{X} - \mu_0}{s\{\overline{X}\}}$$

(c) **Two-sided**

$H_0: \mu = \mu_0$
$H_1: \mu \neq \mu_0$

If $|z^*| \leq z(1-\alpha/2)$, conclude H_0
If $|z^*| > z(1-\alpha/2)$, conclude H_1

$$z^* = \frac{\overline{X} - \mu_0}{s\{\overline{X}\}}$$

11.4 *P*-VALUES

Nature of *P*-Value

The *P*-value of a statistical test is a probability number that measures the extent to which the sample data are consistent with conclusion H_0. It represents a compact summary of the sample findings in a statistical test, and it is frequently used in published reports of statistical test results and in the output of statistical computer packages, as we shall illustrate.

Figure 11.6a shows a statistical decision rule for a one-sided upper-tail test about the population mean μ. The sampling distribution of $\overline{X}$ when $\mu = \mu_0$ is also shown. Figures 11.6b, 11.6c, and 11.6d show the *P*-value that corresponds to the observed value of the sample mean $\overline{X}$ and the associated standardized test statistic z^* for three different possible sample outcomes. The *P*-value is the shaded area under the sampling distribution in each figure. Notice how this shaded area decreases the farther *in the direction of the rejection region* is the observed $\overline{X}$ from μ_0 or, equivalently, the farther is z^* from 0. Thus, the less consistent is the sample evidence (in this case, $\overline{X}$ or z^*) with conclusion H_0 being true ($\mu \leq \mu_0$, in this example), the smaller is the *P*-value.

For a one-sided test concerning the population mean, the *P*-value is defined as follows with respect to the standardized test statistic z^*.

(11.12)   The *P-value* of a one-sided statistical test for μ is the probability that, if $\mu = \mu_0$, the standardized test statistic z^* might have been more extreme in the direction of the rejection region than was actually observed.

Calculation of *P*-Value

In calculating the *P*-value of a test, we distinguish between one-sided and two-sided test alternatives. The procedures to be explained are applicable when the sample size is reasonably large.

One-Sided *P*-Value. The *P*-values illustrated in Figure 11.6 are referred to as *one-sided P-values* because they relate to a one-sided test. We now show how one-sided *P*-values are calculated.

EXAMPLE 1

Upper-Tail Test. For the machine component example, Figure 11.7a contains the decision rule reproduced from Figure 11.2. Recall that $\overline{X} = 1121$ and $z^* = 0.57$ were the sample outcomes. These observed values are located on their respective scales in Figure 11.7b. The one-sided *P*-value in this case is the probability that, if $\mu = \mu_0 = 1100$, z^* might have been farther in the direction of the rejection region than actually was observed. In other words, the *P*-value is the probability that, if $\mu = \mu_0 = 1100$, z^* might have been larger than its actual

FIGURE 11.6   The *P*-value of a test for μ measures the consonance of the observed sample mean $\overline{X}$ with the value μ_0

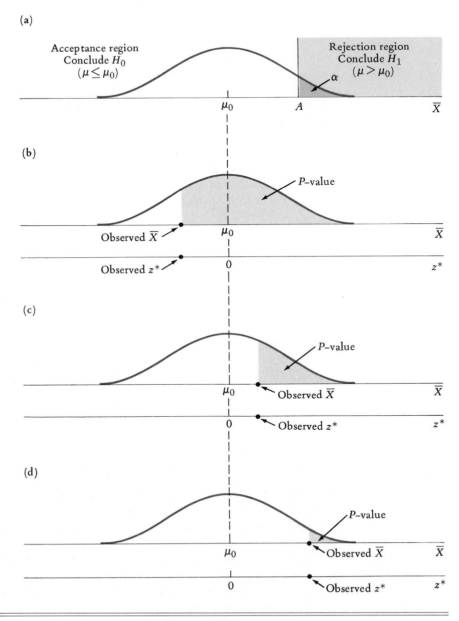

FIGURE 11.7   Calculation of the one-sided *P*-value for the machine component example

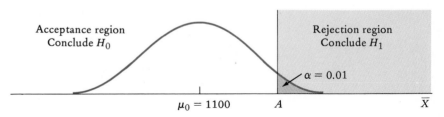

(a)

Acceptance region
Conclude H_0

Rejection region
Conclude H_1

$\alpha = 0.01$

$\mu_0 = 1100$ A $\overline{X}$

(b)

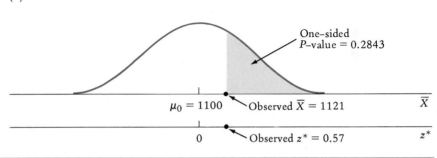

One–sided
P–value = 0.2843

$\mu_0 = 1100$ Observed $\overline{X} = 1121$ $\overline{X}$

0 Observed $z^* = 0.57$ z^*

value of 0.57. This probability is shown by the shaded area in Figure 11.7b and can be seen from Table C.1 to equal $P(Z > z^*) = P(Z > 0.57) = 0.2843$. Thus, the one-sided *P*-value is 0.2843 here.

   This reasonably large *P*-value indicates that the sample outcome is quite consistent with conclusion H_0.   ☐

EXAMPLE 2

Lower-Tail Test.   For the bank service example, Figure 11.8a contains the decision rule reproduced from Figure 11.3. Recall that the sample outcomes were $\overline{X} = 115.8$ and $z^* = -2.36$. The one-sided *P*-value is the probability that, if $\mu = \mu_0 = 123$, z^* might have been farther in the direction of the rejection region than actually was found. Thus, the *P*-value is the probability that, if $\mu = \mu_0 = 123$, z^* might have been smaller than its actual value of -2.36. This probability is shown by the shaded area in Figure 11.8b and equals $P(Z < z^*) = P(Z < -2.36) = 0.0091$. Hence, the one-sided *P*-value here is 0.0091. This number is sufficiently small so that there is little doubt that the sample outcome is not consistent with conclusion H_0.   ☐

Two-Sided *P*-Value.   In a two-sided test, the extent to which $\overline{X}$ is consistent with μ_0 is judged by the distance of $\overline{X}$ from μ_0 or, equivalently, by the distance of z^* from 0,

FIGURE 11.8   Calculation of the one-sided P-value for the bank service example

(a)

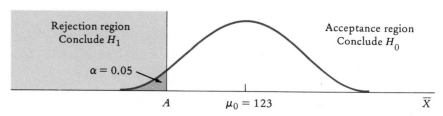

(b)

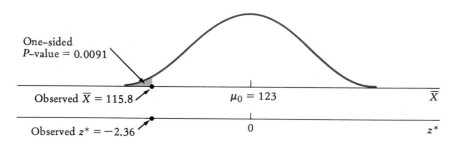

irrespective of the direction. Hence, in a two-sided test, the calculation of the P-value considers extreme departures in either direction. This calculation is done in two steps:

1. Obtain the one-sided P-value for the direction toward the rejection region in which $\overline{X}$ and z^* happen to fall.
2. Double this one-sided P-value.

The resulting number is called a *two-sided P-value*.

EXAMPLE

For the art valuation example, Figure 11.9a contains the decision rule reproduced from Figure 11.4. Recall that $\overline{X} = 733$ and $z^* = 2.33$ were the sample outcomes. To calculate the two-sided P-value here, we first obtain the one-sided P-value corresponding to $z^* = 2.33$. This one-sided P-value is shown by the shaded area in Figure 11.9b and equals $P(Z > z^*) = P(Z > 2.33) = 0.0099$. Next, we double this one-sided P-value to obtain $2(0.0099) = 0.0198$. Thus, the two-sided P-value for this test is 0.0198. This number is fairly small but not negligible, so the sample outcome may or may not be consistent with conclusion H_0. $\square$

Definitions.   We present now the calculational formulas for obtaining the P-values for the three types of tests for μ.

FIGURE 11.9 **Calculation of the two-sided *P*-value for the art valuation example**

(a)

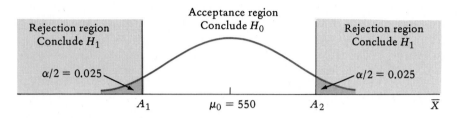

(b)

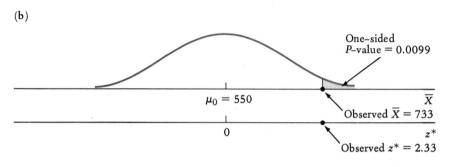

Two–sided P–value = 2(0.0099) = 0.0198

<table>
<tr><td>(11.13a)</td><td>One-Sided Upper-Tail Alternatives (H_0: $\mu \leq \mu_0$, H_1: $\mu > \mu_0$):

One-sided P-value = $P(Z > z^*)$</td></tr>
<tr><td>(11.13b)</td><td>One-Sided Lower-Tail Alternatives (H_0: $\mu \geq \mu_0$, H_1: $\mu < \mu_0$):

One-sided P-value = $P(Z < z^*)$</td></tr>
<tr><td>(11.13c)</td><td>Two-Sided Alternatives (H_0: $\mu = \mu_0$, H_1: $\mu \neq \mu_0$):

Two-sided P-value = $\begin{cases} 2P(Z > z^*) & \text{if } z^* \geq 0 \\ 2P(Z < z^*) & \text{if } z^* < 0 \end{cases}$</td></tr>
</table>

Testing with *P*-Values

As noted earlier, the *P*-value of a test for μ measures the consistency between the sample outcome and the value of μ_0 postulated in H_0. A large *P*-value indicates that μ_0

is plausible and, hence, H_0 should be concluded. A small P-value indicates that μ_0 is not plausible and, hence, H_1 should be concluded. Indeed, the P-value can be used directly to choose between H_0 and H_1, according to whether or not the P-value is larger or smaller than the specified α risk for conducting the test. The result obtained from the test based on the P-value is mathematically equivalent to the result provided by the corresponding decision rule based on the standardized test statistic. The decision rule based on the P-value has the following form.

(11.14)

If P-value $\geq \alpha$, conclude H_0.

If P-value $< \alpha$, conclude H_1.

This decision rule applies whether the test is two-sided or one-sided, using a two-sided or one-sided P-value as appropriate.

EXAMPLE 1

Upper-Tail Test.   Refer to Figure 11.7 for the machine component example, where the α risk is specified to be 0.01 and the one-sided P-value for the observed sample outcome is 0.2843. Since $0.2843 \geq 0.01$, decision rule (11.14) states that H_0 is the appropriate conclusion. This conclusion is identical to the one drawn earlier on the basis of the standardized test statistic, as it must necessarily be.   ☐

EXAMPLE 2

Lower-Tail Test.   Refer to Figure 11.8 for the bank service example, where $\alpha = 0.05$ and one-sided P-value = 0.0091. Since $0.0091 < 0.05$, decision rule (11.14) states that H_1 is the appropriate conclusion, as was found earlier.   ☐

EXAMPLE 3

Two-Sided Test.   Refer to Figure 11.9 for the art valuation example, where $\alpha = 0.05$ and two-sided P-value = 0.0198. Since $0.0198 < 0.05$, decision rule (11.14) states that H_1 is the appropriate conclusion, as was found earlier.   ☐

Comments

1.   The equivalence of decision rule (11.14) based on the P-value and the corresponding decision rule based on the standardized test statistic z^* can be seen from Figures 11.7, 11.8, and 11.9. Note in these figures that the P-value is less than α only when the observed z^* value falls in the rejection region.

2.   In later chapters, we will show that P-values are applicable in statistical tests other than those for a population mean.

3.   In interpreting P-values presented in statistical reports and computer output, one must establish whether the reported P-value is a one-sided or a two-sided P-value. Figure 11.10 (p. 337) shows computer output for a test to be discussed shortly, in which the reported P-value is clearly labeled to be a two-sided one. Other labels for the P-value that are often found in computer output include P, PROBABILITY, and PROB.

11.5   SMALL-SAMPLE TESTS

Construction of Decision Rule

The test procedures for μ discussed so far have assumed that the sample size is reasonably large. This assumption has permitted us to invoke central limit theorem extension (10.13), which states that the sampling distribution of $(\overline{X} - \mu)/s\{\overline{X}\}$ is approximately standard normal when the sample size is large. When, however, the sample size is not large, approximate normality may no longer hold.

Assumption.   We shall now consider the construction of decision rules for small samples when the population is normal or does not depart too markedly from normality.

Test Statistic.   Theorem (10.14) is relevant for the assumed condition. It states that when the population is normal, $(\overline{X} - \mu)/s\{\overline{X}\}$ is distributed as $t(n - 1)$. Furthermore, in view of the robustness of this statistic, we know that it is approximately distributed as $t(n - 1)$ as long as the population does not depart too markedly from a normal distribution. This fact means that the same test statistic (11.5) used for large samples may also be used for small samples when one is making tests concerning the population mean μ for populations that do not depart too markedly from normality. The difference in testing is that an appropriate t percentile is now used in place of the z percentile in constructing the decision rule.

We shall now use t^* to denote the earlier test statistic (11.5) to remind us that the tests for small samples are based on the t distribution.

(11.15)

> Tests concerning the population mean μ, when the population is normal or the departure is not too marked, are based on the standardized test statistic:
>
> $$t^* = \frac{\overline{X} - \mu_0}{s\{\overline{X}\}}$$
>
> *where:*
>
> $$s\{\overline{X}\} = \frac{s}{\sqrt{n}}$$
>
> α risk is controlled at $\mu = \mu_0$
>
> When $\mu = \mu_0$, t^* follows the t distribution with $n - 1$ degrees of freedom.

The appropriate decision rules are constructed in the same fashion as those for large samples. Table 11.2 shows the form of the decision rule for each of the three types of tests. Since the only difference between these tests and those for the large-sample case is the use of t percentiles in place of z percentiles, we proceed directly to an illustration.

TABLE 11.2   Summary of decision rules for tests of μ — Small-sample case, population normal or departure not too marked

Test	Alternatives	Decision Rule
(a) One-sided Upper-tail	H_0: $\mu \leq \mu_0$ H_1: $\mu > \mu_0$	If $t^* \leq t(1 - \alpha; n - 1)$, conclude H_0 If $t^* > t(1 - \alpha; n - 1)$, conclude H_1 where: $t^* = \dfrac{\overline{X} - \mu_0}{s\{\overline{X}\}}$
(b) One-sided Lower-tail	H_0: $\mu \geq \mu_0$ H_1: $\mu < \mu_0$	If $t^* \geq t(\alpha; n - 1)$, conclude H_0 If $t^* < t(\alpha; n - 1)$, conclude H_1 where: $t^* = \dfrac{\overline{X} - \mu_0}{s\{\overline{X}\}}$
(c) Two-sided	H_0: $\mu = \mu_0$ H_1: $\mu \neq \mu_0$	If $|t^*| \leq t(1 - \alpha/2; n - 1)$, conclude H_0 If $|t^*| > t(1 - \alpha/2; n - 1)$, conclude H_1 where: $t^* = \dfrac{\overline{X} - \mu_0}{s\{\overline{X}\}}$

EXAMPLE

In the lake acidity example, the scientist wished to test the mean pH level for a lake to see whether it differs from neutral, that is, from 7.0.

Step 1.   The test alternatives are:

H_0: $\mu = 7.0$

H_1: $\mu \neq 7.0$

Hence, $\mu_0 = 7.0$.

Step 2.   The scientist wishes to control the α risk of the test at 0.01 when $\mu = \mu_0 = 7.0$.

Step 3.   Past experience indicates that pH levels in lake water tend to be approximately normally distributed. The scientist has drawn a random sample of $n = 18$ water samples from the lake; thus, a small-sample testing procedure is required. The appropriate decision rule is the two-sided one in Table 11.2c. For $\alpha = 0.01$ and $n = 18$, we require $t(1 - \alpha/2; n - 1) = t(0.995; 17) = 2.898$. Hence, the decision rule is as follows:

If $|t^*| \leq 2.898$, conclude H_0.
If $|t^*| > 2.898$, conclude H_1.

Step 4.   The scientist's sample data are given in the computer output in Figure 11.10. The output contains the sample statistics $n = 18$, $\overline{X} = 6.8861$, $s =$

FIGURE 11.10 **Sample data, statistics, and test results for the lake acidity example—Small-sample case, population normal or departure not too marked**

DATA

OBS.	VALUE	OBS.	VALUE	OBS.	VALUE
1	6.97	7	6.89	13	7.05
2	6.70	8	6.92	14	6.90
3	6.84	9	6.91	15	6.94
4	6.83	10	6.60	16	6.99
5	6.95	11	6.94	17	6.95
6	6.84	12	6.89	18	6.84

SAMPLE STATISTICS

NO. OF OBSERVATIONS	18
MEAN	6.8861
STD. DEVIATION	.10478
STD. ERROR OF MEAN	.024696

TEST RESULTS

HYPOTHESIZED MEAN	7.0
STD. TEST STATISTIC	-4.612
DF	17
TWO-SIDED P-VALUE	.0002488

0.10478, and $s\{\overline{X}\} = 0.10478/\sqrt{18} = 0.024696$. The standardized test statistic (11.15) is also given in the output and was calculated as follows:

$$t^* = \frac{\overline{X} - \mu_0}{s\{\overline{X}\}} = \frac{6.8861 - 7.0}{0.024696} = -4.612$$

Since $|t^*| = |-4.612| = 4.612$ and $4.612 > 2.898$, the scientist should conclude H_1—that the lake water is not neutral. Since the sample mean is, in fact, less than 7.0, the indication is that the lake water is acidic. $\square$

Testing with *P*-Values

The *P*-value may be computed for a small-sample test based on t^* just as for a large-sample test based on z^*, except that the *t* distribution is now used. Because Table C.3 for the *t* distribution provides only a limited number of percentiles, it usually only permits one to obtain bounds for the *P*-value rather than the actual *P*-value itself. Many computer packages and programmable calculators have built-in programs for calculating the exact *P*-value in these cases.

EXAMPLE

For the lake acidity example, a two-sided *P*-value is required. Recall that $n = 18$ and $t^* = -4.612$. The one-sided *P*-value corresponding to $t^* = -4.612$ is the tail area below -4.612 in the t distribution with $n - 1 = 17$ degrees of freedom. From Table C.3, we see that the smallest percentile available is $t(0.0005; 17) = -3.965$. Since $t^* = -4.612$ is smaller than -3.965, we know that the one-sided P-value is smaller than 0.0005 and thus the two-sided P-value is smaller than $2(0.0005) = 0.001$. The computer output of Figure 11.10 shows that the exact two-sided P-value is 0.0002488. Since the scientist fixed α at 0.01, decision rule (11.14) leads to conclusion H_1 because $0.0002488 < 0.01$, the same conclusion as that drawn earlier on the basis of the standardized test statistic t^*. ☐

11.6 POWER OF TEST

The decision rules for tests of μ have been set up to control the α risk of the test when $\mu = \mu_0$. Recall that the test alternatives are always formulated so that the α risk is the more important risk to control. Still, β risks, that is, the risks of making a Type II error, are also of concern in testing and, if possible, should be kept reasonably small. We now consider how to evaluate the β risks of a statistical test. In the following section, we discuss how to control the β risks by an appropriate choice of sample size.

Assumptions

Throughout this section, we shall assume the following:

1. The population standard deviation σ is known.
2. The random sample size is large.

These assumptions will ensure that the results to be obtained from the normal distribution are exact or nearly exact. We comment subsequently on the effect when the two assumptions do not hold.

For tests in which the assumptions apply, that is, the population standard deviation σ is known and the sample size is large, the standardized test statistic has the following form.

(11.16)

$$z^* = \frac{\overline{X} - \mu_0}{\sigma\{\overline{X}\}}$$

where:

$$\sigma\{\overline{X}\} = \frac{\sigma}{\sqrt{n}}$$

α risk is controlled at $\mu = \mu_0$

Rejection Probabilities

To evaluate the β risks associated with any decision rule, we first consider the probability that the rule will lead to conclusion H_1. This probability depends on the true value of the population mean μ. We denote this probability by $P(H_1; \mu)$.

(11.17)

> The probability of concluding H_1 when the population mean is μ, denoted by $P(H_1; \mu)$, is called the *rejection probability* of the statistical decision rule at μ. A graph of the probability $P(H_1; \mu)$ for different possible values of μ is called the *rejection probability curve* of the decision rule.

We have already encountered a special rejection probability in constructing decision rules for tests of μ, but not by this name. In each of the decision rules summarized in Figure 11.5, the probability of concluding H_1 when $\mu = \mu_0$ is controlled at α. This probability, $P(H_1; \mu_0) = \alpha$, is a rejection probability.

We consider now how rejection probabilities are evaluated and how they, in turn, enable us to measure the β risks.

One-Sided Tests

We explain the evaluation of β risks for one-sided tests first and then take up two-sided tests. In the following example, we consider a one-sided upper-tail test. The evaluation of β risks for one-sided lower-tail tests proceeds in an analogous fashion.

Calculation of Rejection Probabilities. The first step in evaluating the β risks of a statistical decision rule is to calculate a number of rejection probabilities. We illustrate these calculations by an example.

EXAMPLE

Data Entry. A marketing research company conducts regular consumer expenditures surveys, the findings of which are sold on a syndicated basis to many consumer product firms. The company has recently restructured the questionnaire used in the survey, making it longer but also more streamlined to facilitate the computer data entry of questionnaire responses. In the past, data entry has required a mean time of 23.0 minutes per questionnaire. The data entry operators are now experienced with the new questionnaire. The company wishes to test whether or not the mean data entry time for the new questionnaire is greater than $\mu_0 = 23.0$ minutes, on the basis of the observed data entry times for a random sample of 60 new questionnaires. The standard deviation of times for the data entry process has been $\sigma = 6.5$ minutes in the past, and the company believes it will remain the same for the new questionnaire.

The test alternatives are:

$H_0: \mu \leq 23.0$

$H_1: \mu > 23.0$

The α risk is to be controlled at $\alpha = 0.10$ when $\mu = \mu_0 = 23.0$. Hence, the decision rule has the form shown in Figure 11.11a. Note that the rejection probability at $\mu_0 = 23.0$ is $P(H_1; \mu_0 = 23.0) = 0.10$ because the α risk is controlled at 0.10 there.

In (11.7) we have shown how the value of the action limit A in a one-sided upper-tail test is calculated. To use that formula in this example, we replace $s\{\overline{X}\}$

FIGURE 11.11   **Calculation of rejection probabilities at $\mu = 23.0, 24.5,$ and 25.0 for the data entry example — One-sided upper-tail test**

(a)

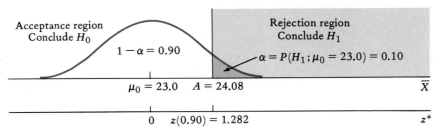

(b)

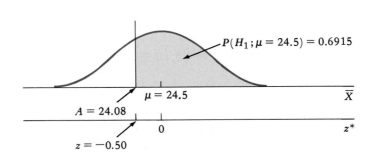

(c)

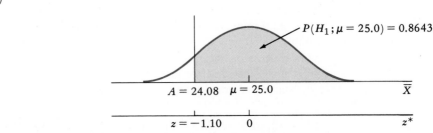

in the formula by $\sigma\{\overline{X}\}$, as the latter is known here. Specifically, because $\sigma = 6.5$ and $n = 60$, we have $\sigma\{\overline{X}\} = \sigma/\sqrt{n} = 6.5/\sqrt{60} = 0.839$. Furthermore, we have $z(1 - \alpha) = z(0.90) = 1.282$. Thus:

$$A = \mu_0 + z(1 - \alpha)\sigma\{\overline{X}\} = 23.0 + 1.282(0.839) = 24.08$$

The decision rule for the test may then be written in terms of $\overline{X}$ as follows:

If $\overline{X} \leq 24.08$, conclude H_0.
If $\overline{X} > 24.08$, conclude H_1.

Using this form of the decision rule facilitates the evaluation of the rejection probability at different values of μ. We begin with $\mu = 24.5$.

1. $P(H_1; \mu = 24.5)$. If the population mean is $\mu = 24.5$, the sampling distribution of $\overline{X}$ will be centered at 24.5, as shown in Figure 11.11b. The probability of concluding H_1 when $\mu = 24.5$, that is, $P(H_1; \mu = 24.5)$, is the shaded area in the rejection region in Figure 11.11b. Since the action limit $A = 24.08$ is located $z = (24.08 - 24.5)/0.839 = -0.50$ standard deviation from $\mu = 24.5$ (recall that $\sigma\{\overline{X}\} = 0.839$), it follows that the shaded area equals:

$$P(H_1; \mu = 24.5) = P(Z > -0.50) = 0.6915$$

2. $P(H_1; \mu = 25.0)$. As a second example, consider the case when the population mean is $\mu = 25.0$. The sampling distribution of $\overline{X}$ will be centered at 25.0, as shown in Figure 11.11c. The action limit $A = 24.08$ is located $z = (24.08 - 25.0)/0.839 = -1.10$ standard deviations from $\mu = 25.0$. Hence, the rejection probability is:

$$P(H_1; \mu = 25.0) = P(Z > -1.10) = 0.8643$$

This probability corresponds to the shaded area in Figure 11.11c. ☐

Rejection Probability Curve.   Once rejection probabilities have been calculated for a number of possible values of μ, the results are then summarized in a rejection probability curve.

EXAMPLE

The rejection probability curve for the decision rule in the data entry example is shown in Figure 11.12. The points previously determined:

$$P(H_1; \mu_0 = 23.0) = 0.10$$

$$P(H_1; \mu = 24.5) = 0.6915$$

$$P(H_1; \mu = 25.0) = 0.8643$$

are shown explicitly on the curve to illustrate the plotting procedure. As one would expect, the rejection probability increases steadily toward 1.0 as μ in-

FIGURE 11.12   Rejection probability curve for the data entry example — One-sided upper-tail test

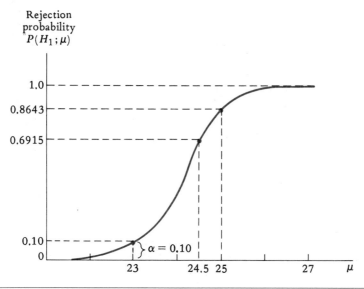

creases. This pattern reflects the reasonable result that the larger the mean data entry time for the new questionnaire, the more probable it is that one will conclude H_1: $\mu > 23.0$ in the test. We see, for instance, that if μ is actually 27.0, the test is almost certain to lead to conclusion H_1. □

Calculation of Error Risks.   The rejection probability curve gives the probability of concluding H_1 for different values of μ. The probabilities of making either a Type I error (α risk) or a Type II error (β risk) at different values of μ can be calculated from this curve.

EXAMPLE

Refer to Figure 11.13, which contains the same rejection probability curve for the data entry example previously shown in Figure 11.12. Consider any μ value such that $\mu \le \mu_0$. Since H_0 is true in this case, the rejection probability at such a value of μ equals the probability of a Type I error and, hence, is an α risk. (Recall that the rejection probability here is the probability of concluding H_1 when, in fact, H_0 is true). When $\mu > \mu_0$, on the other hand, H_1 is true and $P(H_1; \mu)$ is then the probability of a correct decision. Its complement, $1 - P(H_1; \mu)$, now equals the probability of a Type II error and, hence, is a β risk.

   By this reasoning, we are able to assess the error risk at any value of μ, whether it be an α or a β risk. Referring to Figure 11.13, we see that the α risk at $\mu_0 = 23.0$ is 0.10 — indeed, the decision rule was constructed so that this condi-

FIGURE 11.13   **Determination of error risks from the rejection probability curve for the data entry example — One-sided upper-tail test**

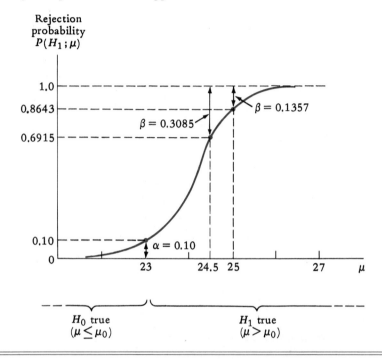

tion would hold. At $\mu = 24.5$, on the other hand, the error probability is a β risk and equals $1 - 0.6915 = 0.3085$. At $\mu = 25.0$, the error probability is also a β risk and equals $1 - 0.8643 = 0.1357$. This last error risk implies that if the mean data entry time for the new questionnaire is $\mu = 25.0$, then the probability is 0.1357 that the test will erroneously lead to the conclusion that the mean data entry time for the new questionnaire is no larger than it was for the old questionnaire.

☐

Comments   1. In one-sided tests, the α risk is a maximum at $\mu = \mu_0$ (as is illustrated by Figure 11.13, where $\alpha = 0.10$ at $\mu_0 = 23.0$). Hence, controlling the α risk at $\mu = \mu_0$ guarantees that the probability of a Type I error at any other value of μ for which H_0 is true is smaller than α. The β risk in a one-sided test is a maximum when μ is close to μ_0. In Figure 11.13, for instance, the β risk approaches $1 - \alpha = 0.90$ when μ is just slightly larger than $\mu_0 = 23.0$. As a practical matter, however, β risks are only a major concern at values of μ some distance from μ_0, where a Type II error would be more costly.

2. Provided the sample size is large, as it was in the data entry example, the standardized statistic $(\overline{X} - \mu)/\sigma\{\overline{X}\}$ is approximately standard normal, and hence, calculations of rejection

probabilities using the normal distribution are reasonably accurate. If the population standard deviation σ is unknown, then $s\{\overline{X}\}$ may be used in place of $\sigma\{\overline{X}\}$ to calculate approximate rejection probabilities with the normal distribution as long as the sample size is large. Calculations of rejection probabilities for decision rules based on small samples can also be made, but they are more complex and we do not consider them here.

3. The rejection probability $P(H_1; \mu)$ at a value of μ for which alternative H_1 is true is often called the *power* of the decision rule at μ because it measures the probability of correctly concluding H_1. In this terminology, the rejection probability curve is often called the *power curve* because power values can be read directly from this curve.

4. For some applications, such as quality control, interest centers on the probability of concluding H_0 at different values of μ. This probability is called the *acceptance probability* or *operating characteristic* of the decision rule at μ and is denoted by $P(H_0; \mu)$. A plot of $P(H_0; \mu)$ against μ is called the *acceptance probability curve* or *operating characteristic curve*. The rejection and acceptance probabilities at a given value of μ are complementary probabilities. Hence, any acceptance probability may be obtained from the corresponding rejection probability by using the complementation theorem (4.19).

(11.18)
$$P(H_0; \mu) = 1 - P(H_1; \mu)$$

A quality control application of the operating characteristic curve is given in Chapter 14.

Two-Sided Tests

The calculations of rejection probabilities and error risks for a two-sided test follow the same principles as for a one-sided test but involve slightly different procedures, which we now illustrate.

EXAMPLE

Growth Pellets.   An agricultural manufacturer produces growth hormone pellets for cattle in large batches. Product specifications require the pellets to have a mean hormone content of $\mu_0 = 1280$ milligrams per pellet. A random sample of 80 pellets is selected from each batch to test whether the batch meets this specification. The process standard deviation of the hormone content per pellet is known from extensive past experience to be $\sigma = 110$ milligrams per pellet.

The test alternatives are:

$H_0: \mu = 1280$

$H_1: \mu \neq 1280$

The α risk is to be controlled at 0.05 when $\mu = \mu_0 = 1280$. Hence, the decision rule has the form shown in Figure 11.14a. Note that the rejection probability at $\mu = \mu_0 = 1280$ equals the α value for the decision rule, that is, $P(H_1; \mu_0 = 1280) = \alpha = 0.05$.

FIGURE 11.14   Calculation of rejection probabilities at μ = 1280 and 1290 for the growth pellets example — Two-sided test

(a)

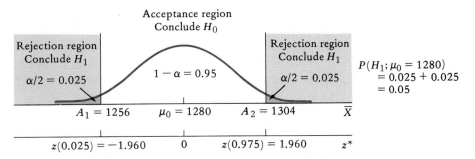

(b)

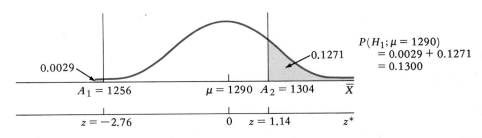

In (11.11), we have shown how the action limits A_1 and A_2 in a two-sided test are calculated. To use (11.11) here, we replace $s\{\overline{X}\}$ by $\sigma\{\overline{X}\}$ because the latter is known — specifically, σ = 110, n = 80, and $\sigma\{\overline{X}\} = \sigma/\sqrt{n} = 110/\sqrt{80} =$ 12.30. Furthermore, we have $z(\alpha/2) = z(0.025) = -1.960$ and $z(1 - \alpha/2) = z(0.975) = 1.960$. Thus:

$$A_1 = \mu_0 + z(\alpha/2)\sigma\{\overline{X}\} = 1280 - 1.960(12.30) = 1256$$

$$A_2 = \mu_0 + z(1 - \alpha/2)\sigma\{\overline{X}\} = 1280 + 1.960(12.30) = 1304$$

Since A_1 and A_2 define the boundaries of the acceptance and rejection regions, the decision rule for the test may be stated in terms of $\overline{X}$ as follows:

If $1256 \leq \overline{X} \leq 1304$, conclude H_0.
If $\overline{X} < 1256$ or $\overline{X} > 1304$, conclude H_1.

Using this form of the decision rule makes it convenient to evaluate the rejection probability of the rule at different values of μ. We shall illustrate one such calculation, at $\mu = 1290$.

If the population mean is 1290, then the sampling distribution of $\overline{X}$ will be centered at 1290, as shown in Figure 11.14b. The probability of concluding H_1 when $\mu = 1290$, that is, $P(H_1; \mu = 1290)$, is the sum of the two tail areas lying in the two parts of the rejection region for the test. Since action limit A_1 corresponds to $z = (1256 - 1290)/12.30 = -2.76$ and action limit A_2 corresponds to $z = (1304 - 1290)/12.30 = 1.14$, the tail areas are $P(Z < -2.76) = 0.0029$ and $P(Z > 1.14) = 0.1271$, respectively. Adding the two tail probabilities, we obtain the rejection probability at $\mu = 1290$ as $P(H_1; \mu = 1290) = 0.0029 + 0.1271 = 0.1300$.

The entire rejection probability curve is shown in Figure 11.15. Note that the curve is symmetrical about $\mu_0 = 1280$ and that it increases steadily toward 1.0 as μ departs from μ_0 in either direction.

In evaluating the error risks inherent in the decision rule, we need to note that, in two-sided tests, an α risk exists only at $\mu = \mu_0$. Thus, the α risk only occurs here at $\mu_0 = 1280$ and may be read directly from the rejection probability curve in Figure 11.15, as indicated there. The β risk at any value of μ other than μ_0 is the complementary probability of the rejection probability at that value of μ. Referring to Figure 11.15, we see that the rejection probability at $\mu = 1290$ is 0.1300. Hence, the β risk when $\mu = 1290$ equals $1 - 0.1300 = 0.8700$.   □

FIGURE 11.15   **Rejection probability curve for the growth pellets example — Two-sided test**

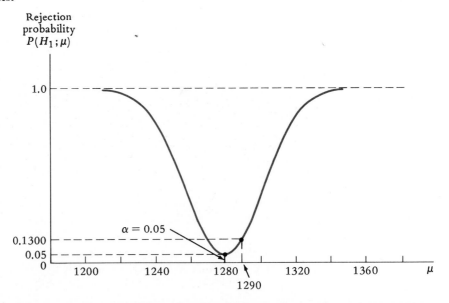

Trade-off Between α and β Risks

We have shown in Section 11.3 how to construct a decision rule for a test when the sample size is given so that the α risk is controlled at a predetermined level for $\mu = \mu_0$. In this section, we have shown how to construct the rejection probability curve for a decision rule and how to assess from that curve the error risks implicit in the rule. We now take this progression of ideas one step further and ask what trade-off between α and β risks occurs when the control on the α risk is changed. We return to the data entry example to illustrate the general nature of this trade-off.

EXAMPLE

Figure 11.16 contains two rejection probability curves for the data entry example. One is the rejection probability curve presented initially in Figure 11.12 for the decision rule where the α risk was controlled at 0.10. The other is a rejection probability curve constructed in the same manner but with the α risk controlled at 0.01. Observe that this latter curve has the same shape as the former but is shifted rightward. The effect of this shift on the error risks is to decrease the α risks for all $\mu \leq \mu_0$ and to increase the β risks for all $\mu > \mu_0$. $\square$

FIGURE 11.16   Rejection probability curves for the decision rules controlling the α risks at 0.10 and 0.01 in the data entry example

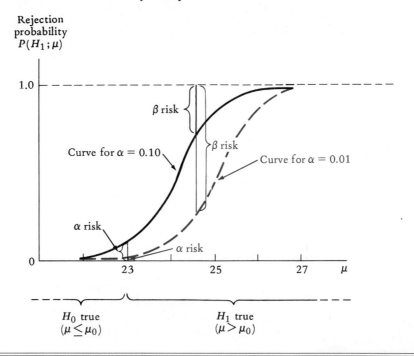

The preceding illustration for a one-sided upper-tail test shows that, for a given sample size, a decrease in α risks is associated with an increase in β risks and vice versa. This same trade-off between the two types of error risks occurs in all statistical tests when n is predetermined. The trade-off reflects an important statistical principle.

(11.19)

> For a given random sample size, one type of error risk can be reduced only at the expense of increasing the other type.

The reference to a given sample size in the principle is important because it is intuitively clear that if the sample size could be increased, then the larger sample would provide more precise information about the true value of μ and, hence, would offer the possibility of reducing both types of error risks simultaneously. We take up, in the next section, planning of sample size to permit control of both types of error risks.

11.7   PLANNING OF SAMPLE SIZE

Need for Planning

Simultaneous control of both α and β risks in a test can be achieved by a planned choice of an appropriate sample size. Planning of the sample size is desirable because a sample size that is too small may entail unsatisfactorily high β risks (the α risk being fixed at the desired level), while a sample size that is too large will be uneconomical.

We now discuss how one can determine the sample size for a test that will keep both α and β risks at prespecified levels. The objective is similar to that of selecting the appropriate sample size to achieve a specified precision in a confidence interval for μ.

Assumptions

The planning procedure we shall use assumes the following:

1. The random sample size ultimately determined is reasonably large.
2. The population is either infinite or, if finite, is large relative to the resulting sample size.

Planning Procedure

The planning procedure requires specifications of the α and β risks to be controlled. Moreover, because the sample size is being planned, we do not yet have the sample standard deviation s available to compute $s\{\overline{X}\}$. Since the population standard deviation σ is generally unknown, we shall therefore also need to specify a planning value for σ to find the needed sample size. Recall that use of a planning value for σ is also necessary for planning the sample size for estimation purposes.

The planning procedure involves the following four steps:

Step 1.   Specify the desired α risk at $\mu = \mu_0$.

Step 2.   Specify the value of μ at which the β risk is to be controlled (denoted by μ_1) and the desired level of the β risk at $\mu = \mu_1$.

Step 3.   Select a planning value for σ.

Step 4.   Calculate the required sample size n by using formula (11.20), which is presented later.

Once the required sample size is determined and the sample results are obtained, then the test is conducted exactly as described in the preceding sections of this chapter.

EXAMPLE

In the machine component example, assume that the sample size has not yet been determined.

Step 1.   The manufacturer wants to control the α risk at 0.01 when $\mu = \mu_0 = 1100$, as before. Figure 11.17a shows the appropriate decision rule for controlling the α risk. Note that the action limit corresponds to $z(0.99) = 2.326$ on the z^* scale. We shall now denote the z value associated with the control of the α risk by z_0. In this example, $z_0 = z(0.99) = 2.326$.

Step 2.   The manufacturer can control the β risk at any appropriate value of μ where H_1 is true. We denote this selected μ value by μ_1. The manufacturer here has decided that when the mean service life of the modified component is $\mu_1 = 1250$, that is, when it is substantially higher than that of the original component, there should be only a 0.10 chance of concluding H_0 on the basis of the test. This specification is equivalent to stating that the probability of concluding H_1 is to be $1 - \beta = 1 - 0.10 = 0.90$ when $\mu_1 = 1250$. Figure 11.17b pertains to this control of the β risk. It shows the sampling distribution of $\overline{X}$ centered at $\mu_1 = 1250$ where the β risk is to be controlled. The requirement that β is to be 0.10 dictates that the left-tail area in the acceptance region be equal to 0.10 and, hence, that the action limit correspond to $z(0.10) = -1.282$ on the z^* scale. We denote this z value associated with the control of the β risk by z_1. Hence, $z_1 = z(0.10) = -1.282$ here.

Step 3.   The manufacturer believes that $\sigma = 250$ hours is a reasonable planning value for the population standard deviation.

Step 4.   Rather than employing formula (11.20), we will determine the required sample size by using Figures 11.17a and 11.17b together. We see that the required sample size n must be such that the interval on the $\overline{X}$ scale from $\mu_0 = 1100$ to A is equal to 2.326 standard deviations. Furthermore, the interval on the $\overline{X}$ scale from A to $\mu_1 = 1250$ must equal 1.282 standard deviations (note that we ignore the sign of $z_1 = -1.282$ because we are concerned only with the length of the interval). Thus, the whole interval length from μ_0 to μ_1, which equals $1250 - 1100 = 150$, must be equivalent to a total of $2.326 + 1.282 = 3.608$

FIGURE 11.17   **Determination of the required sample size to control α and β risks in the machine component example**

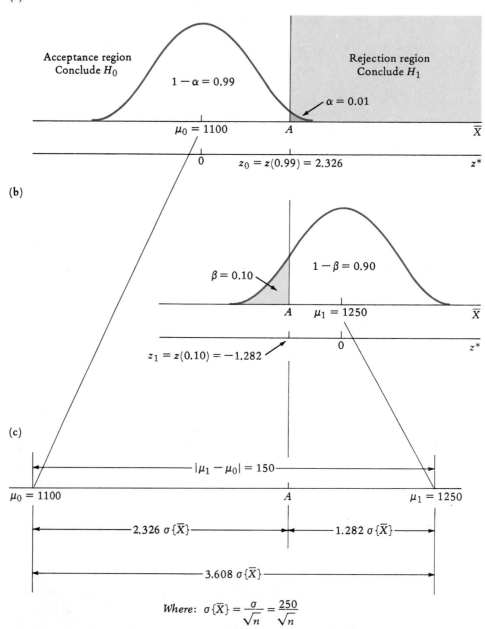

(a)

Acceptance region
Conclude H_0

$1 - \alpha = 0.99$

Rejection region
Conclude H_1

$\alpha = 0.01$

$\mu_0 = 1100$ A $\overline{X}$

0 $z_0 = z(0.99) = 2.326$ z^*

(b)

$\beta = 0.10$

$1 - \beta = 0.90$

A $\mu_1 = 1250$ $\overline{X}$

$z_1 = z(0.10) = -1.282$ 0 z^*

(c)

$|\mu_1 - \mu_0| = 150$

$\mu_0 = 1100$ A $\mu_1 = 1250$

$2.326\,\sigma\{\overline{X}\}$ $1.282\,\sigma\{\overline{X}\}$

$3.608\,\sigma\{\overline{X}\}$

Where: $\sigma\{\overline{X}\} = \dfrac{\sigma}{\sqrt{n}} = \dfrac{250}{\sqrt{n}}$

standard deviations. From the planning value $\sigma = 250$, the standard deviation of the sampling distribution of $\bar{X}$ is $\sigma\{\bar{X}\} = \sigma/\sqrt{n} = 250/\sqrt{n}$, where n is the required sample size. Thus, we have the following equation:

$$150 = 3.608 \left(\frac{250}{\sqrt{n}} \right)$$

Solving this equation for n gives $n = 36$. Figure 11.17c shows pictorially the reasoning that leads to the equation.

The sample size $n = 36$ was, in fact, the one used earlier in the manufacturer's test. It was used precisely because it was planned in this manner. Once the sample size is determined and the actual sample results are in hand, the test procedure is applied in the manner described earlier, using the sample standard deviations s in lieu of the planning value for σ. For this example, for instance, recall that the sample results were $\bar{X} = 1121$ and $s = 222$. Thus, z^* for the test was calculated as:

$$z^* = \frac{\bar{X} - \mu_0}{s\{\bar{X}\}} = \frac{1121 - 1100}{37.0} = 0.57$$

where $s\{\bar{X}\} = s/\sqrt{n} = 222/\sqrt{36} = 37.0$.   □

Comments
1. The α and β risk specifications in the machine component example are equivalent to the specification of two rejection probabilities, namely:

   $P(H_1; \mu_0 = 1100) = \alpha = 0.01$

   $P(H_1; \mu_1 = 1250) = 1 - \beta = 0.90$

   These two specifications fix two points of the rejection probability curve.
2. The procedure described for planning the sample size to control both types of error risks is appropriate if the resulting sample size is reasonably large. Modifications in the procedure are needed if the required sample size turns out to be small.
3. If the planning value σ is reasonably accurate, the actual β risk at $\mu = \mu_1$ for the test will be close to the planned level. This is the case in the machine component example, where the planning value for σ was 250 and the sample standard deviation, estimating the true standard deviation σ, was $s = 222$.

General Sample Size Formula

The reasoning process we have just followed for determining the required sample size in the machine component example can also be applied to lower-tail and two-sided tests. A general sample size formula can be obtained that applies to all three types of tests.

(11.20)   The necessary random sample size to control both α and β risks, for a given planning value of σ, is:

$$n = \frac{\sigma^2(|z_1| + |z_0|)^2}{|\mu_1 - \mu_0|^2}$$

where:

σ is the planning value for the population standard deviation

μ_0 and μ_1 are the μ values where the α and β risks are controlled, respectively

z_0 and z_1 are the z values associated with the specified α and β risks, respectively, and are defined as follows for each type of test.

One-Sided Upper-Tail Test $(H_0\colon \mu \le \mu_0, H_1\colon \mu > \mu_0)$:

$$z_0 = z(1 - \alpha)   \qquad   z_1 = z(\beta)$$

One-Sided Lower-Tail Test $(H_0\colon \mu \ge \mu_0, H_1\colon \mu < \mu_0)$:

$$z_0 = z(\alpha)   \qquad   z_1 = z(1 - \beta)$$

Two-Sided Test $(H_0\colon \mu = \mu_0, H_1\colon \mu \ne \mu_0)$:

$$z_0 = z(1 - \alpha/2)   \qquad   z_1 = z(\beta)$$

We illustrate the use of formula (11.20) by two examples.

EXAMPLE 1

One-Sided Upper-Tail Test.   For the machine component example in Figure 11.17, we require, for use of formula (11.20):

$$z_0 = z(1 - \alpha) = z(0.99) = 2.326   \quad \text{so } |z_0| = 2.326$$

$$z_1 = z(\beta) = z(0.10) = -1.282   \quad \text{so } |z_1| = 1.282$$

$$|\mu_1 - \mu_0| = |1250 - 1100| = 150   \quad \text{and} \quad \sigma = 250$$

Substituting into (11.20) yields:

$$n = \frac{(250)^2(2.326 + 1.282)^2}{(150)^2} = 36$$

EXAMPLE 2

Two-Sided Test.   Suppose the sample size in the art valuation example had not yet been determined.

Step 1.   As before, the α risk is to be controlled at 0.05 when $\mu_0 = 550$ (see Figure 11.4).

Step 2.   The β risk is to be controlled at 0.10 when $\mu_1 = 850$. (By symmetry, the same sample size is obtained if the β risk is controlled at $\mu_1 = 250$.)

Step 3.   A planning value of $\sigma = 925$ is deemed appropriate.

Step 4.   Referring to formula (11.20) for a two-sided test, we see that:

$$z_0 = z(1 - \alpha/2) = z(0.975) = 1.960 \quad \text{so } |z_0| = 1.960$$

$$z_1 = z(\beta) = z(0.10) = -1.282 \quad \text{so } |z_1| = 1.282$$

$$|\mu_1 - \mu_0| = |850 - 550| = 300$$

Substituting into (11.20), we obtain:

$$n = \frac{(925)^2(1.282 + 1.960)^2}{(300)^2} = 100$$

Thus, 100 subjects are needed in the study to achieve the desired control over the α and β risks of the test.

The sample size $n = 100$ was, in fact, the one used in the test we considered earlier. Recall that the sample results were $\overline{X} = 733$ and $s = 787$. The test statistic was therefore calculated as follows:

$$z^* = \frac{\overline{X} - \mu_0}{s\{\overline{X}\}} = \frac{733 - 550}{78.7} = 2.33$$

where $s\{\overline{X}\} = s/\sqrt{n} = 787/\sqrt{100} = 78.7$.   □

Comments

1. Formula (11.20) describes how the required sample size n is affected by the error risk specifications and by the levels of μ at which the risks are controlled. The smaller are the desired α and β risks, the larger are $|z_0|$ and $|z_1|$, respectively, and consequently the larger is the required n. Likewise, the closer μ_1 is to μ_0, the smaller is $|\mu_1 - \mu_0|$ and the larger is the required n. Finally, the larger is the population standard deviation (as anticipated by the planning value σ), the larger is the required sample size n to achieve the desired error control.

2. The derivation of formula (11.20) can be explained in terms of the example in Figure 11.17. The length of the interval from μ_0 to μ_1 may be represented by $|\mu_1 - \mu_0|$. The absolute value is used here because we wish to measure a length without regard to its sign. The length of the interval from μ_0 to A equals $|z_0|$ standard deviations, that is, it equals $|z_0|\sigma\{\overline{X}\}$. Likewise, the length of the interval from A to μ_1 equals $|z_1|$ standard deviations, that is, it equals $|z_1|\sigma\{\overline{X}\}$. Again, absolute values of z_0 and z_1 are used because we wish to measure lengths without regard to signs. Thus, we can see from Figure 11.17c that we have the equality:

$$|\mu_1 - \mu_0| = |z_1|\sigma\{\overline{X}\} + |z_0|\sigma\{\overline{X}\} = (|z_1| + |z_0|)\sigma\{\overline{X}\}$$

Finally, we have that $\sigma\{\overline{X}\} = \sigma/\sqrt{n}$. Making this substitution in the preceding equation gives:

$$|\mu_1 - \mu_0| = (|z_1| + |z_0|)\left(\frac{\sigma}{\sqrt{n}}\right)$$

Solving this equation for n gives formula (11.20).

11.8   OPTIONAL TOPIC—RELATION BETWEEN CONFIDENCE INTERVALS AND TESTS FOR μ

Statistical testing is closely connected with interval estimation. Indeed, tests of μ can be constructed on the basis of an appropriate confidence interval. One-sided tests correspond to one-sided confidence intervals, and two-sided tests correspond to two-sided intervals. The correspondence of tests and confidence intervals for μ is set out formally in Table 11.3. The α risk of the test and the confidence coefficient of the confidence interval are related as follows: If the confidence coefficient is $1 - \alpha$, then the corresponding α risk is α. This correspondence holds for both large and small samples. Finally, note that a lower one-sided confidence interval corresponds to an upper-tail test and that an upper one-sided confidence interval corresponds to a lower-tail test.

> **EXAMPLE 1**

Large-Sample Two-Sided Test.   In the accounts receivable example of Chapter 10, we considered an auditor who sampled the accounts receivable of a freight company in order to estimate the mean audit amount per account (μ). The accounting records of the company show that the mean book amount of the accounts receivable is \$47.86. The auditor wishes to test whether the mean audit amount agrees with the company books; that is, he wishes to test the following two-sided alternatives:

$$H_0: \mu = 47.86$$

$$H_1: \mu \neq 47.86$$

where $\mu_0 = 47.86$.

From a random sample of 100 accounts, the following 95 percent confidence interval for μ was constructed:

$$26.29 \leq \mu \leq 40.09$$

Referring to Table 11.3c, we see that the auditor should conclude H_0 if μ_0 lies in the confidence interval and H_1 if it does not. Here, $\mu_0 = 47.86$ does not lie in

---

TABLE 11.3   Correspondences of confidence intervals and tests for μ

Test	Alternatives	Confidence Interval	Decision Rule Based on Confidence Interval
(a) One-sided Upper-tail test	$H_0: \mu \leq \mu_0$ $H_1: \mu > \mu_0$	Lower one-sided $\mu \geq L$	If $\mu_0 \geq L$, conclude H_0 If $\mu_0 < L$, conclude H_1
(b) One-sided Lower-tail test	$H_0: \mu \geq \mu_0$ $H_1: \mu < \mu_0$	Upper one-sided $\mu \leq U$	If $\mu_0 \leq U$, conclude H_0 If $\mu_0 > U$, conclude H_1
(c) Two-sided test	$H_0: \mu = \mu_0$ $H_1: \mu \neq \mu_0$	Two-sided $L \leq \mu \leq U$	If $L \leq \mu_0 \leq U$, conclude H_0 If $\mu_0 < L$ or $\mu_0 > U$, conclude H_1

the interval. Hence, H_1 should be concluded. Because the confidence level of the interval is 95 percent, the corresponding α risk for the test is 0.05 when $\mu = \mu_0$.

□

EXAMPLE 2

Small-Sample Lower-Tail Test. In the lake acidity example, the following test alternatives are of interest when there is concern with whether the lake is acidic ($\mu < 7.0$):

$$H_0: \mu \geq 7.0$$

$$H_1: \mu < 7.0$$

where $\mu_0 = 7.0$.

The sample results were $n = 18$, $\overline{X} = 6.8861$, and $s\{\overline{X}\} = 0.024696$. According to Table 11.3, an upper confidence interval is required here. A 99.5 percent upper confidence limit for the mean pH of the lake, based on this small sample, is $U = 6.8861 + 2.898(0.024696) = 6.958$, where $t(0.995; 17) = 2.898$. Thus, the corresponding upper confidence interval is $\mu \leq 6.958$.

According to the decision rule in Table 11.3b, we conclude H_1 in this case because $\mu_0 = 7.0$ lies above the upper confidence limit, that is, $\mu_0 = 7.0 > 6.958$. The α risk for this test is 0.005 when $\mu = \mu_0$ because the confidence level of the upper confidence interval is 99.5 percent.

□

Comments

To illustrate the mathematical correspondence between a confidence interval and the associated decision rule, consider a two-sided test. Table 11.3c states that H_0 is concluded if $L \leq \mu_0 \leq U$. Now, for a two-sided confidence interval based on a large sample, L and U are given in (10.11) as $\overline{X} - zs\{\overline{X}\}$ and $\overline{X} + zs\{\overline{X}\}$, respectively, where $z = z(1 - \alpha/2)$. Rearranging the inequalities as shown in the following steps, we obtain the desired result:

1. $L \leq \mu_0 \leq U$

2. $\overline{X} - zs\{\overline{X}\} \leq \mu_0 \leq \overline{X} + zs\{\overline{X}\}$

3. $-z \leq \dfrac{\overline{X} - \mu_0}{s\{\overline{X}\}} \leq z$

4. $-z(1 - \alpha/2) \leq z^* \leq z(1 - \alpha/2)$ since $z^* = \dfrac{\overline{X} - \mu_0}{s\{\overline{X}\}}$

5. $|z^*| \leq z(1 - \alpha/2)$

This last inequality defines the acceptance region of a decision rule for a two-sided test.

PROBLEMS

11.1 For each of the following test situations: (1) specify the population mean μ; (2) give the value of the standard μ_0 against which the population mean is being compared; (3) state the alternatives

H_0 and H_1; (4) indicate whether the alternatives are one-sided upper-tail, one-sided lower-tail, or two-sided; and (5) describe the Type I and Type II errors that are possible.

    a. The mean donation per contributor to the Community Appeal was \$11.83 prior to the initiation of a new public relations campaign. A random sample of donations received while the campaign is in effect is to be used to determine whether the mean donation now exceeds \$11.83. Concluding that the campaign increases the mean donation, when in fact it does not, is the more serious error.

    b. The mean duration of marriages that ended in divorce or annulment during the past few years in a certain state was 8.1 years. A sociologist wishes to test whether new divorce legislation has changed the mean duration, based on a random sample of divorce records accumulated since the legislation was enacted. Concluding that a change in the mean duration has occurred, when in fact it has not, is the more serious error.

11.2 For each of the following test situations: (1) specify the population mean μ; (2) give the value of the standard μ_0 against which the population mean is being compared; (3) state the alternatives H_0 and H_1; (4) indicate whether the alternatives are one-sided upper-tail, one-sided lower-tail, or two-sided; and (5) describe the Type I and Type II errors that are possible.

    a. An engineer has developed a leaching process for reducing the amount of sulfur in coal. Sixty-four 1-kilogram samples are to be selected randomly from coal treated by this process to test whether the mean sulfur content of the treated coal is less than 10 grams per kilogram. Concluding that the process reduces the mean sulfur content to less than 10 grams per kilogram, when in fact it does not, is the more serious error.

    b. Bilingual children achieve a mean score of 60 on a mathematics proficiency test when mathematics is taught to them in their primary language. A random sample of 400 children has been taught the subject in their second language. An equivalent proficiency test will now be administered to see whether the mean score differs from 60. Concluding that the mean score differs from 60, when in fact it does not, is the more serious error.

11.3 A university has the option to switch to a new billing schedule for local telephone calls. This new schedule will yield a cost saving if the mean duration of local calls is less than 6.0 minutes but will be more costly otherwise. A random sample of local calls will be timed in order to decide whether to retain the present schedule or switch to the new one.

    a. Define the parameter μ here.

    b. Specify H_0 and H_1 if the α risk is to be the risk of incurring a cost increase because of an incorrect conclusion about μ.

    c. Specify H_0 and H_1 if the more serious error is to fail to obtain a cost saving when, in fact, the mean duration of calls is 5.5 minutes.

11.4 A firm has developed a diagnostic product for use by physicians in private practice. A decision must now be made whether or not to undertake a promotional campaign for the product. Such a campaign would be beneficial only if the mean number of units ordered per physician is greater than 5.0. Office demonstrations will be conducted with a random sample of physicians in the target market in order to decide whether or not to undertake the campaign.

    a. Define the parameter μ here.

    b. Specify H_0 and H_1 if the more costly error is to undertake the promotional campaign when it would not be beneficial.

    c. Specify H_0 and H_1 if the more costly error is to fail to undertake the promotional campaign when, in fact, the mean number of units ordered per physician is 5.5.

11.5 Refer to Table 11.1. In a statistical test, can one make both a Type I error and a Type II error simultaneously? Explain.

11.6 A machine fills cartons with soap powder and seals the cartons. The machine is set for a mean fill of 4.500 kilograms. However, the mean fill μ can drift upward or downward from 4.500 kilograms in the course of operations. The machine operator is given the following instructions by the quality assurance engineer: "The mean fill should be 4.500 kilograms. When the machine has been operating for an hour, empty the next ten cartons that are filled and weigh the soap powder. If the mean weight per carton for the ten cartons is between 4.480 and 4.520 kilograms, let the machine keep operating. If the mean weight is below 4.480 or above 4.520 kilograms, shut down the machine and call me."

   a. Specify H_0 and H_1 here.
   b. For the engineer's statistical decision rule, give (1) the test statistic, (2) the action limits, (3) the acceptance region, (4) the rejection region.
   c. What constitute the Type I and Type II errors here?

11.7 A buying consortium receives shipments of fresh turkeys from a local turkey farm. One of the clauses in the contract states that the mean dressed weight of the turkeys in a shipment should exceed 5.0 kilograms. In a test to determine whether this provision is being met in a large shipment of turkeys received for the Thanksgiving holiday, a random sample of 60 turkeys is selected and the birds are weighed. Consider the decision rule for a one-sided upper-tail test here, in which the α risk is controlled at 0.05 when $\mu = 5.0$. If, in fact, $\mu = 5.0$, what is the probability (prior to selecting the sample of turkeys from the shipment) that the sample mean dressed weight will fall in (1) the rejection region, (2) the acceptance region of the decision rule?

11.8 Refer to Figure 11.1. Explain why the acceptance region of each of these decision rules includes the neighborhood of μ_0.

11.9 Refer to Figure 11.2. Using the definition of z^*, show that $\mu_0 = 1100$ on the $\overline{X}$ axis corresponds to 0 on the z^* axis.

*11.10 Commercial Loans. The mean size of commercial loans made by a bank has been $60 thousand in the past. A recent change in the bank's credit policy allows larger amounts to be borrowed under the same terms. The credit manager now wishes to test whether the mean size of commercial loans made since the policy change is larger than $60 thousand ($H_1$) or not ($H_0$). The manager wishes to control the α risk at 0.01 when $\mu = 60$. A random sample of $n = 144$ loans made since the policy change yields the following results (in $000): $\overline{X} = 68.1$, $s = 45.0$.

   a. Conduct the test. State the alternatives, the decision rule, the value of the standardized test statistic, and the conclusion.
   b. If the α risk were controlled at 0.05 instead of 0.01, would your conclusion be different? Explain.

11.11 Refer to **Airline Reservations** Problem 10.12. It is known from extensive experience that the mean number of no-shows for other commuter flights of this airline is 1.320. Using the sample results in Problem 10.12, test whether the mean number of no-shows on the 4 P.M. commuter flight to New York City exceeds 1.320 (H_1) or not (H_0). Control the α risk at 0.05 when $\mu = 1.320$. State the alternatives, the decision rule, the value of the standardized test statistic, and the conclusion.

*11.12 Customs Declarations. Travelers returning from abroad are asked to declare the value of goods they are bringing back to their country. Customs authorities wish to test whether the mean reporting error is negative (H_1) or not (H_0). The reporting error is the difference between the declared value and the actual value. The authorities want to control the α risk at 0.001 when $\mu = 0$. A random audit of the personal effects of $n = 300$ travelers yielded the following results: $\overline{X} = -\$35.41$, $s = \$45.94$.

a. Conduct the test. State the alternatives, the decision rule, the value of the standardized test statistic, and the conclusion.

b. Why is the actual α risk of the test conducted in **a** only approximately at the specified level of 0.001?

11.13 Refer to **Trade Association** Problem 10.11. Shortly before enactment of the legislation, a full enumeration of all the member firms in the association showed that the mean number of hourly paid employees per firm was 9.16. It is desired to test whether the current mean number of hourly paid employees is less than 9.16 (H_1) or not (H_0), controlling the α risk at 0.05 when $\mu = 9.16$.

a. Using the sample results from Problem 10.11, conduct the test. State the alternatives, the decision rule, the value of the standardized test statistic, and the conclusion.

b. If, in fact, μ is actually 9.25, is your conclusion in **a** correct? If it is not correct, what type of error has been made (Type I or II)? Explain.

*11.14 **Summer Book Loans.**   In past summers in a large library system, the mean number of books borrowed per cardholder was 8.50. The library administration would like to test whether the mean number of books borrowed per cardholder this summer under modified loan arrangements differs from the level of past summers (H_1) or not (H_0). A random sample of 100 cardholders shows the following results for borrowing this summer: $\overline{X} = 9.34$, $s = 3.31$.

a. Conduct the test, controlling the α risk at 0.05 when $\mu = 8.50$. State the alternatives, the decision rule, the value of the standardized test statistic, and the conclusion.

b. Why is the actual α risk of the test conducted in **a** only approximately at the specified level of 0.05?

11.15 **Follow-up Purchases.**   Financial analysts in a firm operating a large chain of retail computer stores have assumed for planning purposes that customers buying a microcomputer from a store in the chain will spend an additional mean amount of $550 in follow-up purchases in the chain within 12 months following the computer purchase. A random sample of 81 customers showed the following results for follow-up purchases: $\overline{X} = \$446$, $s = \$380$. A test is to be conducted to determine whether the mean follow-up purchases are $550 ($H_0$) or not ($H_1$). The α risk is to be controlled at 0.01 when $\mu = 550$.

a. Conduct the appropriate test. State the alternatives, the decision rule, the value of the standardized test statistic, and the conclusion.

b. If, in fact, $\mu = 500$, is your conclusion in **a** correct? If it is not correct, what type of error has been made (Type I or II)? Explain.

11.16 **Auto Parts Producer.**   A sales analyst in a firm producing auto parts laboriously determined, from a study of all sales invoices for the previous fiscal year, that the mean profit contribution per invoice was $16.50. For the current fiscal year, the analyst will take a random sample of sales invoices to test whether the mean profit contribution this year has changed from $16.50 ($H_1$) or not ($H_0$). A random sample of 225 invoices yielded $\overline{X} = \$17.14$ and $s = \$18.80$. The α risk is to be controlled at 0.05 when $\mu = 16.50$.

a. Conduct the test. State the alternatives, the decision rule, the value of the standardized test statistic, and the conclusion.

b. What constitute Type I and Type II errors here? Given the conclusion in **a**, is it possible that a Type I error has been made in this test? Is a Type II error possible here? Explain.

11.17 In a tasting session, a random sample of 100 subjects from a target consumer population tasted a food item, and each subject individually gave it a rating from 1 (very poor) to 10 (very good). It is desired to test H_0: $\mu \leq 6.0$ versus H_1: $\mu > 6.0$, where μ denotes the mean rating for the food item in the target population. A computer analysis of the sample results gives a one-sided P-value of 0.0068.

a. Does the sample mean lie above or below $\mu_0 = 6.0$?

b. What must be the value of $z*$ in this test situation?

c. The sample standard deviation is $s = 1.76$. What must be the sample mean $\bar{X}$?

d. Does the magnitude of the P-value indicate that the sample results are inconsistent with conclusion H_0? Explain.

11.18 For each of the following large-sample test situations: (1) obtain the P-value, (2) state whether the P-value is one-sided or two-sided, and (3) comment on whether the sample data are consistent with conclusion H_0.

a. One-sided lower-tail alternatives, $z* = -0.32$.

b. Two-sided alternatives, $z* = 2.86$.

c. One-sided upper-tail alternatives, $z* = -0.86$.

11.19 For each of the following large-sample test situations: (1) obtain the P-value, (2) state whether the P-value is one-sided or two-sided, and (3) comment on whether the sample data are consistent with conclusion H_0.

a. Two-sided alternatives, $z* = -0.63$.

b. One-sided upper-tail alternatives, $z* = 3.01$.

c. One-sided lower-tail alternatives, $z* = -2.32$.

*11.20 Refer to **Commercial Loans** Problem 11.10.

a. Calculate the P-value of the test. Interpret its meaning here.

b. Conduct the required test using the P-value in **a**. Is the test conclusion identical to the one drawn on the basis of the decision rule for $z*$? Must this necessarily be the case? Explain.

11.21 Refer to **Airline Reservations** Problems 10.12 and 11.11.

a. Calculate the P-value of the test. Interpret its meaning here.

b. Conduct the required test using the P-value in **a**. Is the test conclusion identical to the one drawn on the basis of the decision rule for $z*$? Must this necessarily be the case? Explain.

*11.22 Refer to **Customs Declarations** Problem 11.12.

a. Calculate the P-value of the test. Is it a one-sided or a two-sided P-value?

b. Conduct the required test using the P-value in **a**. Is the test conclusion identical to the one drawn on the basis of the decision rule for $z*$? Must this necessarily be the case? Explain.

11.23 Refer to **Trade Association** Problems 10.11 and 11.13.

a. Calculate the P-value of the test. Interpret its meaning here.

b. Conduct the required test using the P-value in **a**. Is the test conclusion identical to the one drawn on the basis of the decision rule for $z*$? Must this necessarily be the case? Explain.

*11.24 Refer to **Summer Book Loans** Problem 11.14.

a. Calculate the P-value of the test. Is it a one-sided or a two-sided P-value?

b. Conduct the required test using the P-value in **a**. Is the test conclusion identical to the one drawn on the basis of the decision rule for $z*$? Must this necessarily be the case? Explain.

11.25 Refer to **Follow-up Purchases** Problem 11.15.

a. Calculate the P-value of the test. Interpret its meaning here.

b. Conduct the required test using the P-value in **a**. Is the test conclusion identical to the one drawn on the basis of the decision rule for $z*$? Must this necessarily be the case? Explain.

11.26 Refer to **Auto Parts Producer** Problem 11.16.

a. Calculate the P-value of the test. Is it a one-sided or a two-sided P-value?

b. Conduct the required test using the P-value in **a**. Is the test conclusion identical to the one drawn on the basis of the decision rule for $z*$? Must this necessarily be the case? Explain.

*11.27 Refer to **Nutrition Study** Problem 10.16. The dietary regimen is intended to yield a mean blood
sugar level of 110 milligrams of sugar per 100 cubic centimeters of blood. It is desired to test
whether this target is being met (H_0) or not (H_1), controlling the α risk at 0.05 when $\mu = 110$.
   a. Conduct the test. State the alternatives, the decision rule, the value of the standardized test
      statistic, and the conclusion.
   b. Obtain bounds for the P-value of this test. Is the P-value one-sided or two-sided? Is the
      P-value consistent with the test result in **a**? Explain.

11.28 A random sample of 21 children from the same age group participated in a special language
skills program designed to build vocabulary. Measurements at the end of the program showed
the children's vocabularies had a mean of $\overline{X} = 1060.1$ words and a standard deviation of
$s = 109.0$ words. It is known that the mean vocabulary of children in this age group ordinarily
is 1000 words. An analyst wishes to test whether the mean vocabulary of children who have
taken the special program is greater than 1000 words (H_1) or not (H_0). Assume that the vocabulary size of children who have taken the program is approximately normally distributed.
   a. Conduct the test, controlling the α risk at 0.01 when $\mu = 1000$. State the alternatives, the
      decision rule, the value of the standardized test statistic, and the conclusion.
   b. Obtain bounds for the P-value of this test. Is the P-value one-sided or two-sided? Is the
      P-value consistent with the test result in **a**? Explain.

11.29 A large bakery has installed new washing equipment for baking pans. A washing cycle with the
old equipment required a mean time of 35 minutes. Eight trial cycles with the new equipment
had mean cycle time $\overline{X} = 32.1$ minutes and standard deviation $s = 4.2$ minutes. Assume that
the cycle times for the new equipment have an approximately normal distribution and that the
eight trials constitute a random sample from this population. It is desired to test whether the
mean cycle time with the new equipment is shorter than with the old equipment (H_1) or not (H_0).
   a. Conduct the appropriate test, controlling the α risk at 0.05 when $\mu = 35$. State the alterna-
      tives, the decision rule, the value of the standardized test statistic, and the conclusion.
   b. Obtain bounds for the P-value of this test. Is the P-value one-sided or two-sided? Is the
      P-value consistent with the test result in **a**? Explain.

11.30 Safety regulations governing the evacuation of a high-risk area of a chemical plant call for the
mean evacuation time to be less than 5 minutes. Four safety drills have produced the following
evacuation times (in minutes): 3.8, 4.6, 5.3, 4.8. Assume that the evacuation times for the four
drills constitute a random sample from a normal population.
   a. Test whether the regulation governing mean evacuation time is being met (H_1) or not (H_0).
      Control the α risk at 0.01 when $\mu = 5.0$. State the alternatives, the decision rule, the value
      of the standardized test statistic, and the conclusion.
   b. Obtain bounds for the P-value of this test. Is the P-value one-sided or two-sided? Is the
      P-value consistent with the test result in **a**? Explain.

11.31 A manufacturer sells a safety light that turns on automatically when the electricity fails in an in-
stallation. The manufacturer claims that these lights, when fully charged, have a mean operating
life before recharging is required of more than 300 minutes. In a test of this claim, 12 fully
charged lights were selected at random, and the operating life of each was measured. The sample results (in minutes) were as follows:

   290   331   329   364   332   333   346   356   352   272   316   347

Assume that the operating lives are approximately normally distributed. It is desired to test
whether the mean operating life exceeds 300 minutes (H_1) or not (H_0), controlling the α risk at
0.025 when $\mu = 300$.

a. Conduct the appropriate test. State the alternatives, the decision rule, the value of the standardized test statistic, and the conclusion.

b. Obtain bounds for the P-value of this test. Is the P-value one-sided or two-sided? Is the P-value consistent with the test result in **a**? Explain.

c. A subsequent check showed that the two sample lights that had operating lives of 290 and 272 minutes were not fully charged for the study. Would your test conclusion in **a** change if these two observations were discarded from the sample? Explain.

d. Construct a stem-and-leaf display of the 12 sample observations. Do the 2 observations mentioned in **c** appear to be outliers? Does the assumption of normality appear to be plausible for the other 10 observations? Comment.

*11.32 **New Product.** A market study for a new industrial product indicates that the product should be launched by the firm only if the mean number of units purchased per customer in the first solicitation of the firm's customers is more than 2.0. Initial orders for the product are to be solicited from a random sample of 100 of the firm's many customers. Assume that the standard deviation of the numbers of units purchased in initial orders is $\sigma = 1.6$ units.

a. The alternatives to be tested are H_0: $\mu \le 2.0$, H_1: $\mu > 2.0$, and the α risk is to be controlled at 0.05 when $\mu = 2.0$. Verify that the following decision rule for $\bar{X}$ is appropriate for this test:

If $\bar{X} \le 2.263$, conclude H_0.
If $\bar{X} > 2.263$, conclude H_1.

b. Calculate the rejection probabilities at $\mu = 2.15$ and $\mu = 2.45$ for the decision rule in **a**, and complete the following table:

μ:	1.85	2.00	2.15	2.30	2.45
$P(H_1; \mu)$:	0.005	0.050	____	0.591	____

c. Sketch the rejection probability curve for the decision rule in **a**.

d. What is the incorrect conclusion when $\mu = 2.45$? What is the probability that the decision rule in **a** will lead to the incorrect conclusion when $\mu = 2.45$? Is this probability an α or β risk?

*11.33 Refer to **New Product** Problem 11.32.

a. Construct the decision rule for $\bar{X}$ if the α risk is controlled at 0.01 rather than at 0.05 when $\mu = 2.0$.

b. Obtain the rejection probability at $\mu = 2.45$ for the decision rule in **a** where the α risk is controlled at 0.01. Is the error risk at $\mu = 2.45$ smaller or larger with this decision rule than with the decision rule where α is controlled at 0.05? Is this finding consistent with the principle stated in (11.19)? Explain.

11.34 **Color Graphics.** The developer of a decision support software package wishes to test whether a color graphics enhancement for the package is considered by users to be beneficial, on balance, given its list price of $800. A random sample of 144 users of the package will be invited to try out the enhancement and rate it on a scale ranging from -5 (completely useless) to $+5$ (very beneficial). The test alternatives are H_0: $\mu \le 0$, H_1: $\mu > 0$, where μ denotes the mean rating of users. The α risk of the test is to be controlled at 0.01 when $\mu = 0$. Assume that the standard deviation of users' ratings is $\sigma = 2.3$.

a. Verify that the following decision rule for $\overline{X}$ is appropriate for this test:

If $\overline{X} \leq 0.446$, conclude H_0.
If $\overline{X} > 0.446$, conclude H_1.

b. Calculate the rejection probabilities at $\mu = 0.25$ and $\mu = 0.75$ for the decision rule in **a**, and complete the following table:

μ:	0	0.25	0.50	0.75
$P(H_1; \mu)$:	0.010	_____	0.610	_____

c. Sketch the rejection probability curve for the decision rule in **a**.
d. What is the incorrect conclusion when $\mu = 0.50$? What is the probability that the decision rule in **a** will lead to the incorrect conclusion when $\mu = 0.50$? Is this probability an α or β risk?

11.35 Trade Agreement. A trade agreement governing the movement of agricultural products between two countries stipulates that the mean weight of boxes of butter must be 25.00 kilograms. A large shipment of boxes of butter is to be tested to determine whether it meets this requirement by selecting a random sample of 64 boxes and using the following decision rule:

If $24.95 \leq \overline{X} \leq 25.05$, conclude H_0 ($\mu = 25.00$).
If $\overline{X} < 24.95$ or $\overline{X} > 25.05$, conclude H_1 ($\mu \neq 25.00$).

The standard deviation for weights of boxes of butter is $\sigma = 0.20$ kilogram.
a. Calculate the rejection probabilities at $\mu = 24.90, 25.00, 25.10$ for this decision rule, and complete the following table:

μ:	24.90	24.95	25.00	25.05	25.10
$P(H_1; \mu)$:	_____	0.500	_____	0.500	_____

b. Sketch the rejection probability curve, using the tabulated values in **a**.
c. What is the α risk at $\mu = 25.00$ for this decision rule? What is the error risk at $\mu = 25.10$? Does the latter relate to a Type I or a Type II error?

11.36 Refer to **Trade Agreement** Problem 11.35.
a. Obtain the decision rule for $\overline{X}$ if the α risk is to be controlled at 0.01 when $\mu = 25.00$. Also, obtain the decision rule for $\alpha = 0.10$.
b. For each decision rule in **a**, obtain the β risk when $\mu = 25.10$. Are the α and β risks for the two decision rules consistent with the principle stated in (11.19)? Explain.

11.37 Foreign Applicants. An education researcher is interested in the effect of culture on the performance of applicants who take a North American graduate school admission test. North American applicants' scores are known to have a mean of 550 points. A random sample of 225 applicants from an English-speaking country abroad is scheduled to take the test. The researcher will use the sample results to test H_0: $\mu = 550$ versus H_1: $\mu \neq 550$, where μ denotes the mean test score of applicants from the English-speaking country abroad. The standard deviation of test scores for the applicants from this country abroad is anticipated to be $\sigma = 45$ points. The following decision rule for $\overline{X}$ is to be employed:

If $544 \leq \overline{X} \leq 556$, conclude H_0.

If $\overline{X} < 544$ or $\overline{X} > 556$, conclude H_1.

a. Calculate the rejection probabilities at $\mu = 540, 545, 550, 555, 560$ for this rule.

b. Sketch the rejection probability curve, using the values in a.

c. What is the α risk at $\mu = 550$? Does an α risk exist at any other value of μ? Explain.

d. If applicants from this foreign country have a mean score that is 5 points above that of North American applicants, what is the probability that the decision rule in a will fail to detect this difference? Is this probability an α or β risk?

*11.38 Refer to **Commercial Loans** Problem 11.10. Use the sample results to estimate the β risk for the test when $\mu = 70$. Interpret this risk in the context of this problem.

11.39 Refer to **Airlines Reservations** Problems 10.12 and 11.11. Use the sample results to estimate the β risk for the test when $\mu = 1.60$. Interpret this risk in the context of this problem.

*11.40 Refer to **Customs Declarations** Problem 11.12. Use the sample results to estimate the β risk for the test when $\mu = -6.0$. Interpret this risk in the context of this problem.

11.41 Refer to **Trade Association** Problems 10.11 and 11.13. Use the sample results to estimate the β risk for the test when $\mu = 8.0$. Interpret this risk in the context of this problem.

*11.42 Refer to **Summer Book Loans** Problem 11.14. Use the sample results to estimate the β risk for the test when $\mu = 10.0$. Interpret this risk in the context of this problem.

11.43 Refer to **Follow-up Purchases** Problem 11.15. Use the sample results to estimate the β risk for the test when $\mu = 450$. Interpret this risk in the context of this problem.

11.44 Refer to **Auto Parts Producer** Problem 11.16. Use the sample results to estimate the β risk for the test when $\mu = 19.50$. Interpret this risk in the context of this problem.

*11.45 Refer to **New Product** Problem 11.32. Suppose that the sample size for the study is still under review.

a. It is desired to control the α risk at 0.05 when $\mu = 2.0$ and the β risk at 0.10 when $\mu = 2.45$. Using $\sigma = 1.6$ as the planning value, calculate the required sample size.

b. A random sample of the size determined in a has been selected, and the results are $\overline{X} = 2.34$ and $s = 1.71$. Conduct the required test, stating the decision rule, the value of the standardized test statistic, and the conclusion.

c. What would be the effect on the sample size in a if the β risk were to be controlled at 0.10 when $\mu = 2.30$ instead of when $\mu = 2.45$?

11.46 Refer to **Color Graphics** Problem 11.34. Suppose that the sample size for the test is still under review.

a. It is desired to control the α risk at 0.01 when $\mu = 0$ and the β risk at 0.05 when $\mu = 0.75$. Using $\sigma = 2.3$ as the planning value, calculate the required sample size.

b. A random sample of the size determined in a has been selected, and the results are $\overline{X} = 0.545$ and $s = 2.12$. Conduct the required test, stating the decision rule, the value of the standardized test statistic, and the conclusion.

11.47 Refer to **Trade Agreement** Problem 11.35. Suppose that the sample size for the test is still under review.

a. It is desired to control the α risk at 0.05 when $\mu = 25.00$ and the β risk at 0.05 when $\mu = 24.92$. Using $\sigma = 0.20$ as the planning value, calculate the required sample size.

b. A random sample of the size determined in a has been selected, and the results are $\overline{X} = 25.03$ and $s = 0.183$. Conduct the required test, stating the decision rule, the value of the standardized test statistic, and the conclusion.

c. The value of the sample standard deviation s is smaller than the planning value for σ used in calculating the required sample size. What does this difference imply about the magnitude of the actual β risk for the test at $\mu = 24.92$ in relation to the target β risk of 0.05?

11.48 Refer to **Foreign Applicants** Problem 11.37. Suppose that the sample size for the study is still under review.
   a. It is desired to control the α risk at 0.05 when $\mu = 550$ and the β risk at 0.025 when $\mu = 540$. Using $\sigma = 45$ as the planning value, calculate the required sample size.
   b. A random sample of the size determined in **a** has been selected, and the results are $\overline{X} = 527.2$ and $s = 43.8$. Conduct the required test, stating the decision rule, the value of the standardized test statistic, and the conclusion.
   c. What would be the effect on the required sample size in **a** if the planning value used for σ were 40 instead of 45?

*11.49 Refer to **Airline Reservations** Problem 10.12. Use the 95 percent confidence interval to test whether the mean number of persons failing to keep their reservations is 1.0 (H_0) or not (H_1). State your conclusion. What is the implied α risk of this test?

11.50 Refer to **Nutrition Study** Problem 10.16. Use the 95 percent confidence interval to test whether the mean blood sugar level is 110 (H_0) or not (H_1). State your conclusion. What is the implied α risk of this test?

11.51 Refer to **Customs Declarations** Problem 11.12.
   a. What type of confidence interval will provide the relevant information for the test? What should be the confidence coefficient?
   b. Using the sample results, construct the appropriate confidence interval. Explain how this confidence interval leads to the same conclusion as the test.

11.52 Refer to **Summer Book Loans** Problem 11.14.
   a. What type of confidence interval will provide the relevant information for the test? What should be the confidence coefficient?
   b. Using the sample results, construct the appropriate confidence interval. Explain how this confidence interval leads to the same conclusion as the test.

EXERCISES

11.53 Demonstrate the equivalence of decision rules (11.6) and (11.14) for testing one-sided upper-tail alternatives when n is large.

11.54 Consider the standardized test statistic $(\overline{X} - \mu_0)/s\{\overline{X}\}$. What must be assumed about the sample size or sampled population in order for this statistic to have (1) an approximate standard normal distribution when $\mu = \mu_0$, (2) a $t(n - 1)$ distribution when $\mu = \mu_0$?

11.55 Consider a test of the alternatives H_0: $\mu \le 0$ versus H_1: $\mu > 0$, controlling the α risk at 0.01 when $\mu = 0$. Assume the population standard deviation is $\sigma = 10$.
   a. Calculate the β risk at $\mu = 2.0$ for each of the sample sizes $n = 100, 200, 400$.
   b. What do the results in **a** imply about the effect of increasing the sample size on the β risk at a given μ, for a fixed level of the α risk?

11.56 Demonstrate the equivalence of the decision rules in Table 11.2a and Table 11.3a for testing one-sided upper-tail alternatives when the sampled population is normal.

STUDIES

11.57 (Computer needed.) Refer to the **Power Cells** data set. Assume that the data on number of cycles before failure for the $n = 54$ cells constitute a random sample from a single population with mean μ.

   a. Test the alternatives H_0: $\mu = 135$ versus H_1: $\mu \neq 135$, controlling the α risk at 0.01 when $\mu = 135$. State the decision rule, the value of the standardized test statistic, and the conclusion.

   b. Compute the P-value of the test in **a**. Verify that it is consistent with the conclusion in **a**.

   c. Estimate the β risk for the test in **a** when $\mu = 160$.

   d. Why might the numbers of cycles for the cells in this experiment not be observations drawn from the same population? Would it be meaningful then to consider a test on μ here? Explain.

11.58 A manufacturer has developed a new drive belt for a machine. The original drive belt is known to have a mean operating life of 3500 hours. The manufacturer wants to place a random sample of the new belts on a forced-life test to establish if the mean operating life of the new belt exceeds that of the original belt (H_1) or not (H_0). The manufacturer is considering several possible sample sizes. The α risk of the test is to be controlled at 0.01 when $\mu = 3500$.

   a. Evaluate the rejection probability at $\mu = 3450, 3500, 3550, 3600, 3650, 3700$ for each of the following values of n: (1) 50, (2) 100, (3) 150. Sketch the rejection probability curves on the same graph. Assume that a reasonable planning value for the standard deviation of the operating lives of the new belts is $\sigma = 500$ hours.

   b. The manufacturer definitely wants to conclude that the new belt has a longer mean operating life if, in fact, its mean life is 5 percent longer than the mean life of the original belt. From the curves in **a**, what is the power of the decision rule for each of the three sample sizes when the mean life of the new belt is 5 percent longer than that of the original?

   c. In the end, the manufacturer placed a random sample of $n = 100$ new belts on a forced-life test and obtained these results (in hours): $\overline{X} = 3650.2$, $s = 531.4$. What should the manufacturer conclude from these results? State the decision rule and the standardized test statistic.

11.59 A market survey is being designed to assess consumer reactions to a new household product. The main factors to be taken into account in the design are (1) the degree (X_1) to which a respondent prefers the new product to its principal competitor, and (2) the respondent's subjective assessment of the dollar value of the new product (X_2). The preference variable X_1 will be measured on an 11-point scale ranging from -5 (new product much worse than that of principal competitor) to $+5$ (new product much better than that of principal competitor). The test alternatives of interest in connection with the preference variable are H_0: $\mu_1 \leq 0$, H_1: $\mu_1 > 0$, where μ_1 denotes the mean of X_1. The mean subjective dollar value, denoted by μ_2, is to be estimated by a 99 percent confidence interval. In planning the sample size for the survey, it is desired that the test on the mean preference control each of the α and β risks at 0.01 or less when $\mu_1 = 0$ and $\mu_1 = +1$, respectively. It is also desired that the half-width of the confidence interval for μ_2 not exceed \$5. Reasonable planning values for the standard deviations of X_1 and X_2 are, respectively, $\sigma_1 = 2.5$ and $\sigma_2 = \$20$.

   a. What number of respondents is adequately large to satisfy both the testing and the estimation requirements?

   b. Consider the following two events: (1) The test on μ_1 leads to the incorrect conclusion; (2) the confidence interval for μ_2 is incorrect. In advance of conducting the survey, are these two events necessarily statistically independent? Explain why or why not.

c. The survey was conducted using a sample size of $n = 140$, and the following sample results were obtained.

Variable	Sample Mean	Sample Standard Deviation
Preference	1.32	2.38
Dollar value	$183.40	$18.10

Construct the desired interval estimate for μ_2, and conduct the test for μ_1, controlling the α risk at 0.01 when $\mu_1 = 0$. For the test, state the decision rule, the value of the standardized test statistic, and the conclusion. Interpret both your interval estimate and your test conclusion.

12 INFERENCES FOR POPULATION PROPORTION

In Chapters 10 and 11, we considered inferences about a population mean. In this chapter, we take up inferences about a population proportion.

12.1 POPULATION PROPORTION

In many applications of sampling, the characteristic of interest in the population elements is qualitative with two possible outcomes.

EXAMPLES

1. A government policymaker is concerned with the outlook by U.S. households about economic conditions. The relevant population consists of all U.S. households, and the characteristic of interest is whether or not the household expects an economic upturn within the next three months. The population parameter is taken to be the proportion p of U.S. households expecting an economic upturn within three months. (The parameter could equally well have been the proportion of households that do not expect an economic upturn within three months.)

2. In the processing of insurance applications, a characteristic of concern is whether or not the prepared application provides complete information. The population of interest in this case is an infinite one, because it pertains to the process of preparing insurance applications. The population proportion p of relevance is the proportion of all applications in the infinite population that are complete. (The proportion of applications that are incomplete could also have been considered to be the population proportion of interest.) Equivalently, the act of preparing an insurance application may be viewed as a Bernoulli random trial where the random variable X equals 1 when the application is complete and 0 otherwise, and the outcome 1 occurs with probability p. □

12.2 SAMPLE STATISTICS

When the characteristic of interest is qualitative with two possible outcomes, a sample statistic of interest is the *number of occurrences* among the n sample observations consisting of the particular outcome reflected in the population proportion. This number of occurrences is denoted by X.

1. p is the population proportion of households that expect an economic upturn shortly; X then is the number of households in the sample that expect an economic upturn shortly.

2. p is the probability that an insurance application is complete; X then is the number of complete applications in the sample.

Another sample statistic of interest is the *sample proportion,* denoted by $\bar{p}$. It is defined as follows.

(12.1)

$$\bar{p} = \frac{X}{n}$$

Thus, $\bar{p}$ in the first example represents the proportion of households in the sample that expect an economic upturn, and in the second example, it is the proportion of complete insurance applications in the sample.

12.3   SAMPLING DISTRIBUTIONS OF X AND $\bar{p}$

Binomial Probability Distribution as Sampling Distribution

When a simple random sample of size n is selected from an infinite population, we already know the exact sampling distribution of the number of occurrences X. Recall from Chapter 6 that when n Bernoulli trials are conducted, we have n Bernoulli random variables $X_1, X_2, \ldots, X_n$, each of which can take on the values 1 and 0. If the n trials are independent and $P(X_i = 1) = p$ for all trials, the sum $X = X_1 + X_2 + \cdots + X_n$ is a binomial random variable whose probability function is given by (6.3). The sum X is, of course, the number of occurrences in the sample that consist of the outcome denoted by 1.

Since the conditions of independence and constant probability p for each trial are precisely the conditions (8.11) for simple random sampling from an infinite population, the number of occurrences X in a simple random sample from an infinite population is a binomial random variable.

(12.2)

When a simple random sample of size n is selected from an infinite population (that is, a Bernoulli process), the sum of the n Bernoulli variables, denoted by X:

$$X = X_1 + X_2 + \cdots + X_n$$

is a binomial random variable whose sampling distribution is given by the *binomial probability function:*

$$P(x) = \binom{n}{x} p^x (1 - p)^{n-x}$$

where:

$P(x) = P(X = x)$

$$\binom{n}{x} = \frac{n!}{x!\,(n - x)!}$$

$x = 0, 1, \ldots, n$

Since the sample proportion $\bar{p}$, as defined in (12.1), is a constant multiple of the number of occurrences X, probabilities for $\bar{p}$ correspond to those for X. Thus, when $n = 3$, the probability that $\bar{p} = 2/3$ is the same as the probability that $X = 2$. Consequently, Table C.5 for the binomial probability distribution may be used to obtain the sampling distributions of both X and $\bar{p}$ for simple random sampling from an infinite population.

EXAMPLES

1. Table 12.1 shows the sampling distributions of X and $\bar{p}$ when $n = 6$ and $p = 0.3$. Figure 12.1 shows these distributions graphically, the only difference between the sampling distributions of X and $\bar{p}$ being in the scaling on the horizontal axis. The probabilities were obtained from Table C.5.

TABLE 12.1 **Sampling distributions of X and $\bar{p}$ for $n = 6$ and $p = 0.3$.** Since $\bar{p}$ is a constant multiple of X, the corresponding probabilities are the same.

Sampling Distribution of X		Sampling Distribution of $\bar{p}$	
x	$P(x)$	$\bar{p}$	$P(\bar{p})$
0	0.1176	0	0.1176
1	0.3025	1/6	0.3025
2	0.3241	2/6	0.3241
3	0.1852	3/6	0.1852
4	0.0595	4/6	0.0595
5	0.0102	5/6	0.0102
6	0.0007	1	0.0007
Total	1.00	*Total*	1.00

FIGURE 12.1   **Sampling distributions of X and $\bar{p}$ for $n = 6$ and $p = 0.3$**

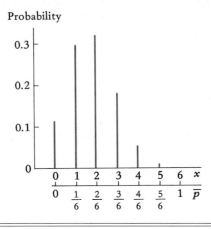

2.   For the preceding example, we wish to find $P(\bar{p} \leq 0.5)$. Using Table 12.1 for $n = 6$, $p = 0.3$, we find the desired probability to be:

$$P(\bar{p} \leq 0.5) = P(X \leq 3) = 0.1176 + 0.3025 + 0.3241 + 0.1852$$

$$= 0.9294$$

Thus, the probability is 0.9294 that $\bar{p}$ is 0.5 or less.

3.   For $n = 12$ and $p = 0.5$, we wish to find $P(X \geq 2)$. We obtain, using Table C.5:

$$P(X \geq 2) = 1 - P(X \leq 1) = 1 - (0.0002 + 0.0029) = 0.9969$$

4.   For $n = 8$ and $p = 0.2$, we wish to find $P(\bar{p} \leq 0.25)$. We obtain:

$$P(\bar{p} \leq 0.25) = P(X \leq 2) = 0.1678 + 0.3355 + 0.2936 = 0.7969$$ □

Comment   The probability function for $\bar{p}$ can be written explicitly by replacing x in (12.2) by $n\bar{p}$.

(12.3)
$$P(\bar{p}) = \binom{n}{n\bar{p}} p^{n\bar{p}}(1 - p)^{n(1-\bar{p})}$$

where:

$$\bar{p} = 0, \frac{1}{n}, \frac{2}{n}, \dots, 1$$

Characteristics of Sampling Distributions of X and $\bar{p}$

Mean and Variance. The mean and the variance of the sampling distribution of X correspond to those of the binomial probability distribution given in (6.5) and (6.6), respectively. We now give the mean, the variance, and the standard deviation of both the sampling distribution of X and the sampling distribution of $\bar{p}$.

(12.4)

For simple random sampling from an infinite population (that is, for a Bernoulli process):

Sampling Distribution of X

$E\{X\} = np$

$\sigma^2\{X\} = np(1 - p)$

$\sigma\{X\} = \sqrt{np(1 - p)}$

Sampling Distribution of $\bar{p}$

$E\{\bar{p}\} = p$

$\sigma^2\{\bar{p}\} = \dfrac{p(1 - p)}{n}$

$\sigma\{\bar{p}\} = \sqrt{\dfrac{p(1 - p)}{n}}$

EXAMPLES

1. When $n = 6$ and $p = 0.3$:

$E\{X\} = 6(0.3) = 1.8 \qquad \sigma^2\{X\} = 6(0.3)(0.7) = 1.26 \qquad \sigma\{X\} = \sqrt{1.26} = 1.12$

$E\{\bar{p}\} = 0.3 \qquad \sigma^2\{\bar{p}\} = \dfrac{0.3(0.7)}{6} = 0.035 \qquad \sigma\{\bar{p}\} = \sqrt{0.035} = 0.187$

2. When $n = 100$ and $p = 0.8$:

$E\{X\} = 100(0.8) = 80 \qquad \sigma^2\{X\} = 100(0.8)(0.2) = 16 \qquad \sigma\{X\} = \sqrt{16} = 4$

$E\{\bar{p}\} = 0.8 \qquad \sigma^2\{\bar{p}\} = \dfrac{0.8(0.2)}{100} = 0.0016 \qquad \sigma\{\bar{p}\} = \sqrt{0.0016} = 0.04$

$\square$

Shape of Distribution. We know from Chapter 6 that the binomial probability distribution is skewed to the right if $p < 0.5$, skewed to the left if $p > 0.5$, and is symmetrical if $p = 0.5$. Hence, the sampling distribution of $\bar{p}$ has these same characteristics and is symmetrical only when $p = 0.5$.

Comments

1. To derive the mean and variance of $\bar{p}$, we utilize the fact that $\bar{p} = X/n$ and the known results for the mean and the variance of the random variable X:

$$E\{\bar{p}\} = E\left\{\frac{X}{n}\right\} = \frac{1}{n}E\{X\} = \frac{np}{n} = p \qquad \text{by (5.11b)}$$

$$\sigma^2\{\bar{p}\} = \sigma^2\left\{\frac{X}{n}\right\} = \frac{1}{n^2}\sigma^2\{X\} = \frac{np(1 - p)}{n^2} = \frac{p(1 - p)}{n} \qquad \text{by (5.12b)}$$

2. The sample proportion $\bar{p}$ is actually a special case of a sample mean $\bar{X}$. Recall that the binomial variable is $X = X_1 + X_2 + \cdots + X_n$, where $X_i = 0, 1$. Hence:

$$\bar{p} = \frac{X}{n} = \frac{X_1 + X_2 + \cdots + X_n}{n} = \bar{X}$$

For example, for $n = 4$ and $X_1 = 0$, $X_2 = 1$, $X_3 = 1$, and $X_4 = 0$, we have:

$$\bar{p} = \frac{0 + 1 + 1 + 0}{4} = 0.5 = \bar{X}$$

All of the earlier results for $\bar{X}$ apply to $\bar{p}$. Thus, we know from Chapter 9 that $E\{\bar{X}\} = E\{X_i\} = \mu$. But for a Bernoulli variable, $E\{X_i\} = 1(p) + 0(1 - p) = p$. Hence, $E\{\bar{X}\} = p$ when the X_i are Bernoulli variables with $E\{X_i\} = p$.

Central Limit Theorem

While the sampling distribution of $\bar{p}$ (or X) is skewed if $p \neq 0.5$, the skewness decreases as the sample size increases. Figure 12.2 illustrates this fact, showing the sampling distribution of $\bar{p}$ when $p = 0.4$ for sample sizes $n = 5$ and $n = 50$. The sampling distribution of $\bar{p}$ for $n = 50$ in Figure 12.2b possesses very little skewness indeed. In fact, the outline of this sampling distribution appears to be approximately normal. This brings us to another encounter with the central limit theorem.

(12.5)   The sampling distributions of X and $\bar{p}$ are approximately normal when the simple random sample size is sufficiently large. As a working rule, the normal approximation is adequate when both $np \geq 5$ and $n(1 - p) \geq 5$.

Thus, the binomial probability distribution can be approximated by a normal probability distribution when the random sample size is sufficiently large.

EXAMPLES

1. When $n = 50$ and $p = 0.4$, we have $np = 50(0.4) = 20$ and $n(1 - p) = 50(0.6) = 30$. Since both quantities exceed 5, our working rule states that the normal distribution can be used here to approximate the sampling distributions of X and $\bar{p}$.
2. When $n = 15$ and $p = 0.8$, we have $np = 15(0.8) = 12$ and $n(1 - p) = 15(0.2) = 3$. Since the latter quantity is less than 5, our working rule states that the normal distribution cannot be used here to approximate the sampling distributions of X and $\bar{p}$.   □

Comments

1. When a finite population of size N is sampled, the sampling distributions of X and $\bar{p}$ are given exactly by the hypergeometric probability function (6.15) with $C = Np$. How-

ever, as long as the sample size is a small fraction of the population size, the binomial probability function provides a good approximation to the hypergeometric distribution. If, moreover, the sample size itself is large, then the normal approximation to the sampling distributions of X and $\bar{p}$ for infinite populations described in (12.5) can be employed as well for large finite populations.

   In Chapter 14 we consider the case when the sample size is not small relative to the population size.

2. As we noted earlier, the sample proportion $\bar{p}$ is a special case of the sample mean $\bar{X}$. Hence, central limit theorem (12.5) for $\bar{p}$ is a special case of central limit theorem (9.5) for $\bar{X}$.

FIGURE 12.2   Sampling distributions of $\bar{p}$ for $n = 5$ and $n = 50$ when $p = 0.4$

(a) $n = 5$

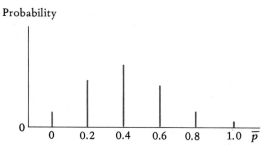

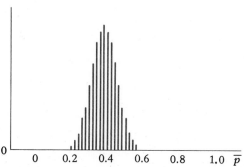
(b) $n = 50$

Normal Approximation to Binomial Distribution

Use of Table C.1 for the standard normal distribution for approximating the binomial distribution requires the following standardized variables.

(12.6)

$$Z = \frac{X - E\{X\}}{\sigma\{X\}} = \frac{X - np}{\sqrt{np(1 - p)}}$$

$$Z = \frac{\bar{p} - E\{\bar{p}\}}{\sigma\{\bar{p}\}} = \frac{\bar{p} - p}{\sqrt{p(1 - p)/n}}$$

We illustrate the theory developed about the sampling distributions of X and $\bar{p}$ by presenting two examples.

EXAMPLE 1

Probability Statement About X Using Correction for Continuity.   An analyst of the insurance company mentioned earlier will select a random sample of 50 insurance applications and is concerned about how many of them will be complete. While, ordinarily, we do not know the population proportion (that is, the probability that $X_i = 1$), let us assume here that we know it to be $p = 0.7$. We wish to find the probability that 30 or more applications in the sample of 50 are complete.

We know from theorem (12.4) that the mean of the sampling distribution of X is $E\{X\} = np = 50(0.7) = 35$, and that the standard deviation is $\sigma\{X\} = \sqrt{np(1 - p)} = \sqrt{50(0.7)(0.3)} = 3.24$. Furthermore, our working rule in (12.5) indicates that the central limit theorem is applicable because both $np = 50(0.7) = 35 \geq 5$ and $n(1 - p) = 50(0.3) = 15 \geq 5$.

In using the normal distribution as an approximation to the binomial distribution, we need to recognize that we are utilizing a continuous probability distribution to approximate a discrete one. Figure 12.3 illustrates the basic problem. Figure 12.3a shows the binomial probability that $X = 30$. Since the normal distribution is continuous and area under the normal curve corresponds to probability, we convert the discrete binomial probability to an area. We do so by treating $X = 30$ as extending from 29.5 to 30.5 and constructing a rectangle to represent the probability at $X = 30$, as shown in Figure 12.3b. Note that the area of the rectangle in Figure 12.3b is the same as the discrete probability in Figure 12.3a, since the height of the rectangle is the same and its width is 1.

Thus, to find $P(X \geq 30)$, we will actually ascertain $P(X \geq 29.5)$ by the normal approximation since $X = 30$ is assumed to go down to 29.5 when discrete probability is converted to area. The desired probability is shown by the shaded area in Figure 12.3c.

The z value is, by (12.6):

$$z = \frac{29.5 - 35}{3.24} = -1.70$$

We thus find that $P(X \geq 29.5) = P(Z \geq -1.70) = 0.955$. Incidentally, the true probability according to the binomial distribution is 0.952, so the normal approximation here is very close.   □

FIGURE 12.3   **Normal approximation to the binomial distribution for $n = 50$, $p = 0.7$, using the correction for continuity.**   Here, $X = 30$ is treated as extending from 29.5 to 30.5.

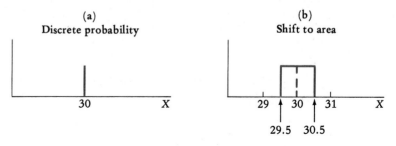

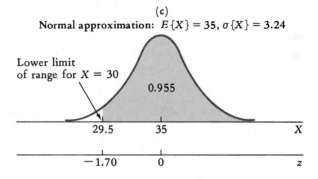

Using 29.5 rather than 30 in finding $P(X \geq 30)$ by means of the normal approximation involves a *correction for continuity*. The correction is so named because it arises when a continuous distribution is used to approximate a discrete one. The correction for continuity becomes trivial when the sample size is large, and we shall ordinarily omit it, as in the next example.

<table>
<tr><td>EXAMPLE 2</td></tr>
</table>

Probability Statement About $\bar{p}$ Omitting Correction for Continuity.   Consider again the insurance application example where the analyst is selecting a random sample of applications. Suppose the sample size now is $n = 200$ and that, as before, $p = 0.7$. The analyst wishes to find the probability that $\bar{p}$ falls within 5 percent points (that is, 0.05) of the population proportion p.

Figure 12.4 presents the normal approximation to the sampling distribution of $\bar{p}$. The mean and the standard deviation of this distribution are, by (12.4):

$$E\{\bar{p}\} = 0.7 \qquad \sigma\{\bar{p}\} = \sqrt{\frac{0.7(0.3)}{200}} = 0.032$$

FIGURE 12.4   **Normal approximation to the binomial distribution for** $n = 200$, $p = 0.7$, **omitting the correction for continuity**

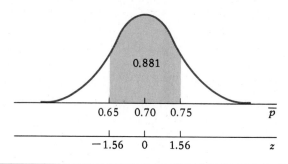

The desired probability is shaded in Figure 12.4. Because of the large sample size, we do not utilize the correction for continuity. The z values are:

$$z = \frac{0.65 - 0.70}{0.032} = -1.56 \qquad z = \frac{0.75 - 0.70}{0.032} = 1.56$$

The desired probability therefore is:

$$P(0.65 \leq \bar{p} \leq 0.75) = P(-1.56 \leq Z \leq 1.56) = 0.881 \qquad \square$$

Poisson Approximation to Binomial Distribution

When p is small and n is large, the binomial probability distribution can also be approximated by the Poisson probability distribution (6.8) with $\lambda = np$. For many practical purposes, a useful approximation is obtained when n is reasonably large and $np < 5$.

EXAMPLE

The probability of a computer breakdown in a day is $p = 0.01$. We wish to find the probability that, in a random sample of $n = 20$ days, there are no days with a breakdown. We assume that the total number of days with breakdowns (X) is a binomial random variable. We have $\lambda = np = 20(0.01) = 0.20$. From the Poisson Table C.6, we find, for $\lambda = 0.20$, that $P(0) = 0.8187$. Hence, the probability is approximately 0.8187 that there will be no days with computer breakdowns in the 20 days. The exact binomial probability from Table C.5 is $P(0) = 0.8179$, so the Poisson approximation is quite close even though n is not especially large here.

$\square$

12.4   ESTIMATION OF POPULATION PROPORTION

We frequently desire to estimate a population proportion by means of a confidence interval.

EXAMPLE

Mail Orders. An executive of a retail firm wished to estimate the proportion p of the firm's current credit card customers who make regular use of the firm's mail-order services. A file check on a random sample of 1240 credit card customers showed that $X = 372$ of these customers used the firm's mail-order services regularly. Hence, the sample proportion is $\bar{p} = 372/1240 = 0.30$. A confidence interval for the population proportion is now desired. □

Confidence Interval for p — Large Sample

We consider now the construction of a confidence interval for the population proportion p when the sample size is large. When n is large, we know from central limit theorem (12.5) that the sampling distribution of $\bar{p}$ is approximately normal. We also know from (12.4) that the mean of the sampling distribution of $\bar{p}$ is $E\{\bar{p}\} = p$ and that the variance is $\sigma^2\{\bar{p}\} = p(1 - p)/n$.

Estimated Standard Deviation of $\bar{p}$. To obtain confidence limits for p when the sample size is large, we need to estimate $\sigma\{\bar{p}\}$. This estimation can be done easily since, as we just noted, $\sigma^2\{\bar{p}\}$ depends only on p and n, and the sample proportion $\bar{p}$ is an unbiased point estimator of p. The estimated standard deviation of $\bar{p}$ is denoted by $s\{\bar{p}\}$ and is defined as follows for infinite or large finite populations.

(12.7)

The estimators of $\sigma^2\{\bar{p}\}$ and $\sigma\{\bar{p}\}$ are, respectively:

$$s^2\{\bar{p}\} = \frac{\bar{p}(1 - \bar{p})}{n - 1} \qquad s\{\bar{p}\} = \sqrt{\frac{\bar{p}(1 - \bar{p})}{n - 1}}$$

It can be shown that $s^2\{\bar{p}\}$ is an unbiased estimator of $\sigma^2\{\bar{p}\}$ for infinite populations.

Development of Confidence Interval. We utilize an extension of the central limit theorem for developing an approximate confidence interval for p when n is large.

(12.8)

For infinite or large finite populations, $(\bar{p} - p)/s\{\bar{p}\}$ follows approximately a standard normal distribution when the random sample size is sufficiently large.

Using the same reasoning as in finding the confidence limits for μ, we obtain analogous confidence limits for p.

(12.9)

The confidence limits for the population proportion p with approximate confidence coefficient $1 - \alpha$, when the random sample size is reasonably large, are:

$$\bar{p} \pm zs\{\bar{p}\}$$

continues

where:

$z = z(1 - \alpha/2)$

$s\{\bar{p}\}$ is given by (12.7)

EXAMPLE

In the mail orders example, the sample results are $n = 1240$ and $\bar{p} = 0.30$. A 99 percent confidence interval for the population proportion p of credit card customers who make regular use of the mail-order services is desired. The population is large relative to the sample size, and we calculate by (12.7):

$$s^2\{\bar{p}\} = \frac{0.30(1 - 0.30)}{1240 - 1} = 0.0001695 \qquad s\{\bar{p}\} = \sqrt{0.0001695} = 0.0130$$

For 99 percent confidence limits, we require $z(1 - \alpha/2) = z(0.995) = 2.576$. Thus, the limits are $0.30 \pm 2.576(0.0130)$, or 0.30 ± 0.03, and the 99 percent confidence interval is:

$$0.27 \leq p \leq 0.33$$

We can report, with 99 percent confidence, that between 27 and 33 percent of current credit card customers make regular use of the firm's mail-order services. ☐

Comment

One-sided confidence intervals for a population proportion can be obtained by using the appropriate limit in (12.9), with $z = z(1 - \alpha)$. For example, suppose we wish to obtain a 90 percent lower confidence interval for p in the mail orders example. We have $\bar{p} = 0.30$ and $s\{\bar{p}\} = 0.0130$. We require $z(1 - \alpha) = z(0.90) = 1.282$, so the lower confidence limit is $0.30 - 1.282(0.0130) = 0.283$, and the desired confidence interval is:

$$p \geq 0.283$$

Planning of Sample Size

Often, the sample size for estimating a population proportion p is not predetermined and can therefore be planned to provide a specified precision for a given confidence coefficient.

EXAMPLE

Worker Location. A commission is studying the spatial distribution of the labor force in a large metropolitan area. One parameter of interest is the proportion of workers (p) whose places of employment are within 15 miles of their residences. It is desired to select a random sample of workers of sufficient size to provide a 95 percent confidence interval for p with a half-width of $h = 0.02$. ☐

Assumptions. The planning procedure to be described assumes the following:

1. The random sample size ultimately determined is reasonably large.

2. The population is either infinite or, if finite, is large relative to the resulting sample size.

Planning Procedure. As usual, the desired half-width of the confidence interval is denoted by h, and $z = z(1 - \alpha/2)$ is the standard normal value corresponding to the specified confidence coefficient. Then, from (12.9), we see that $h = z\sigma\{\bar{p}\}$. We use the notation for the true standard deviation $\sigma\{\bar{p}\}$ here because we are in the planning stage. We know from (12.4) that $\sigma^2\{\bar{p}\} = p(1 - p)/n$. Hence, it follows that:

$$h = z\sigma\{\bar{p}\} = z\sqrt{\frac{p(1 - p)}{n}}$$

In solving this equation for n, we obtain an expression involving the population proportion p, which is unknown. Hence, we require a planning value of p. Often, this planning value can be based on past experience with the same or a similar problem.

(12.10)

The needed random sample size, based on the desired half-width h of the confidence interval and the specified confidence coefficient $1 - \alpha$, for a given planning value of p, is:

$$n = \frac{z^2 p(1 - p)}{h^2}$$

where:

$$z = z(1 - \alpha/2)$$

EXAMPLE

In the worker location example, we have $h = 0.02$ and $z(1 - \alpha/2) = z(0.975) = 1.960$ because a 95 percent confidence interval is desired. Suppose it is expected that p is in the neighborhood of 0.9, so the planning value $p = 0.9$ will be used. Substituting into (12.10) yields:

$$n = \frac{(1.960)^2(0.9)(0.1)}{(0.02)^2} = 864$$

A random sample of 864 workers from the labor force should provide an interval estimate of p with approximately the desired precision for the specified confidence coefficient if the planning value of p is reasonably accurate. □

Comment

In (12.10), the term $p(1 - p)$ takes on its largest value when $p = 0.5$, and it becomes smaller as p approaches 0 or 1. This implies that if no reliable planning value of p can be specified and the sample size must not be too small, the planning value of p should be set equal to 0.5 in computing n from (12.10). This sample size may be quite conservative (that is, much larger than necessary), because the farther p lies from 0.5 (toward either 0 or 1), the smaller is the sample size actually needed to provide a given degree of precision. In the worker location example, $n = 864$ is adequate if the specified planning value is $p = 0.9$, whereas $n = 2401$ would be obtained if $p = 0.5$ were specified as the planning value.

12.5   STATISTICAL TESTS FOR p

We now consider statistical tests concerning the population proportion p when the simple random sample size is large.

Construction and Application of Statistical Decision Rule — Large Sample

Tests about a population proportion p are conducted in a similar manner as tests on the population mean μ. The standardized test statistic has the usual form:

$$z^* = \frac{\bar{p} - p_0}{\sigma\{\bar{p}\}}$$

When $p = p_0$, the level where the α risk is controlled, this standardized test statistic is approximately a standard normal variable for large samples, as will be explained in the following example. Hence, the statistical decision rule is constructed in the usual fashion by means of the standard normal distribution.

(12.11)

Tests concerning the population proportion p, when the random sample size is sufficiently large, are based on the standardized test statistic:

$$z^* = \frac{\bar{p} - p_0}{\sigma\{\bar{p}\}}$$

where:

$$\sigma\{\bar{p}\} = \sqrt{\frac{p_0(1 - p_0)}{n}}$$

α risk is controlled at $p = p_0$

When $p = p_0$, z^* follows approximately the standard normal distribution.

We shall illustrate the use of the test statistic in (12.11) by means of an example involving a one-sided lower-tail test.

EXAMPLE

Fish Preference.   A British food organization currently uses cod for fish and chips, the fish traditionally used in making the food item. Whiting, however, is a cheaper and more plentiful fish. Management wishes to switch to whiting, to take advantage of the cost savings and better availability, if 50 percent or more of consumers prefer whiting to cod.

Step 1. Let *p* denote the proportion of consumers who prefer whiting; the test alternatives are then defined in terms of $p_0 = 0.5$:

$H_0: p \geq 0.5$

$H_1: p < 0.5$

Thus, the α risk is to be controlled at $p_0 = 0.5$.

Step 2. Management wishes to control the α risk at 0.05 when $p = p_0 = 0.5$. Figure 12.5 illustrates the situation. The appropriate type of decision rule is one where a small value of $\bar{p}$ leads to conclusion H_1 ($p < 0.5$), while a value of $\bar{p}$ in the neighborhood of $p_0 = 0.5$ or larger leads to conclusion H_0 ($p \geq 0.5$). The sampling distribution of $\bar{p}$ is centered at $p_0 = 0.5$ where the α risk is to be controlled. The tail area in the rejection region, corresponding to the α risk, must be equal to 0.05.

Step 3. Subjects in a random sample of $n = 265$ consumers are each to be given a blind taste test of both types of fish prepared in the same fashion; $\bar{p}$ will denote the sample proportion who prefer whiting over cod. Because the sample size is large, the sampling distribution of $\bar{p}$ when $p = p_0 = 0.5$ is approximately normal, with mean $p_0 = 0.5$ and standard deviation:

$$\sigma\{\bar{p}\} = \sqrt{\frac{p_0(1 - p_0)}{n}} = \sqrt{\frac{0.5(1 - 0.5)}{265}} = 0.03071$$

Hence, the standardized test statistic in (12.11) is approximately a standard normal variable when $p = p_0$, and the action limit for the decision rule in Figure 12.5

FIGURE 12.5   Illustration of the statistical decision rule for the fish preference example — Large-sample, one-sided lower-tail test for *p*

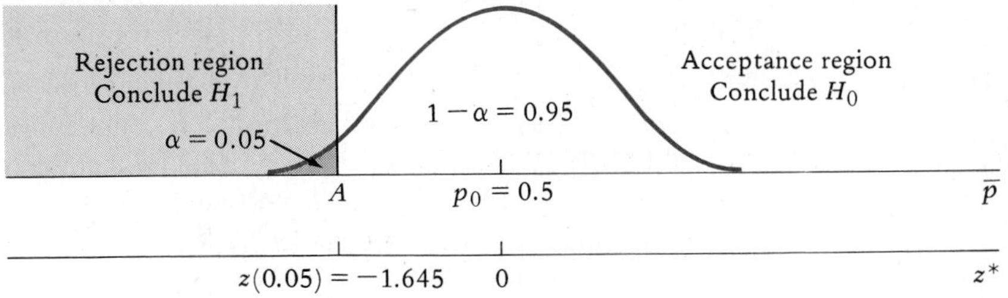

Where: $z^* = \dfrac{\bar{p} - p_0}{\sigma\{\bar{p}\}} = \dfrac{\bar{p} - 0.5}{0.03071}$

$\sigma\{\bar{p}\} = \sqrt{\dfrac{p_0(1 - p_0)}{n}} = \sqrt{\dfrac{0.5(0.5)}{265}} = 0.03071$

---

corresponds to $z(\alpha) = z(0.05) = -1.645$. Thus, the appropriate decision rule for the test is as follows:

If $z^* \geq -1.645$, conclude H_0.
If $z^* < -1.645$, conclude H_1.

Step 4. In the taste test, 144 of the 265 consumers preferred the whiting. Hence, $\bar{p} = 144/265 = 0.5434$. We noted in the previous step that $p_0 = 0.5$ and $\sigma\{\bar{p}\} = 0.03071$. The standardized test statistic therefore is:

$$z^* = \frac{\bar{p} - p_0}{\sigma\{\bar{p}\}} = \frac{0.5434 - 0.5}{0.03071} = 1.41$$

Since $z^* = 1.41 \geq -1.645$, the decision rule leads us to conclude H_0—that at least 50 percent of consumers prefer whiting to cod. $\square$

Comments

1. Note that the calculation of z^* in (12.11) uses the standard deviation $\sigma\{\bar{p}\}$ and not its estimate $s\{\bar{p}\}$. The reason is that p_0, the value of p at which the α risk is controlled, is given. Hence, the sampling distribution of $\bar{p}$ when $p = p_0$ has a known standard deviation since $\sigma\{\bar{p}\}$ depends only on n and p.

2. The test in the fish preference example can also be conducted by using a P-value. The one-sided P-value in this example is equal to $P(Z < z^*) = P(Z < 1.41) = 0.9207$. Since $\alpha = 0.05$ and $0.9207 \geq 0.05$, we conclude H_0. This conclusion is necessarily the same as that using test statistic z^*.

Power of Test — Large Sample

To examine the exposure to β risks of a decision rule concerning the population proportion p, we need to consider the rejection probability curve for the decision rule. Rejection probability curves for tests on p are obtained in similar fashion as those for tests on μ discussed in Section 11.6. The procedure to be explained continues to assume that the simple random sample size is large.

Calculation of Rejection Probabilities. We first consider how to calculate a rejection probability $P(H_1; p)$, that is, the probability that the decision rule will lead to conclusion H_1 for a given value of the population proportion p. Since there are no new principles involved, we illustrate the calculation of rejection probabilities by an example. Note that the value of the population proportion p for which the rejection probability is calculated determines the value of the standard deviation of the sampling distribution of $\bar{p}$.

EXAMPLE

In the fish preference example, management wishes to know the rejection probability for the decision rule illustrated in Figure 12.5 when $p = 0.35$. Figure 12.6

FIGURE 12.6   **Calculation of the rejection probability at $p = 0.35$ for the fish preference example — One-sided lower-tail test**

(a)

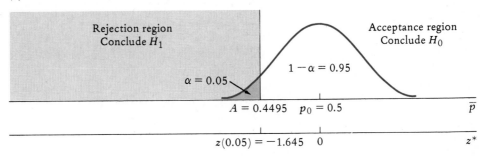

(b)

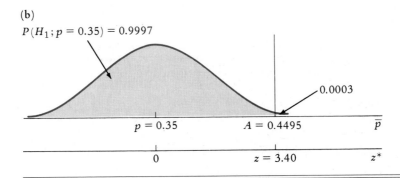

shows the method of calculation. Figure 12.6a is a reproduction of Figure 12.5. We first need to determine the position of the action limit A that controls the α risk at 0.05 when $p = p_0 = 0.5$. For $p_0 = 0.5$, $\sigma\{\bar{p}\} = \sqrt{0.5(0.5)/265} = 0.03071$ (as was calculated earlier). Since $z(0.05) = -1.645$, it follows that the action limit in Figure 12.6a equals:

$$A = p_0 + z(0.05)\sigma\{\bar{p}\} = 0.5 - 1.645(0.03071) = 0.4495$$

When the population proportion is $p = 0.35$, the sampling distribution of $\bar{p}$ is centered at $p = 0.35$, as shown in Figure 12.6b. The standard deviation of this distribution is:

$$\sigma\{\bar{p}\} = \sqrt{\frac{p(1-p)}{n}} = \sqrt{\frac{0.35(1-0.35)}{265}} = 0.02930$$

Observe that $\sigma\{\bar{p}\}$ must be recomputed for each different value of p because $\sigma\{\bar{p}\}$ depends mathematically on p. The area under the sampling distribution that lies in the rejection region is the desired rejection probability. This area is shaded in Figure 12.6b. We denote this area by $P(H_1; p = 0.35)$. Since the action limit $A = 0.4495$ is located $z = (0.4495 - 0.35)/0.02930 = 3.40$ standard deviations from $p = 0.35$, it follows that the rejection probability at $p = 0.35$ for the decision rule is:

$$P(H_1; p = 0.35) = P(Z < 3.40) = 0.9997$$

Thus, management is assured that the decision rule would almost certainly lead to conclusion H_1 if p were, in fact, equal to 0.35.   ☐

Rejection Probability Curve.   The rejection probability curve for a decision rule concerning the population proportion p is plotted in the usual fashion after a number of rejection probabilities have been calculated.

EXAMPLE

In the fish preference example, a number of rejection probabilities for other values of p were also calculated. In each case, $\sigma\{\bar{p}\}$ was recomputed because $\sigma\{\bar{p}\}$ varies with the value of p. These rejection probabilities are plotted as the rejection probability curve shown in Figure 12.7. Note that the point corresponding to $P(H_1; p = 0.35) = 0.9997$ is shown on the curve. The point at which the α risk is controlled is also shown on the curve. Since $\alpha = 0.05$ at $p_0 = 0.5$, this point

FIGURE 12.7   **Rejection probability curve for the fish preference example — One-sided lower-tail test**

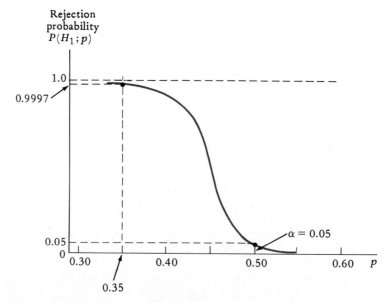

corresponds to $P(H_1; p = 0.5) = 0.05$. From the rejection probability curve, we can see that, for values of p less than 0.40 or thereabouts, the probabilities of concluding H_1 (and, hence, of rejecting H_0) are quite high, indicating that the test is quite powerful for p values in this range. ☐

Comments

1. Rejection probabilities for one-sided upper-tail and two-sided tests are calculated in an analogous fashion.
2. In a test for μ, one must have knowledge of the population standard deviation σ in order to assess the power of the test. In contrast, in a test for p, the rejection probability curve can be constructed on the basis of the sample size n alone, as we have just illustrated, since the standard deviation $\sigma\{\bar{p}\}$ for a given value of p depends only on p and n.

Planning of Sample Size

We now describe how to determine the sample size for a test of p that will keep both α and β risks at prespecified levels. The procedure is similar to that in planning the sample size for a test of the population mean μ.

Assumptions. The planning procedure assumes the following:

1. The random sample size ultimately determined is reasonably large.
2. The population is either infinite or, if finite, is large relative to the resulting sample size.

Planning Procedure. We shall continue to denote the value of p where the α risk is controlled by p_0. The value of p where the β risk is controlled shall be denoted by p_1. We shall denote the z values associated with the α and β risks by z_0 and z_1, respectively, as we did previously in planning the sample size for tests on μ.

The sample size formula, which is applicable to all three types of tests, follows.

(12.12)

The necessary random sample size to control both the α and β risks is:

$$n = \frac{\left[|z_1|\sqrt{p_1(1 - p_1)} + |z_0|\sqrt{p_0(1 - p_0)}\right]^2}{|p_1 - p_0|^2}$$

where:

p_0 and p_1 are the values of p where the α and β risks are controlled, respectively

z_0 and z_1 are the z values associated with the α and β risk specifications, respectively, and are defined as follows for each type of test.

continues

One-Sided Upper-Tail Test (H_0: $p \leq p_0$, H_1: $p > p_0$):

$$z_0 = z(1 - \alpha) \qquad z_1 = z(\beta)$$

One-Sided Lower-Tail Test (H_0: $p \geq p_0$, H_1: $p < p_0$):

$$z_0 = z(\alpha) \qquad z_1 = z(1 - \beta)$$

Two-Sided Test (H_0: $p = p_0$, H_1: $p \neq p_0$):

$$z_0 = z(1 - \alpha/2) \qquad z_1 = z(\beta)$$

We illustrate the use of formula (12.12) by an example.

EXAMPLE

In the fish preference example, suppose that the sample size has not yet been determined.

Step 1.   We continue to specify that the α risk is to be controlled at 0.05 when $p_0 = 0.5$ (see Figure 12.8a).

Step 2.   Management has decided that if only 40 percent of consumers prefer whiting, the probability of concluding H_0 ($p \geq 0.5$) in the test is to be controlled at 0.05. Thus, the β risk is to be controlled at 0.05 when $p_1 = 0.4$ (see Figure 12.8b).

Step 3.   Referring to formula (12.12) for a one-sided lower-tail test, we see that:

$$z_0 = z(\alpha) = z(0.05) = -1.645 \quad \text{so} \quad |z_0| = 1.645$$

$$z_1 = z(1 - \beta) = z(1 - 0.05) = z(0.95) = 1.645 \quad \text{so} \quad |z_1| = 1.645$$

$$|p_1 - p_0| = |0.4 - 0.5| = 0.1$$

$$\sqrt{p_1(1 - p_1)} = \sqrt{0.4(0.6)} = 0.490$$

$$\sqrt{p_0(1 - p_0)} = \sqrt{0.5(0.5)} = 0.500$$

Substituting into (12.12), we obtain:

$$n = \frac{[1.645(0.490) + 1.645(0.500)]^2}{(0.1)^2} = 265$$

This is, in fact, the sample size used in the study. □

Comment

The derivation of the sample size formula in (12.12) follows the same pattern as for tests on the population mean μ. We illustrate the derivation in terms of the fish preference example. Figure 12.8a shows the appropriate decision rule and the control of the α risk, while Figure 12.8b shows the control of the β risk.

FIGURE 12.8 Determination of the required sample size to control the α and β risks in the fish preference example — One-sided lower-tail test for p

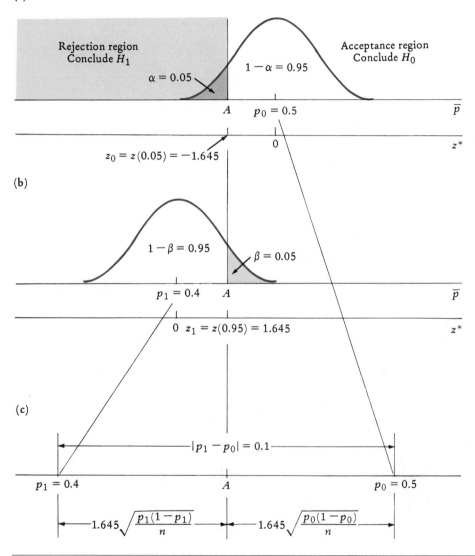

(a)

Rejection region
Conclude H_1

$\alpha = 0.05$

$1 - \alpha = 0.95$

Acceptance region
Conclude H_0

A $p_0 = 0.5$ $\bar{p}$

0 z^*

$z_0 = z(0.05) = -1.645$

(b)

$1 - \beta = 0.95$ $\beta = 0.05$

$p_1 = 0.4$ A $\bar{p}$

0 $z_1 = z(0.95) = 1.645$ z^*

(c)

$|p_1 - p_0| = 0.1$

$p_1 = 0.4$ A $p_0 = 0.5$

$1.645\sqrt{\dfrac{p_1(1-p_1)}{n}}$ $1.645\sqrt{\dfrac{p_0(1-p_0)}{n}}$

The value of $\sigma\{\bar{p}\}$ depends only on p and n. For the sampling distribution of $\bar{p}$ in Figure 12.8a where the α risk is controlled, we have $p = p_0 = 0.5$ and thus:

$$\sigma\{\bar{p}\} = \sqrt{\frac{p_0(1 - p_0)}{n}} = \sqrt{\frac{0.5(0.5)}{n}}$$

For the sampling distribution in Figure 12.8b where the β risk is controlled, we have $p = p_1 = 0.4$ and thus:

$$\sigma\{\bar{p}\} = \sqrt{\frac{p_1(1 - p_1)}{n}} = \sqrt{\frac{0.4(0.6)}{n}}$$

Note that no planning value is required here to calculate $\sigma\{\bar{p}\}$, unlike the case when planning the sample size for tests concerning μ.

Looking at Figures 12.8a and 12.8b together, we can see that the interval from p_1 to p_0 has length $|p_1 - p_0| = |0.4 - 0.5| = 0.1$. This interval has two components. First, the component between p_1 and the action limit A has a length equal to:

$$|z_1|\sigma\{\bar{p}\} = |1.645|\sqrt{\frac{p_1(1 - p_1)}{n}} = 1.645\sqrt{\frac{0.4(0.6)}{n}}$$

Second, the component between the action limit A and p_0 has a length equal to:

$$|z_0|\sigma\{\bar{p}\} = |-1.645|\sqrt{\frac{p_0(1 - p_0)}{n}} = 1.645\sqrt{\frac{0.5(0.5)}{n}}$$

The sum of these two components equals $|p_1 - p_0|$, as shown in Figure 12.8c. Thus, we have the equation:

$$|0.4 - 0.5| = 1.645\sqrt{\frac{0.4(0.6)}{n}} + 1.645\sqrt{\frac{0.5(0.5)}{n}}$$

Rearranging this equation and solving for the required sample size yields:

$$n = \frac{[1.645\sqrt{0.4(0.6)} + 1.645\sqrt{0.5(0.5)}]^2}{|0.4 - 0.5|^2} = 265$$

OPTIONAL TOPIC — CONFIDENCE INTERVAL FOR p — SMALL SAMPLE

When the sample size is small, so that the large-sample confidence limits (12.9) for the population proportion p are not applicable, exact confidence limits for p can be utilized. These limits are exact when the population is infinite, and they can also be used for a finite population as long as the sample size is not large relative to the population size.

EXAMPLE

Bond Issues.   A financial analyst studied new corporate bond issues sold during the past five years. One characteristic of interest is the proportion p of the issues sold at a premium—that is, where the issue price exceeds the face value of the bond. The analyst selected a random sample of $n = 15$ new issues during the period and determined that $X = 6$ of these were sold at a premium. Hence, the sample proportion is $\bar{p} = 6/15 = 0.40$. The analyst now wants to construct an interval estimate for the population proportion p. $\square$

Exact Confidence Limits for *p*

The exact confidence limits for the population proportion p involve the number of occurrences X, the random sample size n, and percentiles of the F distribution. (If a review of the F distribution is needed, the reader should turn to Appendix B, Section B.3, before proceeding.)

(12.13)

The confidence limits L and U for the population proportion p with confidence coefficient $1 - \alpha$, when the population is infinite or large relative to n, are:

$$L = \frac{X}{X + (n - X + 1)F_1} \qquad U = \frac{(X + 1)F_2}{(X + 1)F_2 + (n - X)}$$

where:

$$F_1 = F[1 - \alpha/2; 2(n - X + 1), 2X]$$
$$F_2 = F[1 - \alpha/2; 2(X + 1), 2(n - X)]$$

EXAMPLE

In the bond issues example, the financial analyst wishes to construct a 90 percent confidence interval for p. The sample results are $n = 15$ and $X = 6$, and $1 - \alpha = 0.90$. Consulting Table C.4 for the F distribution, we find:

$$F_1 = F[0.95; 2(15 - 6 + 1), 2(6)] = F(0.95; 20, 12) = 2.54$$
$$F_2 = F[0.95; 2(6 + 1), 2(15 - 6)] = F(0.95; 14, 18) = 2.29$$

where the latter value is obtained by interpolation. Substituting into (12.13) gives:

$$L = \frac{6}{6 + (15 - 6 + 1)(2.54)} = 0.19$$

$$U = \frac{(6 + 1)(2.29)}{(6 + 1)(2.29) + (15 - 6)} = 0.64$$

Therefore, a 90 percent confidence interval for the proportion of bond issues sold at a premium is:

$$0.19 \leq p \leq 0.64$$

The analyst can conclude, with 90 percent confidence, that the population proportion of new issues selling at a premium is between 19 and 64 percent. These limits
are quite wide, because the sample size is so small. Note that the point estimate
$\bar{p} = 0.40$ gives no indication of this lack of precision.   ☐

Comments   1.   Unlike the large-sample confidence limits for p in (12.9), the exact confidence limits for
                       p in (12.13) are not equally spaced about $\bar{p}$. In the example, the limits 0.19 and 0.64 are
                       not equidistant from $\bar{p} = 0.40$.
                 2.   The confidence limits (12.13) are based on two facts. First, the sampling distribution of
                       $\bar{p}$ is given by the binomial probability function (12.2) when the population is infinite.
                       Second, there exists a mathematical relation between the binomial distribution and the
                       F distribution.

PROBLEMS

12.1   **Information Specialists.**   A random sample of $n = 9$ information specialists will be selected
to ascertain whether they use computer graphics on a regular basis. Let X_i be a Bernoulli random
variable coded 1 if the ith specialist answers yes and coded 0 otherwise.
   a. What does the Bernoulli parameter p denote here?
   b. Let $\bar{p}$ and X be defined as in (12.1) and (12.2). (1) What does X denote here, and what possible values can it take on? (2) What does $\bar{p}$ denote here, and what possible values can it
   take on?
   c. Specify the relation between X and $\bar{p}$ here.

12.2   **Invoice Payments.**   Consider a random sample of $n = 12$ invoices. Let X_i be a Bernoulli random variable coded 0 if the ith invoice is paid within 10 days and coded 1 otherwise.
   a. What does the Bernoulli parameter p denote here?
   b. Let $\bar{p}$ and X be defined as in (12.1) and (12.2). (1) What does X denote here, and what possible values can it take on? (2) What does $\bar{p}$ denote here, and what possible values can it
   take on?
   c. Specify the relation between X and $\bar{p}$ here.

12.3   **Customers.**   Consider a random sample of $n = 15$ persons who enter a department store. Let
X denote the number of persons in the sample who make a purchase.
   a. (1) What does $\bar{p}$, as defined in (12.1), represent in this situation? (2) What values can X
   take on here? (3) What values can $\bar{p}$ take on here?
   b. Is X a binomial random variable here? Explain.

****12.4**   Refer to **Information Specialists** Problem 12.1. Assume that X is a binomial random variable
 with $n = 9$ and $p = 0.9$.
   a. Obtain the sampling distributions of X and $\bar{p}$.
   b. Plot the two sampling distributions in one graph, as in Figure 12.1.
   c. Obtain (1) $P(X = 9)$, (2) $P(\bar{p} = 1)$, (3) $P(\bar{p} \geq 7/9)$.
   d. Obtain the means and the standard deviations of the sampling distributions of X and $\bar{p}$, using (12.4).

12.5 Refer to **Invoice Payments** Problem 12.2. Assume that X is a binomial random variable with $n = 12$ and $p = 0.2$.
   a. Obtain the sampling distributions of X and $\bar{p}$.
   b. Plot the two sampling distributions in one graph, as in Figure 12.1.
   c. Obtain (1) $P(X \leq 2)$, (2) $P(\bar{p} = 1/12)$, (3) $P(\bar{p} \geq 1/12)$.
   d. Obtain the means and the standard deviations of the sampling distributions of X and $\bar{p}$, using (12.4).

12.6 Refer to **Customers** Problem 12.3. Assume that X is a binomial random variable with $n = 15$ and $p = 0.3$.
   a. Obtain the sampling distributions of X and $\bar{p}$.
   b. Plot the two sampling distributions in one graph, as in Figure 12.1.
   c. Obtain (1) $P(X = 10)$, (2) $P(\bar{p} \leq 0.20)$, (3) $P(\bar{p} = 0.40)$.
   d. Obtain the means and the standard deviations of the sampling distributions of X and $\bar{p}$, using (12.4).

*12.7 For each of the following cases, indicate whether working rule (12.5) permits the use of a normal approximation to the binomial distribution.
   a. $n = 100$, $p = 0.02$
   b. $n = 80$, $p = 0.15$
   c. $n = 40$, $p = 0.90$
   d. $n = 20$, $p = 0.60$

12.8 A defense consultant will make random spot checks to estimate the proportion of time a scrambling device for military telephone messages is in use. Sixteen moments in time will be selected at random for the spot checks. Will working rule (12.5) permit use of a normal approximation to the sampling distribution of $\bar{p}$ if the device is actually in use (1) 20 percent of the time, (2) 80 percent of the time, (3) 50 percent of the time? Explain.

*12.9 The probability that a robotized machine is in an unproductive state is $p = 0.20$. Let X denote the number of times the machine is in an unproductive state when it is observed at 50 random moments in time.
   a. Verify that working rule (12.5) permits use of a normal approximation here.
   b. Obtain the mean and the standard deviation of the sampling distribution of X.
   c. Using a normal approximation and a correction for continuity, obtain (1) $P(X > 15)$, (2) $P(X \leq 4)$, (3) $P(X = 10)$.
   d. Using your results in c, obtain (1) $P(\bar{p} > 0.30)$, (2) $P(\bar{p} \leq 0.08)$, (3) $P(\bar{p} = 0.20)$.

12.10 The probability of a delayed shipment is $p = 0.10$. Let X denote the number of delayed shipments in a random sample of $n = 60$ shipments.
   a. Verify that working rule (12.5) permits use of a normal approximation here.
   b. Obtain the mean and the standard deviation of the sampling distribution of X.
   c. Using a normal approximation and a correction for continuity, obtain (1) $P(X \leq 1)$, (2) $P(X > 6)$, (3) $P(X = 6)$.
   d. Using your results in c, obtain (1) $P(\bar{p} \leq 1/60)$, (2) $P(\bar{p} > 0.10)$, (3) $P(\bar{p} = 0.10)$.

*12.11 From a large population of residential water meters, a random sample of 250 meters is to be selected to estimate the proportion due for replacement within the next five years. Suppose that $p = 0.40$.
   a. Find the mean and the standard deviation of the sampling distribution of $\bar{p}$.
   b. Using a normal approximation without a correction for continuity, find the probability that (1) $\bar{p}$ is less than 0.35, (2) $\bar{p}$ lies within ± 0.05 of the population proportion p.

c. Calculate the probability in **b**(2) for a sample size of 1000. What is the effect of the larger sample size?

12.12 A nationwide survey of 225 food stores selected at random is to be conducted to estimate the proportion of food stores carrying fertile eggs. Suppose that $p = 0.30$.

    a. Find the mean and the standard deviation of the sampling distribution of $\bar{p}$.

    b. Using a normal approximation without a correction for continuity, find (1) the probability that $\bar{p}$ will exceed 0.35, (2) the interval centered about p within which $\bar{p}$ will fall with probability 0.95.

    c. Calculate the interval in **b**(2) for a sample of size 500. What is the effect of the larger sample size?

12.13 A public opinion poll of 1000 persons randomly selected is to be conducted to estimate the proportion of the population who believe business conditions will worsen in the next 12 months. Assume that the binomial distribution is applicable and that it can be approximated by a normal distribution. Ignore the correction for continuity. What is the probability that the proportion of persons polled who believe business conditions will worsen is within ±0.04 of the population proportion p if (1) $p = 0.20$, (2) $p = 0.50$, (3) $p = 0.80$? For which of these three values of p is the calculated probability smallest?

*12.14 A bank has many small consumer loans outstanding. One percent of these loans are technically in default. Using a Poisson approximation to the binomial distribution, find the probability that an auditor who examines a random sample of 200 small consumer loans will find (1) exactly one in default, (2) five or more in default.

12.15 A computer simulation model of a complex communications network is run for 700 independent trials, each trial being a simulation of the network's operation for a hypothetical 10-year period. The probability of a major network failure on any one trial is $p = 0.005$. Using a Poisson approximation to the binomial distribution, find the probability that (1) exactly 2 failures occur among the 700 trials, (2) 10 or more failures occur among the 700 trials.

*12.16 Savings Transfers. A savings bank noted that 180 of the 225 new accounts opened last month involved a transfer of savings from an account at another bank. It is desired to estimate the process proportion p of new accounts at this bank that involve transfers from another bank. Assume that the data for the 225 new accounts constitute a random sample from the process.

    a. Calculate $\bar{p}$ and $s\{\bar{p}\}$. What does the latter estimate here?

    b. Obtain a 95 percent confidence interval for the process proportion p. Interpret your confidence interval.

    c. An executive of this savings bank had claimed that three-quarters of all new accounts involve a transfer from another bank. Does the confidence interval in **b** suggest that the executive's claim is plausible? Comment.

12.17 In a sample survey of $n = 400$ persons over the age of 40 randomly selected, 65 percent were found to use some hair-conditioning product regularly. It is desired to estimate the population proportion p of persons over the age of 40 who use some hair-conditioning product regularly.

    a. Estimate the standard deviation of the sampling distribution of $\bar{p}$.

    b. Construct a 99 percent confidence interval for the population proportion p. Interpret your confidence interval.

    c. Why is the confidence coefficient of the interval constructed in **b** only approximately 99 percent?

12.18 Chronic Illness. In a random sample of $n = 1000$ workers from a certain occupation, it was found that 153 of them had some chronic illness. Construct a 90 percent confidence interval for the population proportion of workers with a chronic illness. Interpret your confidence interval.

12.19 In a shipment of 140 ornamental glass globes received from a new vendor, 28 were flawed. Assume that the globes in the shipment constitute a random sample from the vendor's current production process. Construct a 95 percent confidence interval for the process proportion of flawed globes. Interpret your confidence interval.

12.20 Refer to **Chronic Illness** Problem 12.18. Construct a 90 percent upper confidence interval for the population proportion. Interpret your confidence interval.

*12.21 A sample survey of persons with a certain type of arthritis is to be conducted to estimate the population proportion of persons who are taking a new antiarthritic drug. A 90 percent confidence interval is desired with a half-width of 0.02. Preliminary sales data for the new drug suggest that a planning value of $p = 0.10$ is reasonable.
   a. Determine the required sample size.
   b. What would be the required sample size if a planning value of $p = 0.12$ were used? Is this required sample size substantially different from that determined in **a**? Explain.

12.22 A graduate business school wishes to estimate the proportion of its graduates who have reached executive positions. A random sample of graduates is to be selected. The population proportion is to be estimated, using a 90 percent confidence interval with a half-width of 0.05. A reasonable planning value, based on information from other graduate schools, is $p = 0.6$.
   a. How many graduates should be selected for the sample?
   b. If the desired half-width were to be 0.025 instead of 0.05, what sample size would be required? Is this required sample size substantially different from that determined in **a**? Explain.

12.23 A lathe operator is to be observed at random times to estimate the proportion of times she is in a productive state. It is desired to estimate this process proportion within ± 0.05, with a 95 percent confidence coefficient. It is expected that the process proportion is 0.8 or greater.
   a. What sample size will be large enough to ensure that the estimate has the desired precision?
   b. It was finally decided to take $n = 250$ random observations. The lathe operator was found to be in a productive state in 225 of these observations. Construct a 95 percent confidence interval for the process proportion. Interpret your confidence interval.

12.24 As part of a study of postsecondary education, a random sample of women in the graduating classes of colleges and universities is to be selected to estimate the proportion who expect to enter the labor force upon graduation. It is desired to estimate the population proportion within ± 0.03, with a confidence coefficient of 99 percent. No reliable planning value for p is available.
   a. What sample size will be large enough to ensure that the estimate has the desired precision?
   b. It was finally decided to include 1225 women in the study. Of these, 950 stated that they expected to enter the labor force upon graduation. Construct a 99 percent confidence interval for the population proportion. Does your confidence interval provide an estimate of the proportion of women in the graduating classes who actually will enter the labor force upon graduation? Discuss.

12.25 If the planning value p used in formula (12.10) for calculating the needed sample size is incorrect, does this affect the validity of the confidence interval obtained from the sample results? Does it affect the precision of the confidence interval? Comment.

12.26 Explain why the estimated standard deviation $s\{\bar{p}\}$ in (12.7) is not used for calculating the standardized test statistic z^* in (12.11).

*12.27 **Serviceable Items.** A large stock of items has been stored under unfavorable conditions. If fewer than 70 percent of the items are serviceable, the stock should be scrapped. It is desired to test whether or not the proportion of items that are serviceable is less than 0.70. The α risk is to be controlled at 0.01 when $p = 0.70$. A random sample of 100 items from the stock has been selected, and 65 of these items were found to be serviceable.

    a. Conduct the appropriate test. State the alternatives, the decision rule, the value of the test statistic, and the conclusion.

    b. Why is the α risk controlled only approximately at 0.01 when $p = 0.70$?

    c. Calculate the P-value of the test. Is it a one-sided or a two-sided P-value?

12.28 Refer to **Chronic Illness** Problem 12.18. In a related occupation, 11 percent of the workers have a chronic illness. It is desired to test whether or not the population proportion of workers in this occupation with a chronic illness is greater than 0.11. The α risk is to be controlled at 0.05 when $p = 0.11$.

    a. Conduct the appropriate test. State the alternatives, the decision rule, the value of the test statistic, and the conclusion.

    b. Calculate the P-value of the test. Is the P-value consistent with the test result in **a**? Explain.

12.29 **Divorces and Annulments.**   Nationally, the proportion of marriages ending in divorce or annulment that involve no children is 0.40. In one state, a random sample of 1000 divorces and annulments showed that 437 involved no children. It is desired to test whether or not the state proportion is equal to 0.40. The α risk is to be controlled at 0.05 when $p = 0.40$.

    a. Conduct the appropriate test. State the alternatives, the decision rule, the value of the test statistic, and the conclusion.

    b. What is a Type II error in this test situation? Is it possible that the conclusion in **a** has resulted in a Type II error? Explain.

    c. Calculate the P-value of the test. Conduct the required test using the P-value.

12.30 **Shoes.**   In the past, 8 percent of the shoes produced in a plant have had small defects and were classified as irregular. Management has modified the production process and now wishes to test whether or not the process proportion of shoes that are irregular has changed from the past level. A random sample of 400 shoes produced under the modified process was selected, and 41 were classified as irregular. The α risk is to be controlled at 0.02 when $p = 0.08$.

    a. Conduct the appropriate test. State the alternatives, the decision rule, the value of the test statistic, and the conclusion.

    b. Calculate the P-value of the test. Is the P-value consistent with the test result in **a**? Explain.

*12.31 Refer to **Serviceable Items** Problem 12.27.

    a. Restate the decision rule in terms of $\bar{p}$. What is the value of the action limit?

    b. Obtain the rejection probability $P(H_1; p)$ for each of the following values of p: (1) 0.50, (2) 0.60, (3) 0.70. Why is it necessary to recalculate $\sigma\{\bar{p}\}$ for each given value of p?

    c. Plot the rejection probability curve for the decision rule. What is the probability that the decision rule will lead to the correct conclusion if $p = 0.60$?

12.32 Refer to **Chronic Illness** Problems 12.18 and 12.28.

    a. Restate the decision rule in Problem 12.28**a** in terms of $\bar{p}$.

    b. Obtain the rejection probability $P(H_1; p)$ for each of the following values of p: (1) 0.09, (2) 0.11, (3) 0.13.

    c. Plot the rejection probability curve for the decision rule. What is the risk of an incorrect decision if $p = 0.13$?

12.33 Refer to **Divorces and Annulments** Problem 12.29.

    a. Restate the decision rule in terms of $\bar{p}$.

    b. Obtain the rejection probability $P(H_1; p)$ for each of the following values of p: (1) 0.35, (2) 0.40, (3) 0.45. Why is it necessary to recalculate $\sigma\{\bar{p}\}$ for each given value of p?

    c. Plot the rejection probability curve for the decision rule. What is the β risk at $p = 0.45$ for this decision rule?

12.34 Refer to **Shoes** Problem 12.30.

    a. Restate the decision rule in terms of $\bar{p}$.

    b. Obtain the rejection probability $P(H_1; p)$ for each of the following values of p: (1) 0.03, (2) 0.08, (3) 0.13.

    c. Plot the rejection probability curve for the decision rule. Is the probability that the test leads to a correct conclusion when p exceeds 0.13 reasonably large?

*12.35 An accountant will audit a large population of sales invoices by examining a simple random sample of invoices. If 1 percent or less of the invoices in the population have errors, the auditor will consider the quality of the population as satisfactory (H_0); otherwise, he will consider the quality as unsatisfactory (H_1). The auditor wishes to control the α risk at 0.05 when $p = 0.01$ and the β risk at 0.05 when $p = 0.03$.

    a. What random sample size should be employed?

    b. A random sample of the size determined in **a** was selected, and the proportion of invoices with errors was found to be 0.028. What conclusion should be reached?

12.36 A field crop has been sprayed for stem rust. An agronomist plans to examine a random sample of plants from the field for rust. Let p denote the proportion of plants in the field with rust. The test alternatives of interest to the agronomist are $H_0: p \geq 0.15$, $H_1: p < 0.15$. The agronomist wishes to control the α risk at 0.05 when $p = 0.15$ and the β risk at 0.05 when $p = 0.10$.

    a. What number of plants should be selected for the agronomist's random sample?

    b. The agronomist used the sample size determined in **a** and obtained $\bar{p} = 0.121$. What is the appropriate test conclusion?

    c. Why are the α and β risks only approximately equal to 0.05 at the specified levels of p in the test performed in **b**?

12.37 A political analyst is planning a poll of eligible voters to ascertain their voting intentions for a forthcoming election between the incumbent candidate A and the opposition candidate B. Let p denote the population proportion of voters who would currently express an intention to vote for A. The analyst wishes to test $H_0: p = 0.50$ versus $H_1: p \neq 0.50$. She wishes to employ a sample size that will control the α risk at 0.02 when $p = 0.50$ and the β risk at 0.02 when $p = 0.45$.

    a. What number of respondents should be selected for the random sample?

    b. The analyst used the sample size determined in **a** for the poll and obtained $\bar{p} = 0.463$. What conclusion should be reached?

12.38 A purchaser of a large shipment of tiles wishes to control the β risk of accepting the shipment at 0.01 when the proportion of damaged tiles is $p = 0.07$. At the same time, the vendor wishes to control the α risk of having the shipment rejected at 0.025 when the proportion of damaged tiles is $p = 0.03$. A random sample of tiles will be selected from the shipment by the purchaser, on the basis of which a decision will be made whether to accept ($H_0: p \leq 0.03$) or reject ($H_1: p > 0.03$) the shipment.

    a. What random sample size is necessary to meet both the purchaser's and the vendor's specifications?

    b. A random sample of the size determined in **a** was selected and yielded $\bar{p} = 0.084$. What conclusion should be reached?

*12.39 An auditor examined a simple random sample of 15 accounts from a large population of accounts receivable. He discovered that one account in the sample was in error. Estimate the error rate in the population with an exact 98 percent confidence interval. Interpret your confidence interval.

12.40 A political scientist conducted in-depth interviews with a random sample of 16 lawyers and learned that 11 favor the present system of judicial appointment. Construct an exact 90 percent confidence interval for the population proportion of lawyers in favor of the present system of appointment.

12.41 Refer to **Savings Transfers** Problem 12.16.
   a. Obtain an exact 95 percent confidence interval for the process proportion p. [*Hint: $F(0.975$; $92, 360) = 1.36$ and $F(0.975; 362, 90) = 1.41$.*]
   b. Compare the exact confidence interval in **a** with that in Problem 12.16**b** based on the normal approximation to the binomial distribution. Is the normal approximation adequate here?

EXERCISES

12.42 The probability that a garnet, when cut and polished, has a flaw that classes it as a "second" is $p = 0.10$. Assume that the production process is a Bernoulli process, and consider a sample of $n = 3$ garnets. Let X denote the number of seconds in the sample.
   a. Obtain the sampling distribution of X by direct enumeration of all possible sample outcomes. Verify that this distribution is the same as that given by (12.2).
   b. Calculate the mean and the standard deviation of the sampling distribution in **a**. Verify that these results agree with those given by (12.4).

12.43 Let X be a binomial random variable with $n = 300$ and $p = 0.995$. Use the Poisson approximation to calculate $P(X = 296)$.

12.44 Show that $s^2\{\bar{p}\}$ as defined in (12.7) is an unbiased estimator of $\sigma^2\{\bar{p}\}$ as defined in (12.4). (*Hint: $E\{\bar{p}^2\} = p^2 + \sigma^2\{\bar{p}\}$.*)

12.45 A market researcher must determine a random sample size for a consumer survey that will be used to estimate the proportion of consumers who have tried a new dairy product. He wishes to estimate the population proportion p with a *relative* margin of sampling error of 20 percent (that is, within $\pm 0.20p$), using a 95 percent confidence coefficient. A pilot study suggests that $p = 0.06$ is a reasonable planning value. What sample size is required?

12.46 A survey of $n = 15$ small manufacturers shows that $X = 3$ have increased the size of their labor force during the last 12 months. Assume the binomial probability distribution applies.
   a. In Table C.5, for which value of p is the likelihood of obtaining $X = 3$ the greatest when $n = 15$?
   b. The maximum likelihood estimator of p here can be shown to be $\bar{p} = X/n$. Is your finding in **a** consistent with this fact?

12.47 (Calculus needed.) If X is a binomial random variable with parameters n and p, show that the sample proportion $\bar{p} = X/n$ is the maximum likelihood estimator of p.

12.48 A municipal airport reports that it has handled 6000 aircraft landings in the past year without a mishap. Assuming that the aircraft landings constitute a Bernoulli process, obtain a 95 percent upper confidence interval for the probability that a landing at the airport will involve a mishap. [*Hint: Use $F(0.95; 2, \infty)$.*]

STUDIES

12.49 An auditor will verify all accounts in a population on a 100 percent basis if she finds that one or more accounts in a random sample of 120 accounts are in error. Assume that the binomial probability distribution is applicable.

   a. Find the probability that the auditor will verify all accounts on a 100 percent basis if the population error rate p is (1) 0.01, (2) 0.05, (3) 0.10. Assuming that a 5 percent error rate is serious, is the auditor's sampling plan effective when $p = 0.05$?

   b. Find the probability of no account in error in the sample if $p = 0.01$ by means of the Poisson approximation. How good is the approximation in this instance?

12.50 A trade association of small businesses plans to select a random sample of member firms to estimate (1) the proportion of firms with no full-time employees, (2) the proportion of firms in which the owner works more than 60 hours per week, and (3) the proportion of firms that carry no liability insurance. The random sample size should be adequate to provide 95 percent confidence intervals for each of these three population proportions with half-widths of 0.02 or smaller. Planning values for the three population proportions deemed appropriate by the research director of the trade association are 0.4, 0.7, and 0.2, respectively.

   a. What sample size should the trade association employ so that the precision requirements for all three estimates can be met? Which planning value governed the choice of sample size? Explain why it determined the sample size.

   b. The trade association finally decided to sample only 1000 firms to expedite the completion of the study. The final report will contain many other estimates of population proportions besides the three mentioned previously. To simplify the presentation of the precisions of the many estimates, calculate the half-width $zs\{\bar{p}\}$ of a 95 percent confidence interval for this survey for $\bar{p} = 0.01, 0.10, 0.20, 0.50, 0.80, 0.90, 0.99$. Plot the half-widths as a function of the sample proportion $\bar{p}$. Prepare a brief statement that explains how the graph is to be used.

13 COMPARISONS OF TWO POPULATIONS AND OTHER INFERENCES

Thus far, we have discussed statistical inferences about a population mean and a population proportion. In this chapter, we take up comparative studies where a comparison of two populations is of interest. Specifically, we describe the construction of interval estimates and tests for the difference between two population means and the difference between two population proportions. In addition, we consider, as an optional topic, inferences about the population variance.

13.1 COMPARATIVE STUDIES

Nature of Comparative Studies

The general purpose of comparative studies is to establish similarities or to detect and measure differences between populations. The populations involved in comparative studies may be of two basic types: (1) *existing* populations and (2) *hypothetical* populations related to an experiment.

EXAMPLES

1. An economist wishes to compare the proportions of families with incomes below the poverty level in two regions of the country, from sample data for each region. Specifically, he wishes to make inferences about the proportions p_1 and p_2 of the two populations from data based on the two samples. The populations here are existing populations.

2. **Social Behavior.** A social psychologist wants to compare persons' attitudes toward a particular social behavior before and after they view an informational film about this behavior, based on a study of a sample of subjects. In this case, persons' attitudes "before" and "after" constitute two hypothetical populations related to an experiment in which the stimulus is the informational film. Using an established questionnaire to measure this attitude, the psychologist wishes to determine if the film causes a shift in the mean attitude level of subjects, where μ_1 and μ_2 denote the before and after stimulus means. That is, the psychologist wishes to ascertain whether or not μ_1 and μ_2 are equal and, if not, how great the difference is and in what direction.

Design Considerations — Independent Versus Matched Samples

In comparative sample studies, there is often the opportunity to choose between two alternative designs, called *independent samples* and *matched samples*. Clearly, one would like to choose the design that, for given cost, leads to smaller sampling errors. The following example highlights the differences between the two designs.

EXAMPLE

It is desired to obtain an estimate of the weight loss in a shipment of bananas during transit. Contrast the following two procedures.

1. *Independent samples.* A random sample of banana bunches is selected from the lot and weighed before loading. After shipment, an independent random sample of bunches is selected and weighed during the unloading. The difference in the two sample mean weights per bunch is used as the estimate of weight loss per bunch.

2. *Matched samples.* Again, a random sample of bunches is selected and weighed before loading. After transit, the same bunches selected before loading are weighed again, and the difference in weight for each bunch is noted. The mean of these differences is used as an estimate of the weight loss per bunch.

Often, the matched-samples method will lead to a smaller sampling error. To illustrate, Table 13.1 contains hypothetical data on weights for a population of 10 banana bunches, giving the weights before and after shipment. In this idealized example, observe that each of the 10 bunches experiences the same weight loss of 2.0 kilograms. If independent samples of size 2, say, are drawn before and after transit, the uniform weight loss per bunch is hidden by the variation in the weights of bunches. The first sample might consist of bunches 1 and 8, with a mean

TABLE 13.1   Hypothetical weights for a population of banana bunches before and after shipment

Bunch	Weight Before Shipment (kilograms)	Weight After Shipment (kilograms)
1	34	32
2	29	27
3	31	29
4	28	26
5	23	21
6	32	30
7	27	25
8	23	21
9	28	26
10	23	21

weight of 28.5 kilograms, while the second sample might contain bunches 4 and 10, with a mean weight of 23.5 kilograms. The sample difference of 5.0 kilograms is far greater than the true difference of 2.0 kilograms. In contrast, the use of matched samples will always reveal exactly the 2.0-kilogram weight loss per bunch.   □

The example just presented is an extreme and simplified one, but it does illustrate the following general principle: If the observations before and after are positively related, the method involving matched observations will lead to smaller sampling errors than the use of independent samples. Note from Table 13.1 that the observations in the banana shipment example are positively related. The heaviest bunches before shipment are still the heaviest bunches after shipment, and similarly for the lightest bunches. Weighing the same sample of banana bunches before and after shipment eliminates variation in weights between banana bunches as a source of sampling error, and only the variation in the amounts of weight loss remains. On the other hand, the use of independent samples involves both of these sources of variation.

Matched samples need not always involve the same sample elements; instead, elements with similar characteristics may be paired in advance of the experiment. In a pharmacological research study involving the reaction of experimental animals to two drug dosages, two animals from the same litter and reared under identical conditions are randomly assigned the two dosages under study. By comparing the effects of the two drug dosages on the animals from the same litter, the effects of heredity, physiological condition, and environment are virtually eliminated from the sampling error.

In spite of the greater precision provided by matched samples when a positive relation exists between the paired observations, independent samples are used frequently. The cost or, sometimes, the physical impossibility of obtaining matched observations often necessitates the use of independent samples. For instance, in the banana shipment example, logistical difficulties may make it infeasible to tag banana bunches before shipment so that they can be identified for subsequent reweighing.

13.2 INFERENCES ABOUT DIFFERENCE BETWEEN TWO POPULATION MEANS—INDEPENDENT SAMPLES

As examples in the previous section have illustrated, we often want to estimate or test the difference between two population means. In this section, we discuss inferences of this nature for the case of two independent samples. First, we consider inferences when the populations are normal or approximately normal. Then, we take up inferences for any populations when the sample sizes are large.

Confidence Interval for $\mu_2 - \mu_1$ —Normal Populations with Equal Variances

Assumptions.   The procedure to be described assumes the following:

1.   The two populations are normal or the departures are not too marked. The population means are denoted by μ_1 and μ_2.

2. The two populations have the same variance, denoted by σ^2.
3. Independent random samples of sizes n_1 and n_2 are drawn from the two populations, respectively.

EXAMPLE

Admission Test Scores.   An admissions officer of a small liberal arts college wishes to compare the mean admission test scores of applicants educated in rural high schools (population 1, X variable) and of applicants educated in urban high schools (population 2, Y variable). Independent random samples of sizes n_1 and n_2 from the two populations, respectively, will be used. It is reasonable to assume here that admission test scores in both populations are normally distributed with means μ_1 and μ_2, respectively, and common variance σ^2. Interest is in estimating the difference in mean test scores, $\mu_2 - \mu_1$, by a confidence interval.   ☐

Point Estimator of $\mu_2 - \mu_1$.   A point estimator of $\mu_2 - \mu_1$ is the difference between the two sample means.

(13.1)

A point estimator of $\mu_2 - \mu_1$ is:

$$\overline{Y} - \overline{X}$$

where:
$\overline{Y}$ is the sample mean of the Y observations from population 2
$\overline{X}$ is the sample mean of the X observations from population 1

Let us now study the properties of the sampling distribution of $\overline{Y} - \overline{X}$.

Sampling Distribution of $\overline{Y} - \overline{X}$.   Since the populations sampled are assumed to be normal, we know from (9.7) that both $\overline{X}$ and $\overline{Y}$ have normal sampling distributions. The means of these sampling distributions are μ_1 and μ_2, from (9.3), and the variances are σ^2/n_1 and σ^2/n_2, from (9.4).

We turn now to the properties of the sampling distribution of $\overline{Y} - \overline{X}$. The following theorem tells us that this difference is also normally distributed.

(13.2)

Any linear combination of independent normal random variables is also normally distributed.

The difference $\overline{Y} - \overline{X}$ is such a linear combination of independent normal random variables here. Remember that $\overline{X}$ and $\overline{Y}$ are independent because the two samples are independent.

Theorem (5.14a) tells us that the expected value of $\overline{Y} - \overline{X}$ is:

$$E\{\overline{Y} - \overline{X}\} = E\{\overline{Y}\} - E\{\overline{X}\} = \mu_2 - \mu_1$$

And theorem (5.14b) enables us to ascertain the variance:

$$\sigma^2\{\overline{Y} - \overline{X}\} = \sigma^2\{\overline{Y}\} + \sigma^2\{\overline{X}\} = \frac{\sigma^2}{n_2} + \frac{\sigma^2}{n_1} = \sigma^2\left(\frac{1}{n_2} + \frac{1}{n_1}\right)$$

We now summarize these results.

(13.3)

When two independent random samples of sizes n_1 and n_2 are selected from normal populations with means μ_1 and μ_2 and common variance σ^2, the sampling distribution of $\overline{Y} - \overline{X}$ has the following properties:

Mean: $E\{\overline{Y} - \overline{X}\} = \mu_2 - \mu_1$

Variance: $\sigma^2\{\overline{Y} - \overline{X}\} = \sigma^2\left(\dfrac{1}{n_2} + \dfrac{1}{n_1}\right)$

Functional form: Normal distribution

Development of Confidence Interval. To construct a confidence interval for $\mu_2 - \mu_1$, we require an estimate of the common population variance σ^2. Let s_1^2 and s_2^2 denote the variances of the two samples. It can be shown that the following weighted average of s_1^2 and s_2^2 is the best unbiased estimator of σ^2.

(13.4)

$$s_c^2 = \frac{(n_1 - 1)s_1^2 + (n_2 - 1)s_2^2}{(n_1 - 1) + (n_2 - 1)}$$

The estimator s_c^2 is called a *pooled* or *combined estimator* of σ^2.

With s_c^2 as an unbiased estimator of σ^2, it can be seen from (13.3) that the following is an unbiased estimator of $\sigma^2\{\overline{Y} - \overline{X}\}$.

(13.5)

$$s^2\{\overline{Y} - \overline{X}\} = s_c^2\left(\frac{1}{n_2} + \frac{1}{n_1}\right)$$

where:
s_c^2 is given by (13.4)

A statistical theorem tells us the following.

(13.6)

For two independent random samples from normal populations with common variance:

$$\frac{(\overline{Y} - \overline{X}) - (\mu_2 - \mu_1)}{s\{\overline{Y} - \overline{X}\}} = t(n_1 + n_2 - 2)$$

where:
$s\{\overline{Y} - \overline{X}\}$ is given by (13.5)

Hence, the situation is parallel to that for interval estimation of μ for a normal population.

(13.7)

> The confidence limits for $\mu_2 - \mu_1$ with confidence coefficient $1 - \alpha$, when the populations are normal (or do not depart too markedly from normality) with common variance σ^2 and the random samples are independent, are:
>
> $$(\overline{Y} - \overline{X}) \pm ts\{\overline{Y} - \overline{X}\}$$
>
> *where:*
>
> $t = t(1 - \alpha/2; n_1 + n_2 - 2)$
> $s\{\overline{Y} - \overline{X}\}$ is given by (13.5)

The t value in this case is based on $n_1 + n_2 - 2$ degrees of freedom because the common variance σ^2 is estimated from two sample variances having a total number of degrees of freedom of $(n_1 - 1) + (n_2 - 1) = n_1 + n_2 - 2$.

EXAMPLE

Independent random samples of $n_1 = 15$ rural applicants and $n_2 = 17$ urban applicants were selected in the admission test scores example. The results were as follows:

$$n_1 = 15 \qquad \overline{X} = 495 \qquad s_1 = 55$$
$$n_2 = 17 \qquad \overline{Y} = 545 \qquad s_2 = 50$$

A 95 percent confidence interval for $\mu_2 - \mu_1$ is desired.
From these sample results, we have:

$$\overline{Y} - \overline{X} = 545 - 495 = 50$$

$$s_c^2 = \frac{(15 - 1)(55)^2 + (17 - 1)(50)^2}{(15 - 1) + (17 - 1)} = 2745$$

$$s^2\{\overline{Y} - \overline{X}\} = 2745\left(\frac{1}{17} + \frac{1}{15}\right) = 344.47$$

$$s\{\overline{Y} - \overline{X}\} = \sqrt{344.47} = 18.56$$

For a 95 percent confidence interval, we require $t(1 - \alpha/2; n_1 + n_2 - 2) = t(0.975; 30) = 2.042$. Hence, the desired confidence limits are $50 \pm 2.042(18.56)$, and the resulting confidence interval is:

$$12.1 \leq \mu_2 - \mu_1 \leq 87.9$$

Therefore, with 95 percent confidence, it can be stated that the mean admission test score for urban applicants for this college exceeds the mean score for rural applicants by between 12.1 and 87.9 points. Clearly, larger samples would be required to measure this difference more precisely. □

Confidence Interval for $\mu_2 - \mu_1$
—Large Samples from Arbitrary Populations

We now turn to a procedure for estimating $\mu_2 - \mu_1$ when the sample sizes are large and there need be no restrictions on the nature of the populations.

Assumptions.   The procedure to be described assumes the following:

1.   There are no restrictions on either population other than that the sample mean be approximately normally distributed for large samples.
2.   Independent random samples of sizes n_1 and n_2 are selected, both of which are reasonably large.

EXAMPLE

Family Income.   An economist wishes to estimate the difference in mean family incomes for households in two socioeconomic groups, neither of which contains households with extremely high incomes. Large independent random samples of n_1 and n_2 households are drawn from the respective groups and their incomes ascertained. The difference in the sample means, $\overline{Y} - \overline{X}$, will again be used to construct an interval estimate for $\mu_2 - \mu_1$ (the difference in mean family incomes for the two groups).   ☐

Sampling Distribution of $\overline{Y} - \overline{X}$.   From the assumptions made in this case, we can determine the properties of the sampling distribution of $\overline{Y} - \overline{X}$.

(13.8)

> When large independent random samples are selected from almost any two populations, the sampling distribution of $\overline{Y} - \overline{X}$ has the following properties:
>
> *Mean:* $E\{\overline{Y} - \overline{X}\} = \mu_2 - \mu_1$
>
> *Variance:* $\sigma^2\{\overline{Y} - \overline{X}\} = \sigma^2\{\overline{Y}\} + \sigma^2\{\overline{X}\}$
>
> *Functional form:*   Approximately normal distribution

Because the samples are drawn independently, $\overline{Y} - \overline{X}$ has a variance by (5.14b) equal to the sum of the variances of $\overline{Y}$ and $\overline{X}$. Also, since both sample sizes are large, central limit theorem (9.5) guarantees that $\overline{Y}$ and $\overline{X}$ are each approximately normal, and consequently, by (13.2) their difference $\overline{Y} - \overline{X}$ is also approximately normal.

Development of Confidence Interval.   To construct a confidence interval for $\mu_2 - \mu_1$, we estimate $\sigma^2\{\overline{Y} - \overline{X}\}$ with the following unbiased estimator.

(13.9)

> $s^2\{\overline{Y} - \overline{X}\} = s^2\{\overline{Y}\} + s^2\{\overline{X}\}$
>
> *where:*
> $s^2\{\overline{Y}\}$ and $s^2\{\overline{X}\}$ are given by (10.8)

Because we are dealing with large random samples, we can draw a parallel between this case and the case of interval estimation for μ based on a large random sample.

(13.10)

The confidence limits for $\mu_2 - \mu_1$ with approximate confidence coefficient $1 - \alpha$, when the random samples are independent and reasonably large, are:

$$(\bar{Y} - \bar{X}) \pm zs\{\bar{Y} - \bar{X}\}$$

where:

$z = z(1 - \alpha/2)$

$s\{\bar{Y} - \bar{X}\}$ is given by (13.9)

EXAMPLE

In the family income example, samples of $n_1 = 200$ and $n_2 = 250$ families were selected from the respective populations. The results were as follows:

$n_1 = 200$ $\bar{X} = \$15,530$ $s_1 = \$5160$

$n_2 = 250$ $\bar{Y} = \$16,910$ $s_2 = \$5840$

A 95 percent confidence interval for $\mu_2 - \mu_1$ is desired.
We have:

$$\bar{Y} - \bar{X} = 16,910 - 15,530 = 1380$$

$$s^2\{\bar{Y} - \bar{X}\} = \frac{(5160)^2}{200} + \frac{(5840)^2}{250} = 269,550$$

$$s\{\bar{Y} - \bar{X}\} = \sqrt{269,550} = 519$$

We require $z(1 - \alpha/2) = z(0.975) = 1.960$. Hence, the desired confidence limits are $1380 \pm 1.960(519)$, and the resulting confidence interval is:

$$363 \le \mu_2 - \mu_1 \le 2397$$

Therefore, it can be concluded, with 95 percent confidence, that mean family income in the second group exceeds that in the first group by between \$363 and \$2397.   □

Statistical Tests for $\mu_2 - \mu_1$

For both the case of normal populations and that of arbitrary populations with large sample sizes, decision rules for making statistical tests concerning $\mu_2 - \mu_1$ are constructed in the same manner as for tests concerning a single population mean. The standardized test statistics are as follows when the α risk is to be controlled at $\mu_2 = \mu_1$ or, equivalently, at $\mu_2 - \mu_1 = 0$.

(13.11)

Tests concerning $\mu_2 - \mu_1$, when the two random samples are independent, are based on the following standardized test statistics.

(a) Normal Populations with
Equal Variances

$$t^* = \frac{(\overline{Y} - \overline{X}) - 0}{s\{\overline{Y} - \overline{X}\}}$$

where:

$s\{\overline{Y} - \overline{X}\}$ is given by (13.5)
α risk is controlled at $\mu_2 - \mu_1 = 0$

When $\mu_2 - \mu_1 = 0$, t^* follows the
t distribution with $n_1 + n_2 - 2$
degrees of freedom.

(b) Large Samples from
Arbitrary Populations

$$z^* = \frac{(\overline{Y} - \overline{X}) - 0}{s\{\overline{Y} - \overline{X}\}}$$

where:

$s\{\overline{Y} - \overline{X}\}$ is given by (13.9)
α risk is controlled at $\mu_2 - \mu_1 = 0$

When $\mu_2 - \mu_1 = 0$, z^* follows
approximately the standard normal
distribution.

We present now an example to illustrate the use of these test statistics.

EXAMPLE

In the family income example, it is desired to test whether both socioeconomic groups have the same mean family income.

Step 1.   The alternatives here are as follows:

H_0: $\mu_2 - \mu_1 = 0$

H_1: $\mu_2 - \mu_1 \neq 0$

Step 2.   The α risk is to be controlled at 0.05 when $\mu_2 - \mu_1 = 0$, that is, when $\mu_2 = \mu_1$.

Step 3.   Figure 13.1 shows the appropriate type of decision rule. The test is two-sided, and the sampling distribution of $\overline{Y} - \overline{X}$ is centered at $\mu_2 - \mu_1 = 0$ where the α risk is to be controlled. Because the sample sizes are large, the standard normal distribution is utilized for constructing the decision rule. Since the α risk is to be controlled at 0.05, the action limits in the decision rule of Figure 13.1 correspond to $z(\alpha/2) = z(0.025) = -1.960$ and $z(1 - \alpha/2) = z(0.975) = 1.960$. Thus, the decision rule is as follows:

If $|z^*| \leq 1.960$, conclude H_0.
If $|z^*| > 1.960$, conclude H_1.

Step 4.   From the sample results given earlier, we have $\overline{Y} - \overline{X} = 1380$ and $s\{\overline{Y} - \overline{X}\} = 519$. Substituting these numbers in (13.11b) for test statistic z^* gives:

$$z^* = \frac{(\overline{Y} - \overline{X}) - 0}{s\{\overline{Y} - \overline{X}\}} = \frac{1380 - 0}{519} = 2.66$$

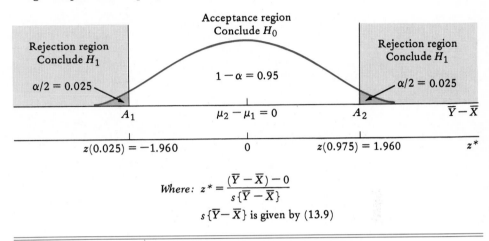

FIGURE 13.1 **Illustration of the statistical decision rule for the family income example— Large independent samples, two-sided test for $\mu_2 - \mu_1$**

Where: $z^* = \dfrac{(\overline{Y} - \overline{X}) - 0}{s\{\overline{Y} - \overline{X}\}}$

$s\{\overline{Y} - \overline{X}\}$ is given by (13.9)

Since $|z^*| = |2.66| = 2.66 > 1.960$, we conclude H_1—that the two groups do not have equal mean incomes. Recall that the 95 percent confidence interval for $\mu_2 - \mu_1$ obtained earlier for this example indicated also that the two groups do not have equal mean incomes. $\square$

13.3 INFERENCES ABOUT DIFFERENCE BETWEEN TWO POPULATION MEANS—MATCHED SAMPLES

When matched samples are employed for making inferences about $\mu_2 - \mu_1$, it turns out that the procedures simplify to those for a single population mean μ. The following example illustrates this result.

EXAMPLE

In the social behavior example, a social psychologist wishes to compare persons' attitudes toward a particular social behavior before and after they view a film about the behavior. For each subject, an attitude measurement is made before the film is viewed (X), and a second measurement is made after the film is viewed (Y). Letting μ_1 and μ_2 denote the population mean values of X and Y, respectively, the psychologist wishes to estimate $\mu_2 - \mu_1$.

We shall let X_i and Y_i denote the "before" and "after" measurements, respectively, for the ith subject, so (X_i, Y_i) is a "matched pair" of observations. To measure the change in attitude of the ith subject, the psychologist will use the following difference.

(13.12)
$$D_i = Y_i - X_i$$

If n subjects are studied, there will be n differences $D_1, D_2, \ldots, D_n$. □

The differences $D_1, D_2, \ldots, D_n$ may be thought of as a random sample of observations from a population of differences. We denote the mean and variance of this population of differences by μ_D and σ_D^2, respectively.

(13.13)
$$E\{D_i\} = \mu_D \qquad \sigma^2\{D_i\} = \sigma_D^2$$

Using definition (13.12), we can express μ_D in terms of μ_1 and μ_2:

$$E\{D_i\} = E\{Y_i - X_i\} = E\{Y_i\} - E\{X_i\} = \mu_2 - \mu_1$$

Hence, we have the following identity.

(13.14)
$$\mu_D = \mu_2 - \mu_1$$

Thus, we shall estimate μ_D by the earlier procedures for a single population mean, but we are really estimating $\mu_D = \mu_2 - \mu_1$, the difference between the means of the two populations.

The procedure, therefore, is to first calculate the sample mean of the observed differences $D_1, D_2, \ldots, D_n$, to be denoted by $\overline{D}$. Next, we calculate the sample variance of the D_i, to be denoted by s_D^2 as a reminder that it is a variance of differences. Finally, we calculate $s^2\{\overline{D}\}$, the estimated variance of $\overline{D}$.

(13.15)
$$\overline{D} = \frac{\sum_{i=1}^{n} D_i}{n}$$

(13.16)
$$s_D^2 = \frac{\sum_{i=1}^{n} (D_i - \overline{D})^2}{n - 1}$$

(13.17)
$$s^2\{\overline{D}\} = \frac{s_D^2}{n}$$

Then, we proceed as earlier, depending on whether the population of differences is approximately normal or the population of differences is arbitrary and the number of differences (n) is large.

Confidence Interval for $\mu_2 - \mu_1$ —Normal Population of Differences

When the differences D_i are normally distributed, or when their distribution does not depart from normality too markedly, a confidence interval for $\mu_2 - \mu_1$ is equivalent to that in (10.15).

(13.18)

> The confidence limits for $\mu_2 - \mu_1$ with confidence coefficient $1 - \alpha$, when the population of differences is normal or does not depart too markedly from normality, are:
>
> $$\overline{D} \pm ts\{\overline{D}\}$$
>
> *where:*
>
> $t = t(1 - \alpha/2; n - 1)$
>
> $\overline{D}$ and $s\{\overline{D}\}$ are given by (13.15) and (13.17), respectively

EXAMPLE

In the social behavior example, $n = 10$ subjects were studied, and the attitude measurements shown in Table 13.2 were obtained. It is desired to estimate $\mu_2 - \mu_1$ with a 90 percent confidence interval. It is reasonable here to assume that the population of differences is approximately normally distributed.

For the sample data, we have $\overline{D} = 7.64$ and $s_D = 12.57$. Therefore:

$$s^2\{\overline{D}\} = \frac{s_D^2}{n} = \frac{(12.57)^2}{10} = 15.80 \qquad s\{\overline{D}\} = \sqrt{15.80} = 3.97$$

We require $t(0.95; 9) = 1.833$. The confidence limits equal $7.64 \pm 1.833(3.97)$, and the desired confidence interval is:

$$0.36 \leq \mu_2 - \mu_1 \leq 14.92$$

TABLE 13.2 **Sample data on attitudes before and after viewing an informational film—Social behavior example**

Subject i	Before X_i	After Y_i	$D_i = Y_i - X_i$	Subject i	Before X_i	After Y_i	$D_i = Y_i - X_i$
1	41.0	46.9	+5.9	6	22.5	56.8	+34.3
2	60.3	64.5	+4.2	7	67.5	60.7	− 6.8
3	23.9	33.3	+9.4	8	50.3	57.3	+ 7.0
4	36.2	36.0	−0.2	9	50.9	65.4	+14.5
5	52.7	43.5	−9.2	10	24.6	41.9	+17.3

$$\overline{D} = 7.64 \qquad s_D = 12.57$$

The psychologist can assert, with 90 percent confidence, that the mean attitude measurement after viewing the film exceeds the mean attitude measurement before viewing by between 0.36 and 14.92 units on the measurement scale. Thus, a change in mean attitude occurred after viewing the film, but the small-scale study does not make clear whether the change is small or large. □

Comment   Matched sampling was highly effective here. To see why, we need to consider $\sigma^2\{D_i\}$. Since $D_i = Y_i - X_i$, we can use (5.30) to find:

$$\sigma^2\{D_i\} = \sigma^2\{Y_i - X_i\} = \sigma^2\{Y_i\} + \sigma^2\{X_i\} - 2\sigma\{X_i, Y_i\}$$

This variance has been denoted by σ_D^2 earlier—that is, $\sigma^2\{D_i\} = \sigma_D^2$. Thus, when $\sigma\{X_i, Y_i\}$, the covariance between X_i and Y_i, is positive, σ_D^2 will be smaller than when X_i and Y_i are independent. Since $\sigma^2\{\overline{D}\} = \sigma_D^2/n$, it follows that $\sigma^2\{\overline{D}\}$ will be smaller when the X_i and Y_i observations are positively related than when they are independent. Here, the "before" and "after" attitude measurements are positively related, leading to a more precise estimate than if independent samples had been employed. Matched-sample designs seek to assure a high positive relation between X and Y to attain this improvement in precision.

Confidence Interval for $\mu_2 - \mu_1$ —Large Sample from Arbitrary Population of Differences

When the sample of differences is reasonably large, the nature of the population of differences does not matter, and the confidence interval for $\mu_2 - \mu_1$ is equivalent to that in (10.11).

(13.19)

The confidence limits for $\mu_2 - \mu_1$ with approximate confidence coefficient $1 - \alpha$, when the sample of differences is reasonably large, are:

$$\overline{D} \pm zs\{\overline{D}\}$$

where:

$z = z(1 - \alpha/2)$

$\overline{D}$ and $s\{\overline{D}\}$ are given by (13.15) and (13.17), respectively

Statistical Tests for $\mu_2 - \mu_1$

Like the construction of confidence intervals for $\mu_2 - \mu_1$, tests on $\mu_2 - \mu_1$ for the case of matched samples parallel those for a single population mean. Here, $\overline{D}$ corresponds to $\overline{X}$, $\mu_D = \mu_2 - \mu_1$ corresponds to μ, and σ_D corresponds to σ. The standardized test statistics are as follows when the α risk is to be controlled at $\mu_2 = \mu_1$ or, equivalently, at $\mu_D = \mu_2 - \mu_1 = 0$.

(13.20)

Tests concerning $\mu_2 - \mu_1$, when the two samples are matched, are based on the following standardized test statistics.

(a) Normal Population of Differences   *(b) Large Sample of Differences*

$$t^* = \frac{\overline{D} - 0}{s\{\overline{D}\}}$$ $$z^* = \frac{\overline{D} - 0}{s\{\overline{D}\}}$$

where:

$\overline{D}$ and $s\{\overline{D}\}$ are given by (13.15) and (13.17), respectively
α risk is controlled at $\mu_D = \mu_2 - \mu_1 = 0$

When $\mu_D = 0$, t^* follows the t distribution with $n - 1$ degrees of freedom.   When $\mu_D = 0$, z^* follows approximately the standard normal distribution.

EXAMPLE

In the social behavior example, the psychologist wishes to test whether or not viewing the film leads to an upward shift in the mean attitude level.

Step 1.   The test alternatives here are as follows:

H_0: $\mu_2 - \mu_1 \leq 0$

H_1: $\mu_2 - \mu_1 > 0$

Step 2.   The α risk is to be controlled at 0.05 when $\mu_2 - \mu_1 = 0$.
Step 3.   The population of differences can be assumed to be approximately normal, and a one-sided upper-tail test is desired. Hence, the appropriate type of decision rule is the one given in Table 11.2a, based on the t distribution. Since $\alpha = 0.05$ and $n = 10$, we require $t(1 - \alpha; n - 1) = t(0.95; 9) = 1.833$, and the decision rule for the test is as follows:

If $t^* \leq 1.833$, conclude H_0.
If $t^* > 1.833$, conclude H_1.

Step 4.   We found earlier that $\overline{D} = 7.64$ and $s\{\overline{D}\} = 3.97$. Hence, test statistic (13.20a) here is:

$$t^* = \frac{\overline{D} - 0}{s\{\overline{D}\}} = \frac{7.64 - 0}{3.97} = 1.92$$

Since $t^* = 1.92 > 1.833$, we conclude H_1—that there has been an upward shift in the mean attitude level after viewing the film.   ☐

INFERENCES ABOUT DIFFERENCE BETWEEN TWO POPULATION PROPORTIONS

We consider now inferences about the difference between two population proportions.

EXAMPLE

Market Survey.   A market survey organization carried out a product taste study with consumers in two regions. Independent random samples of $n_1 = 400$ consumers and $n_2 = 300$ consumers were selected in the two regions. Each person was asked to indicate which of two servings of product had a better taste. Unknown to the subject, one serving was a new high-protein breakfast cereal and the other was an existing cereal. In the first region, proportion $\bar{p}_1 = 0.55$ of the sample persons preferred the new cereal. In the second region, the proportion was $\bar{p}_2 = 0.65$. It is now desired to construct an interval estimate of $p_2 - p_1$, the difference in the population proportions of consumers in the two regions who prefer the new cereal.   □

Assumptions

Throughout this section, we assume the following:

1.   Independent random samples are selected from the two populations.
2.   The two random samples are each reasonably large.

Sampling Distribution of $\bar{p}_2 - \bar{p}_1$

Estimating the difference $p_2 - p_1$ is very similar to estimating the difference between two population means. A point estimator of $p_2 - p_1$ is the difference in sample proportions.

(13.21)

$$\bar{p}_2 - \bar{p}_1$$

We now summarize the properties of the sampling distribution of $\bar{p}_2 - \bar{p}_1$ when large and statistically independent random samples are drawn from the two populations.

(13.22)

When large independent random samples are selected from two populations, the sampling distribution of $\bar{p}_2 - \bar{p}_1$ has the following properties:

Mean: $E\{\bar{p}_2 - \bar{p}_1\} = p_2 - p_1$

Variance: $\sigma^2\{\bar{p}_2 - \bar{p}_1\} = \sigma^2\{\bar{p}_2\} + \sigma^2\{\bar{p}_1\}$

Functional form: Approximately normal distribution

Observe that the variance of $\bar{p}_2 - \bar{p}_1$ is the sum of the variances of $\bar{p}_1$ and $\bar{p}_2$ by (5.14b), because the samples are independent. In addition, since the individual sample sizes are large, $\bar{p}_1$ and $\bar{p}_2$ are each approximately normal by (12.5), and by (13.2), their difference $\bar{p}_2 - \bar{p}_1$ is therefore also approximately normal.

Confidence Interval for $p_2 - p_1$

To construct an interval estimate for $p_2 - p_1$, we need an unbiased estimator of $\sigma^2\{\bar{p}_2 - \bar{p}_1\}$. We obtain it by using $s^2\{\bar{p}_1\}$ and $s^2\{\bar{p}_2\}$, as given in (12.7), as estimators of the true variances $\sigma^2\{\bar{p}_1\}$ and $\sigma^2\{\bar{p}_2\}$. The resulting estimator is denoted by $s^2\{\bar{p}_2 - \bar{p}_1\}$.

(13.23)

$$s^2\{\bar{p}_2 - \bar{p}_1\} = s^2\{\bar{p}_2\} + s^2\{\bar{p}_1\}$$

where:

$s^2\{\bar{p}_2\}$ and $s^2\{\bar{p}_1\}$ are given by (12.7)

The confidence interval for $p_2 - p_1$ takes the usual form for the large-sample case.

(13.24)

The confidence limits for $p_2 - p_1$ with approximate confidence coefficient $1 - \alpha$, when the two random samples are independent and reasonably large, are:

$$(\bar{p}_2 - \bar{p}_1) \pm zs\{\bar{p}_2 - \bar{p}_1\}$$

where:

$z = z(1 - \alpha/2)$
$s\{\bar{p}_2 - \bar{p}_1\}$ is given by (13.23)

EXAMPLE

For the market survey example, we have the following:

$$n_1 = 400 \qquad \bar{p}_1 = 0.55$$
$$n_2 = 300 \qquad \bar{p}_2 = 0.65$$

A 90 percent confidence interval for $p_2 - p_1$ is desired.
We have $\bar{p}_2 - \bar{p}_1 = 0.65 - 0.55 = 0.10$, and:

$$s^2\{\bar{p}_1\} = \frac{\bar{p}_1(1 - \bar{p}_1)}{n_1 - 1} = \frac{0.55(0.45)}{400 - 1} = 0.0006203$$

$$s^2\{\bar{p}_2\} = \frac{\bar{p}_2(1 - \bar{p}_2)}{n_2 - 1} = \frac{0.65(0.35)}{300 - 1} = 0.0007609$$

$$s^2\{\bar{p}_2 - \bar{p}_1\} = 0.0007609 + 0.0006203 = 0.0013812$$

$$s\{\bar{p}_2 - \bar{p}_1\} = \sqrt{0.0013812} = 0.0372$$

We require $z(0.95) = 1.645$, and the confidence limits are $0.10 \pm 1.645(0.0372)$. The resulting confidence interval is:

$$0.039 \leq p_2 - p_1 \leq 0.161$$

We conclude, with 90 percent confidence, that the proportion of consumers in the second region who prefer the new cereal exceeds the proportion for the first region by between 3.9 and 16.1 percent points. ☐

Statistical Tests for $p_2 - p_1$

Statistical tests regarding the difference between two population proportions for the large-sample case are similar to those for the difference between two population means. We illustrate the procedure for a two-sided test.

EXAMPLE

In the market survey example, it is desired to test whether or not the proportions of consumers preferring the new high-protein cereal in the two regions are equal.

Step 1.   The test alternatives here are:

H_0: $p_2 - p_1 = 0$

H_1: $p_2 - p_1 \neq 0$

Step 2.   The α risk at $p_2 - p_1 = 0$ is to be controlled at 0.10.

Step 3.   Figure 13.2 shows the appropriate type of decision rule for the large-sample case. The test is two-sided, and the sampling distribution of $\bar{p}_2 - \bar{p}_1$ is centered at $p_2 - p_1 = 0$, where the α risk is to be controlled. To calculate the standardized test statistic z^*, we require an estimate of $\sigma\{\bar{p}_2 - \bar{p}_1\}$ for the case where $p_2 - p_1 = 0$, that is, where $p_2 = p_1$. Let p denote the common value of p_2 and p_1 when $p_2 = p_1$. To estimate p from the two samples, we employ the following *pooled* estimator of p.

(13.25)

$$\bar{p}' = \frac{n_1\bar{p}_1 + n_2\bar{p}_2}{n_1 + n_2}$$

Note that $\bar{p}'$ is simply a weighted average of $\bar{p}_1$ and $\bar{p}_2$, using the sample sizes as weights.

When $p_1 = p_2 = p$, we know that:

$$\sigma^2\{\bar{p}_2 - \bar{p}_1\} = \frac{p_2(1 - p_2)}{n_2} + \frac{p_1(1 - p_1)}{n_1} = p(1 - p)\left(\frac{1}{n_2} + \frac{1}{n_1}\right)$$

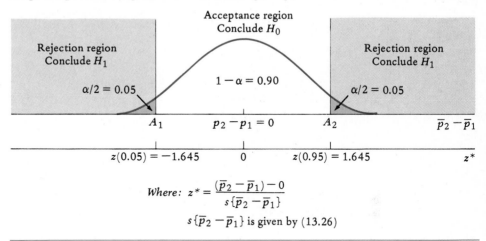

FIGURE 13.2 Illustration of the statistical decision rule for the market survey example — Large independent samples, two-sided test for $p_2 - p_1$

Hence, a sample estimator of $\sigma^2\{\bar{p}_2 - \bar{p}_1\}$ when $p_1 = p_2$, based on the pooled estimator $\bar{p}'$, is as follows.

(13.26)

$$s^2\{\bar{p}_2 - \bar{p}_1\} = \bar{p}'(1 - \bar{p}')\left(\frac{1}{n_2} + \frac{1}{n_1}\right)$$

where:

$\bar{p}'$ is given by (13.25)

When the sample sizes are large, the standardized test statistic:

$$z^* = \frac{(\bar{p}_2 - \bar{p}_1) - 0}{s\{\bar{p}_2 - \bar{p}_1\}}$$

has an approximate standard normal distribution when $p_2 - p_1 = 0$. The action limits in the decision rule of Figure 13.2 therefore correspond to $z(\alpha/2) = z(0.05) = -1.645$ and $z(1 - \alpha/2) = z(0.95) = 1.645$. Thus, the decision rule is as follows:

If $|z^*| \leq 1.645$, conclude H_0.
If $|z^*| > 1.645$, conclude H_1.

Step 4.   Recall that the sample results in this example were as follows:

$$n_1 = 400 \qquad \bar{p}_1 = 0.55$$

$$n_2 = 300 \qquad \bar{p}_2 = 0.65$$

Substituting these values in (13.25) and (13.26) gives:

$$\bar{p}' = \frac{n_1\bar{p}_1 + n_2\bar{p}_2}{n_1 + n_2} = \frac{400(0.55) + 300(0.65)}{400 + 300} = 0.593$$

$$s^2\{\bar{p}_2 - \bar{p}_1\} = \bar{p}'(1 - \bar{p}')\left(\frac{1}{n_2} + \frac{1}{n_1}\right)$$

$$= 0.593(0.407)\left(\frac{1}{300} + \frac{1}{400}\right) = 0.001408$$

$$s\{\bar{p}_2 - \bar{p}_1\} = \sqrt{0.001408} = 0.0375$$

Finally, we obtain for the standardized test statistic:

$$z* = \frac{(\bar{p}_2 - \bar{p}_1) - 0}{s\{\bar{p}_2 - \bar{p}_1\}} = \frac{(0.65 - 0.55) - 0}{0.0375} = 2.67$$

Since $|z*| = |2.67| = 2.67 > 1.645$, we conclude H_1—that the proportions of consumers preferring the new cereal in the two regions differ.   □

Test Statistic.   The test statistic developed here is appropriate for all large-sample tests concerning $p_2 - p_1$ when the α risk is to be controlled at $p_2 = p_1$ or, equivalently, at $p_2 - p_1 = 0$.

(13.27)

Tests concerning $p_2 - p_1$, when the two random samples are large and independent, are based on the standardized test statistic:

$$z* = \frac{(\bar{p}_2 - \bar{p}_1) - 0}{s\{\bar{p}_2 - \bar{p}_1\}}$$

where:

$s\{\bar{p}_2 - \bar{p}_1\}$ is given by (13.26)

α risk is controlled at $p_2 - p_1 = 0$

When $p_2 - p_1 = 0$, $z*$ follows approximately the standard normal distribution.

13.5 OPTIONAL TOPIC—INFERENCES ABOUT POPULATION VARIANCE AND RATIO OF TWO POPULATION VARIANCES

Interest in Population Variance

On various previous occasions, we have used the sample variance s^2 as an unbiased point estimator of the population variance σ^2. In some applications, we are interested in constructing an interval estimate for σ^2 or in testing the value of σ^2.

EXAMPLE

Investment Return. The population variance is often of interest in studies where the extent of variability in the population is of concern. Consider the following quarterly rates of return (dividends plus price change as a percentage of opening price) on a common stock investment over a two-year period.

Quarter:	1	2	3	4	5	6	7	8
Rate of Return:	4.8	2.8	9.9	7.6	9.5	6.0	8.4	−5.0

Suppose that these eight observations can be considered as a random sample from the infinite population of all quarterly rates of return for the stock and that this population is approximately normal. The population variance σ^2 may be viewed as a measure of uncertainty for the common stock investment, and we are therefore interested in constructing an interval estimate for σ^2. ☐

Assumption. We consider here inferences about the population variance σ^2 for the special case *when the population is normal or approximately normal*. In Chapter 15, we consider inferences about the population variance when the population is not normal or approximately normal.

Confidence Interval for σ^2

Sampling Distribution of s^2. Statistical theory provides us with the following important theorem.

(13.28)

If a random sample of size n is selected from a normal population with variance σ^2, then:

$$\frac{(n-1)s^2}{\sigma^2} = \chi^2(n-1)$$

where:

s^2 is the sample variance defined in (8.12b)

(If a review of the χ^2 distribution is needed, the reader should turn to Appendix B, Section B.1, before proceeding.)

Figure 13.3a contains a representative χ^2 distribution that illustrates theorem (13.28). A χ^2 distribution with v degrees of freedom has its mean at v. Here, $v = n - 1$, so the distribution of $\chi^2(n - 1)$ is located around $n - 1$. Since a χ^2 distribution is always right-skewed, it is possible for s^2 to be substantially larger than σ^2 on some occasions.

Confidence Limits.   The confidence limits for σ^2 are based on theorem (13.28) and make use of the percentiles of the χ^2 distribution shown in Figure 13.3b.

(13.29)

> The confidence limits L and U for the population variance σ^2 with confidence coefficient $1 - \alpha$, when the population is normal or approximately so, are:
>
> $$L = \frac{(n - 1)s^2}{\chi^2(1 - \alpha/2; n - 1)} \qquad U = \frac{(n - 1)s^2}{\chi^2(\alpha/2; n - 1)}$$

FIGURE 13.3   Sampling distribution of $(n - 1)s^2/\sigma^2$ for a normal population

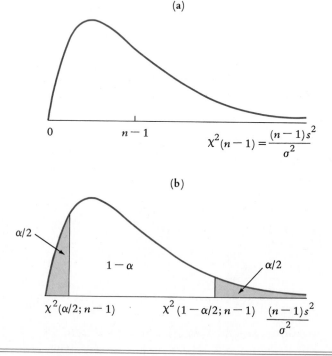

(a)

$$\chi^2(n - 1) = \frac{(n - 1)s^2}{\sigma^2}$$

(b)

EXAMPLE

For the investment return example, $n = 8$ and $s^2 = 23.78$ (calculation not shown). A 95 percent confidence interval is desired; hence, we require from Table C.2:

$$\chi^2(1 - \alpha/2; n - 1) = \chi^2(0.975; 7) = 16.01$$
$$\chi^2(\alpha/2; n - 1) = \chi^2(0.025; 7) = 1.69$$

Substituting these values into (13.29) yields:

$$L = \frac{(8 - 1)23.78}{16.01} = 10.4 \qquad U = \frac{(8 - 1)23.78}{1.69} = 98.5$$

Thus, the 95 percent confidence interval for σ^2 is:

$$10.4 \le \sigma^2 \le 98.5$$

Note that the interval is very wide. The magnitude of σ^2 cannot be ascertained precisely here with such a small sample size. □

Comments

1. Observe that the confidence limits for σ^2 are not equally spaced about s^2. This result reflects the right-skewness of the χ^2 sampling distribution.

2. A confidence interval for σ is obtained by taking the square roots of the confidence limits for σ^2 in (13.29). For the investment return example, the 95 percent confidence interval for σ is:

$$3.2 = \sqrt{10.4} \le \sigma \le \sqrt{98.5} = 9.9$$

Hence, the standard deviation of the common stock's quarterly rate of return is between 3.2 and 9.9 percent points, with 95 percent confidence.

3. For large n, the χ^2 percentiles in (13.29) may be approximated by standard normal percentiles, using the approximation given in (B.4) in Appendix B.

4. One-sided confidence limits for σ^2, with confidence coefficient $1 - \alpha$, can be obtained from (13.29) by replacing $1 - \alpha/2$ by $1 - \alpha$ in the appropriate percentile.

5. The confidence limits in (13.29) are obtained by means of theorem (13.28), which tells us that $(n - 1)s^2/\sigma^2$ follows a $\chi^2(n - 1)$ distribution. It is a consequence of this theorem and the definition of percentiles for a χ^2 distribution that:

$$P\left[\chi^2(\alpha/2; n - 1) \le \frac{(n - 1)s^2}{\sigma^2} \le \chi^2(1 - \alpha/2; n - 1) \right] = 1 - \alpha$$

This probability statement is illustrated in Figure 13.3b. Observe that α is divided equally between the two tails, as has been our practice in constructing two-sided confidence intervals. Rearranging the inequalities, we obtain:

$$P\left[\frac{(n - 1)s^2}{\chi^2(1 - \alpha/2; n - 1)} \le \sigma^2 \le \frac{(n - 1)s^2}{\chi^2(\alpha/2; n - 1)} \right] = 1 - \alpha$$

yielding the $1 - \alpha$ confidence limits for σ^2 in (13.29).

Statistical Tests for σ^2

Statistical tests for σ^2 may be constructed by using theorem (13.28). All tests are based on the same test statistic, denoted by X^2. As will be explained in the following example, this test statistic follows the χ^2 distribution with $n - 1$ degrees of freedom when $\sigma^2 = \sigma_0^2$, the value at which the α risk is controlled. Hence, percentiles of the $\chi^2(n - 1)$ distribution are utilized in constructing the decision rule.

(13.30)

Tests concerning the population variance σ^2, when the population is normal or approximately normal, are based on the test statistic:

$$X^2 = \frac{(n - 1)s^2}{\sigma_0^2}$$

where:

α risk is controlled at $\sigma^2 = \sigma_0^2$

When $\sigma^2 = \sigma_0^2$, X^2 follows the χ^2 distribution with $n - 1$ degrees of freedom.

We illustrate the use of test statistic X^2 with a two-sided test.

EXAMPLE

An investment analyst has developed a model according to which σ^2 for the common stock in the investment return example, given the nature of the business, the risks involved, and other related factors, should equal 5. We wish to test whether or not the model is correct, that is, whether or not $\sigma^2 = 5$.

Step 1. The test alternatives here are:

$H_0: \sigma^2 = 5$

$H_1: \sigma^2 \neq 5$

Step 2. The α risk is to be controlled at 0.05 when $\sigma^2 = \sigma_0^2 = 5$.
Step 3. Figure 13.4 shows the appropriate type of decision rule for this test. From theorem (13.28) we know that test statistic X^2 in (13.30) follows a $\chi^2(n - 1)$ distribution when $\sigma^2 = \sigma_0^2$. For this example, $n - 1 = 8 - 1 = 7$. Hence, the sampling distribution of X^2 is the $\chi^2(7)$ distribution whose mean is $n - 1 = 7$, as shown in Figure 13.4.

If s^2 is substantially smaller or larger than σ_0^2, then one should conclude H_1, that is, conclude $\sigma^2 \neq \sigma_0^2$. Correspondingly, because $X^2 = (n - 1)s^2/\sigma_0^2$, we should conclude H_1 if X^2 is either much smaller or much larger than $n - 1$. The appropriate type of decision rule is therefore a two-sided one. The action limits are placed so that each tail area equals $\alpha/2$. Since $\alpha = 0.05$ here, the action limits are the following χ^2 percentiles:

FIGURE 13.4 **Illustration of the statistical decision rule for the investment return example—Population normal or nearly normal, two-sided test for σ^2**

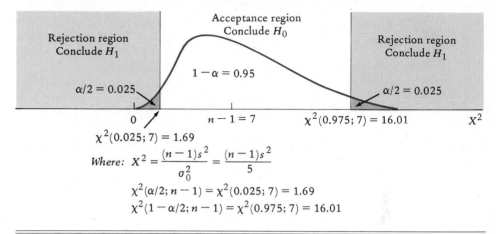

$$\chi^2(\alpha/2; n - 1) = \chi^2(0.025; 7) = 1.69$$
$$\chi^2(1 - \alpha/2; n - 1) = \chi^2(0.975; 7) = 16.01$$

The decision rule for the test is therefore as follows:

If $1.69 \leq X^2 \leq 16.01$, conclude H_0.
If $X^2 < 1.69$ or $X^2 > 16.01$, conclude H_1.

Step 4. Since $s^2 = 23.78$, the test statistic in (13.30) equals:

$$X^2 = \frac{(n - 1)s^2}{\sigma_0^2} = \frac{(8 - 1)(23.78)}{5} = 33.3$$

Because $X^2 = 33.3 > 16.01$, H_1 is concluded. Thus, the variance of quarterly returns for the stock does not equal 5, the magnitude projected by the model. If the model were to perform as badly for other stocks, then the analyst should abandon it. □

Comparison of Two Population Variances

In some comparative studies, interest lies in the variabilities of two populations. For instance, a manufacturer may wish to know whether two versions of a product have the same variability in shelf life. We now discuss interval estimates and tests for the ratio of two population variances.

EXAMPLE

Property Appraisal.   A behavioral scientist is interested in determining whether group discussion tends to affect the closeness of judgments by property appraisers. She obtained the cooperation of 16 real estate appraisers. Randomly, she selected 6 and asked them to make individual appraisals of a parcel of commercial property after all 6 had participated in a group discussion of relevant appraisal factors. The other 10 appraisers were also asked to make individual appraisals of the same property but without any prior group discussion.

As a measure of the closeness of agreement among appraisers, the scientist is using the variance. We shall let σ_1^2 denote the variance of appraisals made with group discussion and σ_2^2 the variance of appraisals made without group discussion. The scientist wants to compare the magnitudes of σ_1^2 and σ_2^2. Statistical theory provides a ready way for her to do so by computing an interval estimate of the ratio σ_2^2/σ_1^2.   □

Assumptions.   We consider here inferences for σ_2^2/σ_1^2 under the following assumptions:

1.   The populations are normal or approximately normal.
2.   The two random samples are independent.

Confidence Interval for σ_2^2/σ_1^2

To construct a confidence interval for σ_2^2/σ_1^2, we employ the following important statistical theorem.

(13.31)

> If independent random samples of sizes n_1 and n_2 are selected from normal populations with variances σ_1^2 and σ_2^2, respectively, then:
>
> $$\frac{s_1^2/\sigma_1^2}{s_2^2/\sigma_2^2} = F(n_1 - 1, n_2 - 1)$$
>
> *where:*
>
> s_1^2 and s_2^2 are the two sample variances, respectively

(If a review of the F distribution is needed, the reader should turn to Appendix B, Section B.3, before proceeding.)

The confidence limits for σ_2^2/σ_1^2 are based on theorem (13.31) and make use of percentiles of the F distribution.

(13.32)

> The confidence limits L and U for σ_2^2/σ_1^2 with confidence coefficient $1 - \alpha$, when the populations are normal or approximately so and the random samples are independent, are:
>
> $$L = F(\alpha/2; n_1 - 1, n_2 - 1)\frac{s_2^2}{s_1^2} \qquad U = F(1 - \alpha/2; n_1 - 1, n_2 - 1)\frac{s_2^2}{s_1^2}$$

Note that the numerator degrees of freedom for the F distribution are those of s_1^2, and the denominator degrees of freedom are those of s_2^2.

EXAMPLE

The data for the property appraisal example are presented in Table 13.3. A 90 percent confidence interval for σ_2^2/σ_1^2 is desired. The scientist states that it is reasonable to assume that the respective populations of appraisals are approximately normal.

We note first that $s_2^2/s_1^2 = 194.2/77.9 = 2.49$. For a 90 percent confidence interval, we require from Table C.4:

$$F(1 - \alpha/2; n_1 - 1, n_2 - 1) = F(0.95; 5, 9) = 3.48$$

$$F(\alpha/2; n_1 - 1, n_2 - 1) = F(0.05; 5, 9) = \frac{1}{4.77} = 0.210$$

TABLE 13.3 **Experimental data on appraisal values for a commercial property — Property appraisal example**

Appraisals With Group Discussion ($000)	Appraisals Without Group Discussion ($000)
97	118
111	109
102	84
99	85
88	100
111	121
	115
	93
	91
	112
$n_1 = 6$	$n_2 = 10$
$s_1^2 = 77.9$	$s_2^2 = 194.2$

Therefore, the confidence limits are $0.210(2.49) = 0.523$ and $3.48(2.49) = 8.67$, and the desired confidence interval is:

$$0.523 \leq \frac{\sigma_2^2}{\sigma_1^2} \leq 8.67$$

Since the confidence interval is so wide and straddles 1, no clear conclusion about the effect of group discussions on the variability of appraisals is possible. The scientist will have to enlarge her study to draw any firmer conclusions. ☐

Comments

1. A confidence interval for the ratio of two standard deviations, σ_2/σ_1, is obtained by taking the square roots of the confidence limits for the variance ratio in (13.32). For the property appraisal example, the 90 percent confidence interval is:

    $$0.72 = \sqrt{0.523} \leq \frac{\sigma_2}{\sigma_1} \leq \sqrt{8.67} = 2.94$$

2. To obtain the confidence limits in (13.32), we use theorem (13.31) and the definition of percentiles of the F distribution to make the following probability statement:

    $$P\left[F(\alpha/2; n_1 - 1, n_2 - 1) \leq \frac{s_1^2/\sigma_1^2}{s_2^2/\sigma_2^2} \leq F(1 - \alpha/2; n_1 - 1, n_2 - 1) \right] = 1 - \alpha$$

    Observe that α is equally divided between the two tails, as has been our practice in constructing two-sided confidence intervals. Rearranging the inequalities, we obtain:

    $$P\left[F(\alpha/2; n_1 - 1, n_2 - 1)\frac{s_2^2}{s_1^2} \leq \frac{\sigma_2^2}{\sigma_1^2} \leq F(1 - \alpha/2; n_1 - 1, n_2 - 1)\frac{s_2^2}{s_1^2} \right] = 1 - \alpha$$

    yielding the $1 - \alpha$ confidence limits for σ_2^2/σ_1^2.

Statistical Tests for σ_2^2/σ_1^2

A variety of statistical tests concerning σ_2^2/σ_1^2 may be constructed. Here, we consider tests where the α risk is controlled at $\sigma_2^2 = \sigma_1^2$ or, equivalently, at $\sigma_2^2/\sigma_1^2 = 1$. All tests are based on the same test statistic, denoted by F^*, which follows the F distribution when $\sigma_2^2 = \sigma_1^2$. Hence, percentiles of the F distribution are used in constructing the decision rule.

(13.33)

Tests concerning σ_2^2/σ_1^2, when the populations are normal or approximately normal and the random samples are independent, are based on the test statistic:

$$F^* = \frac{s_2^2}{s_1^2}$$

where:

α risk is controlled at $\sigma_2^2 = \sigma_1^2$ or, equivalently, at $\sigma_2^2/\sigma_1^2 = 1$

When $\sigma_2^2 = \sigma_1^2$, F^* follows the F distribution with $n_2 - 1$ degrees of freedom in the numerator and $n_1 - 1$ degrees of freedom in the denominator.

We illustrate the use of test statistic F^* with a one-sided upper-tail test.

EXAMPLE

In the property appraisal example, the behavioral scientist had hypothesized that group discussion would reduce the variability of appraisals—that is, it would make σ_1^2 smaller than σ_2^2. The appropriateness of this hypothesis is to be tested.

Step 1. The test alternatives here are:

$$H_0: \frac{\sigma_2^2}{\sigma_1^2} \leq 1$$

$$H_1: \frac{\sigma_2^2}{\sigma_1^2} > 1$$

Note that alternative H_1 is equivalent to $\sigma_2^2 > \sigma_1^2$.

Step 2. The scientist wishes to control the α risk at 0.05 when $\sigma_2^2/\sigma_1^2 = 1$ or, equivalently, when $\sigma_2^2 = \sigma_1^2$.

Step 3. Figure 13.5 shows the appropriate type of decision rule for this test. The sampling distribution of F^* in Figure 13.5 is the $F(n_2 - 1, n_1 - 1)$ distribu-

FIGURE 13.5 Illustration of the statistical decision rule for the property appraisal example—Independent samples, populations normal or nearly normal, upper-tail test for σ_2^2/σ_1^2

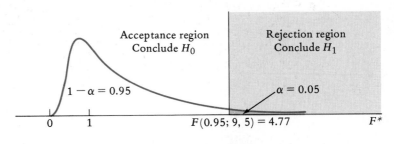

$$F(1 - \alpha; n_2 - 1, n_1 - 1) = F(0.95; 9, 5) = 4.77$$

tion because theorem (13.31) tells us that test statistic F^* follows this F distribution when $\sigma_1^2 = \sigma_2^2$, the case for which the α risk is to be controlled. [Note that when $\sigma_1^2 = \sigma_2^2$, the denominators in the two ratios in (13.31) cancel, yielding the test statistic F^* with the roles of n_1 and n_2 reversed.]

If s_2^2 is substantially larger than s_1^2, the scientist should conclude H_1, that is, conclude $\sigma_2^2/\sigma_1^2 > 1$. But s_2^2 being much larger than s_1^2 is equivalent to $F^* = s_2^2/s_1^2$ being large. Thus, large values of F^* lead to conclusion H_1. The appropriate type of decision rule is therefore one-sided upper-tail, as shown in Figure 13.5. Since $\alpha = 0.05$ here, the action limit must correspond to the 95th F percentile:

$$F(1 - \alpha; n_2 - 1, n_1 - 1) = F(0.95; 9, 5) = 4.77$$

The decision rule for the test is therefore as follows:

If $F^* \leq 4.77$, conclude H_0.
If $F^* > 4.77$, conclude H_1.

Step 4. For the two samples, we have $s_1^2 = 77.9$ and $s_2^2 = 194.2$. Hence:

$$F^* = \frac{194.2}{77.9} = 2.49$$

Since $F^* = 2.49 \leq 4.77$, the scientist should conclude H_0—that group discussion does not reduce the variability of appraisals.   □

Need for Normality

The inferential procedures for a population variance and the comparison of two population variances described in this section are exact when the underlying population is

FIGURE 13.6   Logarithmic transformation making skewed distribution more normal

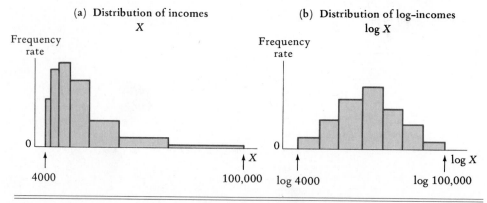

normal. These procedures yield approximately correct results when departures from normality are small, but under moderate and large departures, the results can be quite unreliable.

Fortunately, when nonnormal populations are encountered, several alternative techniques are available. One is to use mathematical transformations on the sample data. For example, a logarithmic transformation is frequently employed on income and other right-skewed data because such data in the form of log X_i generally are more normal than the untransformed data X_i. Figure 13.6 illustrates this point. Figure 13.6a shows a histogram of the income distribution for persons under 35 years old in a community. Figure 13.6b presents the same data on a log X scale. Clearly, the logarithmic transformation has resulted here in a much more symmetric distribution.

A second alternative is to employ an inferential procedure that does not require normality of the population. One such procedure is the jackknife procedure discussed in Chapter 15.

PROBLEMS

13.1 A company with a large number of salespersons wished to compare the mean automobile expenses per salesperson for the fiscal year just ended with the mean expenses for the preceding fiscal year. A random sample of salespersons was selected, and the sample mean automobile expenses were obtained for each of the two years. The difference in sample means was then used by the company for its desired comparison.

   a. Did the design here use matched samples or independent samples? Explain.
   b. Automobile expenses tend to differ among salespersons because of differences in their territories. Does this fact imply that the type of design identified in **a** offers an advantage in estimating the difference in mean automobile expenses per salesperson for the two years? Explain.

13.2 In each of the following studies, indicate whether independent samples or matched samples are employed.

   a. Eighty technician trainees are randomly divided into two equal groups. One group receives field training only, while the second group receives a mixture of field and classroom training. The mean skill level of each group of trainees is measured at the end of the training period to compare the effectiveness of the two training systems.
   b. For a comparison of the color permanence of two industrial paints, ten wood panels are each painted half with one paint and half with the other paint. The mean amount of color fading per panel is measured for each paint after exposure to intense white light at a fixed temperature for two months.

*13.3 **Weather Environments.** Annual snowfall in a northern city averaged 284 centimeters in the past 21 years (period 2), with a standard deviation of 66 centimeters. In the 21 years before that (period 1), the average annual snowfall was 264 centimeters, with a standard deviation of 65 centimeters. Assume that the two sets of observations constitute independent random samples from normal populations with the same variance.

   a. Calculate s_c^2. What parameter does this statistic estimate?
   b. Estimate $\mu_2 - \mu_1$, the difference in mean annual snowfalls for the weather environments prevailing in these two periods, using a 90 percent confidence interval. Interpret your confidence interval.

13.4 **Tread Life.** An operations analyst wished to estimate the difference in mean tread life for a certain make of automobile tire when it is inflated to the standard pressure and when it is inflated to a higher pressure to improve gas mileage. She selected two independent random samples of 15 tires each from the production process, inflated the tires in sample 1 to the standard pressure and the tires in sample 2 to the higher pressure, and conducted tread life tests for all tires. The results, expressed in thousands of miles of tread life, follow. The size of sample 1 turned out to be $n_1 = 14$ because one of the tires was defective and the testing could not be completed.

Standard Pressure	Higher Pressure
$n_1 = 14$	$n_2 = 15$
$\overline{X} = 43.0$	$\overline{Y} = 40.7$
$s_1 = 1.1$	$s_2 = 1.3$

Assume that both populations are normal with the same variance.
   a. Identify the two populations here, taking into account the exclusion of the defective tire from the study.
   b. Estimate $\mu_2 - \mu_1$, the difference in mean tread life for the two tire pressures, using a 95 percent confidence interval. Interpret your confidence interval.

13.5 **Pain Relief.** In a small-scale experiment to compare the pain-relieving effectiveness of two new medications for arthritis sufferers, 27 volunteers were divided at random into two groups of sizes $n_1 = 14$ and $n_2 = 13$. The volunteers in each group were given small but comparable doses of medication. In the group receiving medication 1, the mean hours of relief were $\overline{X} = 6.4$ and the standard deviation was $s_1 = 1.4$. The comparable results for the group given medication 2 were $\overline{Y} = 9.3$ and $s_2 = 1.4$. Assume that both populations are normal with the same variance. Estimate $\mu_2 - \mu_1$, the difference in mean hours of relief for the two medications, using a 95 percent confidence interval. Interpret your confidence interval.

*13.6 **Professional Salaries.** In a study of salaries in a large corporation, random samples of 150 professional employees were selected independently from each of two large departments. The following salary statistics were obtained.

Accounting Department	Engineering Department
$n_1 = 150$	$n_2 = 150$
$\overline{X} = \$37,250$	$\overline{Y} = \$39,212$
$s_1 = \$5541$	$s_2 = \$5356$

   a. Construct a 90 percent confidence interval for $\mu_2 - \mu_1$, the difference in mean salaries for the two departments. Interpret your confidence interval.
   b. Why is the confidence coefficient for the confidence interval in **a** only approximately 90 percent?

13.7 **Electricity Use.** A simple random sample of 400 households from a large community was selected to estimate the mean electricity use per household during February of last year. Another simple random sample of 450 households was selected, independently of the first, to estimate mean electricity use during February of this year. The sample results (expressed in kilowatthours) were as follows:

Last Year	This Year
$n_1 = 400$	$n_2 = 450$
$\bar{X} = 1252$	$\bar{Y} = 1330$
$s_1 = 257$	$s_2 = 251$

Construct a 99 percent confidence interval for $\mu_2 - \mu_1$. Can you conclude that the mean use per household has changed between the two years? If so, was the change small or large? Discuss.

13.8 Business Directory. Two versions of a business advertising directory were distributed to two independent random samples of 300 households in a large community, one version being distributed to each sample. The two versions differed in the layouts of their contents. Subsequently, each sample household was asked to rate the usefulness of its version of the directory on a 9-point scale from 1 (not useful) to 9 (very useful). The sample results were as follows:

Version 1	Version 2
$n_1 = 300$	$n_2 = 300$
$\bar{X} = 5.83$	$\bar{Y} = 5.21$
$s_1 = 2.04$	$s_2 = 2.19$

Construct a 95 percent confidence interval for $\mu_2 - \mu_1$. Interpret your confidence interval.

*13.9 Refer to **Weather Environments** Problem 13.3. It is desired to test whether or not the mean annual snowfall is the same in the two weather environments. The α risk is to be controlled at 0.10 when $\mu_2 - \mu_1 = 0$.
  a. Conduct the test. State the alternatives, the decision rule, the value of the test statistic, and the conclusion.
  b. What is the P-value of this test?

13.10 Refer to **Tread Life** Problem 13.4. The operations analyst wishes to test whether or not the mean tread life for tires inflated to the higher pressure is less than that for tires inflated to the standard pressure. The α risk is to be controlled at 0.025 when $\mu_2 - \mu_1 = 0$.
  a. Conduct the test. State the alternatives, the decision rule, the value of the test statistic, and the conclusion.
  b. What is the P-value of this test?

13.11 Refer to **Pain Relief** Problem 13.5. The experimenter wishes to test whether or not the two medications give the same mean hours of relief. The α risk is to be controlled at 0.05 when $\mu_2 - \mu_1 = 0$.
  a. Conduct the test. State the alternatives, the decision rule, the value of the test statistic, and the conclusion.
  b. What is the P-value of this test?

*13.12 Refer to **Professional Salaries** Problem 13.6. It is desired to test whether or not the mean salaries in the engineering and accounting departments are equal.
  a. Conduct the test, controlling the α risk at 0.10 when $\mu_2 - \mu_1 = 0$. State the alternatives, the decision rule, the value of the test statistic, and the conclusion.
  b. What is the P-value of this test?

13.13 Refer to **Electricity Use** Problem 13.7.
  a. Test whether or not the mean household use of electricity increased from February of last year to February of this year. Control the α risk at 0.005 when $\mu_2 - \mu_1 = 0$. State the alternatives, the decision rule, the value of the test statistic, and the conclusion.
  b. What is the P-value of this test? Is it consistent with the conclusion reached in **a**?

13.14 Refer to **Business Directory** Problem 13.8.

   a. Test whether or not the mean ratings for the two directories are the same, controlling the α risk at 0.05 when $\mu_2 - \mu_1 = 0$. State the alternatives, the decision rule, the value of the test statistic, and the conclusion.

   b. What is the P-value of this test?

*13.15 **Oil Additive.**   In a test involving highway cruising at 55 mph, eight new cars of a certain make were each driven 1000 miles without an oil additive and 1000 miles with an oil additive. The order of the two runs was randomized. Let X_i and Y_i denote, respectively, the miles per gallon for the ith car without and with the oil additive. Data for the differences $D_i = Y_i - X_i$ follow.

i:	1	2	3	4	5	6	7	8
D_i:	1.87	1.71	2.38	2.19	1.89	1.96	1.76	2.00

Assume that the population of differences is normal.

   a. Construct a 90 percent confidence interval for μ_D, the mean difference per car in miles per gallon with and without the additive. Interpret your confidence interval.

   b. Does the confidence interval indicate that the oil additive has an effect?

13.16 **Achievement Scores.**   An educational psychologist is studying differences in mathematics achievement scores produced by two tests (A and B). Eighteen subjects randomly selected were each given both tests, half taking test A first and the remainder taking test B first. The results were as follows, where X and Y denote the scores for tests A and B, respectively.

Subject	Test Score		Subject	Test Score		Subject	Test Score	
i	Y_i	X_i	i	Y_i	X_i	i	Y_i	X_i
1	81	93	7	103	94	13	123	97
2	103	70	8	102	87	14	98	73
3	88	79	9	111	119	15	124	99
4	85	91	10	114	111	16	113	108
5	84	83	11	109	95	17	80	77
6	100	106	12	91	93	18	107	98

Assume that the population of test score differences $D_i = Y_i - X_i$ is approximately normal.

   a. Construct a 95 percent confidence interval for $\mu_2 - \mu_1$, the mean difference in scores for the two test procedures. Interpret your confidence interval.

   b. Can you conclude from your interval estimate in **a** that the mean scores for the two test procedures differ? Explain.

13.17 **Diagnostic Device.**   In an experiment, two equally proficient specialists each diagnosed 20 representative types of problems that tend to arise in complex electronic circuits. For each problem, one of the specialists (chosen at random) used a newly developed diagnostic procedure utilizing a microprocessor chip, while the other specialist used the standard procedure. The difference in diagnosis time (standard procedure minus new procedure) was obtained for each of the 20 problems. The sample mean difference in diagnosis time per problem was $\overline{D} = 87$ minutes. The standard deviation of the time differences for the 20 problems was $s_D = 28$ minutes. Assume that the population of time differences is normal.

   a. Construct a 99 percent confidence interval for the mean diagnosis time difference per problem. Interpret your confidence interval.

b. If it had been found that the problem took longer to diagnose with the standard procedure than with the new procedure in 17 of the 20 diagnoses, would this fact imply that the assumption of normality might not apply? That the confidence interval in **a** might not be valid? Comment.

*13.18 **Bond Yields.** A financial analyst investigated whether bond yields are affected by the presence of a certain sinking fund provision in the bond indenture. He gathered data on 160 pairs of bond issues selected randomly from bond issues in the past ten years. Each pair was identical with respect to coupon rate, year of maturity, and investment quality rating, but only one bond issue in each pair had the sinking fund provision. Taking the difference $D_i = Y_i - X_i$ for each pair between the yield of the bond issue with the provision (Y_i) and the yield of the bond issue without it (X_i), the analyst obtained a mean difference in yields of $\overline{D} = -0.18$ percent point and a standard deviation of differences of $s_D = 0.46$ percent point.

   a. Obtain a 98 percent confidence interval for the mean difference in yields with and without the provision. Interpret your confidence interval.

   b. The analyst would obtain the same mean yield difference if he simply calculated the mean yield of the 160 issues with the provision, then calculated the mean yield of the 160 issues without the provision, and, finally, took the difference between the two means. What is the likely benefit of matching pairs of bond issues?

13.19 **Property Appraisals.** A municipality requires that each residential property seized for nonpayment of taxes be appraised independently by two licensed appraisers before it is disposed. In the past 24 months, appraisers Smith and Jones independently appraised 50 such properties. The difference in appraisals $D_i = Y_i - X_i$ was calculated for each sample property, where X_i and Y_i denote Smith's and Jones' appraised values, respectively. The mean and the standard deviation of the 50 differences were $\overline{D} = \$1.21$ thousand and $s_D = \$2.61$ thousand, respectively.

   a. Construct a 95 percent confidence interval for the mean difference in appraisal values for these two appraisers.

   b. To what population of differences does your confidence interval apply?

13.20 **Isotope Dating.** An archeologist obtained a random sample of 130 artifacts, which he dated by using two different destructive isotope procedures. Two small fragments were broken from each artifact and subjected to the two dating procedures. The age differences for the sample of 130 artifacts had a mean and a standard deviation of $\overline{D} = 53$ and $s_D = 680$ years, respectively.

   a. Construct a 90 percent confidence interval for the mean difference in ages according to the two isotope-dating procedures. Interpret your confidence interval.

   b. Does your confidence interval in **a** indicate that the two dating procedures, on the average, furnish different ages? Discuss.

*13.21 Refer to **Oil Additive** Problem 13.15.

   a. Test whether or not the oil additive increases the mean gas mileage. Control the α risk at 0.05 when $\mu_D = 0$. State the alternatives, the decision rule, the value of the test statistic, and the conclusion.

   b. What is the P-value of the test?

13.22 Refer to **Achievement Scores** Problem 13.16.

   a. Test whether or not the mean achievement score for test B is higher than that for test A. Control the α risk at 0.025 when $\mu_2 - \mu_1 = 0$. State the alternatives, the decision rule, the value of the test statistic, and the conclusion.

   b. What is the P-value of the test?

   c. If all the subjects had taken test A first and then test B, could one conclude that a difference in mean test scores is necessarily due to differences in the two tests? Explain.

13.23 Refer to **Diagnostic Device** Problem 13.17.
     a. Test whether or not the mean diagnosis times for the new and standard procedures differ. Control the α risk at 0.01 when $\mu_D = 0$. State the alternatives, the decision rule, the value of the test statistic, and the conclusion.
     b. What is the P-value of the test?

*13.24 Refer to **Bond Yields** Problem 13.18. The analyst hypothesized that the sinking fund provision reduces mean bond yields.
     a. Test whether or not this hypothesis is true. Control the α risk at 0.01 when $\mu_D = 0$. State the alternatives, the decision rule, the value of the test statistic, and the conclusion.
     b. What is the P-value of the test?

13.25 Refer to **Property Appraisals** Problem 13.19. An observer who has not seen the actual comparative appraisal data suspects that Jones' appraised values are higher, on the average, than Smith's.
     a. Test whether or not this hypothesis is true. Control the α risk at 0.025 when $\mu_D = 0$. State the alternatives, the decision rule, the value of the test statistic, and the conclusion.
     b. What is the P-value of the test?

13.26 Refer to **Isotope Dating** Problem 13.20.
     a. Test whether or not the ages given to an artifact by the two dating procedures, on the average, are the same. Control the α risk at 0.10 when $\mu_D = 0$. State the alternatives, the decision rule, the value of the test statistic, and the conclusion.
     b. What is the P-value of the test? Is this P-value consistent with the conclusion reached in **a**?
     c. If $\mu_D = 0$, does this imply that the two dating procedures give the same age to an artifact? Explain.

13.27 A researcher selected a random sample of 50 families from the population of two-parent families having one or more children enrolled in schools in a large school district. She asked each parent separately whether he or she favors a six-week extension in the school year to improve academic coverage. The researcher now wishes to test whether the proportions of fathers and mothers in the population who favor the extension are the same. Has the researcher obtained independent random samples of fathers and mothers from the population? Discuss.

*13.28 **Divorces and Annulments.** A random sample of $n_1 = 1000$ recent divorces and annulments in a state contained 437 that involved no children. In a neighboring state, an independent random sample of $n_2 = 1200$ divorces and annulments contained 538 that involved no children. Let p_1 and p_2 denote, respectively, the proportions of divorces and annulments involving no children in the first and second states.
     a. Calculate $\bar{p}_2 - \bar{p}_1$ and $s\{\bar{p}_2 - \bar{p}_1\}$. What does each of these statistics estimate?
     b. Construct a 90 percent confidence interval for $p_2 - p_1$.
     c. Does it appear from your confidence interval in **b** that the two state proportions are close to each other? Explain.

13.29 **Employee Stability.** The following data show the number of business school graduates hired by a firm three years ago who are still with the firm, classified by degree.

Degree	Number Hired	Number Remaining
1. Bachelor's	210	141
2. Master's	150	48

Assume that the numbers remaining, 141 and 48, are outcomes of independent binomial random variables. Construct a 95 percent confidence interval for $p_2 - p_1$, the difference in the probabilities of master's and bachelor's degree graduates remaining with the firm for three years. Interpret your confidence interval.

13.30 Environmental Amendment. A public opinion research institute, taking independent random samples of $n_1 = 485$ males and $n_2 = 485$ females in a state, asked each respondent whether he or she favored the continuation of an environmental amendment in the state constitution. It was found that 335 of the males and 354 of the females were in favor of continuation.

    a. Construct a 90 percent confidence interval for $p_2 - p_1$, the difference in the proportions of females and males in the state who favor continuation of the amendment.

    b. Does your confidence interval in **a** indicate that the proportions of females and males in the state who are in favor of continuation are different or the same? If the proportions differ, might this difference be large? Comment.

13.31 Microwave Ovens. A manufacturer of consumer products obtained data on breakdowns in two makes of portable turntables designed to rotate food in microwave ovens. In the sample consisting of $n_1 = 197$ turntables of make 1, it was found that 53 broke down within one year after the expiration of the 90-day warranty. The comparable figure in the sample consisting of $n_2 = 290$ units of make 2 was 38. Assume that the samples are independent random samples from their respective populations. Obtain a 99 percent confidence interval for $p_2 - p_1$, the difference in the proportions of units that break down within one year after the expiration of the 90-day warranty for the two makes. Interpret your confidence interval.

*13.32 Refer to **Divorces and Annulments** Problem 13.28. An analyst wishes to test whether or not the two state proportions are equal.

    a. Obtain $\bar{p}'$. What does this statistic estimate here?

    b. Conduct the test, controlling the α risk at 0.10 when $p_2 - p_1 = 0$. State the alternatives, the decision rule, the value of the test statistic, and the conclusion.

    c. What is the P-value of the test?

13.33 Refer to **Employee Stability** Problem 13.29.

    a. Test whether or not the probability of a bachelor's degree graduate remaining with the firm for three years or longer exceeds that of a master's degree graduate. Control the α risk at 0.025 when $p_2 - p_1 = 0$. State the alternatives, the decision rule, the value of the test statistic, and the conclusion.

    b. What is the P-value of the test?

13.34 Refer to **Environmental Amendment** Problem 13.30.

    a. Test whether or not the proportions of females and males favoring continuation of the amendment are the same. Control the α risk at 0.10 when $p_2 - p_1 = 0$. State the alternatives, the decision rule, the value of the test statistic, and the conclusion.

    b. What is the P-value of the test?

13.35 Refer to **Microwave Ovens** Problem 13.31.

    a. Test whether or not the proportion breaking down within one year for make 1 is larger than the proportion for make 2. Control the α risk at 0.005 when $p_2 - p_1 = 0$. State the alternatives, the decision rule, the value of the test statistic, and the conclusion.

    b. What is the P-value of the test?

*13.36 Active Ingredient. A pharmaceutical firm produces tablets that are supposed to contain a consistent amount of active ingredient. A random sample of 41 tablets just taken from the production process has a standard deviation $s = 1.09$ milligrams of active ingredient per tablet. Assume that the distribution of amounts of active ingredient per tablet is normal.

    a. Construct a 95 percent confidence interval for the process variance of amounts of active ingredient in the tablets.

   b. Convert your interval estimate in **a** into a 95 percent confidence interval for the process standard deviation. Interpret your confidence interval.

   c. Why are the confidence limits in **a** not equally spaced about s^2?

13.37 **Fish Hatchery.** The standard deviation of the lengths of 25 mature trout raised on an experimental diet in a hatchery was $s = 4.35$ centimeters. Assume that the distribution of lengths of trout is normal.

   a. Construct a 99 percent confidence interval for the population variance.

   b. Convert your interval estimate in **a** into a 99 percent confidence interval for the population standard deviation. Interpret your confidence interval.

   c. Why are the confidence limits in **a** not equally spaced about s^2?

13.38 Refer to **Oil Additive** Problem 13.15.

   a. Construct a 95 percent confidence interval for the variance σ_D^2 of the population of differences.

   b. Convert your confidence interval in **a** into a confidence interval for σ_D. Interpret this confidence interval.

13.39 **Heat Diffusion.** A scientist studied heat diffusion in metals by measuring the temperature at one end of a metal rod 10 seconds after applying a standard heat source at the opposite end of the rod. The following temperature readings were obtained in five replications of the experiment.

Replication:	1	2	3	4	5
Temperature (°C):	216	221	220	218	225

The scientist wishes to obtain an interval estimate of the population standard deviation of the temperature readings. Assume that the five readings constitute a random sample from a normal population.

   a. Construct a 99 percent confidence interval for σ^2.

   b. Convert the confidence interval in **a** into a confidence interval for σ. Interpret this confidence interval.

*13.40 Refer to **Active Ingredient** Problem 13.36. Quality standards require that the process variance for the amount of active ingredient be 1.10 or less. Test the alternatives H_0: $\sigma^2 \leq 1.10$ versus H_1: $\sigma^2 > 1.10$, controlling the α risk at 0.025 when $\sigma^2 = 1.10$. State the decision rule, the value of the test statistic, and the conclusion.

13.41 Refer to **Fish Hatchery** Problem 13.37. The variance of lengths of trout raised on a standard diet is 16.32. It is desired to test whether or not the population variance of the lengths of trout raised on the experimental diet equals 16.32. The α risk is to be controlled at 0.01 when $\sigma^2 = 16.32$. Conduct the test. State the alternatives, the decision rule, the value of the test statistic, and the conclusion.

13.42 Refer to **Oil Additive** Problem 13.15. Test whether or not the variance σ_D^2 of the population of differences is less than 0.063. Control the α risk at 0.025 when $\sigma_D^2 = 0.063$. State the alternatives, the decision rule, the value of the test statistic, and the conclusion.

13.43 Refer to **Heat Diffusion** Problem 13.39. Theoretical calculations show that the population variance of temperature readings in this experiment should be 4.00. Test whether or not the theory is supported by the experimental findings. Control the α risk at 0.01 when $\sigma^2 = 4.00$. State the alternatives, the decision rule, the value of the test statistic, and the conclusion.

*13.44 Refer to **Active Ingredient** Problem 13.36. A second random sample of $n_2 = 31$ tablets selected independently from the production process a week later has a standard deviation of

$s_2 = 1.21$ milligrams of active ingredient per tablet. Assume that the distribution of amounts of active ingredient per tablet is still normal.

    a. Construct a 98 percent confidence interval for σ_2^2/σ_1^2, the ratio of the process variances for the two weeks.

    b. Convert the interval estimate in **a** into a 98 percent confidence interval for the ratio of the process standard deviations. Interpret this confidence interval.

13.45 Technicians. Two technicians have made measurements of impurity levels in specimens from a standard solution. One technician measured 11 specimens and the other measured 9 specimens. It is desired to test whether or not the measurements of impurity levels have the same variance for both technicians. The technicians' sets of measurements can be assumed to be independent random samples from normal populations. The sample results (in parts per million) are as follows:

Technician (i)	n_i	s_i
1	11	38.6
2	9	21.7

    a. Construct a 90 percent confidence interval for σ_2^2/σ_1^2, the ratio of the variances of the technicians' measurements for specimens from this standard solution.

    b. Does your confidence interval in **a** indicate that the measurements made by one technician differ in variability from those made by the other? Explain.

    c. If the measurements by one of the technicians were not close to normally distributed because of occasional large measurement errors, might the actual confidence level of the interval in **a** differ substantially from the stipulated level of 0.90? Explain.

13.46 The profit outcomes (in $000) of two independent plays of a business game by team 1 were 68 and 96. Two independent plays of the game by team 2 yielded profit outcomes of 181 and -13 (a loss). Assume that profit outcomes for each team are normally distributed.

    a. Construct a 90 percent confidence interval for σ_2^2/σ_1^2, the ratio of variances of profit outcomes for this game for the two teams.

    b. Convert the confidence interval in **a** to one for the ratio of standard deviations. Is this sparse evidence sufficient to conclude that the profit outcomes for the two teams differ in their variability? Comment.

*13.47 Refer to **Active Ingredient** Problems 13.36 and 13.44. Test whether or not the process variance has increased between the first and second weeks, controlling the α risk at 0.01 when $\sigma_1^2 = \sigma_2^2$. State the alternatives, the decision rule, the value of the test statistic, and the conclusion.

13.48 Refer to **Technicians** Problem 13.45. Test whether or not the measurement variances of the two technicians are equal, controlling the α risk at 0.10 when $\sigma_1^2 = \sigma_2^2$. State the alternatives, the decision rule, the value of the test statistic, and the conclusion.

13.49 Refer to **Fish Hatchery** Problem 13.37. It was hypothesized that the variability in lengths of mature trout could be reduced by increasing the protein in the diet. The standard deviation of the lengths of $n_2 = 25$ mature trout raised on a higher-protein diet was found to be $s_2 = 2.76$ centimeters. Assume that the population of trout lengths here also is normally distributed.

    a. Test whether or not the higher-protein diet leads to a smaller variance, controlling the α risk at 0.05 when $\sigma_1^2 = \sigma_2^2$. State the alternatives, the decision rule, the value of the test statistic, and the conclusion.

    b. Describe the nature of a Type II error in the context of the test in **a**. Is it possible that your conclusion in **a** is, in fact, incorrect in the sense of a Type II error? Explain.

EXERCISES

13.50 A panel consists of 300 grain farmers who are surveyed annually about cropping intentions for different grains. Each panel member serves for two years, with half of the panel rotated off each year and replaced by a random sample of farmers who were not on the panel in the previous year. A key use of the panel is to estimate the change in mean acreage per farm that is intended to be planted in wheat between one year and the next.

    a. Denote this change in mean intended wheat acreage per farm by $\mu_2 - \mu_1$. Would either of the interval estimation procedures in (13.10) or (13.19) be appropriate in this situation if used alone? Comment.

    b. If the matched half of the panel data and the independent half of the panel data in any two consecutive years were used to obtain separate sample estimates by means of (13.19) and (13.10), respectively, would the resulting interval estimates be estimating the same population quantity? Comment.

13.51 Demonstrate that (1) s_c^2 in (13.4) is an unbiased estimator of σ^2, (2) $(n_1 + n_2 - 2)s_c^2/\sigma^2$ has a $\chi^2(n_1 + n_2 - 2)$ distribution.

13.52 Let T_1 and T_2 be unbiased estimators of total health expenditures by U.S. families in two successive years, which we denote by τ_1 and τ_2, respectively.

    a. Show that $T_2 - T_1$ is an unbiased estimator of the difference $\tau_2 - \tau_1$.

    b. Show that the more positive the covariance of T_1 and T_2 (other factors being unchanged), the more efficient is $T_2 - T_1$ as an estimator of $\tau_2 - \tau_1$.

    c. Explain the relevance of the results in **a** and **b** to the use of matched samples to estimate the change in total health expenditures between the two years.

13.53 Explain why a pooled estimator is employed in testing whether or not $p_2 - p_1 = 0$, using (13.27), whereas it is not employed in constructing a confidence interval for $p_2 - p_1$, using (13.24).

13.54 Examine whether the confidence limits (13.29) for the population variance σ^2 become more symmetrical about the point estimator s^2 as the sample size increases by considering the case $s^2 = 100$, $1 - \alpha = 0.95$, for $n = 5, 11, 21, 51, 101$. What do you conclude?

13.55 Refer to **Active Ingredient** Problem 13.36. Specifications require that at least 95 percent of tablets have an amount of active ingredient within ± 2.0 milligrams of the mean amount. Test whether or not the process meets this specification, controlling the α risk at 0.01. State the alternatives, the decision rule, the value of the test statistic, and the conclusion. [*Hint:* What value of σ satisfies $2/\sigma = z(0.975)$?]

STUDIES

13.56 Refer to the **Power Cells** data set. Pair each cell in the group 1 to 18 with the corresponding cell from the group 19 to 36 that has the same discharge rate, depth of discharge, and ambient temperature (that is, the matched pairs are 1 and 19, 2 and 20, . . . , 18 and 36). The experimental conditions for the two cells in a matched pair differ only in the charge rate (0.4 for cells 1 to 18, 1.0 for cells 19 to 36). Let Y_i and X_i denote the numbers of cycles before failure for the low charge-rate cell and the high charge-rate cell in a pair, respectively.

    a. Obtain the differences $D_i = Y_i - X_i$ for the 18 matched pairs of cells.

b. On the assumption that the differences in **a** constitute a random sample from an approximately normal distribution, obtain a 95 percent confidence interval for μ_D, the mean difference in number of cycles before failure for the two charge rates.

c. Use your confidence interval in **b** to test the alternatives H_0: $\mu_D = 0$ versus H_1: $\mu_D \neq 0$. What is your conclusion?

d. Let s_2^2 denote the sample variance of the numbers of cycles before failure for cells 1 to 18. Let s_1^2 be the comparable statistic for cells 19 to 36. Why would it be inappropriate to employ (13.33) to test if the corresponding population variances are equal?

13.57 A labor economist is studying the durations of the most recent strikes in the vehicles and construction industries to see whether strikes in the two industries are equally difficult to settle. To achieve approximate normality and equal variances, the economist decided to work with the logarithms (to base 10) of the duration data (expressed in days), and obtained the following sample statistics.

Industry	Number of Strikes	Mean Log-Duration	Standard Deviation of Log-Duration
1. Vehicles	13	0.593	0.294
2. Construction	15	0.973	0.349

The economist believes it is reasonable to view the data as independent random samples.

a. Construct a 90 percent confidence interval for the difference in the mean log-durations of strikes in the two industries. Interpret your interval estimate.

b. What do the antilogarithms of the two confidence limits in **a** represent? [*Hint:* Recall formula (3.25) and the properties of logarithms.]

c. Test whether strikes in the two industries have the same mean log-durations, controlling the α risk at 0.10. State the alternatives, the decision rule, the value of the test statistic, and the conclusion.

d. Test the economist's assumption that the log-durations of strikes in the two industries have equal variances, controlling the α risk at 0.10 when $\sigma_1^2 = \sigma_2^2$. State the alternatives, the decision rule, the value of the test statistic, and the conclusion.

13.58 Sixty-two persons who were seriously overweight were randomly assigned to one of two weight reduction regimens. During the study period, one person in regimen 2 moved out of town. All other persons remained in the study. At the end of the study period, the weight losses were ascertained. The data on weight losses (in kilograms) follow.

Regimen 1						Regimen 2					
15.2	14.4	16.3	13.6	11.7	10.2	16.2	18.6	12.6	17.9	18.2	20.0
12.6	12.6	14.3	16.7	13.0	17.1	17.3	18.3	19.7	15.1	16.8	16.8
11.7	12.9	14.4	14.2	14.5	14.8	15.5	20.5	16.7	16.4	14.8	18.3
12.7	15.1	11.1	10.0	12.1	14.7	16.6	17.9	18.8	16.9	18.2	17.3
11.9	17.2	11.3	13.3	14.6	12.9	16.2	18.2	18.1	15.2	16.5	19.2
13.7											

a. Construct a frequency distribution for each sample, and plot the frequency polygons on the same graph. Do the two sets of data appear to have about the same amount of variability? Do the distributions suggest that the populations are normal? Discuss.

b. Assume that the two populations are normal. Test whether or not the two population variances are equal, controlling the α risk at 0.02 when $\sigma_1^2 = \sigma_2^2$. State the alternatives, the decision rule, the value of the test statistic, and the conclusion.

c. Assume that the two populations are normal with equal variances. Test whether or not the mean weight loss for regimen 2 exceeds that for regimen 1, as had been expected. Control the α risk at 0.05 when $\mu_2 - \mu_1 = 0$. State the alternatives, the decision rule, the value of the test statistic, and the conclusion. [*Hint:* $t(0.95; 59) = 1.671.$]

d. Estimate the difference in the mean weight losses, using a 90 percent confidence interval. Interpret your interval estimate.

e. Obtain a point estimate (which relies on the normality of the two populations) of the probability of a weight loss of 16.0 kilograms or more for each of the two regimens. Does there appear to be a substantial difference between the two probabilities? Comment.

QUALITY CONTROL AND OTHER APPLICATIONS OF SAMPLING

Quality control is an important area of application of the statistical inference procedures discussed in preceding chapters. We begin this chapter by taking up two important quality control methods, namely, acceptance sampling and control charts.

Sampling methods are also used in many other areas of application. Later in this chapter, we consider simple random sampling when the population is finite and the sample size is not small relative to the population size. We also describe other important probability sampling procedures that are widely used, namely, stratified, cluster, and systematic sampling methods.

ACCEPTANCE SAMPLING

Acceptance sampling is an application of statistical testing to quality control. The choice is between *accepting* or *rejecting* a lot of items, such as an incoming shipment of material, an outgoing shipment of finished product, or a batch of clerical work. The lots constitute finite populations; hence, they could be given 100 percent inspection unless the testing is destructive. However, 100 percent inspection can be prohibitively expensive and usually would not detect all defects and errors anyway in view of inspection fatigue and other sources of error. Hence, sampling is frequently used to decide whether the lot should be accepted (H_0) or rejected (H_1). The latter decision may lead to 100 percent inspection of the items, a return of the lot to the vendor, complete verification of the batch of clerical work, or some other action of this nature.

Single Sampling Plans

In acceptance sampling, the quality of an item often is expressed as satisfactory or defective. A *single sampling plan* for such a quality characteristic specifies the random sample size and the action limit for the number of defective items in the sample.

EXAMPLE

Bushings. A firm receives bushings machined to fine tolerances in lots of 10,000 from a supplier. The alternative conclusions for any particular lot are:

$$H_0: p \leq 0.04 \quad \text{(accept lot)}$$
$$H_1: p > 0.04 \quad \text{(reject lot)}$$

where p denotes the proportion of defective bushings in the lot. Here $p_0 = 0.04$ denotes the maximum acceptable proportion of defectives. In acceptance sampling terminology, p_0 is referred to as the *acceptable quality level,* or AQL. In our earlier testing terminology, p_0 is the value of p at which the α risk is to be controlled.

The firm and the supplier have reached an agreement that the following single sampling plan, stated in condensed form in Table 14.1a, be employed:

Select a random sample of 125 bushings from the lot. If 10 or fewer bushings in the sample are defective, accept the lot; if 11 or more are defective, reject the lot.

The maximum number of defectives that leads to acceptance of the lot is 10 here. This value is called the *acceptance number* of the sampling plan. The minimum number of defectives leading to rejection, here 11, is called the *rejection number.*

$\square$

This sampling plan is equivalent to a decision rule for a test of a population proportion. Let X denote the number of defective bushings in the sample of $n = 125$. The plan then corresponds to the following decision rule:

If $X \leq 10$, conclude H_0 (accept lot).
If $X > 10$, conclude H_1 (reject lot).

TABLE 14.1   **Matched single and multiple sampling plans for the bushings example**

(a)
Single sampling plan

Sample Size	Acceptance Number	Rejection Number
125	10	11

(b)
Multiple sampling plan

Sample	Sample Size	Cumulative Sample Size	Acceptance Number	Rejection Number
First	32	32	0	5
Second	32	64	3	8
Third	32	96	6	10
Fourth	32	128	8	13
Fifth	32	160	11	15
Sixth	32	192	14	17
Seventh	32	224	18	19

SOURCE: Military Standard 105D Tables (Ref. 14.1).

Equivalently, since $\bar{p} = X/n$ and $10/125 = 0.08$, the plan corresponds to the following decision rule in terms of $\bar{p}$:

If $\bar{p} \le 0.08$, conclude H_0 (accept lot).
If $\bar{p} > 0.08$, conclude H_1 (reject lot).

Operating Characteristics. In acceptance sampling applications, a characteristic of the sampling plan that is of particular interest is the probability of accepting the lot for different values of p. These probabilities, which we denote by $P(H_0; p)$, are approximately binomial probabilities when the lot size is large relative to the sample size (as is the case in the bushings example) and, hence, are readily computed. The plot of $P(H_0; p)$ for different values of p is called the *operating characteristic curve* of the sampling plan. The operating characteristic curve is the complement of the corresponding rejection probability curve, as can be seen from (11.18).

EXAMPLE

Figure 14.1 shows the operating characteristic curve of the sampling plan in the bushings example. The following are two probabilities on this curve corresponding to p values of 0.04 and 0.16, respectively; recall that H_0 (accept lot) is concluded if $X \le 10$ for the sample of $n = 125$ bushings:

$$P(H_0; p = 0.04) = P(X \le 10; p = 0.04) = 0.9881$$

$$P(H_0; p = 0.16) = P(X \le 10; p = 0.16) = 0.0066$$

FIGURE 14.1 **Operating characteristic curve of the sampling plan in the bushings example**

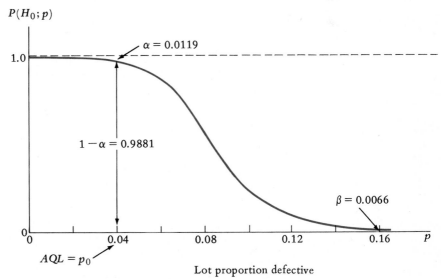

These probability numbers were obtained from a computer routine that calculates cumulative binomial probabilities. Alternatively, the probabilities could have been approximated by the normal distribution. □

The α risk here is the probability of rejecting an acceptable shipment and, for this reason, is often called the *producer's* or *supplier's risk*. On the other hand, the β risk is the probability of accepting an unacceptable shipment and, hence, is often called the *buyer's* or *purchaser's risk*.

EXAMPLE

Referring to the bushings example in Figure 14.1, we see that the supplier's risk when $p = p_0 = 0.04$ equals:

$$\alpha = 1 - P(H_0; p = 0.04) = 1 - 0.9881 = 0.0119$$

We also see that the β risk when $p = 0.16$ equals:

$$\beta = P(H_0; p = 0.16) = 0.0066$$

Thus, the sampling plan exposes the firm to a risk of 0.0119 of rejecting a lot in which the proportion defective is 0.04, and to a risk of 0.0066 of accepting a lot in which the proportion defective is 0.16. □

Multiple Sampling Plans

A *multiple sampling plan* involves sampling in two or more stages. At each stage, a decision is made whether to accept or reject the lot or to continue sampling. The use of multiple sampling plans usually results in smaller total sample sizes than corresponding single sampling plans with the same α and β risks.

EXAMPLE

Table 14.1b contains a multiple sampling plan for the bushings example that has approximately the same operating characteristic curve as the single sampling plan in Table 14.1a. With this multiple sampling plan, an initial sample of 32 bushings is selected. If 0 bushings are defective (the acceptance number), the lot is accepted. If 5 (the rejection number) or more are defective, the lot is rejected. If the number of defectives is 1, 2, 3, or 4, sampling continues and a second sample of 32 bushings is selected. We now consider the *combined* sample of 64 bushings. If 3 or fewer bushings in the combined sample are defective, we accept the lot. If 8 or more are defective, we reject it. If the number of defectives is 4, 5, 6, or 7, a third sample of 32 is selected. This process continues, if necessary, until the seventh sample of 32 is selected. At this stage, the lot is either accepted (18 or fewer defectives in all seven samples combined) or rejected (19 or more defectives).

It is intuitively clear that the total sample size depends on the lot proportion defective p. The multiple sampling plan will detect exceptionally good or bad lots at the first stage, with a resulting sample size of 32 compared with the sample size of 125 used in the single sampling plan. Lots of intermediate quality may require two or more samples before a final decision is reached. Only infrequently, however, will the combined sample size exceed 128 — that is, go beyond the fourth sample. □

Comments
1. When multiple sampling plans provide for sampling one item at each stage, so that an accept, reject, or continue decision is made with each sample observation, they are called *sequential sampling plans*.
2. Extensive tables of sampling plans have been published. One set is the *Military Standard 105D Tables* (Ref. 14.1) from which the sampling plans in Table 14.1 have been extracted. These tables enable the user to select readily a plan to fit the circumstances.

14.2 CONTROL CHARTS

Control charts are a widely used application of statistical tests to the control of quality in ongoing processes, such as manufacturing processes (for example, manufacture of an automobile engine) and processes of service activities (for instance, processing of credit card applications). Control charts can be designed for the quality control of a variety of process characteristics, such as the process mean and the process proportion defective. Charts other than control charts are also employed. We begin with a discussion of control charts for the process mean and then take up other control charts.

Control Chart for Process Mean

Control charts for the process mean can be very helpful in detecting when a change in a process occurs that may require managerial action. We shall utilize the following example to discuss control charts for the process mean.

EXAMPLE

Drug Tablets. In the production of a certain tablet by a pharmaceutical firm, the weight of the tablet is an important quality characteristic to be controlled. Management wishes to know when a change in the process mean weight takes place so that appropriate action can be taken. □

Constructing a Control Chart. To set up a control chart for tablet weight, we must obtain information about the process mean μ and standard deviation σ of the weight of tablets. Extensive past experience indicates that the distribution of tablet weights is approximately normal, with $\mu = 0.441$ gram and $\sigma = 0.008$ gram. These levels are considered satisfactory by management.

Random samples of $n = 5$ tablets will be taken from the process periodically and decisions made, on the basis of the sample mean $\overline{X}$, whether or not the process mean has remained stable. The sampling distribution of $\overline{X}$ will be approximately normal here, even though the sample size is small, because of the approximate normality of the population. Furthermore, we know from (9.3) and (9.4) that the mean and the standard deviation of $\overline{X}$ are $E\{\overline{X}\} = \mu = 0.441$ gram and $\sigma\{\overline{X}\} = \sigma/\sqrt{n} = 0.008/\sqrt{5} = 0.00358$ gram, respectively.

The test to be performed for each sample of $n = 5$ tablets from the process involves the following two-sided alternatives.

(14.1)

H_0: $\mu = \mu_0$ (let process alone)

H_1: $\mu \neq \mu_0$ (look for cause of change in μ)

Here, μ_0 denotes the process mean during the past. We have $\mu_0 = 0.441$ gram, as noted earlier.

The appropriate type of decision rule is the one in Figure 11.5c. For purposes of setting up the control chart, however, the decision rule is expressed in terms of $\overline{X}$, as follows.

(14.2)

If $LCL \leq \overline{X} \leq UCL$, conclude H_0 (let process alone).
If $\overline{X} < LCL$ or $\overline{X} > UCL$, conclude H_1 (look for a cause of change in μ).

where:

$LCL = \mu_0 - 3\sigma\{\overline{X}\}$
$UCL = \mu_0 + 3\sigma\{\overline{X}\}$

Here, LCL and UCL correspond to the lower and upper action limits A_1 and A_2, respectively, in Figure 11.5c. Limits LCL and UCL are called the *lower control limit* and *upper control limit,* respectively. For our example, where $\mu_0 = 0.441$ and $\sigma\{\overline{X}\} = 0.00358$, the control limits are:

$$LCL = 0.441 - 3(0.00358) = 0.430 \text{ gram}$$

$$UCL = 0.441 + 3(0.00358) = 0.452 \text{ gram}$$

Because the sampling distribution of $\overline{X}$ is approximately normal, we know that the probability of $\overline{X}$ falling below the LCL or above the UCL when $\mu = \mu_0$ is negligibly small; in fact, the probability that $\overline{X}$ will be beyond 3 standard deviations from μ_0 when $\mu = \mu_0$ is only 0.003. Thus, decision rule (14.2) has a small α risk; that is, it is unlikely to lead one to look for a cause of change in the process mean when, in fact, no change has occurred.

Comment   We have shown the control limits as being located $z = 3$ standard deviations above and below μ_0 because this range is customary in the United States and Canada. A z value other than 3 is sometimes used, with a consequent change in the α risk of the test.

Plotting Procedure. To facilitate use of decision rule (14.2), as well as for other reasons to be discussed shortly, the rule is usually presented graphically as a *control chart*. Figure 14.2 presents the control chart for the drug tablets example. Note that the level of the process mean μ_0 and the control limits LCL and UCL are plotted as horizontal lines. The vertical axis represents the sample mean weight of tablets $\overline{X}$, and the horizontal axis represents the time over which the periodic samples — in this instance, samples of $n = 5$ tablets each — are taken. Points representing $\overline{X}$ for each sample are plotted on the control chart. As long as the points stay within the control limits, the process is to be let alone. If a point falls outside the control limits, however, it is concluded that the process mean has changed as the result of an assignable cause that was not present previously. A search for the cause is then undertaken, on the basis of which appropriate remedial action is initiated. Points inside the control limits indicate that the process is *in control,* and points that fall outside the control limits indicate that the process is *out of control*.

For the drug tablets example, Table 14.2 gives the sample mean weights for each of 30 samples taken hourly during three days. These $\overline{X}$ values have been plotted on the

FIGURE 14.2   Control chart for the process mean weight for the drug tablets example

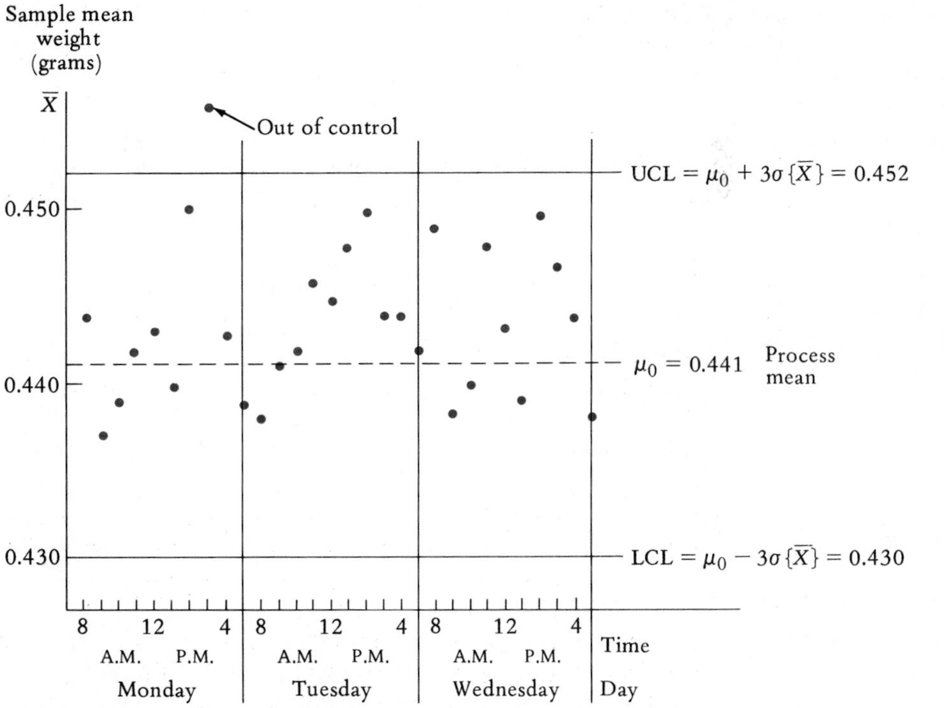

TABLE 14.2   **Mean weights of $n = 5$ tablets in 30 hourly samples for the drug tablets example**

Sample	Day and Hour	Sample Mean Weight ($\overline{X}$; grams)	Sample	Day and Hour	Sample Mean Weight ($\overline{X}$; grams)
1	Mon., 8 A.M.	0.444	16	Tues., 1 P.M.	0.448
2	9	0.437	17	2	0.450
3	10	0.439	18	3	0.444
4	11	0.442	19	4	0.444
5	12	0.443	20	5	0.442
6	1 P.M.	0.440	21	Wed., 8 A.M.	0.449
7	2	0.450	22	9	0.438
8	3	0.456	23	10	0.440
9	4	0.443	24	11	0.448
10	5	0.439	25	12	0.443
11	Tues., 8 A.M.	0.438	26	1 P.M.	0.439
12	9	0.441	27	2	0.450
13	10	0.442	28	3	0.447
14	11	0.446	29	4	0.444
15	12	0.445	30	5	0.438

control chart in Figure 14.2 in the prescribed manner. Note that the process was out of control on Monday at 3 P.M.

Information Provided by Control Chart.   One important type of information furnished by the control chart is the approximate time when a process goes out of control. This information often provides a valuable clue to the assignable cause. For instance, a process may have gone out of control at about the same time that raw material from a new supplier was put into the process. In another instance, a point may have fallen outside the control limits shortly after an operator adjusted the machine.

The periodic sampling of the process assures management that the process will not operate too long after it has gone out of control without this condition being detected. If inspection were undertaken only after the production of large batches of items, the process might turn out many defective items before the condition is discovered.

The control chart furnishes other valuable clues in addition to indicating when a process is out of control. Through the plotting of sample results over a period of time, one can study trends or other changes in the quality of performance. For instance, Figure 14.2 indicates that many of the sample means exceeded $\mu_0 = 0.441$, suggesting a modest upward shift in the process mean and perhaps some reduction in process variability.

Trends and other patterns in the sequence of plotted points may give an early warning of an impending out-of-control situation, and therefore, remedial action might be taken before the process actually goes out of control. Statistical tests should be used, of course, to decide whether a pattern in plotted points (such as a trend) is a real one or simply a chance phenomenon. In the case of trends, we present a test in Chapter 15 that is useful for determining whether or not runs of points upward or downward on a control chart are due to chance.

When an assignable cause of a process change has been discovered and appropriate action taken, changes in the process mean μ, the process standard deviation σ, or both, frequently result. The control chart must then be revised to reflect the new conditions.

Control Chart for Process Standard Deviation

The quality of a product depends not only on the process mean μ but also on the process variability σ, because quality is synonymous with the product characteristic being uniformly close to product specifications or standards. If the process standard deviation increases, for instance, then the product characteristic becomes more variable, and a larger proportion of items will not meet specifications. Hence, it is usually important to control simultaneously both the process mean and the process variability. Process variability is controlled by use of another control chart that is similar to the one described for controlling the process mean. The sample statistic plotted usually is the sample range or the sample standard deviation.

Control Charts for Other Process Characteristics

Control charts also can be constructed for other process characteristics that are related to quality. For instance, the process proportion of items that have a defect or some other unsatisfactory attribute is often an important quality characteristic. We now briefly illustrate two additional types of control charts.

Proportion-Defective Chart.   A proportion-defective chart is a control chart for a process proportion p. The chart is a plot of the sample proportion defective $\bar{p}$ for successive samples. We shall describe the construction of the chart for the case where the samples are sufficiently large that the normal approximation to the sampling distribution of $\bar{p}$ applies. The method of construction is the same in principle as that for the process mean chart. Let p_0 denote the process proportion defective. Using the theory in Chapter 12, we then compute the control limits as follows.

(14.3)
$$\text{LCL} = p_0 - 3\sqrt{\frac{p_0(1 - p_0)}{n}}$$

$$\text{UCL} = p_0 + 3\sqrt{\frac{p_0(1 - p_0)}{n}}$$

EXAMPLE

A mail-order house fills thousands of orders each day. Management wishes to control the proportion of orders that are filled incorrectly. Extensive past experience shows the process proportion to be $p_0 = 0.07$ when the process is under control. The process proportion will be monitored by drawing random samples of $n = 150$ orders from the orders filled each day and noting the sample proportion filled incorrectly.

Figure 14.3 shows a control chart on which are plotted the sample proportions for 18 consecutive days. The lower and upper control limits for this chart were calculated from (14.3) as follows:

$$\text{LCL} = 0.07 - 3\sqrt{\frac{0.07(1 - 0.07)}{150}} = 0.0075$$

$$\text{UCL} = 0.07 + 3\sqrt{\frac{0.07(1 - 0.07)}{150}} = 0.1325$$

The chart shows that no out-of-control points occurred during this span of time.

$\square$

Comments	
1.	The lower control limit for the proportion-defective chart is defined to be zero if the LCL computed according to formula (14.3) turns out to be negative.
2.	The lower control limit can be omitted from a proportion-defective chart when there is no interest in having an out-of-control signal for a very small sample proportion defective. Usually, however, out-of-control points falling below the lower control limit can be helpful, indicating the presence of conditions that are favorable to permanently reducing the process proportion defective p_0. In the mail-order example, for instance, the process proportion of incorrectly filled orders may decrease because of the institution of a training program for the order clerks, the use of a clearer and more detailed order form, or some other favorable cause.

Number-of-Defects Chart.   Some quality control applications involve monitoring the number of events that occur in a unit of activity. Examples include the number of defects in a kilometer of optical fiber and the number of worker accidents per million man-hours worked in a plant. A simple control chart based on the Poisson distribution is useful in these applications.

EXAMPLE

The number of defective castings produced per hour (X) in a metal casting process follows a Poisson distribution. Extensive past experience has shown that the process mean number of defective castings produced per hour is $\lambda_0 = 1.5$ when the process is under control. The control chart in Figure 14.4 shows a plot of X, the number of defective castings in each of 30 consecutive hours. This particular control chart contains only an upper control limit (UCL) because management here is primarily concerned with the process when the number of defective castings per hour is too large.

FIGURE 14.3   **Control chart for the process proportion of incorrectly filled orders**

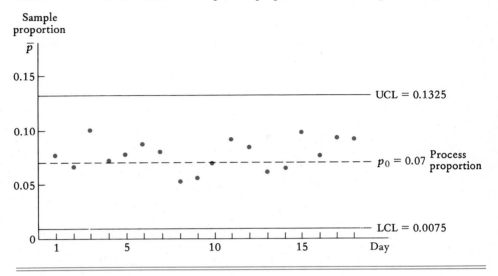

FIGURE 14.4   **Control chart for the process mean number of defective castings produced per hour**

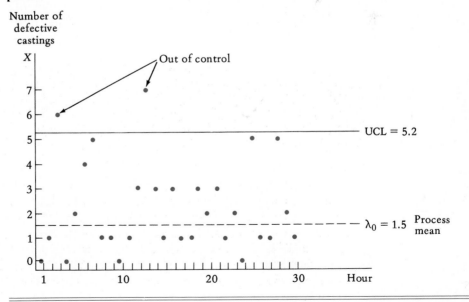

The upper control limit in Figure 14.4 is calculated in the usual fashion as being three standard deviations above the process mean:

$$\text{UCL} = \lambda_0 + 3\sqrt{\lambda_0} = 1.5 + 3\sqrt{1.5} = 5.2$$

Recall from (6.9) and (6.10) that the standard deviation of a Poisson variable is equal to the square root of the mean.

The upper control limit in Figure 14.4 has the property that there is only a small probability that X will exceed the control limit when the process mean number of defectives is equal to λ_0. When $\lambda_0 = 1.5$, the probability that X will exceed $\text{UCL} = 5.2$ is the Poisson probability $P(X \geq 6)$ for parameter value $\lambda = 1.5$. (Remember that the number of defective castings is an integer.) Referring to Table C.6, we find this probability equals:

$$P(X \geq 6) = P(X = 6) + P(X = 7) + P(X = 8) + \cdots$$

$$= 0.0035 + 0.0008 + 0.0001 + \cdots = 0.0044$$

Hence, the probability is only 0.0044 that a point on the control chart will be out of control when, in fact, the process mean number of defective castings is $\lambda_0 = 1.5$.

Figure 14.4 shows that the process was out of control during hours 3 and 13.

□

Cusum Charts

An alternative to conventional control charts is the *cumulative sum chart*, or *cusum chart* for short. The cusum chart, like control charts, is useful for controlling process characteristics, especially for detecting moderate departures of the process characteristic from its standard that are persistent over time. We shall now consider the cusum chart for a process mean to illustrate its use.

Let $\overline{X}_i$ be the sample mean for the ith sample drawn from a process; then, $\overline{X}_i - \mu_0$ is a measure of the departure of the sample mean from the process mean μ_0 at that time point. The sum of these departures for all samples from the first one to the current one (k, say) is the basis of the cusum chart. This sum, denoted by C_k, may be written as follows.

(14.4)
$$C_k = \sum_{i=1}^{k} (\overline{X}_i - \mu_0)$$

where:

$\overline{X}_i$ is the sample mean for the ith sample

μ_0 is the process mean

Plotting the values of C_k against their time points k gives a cusum chart for the process mean.

TABLE 14.3 **Fragment of the cusum calculations for the drug tablets example data in Table 14.2 ($\mu_0 = 0.441$)**

Sample k	Sample Mean $\overline{X}_k$	Sample Departure $\overline{X}_k - \mu_0$	Cumulative Sum $C_k = \sum\limits_{i=1}^{k} (\overline{X}_i - \mu_0)$
1	0.444	0.003	0.003
2	0.437	−0.004	−0.001
3	0.439	−0.002	−0.003
4	0.442	0.001	−0.002
⋮	⋮	⋮	⋮
29	0.444	0.003	0.082
30	0.438	−0.003	0.079

EXAMPLE

Table 14.3 shows a fragment of the cusum calculations for the drug tablets example. Recall that the process mean is $\mu_0 = 0.441$ gram in this example. The mean weights for the samples of tablets are given in Table 14.2. The detailed calculations for the first three cumulative sums are:

$$C_1 = (0.444 - 0.441) = 0.003$$

$$C_2 = (0.444 - 0.441) + (0.437 - 0.441) = 0.003 - 0.004 = -0.001$$

$$C_3 = (0.444 - 0.441) + (0.437 - 0.441) + (0.439 - 0.441)$$

$$= 0.003 - 0.004 - 0.002 = -0.003$$

Figure 14.5 contains the cusum chart. The plotted points have been connected by lines to facilitate visual interpretation. When the process mean remains in control, the consecutive values of C_k will fluctuate about 0. In Figure 14.5, however, the cumulative sum shows a steady trend upward, starting midday Monday and persisting through to the end of Wednesday. The trend confirms what we had noted earlier in the control chart in Figure 14.2, namely, that the sample mean weights $\overline{X}$ tended to be larger than μ_0 throughout most of the period. Note that the cusum chart gives a reasonably early and clear signal that the process mean is out of control by the end of Monday or early Tuesday.   □

Formal methods exist for testing when a process has gone out of control in a cusum plot. These methods are covered by specialized texts in quality control (see, for example, Ref. 14.2).

14.3 SIMPLE RANDOM SAMPLING FROM FINITE POPULATIONS

Throughout the text up to this point, we have assumed the sampling fraction n/N is small whenever the population sampled is finite. There are instances, however, when

FIGURE 14.5   Cusum chart for the cumulative sum of departures of sample mean weights from the process mean ($\mu_0 = 0.441$) for the drug tablets example

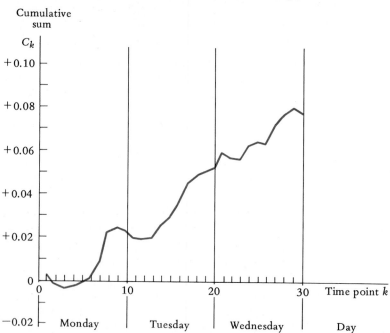

this condition does not hold, for example, when a sample of 5 generators in a population of 15 is sampled to study the working condition of an important safety feature. In cases where the sampling fraction n/N is not small, recognition of the finite nature of the population is essential.

Estimation of Population Mean

The formula for the estimated standard deviation of $\overline{X}$ given by (10.8), that is, $s\{\overline{X}\} = s/\sqrt{n}$, is exact only for the case of an infinite population. When the population is finite and the sampling fraction n/N is small, formula (10.8) still applies approximately. However, when the sampling fraction n/N is not small, the estimated standard deviation $s\{\overline{X}\}$ must be modified as follows.

(14.5)

For finite populations, the estimated variance $s^2\{\overline{X}\}$ and standard deviation $s\{\overline{X}\}$ are:

$$s^2\{\overline{X}\} = \left(1 - \frac{n}{N}\right)\frac{s^2}{n} \qquad s\{\overline{X}\} = \sqrt{1 - \frac{n}{N}}\,\frac{s}{\sqrt{n}}$$

The factor $\sqrt{1 - (n/N)}$ is called the *finite population correction* because the standard deviation $s/\sqrt{n}$ for an infinite population is "corrected" by this factor to yield the proper estimate when the population is finite.

Two characteristics of the finite population correction are noteworthy:

1. For a given sample size n, the finite population correction $\sqrt{1 - (n/N)}$ is nearly equal to 1 when N is large relative to n. It is for this reason that a finite population can be treated as an infinite one when the sampling fraction n/N is small, since $s\{\overline{X}\}$ then equals $s/\sqrt{n}$ for all practical purposes. A working rule often employed is as follows:

    > A finite population can be treated as infinite when $n/N \leq 0.05$, that is, when no more than 5 percent of the population is sampled.

    In all applications considered thus far in the text, the sampling fraction n/N has been small enough for the correction to be ignored.

2. For a given finite population, the finite population correction approaches 0 as n approaches N. Hence, $s\{\overline{X}\} = 0$ when $n = N$. This result is intuitively reasonable since a census of the population is taken when $n = N$, and hence there is no error due to sampling present in $\overline{X}$, that is, $\overline{X} = \mu$ then.

Confidence Interval for μ.   Whenever the sample size is not an exceptionally large portion of the finite population size, the estimation and testing procedures for a population mean μ recognizing the finite nature of the population are the same as when the population can be treated as infinite, except that (14.5) is used to obtain $s\{\overline{X}\}$. We now illustrate the construction of a confidence interval for μ when the sampling fraction n/N is large.

EXAMPLE

A chemical company serves $N = 810$ dry-cleaning establishments. It selected a simple random sample of $n = 140$ of these establishments to estimate the mean annual use per establishment (μ) that would be made of a new solvent product. The survey results gave $\overline{X} = 410$ gallons and $s = 215$ gallons. A 95 percent confidence interval for μ is desired.

The sampling fraction n/N in this case is $140/810 = 0.1728$, which is large enough to make the finite population correction appropriate but not so large as to affect the estimation procedure. Hence, the confidence interval for μ is of the same form as (10.11), except that $s\{\overline{X}\}$ is computed by using (14.5).

We have, from (14.5):

$$s\{\overline{X}\} = \sqrt{1 - \frac{140}{810}} \frac{215}{\sqrt{140}} = 16.526$$

For 95 percent confidence, $z(0.975) = 1.960$ is required. Hence, the confidence limits are $410 \pm 1.960(16.526)$, and the desired confidence interval is $377.6 \leq \mu \leq 442.4$.   □

Planning of Sample Size.   When planning a sample from a finite population, the finite nature of the population must be taken into account when the resulting sample size is not a small fraction of the population size. The appropriate planning formula for a $1 - \alpha$ confidence interval for μ is the following modification of (10.18).

(14.6)
$$n = \frac{z^2\sigma^2}{h^2 + \dfrac{z^2\sigma^2}{N}}$$

where:

$z = z(1 - \alpha/2)$
h is the specified half-width
σ is the planning value for the population standard deviation

The planning formula (14.6) still assumes that the resulting sample size will be large.

EXAMPLE

A random sample from a population of $N = 900$ Canadian general hospitals is to be taken to obtain a confidence interval for the mean number of pediatric beds per hospital, with a half-width of $h = 5$ and a 95 percent confidence coefficient. Results of a previous study suggest that an appropriate planning value of the population standard deviation is $\sigma = 40$ beds. Since $z(0.975) = 1.960$, the needed sample size is:

$$n = \frac{(1.960)^2(40)^2}{(5)^2 + \dfrac{(1.960)^2(40)^2}{900}} = 193$$

Note that $n/N = 193/900 = 0.214$. Thus, over 21 percent of all hospitals must be sampled to obtain an estimate with the desired precision. $\square$

Comment

As the population size N gets large, the second term in the denominator of (14.6) approaches zero, and (14.6) simplifies to (10.18).

Estimation of Population Proportion

Confidence Interval for p.   When estimating a population proportion p where the finite nature of the population must be recognized, the estimated standard deviation $s\{\bar{p}\}$ needs to include the finite population correction.

(14.7)

For finite populations, the estimated variance $s^2\{\bar{p}\}$ and standard deviation $s\{\bar{p}\}$ are:

$$s^2\{\bar{p}\} = \left(1 - \frac{n}{N}\right)\frac{\bar{p}(1-\bar{p})}{n-1} \qquad s\{\bar{p}\} = \sqrt{1 - \frac{n}{N}}\sqrt{\frac{\bar{p}(1-\bar{p})}{n-1}}$$

The construction of the confidence interval for p proceeds as for the case where the population is infinite, using (12.9), except that $s\{\bar{p}\}$ is obtained from (14.7).

EXAMPLE

For a population of $N = 1200$ railroad tank cars, a simple random sample of $n = 200$ cars revealed that 33 had defective discharge valves. A 90 percent confidence interval for the population proportion of tank cars with defective valves is desired. We require $z(0.95) = 1.645$, $\bar{p} = 33/200 = 0.165$, and:

$$s\{\bar{p}\} = \sqrt{1 - \frac{200}{1200}}\sqrt{\frac{0.165(1 - 0.165)}{200 - 1}} = 0.0240$$

The confidence limits therefore are $0.165 \pm 1.645(0.0240)$, and the desired confidence interval is $0.13 \le p \le 0.20$. □

Planning of Sample Size. When the required sample size for estimating a population proportion p will represent a fraction of the population that is not small, the planning formula in (12.10) must be modified as follows.

(14.8)

$$n = \frac{z^2 p(1-p)}{h^2 + \dfrac{z^2 p(1-p)}{N}}$$

where:
$z = z(1 - \alpha/2)$
p is the planning value for the population proportion
h is the specified half-width

Planning formula (14.8) still assumes that the resulting sample size will be large.

EXAMPLE

Union leaders of a local with $N = 850$ members want to survey members' opinions on a proposed change in their labor contract. It is desired to estimate the proportion p supporting the change by a confidence interval with half-width $h = 0.05$ and confidence coefficient 0.99. A planning value of $p = 0.5$ is proposed. Since $z(0.995) = 2.576$, the number of members needed to be included in the sample is:

$$n = \frac{(2.576)^2(0.5)(1 - 0.5)}{(0.05)^2 + \dfrac{(2.576)^2(0.5)(1 - 0.5)}{850}} = 373$$

□

Efficiency of a Sampling Procedure

Simple random sampling is not always the best sampling procedure to employ. Other procedures often are more efficient. The concept of efficiency of a sampling procedure is usually defined in terms of the precision of the estimator, as measured by the width of the confidence interval for a specified confidence coefficient. Every probability sampling procedure can be evaluated in terms of the cost of attaining a given statistical precision, or, alternatively, in terms of the precision furnished for a given cost.

(14.9)

> A probability sampling procedure is said to be more *efficient* than simple random sampling if it offers, at a given level of confidence, the same precision at less cost or greater precision at the same cost.

We now consider a number of probability sampling procedures that, under certain conditions, are more efficient than simple random sampling.

14.4   STRATIFIED SAMPLING

Stratified sampling is a sampling procedure that is widely used.

EXAMPLE

Teachers Association.   A state teachers association studied the educational qualifications of its membership, which consists of 50,000 elementary school, high school, and college teachers. A major characteristic of interest was the mean number of years of formal education (including teacher training) of the association's teachers. The analyst conducting the study knew that the number of years of education tends to vary for teachers by level of school, being greater for teachers in higher-level educational institutions. Hence, the analyst used a sample design in which independent simple random samples of teachers are selected from among the member teachers in each of the three school levels. Column 1 of Table 14.4

TABLE 14.4   Illustration of stratified sampling — Teachers association example

| | | (1) | (2) |
| | | Number of Members | |
Stratum	Level of School	In Population	In Sample
1	Elementary	25,000	150
2	High school	20,000	120
3	College	5,000	30
	Total	50,000	300

shows the total number of member teachers in each school level, and column 2 shows the sample size for each group. ☐

Definition

The sampling procedure used here is called stratified random sampling.

(14.10)

> With *stratified random sampling,* the population is divided into a number of mutually exclusive subpopulations, or *strata,* and independent simple random samples are selected from the strata.

In the example, the 25,000 teachers in elementary schools represent a stratum of the population of 50,000 teachers.

Advantages of Stratified Sampling

There are three major advantages of stratified sampling: (1) efficiency, (2) information about subpopulations, and (3) feasibility.

Efficiency. Frequently, a stratified sample will be much more efficient than a simple random sample. We illustrate this in Table 14.5a, which contains a hypothetical population of nine teachers. The number of years of education is given for each. The population variance is 2.61. This is the variability affecting the precision of the sample mean $\overline{X}$ with simple random sampling.

Suppose we stratify the teachers by level of school, as in Table 14.5b, which also shows the strata variances. Note that the variability of years of education is much smaller within each stratum than in the entire population. Consequently, a stratified random sample of given size will yield a more precise estimate than a simple random sample of the same size, since much smaller variability is encountered within each stratum.

Thus, for stratified sampling to be efficient, the strata must be designed to contain relatively homogeneous elements. Homogeneity is accomplished when the basis of stratification is related to the characteristic under study. In the example, level of school in which members teach is related to number of years of education. Similarly, in a survey of firms to estimate mean charitable contributions per firm, a stratification by size of firm is likely to be highly effective in view of the positive relation between size of firm and amount of contributions.

While the selection of a stratified random sample may cost somewhat more than the selection of a simple random sample of equal size (for example, a separate frame is required for each stratum), this increase in cost is usually more than balanced by the substantially improved precision. Thus, a stratified sample that is designed to yield a specified precision usually is significantly smaller than the corresponding simple random sample and consequently costs less.

TABLE 14.5   Illustration of the effects of stratification—Teachers association example

(a)
Population

Teacher	Level of School	Years of Education
1	Elementary	15.5
2	College	19.0
3	High school	16.5
4	Elementary	16.0
5	High school	18.0
6	College	20.5
7	High school	17.5
8	College	19.5
9	Elementary	16.5

Variance = 2.61

(b)
Strata

Stratum 1 Elementary		Stratum 2 High School		Stratum 3 College	
Teacher	Years of Education	Teacher	Years of Education	Teacher	Years of Education
1	15.5	3	16.5	2	19.0
4	16.0	5	18.0	6	20.5
9	16.5	7	17.5	8	19.5
Variance = 0.17		Variance = 0.39		Variance = 0.39	

Information About Subpopulations.   Stratified sampling can provide estimates of strata characteristics in addition to estimates of overall population characteristics. For example, a study by a university of the effects of tuition increases is to provide separate information for undergraduate and graduate students, as well as information for all students. Stratification by student class status will permit precise estimates for undergraduate and graduate students separately, as well as for all students.

Feasibility.   At times, stratified sampling is the most feasible type of sampling. Consider the sampling of welfare claims in a state. In the one large city of the state, the claims records are in computerized files, whereas elsewhere in the state, the records are kept locally in file cabinets. Administrative considerations here require separate sampling of claims in the city and elsewhere in the state.

Estimation of Population Mean

Notation.   The total number of strata into which the population is divided is denoted by k. The size of the jth stratum (that is, the number of population elements in it) is denoted by N_j. The total population size is denoted as usual by N.

(14.11)
$$N = \sum_{j=1}^{k} N_j$$

Analogously, n_j denotes the size of the sample selected from the jth stratum, and n denotes the total sample size.

(14.12)
$$n = \sum_{j=1}^{k} n_j$$

Finally, $\overline{X}_j$ and s_j^2 denote the mean and the variance, respectively, of the sample drawn from the jth stratum.

In the teachers association example (see Table 14.4), $k = 3$ strata are employed. The stratum size and the sample size for the first stratum (elementary schools) are $N_1 = 25{,}000$ and $n_1 = 150$, respectively. The total population and sample sizes are $N = 50{,}000$ and $n = 300$.

Development of Confidence Interval.   We require a point estimator of the population mean μ.

(14.13)
The stratified sample estimator of the population mean μ, denoted by $\overline{X}_{st}$, is:

$$\overline{X}_{st} = \left(\frac{N_1}{N}\right)\overline{X}_1 + \left(\frac{N_2}{N}\right)\overline{X}_2 + \cdots + \left(\frac{N_k}{N}\right)\overline{X}_k = \sum_{j=1}^{k} \left(\frac{N_j}{N}\right)\overline{X}_j$$

The estimator $\overline{X}_{st}$ is an unbiased estimator of μ. Note that $\overline{X}_{st}$ is simply a weighted average of the strata sample means, with the weights being the proportions of the total population in each stratum.

We can use the theory of random variables developed in Chapter 5 to obtain $\sigma^2\{\overline{X}_{st}\}$, the variance of the sampling distribution of $\overline{X}_{st}$. Since the $\overline{X}_j$ are based on independent random samples, we know from (5.15b) that the variances of the terms in $\overline{X}_{st}$ are additive. Furthermore, by (5.12b), the variance of $(N_j/N)\overline{X}_j$ is $(N_j/N)^2\sigma^2\{\overline{X}_j\}$. Thus:

(14.14)
$$\sigma^2\{\overline{X}_{st}\} = \left(\frac{N_1}{N}\right)^2 \sigma^2\{\overline{X}_1\} + \left(\frac{N_2}{N}\right)^2 \sigma^2\{\overline{X}_2\} + \cdots + \left(\frac{N_k}{N}\right)^2 \sigma^2\{\overline{X}_k\}$$

We must still estimate the variances $\sigma^2\{\overline{X}_j\}$. For this purpose, we employ $s^2\{\overline{X}_j\}$ given by (14.5). Hence, the following is a point estimator of $\sigma^2\{\overline{X}_{st}\}$.

(14.15)
$$s^2\{\overline{X}_{st}\} = \left(\frac{N_1}{N}\right)^2 s^2\{\overline{X}_1\} + \left(\frac{N_2}{N}\right)^2 s^2\{\overline{X}_2\} + \cdots + \left(\frac{N_k}{N}\right)^2 s^2\{\overline{X}_k\}$$

$$= \sum_{j=1}^{k} \left(\frac{N_j}{N}\right)^2 s^2\{\overline{X}_j\}$$

As usual, the finite population correction is omitted for any stratum for which the sampling fraction n_j/N_j does not exceed 5 percent.

To complete the construction of the interval estimate of μ, we note that the sampling distribution of $(\overline{X}_{st} - \mu)/s\{\overline{X}_{st}\}$ is approximately normal when the sample size n is reasonably large. Thus, the confidence limits for μ have the usual large-sample form.

(14.16)
The confidence limits for μ with approximate confidence coefficient $1 - \alpha$, when stratified random sampling is employed and the sample size is reasonably large, are:

$$\overline{X}_{st} \pm zs\{\overline{X}_{st}\}$$

where:

$z = z(1 - \alpha/2)$

$\overline{X}_{st}$ is given by (14.13)

$s\{\overline{X}_{st}\}$ is given by (14.15)

EXAMPLE

For the teachers association example, a 95 percent confidence interval for μ, the mean number of years of education, is desired. Table 14.6 contains the sample results and shows the calculation of $\overline{X}_{st}$. We find that $\overline{X}_{st} = 16.25$ years. This table also shows the calculation of $s^2\{\overline{X}_{st}\}$. Since the strata sizes N_j are large relative to the sample sizes n_j, we have ignored the finite population corrections and used s_j^2/n_j for calculating $s^2\{\overline{X}_j\}$. Since $s^2\{\overline{X}_{st}\} = 0.01545$, it follows that $s\{\overline{X}_{st}\} = \sqrt{0.01545} = 0.1243$ year.

For a 95 percent confidence coefficient, we require $z(0.975) = 1.960$. Hence, the 95 percent confidence limits for μ are $16.25 \pm 1.960(0.1243)$, and the confidence interval is:

$$16.01 \leq \mu \leq 16.49$$

It can be asserted, with 95 percent confidence, that the mean number of years of education of the member teachers is between 16.0 and 16.5. □

Comment	Stratified sampling can also be used to estimate a population proportion p. For example, an economically depressed district of a city might be stratified into groups of contiguous city blocks and a sample of households drawn from each stratum to estimate the proportion of households in the district with inadequate living space. The procedures for estimating the population proportion p are analogous to those for estimating the population mean μ.

Planning Strata Sample Sizes

An important design consideration in stratified sampling is the allocation of the total sample size to the individual strata.

Proportional Allocation. Note from Table 14.6 that the sample size is the same proportion (0.6 percent) of the stratum size for all strata. This method of allocation is called proportional allocation.

(14.17)

With *proportional allocation*, the sample size for the jth stratum is:

$$n_j = \left(\frac{N_j}{N}\right) n$$

Proportional allocation is frequently used because it is simple and often quite effective.

Optimal Allocation. This approach allocates the total sample size to the individual strata so as to minimize $\sigma^2\{\overline{X}_{st}\}$, thereby providing an estimate of μ that is as precise as

TABLE 14.6 **Stratified sample results for the teachers association example**

j	Level of School	N_j	n_j	$\overline{X}_j$	s_j^2	$\left(\dfrac{N_j}{N}\right)\overline{X}_j$	$\left(\dfrac{N_j}{N}\right)^2\left(\dfrac{s_j^2}{n_j}\right)$
1	Elementary	25,000	150	14.8	6.40	7.40	0.01067
2	High school	20,000	120	17.3	2.69	6.92	0.00359
3	College	5,000	30	19.3	3.57	1.93	0.00119
	Total	50,000	300			16.25	0.01545
		↑	↑			↑	↑
		N	n			$\overline{X}_{st}$	$s^2\{\overline{X}_{st}\}$

possible for the given total sample size. A statistical theorem provides the strata sample sizes for optimal allocation.

(14.18)

With *optimal allocation,* the sample size for the jth stratum is:

$$n_j = \left(\frac{N_j \sigma_j}{\sum\limits_{j=1}^{k} N_j \sigma_j} \right) n$$

where:

σ_j is the standard deviation for the jth stratum

Note that (14.18) involves the strata standard deviations σ_j. Since these parameters are generally unknown, planning values for the σ_j are required. Approximate information about the order of magnitude of the σ_j is usually adequate for planning purposes.

EXAMPLE

The analyst for the teachers association considered optimal allocation for the stratified sample. Table 14.7 shows the planning values for the σ_j, based on earlier surveys, together with the calculations for optimally allocating the total sample size $n = 300$ to the three strata. We see that the strata sample sizes for optimal allocation differ somewhat from those for proportional allocation in Table 14.6.

In the end, the analyst did not utilize the optimal allocation sample sizes. The reason was that the sample was intended to provide information about various other characteristics besides the mean number of years of education. It turned out that the optimal allocations for estimating other major characteristics differed markedly from the one here, because the relative magnitudes of the σ_j for the other characteristics were not the same as for years of education. The analyst then decided that proportional allocation represented a reasonable compromise. □

TABLE 14.7   **Optimal allocation of total sample size in stratified sampling for the teachers association example**

j	Level of School	N_j	Planning Value of σ_j	$N_j \sigma_j$	$\dfrac{N_j \sigma_j}{\sum N_j \sigma_j}$	n_j
1	Elementary	25,000	2.5	62,500	0.6098	183
2	High school	20,000	1.5	30,000	0.2927	88
3	College	5,000	2.0	10,000	0.0976	29
	Total	50,000		$\sum N_j \sigma_j = 102,500$	1.000	300

14.5   CLUSTER SAMPLING

Single-Stage Cluster Sampling

Sometimes, it is efficient to sample population elements in groups rather than individually.

EXAMPLE

Inspection.   An inspector needs to examine a shipment to determine the extent of deterioration, if any, of the parts in the shipment. The shipment consists of 2000 sealed cartons, each containing 5 parts. If a simple random sample of 100 parts is to be examined, it is conceivable that as many as 100 cartons may have to be unsealed. Since unsealed cartons are difficult to store and handle, and items in unsealed cartons may be stolen, the inspector wishes to open fewer cartons. His sampling procedure calls for the selection of 20 of the 2000 cartons at random and inspection of all the parts in each selected carton.   □

Definition.   The procedure just described illustrates single-stage cluster sampling. Each carton represents a group or cluster of elements, and a probability sample of these clusters is selected. All elements in the selected clusters are then included in the sample.

(14.19)

> With *single-stage cluster sampling*, the population is divided into *clusters* of elements, a probability sample of clusters is selected, and all elements in the selected clusters are included in the sample.

Efficiency.   In general, cluster sampling requires a larger number of elements to yield a specified precision than simple random sampling. The reason is that the elements within a cluster (persons in a family, parts in a carton) tend to be more homogeneous than is the case in the population at large. Nevertheless, cost considerations under some circumstances make cluster sampling more efficient than simple random sampling:

1.   When the cost of constructing a frame of elements is high, cluster sampling may be more efficient. For example, in a survey of eligible voters in a city, a current listing of city blocks is available but a current listing of eligible voters is not. Here, cluster sampling of city blocks has the advantage of requiring the preparation of voters' lists only for city blocks actually sampled. On the other hand, simple random sampling of eligible voters would require a voters' list for the entire city, which is an expensive undertaking.
2.   Cluster sampling may also be efficient when it is more expensive to sample items scattered throughout the population than items that are "close" to one another. In the inspection example, it is much more costly to open 100 cartons to inspect

100 parts than to inspect the same number of parts in 20 cartons. As another example, a block sample of households in a large city requires substantially less travel time by interviewers than a random sample of households scattered throughout the city.

Multistage Cluster Sampling

In some applications of cluster sampling, it is desirable to sample in two or more stages.

EXAMPLE

Food Expenditures Survey.   In a statewide survey of urban households to obtain data on food expenditures, a probability sample of cities is first selected. Within the chosen cities, probability samples of city blocks are selected, and within the selected blocks, random samples of households are chosen.   ☐

Definition.   The procedure just described illustrates a three-stage cluster sample. Cities are sampled in the first stage and are called primary sampling units. Blocks are sampled in the second stage and are called secondary sampling units. Households within blocks are sampled in a third stage and are called tertiary sampling units.

(14.20)

> With *multistage cluster sampling*, a probability sample of *primary sampling units* is selected. From each chosen primary sampling unit, a probability sample of *secondary sampling units* is selected, and so on, until the final stage of sampling.

Efficiency.   Cluster sampling in more than one stage is efficient when the primary sampling units are large and heterogeneous, such as cities or counties. It is also efficient when the cost of frame construction is low for the primary sampling units and much higher for secondary and other subsequent sampling units. For instance, a frame of cities in a state is readily available, as are frames for blocks in most cities, but the cost of constructing a current frame of households in a state is very large. Also, when the elements within the clusters at any stage of sampling are homogeneous, precision will be enhanced if the budget is devoted to the selection of more clusters rather than to the selection of more elements within a cluster. The reason is that selecting more elements in a cluster will not provide much more information than is provided by a few elements, in view of the homogeneity within clusters. As an extreme example, if all persons in a household have the same preference among presidential candidates, a sample of one person from a household provides full information about that cluster.

Comments

1.   In both single- and multistage cluster sampling, probability sampling procedures other than simple random sampling may be advantageous for selecting the sampling units at the different stages. In the food expenditures survey example, cities might be selected by using probabilities proportional to their population sizes, and blocks within a city might be selected similarly.

2. When the clusters refer to geographical areas such as counties or city blocks, cluster sampling is often referred to as *area sampling*.

3. Cluster sampling can be used in conjunction with stratified sampling. For example, a statewide sample of urban households might involve, in the first stage of sampling, a stratified sample of cities. In the second stage, city blocks in a selected city might be sampled by using stratification based on city districts.

4. Estimation of the population mean or other parameter with multistage cluster sampling requires special formulas given in texts on sample survey methods.

14.6 SYSTEMATIC SAMPLING

Systematic selection of sample elements is widely employed because of its ease and convenience.

EXAMPLE

Audit. An auditor wishes to sample 200 sales receipts from a population of 10,000 receipts issued during the past quarter. The receipts are serialized and stacked in order of the serial numbers. To select the sample quickly, the auditor decides to select every $10,000/200 = 50$th receipt. He chooses one of the first 50 receipts at random (he happens to choose random number 39) and then selects every 50th receipt thereafter. Thus, the sample consists of the 39th, 89th, 139th, ..., 9989th receipts. The actual selection is extremely easy because of the serialized ordering of the population. □

Definition

The sampling procedure described here is called systematic random sampling.

(14.21)

> With *systematic random sampling,* every kth element in the frame is selected for the sample, with the starting point among the first k elements determined at random.

When every kth element is selected, then $100/k$ percent of the population is sampled. The resulting sample is called a *100/k percent systematic sample.* The auditor, for instance, selected a $100/50 = 2$ percent systematic sample.

Efficiency

Systematic sampling has the advantage of being simple to execute and hence is economical and convenient to use. A systematic random sample frequently will also provide more precise estimates than a simple random sample of the same size. This occurs when the frame listing in effect stratifies the population elements. For example, the sales receipts in the audit example are issued consecutively during the quarter.

Thus, a systematic sample will automatically contain receipts issued at different times during the quarter. In essence, the auditor's systematic sample is stratified by time.

Comments	1.	In using systematic sampling, one must avoid a sampling interval that corresponds to any natural period in the frame. Thus, in sampling average monthly stock prices for a 30-year period, one should not sample every 12th month, or some multiple thereof. The resulting price data would always correspond to the same month of the year and might be systematically different from the data for the other months.
	2.	Systematic sampling can be used in combination with other sampling procedures. For example, a stratified sample may involve systematic selection within each stratum.

CITED REFERENCES

14.1   *Military Standard 105D Tables: Sampling Procedures and Tables for Inspection by Attributes.* Office of the Assistant Secretary of Defense, Washington, D.C., 1963.

14.2   A. J. Duncan, *Quality Control and Industrial Statistics,* 5th ed. Homewood, Ill.: Irwin, 1986.

PROBLEMS

*14.1   **Glass Containers.** Consider the following single sampling plan for large shipments of glass containers received from a supplier.

Sample Size	Acceptance Number	Rejection Number
20	1	2

     a. What action should be taken if the sample contains three defectives? If the sample contains one defective?

     b. Express this single sampling plan as a decision rule involving the sample proportion $\bar{p}$.

14.2   **Oil Filters.** Large incoming lots of oil filters received by an engine assembly plant are checked for an excess of defective filters, based on inspection of a random sample of 120 filters from each lot. The action for each lot is determined by the following decision rule for the sample proportion defective $\bar{p}$:

If $\bar{p} \leq 0.025$, conclude H_0 (accept lot).
If $\bar{p} > 0.025$, conclude H_1 (reject lot).

     a. What are the acceptance and rejection numbers of the single sampling plan that correspond to this decision rule and sample size?

     b. What action should be taken if the sample from a lot contains no defective? If the sample contains four defectives?

14.3 **Canned Tuna.** A fish processor ships canned tuna in truckload lots to a retail food chain. The processor follows a zero-defects inspection plan for its outgoing tuna shipments whereby a random sample of 15 cans from a lot is inspected and the lot is released for shipment only if all 15 cans meet quality specifications.

    a. Describe this inspection plan as a single sampling plan in the format of Table 14.1a.

    b. Express a zero-defects inspection plan as a decision rule involving the sample proportion $\bar{p}$.

*14.4 Refer to **Glass Containers** Problem 14.1. The acceptable quality level (AQL) for a shipment is $p_0 = 0.02$.

    a. Use Table C.5 to obtain the probability of accepting, with this sampling plan, a shipment in which the proportion of defective containers is $p = 0.02, 0.10, 0.20$.

    b. Plot the operating characteristic curve of this sampling plan.

    c. What is the probability of an incorrect conclusion with this sampling plan for a shipment in which $p = 0.02$? Is this probability the supplier's risk or the purchaser's risk?

14.5 Refer to **Oil Filters** Problem 14.2. The acceptable quality level (AQL) for a lot is $p_0 = 0.01$.

    a. Use (6.3) to obtain the points on the operating characteristic curve of this sampling plan when $p = 0.01, 0.04, 0.06$.

    b. Plot the operating characteristic curve.

    c. A lot with proportion defective $p = 0.06$ is definitely unacceptable. What is the risk of an erroneous decision with this sampling plan when $p = 0.06$? Is this the supplier's risk or the purchaser's risk?

14.6 Refer to **Canned Tuna** Problem 14.3. Let p denote the proportion of cans in a lot that do not meet quality specifications.

    a. Use Table C.5 to obtain the points on the operating characteristic curve of this zero-defects inspection plan when $p = 0.01, 0.10, 0.20$.

    b. Plot the operating characteristic curve.

    c. The acceptable quality level (AQL) for outgoing lots is $p_0 = 0.01$. What is the risk of not releasing a shipment for which $p = 0.01$? Is this the producer's risk or the buyer's risk?

*14.7 Consider the following multiple sampling plan with two stages for examining incoming lots.

Sample	Sample Size	Cumulative Sample Size	Acceptance Number	Rejection Number
First	15	15	1	3
Second	15	30	3	4

    a. What action should be taken if the first sample of 15 items from a lot contains (1) 0 defective, (2) 3 defectives, (3) 2 defectives?

    b. What action should be taken if the first sample contains 2 defectives and the second contains 2 defectives?

14.8 Refer to **Oil Filters** Problem 14.2. The following multiple sampling plan with three stages has been proposed to replace the current single sampling plan for incoming lots.

Sample	Sample Size	Cumulative Sample Size	Acceptance Number	Rejection Number
First	40	40	0	3
Second	40	80	2	4
Third	40	120	3	4

    a. What action should be taken if the first sample of 40 filters contains (1) 0 defective, (2) 2 defectives, (3) 4 defectives?

    b. What action should be taken if the first sample contains 1 defective and the second contains 3 defectives?

    c. What action should be taken if the first sample contains 2 defectives, the second sample contains 1 defective, and the third sample contains no defective?

14.9 A speaker stated: "Generally, acceptance sampling plans are concerned with inferences about finite populations, whereas control charts are concerned with inferences about processes, that is, infinite populations." Do you agree? Discuss.

*14.10 Melting Points. Extensive past experience with a process for producing alloy filaments has shown that the melting points of the filaments are normally distributed with $\mu = 335°C$ and $\sigma = 21°C$. Samples of nine alloy filaments are taken daily from the production process and the melting point of each is determined.

    a. Construct a control chart for the process mean based on the sample mean $\bar{X}$, using $z = 3$ standard deviation control limits. Give the values of μ_0, LCL, and UCL for your chart.

    b. The sample means for the last 28 operating days are as follows:

Day	$\bar{X}$	Day	$\bar{X}$	Day	$\bar{X}$	Day	$\bar{X}$	Day	$\bar{X}$	Day	$\bar{X}$	Day	$\bar{X}$
1	340	5	351	9	336	13	339	17	325	21	340	25	351
2	328	6	340	10	317	14	355	18	338	22	345	26	344
3	342	7	323	11	337	15	334	19	326	23	330	27	328
4	335	8	339	12	350	16	310	20	328	24	332	28	333

    Plot these $\bar{X}$ values on your control chart. Identify any out-of-control points in the plot.

    c. A worker has suggested that it would be more convenient to postpone plotting the daily sample means until the end of each week and then examine the points for the entire week for out-of-control conditions. Why would such a postponement partially defeat the purpose of the control chart?

14.11 It is known from past experience that the lengths of shafts produced by a process are normally distributed, with $\mu = 280.0$ centimeters and $\sigma = 2.0$ centimeters. The process mean is monitored by taking random samples of four shafts each hour and plotting the sample means $\bar{X}$ on a control chart.

    a. Construct a control chart for the process mean, using $z = 3$ standard deviation control limits. Give the values of μ_0, LCL, and UCL for your chart.

    b. The first five samples taken today yielded the following values for the lengths (in centimeters) of the individual shafts.

Sample	Observation			
	1	2	3	4
1	279.3	280.1	279.4	281.0
2	280.2	277.5	282.4	280.5
3	277.0	278.6	280.3	278.5
4	277.6	279.2	277.3	280.4
5	275.8	274.7	277.9	276.0

    Calculate and plot the sample means on your control chart. Do any of the sample points indicate that the process is out of control with respect to its mean level? Comment.

14.12 Product Weight. The weight of product placed in containers by a machine is normally distributed with standard deviation $\sigma = 1.5$ grams. The mean fill per container depends on the setting of the machine but tends to drift during the filling operation. The desired mean fill is $\mu_0 = 375.0$ grams. A random sample of six containers is selected every hour during filling, and the sample mean $\overline{X}$ is calculated.

    a. Construct a control chart for the process mean fill μ, using $z = 3$ standard deviation control limits. Give the values of the upper and lower control limits.

    b. The sample means for hours 25 through 32 are as follows:

Hour:	25	26	27	28	29	30	31	32
$\overline{X}$:	374.2	375.1	374.9	376.3	376.2	376.6	376.7	377.2

    Plot these means on your control chart. Was the mean fill in control during these hours? Does your plot provide any other information about the process? Explain.

    c. If the sample size were increased to $n = 12$ and the control limits recalculated, would this change affect the probability of obtaining an out-of-control point when (1) the mean fill is in control at $\mu_0 = 375.0$, (2) the mean fill is out of control at $\mu = 373.0$? Make appropriate calculations, assuming that the process standard deviation remains at $\sigma = 1.5$ grams.

14.13 Refer to **Melting Points** Problem 14.10.

    a. Under what conditions should the control limits constructed in Problem 14.10a be changed?

    b. What process characteristic in addition to the mean would you also be interested in controlling with respect to filament melting points? Discuss.

14.14 The regulation of a serious medical disorder requires that the patient ingest regular minute doses of a drug. The drug is administered in the form of tablets that contain small amounts of the drug and an inert binding substance. Why would the pharmaceutical company manufacturing this tablet be concerned with controlling the process standard deviation, as well as the process mean, of the amount of the drug per tablet?

*14.15 A credit card company bills its credit card holders monthly. Extensive past experience has shown that 4 percent of the bills contain one or more billing errors. Random samples of 300 bills are selected monthly and checked for errors.

    a. Construct a control chart for the process proportion of bills containing one or more billing errors, using $z = 3$ standard deviation control limits. Give the values of p_0, LCL, and UCL for your chart.

    b. The numbers of bills with one or more billing errors found in the samples for the past 12 months are, chronologically, as follows:

Month:	1	2	3	4	5	6	7	8	9	10	11	12
Number:	10	15	10	13	13	10	11	13	14	8	9	3

    Calculate the sample proportions $\overline{p}$ for the past 12 months, and plot them on your control chart. Was billing quality in control during the past 12 months? Does your plot provide any other information about the process? Discuss.

    c. Why might management wish to use a control chart with upper and lower control limits here instead of a chart with an upper limit only?

14.16 In the past, 20 percent of the figurines manufactured by an automatic process were found on later inspection to require additional finishing. A control chart will now be instituted to monitor the process on a daily basis by inspecting a random sample of $n = 50$ figurines taken from each day's production and noting the proportion that require additional finishing.

    a. Construct a control chart for the process proportion of figurines requiring additional finishing, using $z = 3$ standard deviation control limits. Give the values of p_0, LCL, and UCL for your chart.

    b. The numbers of figurines that required additional finishing in the samples for the first 15 operating days are, chronologically, as follows:

Day:	1	2	3	4	5	6	7	8	9	10	11	12	13	14	15
Number:	9	12	8	11	13	10	10	7	11	9	11	14	13	17	20

Calculate the sample proportions $\bar{p}$ for these 15 days, and plot them on your control chart. Do any of the sample points indicate that the process was out of control during these 15 days?

14.17 In the control of a process for manufacturing chinaware, 100 pieces are selected at random and inspected at the end of each operating day. Each inspected piece is classified as "regular" (no flaws) or "second" (one or more flaws). Under normal conditions, 25 percent of the pieces are seconds.

    a. Construct a control chart for the process proportion of seconds, using $z = 3$ standard deviation control limits. Give the values of p_0, LCL, and UCL for your chart.

    b. During the past three weeks, the numbers of seconds in the daily samples have been as follows:

	Day of Week				
Week	M	T	W	Th	F
1	32	24	21	20	24
2	37	26	32	23	27
3	39	20	24	25	28

Calculate the sample proportions $\bar{p}$ for these days, and plot them on your control chart. Was the process in control with respect to the proportion of seconds during these three weeks? Does the plot provide any other information about the process? Discuss.

*14.18 The number of false fire alarms turned in weekly in a city follows a Poisson distribution, with mean $\lambda = 5.5$. A control chart is to be set up to monitor the process mean number of false alarms received per week.

    a. Construct a number-of-defects control chart for the process mean, using a $z = 3$ standard deviation upper control limit. Give the values of λ_0 and UCL for your chart.

    b. The numbers of false alarms turned in weekly for the past 20 weeks were as follows:

Week:	1	2	3	4	5	6	7	8	9	10	11	12	13	14	15	16	17	18	19	20
Number:	3	6	5	7	4	9	3	7	5	2	8	4	2	6	5	5	11	8	10	11

Plot these numbers on your control chart. Was the process in control with respect to the mean number of false alarms? Does the control chart provide any other information about the process? Discuss.

    c. For your control chart in **a**, what is the probability of obtaining an out-of-control point when (1) $\lambda = 5.5$, (2) $\lambda = 15$?

14.19 An automatic process for producing fluorescent tubes includes an immediate check of each tube to determine whether it is defective. When the process is operating satisfactorily, the number of defective tubes in an hour's production follows a Poisson distribution with $\lambda = 11.0$.

    a. Construct a number-of-defects control chart for the process mean number of defective tubes per hour, using $z = 3$ standard deviation upper and lower control limits. Give the values of λ_0, LCL, and UCL for your chart.

    b. The numbers of defective tubes per operating hour in the past ten operating hours were as follows:

Hour:	1	2	3	4	5	6	7	8	9	10
Number:	10	7	12	11	15	13	16	4	11	18

    Plot these numbers on your control chart. Was the process in control with respect to the process mean? Explain.

    c. For your control chart in **a**, what is the probability when $\lambda = 11.0$ of obtaining a point above the upper control limit? Of obtaining a point below the lower control limit?

14.20 A cusum plot for a process mean initially fluctuated about the zero line of the chart and then commenced a lengthy, persistent downward drift. What does this pattern indicate about the relation, during the time period under consideration, of the process mean to the standard (μ_0) upon which the chart is based?

*14.21 Refer to **Melting Points** Problem 14.10.

    a. Calculate the daily cumulative sums C_k for $k = 1, 2, \ldots, 28$, and plot these sums on a chart, as in Figure 14.5.

    b. Does your cusum plot provide any evidence that the process mean was out of control at any time during this period? Comment.

14.22 Refer to **Product Weight** Problem 14.12. The cumulative sum for sample means from hours 1 through 24 is $C_{24} = 1.5$ (data not shown).

    a. Calculate the cumulative sums C_k for $k = 25, \ldots, 32$, and plot these sums on a chart, as in Figure 14.5.

    b. Does your cusum plot give an earlier indication that the process mean may differ from $\mu_0 = 375.0$ than the $\overline{X}$ control chart in Problem 14.12? Explain.

*14.23 What is the ratio of the value of $s\{\overline{X}\}$ calculated without the finite population correction to the exact value calculated according to (14.5) when 3 percent of a finite population is sampled, that is, when $n/N = 0.03$?

14.24   **a.** Calculate the finite population correction $\sqrt{1 - (n/N)}$ for each of the following three cases.

N:	1000	10,000	100,000
n:	500	500	500

    b. Use the results in **a** to calculate $s\{\overline{X}\}$ for each of the three cases when $s = 4$. Is there much of a change in $s\{\overline{X}\}$ when N goes from 1000 to 10,000? From 10,000 to 100,000?

*14.25 A simple random sample of 50 bank branches was selected from a population of 400 branches to estimate the mean number of savings certificates issued per branch during the past month. The sample results were $\overline{X} = 200$ and $s = 15$. Construct a 95 percent confidence interval for the population mean. Interpret your confidence interval.

14.26 **Expense Claims.** An auditor selected a random sample of 100 expense claims from among the 500 filed by company executives during the first quarter of the year to ascertain the expense amount supported by acceptable receipts for each claim. The sample results were $\overline{X} = \$418$ and $s = \$180$. Construct a 90 percent confidence interval for the mean supported expenses per claim for all claims filed during the quarter. Interpret your confidence interval.

14.27 A random sample of 80 full-time classroom teachers was selected from the 427 teachers employed by a school system, and the number of days of accrued sick leave for each teacher in the sample was determined from personnel files. The sample results (in days) were $\overline{X} = 58$ and $s = 12$. Construct a 99 percent confidence interval for the population mean number of days of accrued sick leave per teacher. Interpret your confidence interval.

*14.28 An analyst requires a 95 percent confidence interval for the mean number of days to expiration in a population of 1200 electric motor warranties. The confidence interval is to have a half-width of 2.5 days. A reasonable planning value for the population standard deviation is $\sigma = 45$ days.
   a. Determine the required sample size.
   b. Recompute the required sample size as though the population were infinite. Does the treatment of the population as infinite increase the required sample size substantially here? Comment.

14.29 Refer to **Expense Claims** Problem 14.26. The auditor is now planning to review the 650 expense claims filed by company executives in the second quarter of the year. What random sample size is required if the mean supported expenses per claim is to be estimated with a 90 percent confidence interval having a half-width of $25? Assume the sample standard deviation from the first quarter's examination is a suitable planning value for σ.

14.30 A plant scientist requires a 99 percent confidence interval for the mean area per leaf (in square centimeters) for a plant, based on the surface areas of a random sample of leaves selected from the 120 leaves on the plant. The desired half-width of the confidence interval is 2 square centimeters per leaf. Previous experience indicates that an appropriate planning value for the standard deviation of the leaf areas is $\sigma = 7$ square centimeters. Obtain the required sample size.

*14.31 In a random sample of 60 garment manufacturers selected from the 300 garment manufacturers operating in a state, 33 were found to pay full-time employees more than the state's minimum wage rate. Construct a 90 percent confidence interval for the proportion of garment manufacturers in the state who pay employees more than the minimum wage rate. Interpret your confidence interval.

14.32 A pharmaceutical company reported that, in a random sample of 250 physicians drawn from the 2000 physicians practicing in one of the company's sales districts, 60 percent write more than 20 prescriptions per working day. Construct a 95 percent confidence interval for the population proportion of physicians who write more than 20 prescriptions per working day. Interpret your confidence interval.

14.33 In a simple random sample of 100 seniors taken from the 427 seniors majoring in management, 32 stated that they intend to continue studying at this university to obtain a graduate degree in management. Obtain a 99 percent confidence interval for the population proportion of all management seniors who intend to obtain a graduate degree in management at this university. Interpret your confidence interval.

*14.34 It is desired to estimate the proportion of the 3820 freshmen entering a university who plan to bring a personal computer with them. What random sample size is needed to estimate this proportion with a 99 percent confidence interval having a half-width of 0.05? Assume that $p = 0.4$ is a suitable planning value.

14.35 A population of 900 firms is to be surveyed to estimate the proportion of firms that have major union contracts that come up for renewal in the next 12 months. What random sample size is needed to estimate the proportion within ± 0.05 with 95 percent confidence? Assume that $p = 0.5$ is a suitable planning value.

14.36 A batch of 2500 drawings has been found in an old building in Brussels. Some of these drawings are by a sixteenth-century master, while the remainder are by apprentices who worked in the master's studio. A random sample of drawings is to be examined carefully to estimate the proportion of drawings that are by apprentices. The 90 percent confidence interval is to have a half-width of 0.03. A planning value of $p = 0.85$ is reasonable.
   a. What simple random sample size is required?
   b. If no reliable planning value for p were available and it were desired to obtain a 90 percent confidence interval for p that is guaranteed to have a half-width that does not exceed 0.03, what planning value should be used to calculate the required sample size?

14.37 Stratified random sampling is to be employed in each of the following cases. Describe a basis of stratification likely to be useful for each.
   a. A sample of personal checking accounts of a bank, to estimate the mean number of transactions per account for May.
   b. A sample of logs in a shipment, to estimate the mean dollar yield of lumber per log.
   c. A sample of school bus drivers from all those licensed in Iowa, to estimate the mean number of years of driving experience per driver.

14.38 Stratified random sampling is to be employed in each of the following cases. Describe a basis of stratification likely to be useful for each.
   a. A sample of the large number of bales of silk in a warehouse damaged by fire, to estimate the mean salvage value per bale.
   b. A sample of the grain in a ship's hold, to estimate the mean protein level of the grain shipment.
   c. A sample of a university's faculty members who are authors of research publications listed in the university's annual report, to estimate the mean research expenditures per member for the year.

*14.39 A stratified random sample of 700 households in a community was selected to estimate mean home improvement expenditures last year. The strata employed were geographic areas. The strata sizes and sample results were as follows:

Stratum (j)	N_j	n_j	$\overline{X}_j$	s_j
1	12,000	300	480	200
2	6,000	150	380	150
3	10,000	250	510	300
Total	28,000	700		

Obtain a 95 percent confidence interval for the mean home improvement expenditures per household in the community. Ignore the finite population correction in your calculations. Interpret your confidence interval.

14.40 A dairy association with 9000 member producers selected a stratified random sample of 400 members to estimate the mean herd size per producer for the entire association. The strata were set up according to the amount of milk shipped in the last calendar year, which was known for each member. The strata sizes and sample results on herd sizes follow.

Stratum (j)	N_j	n_j	$\overline{X}_j$	s_j
1	4500	170	15.7	8.3
2	2700	90	25.1	7.7
3	1800	140	58.2	18.0
Total	9000	400		

Obtain a 99 percent confidence interval for the population mean herd size. Use the finite population correction in your calculations for all strata. Interpret your confidence interval.

14.41 A stratified random sample of 400 stores was selected from a population of 8000 stores to estimate the mean pilferage loss per store for the past fiscal year. The strata were set up according to the floor area of the store, which was known for each store in the population. The strata sizes and sample results follow.

Stratum (j)	N_j	n_j	$\overline{X}_j$	s_j
1	4500	125	2060	220
2	2700	75	3840	270
3	800	200	4200	350
Total	8000	400		

a. Obtain a 90 percent confidence interval for the mean pilferage loss per store in the population. Use the finite population correction in your calculations for all strata. Interpret your confidence interval.

b. Is the validity of the confidence interval in **a** affected by whether or not pilferage loss in a store is related to its floor area? Explain.

14.42 a. What information about the population strata is required for optimal allocation but not for proportional allocation of the total sample size in planning a stratified random sample?

b. Under what special circumstances will optimal allocation and proportional allocation lead to the same allocation of strata sample sizes, for a given total sample size?

*14.43 Invoice Payments.   A company wishes to estimate the mean number of days elapsed between the date an invoice is sent to a customer and the date payment is received. The study is to be confined to the 18,000 invoices issued in the last fiscal quarter. Three strata, according to the amount of the invoice, are to be employed. Strata sizes and the anticipated strata standard deviations are as follows:

Stratum (j)	N_j	Anticipated σ_j (days)
1	9,000	8
2	6,000	5
3	3,000	3
Total	18,000	

The budget permits a sample of 600 invoices to be examined.

a. Determine the strata sample sizes with proportional allocation.

b. Determine the strata sample sizes with optimal allocation.

c. Do the two methods of allocation lead to substantially different strata sample sizes here? Discuss.

14.44 A state contains 100 large municipalities (stratum 1) and 500 small municipalities (stratum 2). It is desired to estimate the mean annual expenditures per municipality for snow clearing in the state last year, based on a total random sample of 60 municipalities drawn from these two strata.
   a. Determine the strata sample sizes with proportional allocation.
   b. Planning values for the stratum standard deviations (in $ million) are $\sigma_1 = 2.60$ and $\sigma_2 = 0.20$, respectively. Determine the strata sample sizes with optimal allocation.
   c. Do the two methods of allocation lead to substantially different strata sample sizes here? Discuss.

14.45 A marketing research firm has grouped the households in a city by locational proximity, with each group consisting of 20 households. A random sample of these groups will be selected by the firm for a household survey. All households in the groups selected for the sample will be interviewed in the survey. Is such a sample a stratified random sample, in which location is used as a basis of stratification, or a single-stage cluster sample, in which the groups constitute clusters of households? Explain.

14.46 Explain how cluster sampling might be employed in each of the following situations.
   a. Sampling of peaches in a fruit orchard, to estimate the percentage of peaches that suffered frost damage.
   b. Sampling of fifth-grade students throughout a state, to estimate the mean reading achievement score.

14.47 Explain how cluster sampling might be employed in each of the following situations.
   a. Sampling of records of patients discharged from acute-care hospitals in a state, to estimate the mean length of hospital stay per patient.
   b. Sampling of travelers departing on flights originating in the United States and Canada, to estimate the mean number of flight segments per trip.

14.48 An auditor must check a company's inventory of raw materials and finished goods located in a large number of warehouses throughout the country. The auditor will perform the check by selecting a sample of inventory items, counting the actual inventory on hand for these items, and comparing these counts with company records. The total error in the book inventory is to be estimated.
   a. Explain how the auditor might employ cluster sampling in this investigation. Does your proposal involve single-stage or multistage cluster sampling?
   b. What advantages does cluster sampling offer here relative to simple random sampling? What disadvantages?

14.49 A 4 percent systematic random sample is to be selected from a computer listing of 6000 employee payroll records. With reference to definition (14.21), what is the value of k here? How many records will be selected for the sample in this case?

14.50 A population of 5000 physicians specializing in plastic surgery is listed in order of age. A systematic random sample of every 50th physician was selected from this frame, to estimate the mean number of years of practice in this specialty. The analyst has evaluated the sample results as if the sample were a simple random one. Will the calculated precision so obtained likely overstate, understate, or be the same as the actual precision of the systematic random sample estimate? Explain.

14.51 In a silviculture experiment, 4000 tree seedlings were planted in rows on a plot. Five years later, a 5 percent systematic sample of the young trees was selected to estimate the mean growth per tree. The successive rows of the plot were used as the population frame.
   a. How many trees were selected for the sample?

b. The southern half of the plot received more rain during this five-year period than the northern half. Does this fact imply that the selected systematic sample will reflect the benefits of stratification with respect to the effects of rainfall on growth? Explain.

EXERCISES

14.52 Consider the following single sampling plan for large lots:

Sample Size	Acceptance Number	Rejection Number
50	2	3

Use the Poisson approximation to the binomial distribution to calculate the probability of accepting, with this sampling plan, a lot containing 3 percent defectives.

14.53 Assume that a process is operating in control with mean μ_0 and variance σ^2. Let $\overline{X}_i$, $i = 1, 2, \ldots, k$, denote the means of independent random samples of size n drawn from this process while it is operating in control.
   a. Show that the cumulative sum C_k, defined in (14.4), has expected value 0 and variance $k(\sigma^2/n)$ under these conditions.
   b. Do the results in a imply that the value of C_k is expected to lie farther from the zero line of the cusum chart as k increases when the process remains in control? Explain.

14.54 In the estimation of a population mean from a stratified random sample, why is the estimated variance (14.15) not applicable when the strata samples are not drawn independently?

14.55 An observer has stated that if the strata standard deviations σ_j are the same for all strata, stratified random sampling offers no greater precision than simple random sampling for the same sample size. Do you agree? Discuss. (*Hint:* If the strata standard deviations are all equal, must the population standard deviation also be the same?)

14.56 Refer to **Invoice Payments** Problem 14.43.
   a. What is the anticipated standard deviation of $\overline{X}_{st}$ with proportional allocation? With optimal allocation?
   b. Is optimal allocation expected to lead to a substantial gain in precision over proportional allocation here? Comment.
   c. If the sample survey were also to estimate the mean delivery charge per invoice, would the strata sample sizes obtained in Problem 14.43b also be optimal for this purpose? Discuss.

STUDIES

14.57 Refer to the sampling plan in Problem 14.7.
   a. Use Table C.5 to obtain the probability that a lot will either be accepted or rejected by the sampling plan in the first stage if the lot proportion defective is $p = 0.01, 0.02, 0.10, 0.40$.
   b. Determine the expected sample size required to reach an accept-reject decision when $p = 0.01, 0.02, 0.10, 0.40$. (*Hint:* For any given lot, the total sample size will be either 15 or 30.) What is the relation between the lot proportion defective and the expected sample size required to reach a decision for this sampling plan?

14.58 A company receives raw material in lots of 8000 units. A random sample of 50 units is selected from each lot, on the basis of which a decision is made to accept (H_0) or reject (H_1) the lot. A rejected lot is inspected 100 percent, all defective units are replaced with acceptable ones, and the supplier is charged for the work and the defective materials. The decision rule used has the following properties:

Lot Proportion Defective p	Acceptance Probability $P(H_0; p)$	Rejection Probability $P(H_1; p)$
0.01	0.91	0.09
0.03	0.56	0.44
0.10	0.03	0.97

   a. Determine the average quality of lots after inspection when the incoming lot proportion defective is $p = 0.01, 0.03, 0.10$. Ignore defective units found in samples from accepted lots.
   b. Explain why the average quality of lots after inspection is better when the lot proportion defective is 0.10 than when it is 0.03. Is it therefore desirable for the supplier to submit poor-quality lots? Explain.
   c. Determine the average number of units inspected per lot when the incoming lot proportion defective is $p = 0.01, 0.03, 0.10$.

14.59 (Computer needed.) Refer to the **Financial Characteristics** data set.
   a. Select a stratified random sample of 64 firms from the population of all firms. Use industry groupings as strata and employ proportional allocation.
   b. Estimate the mean net income per firm in year 1 for the population, using a 95 percent confidence interval. Use the finite population correction in your calculations. Interpret your interval estimate.
   c. How would a systematic random sample of 64 firms be selected here? Why might this systematic sample be considered, for purposes of analysis, as approximately equivalent to the stratified sample selected in **a**?
   d. Use the sample results in **a** to estimate the population variance here. {*Hint:* $\sigma^2 =$
   $\sum_{j=1}^{k} (N_j/N) [\sigma_j^2 + (\mu_j - \mu)^2]$, where μ and σ^2 denote the population mean and variance, and μ_j and σ_j^2 ($j = 1, 2, \ldots, k$) denote the strata means and variances.}
   e. Use the estimate of σ^2 in **d** to estimate the standard deviation of the sample mean $\overline{X}$ based on a simple random sample of 64 firms. How does this estimated standard deviation compare with the value of $s\{\overline{X}_{st}\}$ calculated in **b**? Is stratification based on industry grouping effective here? Discuss.

UNIT FOUR

Estimation and Testing — II

15 NONPARAMETRIC PROCEDURES

In the estimation and testing procedures considered thus far, heavy reliance has been placed either on the assumption that the underlying population is of a particular form (for example, normal) or, when little can be assumed about the nature of the population, on large sample sizes and the central limit theorem. In this chapter, we discuss some procedures that require only minimal assumptions about the population and can be used for small as well as large sample sizes. These procedures are called *nonparametric* or *distribution-free* procedures. They are typically simple to implement, being generally derived from elementary probability considerations and often employing simple statistics based on ranked or ordered sample data.

15.1 SINGLE-POPULATION STUDIES

We first discuss nonparametric procedures that are concerned with the median of a single population. The estimation of the population median is important in many applications, for two major reasons. First, when the population is highly skewed (for instance, in studies of family incomes, store sales, and manufacturers' inventories), the population median is located more in the center of the distribution than the population mean and thus may be a more meaningful measure of position. Second, when the population is symmetrical, the population mean and the population median coincide and thus are equally meaningful measures of position.

Assumptions

The inference procedures for the population median that we now consider require only the following two assumptions:

1. The population is continuous.
2. The sample is a random one.

Recall that a random variable for a continuous population can take on any value in an interval. Figure 5.4 illustrates two continuous populations.

Point Estimation of Population Median

We denote the population median by η (Greek eta). We shall use the sample median Md as defined in (3.8) as the point estimator of η. Recall that the sample median Md is the middle value when all items in the sample are arrayed in ascending or descending order. When the sample size is even, Md is taken to be the mean of the middle two values.

The choice of the sample median Md as a point estimator of η is appropriate for several reasons:

1. For a symmetrical population, it can be shown that Md is an unbiased estimator of η.
2. For any continuous population, whether symmetrical or not, the sample median, when n is odd, has an equal chance of lying above η or below it. In other words, the median of the sampling distribution of Md equals η. The same result holds approximately when n is even. The reason why the result is not exact when n is even is our convention of taking Md as the mean of the middle two values in this case.
3. The sample median is a consistent estimator of η when the population is continuous, but it is not always the most efficient.

EXAMPLE

Case Duration.   An administrator is studying how long it takes to settle claims made under a workers' compensation insurance plan for injuries received from on-the-job accidents. For one category of claims, case durations for a random sample of nine settled cases are as follows:

Case:	1	2	3	4	5	6	7	8	9
Case Duration (months):	4.5	5.8	2.6	4.3	2.7	10.5	5.6	1.9	4.1

Case duration refers to the length of time from the occurrence of the accident to the closing date for the case. We are to estimate the median duration η for claims in this category.

Arraying the nine case durations in ascending order, we see that the sample median is $Md = 4.3$ months.

Value:	1.9	2.6	2.7	4.1	4.3	4.5	5.6	5.8	10.5
Rank:	1	2	3	4	5	6	7	8	9

Therefore, $Md = 4.3$ months is a point estimate of η for the population of case durations under study here.   □

Interval Estimation of Population Median

A confidence interval for the population median is easy to construct. Let us denote the rth smallest sample outcome by L_r and the rth largest by U_r. In the case duration example, for instance, $L_2 = 2.6$ and $U_2 = 5.8$. By an appropriate choice of r, the values L_r and U_r may be used as confidence limits for the population median η.

(15.1)

A confidence interval for the population median η, when the population is continuous, is:

$$L_r \leq \eta \leq U_r$$

where:

confidence coefficient $1 - \alpha$ is given in Table C.10

To understand the basis of (15.1), consider a random sample of size $n = 2$ and denote the two sample observations by X_1 and X_2, as usual. Here, L_1 equals the smaller of X_1 and X_2, and U_1 equals the larger. According to Table C.10, the confidence interval:

$$L_1 \leq \eta \leq U_1$$

has a confidence coefficient of 0.50 for a sample of size $n = 2$.

The reason why the confidence coefficient equals 0.50 here is as follows: For any continuous population, the probability that an observation X_i falls below the median η is $P(X_i < \eta) = 0.5$, and the probability that X_i falls above η is $P(X_i > \eta) = 0.5$. Figure 15.1 illustrates this situation. Remember that, for a continuous population, we have $P(X_i = \eta) = 0$. Now, the confidence interval $L_1 \leq \eta \leq U_1$ will be *incorrect* only if (1) both X_1 and X_2 fall below η (so $U_1 < \eta$) or (2) both X_1 and X_2 fall above η (so $L_1 > \eta$). Because of the independence of the observations in a random sample, the probabilities of these two events are as follows:

1. $P[(X_1 < \eta) \cap (X_2 < \eta)] = 0.5(0.5) = 0.25$
2. $P[(X_1 > \eta) \cap (X_2 > \eta)] = 0.5(0.5) = 0.25$

FIGURE 15.1   Position of population median η in a continuous population

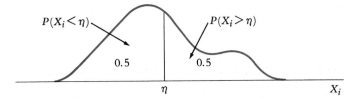

Moreover, these two events are mutually exclusive. Hence, by the complementation theorem (4.19), we obtain:

$$P(L_1 \leq \eta \leq U_1) = 1 - (0.25 + 0.25) = 0.50$$

so that the confidence coefficient here is 0.50.

Clearly, the use of the smallest (L_1) and largest (U_1) sample observations as confidence limits for η for larger sample sizes gives a higher confidence coefficient. Indeed, we are not required to use the extreme sample outcomes but may make a judicious choice of L_r and U_r in constructing the confidence limits for η. The method of obtaining appropriate confidence limits L_r and U_r depends on the sample size. We discuss the case of small samples first and then the case of large samples.

Small Sample.   Table C.10 provides the confidence coefficient $1 - \alpha$ for different L_r and U_r, for sample sizes up to 15. We now illustrate the meaning of this table.

1. If $n = 10$, use of the confidence limits L_1 and U_1 (that is, the two extreme sample observations) involves a confidence coefficient of 0.998.
2. If $n = 11$, use of the confidence limits L_2 and U_2 (that is, the second-smallest and second-largest sample observations) involves a confidence coefficient of 0.988.
3. If $n = 15$, use of the confidence limits L_4 and U_4 (that is, the fourth-smallest and fourth-largest sample observations) involves a confidence coefficient of 0.965.

Table C.10 allows for easy determination of L_r and U_r to control the confidence coefficient near a desired level.

EXAMPLES

1. In the case duration example, a confidence interval for η is desired with a confidence coefficient near 95 percent. Table C.10 shows that, for $n = 9$, the closest we can come to 95 percent is by use of $r = 2$, which involves a 96.1 percent confidence coefficient. Hence, we require L_2 and U_2. We see from the sample results presented earlier that $L_2 = 2.6$ and $U_2 = 5.8$. Hence, our confidence interval is:

    $$2.6 \leq \eta \leq 5.8$$

    Thus, we can conclude, with 96.1 percent confidence, that the median case duration is between 2.6 and 5.8 months.

2. A random sample of $n = 8$ households showed the following family incomes (in $ thousand) in ascending order:

Value:	5.9	6.4	8.3	11.2	14.9	16.3	17.6	31.7
Rank:	1	2	3	4	5	6	7	8

    A confidence interval is desired for the population median family income with confidence coefficient near 90 percent. We see from Table C.10 that, for $n = 8$, use of $r = 2$ involves a 93.0 percent confidence coefficient, which exceeds the desired 90 percent level just a little. Hence, we use $L_2 = 6.4$ and $U_2 = 17.6$ as confidence limits, and the 93 percent confidence interval for the population median family income is:

$$6.4 \leq \eta \leq 17.6$$

The limits are wide here because of the very small sample size and the large variability in the observations. □

Large Sample.   For sample sizes larger than 15, the value of r required for the confidence interval $L_r \leq \eta \leq U_r$ can be obtained by an approximation formula.

(15.2)

> When the sample size n exceeds 15, for a desired $1 - \alpha$ confidence coefficient for the confidence interval $L_r \leq \eta \leq U_r$, select the largest integer r that does not exceed:
>
> $$0.5[n + 1 - z(1 - \alpha/2)\sqrt{n}]$$

EXAMPLES

1. A random sample of 43 tax returns has been selected. It is desired to construct a 95 percent confidence interval for the population median amount of interest income per return. Since $1 - \alpha = 0.95$, we require $z(1 - \alpha/2) = z(0.975) = 1.960$, and we obtain:

    $$0.5(43 + 1 - 1.960\sqrt{43}) = 15.57$$

    The largest integer not exceeding 15.57 is 15. Hence, $r = 15$, and the required confidence limits for η are the 15th-smallest (L_{15}) and the 15th-largest (U_{15}) interest incomes in the sample of 43 returns.

2. A random sample of 75 vouchers has been selected. It is desired to construct a 90 percent confidence interval for the population median amount expended per voucher. Since $1 - \alpha = 0.90$, we require $z(0.95) = 1.645$, and we obtain:

    $$0.5(75 + 1 - 1.645\sqrt{75}) = 30.88$$

    The largest integer not exceeding 30.88 is 30. Hence, $r = 30$, and we use L_{30} and U_{30} (the 30th-smallest and 30th-largest voucher amounts) for the confidence limits for η. □

Evaluation of Confidence Coefficient.   We now give a more general explanation of how the confidence coefficients in Table C.10 are obtained. We return to the case duration example, where a random sample of $n = 9$ cases was studied. Suppose we wish to use L_2 and U_2 as the confidence limits for η. In other words, we plan to use the second-smallest and second-largest sample durations for the confidence limits.

Figure 15.2 shows schematically seven different possible locations for the nine sample observations in relation to the population median η. Note that the confidence

FIGURE 15.2   Several confidence intervals of the form $L_2 \leq \eta \leq U_2$ for a sample of size $n = 9$.   The confidence interval is correct whenever at least two but no more than seven observations fall above η.

Case	Population median η		Interval spans η	Number of observations above η
1	L_2	U_2	Yes	7
2	L_2	U_2	Yes	2
3	L_2	U_2	Yes	4
4	L_2 U_2		No	0
5	L_2 U_2		No	1
6		L_2 U_2	No	8
7		L_2 U_2	No	9

interval $L_2 \leq \eta \leq U_2$ is correct whenever at least two observations fall below η and at least two observations fall above it (cases 1, 2, and 3). The confidence interval is incorrect when none or only one observation is above η (cases 4 and 5) or when eight or nine observations are above η (cases 6 and 7). (Remember that, for a continuous population, the probability is zero that an observation equals the population median.)

Thus, the confidence interval $L_2 \leq \eta \leq U_2$ is correct here whenever at least two but not more than seven observations fall above η. Let B denote the number of sample observations above the median. We can then express the probability that the confidence interval will be correct as follows:

$$P(2 \leq B \leq 7)$$

But B is a binomial random variable with $p = 0.5$ because (1) for each observation X_i, $P(X_i > \eta) = 0.5$, since the population is continuous, and (2) the sample observations are independent, since a random sample is assumed. Hence, we have, by (6.3), for our example where $n = 9$:

$$P(2 \leq B \leq 7) = \sum_{i=2}^{7} \binom{9}{i} (0.5)^i (0.5)^{9-i}$$

Consulting the binomial tables (Table C.5) for $n = 9$ and $p = 0.50$, reproduced in Figure 15.3, we add the probability values associated with outcomes $2, 3, \ldots, 7$ and obtain $0.0703 + 0.1641 + \cdots + 0.0703 = 0.961$. This is the confidence coefficient shown in Table C.10 for $n = 9$ and $r = 2$.

FIGURE 15.3   **Fragment of the binomial table showing the probability that L_2 and U_2 will span η when $n = 9$**

$$p = 0.50$$

n	x	
9	0	0.0020
	1	0.0176
	2	0.0703
	3	0.1641 $\longleftarrow 0.961$
	4	0.2461
	5	0.2461
	6	0.1641
	7	0.0703
	8	0.0176
	9	0.0020

In general, the confidence interval $L_r \leq \eta \leq U_r$ will be correct if at least r and not more than $n - r$ out of the n sample observations fall above the median. Hence, the confidence coefficient is equivalent to the probability $P(r \leq B \leq n - r)$.

(15.3)

The confidence interval $L_r \leq \eta \leq U_r$ for the population median η has a confidence coefficient given by:

$$P(r \leq B \leq n - r) = \sum_{i=r}^{n-r} \binom{n}{i}(0.5)^i(0.5)^{n-i}$$

where:

B is the number of sample observations above η

Comments

1. In general, the confidence limits L_r and U_r will not be equidistant from the sample median Md. Recall that, in the case duration example, we had $Md = 4.3$, $L_2 = 2.6$, and $U_2 = 5.8$.

2. The procedure just described does not usually enable us to achieve exactly a desired confidence coefficient such as 90 or 95 percent. The reason is that the binomial probability distribution, which determines the confidence coefficient, is discrete. However, we can usually get quite close to the desired confidence coefficient when the sample size is not exceedingly small.

3. Formula (15.2) for r when n is not small is obtained from the normal approximation to the binomial probability in (15.3), using the correction for continuity discussed in Chapter 12.

4. The rth-smallest sample value, L_r, is often referred to as the *rth order statistic* of the sample.
5. Confidence intervals for other population percentiles, such as the 25th or 75th percentiles, can be obtained by a procedure similar to the one discussed here for the median.

Sign Test for Population Median

We now describe a simple test for a population median η called a *sign test*. We consider the usual three types of test alternatives.

(15.4)

(15.4a)	(15.4b)	(15.4c)
$H_0: \eta = \eta_0$	$H_0: \eta \leq \eta_0$	$H_0: \eta \geq \eta_0$
$H_1: \eta \neq \eta_0$	$H_1: \eta > \eta_0$	$H_1: \eta < \eta_0$

Here, η_0 denotes the value of the population median at which the α risk is to be controlled.

The sign test is based on the number of sample observations that fall above η_0. It is actually an ordinary test concerning a population proportion p, where p here denotes the probability that an observation falls above η_0, that is, $p = P(X_i > \eta_0)$. Observe from Figure 15.1 that if η is greater than η_0, then $p = P(X_i > \eta_0) > 0.5$; and if η is below η_0, then $p = P(X_i > \eta_0) < 0.5$. When η coincides with η_0, that is, when $\eta = \eta_0$, then $p = 0.5$.

Hence, we can restate the alternatives concerning η in (15.4) in terms of alternatives concerning p.

(15.5)

(15.5a)	(15.5b)	(15.5c)
$H_0: p = 0.5$	$H_0: p \leq 0.5$	$H_0: p \geq 0.5$
$H_1: p \neq 0.5$	$H_1: p > 0.5$	$H_1: p < 0.5$

where:

$p = P(X_i > \eta_0)$

Thus, the sign test is simply a test about a population proportion where the α risk is controlled at $p_0 = 0.5$. We shall let B denote the number of sample observations larger than η_0 and $\bar{p} = B/n$ the proportion of sample observations larger than η_0. When $\eta = \eta_0$ (or $p = p_0 = 0.5$), then B and $\bar{p}$ follow the binomial probability distribution with $p = 0.5$.

Large Sample. For large n, the sign test uses test statistic z^* in (12.11) with $p_0 = 0.5$ and the normal approximation, as described in Chapter 12.

EXAMPLE

Travelers Checks. A major financial institution that issues travelers checks is concerned about the length of time these checks are outstanding, that is, the length of time from issue to redemption. Management wishes to test whether the population median time outstanding for checks issued in the immediately preceding year (η) equals the population median time outstanding for checks issued in earlier years, which is known to be $\eta_0 = 2.1$ months.

  Step 1.   The alternatives here are:

$$H_0\!: \eta = 2.1 \qquad\qquad H_0\!: p = 0.5$$
$$\text{or}$$
$$H_1\!: \eta \neq 2.1 \qquad\qquad H_1\!: p \neq 0.5$$

where $\eta_0 = 2.1$ and $p = P(X_i > 2.1)$.
  Step 2.   The α risk is to be controlled at 0.05 when $\eta = \eta_0 = 2.1$, or equivalently, when $p = p_0 = 0.5$.
  Step 3.   A random sample of $n = 250$ travelers checks issued in the preceding year was selected, and the length of time outstanding was obtained for each check. The appropriate type of decision rule is the two-sided one shown in Figure 15.4.
  The sampling distribution of the standardized test statistic z^* in (12.11) when $p = p_0 = 0.5$ is approximately normal because of the large sample size. Hence,

FIGURE 15.4 Illustration of the statistical decision rule for the travelers checks example — Large-sample, two-sided sign test for η

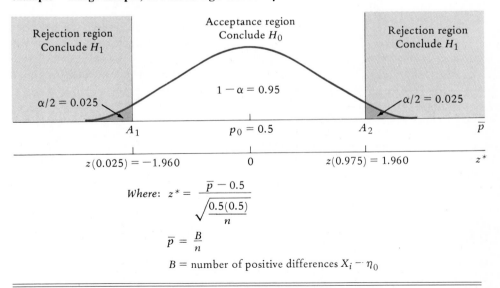

for $\alpha = 0.05$, we require $z(0.975) = 1.960$. Thus, the appropriate decision rule for the test is as follows:

If $|z*| \leq 1.960$, conclude H_0.
If $|z*| > 1.960$, conclude H_1.

Step 4.   The sample data, in part, are as follows:

Check i	Months Outstanding X_i	$X_i - \eta_0$	Sign
1	1.6	-0.5	$-$
2	4.2	$+2.1$	$+$
$\vdots$	$\vdots$	$\vdots$	$\vdots$
250	12.5	$+10.4$	$+$

We can obtain B, the number of observations larger than η_0, by taking the differences $X_i - \eta_0$ and counting the number of plus signs (hence, the name *sign test*). In this sample, it was found that $B = 89$ and hence $\bar{p} = B/n = 89/250 = 0.356$. The standardized test statistic for $p_0 = 0.5$ is therefore:

$$z* = \frac{\bar{p} - p_0}{\sqrt{\dfrac{p_0(1 - p_0)}{n}}} = \frac{0.356 - 0.5}{\sqrt{\dfrac{0.5(1 - 0.5)}{250}}} = -4.55$$

Since $|z*| = 4.55 > 1.960$, conclusion H_1 is drawn — that the population median time outstanding for travelers checks issued in the preceding year is not 2.1 months.   □

Small Sample.   When n is small, the decision rule for the sign test is set up by using the statistic B and its exact sampling distribution (that is, the binomial distribution).

EXAMPLE

In the case duration example, we wish to test whether or not the population median case duration is six months.

Step 1.   The alternatives here are:

H_0: $\eta = 6$ H_0: $p = 0.5$
                    or
H_1: $\eta \neq 6$ H_1: $p \neq 0.5$

where $\eta_0 = 6$ and $p = P(X_i > 6)$.

Step 2.   The α risk is to be as close as possible to 0.05 without exceeding it when $\eta = \eta_0 = 6$, or equivalently, when $p = p_0 = 0.5$.

Step 3.   Figure 15.5 shows the appropriate two-sided type of decision rule. The sampling distribution of B when $p = p_0 = 0.5$, where the α risk is controlled, is a binomial distribution with $n = 9$ and $p = 0.5$. This sampling distri-

---

**FIGURE 15.5   Illustration of the statistical decision rule for the case duration example —
Small-sample, two-sided sign test for η**

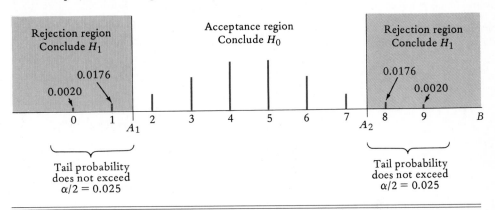

---

bution is also shown in Figure 15.5. Since $\alpha = 0.05$, the probability to the left of
action limit A_1 in Figure 15.5 must not exceed $\alpha/2 = 0.025$, and the probability
to the right of A_2 must not exceed 0.025. The appropriate binomial probabilities
are found in Figure 15.3. We see that only outcomes 0 and 1 can be to the left of
A_1 so that the probability in the left part of the rejection region will not exceed
0.025. By symmetry, only outcomes 8 and 9 can be to the right of A_2. Hence, the
action limits A_1 and A_2 are positioned as shown in Figure 15.5, and the decision
rule, expressed in terms of B, is as follows:

If $2 \le B \le 7$, conclude H_0.
If $B < 2$ or $B > 7$, conclude H_1.

Step 4. In our example, $B = 1$ because only one case duration, namely
10.5, exceeds $\eta_0 = 6$ (see p. 482). Hence, we conclude H_1 — that η does not
equal 6 months.

Note that the exact α risk here can be seen from Figure 15.5 to be
$2(0.0020 + 0.0176) = 0.039.$ □

Comments   1.   The statistic used in the sign test can also be the number of observations *below* η_0 rather
than the number *above*. Either statistic gives the same conclusion.

2.   In theory, sample values should not exactly equal η_0 because the population is assumed
to be continuous. In practice, zero differences $(X_i - \eta_0)$ may occur occasionally because
of rounding. It is then conventional to disregard the zero differences for the sign test and
reduce the sample size accordingly.

3. Confidence interval (15.1) for η can be used to test the two-sided alternatives (15.4a). Then, H_0 is concluded if η_0 lies in the confidence interval, and H_1 is concluded if it does not. In the case duration example, for instance, the 96.1 percent confidence interval for η was earlier found to be $2.6 \leq \eta \leq 5.8$ months. Since $\eta_0 = 6$ months lies outside this interval, H_1 is concluded—that $\eta \neq 6$. The α risk implicit in the test based on the confidence interval is $\alpha = 1 - 0.961 = 0.039$.

15.2   COMPARATIVE STUDIES OF TWO POPULATIONS—INDEPENDENT SAMPLES

In this section, we describe a nonparametric procedure for comparing the positions of two population distributions that have identical shapes, as illustrated in Figure 15.6. The comparison is based on independent random samples.

EXAMPLE

Marketing Instruction.   An instructor, wishing to compare the effectiveness of two different pedagogical approaches to the introductory marketing course, randomly split the total enrollment of 24 students into two equal course sections. One section was given live lectures by the instructor, while the other viewed videotaped lectures prepared by the same instructor. In other respects, the two course sections were handled in identical fashion. The instructor now wants to test whether or not the final grade distributions for students exposed to videotaped lectures (X population) and to live lectures (Y population) are located at the same position. She believes that the type of instruction should not affect the shape of the grade distribution but only its position. Figure 15.6 illustrates the situation for the case when live lectures lead to higher grades.   □

FIGURE 15.6   **Two population distributions of identical shape differing in position by amount δ**

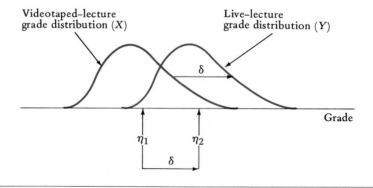

Shift Parameter

The two distributions for X and Y in Figure 15.6 are identical in shape but differ in position. We denote the difference in positions by δ (Greek delta) and call it the *shift parameter*. If $\delta > 0$, as in Figure 15.6, the distribution for Y lies a distance δ to the right of the distribution for X. If $\delta < 0$, the distribution for Y lies a distance δ to the left of the distribution for X. Finally, if $\delta = 0$, the two distributions are at the same position and hence are identical.

Mann-Whitney-Wilcoxon (M–W–W) Test

The Mann-Whitney-Wilcoxon test (the M–W–W test, for short) is a widely used non-parametric procedure for determining whether or not two population distributions of the same shape differ in position. In our subsequent discussion, populations 1 and 2 correspond to the X and Y populations, respectively.

Assumptions

1. The two population distributions are continuous and have the same shape, but they may differ by the distance δ in position.
2. Independent random samples of sizes n_1 and n_2 are drawn from the two populations.

Note that assumption 1 does *not* specify the shape of the distributions.

Alternatives.   The M–W–W test can be used for testing two-sided or one-sided alternatives concerning the shift parameter δ.

(15.6)

(15.6a)	(15.6b)	(15.6c)
$H_0: \delta = 0$	$H_0: \delta \leq 0$	$H_0: \delta \geq 0$
$H_1: \delta \neq 0$	$H_1: \delta > 0$	$H_1: \delta < 0$

Since the two population distributions are assumed to be identical in shape, the shift parameter δ measures the distance between the two population means, $\mu_2 - \mu_1$, or the distance between the two population medians, $\eta_2 - \eta_1$. Figure 15.6 illustrates this fact for the two population medians. Hence, the tests on δ may also be viewed as tests on $\mu_2 - \mu_1$ or on $\eta_2 - \eta_1$.

Test Statistic.   The test statistic for the M–W–W test is simple and is calculated as follows:

1. Combine the n_1 sample observations from the X population and the n_2 sample observations from the Y population, and array them in ascending order.

2. Assign ranks to the observations (starting with 1 for the smallest observation).
3. Sum the ranks for the n_2 sample observations from the Y population and denote this sum by S_2.

EXAMPLE

The following are the sample results when $n_1 = 2$ and $n_2 = 3$ sample observations were selected from two populations, respectively:

Sample from X Population		Sample from Y Population		
$n_1 = 2$		$n_2 = 3$		
$X_1 = 4$	$X_2 = 12$	$Y_1 = 14$	$Y_2 = 10$	$Y_3 = 17$

The array and the ranking of the combined samples are as follows:

Sample Value:	4	10	12	14	17
Rank:	1	2	3	4	5
Population:	X	Y	X	Y	Y

The ranks shown in boxes are for the three sample observations from the Y population. We see that the test statistic is:

$$S_2 = 2 + 4 + 5 = 11$$

Since there were $n_2 = 3$ sample observations from the Y population in this example, the smallest value that S_2 could have assumed is the sum of the first three ranks, namely, $1 + 2 + 3 = 6$. The largest possible value of S_2 is the sum of the last three ranks, namely, $3 + 4 + 5 = 12$.

In general, the smallest possible value of S_2 is the sum of the first n_2 ranks:

$$1 + 2 + \cdots + n_2 = \frac{n_2(n_2 + 1)}{2}$$

The largest possible value of S_2 is the sum of the last n_2 ranks:

$$(n_1 + 1) + (n_1 + 2) + \cdots + (n_1 + n_2) = n_1 n_2 + \frac{n_2(n_2 + 1)}{2}$$

For this example, $n_1 = 2$ and $n_2 = 3$, so:

$$\frac{n_2(n_2 + 1)}{2} = \frac{3(3 + 1)}{2} = 6$$

$$n_1 n_2 + \frac{n_2(n_2 + 1)}{2} = 2(3) + \frac{3(3 + 1)}{2} = 12$$

If the Y distribution lies to the right of the X distribution (so $\delta > 0$, as in Figure 15.6), the Y values will tend to be larger than the X values, and consequently, S_2 will be close to its largest value. Conversely, if the Y distribution is to the left of the X

distribution (so $\delta < 0$), the Y values will tend to be smaller than the X values, and S_2 will tend to be close to its smallest possible value. Finally, if the X and Y distributions are identical (so $\delta = 0$), all possible orderings of the combined samples will be equally likely, and S_2 will tend to have a value midway between the two extremes.

Thus, the appropriate type of decision rule for each set of alternatives in (15.6) is intuitively clear. In a test, for example, of:

$$H_0: \delta = 0$$

$$H_1: \delta \neq 0$$

very small and very large values of S_2 will lead to H_1, and intermediate values will lead to H_0.

Sampling Distribution of S_2 When $\delta = 0$.   To control the α risk when $\delta = 0$, we need to know the sampling distribution of S_2 when $\delta = 0$. As we mentioned earlier, all possible orderings of the combined $n_1 + n_2$ observations are equally likely when $\delta = 0$, and the exact sampling distribution of S_2 can be worked out. It depends only on the sample sizes n_1 and n_2. Tables of the exact sampling distribution of S_2 are available but are not needed unless n_1 and n_2 are very small.

(15.7)

> When independent random samples of sizes n_1 and n_2 are selected from two identical populations (that is, identical shape and $\delta = 0$), the sampling distribution of S_2 has mean and variance:
>
> $$E\{S_2\} = \frac{n_2(n_1 + n_2 + 1)}{2} \qquad \sigma^2\{S_2\} = \frac{n_1 n_2(n_1 + n_2 + 1)}{12}$$
>
> Furthermore, when n_1 and n_2 are sufficiently large, the sampling distribution of S_2 is approximately normal. As a working rule, the normal approximation is adequate when n_1 and n_2 are each 10 or more.

Thus, the sample sizes need not be particularly large for the normal approximation to be applicable.

Standardized Test Statistic.   All tests concerning the shift parameter δ are based on the following standardized test statistic.

(15.8)

> Tests concerning the shift parameter δ, when the sample sizes are sufficiently large, are based on the standardized test statistic:
>
> $$z^* = \frac{S_2 - E\{S_2\}}{\sigma\{S_2\}}$$

continues

where:

$E\{S_2\}$ and $\sigma\{S_2\}$ are given by (15.7)

α risk is controlled at $\delta = 0$

When $\delta = 0$, z^* follows approximately a standard normal distribution.

Given the approximate normality of the sampling distribution of z^* when $\delta = 0$, the construction of the decision rule parallels earlier large-sample tests.

EXAMPLE 1

Two-Sided Test.   In the marketing instruction example, the instructor wishes to test whether or not type of instruction affects the position of the grade distribution.

Step 1.   The alternatives here are two-sided:

H_0: $\delta = 0$

H_1: $\delta \neq 0$

Step 2.   The instructor wants to control the α risk at 0.05 when $\delta = 0$.
Step 3.   Table 15.1a contains the sample results. Note that the grades in each course section have been arrayed from smallest to largest. Also, note that $n_1 = 11$

TABLE 15.1   Sample results for the marketing instruction example

(a)
Final grades (arrayed by course section)

Video ($n_1 = 11$)

| X_i: | 49 | 52 | 56 | 61 | 72 | 74 | 75 | 80 | 87 | 88 | 94 | |

Live ($n_2 = 12$)

| Y_i: | 42 | 57 | 64 | 69 | 71 | 73 | 78 | 78 | 84 | 85 | 91 | 97 |

(b)
Combined array (Y ranks boxed)

Value:	42	49	52	56	57	61	64	69	71	72	73	74
Rank:	1	2	3	4	5	6	7	8	9	10	11	12

Value:	75	78	78	80	84	85	87	88	91	94	97
Rank:	13	14.5	14.5	16	17	18	19	20	21	22	23

(c)
Y ranks

| 1 | 5 | 7 | 8 | 9 | 11 | 14.5 | 14.5 | 17 | 18 | 21 | 23 | $S_2 = 149$ |

because one student withdrew from the video-lecture class. Figure 15.7 shows the appropriate two-sided type of decision rule. The sampling distribution of S_2 when $\delta = 0$ is approximately normal here because $n_1 = 11$ and $n_2 = 12$. To control the α risk at 0.05, we therefore require $z(0.975) = 1.960$. Hence, the decision rule for the test is as follows:

If $|z^*| \leq 1.960$, conclude H_0.
If $|z^*| > 1.960$, conclude H_1.

Step 4. The calculation of test statistic S_2 is shown in Tables 15.1b and 15.1c. Table 15.1b contains the array for the combined samples and shows the rank for each sample observation (the Y ranks are shown in boxes). Note that the one case of ties is handled by assigning the mean rank to each tied observation. Thus, since the two grades of 78 would normally have had ranks 14 and 15, each has been given rank 14.5. Table 15.1c assembles the ranks for the sample from the Y population. The test statistic is $S_2 = 149$.

The minimum and maximum values of S_2 here are:

$$\frac{12(13)}{2} = 78 \qquad 11(12) + \frac{12(13)}{2} = 210$$

so we see that $S_2 = 149$ is not an extreme value.

FIGURE 15.7 **Illustration of the statistical decision rule for the marketing instruction example — Large-sample, two-sided M–W–W test for δ**

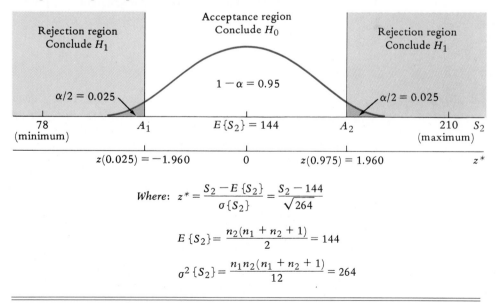

Where: $z^* = \dfrac{S_2 - E\{S_2\}}{\sigma\{S_2\}} = \dfrac{S_2 - 144}{\sqrt{264}}$

$$E\{S_2\} = \frac{n_2(n_1 + n_2 + 1)}{2} = 144$$

$$\sigma^2\{S_2\} = \frac{n_1 n_2(n_1 + n_2 + 1)}{12} = 264$$

The value of the standardized test statistic in (15.8) can now be calculated, as follows:

$$E\{S_2\} = \frac{12(11 + 12 + 1)}{2} = 144$$

$$\sigma^2\{S_2\} = \frac{11(12)(11 + 12 + 1)}{12} = 264 \qquad \sigma\{S_2\} = \sqrt{264} = 16.25$$

$$z^* = \frac{S_2 - E\{S_2\}}{\sigma\{S_2\}} = \frac{149 - 144}{16.25} = 0.31$$

Since $|z^*| = 0.31 \leq 1.960$, we conclude H_0—that $\delta = 0$. Thus, this small-scale experiment does not indicate any difference in students' grades for live and video-taped lectures. □

EXAMPLE 2

One-Sided Upper-Tail Test.   In the marketing instruction example, suppose that the instructor had hypothesized that live lectures would lead to better performance than video lectures and wishes to test this hypothesis.

Step 1.   The alternatives here are the upper-tail type:

H_0: $\delta \leq 0$

H_1: $\delta > 0$

Step 2.   The α risk is to be controlled at 0.05 when $\delta = 0$.
Step 3.   Since S_2 will tend to be large if $\delta > 0$, large values of z^* should lead to conclusion H_1. For $\alpha = 0.05$, we require $z(1 - \alpha) = z(0.95) = 1.645$. Thus, the appropriate decision rule is as follows:

If $z^* \leq 1.645$, conclude H_0.
If $z^* > 1.645$, conclude H_1.

Step 4.   Our earlier calculation gave $z^* = 0.31$. Since $z^* = 0.31 \leq 1.645$, we conclude H_0—that live lectures do not lead to better performance than video lectures. □

Comments

1. The M–W–W test procedure could equally well have been based on the sum of the ranks of the sample from the X population, which we might denote by S_1. However, the expected value of S_1 when $\delta = 0$ is not the same as $E\{S_2\}$ unless $n_1 = n_2$. It is given next.

(15.9)

$$E\{S_1\} = \frac{n_1(n_1 + n_2 + 1)}{2}$$

The variance of S_1 when $\delta = 0$ is the same as that of S_2.

2.   The procedure we have described for handling ties is suitable provided they are not numerous. The decision rule for the M–W–W test must be modified if ties are numerous.

3.   Although the M–W–W test uses only the ranks rather than the values of the observations, for reasonably large samples the M–W–W test is at times almost as powerful as tests that depend on specific knowledge about the form of the underlying populations. For instance, it compares favorably with the t test described in Section 13.2 for comparing the positions of two normal distributions. That is, when both tests have the same α risk, the probability of correctly concluding H_1 when $\delta \neq 0$ is almost as large for the M–W–W test as for the t test.

4.   The statistic S_2 is also called the *Wilcoxon rank sum statistic* because it was originally developed by Wilcoxon. Mann and Whitney independently formulated an alternative test statistic that is a simple function of S_2 and leads to identical tests.

15.3   COMPARATIVE STUDIES OF TWO POPULATIONS—MATCHED SAMPLES

In this section, we describe several nonparametric procedures for comparing two populations based on matched samples. Recall from Chapter 13 that matched samples involve the pairing of sample observations X_i and Y_i from two populations, as, for example, when male and female employees are paired on the basis of their job qualifications in order to study possible salary differentials. Like the procedures discussed in Chapter 13, the nonparametric procedures are based on the differences $D_i = Y_i - X_i$ for matched pairs.

We shall focus in this section on the median of the population of differences, denoted by η_D. When the Y_i values tend to be larger than the X_i values, the D_i values tend to be positive and η_D is positive. Similarly, when the Y_i values tend to be smaller than the X_i values, the D_i values tend to be negative and η_D is negative. Finally, if the X and Y distributions do not differ in position, the D_i values typically tend to be positive and negative with approximately equal frequency and η_D then is near zero. Thus, η_D may be thought of as a parameter that measures how far apart the X and Y distributions are. This result is particularly evident in the case of a symmetrical distribution of differences, because then η_D equals μ_D, the mean of the population of differences, and hence, $\eta_D = \mu_D = \mu_2 - \mu_1$.

Confidence Interval for η_D

As in Chapter 13, nonparametric inferences for matched samples are based on inference procedures for single populations, the population being that of the differences D_i. Thus, the setting up of a confidence interval for the median difference η_D utilizes the procedures of Section 15.1.

Assumptions

1.   The population of differences D_i is continuous.
2.   The n differences are a random sample from the population of differences.

EXAMPLE

Credit Information.   A business researcher studied the impact of information on credit decisions. Forty volunteer credit managers were grouped into $n = 20$ pairs, each pair having similar backgrounds with respect to business experience, education, and so on. Two case descriptions were prepared of a hypothetical business that is requesting trade credit. The first one provided limited information about the business (for example, product line, years in business, total assets). The second case description provided detailed information, including financial and operating data for the past several years. In each pair, a random number was used to decide which manager received description 1 and which one received description 2. Each manager then studied the description and determined the maximum amount of credit he or she would extend to the business.

The sample results are presented in Table 15.2. The credit amounts with limited information are denoted by X, and those with detailed information are denoted by Y. The researcher would like to estimate the median η_D of the population of differences $D_i = Y_i - X_i$, with a 95 percent confidence interval.

TABLE 15.2   **Sample results for the credit information example**

Pair of Managers i	Credit Amount ($000) Limited Information X_i	Detailed Information Y_i	Difference D_i	Rank
1	85	90	$+ 5$	4.5
2	50	60	$+10$	6
3	85	113	$+28$	15
4	70	100	$+30$	16
5	30	45	$+15$	7
6	75	99	$+24$	12
7	67	86	$+19$	9
8	60	85	$+25$	13.5
9	60	85	$+25$	13.5
10	85	87	$+ 2$	3
11	70	50	-20	1
12	85	90	$+ 5$	4.5
13	70	93	$+23$	11
14	77	115	$+38$	18
15	70	62	$- 8$	2
16	50	85	$+35$	17
17	55	72	$+17$	8
18	75	125	$+50$	20
19	81	103	$+22$	10
20	20	60	$+40$	19

The n sample differences are calculated in Table 15.2 and are ranked there. Since $n = 20$ is a reasonably large sample, we use formula (15.2) for finding r. For $1 - \alpha = 0.95$, we require $z(1 - \alpha/2) = z(0.975) = 1.960$. Hence:

$$0.5[n + 1 - z(1 - \alpha/2)\sqrt{n}] = 0.5(20 + 1 - 1.960\sqrt{20}) = 6.12$$

so $r = 6$. We therefore choose L_6 and U_6, the 6th-smallest (rank 6) and 6th-largest (rank 15) sample differences, for our confidence limits. A study of Table 15.2 shows $L_6 = 10$ and $U_6 = 28$. Thus, the desired confidence interval is:

$$10 \leq \eta_D \leq 28$$

The researcher can state, with 95 percent confidence, that detailed information in the case description has resulted in a median increase of between \$10 and \$28 thousand in trade credit extended. These limits clearly indicate that the maximum credit tends to be larger here with detailed than with limited information. □

Sign Test for η_D

The sign test discussed in Section 15.1 can be used to test the median difference η_D with matched samples.

Assumptions

1. The population of differences D_i is continuous.
2. The n differences are a random sample from the population of differences.

EXAMPLE

In the credit information example, the researcher wishes to test whether or not the median difference in credit extended with limited and detailed information is zero.

Step 1. The test alternatives here are:

$$H_0: \eta_D = 0 \qquad H_0: p = 0.5$$
$$\text{or}$$
$$H_1: \eta_D \neq 0 \qquad H_1: p \neq 0.5$$

where $p = P(D_i > 0)$.

Step 2. The α risk is to be controlled at 0.05 when $\eta_D = 0$, or equivalently, when $p = p_0 = 0.5$.

Step 3. The sample size $n = 20$ is large enough so that the sampling distribution of the standardized test statistic z^* in (12.11) when $p = p_0 = 0.5$ is approximately standard normal. Hence, for $\alpha = 0.05$, we require $z(0.975) = 1.960$, and the appropriate decision rule is as follows:

If $|z^*| \leq 1.960$, conclude H_0.
If $|z^*| > 1.960$, conclude H_1.

Step 4. From the sample results in Table 15.2, we see that the number of positive sample differences is $B = 18$, and thus, $\bar{p} = 18/20 = 0.90$. Therefore, the standardized test statistic for $p_0 = 0.5$ is:

$$z^* = \frac{\bar{p} - p_0}{\sqrt{\dfrac{p_0(1 - p_0)}{n}}} = \frac{0.90 - 0.5}{\sqrt{\dfrac{0.5(1 - 0.5)}{20}}} = 3.58$$

Since $|z^*| = 3.58 > 1.960$, we conclude H_1 — that $\eta_D \neq 0$.   ☐

Wilcoxon Signed Rank Test for η_D

When one can assume that the population of differences is symmetrical, the Wilcoxon signed rank test is generally more powerful than the sign test for making inferences about the population median difference η_D.

In practice, and especially in experimental settings, the population of differences between matched pairs frequently will be symmetrical, or approximately so. In the case of experiments where matched subjects are each assigned randomly to two different treatments, a symmetrical distribution is to be expected when the two treatments have no differential effects, because each subject has an equal chance of being assigned to either treatment.

Assumptions

1.   The population of differences D_i is continuous and symmetrical.
2.   The n differences are a random sample from the population of differences.

Alternatives.   The Wilcoxon signed rank test can be employed for both two-sided and one-sided alternatives concerning η_D.

Test Statistic.   The test statistic for the Wilcoxon test is simple and is calculated as follows:

1.   Obtain the absolute differences $|D_i|$ and rank them.
2.   To each rank, attach a plus sign or a minus sign according to whether D_i is positive or negative, respectively.
3.   Sum the signed ranks, and denote the sum by T.

If a difference D_i should happen to equal zero, discard it and reduce the sample size accordingly. When absolute differences are tied, they are assigned the average value of the corresponding ranks, as in the M–W–W test.

EXAMPLE

Table 15.3 shows the calculation of the test statistic T for the credit information example. We repeat in column 1 the differences D_i from Table 15.2. The absolute values of the differences are shown in column 2. The ranks from smallest absolute difference (rank 1) to largest (rank 20) are shown in column 3. The signed ranks are shown in column 4. It can be seen from Table 15.3 that $T = 184$.   ☐

TABLE 15.3 Wilcoxon signed rank test calculations — Credit information example

Pair of Managers i	(1) Difference D_i	(2) Absolute Difference $|D_i|$	(3) Rank	(4) Signed Rank
1	+ 5	5	2.5	+ 2.5
2	+10	10	5	+ 5
3	+28	28	15	+15
4	+30	30	16	+16
5	+15	15	6	+ 6
6	+24	24	12	+12
7	+19	19	8	+ 8
8	+25	25	13.5	+13.5
9	+25	25	13.5	+13.5
10	+ 2	2	1	+ 1
11	−20	20	9	− 9
12	+ 5	5	2.5	+ 2.5
13	+23	23	11	+11
14	+38	38	18	+18
15	− 8	8	4	− 4
16	+35	35	17	+17
17	+17	17	7	+ 7
18	+50	50	20	+20
19	+22	22	10	+10
20	+40	40	19	+19

$$T = 184$$

If all the differences are positive, T is the sum of the ranks $1, 2, \ldots, n$:

$$1 + 2 + \cdots + n = \frac{n(n + 1)}{2}$$

For $n = 20$, $n(n + 1)/2 = 210$. If all the differences are negative, T equals $-n(n + 1)/2$. For $n = 20$, this sum is -210. Finally, if the distribution of differences is symmetrical about 0 (that is, if $\eta_D = 0$), the rank associated with any absolute difference has an equal probability of being positive or negative. In this case, T will tend to have a value midway between the two extremes, that is, near 0. Thus, the appropriate decision rules are clear. For a two-sided test, for example, values of T close to $-n(n + 1)/2$ or close to $n(n + 1)/2$ lead us to conclude H_1 ($\eta_D \neq 0$); otherwise, we conclude H_0 ($\eta_D = 0$).

Sampling Distribution of T When $\eta_D = 0$. To control the α risk at a specified level when $\eta_D = 0$, we need to know the sampling distribution of T when $\eta_D = 0$. As we mentioned earlier, when $\eta_D = 0$, the rank associated with any absolute difference has an equal probability of being positive or negative. Hence, the sampling distribution of T can be determined exactly for all sample sizes. It depends only on the number of differences n. Tables of the exact sampling distribution of T for small sample sizes are available but are not needed unless n is very small.

(15.10)

> When a random sample of n differences is selected from a symmetrical population of differences with $\eta_D = 0$, the sampling distribution of T has mean and variance:
>
> $$E\{T\} = 0 \qquad \sigma^2\{T\} = \frac{n(n + 1)(2n + 1)}{6}$$
>
> Furthermore, when n is sufficiently large, the sampling distribution of T is approximately normal. As a working rule, the normal approximation is adequate when n is 10 or more.

Standardized Test Statistic.   All tests concerning η_D based on the Wilcoxon signed rank method are conducted by means of the following standardized test statistic.

(15.11)

> Wilcoxon signed rank tests concerning η_D, when the number of differences is sufficiently large, are based on the standardized test statistic:
>
> $$z^* = \frac{T - 0}{\sigma\{T\}}$$
>
> *where:*
> $\sigma\{T\}$ is given by (15.10)
> α risk is controlled at $\eta_D = 0$
>
> When $\eta_D = 0$, z^* follows approximately a standard normal distribution.

Given the approximate normality of the sampling distribution of z^* when $\eta_D = 0$, the construction of the decision rule parallels earlier large-sample tests.

EXAMPLE

In the credit information example, the researcher wishes to test whether or not the median difference η_D is zero. It is reasonable to assume here that the distribution of differences when $\eta_D = 0$ is symmetric.

Step 1.   The test alternatives here are:

H_0: $\eta_D = 0$

H_1: $\eta_D \neq 0$

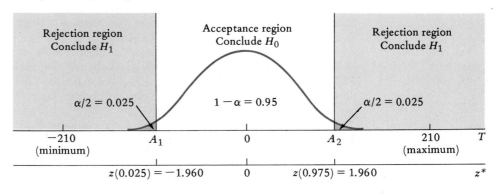

FIGURE 15.8 **Illustration of the statistical decision rule for the credit information example — Large-sample, two-sided Wilcoxon signed rank test for η_D**

Where: $z^* = \dfrac{T - 0}{\sigma\{T\}} = \dfrac{T - 0}{\sqrt{2870}}$

$\sigma^2\{T\} = \dfrac{n(n + 1)(2n + 1)}{6} = 2870$

Step 2. The researcher wants to control the α risk at 0.05 when $\eta_D = 0$.

Step 3. Figure 15.8 shows the appropriate two-sided type of decision rule. The sampling distribution of T when $\eta_D = 0$ is approximately normal in this example because $n = 20$. Hence, to control α at 0.05, we require $z(0.975) = 1.960$, and the decision rule for the test is as follows:

If $|z^*| \leq 1.960$, conclude H_0.
If $|z^*| > 1.960$, conclude H_1.

Step 4. The calculations in Table 15.3 gave $T = 184$. Thus, the standardized test statistic (15.11) here is:

$$\sigma^2\{T\} = \frac{20(20 + 1)[2(20) + 1]}{6} = 2870 \qquad \sigma\{T\} = \sqrt{2870} = 53.57$$

$$z^* = \frac{T - 0}{\sigma\{T\}} = \frac{184 - 0}{53.57} = 3.43$$

Since $|z^*| = 3.43 > 1.960$, conclusion H_1 is indicated — that $\eta_D \neq 0$. ☐

Comments 1. The Wilcoxon signed rank test accommodates tied ranks if they are not too numerous. If they are numerous, modifications must be made in the decision rule.

2.   The Wilcoxon signed rank test may also be used to test if a single population (which is symmetrical and continuous) has a median equal to some specified value, say, $\eta = \eta_0$. From the random sample observations $X_1, X_2, \ldots, X_n$, the differences $D_i = X_i - \eta_0$ are computed, and the test of whether or not $\eta_D = 0$ is applied to these differences in the standard fashion. Clearly, testing whether the differences D_i have a zero median is equivalent to testing whether the X_i have a median equal to η_0.

For example, in the case duration example, suppose that we wish to test H_0: $\eta = 6$ versus H_1: $\eta \neq 6$ and can assume that the distribution of case durations is symmetric. We would then use the differences $D_1 = 4.5 - 6 = -1.5$, $D_2 = 5.8 - 6 = -0.2$, and so on, and employ the Wilcoxon signed rank test for the alternatives H_0: $\eta_D = 0$, H_1: $\eta_D \neq 0$.

3.   Like the M–W–W test, the Wilcoxon signed rank test often compares favorably with parametric tests based on stronger assumptions about the underlying population (such as the t test for a normal population of differences, discussed in Section 13.3). For this reason, the test is widely used in practice.

15.4   OPTIONAL TOPIC — TEST FOR RANDOMNESS

In this section, we take up a nonparametric procedure for examining whether or not a set of observations constitutes a random sample from an infinite population.

Nature of Randomness

We saw from definition (8.11) in Chapter 8 that a random sample from an infinite population is one in which the observations are statistically independent and have the same probability distribution. That is, the process generating the sample data is independent and stationary as defined in (6.1). Figure 15.9a illustrates a sequence of observations generated by an independent and stationary process from the standard normal distribution $N(0, 1)$.

Departures from randomness can occur in many ways, some of which are illustrated in Figure 15.9. The sequence in Figure 15.9b comes from a process with an upward trend, which is inconsistent with the assumption that all outcomes of a random sequence are from the same distribution. In Figure 15.9c, the sequence of observations exhibits increasing variability, which again illustrates that the observations are not from the same distribution. Figures 15.9d and 15.9e contain sequences in which consecutive observations are dependent. In Figure 15.9d, the process involves a cyclical pattern; while in Figure 15.9e, the process alternates regularly.

Tests for randomness are of major importance because the assumption of randomness underlies statistical inference. In addition, tests for randomness are important for time series analysis, discussed in Chapters 24 and 25.

Test of Runs Up and Down

Departures from randomness can take so many forms that no single test for randomness is best for all situations. One of the most common departures from randomness is

FIGURE 15.9   Sequences of observations from random and nonrandom processes

(a) Random normal process   (b) Process involving
                                                   an upward trend

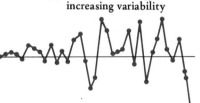   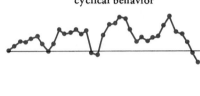

(c) Process involving   (d) Process involving
     increasing variability   cyclical behavior

(e) Process involving regular and
     frequent changes in direction

the tendency of a sequence to persist in its direction of movement. This type of departure is found in Figure 15.9b, where the sequence contains a trend, and in Figure 15.9d, where the persistence of movement produces a pronounced cyclical pattern. We now present a nonparametric test that is particularly effective when a process contains this persistence, as well as when a process contains excessively frequent changes in direction, as in Figure 15.9e.

Assumption.   The test assumes only that all sample observations come from continuous populations.

Alternatives.   The runs-up-and-down test, as already noted, can be used for detecting trend and other forms of persistence in a sequence of sample observations. Here, the alternatives are as follows.

(15.12)

H_0: Sequence generated by a random process.
H_1: Sequence generated by a process containing persistence.

When the test is used to detect departures from randomness consisting of either frequent or infrequent changes in direction, the alternatives are as given next.

(15.13)

H_0: Sequence generated by a random process.
H_1: Sequence generated by a process containing either persistence or frequent changes in direction.

EXAMPLE

Lake Level. Table 15.4 shows the maximum level of a lake each year for a 30-year period. It is desired to test whether the 30 observations have been generated by a random process or whether the process contains a persistent trend. The presence of a trend would have significant environmental policy implications. ☐

Test Statistic. The test statistic is a simple one — namely, the total number of runs up and down. In Table 15.4, the direction of change in the sequence of maximum lake levels is shown by plus and minus signs. Thus, since $X_2 = 6.59$ is less than $X_1 = 6.63$, the change is $-$. A *run* is a succession of plus or minus signs, surrounded by the oppo-

TABLE 15.4   **Determination of the number of runs in the maximum lake level for 30 consecutive years — Lake level example**

Year	Level (meters above 190)	Change		Year	Level (meters above 190)	Change	
1	6.63			16	5.91	$-$	
2	6.59	$-$	} Run 1	17	5.81	$-$	
3	6.46	$-$		18	5.64	$-$	} Run 7
4	6.49	$+$	} Run 2	19	5.51	$-$	
5	6.45	$-$		20	5.31	$-$	
6	6.41	$-$		21	5.36	$+$	} Run 8
7	6.38	$-$	} Run 3	22	5.17	$-$	
8	6.26	$-$		23	5.07	$-$	} Run 9
9	6.09	$-$		24	4.97	$-$	
10	5.99	$-$		25	5.00	$+$	} Run 10
11	5.92	$-$		26	5.01	$+$	
12	5.93	$+$	} Run 4	27	4.85	$-$	
13	5.83	$-$	} Run 5	28	4.79	$-$	} Run 11
14	5.82	$-$		29	4.73	$-$	
15	5.95	$+$	} Run 6	30	4.76	$+$	} Run 12

site sign. Thus, we see in Table 15.4 that the first run consists of two minus signs and the next run consists of one plus sign. Let R denote the total number of runs in the sequence. We have $R = 12$ for the data in Table 15.4.

When a trend or other such form of persistence is present, R will clearly be small. The smallest possible value of R is 1 (a single run). On the other hand, if the process involves frequent changes in direction, R will be large. The largest possible value of R is $n - 1$, when the signs alternate with each observation. Thus, when one is testing the alternatives in (15.12), small values of R will lead to H_1; and when one is testing the alternatives in (15.13), both small and large values of R will lead to H_1.

Sampling Distribution of R When H_0 Holds.   To control the α risk at a specified level, we need to know the sampling distribution of R when the sequence is generated by a random process. This sampling distribution can be determined exactly for all sample sizes, since under H_0 all possible permutations of the observed values are equally likely. Tables of this sampling distribution for small sample sizes are available; but for most practical applications, they are not needed, since a normal approximation is applicable for all but very small sample sizes.

(15.14)

When a sequence of n observations is generated by a random process (that is, stationary and independent) and the distribution is continuous, the sampling distribution of R has mean and variance:

$$E\{R\} = \frac{2n - 1}{3} \qquad \sigma^2\{R\} = \frac{16n - 29}{90}$$

Furthermore, when n is sufficiently large, the sampling distribution of R is approximately normal. As a working rule, the normal approximation is adequate if n is 20 or more.

Standardized Test Statistic.   All tests for randomness using the runs-up-and-down method are conducted by means of the following standardized test statistic.

(15.15)

Runs-up-and-down tests, when the number of observations is sufficiently large, are based on the standardized test statistic:

$$z^* = \frac{R - E\{R\}}{\sigma\{R\}}$$

where:

$E\{R\}$ and $\sigma\{R\}$ are given by (15.14)

When the process is random, z^* follows approximately a standard normal distribution.

Given the approximate normality of the sampling distribution of $z*$ when H_0 holds, the construction of the decision rule parallels earlier large-sample tests.

EXAMPLE

In the lake level example, it is desired to test whether or not persistence is present in the process.

Step 1.   The alternatives here are those in (15.12).

Step 2.   The α risk is to be controlled at 0.01 when the process is random.

Step 3.   Figure 15.10 shows the appropriate type of decision rule. The test is a one-sided lower-tail one because, as noted earlier, evidence of persistence in a process is a small number of runs in the sequence of observations. The sampling distribution of R under H_0 is approximately normal in this example because $n = 30$. Hence, for $\alpha = 0.01$, we require $z(0.01) = -2.326$, and the decision rule for the test is as follows:

If $z* \geq -2.326$, conclude H_0.
If $z* < -2.326$, conclude H_1.

Step 4.   Table 15.4 shows that the number of runs up and down is $R = 12$. We therefore calculate standardized test statistic (15.15) as follows:

$$E\{R\} = \frac{2(30) - 1}{3} = 19.67$$

FIGURE 15.10   Illustration of the statistical decision rule for the lake level example — Large-sample, one-sided, lower-tail test for runs up and down

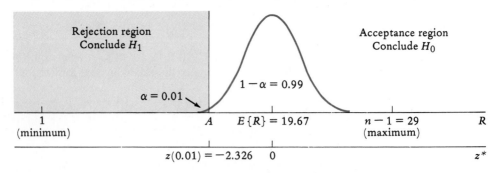

Where: $z* = \dfrac{R - E\{R\}}{\sigma\{R\}} = \dfrac{R - 19.67}{\sqrt{5.01}}$

$E\{R\} = \dfrac{2n - 1}{3} = 19.67$

$\sigma^2\{R\} = \dfrac{16n - 29}{90} = 5.01$

$$\sigma^2\{R\} = \frac{16(30) - 29}{90} = 5.01 \qquad \sigma\{R\} = \sqrt{5.01} = 2.24$$

$$z^* = \frac{R - E\{R\}}{\sigma\{R\}} = \frac{12 - 19.67}{2.24} = -3.42$$

Since $z^* = -3.42 < -2.326$, we conclude H_1—that the lake level is undergoing a persistent change. Thus, it is not appropriate to treat the 30 observations as a random sample from a population. $\qquad\square$

Comment	If consecutive observations are tied, we suggest that the tied observations be treated as a continuation of the existing run. For instance, tied observations following a minus sign should be treated as a minus. If ties are too numerous, the decision rule must be modified.

15.5 OPTIONAL TOPIC—JACKKNIFE ESTIMATION OF POPULATION STANDARD DEVIATION

In Section 13.5, we considered confidence interval estimation of the population variance and standard deviation when the population is normal. We cautioned there that the methodology does not yield reliable results when the population distribution is not close to normal. We now consider a robust procedure for estimating the population standard deviation, that is, a procedure applicable under a wide variety of conditions. This procedure is called the *jackknife procedure*. It is so named because the procedure is a multipurpose estimation tool that is reliable under quite general conditions.

The jackknife procedure for estimating the population standard deviation σ is conceptually quite simple, involving the following three steps.

Step 1.   From the n sample observations, n standard deviations are calculated, each time omitting one sample observation. We shall let s_{-i} denote the sample standard deviation obtained when X_i is omitted. Clearly, each s_{-i} is an estimator of the population standard deviation σ.

Step 2.   Next, n *pseudovalues* J_i are calculated, utilizing the standard deviation s for the full sample and the standard deviations s_{-i} obtained by omitting one X_i observation at a time.

(15.16)
$$J_i = ns - (n - 1)s_{-i} \qquad i = 1, 2, \ldots, n$$

Since both s and s_{-i} are estimators of the population standard deviation σ, each J_i is also an estimator of σ.

Step 3.   Now, we treat the n pseudovalues $J_1, J_2, \ldots, J_n$ as if they were a random sample from a normal population. We compute the mean and standard deviation of the pseudovalues in the usual way—here, we denote them by $\bar{J}$ and s_J.

$$(15.17) \qquad \bar{J} = \frac{\sum\limits_{i=1}^{n} J_i}{n} \qquad s_J = \sqrt{\frac{\sum\limits_{i=1}^{n} (J_i - \bar{J})^2}{n - 1}}$$

Next, we compute the estimated standard deviation of $\bar{J}$, denoted here by $s\{\bar{J}\}$, just as is done for the sample mean $\bar{X}$.

$$(15.18) \qquad s\{\bar{J}\} = \frac{s_J}{\sqrt{n}}$$

The statistic $\bar{J}$ is an estimator of the population standard deviation σ because each J_i is an estimator of σ. To estimate the population standard deviation σ by means of an interval estimate, we employ confidence interval (10.15) based on a sample mean. This interval, which utilizes the t distribution, is appropriate here because the pseudovalues J_i are treated as if they were a random sample from a normal population.

(15.19)

The jackknife confidence limits for σ, with approximate confidence coefficient $1 - \alpha$, are:

$$\bar{J} \pm ts\{\bar{J}\}$$

where:
$t = t(1 - \alpha/2; n - 1)$
$\bar{J}$ is given by (15.17)
$s\{\bar{J}\}$ is given by (15.18)

EXAMPLE

Fire Insurance Claims.   A random sample of seven fire insurance claims under a particular type of residential policy consists of the following claims (in thousands of dollars):

| 6.2 | 3.1 | 1.0 | 9.3 | 4.2 | 0.7 | 11.7 |

It is desired to obtain a 95 percent confidence interval for σ, the population standard deviation for claims. Since the population of claims is quite skewed, confidence limits (13.29) for a normal population are not appropriate, and the jackknife procedure is to be employed. We shall follow the three-step procedure.

Step 1.   The sample mean for the $n = 7$ claims is $\overline{X} = 5.17$. Hence, the sample variance based on the full sample is:

$$s^2 = \frac{1}{7 - 1}[(6.2 - 5.17)^2 + (3.1 - 5.17)^2 + \cdots + (11.7 - 5.17)^2]$$

$$= 17.226$$

so $s = \sqrt{17.226} = 4.15$. This value is shown in the last line of Table 15.5.

Next, we calculate the standard deviations s_{-i} by omitting one of the sample observations at a time. When X_1 is omitted, for instance, the mean of the remaining six observations is 5.00, and we obtain:

$$s^2_{-1} = \frac{1}{6 - 1}[(3.1 - 5.00)^2 + (1.0 - 5.00)^2 + \cdots + (11.7 - 5.00)^2]$$

$$= 20.424$$

so $s_{-1} = \sqrt{20.424} = 4.52$. This standard deviation is shown in Table 15.5, column 2, first line. The other standard deviations s_{-i} are also shown in this table.

Step 2.   Next, we calculate the pseudovalues J_i, using (15.16). For $i = 1$, we obtain:

$$J_1 = ns - (n - 1)s_{-1} = 7(4.15) - 6(4.52) = 1.93$$

This and the other pseudovalues are shown in column 3 of Table 15.5.

Step 3.   Now, we treat the n pseudovalues J_i as if they were a random sample from a normal population. Using (15.17), we obtain:

$$\overline{J} = \frac{1}{7}(1.93 + 2.41 + \cdots + 9.37) = \frac{29.77}{7} = 4.25$$

TABLE 15.5   **Calculations for a jackknife estimate of σ—Fire insurance claims example**

	(1)	(2)	(3)
	Omitted Observation	Sample Standard Deviation	Pseudovalue
i	X_i	s_{-i}	J_i
1	6.2	4.52	1.93
2	3.1	4.44	2.41
3	1.0	4.08	4.57
4	9.3	4.09	4.51
5	4.2	4.52	1.93
6	0.7	4.00	5.05
7	11.7	3.28	9.37
	None	$s = 4.15$ *Total*	29.77

and:

$$s_J^2 = \frac{1}{7 - 1}[(1.93 - 4.25)^2 + (2.41 - 4.25)^2 + \cdots + (9.37 - 4.25)^2]$$

$$= 6.862$$

so $s_J = \sqrt{6.862} = 2.62$. Finally, the estimated standard deviation of $\bar{J}$, using (15.18), is:

$$s\{\bar{J}\} = \frac{s_J}{\sqrt{n}} = \frac{2.62}{\sqrt{7}} = 0.99$$

For $1 - \alpha = 0.95$ and $n - 1 = 6$, we require $t(0.975; 6) = 2.447$. Hence, the confidence limits, using (15.19), are $4.25 \pm 2.447(0.99)$, and the confidence interval is:

$$1.83 \le \sigma \le 6.67$$

We conclude, with 95 percent confidence, that the standard deviation of the population of claims is between $1.8 thousand and $6.7 thousand. □

Comments

1. We noted already that jackknife estimators are robust. In addition, jackknife estimators have the desirable property that they avoid some types of bias that may be present in the original estimator. For example, s is a biased estimator of σ, but $\bar{J}$ has almost no bias. Jackknife estimators are also very useful for obtaining confidence intervals for parameters that must be estimated by very complicated statistics. Since this situation is common in large-scale surveys, jackknife estimation is widely used in analyzing survey data.

2. The appropriateness of confidence intervals based on the jackknife method can sometimes be improved by employing a mathematical transformation to reduce possible skewness in the distribution of the pseudovalues. In our example, for instance, we could have used $\log s_{-i}$ and $\log s$ instead of s_{-i} and s for obtaining the pseudovalues. If the population of claims were an extremely skewed distribution, this transformation would help make the pseudovalues more nearly normally distributed.

3. When n is large, computing n pseudovalues by omitting one observation at a time involves much computation. As a remedy for this situation, the sample can be divided randomly into groups. For example, a sample of size 300 might be divided into 10 groups of size 30 each. To estimate the population standard deviation, one computes 10 sample standard deviations s_{-i}, dropping out one group of observations at a time. Ten pseudovalues would then be calculated according to (15.16), and the confidence interval for σ would be obtained by using (15.19) but now with $n = 10$.

PROBLEMS

*15.1 A random sample of size $n = 14$ has been selected from a continuous population, and a confidence interval of the form (15.1) is to be constructed.
   a. What is the value of r for which this confidence interval has a confidence coefficient closest to 0.99?
   b. What is the exact confidence coefficient when $r = 2$?

15.2 A random sample of size $n = 10$ has been selected from a continuous population, and a confidence interval of the form (15.1) is to be constructed.
   a. What is the value of r for which this confidence interval has a confidence coefficient closest to 0.95?
   b. What is the exact confidence coefficient when $r = 3$?

*15.3 The weights (in kilograms) of 9 smallmouth bass taken from a lake were as follows:

$$2.83 \quad 2.96 \quad 2.47 \quad 3.00 \quad 2.08 \quad 2.54 \quad 2.84 \quad 3.04 \quad 2.31$$

Assume that these observations represent a random sample from a continuous population with median η.
   a. Array the observations, and obtain a point estimate of η.
   b. Construct a confidence interval for η with a confidence coefficient near 0.95. Interpret your confidence interval.

15.4 **Damage Suits.** In 12 recent automobile damage suits where the claims were settled out of court, the damages agreed upon (in $ thousand) were:

$$5.2 \quad 5.5 \quad 3.8 \quad 12.5 \quad 8.3 \quad 2.1 \quad 1.7 \quad 20.0 \quad 4.8 \quad 6.9 \quad 7.5 \quad 10.6$$

Assume that these data represent a random sample from a continuous population with median η.
   a. Array the observations, and obtain a point estimate of η.
   b. Construct a confidence interval for η with a confidence coefficient near 0.99. Interpret your confidence interval.

15.5 A chemist made eight independent measurements of the percent yield of a microchemical reaction, obtaining the following:

$$68 \quad 43 \quad 71 \quad 65 \quad 53 \quad 62 \quad 68 \quad 49$$

Assume that these data represent a random sample from a continuous population with median η.
   a. Array the observations, and obtain a point estimate of η.
   b. Construct a confidence interval for η with a confidence coefficient near 0.90. Interpret your confidence interval.

*15.6 Consider a random sample of size $n = 36$ from a continuous population. Using approximation (15.2), give the value of r for which the interval estimate (15.1) has a confidence coefficient of about 0.95.

15.7 Consider a random sample of size $n = 25$ from a continuous population. Using approximation (15.2), give the value of r for which confidence interval (15.1) has a confidence coefficient of approximately 0.99.

*15.8   **Failure Ages.**   An aluminum company experienced the following failure ages (in days) with a group of 21 electrolytic cells of design A that were installed in one of its smelter potrooms (in order of failure):

518	903	1192	1477	1814	2060	2421
775	1015	1354	1604	1826	2274	2591
888	1189	1367	1708	2040	2330	2716

The 21 observations may be considered a random sample from a continuous population of failure ages.

   a. Construct a 90 percent confidence interval for the population median failure age of cells of design A. Interpret your confidence interval.
   b. When the distribution of failure ages is symmetrical, why does your confidence interval in **a** also serve as a 90 percent confidence interval for the population mean failure age?
   c. The target median failure age for cells is 1800 days. Does your confidence interval in **a** suggest that the target is being met by design A? Explain.

15.9   Refer to the data array of time intervals between oil spills before pollution controls in Figure 3.8. Assume that the 40 observations are a random sample from a continuous population. Construct a 95 percent confidence interval for the population median. Interpret your confidence interval.

15.10 Refer to the **Financial Characteristics** data set. For textile products firms, construct a 99 percent confidence interval for the population median net assets in year 1. Assume that the 52 observations constitute a random sample from a continuous population. Interpret your confidence interval.

*15.11 In fund-raising drives by a charitable foundation, the median contribution from regular contributors has been $62.50 in the recent past. In an experiment, an analyst in the foundation prepared a glossy brochure describing the foundation's work and distributed it to a random sample of 250 regular contributors shortly before the next fund-raising drive. A subsequent follow-up indicated that 140 of the 250 contributors gave more than $62.50, 109 gave less than $62.50, and one contributed exactly $62.50. The analyst now plans to conduct a sign test to determine whether or not the median contribution η of regular contributors who are sent the brochure still is $62.50.

   a. State the test alternatives in terms of η. Then, restate the test alternatives in terms of p, and define p.
   b. Conduct the sign test, controlling the α risk at 0.10 when $\eta = 62.50$. State the decision rule, the value of the test statistic, and the conclusion.
   c. What is the P-value of the test?

15.12 Students enrolled in a community school system achieved a median score of 512 on a chemistry aptitude test during the past few years. Of 800 students who took the aptitude test this year, 493 had scores above 512, and 307 had scores below 512.

   a. Use the sign test to decide whether or not the population median failure age of cells of design A exceeds 1800 days. Control the α risk at 0.05 when $\eta = 1800$. State the alternatives, the decision rule, the value of the test statistic, and the conclusion.
   b. What is the P-value of the test?

15.13 Refer to **Failure Ages** Problem 15.8.

   a. Use the sign test to decide whether or not the population median failure age of cells of design A exceeds 1800 days. Control the α risk at 0.05 when $\eta = 1800$. State the alternatives, the decision rule, the value of the test statistic, and the conclusion.
   b. What is the P-value of the test?

*15.14 A gas industry report states that 50 percent of residential customers pay their gas bills within 14 days after billing. In a recent random sample of 10 residential customers of one gas company, the bills were paid within the following numbers of days after billing (arrayed by magnitude):

| 2 | 4 | 8 | 13 | 15 | 28 | 29 | 35 | 36 | 46 |

  a. Use the sign test to determine whether or not the median payment period for this company equals 14 days. Control the α risk near 0.02 when $\eta = 14$. State the alternatives, the decision rule, the value of the test statistic, and the conclusion.
  b. What is the exact α risk of your test?

15.15 The median score of the general population on a standard psychological test designed to measure extrinsic-reward motivation is 115. A random sample of eight successful sales managers who took the test obtained the following scores (arrayed by magnitude):

| 116 | 120 | 121 | 128 | 131 | 138 | 139 | 145 |

  a. Use the sign test to determine whether or not the median test score of successful managers exceeds that of the general population. Control the α risk near 0.05 when $\eta = 115$. State the alternatives, the decision rule, the value of the test statistic, and the conclusion.
  b. What is the exact α risk of your test?

15.16 Refer to **Damage Suits** Problem 15.4. An analyst wishes to determine whether or not the population median amount awarded equals $13.0 thousand.
  a. Use the sign test to conduct the test, controlling the α risk near 0.01 when $\eta = 13.0$. State the alternatives, the decision rule, the value of the test statistic, and the conclusion.
  b. What is the exact α risk of the test in **a**?
  c. Explain how the confidence interval in Problem 15.4b could be used to conduct the same test as in **a**.

*15.17 In a random sample of $n_1 = 15$ homeowners, each was asked to state the maximum amount (in $ thousand) they would pay for a home security system described to them in a promotional brochure. An independent random sample of $n_2 = 12$ homeowners was asked the same question but was allowed to inspect an installed system in operation, in addition to receiving the brochure. The results follow (arrayed by magnitude).

Brochure Only X_i					Brochure and Inspection Y_i			
2.4	3.3	3.9	4.5	5.3	4.8	5.9	6.4	6.9
3.0	3.5	4.3	4.5	5.4	5.1	6.0	6.5	7.1
3.1	3.6	4.4	4.7	5.5	5.2	6.3	6.7	7.5

  a. In an M–W–W test to determine whether or not homeowners are willing to pay a larger amount for the system if they have inspected an installed system than if they have not, what is the smallest possible value of S_2? What is the largest possible value of S_2? What is the actual value of S_2 based on the study results?
  b. Conduct the required test, controlling the α risk at 0.05 when $\delta = 0$. State the alternatives, the decision rule, the value of the test statistic, and the conclusion.
  c. What is the P-value of the test?
  d. What assumptions, if any, are required to use test procedure (13.11a) that are not required for the M–W–W test?

15.18 Employee Productivity.   The numbers of units produced by $n_1 = 25$ employees randomly selected from shift 1 and by $n_2 = 25$ employees randomly and independently selected from shift 2 are as follows (arrayed by magnitude):

		Shift 1 X_i					Shift 2 Y_i		
169	185	198	202	213	172	186	203	212	231
176	188	198	204	215	175	190	206	218	235
177	189	199	205	220	178	191	207	221	237
182	192	200	208	225	181	196	209	227	246
183	195	201	211	230	184	197	210	229	253

a. Use the M–W–W test to determine whether or not employee productivity in shift 2 is higher than that in shift 1. Control the α risk at 0.05 when $\delta = 0$. State the alternatives, the decision rule, the value of the test statistic, and the conclusion.

b. What is the P-value of the test?

c. Interpret the shift parameter δ here.

15.19 Refer to **Employee Productivity** Problem 15.18.

a. Construct separate stem-and-leaf displays for the two samples. Use as stems 16, 17, and so on. Do the frequency patterns of the displays support the applicability of the M–W–W test here? Comment.

b. Does the difference in positions of the two displays in **a** represent an estimate of δ here? Comment.

15.20 A new method of swathing grain is under experimentation in a dry northern region. The method leaves strips of standing grain to increase snow retention on fields during winter and thereby to increase soil moisture and crop yields in the following growing season. In one experiment, 34 similar fields were divided randomly into two equal groups. Fields in one group were swathed conventionally and fields in the other group were swathed by using the snow-strip technique. Grain yields (in bushels per acre) for the fields in the following year were as follows (arrayed by magnitude):

	Conventional Swathing X_i				Snow-Strip Swathing Y_i		
15.6	19.7	23.5	27.1	21.4	26.3	30.2	33.2
16.1	20.8	24.4	28.3	23.3	27.5	30.3	36.3
17.7	21.0	25.7		25.7	28.0	30.7	
18.6	22.5	26.5		25.9	28.4	31.4	
19.7	22.8	26.6		26.2	29.4	32.9	

a. Use the M–W–W test to determine whether or not snow-strip swathing leads to higher crop yields in the following year. Control the α risk at 0.01 when $\delta = 0$. State the alternatives, the decision rule, the value of the test statistic, and the conclusion.

b. What is the P-value of the test?

c. Under the assumptions of the M–W–W test, $\delta = \mu_2 - \mu_1 = \eta_2 - \eta_1$. Estimate δ by calculating $\overline{Y} - \overline{X}$ and also by calculating the difference between the two sample medians. How do the two estimates compare?

*15.21 Transferred Executives.   A company's personnel policy provides a transferred executive with a guaranteed purchase price for his or her home if it cannot be sold for a larger amount within a fixed period from the transfer notification date. The guaranteed price is the average of two appraisals made independently by company-appointed real estate appraisers. The following are appraisal data (in $ thousand) for a random sample of 15 houses appraised by the same pair of appraisers.

House	Appraiser No. 1 X_i	Appraiser No. 2 Y_i	House	Appraiser No. 1 X_i	Appraiser No. 2 Y_i	House	Appraiser No. 1 X_i	Appraiser No. 2 Y_i
1	140.3	140.6	6	159.7	160.7	11	174.4	175.5
2	154.6	155.1	7	161.5	161.2	12	164.4	167.0
3	153.1	153.5	8	150.1	150.8	13	145.2	146.9
4	165.1	167.8	9	163.7	163.3	14	180.1	182.3
5	154.8	155.2	10	158.6	160.4	15	178.5	177.7

Construct a confidence interval for the median of the population of differences $D_i = Y_i - X_i$, with a confidence coefficient near 90 percent. Interpret your confidence interval.

15.22 **Physical Fitness.** The data that follow show the physical fitness of 24 members of a health club at the time of joining (X_i) and six months later (Y_i). Physical fitness is measured by an index based on a number of physical-fitness tests. The larger the index, the better the physical fitness. Assume that the 24 members constitute a random sample.

Member	Index X_i	Index Y_i	Member	Index X_i	Index Y_i	Member	Index X_i	Index Y_i
1	92	100	9	113	116	17	116	115
2	118	124	10	87	87	18	99	95
3	87	101	11	87	97	19	105	120
4	83	94	12	79	99	20	97	122
5	100	113	13	113	113	21	100	102
6	94	101	14	86	98	22	84	82
7	112	117	15	86	95	23	122	116
8	106	123	16	112	109	24	100	124

a. Construct a 95 percent confidence interval for the median of the population of differences $D_i = Y_i - X_i$. Interpret your confidence interval.
b. Does it appear that physical fitness improves with club membership?

15.23 A random sample of ten boys experienced the following growths in height (in centimeters) during their sixth year of life:

6.0   4.7   4.2   5.0   4.6   4.0   5.4   5.3   6.5   5.5

a. Construct a confidence interval for the population median height change of boys in their sixth year of life. Use a confidence coefficient near 0.98. Interpret your confidence interval.
b. What assumption about the population of height changes must hold for the estimation procedure in a to be valid? Is the assumption satisfied here? Explain.

*15.24 Refer to **Transferred Executives** Problem 15.21.

    a. Use the sign test to determine whether or not the population median difference is zero. Control the α risk near 0.10 when $\eta_D = 0$. State the alternatives, the decision rule, the value of the test statistic, and the conclusion.

    b. What is the exact α risk of the test?

15.25 Refer to **Physical Fitness** Problem 15.22. Use the sign test to determine whether or not the population median difference exceeds 10 points. Control the α risk at 0.025 when $\eta_D = 10$. State the alternatives, the decision rule, the value of the test statistic, and the conclusion.

15.26 Refer to **Achievement Scores** Problem 13.16.

    a. Use the sign test to determine whether or not the population median of test score differences is zero. Control the α risk at 0.05 when $\eta_D = 0$. State the alternatives, the decision rule, the value of the test statistic, and the conclusion.

    b. Since the population of test score differences is approximately normal, is the test in **a** also a test of whether or not the difference in population means $\mu_2 - \mu_1$ is zero? Explain.

*15.27 Refer to **Transferred Executives** Problem 15.21. Assume that the population of differences is symmetrical. It is desired to use the Wilcoxon signed rank test to determine whether or not the population median difference is zero.

    a. What is the lowest possible value of T here? What is the highest possible value of T here? What is the actual value of T based on the sample data?

    b. Conduct the Wilcoxon signed rank test, controlling the α risk at 0.10 when $\eta_D = 0$. State the alternatives, the decision rule, the value of the test statistic, and the conclusion.

    c. What is the P-value of the test?

    d. Make a frequency distribution of the differences in appraisals for the 15 houses and plot it as a histogram. Use class limits: $-1, 0, 1, 2, 3$. Does the histogram support the appropriateness of the Wilcoxon signed rank test here? Explain.

15.28 Refer to **Achievement Scores** Problem 13.16.

    a. Use the Wilcoxon signed rank test to determine whether or not the population median test score difference is zero. Control the α risk at 0.05 when $\eta_D = 0$. State the alternatives, the decision rule, the value of the test statistic, and the conclusion.

    b. Since the population of test score differences is approximately normal, is use of the Wilcoxon signed rank test valid? Explain.

15.29 The differences in ages between husband and wife in a random sample of 12 married couples are to be studied to determine whether or not the population median age difference exceeds zero, using the Wilcoxon signed rank test. The sample differences (husband's age minus wife's age) follow.

i:	1	2	3	4	5	6	7	8	9	10	11	12
D_i:	8.8	-3.6	0.6	2.8	3.6	5.1	-0.4	-0.2	5.4	-0.4	1.5	4.4

Assume that the population of differences is symmetrical.

    a. Conduct the Wilcoxon signed rank test, controlling the α risk at 0.10 when $\eta_D = 0$. State the alternatives, the decision rule, the value of the test statistic, and the conclusion.

    b. What is the P-value of the test?

15.30 Refer to **Failure Ages** Problem 15.8. Assume that the population of failure ages is symmetrical. Cells of competing design B are known to have a median failure age of 1300 days. Use the Wilcoxon signed rank test to determine whether or not the population median failure age of cells of design A exceeds 1300 days. Control the α risk at 0.05 when $\eta = 1300$. State the alternatives, the decision rule, the value of the test statistic, and the conclusion.

*15.31 Refer to **Melting Points** Problem 14.10.

a. How many runs up and down are contained in the sequence of 28 sample means? How many runs would be expected if the process were random?

b. Use the test of runs up and down to determine whether the sequence of sample means was generated by a random process or by a process containing either persistence or frequent changes in direction. Control the α risk at 0.05 when the sequence is generated by a random process. State the alternatives, the decision rule, the value of the test statistic, and the conclusion.

15.32 **Random Number Generator.** The following 50 numbers were generated by a computer program written by a graduate student to generate independent standard normal random numbers.

−0.957	−0.238	−1.029	0.551	−0.298
0.525	−0.869	0.479	0.418	1.064
−1.865	−1.016	2.709	0.074	0.162
−0.273	0.417	−0.057	0.524	−0.129
−0.035	0.056	−0.300	0.479	−1.204
0.371	0.561	−0.594	0.326	1.097
−0.702	−2.357	−1.047	1.114	−0.916
−0.432	1.956	−1.347	1.068	1.222
−0.465	−0.281	0.996	0.772	−1.153
0.120	0.932	−1.023	−0.226	1.298

The numbers were generated in successive rows from left to right.

a. Plot the sequence of numbers. Is any evidence of nonrandomness apparent in your plot? Explain.

b. Conduct a two-sided test of runs up and down on the sequence of numbers. Control the α risk at 0.10 when the sequence is generated by a random process. State the alternatives, the decision rule, the value of the test statistic, and the conclusion.

c. Describe the types of nonrandomness that the two-sided test in **a** is designed to detect. Does the validity of the test in **a** depend on the generated numbers being standard normal? Explain.

d. If runs tests were performed on 100 independent sequences of 50 numbers, each generated by the computer program and α controlled at 0.10 for each test, how many times would the conclusion of nonrandomness be expected if the computer program generates numbers in a random sequence?

15.33 **Marriages.** The numbers of marriages in a city for each month from July of year 1 to June of year 5 follow.

Year 1		Year 2			Year 3			Year 4			Year 5	
J	219	J	121	J	228	J	114	J	231	J	131	J	227	J	140
A	226	F	131	A	226	F	128	A	240	F	135	A	254	F	155
S	185	M	140	S	195	M	151	S	223	M	150	S	204	M	153
O	186	A	176	O	183	A	178	O	185	A	189	O	199	A	190
N	162	M	187	N	165	M	202	N	174	M	221	N	187	M	224
D	180	J	245	D	182	J	262	D	194	J	277	D	181	J	278

a. Plot this monthly time series. Does your plot suggest any departure from randomness? Explain.

    b. Perform a test of runs up and down to determine whether or not persistence is present. Control the α risk at 0.01 when the process is random. State the alternatives, the decision rule, the value of the test statistic, and the conclusion.

*15.34 The durations (in days) of the six most recent transit strikes in a city are:

$$5 \quad 2 \quad 7 \quad 6 \quad 4 \quad 10$$

As a measure of the variability of durations of transit strikes in this city, a 95 percent confidence interval for the standard deviation of the probability distribution of strike durations is to be constructed. Assume the six observations constitute a random sample from this distribution.
    a. Verify that $s = 2.733$.
    b. Obtain the desired confidence interval, using the jackknife procedure. Some of the required calculational results follow.

i:	1	2	3	4	5	6
X_i:	5	2	7	6	4	10
s_{-i}:	3.033	2.302	2.966	____	____	1.924
J_i:	1.229	4.884	1.563	____	____	6.777

15.35 Rates of Return. Rates of return (in percent) earned on the common stock of Polyphase, Inc., during each of the past five quarters are:

$$-5 \quad 10 \quad 2 \quad 19 \quad 15$$

Construct a 90 percent confidence interval for the population standard deviation of quarterly rates of return on this stock, using the jackknife procedure. Assume the five observations constitute a random sample from the population of all quarterly rates of return. Some of the required calculational results follow.

i:	1	2	3	4	5	
X_i:	-5	10	2	19	15	
s_{-i}:	7.326	11.177	10.500	____	____	$s = 9.731$
J_i:	19.354	3.951	6.657	____	____	

15.36 In the last six sales of commercial property in a tax district, the assessed values of the properties, expressed as a percentage of market value, were:

$$38 \quad 46 \quad 22 \quad 49 \quad 27 \quad 34$$

Use the jackknife procedure to construct a 99 percent confidence interval for the population standard deviation of assessed value as a percentage of market value for commercial properties in this tax district. Assume the six percentages constitute a random sample from the population. Some calculational results follow.

i:	1	2	3	4	5	6	
X_i:	38	46	22	49	27	34	
s_{-i}:	11.718	10.416	8.927	9.370	____	____	$s = 10.526$
J_i:	4.569	11.075	18.520	16.306	____	____	

15.37 Refer to **Rates of Return** Problem 15.35.
  a. Use (13.29) to construct a 90 percent confidence interval for σ. What assumption is required for this confidence interval that is not required by the jackknife procedure?
  b. How does your confidence interval in **a** compare with that obtained in Problem 15.35 by the jackknife method?

═══════ ## EXERCISES

15.38 For a random sample from a symmetrical population, explain why $E\{Md\} = \mu = \eta$.

15.39 For an odd-sized random sample from a continuous population, show that η is the median of the sampling distribution of Md. (*Hint:* Does Md have an equal probability of falling on either side of η?)

15.40 Use Table C.5 to evaluate the confidence level of the interval $L_8 \le \eta \le U_8$ when $n = 20$.

15.41 Verify (15.2). [*Hint:* Use the normal approximation to the binomial probability in (15.3), together with the correction for continuity discussed in Section 12.3.]

15.42 Derive (15.9) from (15.7). [*Hint:* $S_1 + S_2 = (n_1 + n_2)(n_1 + n_2 + 1)/2$ for the M–W–W test statistics.]

15.43 Derive $E\{T\}$ and $\sigma^2\{T\}$ in (15.10) when $\eta_D = 0$. [*Hint:* $\sum_{i=1}^{n} i^2 = n(n + 1)(2n + 1)/6$.]

═══════ ## STUDIES

15.44 Refer to **Failure Ages** Problem 15.8. The company has also installed 13 cells of a new design C. The distribution of cell failure ages for this design is known to be approximately normal. All 13 cells were installed on the same day. The first 10 failures among the 13 cells in the sample have occurred, and their failure ages (in days) are (in order of failure):

  808 980 1351 1543 1627 1666   1726 1993 2168 2271

The company wishes to estimate the mean failure age of cells of this design from these incomplete sample data.
  a. Can any of the estimation procedures for the mean of a normal population discussed earlier in the text be employed in this case? Explain.
  b. Since the mean and the median of a normal population coincide, it is decided to construct a confidence interval for the population median failure age, with a confidence coefficient near 90 percent. Obtain this interval estimate and interpret it.
  c. Use the Wilcoxon signed rank test to determine whether or not the median failure age of cells of design C exceeds 1300 days. Control the α risk at 0.05 when $\eta = 1300$. Explain why this test procedure can be applied even though the failure ages of the last three cells are not known yet.

15.45 Ten sales executives selected randomly had the following annual incomes (in $ thousand):

  37.0 45.7 43.1 56.8 94.1 36.9 51.4 38.3 64.1 142.9

An analyst suspects that the population income distribution of sales executives is quite skewed, so inference procedures not depending on approximate normality should be used.

a. Estimate the population median income by an interval estimate, with a confidence coefficient near 95 percent.

b. The analyst wishes to test whether or not the population median income of sales executives exceeds $60 thousand. Conduct a sign test, controlling the α risk near 0.025 when $\eta = 60$. State the alternatives, the decision rule, the value of the test statistic, and the conclusion.

c. The analyst also wishes to estimate the population standard deviation σ. To improve the appropriateness of the jackknife procedure, the analyst believes it would be desirable to use the logarithms of the standard deviations s and s_{-i}, as described in Comment 2, page 514. Construct a 95 percent confidence interval for σ by first obtaining a confidence interval for $\log \sigma$, using the pseudovalues $J_i = n \log(s) - (n - 1) \log(s_{-i})$, and then converting this interval into one for σ.

15.46 (Computer needed.) Refer to the **Financial Characteristics** data set. The percent rates of return on net assets (net income as a percentage of net assets) for crude oil producers (firms 1–60) are of interest. Assume that the 60 firms are a random sample of all crude oil producers.

a. Compute the percent rate of return on net assets for each of years 1 and 2 for the 60 crude oil producers.

b. Obtain a 95 percent confidence interval for the population median percent rate of return of crude oil producers in year 2. Interpret your confidence interval.

c. Use the sign test to determine whether or not the population median of differences $D_i = Y_i - X_i$ in percent rate of return between year 2 (Y_i) and year 1 (X_i) equals 0. Control the α risk at 0.05 when $\eta_D = 0$. State the alternatives, the decision rule, the value of the test statistic, and the conclusion.

d. Construct a box plot of the differences D_i in percent rate of return for crude oil producers. Does the plot suggest that the test in c could have been conducted by means of the Wilcoxon signed rank test? Comment.

16 GOODNESS OF FIT

Frequently, the functional form of a population requires investigation. In this chapter, we present several inference procedures that are widely used for this purpose.

16.1 ALTERNATIVES IN GOODNESS-OF-FIT TESTS

Information about the functional form of the population of interest may be needed for (1) confirming theories or validating models, or (2) determining whether the population assumptions of a statistical inference procedure are appropriate for the given circumstances. We illustrate these types of situations with three examples.

EXAMPLES

1. **Spare Part.** An operations analyst is developing a model of replacement demand for a spare part. She knows that for many spare parts, replacement demand follows a Poisson distribution. The analyst wishes to decide whether the weekly demand for this particular part follows a Poisson distribution. If so, the analyst will utilize Poisson probabilities in the model to ascertain the optimal inventory level. She has collected data on the number of replacement units of the spare part demanded per week for a sample of 52 weeks. She now wants to use these data to test which of the following alternatives should be accepted.

(16.1)

H_0: The probability distribution is Poisson.
H_1: The probability distribution is not Poisson.

2. **Milk Production.** A researcher for a milk producers' association has production data for a sample of 40 milk cows of the same age and breed. He wishes to estimate the variability in output among cows in the relevant population and would like to use confidence limits (13.29), which require the population to be approximately normal. A test of the assumption of normality involves the following alternatives.

(16.2)

> H_0: The probability distribution is normal.
> H_1: The probability distribution is not normal.

3. **Hospital Stays.** Extensive experience in another country has shown that the length of hospital stay (in days) for a certain psychiatric condition has an exponential distribution with parameter value $\lambda = 0.036$. A medical researcher has collected comparable data for 10 cases in this country and wishes to test whether these data also came from an exponential distribution with $\lambda = 0.036$; that is, he wishes to test the following alternatives.

(16.3)

> H_0: The probability distribution is exponential with $\lambda = 0.036$.
> H_1: The probability distribution is not exponential with $\lambda = 0.036$.

Tests for deciding between the alternatives in the three examples are called *goodness-of-fit tests*. The name arises because the alternatives involve the question whether or not a particular probability distribution is a good model for the population sampled, and this question will be judged by whether or not the probability distribution specified in H_0 is a good fit for the sample data.

The alternatives in the three examples differ in some important ways. In (16.1), the alternatives relate to a discrete probability distribution, whereas in (16.2) and (16.3), they relate to continuous probability distributions. Another difference is whether or not the parameter values are specified. The alternatives in (16.1) and (16.2) pertain to a family of probability distributions (Poisson, normal) without specifying the parameter value(s). On the other hand, H_0 in (16.3) specifies not only the family (exponential) but also the parameter value ($\lambda = 0.036$). Thus, conclusion H_1 in (16.1) and (16.2) implies that the probability distribution is not that specified in H_0 (Poisson, normal). On the other hand, conclusion H_1 in (16.3) implies either that the probability distribution is not exponential or that it is exponential but λ is not equal to 0.036.

16.2   CHI-SQUARE TEST

One of the most widely used goodness-of-fit tests is the *chi-square test*.

Assumptions

There are only two assumptions for the chi-square test:

1. The sample is a simple random one from the population.
2. The sample size is reasonably large.

Nature of Test

The chi-square test is based on a comparison of the sample data with the expected outcomes if H_0 is true. We first classify the sample data into k classes. For the ith class, we denote the observed number or frequency of sample values falling into that class by f_i. Of course, the f_i must sum to the total sample size n.

(16.4)

$$\sum_{i=1}^{k} f_i = n$$

Now, given the form of the distribution specified in H_0, we can determine the expected frequency for each class when H_0 is true. The expected frequency for the ith class when H_0 holds is denoted by F_i. As is the case for the observed frequencies, the expected frequencies F_i also must sum to the total sample size n.

(16.5)

$$\sum_{i=1}^{k} F_i = n$$

The two sets of frequencies to be compared are shown in Table 16.1.

EXAMPLE

In the spare part example, the alternatives are as follows:

H_0: The probability distribution is Poisson.
H_1: The probability distribution is not Poisson.

The weekly demand data for a random sample of 52 weeks are presented in Table 16.2a. In Table 16.2b, the 52 sample observations are classified into $k = 8$ classes, and the observed frequency for each class (f_i) is shown. The reason for using these eight classes will be explained shortly.

TABLE 16.1   Basic structure of a chi-square test for goodness of fit

Class	Observed Frequency	Expected Frequency Under H_0	Relative Squared Residual
1	f_1	F_1	$(f_1 - F_1)^2/F_1$
2	f_2	F_2	$(f_2 - F_2)^2/F_2$
$\vdots$	$\vdots$	$\vdots$	
k	f_k	F_k	$(f_k - F_k)^2/F_k$
Total	n	n	$X^2 = \Sigma (f_i - F_i)^2/F_i$

TABLE 16.2   Chi-square test for the goodness of fit of a Poisson distribution — Spare part example

(a)
Weekly replacement demands

Weeks	Demand												
1–13	5	5	2	1	3	3	3	3	8	6	5	3	6
14–26	4	6	5	3	2	6	2	5	3	2	8	2	2
27–39	6	3	6	2	5	4	6	4	5	3	4	2	2
40–52	6	6	3	7	5	0	4	4	10	1	3	4	0

(b)
Frequency distribution and calculations for chi-square test

Class i	Number Demanded X	Observed Frequency f_i	Probability Under H_0	Expected Frequency Under H_0 F_i	Residual $f_i - F_i$	Relative Squared Residual $(f_i - F_i)^2/F_i$
1	0–1	4	0.0916	4.763	−0.763	0.122
2	2	9	0.1465	7.618	1.382	0.251
3	3	11	0.1954	10.161	0.839	0.069
4	4	7	0.1954	10.161	−3.161	0.983
5	5	8	0.1563	8.128	−0.128	0.002
6	6	9	0.1042	5.418	3.582	2.368
7	7	1	0.0595	3.094	−2.094	1.417
8	8 or more	3	0.0511	2.657	0.343	0.044
	Total	52	1.0000	52.000	0.000	$X^2 = 5.256$

The observed frequencies under H_0 are hypothesized to have arisen from a Poisson distribution. We must now obtain the expected frequencies if the population is a Poisson distribution. We know from (6.8) that the Poisson distribution has one parameter — its mean λ. Since H_0 does not specify the value of this parameter, we estimate it by the sample mean $\overline{X}$. For the 52 sample values in Table 16.2a, $\overline{X} = 4.0$ units (calculation not shown). Now, we use the Poisson distribution with $\lambda = 4.0$ in Table C.6 to obtain the probabilities that a sample observation will fall into each of the k classes in Table 16.2b. For instance, the probability that a week's demand X will be 0 or 1 and hence fall in the first class is $P(0) + P(1) = 0.0183 + 0.0733 = 0.0916$, and that X will be 2 and fall in the second class is $P(2) = 0.1465$. These probabilities are shown in Table 16.2b.

Next, we obtain the expected frequencies if H_0 holds (F_i) by multiplying each probability by the sample size n ($n = 52$ in this example). For the first class, we obtain $F_1 = 52(0.0916) = 4.763$, and for the second class, $F_2 = 52(0.1465) = 7.618$. These expected frequencies are also shown in Table 16.2b. We now have the observed frequencies f_i and the expected frequencies F_i based on the assumption that weekly demand is Poisson distributed. At this point, we are ready to calculate the test statistic. □

Test Statistic

The test statistic is based on a comparison of the observed frequencies f_i with the corresponding expected frequencies F_i under H_0. Each difference between the observed frequency and the expected frequency, $f_i - F_i$, is a *residual*, as explained in Chapter 2 (Section 2.4). The closer the observed frequencies are to the expected frequencies, the greater is the weight of evidence in favor of H_0. The farther apart the observed and expected frequencies, the greater is the weight of evidence in favor of H_1. The specific measure of closeness used for each class is the squared residual expressed relative to the expected frequency, that is, $(f_i - F_i)^2/F_i$, as shown in Table 16.1. We shall call this ratio the *relative squared residual*. The test statistic is the sum of the relative squared residuals for all classes and will be denoted by X^2.

(16.6)
$$X^2 = \sum_{i=1}^{k} \frac{(f_i - F_i)^2}{F_i}$$

Note that the farther the observed frequency f_i departs in either direction from the expected frequency F_i, the larger is $(f_i - F_i)^2$ and hence the larger is X^2. On the other hand, if f_i and F_i are identical for all classes, $X^2 = 0$ because each $(f_i - F_i)^2 = 0$. It follows, therefore, that large values of X^2 are indicative of H_1 being true, while small values of X^2 are indicative of H_0 being true. Thus, an upper-tail test as shown in Figure 16.1 is appropriate.

FIGURE 16.1   General form of the statistical decision rule for a chi-square goodness-of-fit test

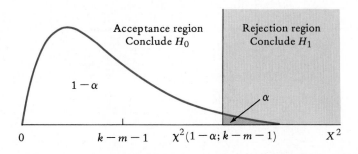

Sampling Distribution of X^2 When H_0 Holds

To control the α risk, we require the sampling distribution of X^2 when H_0 is true. The following theorem applies.

(16.7)

> When the population sampled has the probability distribution specified in H_0 and the sample size n is reasonably large:
>
> $$X^2 \simeq \chi^2(k - m - 1)$$
>
> *where:*
>
> k is the number of classes
> m is the number of parameters estimated from the sample data
>
> As a working rule, n is adequately large when all of the expected frequencies F_i are 2 or more and at least 50 percent of them are 5 or more.

Thus, when H_0 holds, X^2 is distributed approximately as a χ^2 random variable with $k - m - 1$ degrees of freedom, provided none of the F_i values is too small. (If a review of the χ^2 distribution is needed, the reader should turn to Appendix B, Section B.1, before proceeding.)

Decision Rule

Figure 16.1 illustrates the sampling distribution of X^2 when H_0 is true. If the risk of a Type I error is to be controlled at α, then the area in the upper tail must be α, and the action limit must equal $\chi^2(1 - \alpha; k - m - 1)$, that is, the $100(1 - \alpha)$ percentile of the $\chi^2(k - m - 1)$ distribution. Hence, the decision rule for a chi-square test has the following general form.

(16.8)

> When the alternatives are as follows:
>
> H_0: Population has specified form.
> H_1: Population does not have specified form.
>
> and the sample size is sufficiently large for theorem (16.7) to apply, the appropriate decision rule to control the α risk is as follows:
>
> If $X^2 \leq \chi^2(1 - \alpha; k - m - 1)$, conclude H_0.
> If $X^2 > \chi^2(1 - \alpha; k - m - 1)$, conclude H_1.
>
> *where:*
> X^2 is given by (16.6)

The application of the chi-square goodness-of-fit test varies a little depending on whether the probability distribution under consideration is discrete or continuous. We consider each of these cases in turn.

16.3 TESTS INVOLVING DISCRETE PROBABILITY DISTRIBUTIONS

When a chi-square goodness-of-fit test is conducted for a discrete probability distribution, an important concern is whether or not the expected frequencies F_i are large enough to meet the requirements of theorem (16.7). If they are not large enough, we conventionally pool or combine adjacent classes to bring the combined expected frequencies into conformance with the χ^2 approximation requirements. We shall illustrate this procedure with the spare part example.

> **EXAMPLE**

In the spare part example, the operations analyst wishes to test whether or not the probability distribution of replacement demand for the spare part is Poisson. Table 16.2a presented the data. Several of the classes in Table 16.2b were formed by pooling to meet the χ^2 approximation requirements. For example, the first class was formed as follows:

	Original Classes				Pooled Class		
X	f_i	Probability Under H_0	F_i	X	f_i	Probability Under H_0	F_i
0	2	0.0183	0.952	0–1	4	0.0916	4.763
1	2	0.0733	3.812				

Note that if $X = 0$ and $X = 1$ had not been pooled, the expected frequency in the first class would have been less than 2.

Pooling was also employed to form the class $X = 8$ or more in Table 16.2b. A check of the F_i values in Table 16.2b shows that all of them are at least 2 and 62.5 percent of them (5 out of 8) are at least 5. Hence, the χ^2 approximation of theorem (16.7) applies.

Step 1. The test alternatives here are as follows:

H_0: The probability distribution is Poisson.
H_1: The probability distribution is not Poisson.

Step 2. The α risk is to be controlled at 0.05.
Step 3. Since there are $k = 8$ classes (see Table 16.2b) and $m = 1$ parameter was estimated from the sample data (namely, λ), the required χ^2 distribution has $k - m - 1 = 8 - 1 - 1 = 6$ degrees of freedom. For $\alpha = 0.05$, Table C.2

shows $\chi^2(0.95; 6) = 12.59$. The decision rule therefore is as follows:

If $X^2 \leq 12.59$, conclude H_0.
If $X^2 > 12.59$, conclude H_1.

Step 4.   Table 16.2b shows the steps in calculating the test statistic X^2. We see that $X^2 = 5.256 \leq 12.59$. Hence, we conclude H_0—that a Poisson distribution adequately fits the observed frequencies for weekly demand.   □

Comment   Computer packages that perform chi-square goodness-of-fit tests often provide the upper-tail *P*-value for the test. The *P*-value for the test equals $P[\chi^2(k - m - 1) > X^2]$, where X^2 is the observed value of the test statistic. For the spare part example, $X^2 = 5.256$ and $k - m - 1 = 6$. Hence, the *P*-value equals $P[\chi^2(6) > 5.256]$. Computer printout for this example (not shown here) gives this *P*-value as 0.511. Since the α risk is to be controlled at 0.05 and $0.511 \geq 0.05$, it follows from the decision rule in (11.14) that H_0 should be concluded. This conclusion is necessarily the same one reached by using the test statistic X^2 itself.

TESTS INVOLVING CONTINUOUS PROBABILITY DISTRIBUTIONS

When the probability distribution specified in H_0 is continuous (for example, normal), the class intervals used to classify the sample data can be chosen in many ways. To strengthen the discriminating capability of the test, we employ the following procedure: *The classes of the continuous probability distribution specified in H_0 shall be chosen to have equal probabilities; that is, the expected frequencies for all classes shall be equal.*

Figure 16.2 presents the procedure graphically. If, say, $k = 10$ classes are to be used, each class has probability $1/k = 0.10$. The percentile separating classes 1 and

FIGURE 16.2   Partitioning a continuous probability distribution into k classes of equal probability

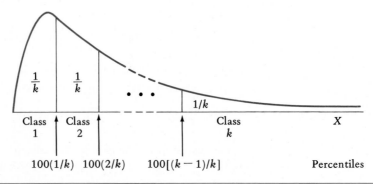

2, then, is the $100(1/k) = 100(1/10) = $ 10th percentile. The percentile separating classes 2 and 3 is the 20th percentile, and so on.

EXAMPLE

In the milk production example, the researcher wishes to determine whether or not the milk production of cows of a certain breed and age is normally distributed. The production data for the sample of 40 cows are presented in Table 16.3a.

Step 1. The alternatives here are as follows:

H_0: The probability distribution is normal.
H_1: The probability distribution is not normal.

Step 2. The researcher wishes to control the α risk at 0.01.

Step 3. To apply the chi-square test, we first must estimate the unknown parameters μ and σ of the normal distribution since they are not specified in H_0. As usual, we employ the sample estimators $\overline{X}$ and s, respectively. For the sample data in Table 16.3a, $\overline{X} = 15.96$ and $s = 2.144$ (calculations not shown).

TABLE 16.3   **Chi-square test for the goodness of fit of a normal distribution — Milk production example**

(a)
Milk production per cow (thousand pounds)

16.93	18.79	14.62	13.98	15.79	12.39	13.20	16.08	13.97	16.16
16.12	17.81	18.74	15.99	13.32	13.63	16.40	13.76	16.58	15.25
18.97	18.36	15.04	18.79	18.08	17.32	16.32	17.54	18.05	14.20
18.04	13.00	13.25	12.43	16.56	14.12	20.55	16.75	13.29	18.23

(b)
Frequency distribution and calculations for chi-square test

Milk Production X	Observed Frequency f_i	Probability Under H_0	Expected Frequency Under H_0 F_i	Residual $f_i - F_i$	Relative Squared Residual $(f_i - F_i)^2/F_i$
Under 13.49	7	0.125	5	2	0.80
13.49–14.51	6	0.125	5	1	0.20
14.52–15.26	3	0.125	5	−2	0.80
15.27–15.95	1	0.125	5	−4	3.20
15.96–16.64	8	0.125	5	3	1.80
16.65–17.39	3	0.125	5	−2	0.80
17.40–18.42	7	0.125	5	2	0.80
18.43 and over	5	0.125	5	0	0.00
Total	40	1.000	40	0	$X^2 = 8.40$

We shall utilize eight class intervals. Since $n = 40$, each class will then have an expected frequency of 5. If we were to use more than eight classes, the expected frequencies F_i all would be less than 5 (remember that they are to be equal for all classes), which would violate our requirement that at least 50 percent of the expected frequencies be 5 or more. If we were to use fewer than eight classes, the test would have less discriminating capability for detecting nonnormality (that is, the risks of Type II errors would be greater).

Figure 16.3 illustrates the construction of the classes. The probability of each class is to be the same; hence, it must be $1/8 = 0.125$. The first class limit corresponds, therefore, to the 12.5th percentile, the second to the 25th percentile, and so on. These percentiles are estimated as follows, using $\overline{X} = 15.96$ and $s = 2.144$:

$$\overline{X} + z(0.125)s = 15.96 - 1.15(2.144) = 13.49$$

$$\overline{X} + z(0.250)s = 15.96 - 0.67(2.144) = 14.52$$

$$\vdots$$

$$\overline{X} + z(0.875)s = 15.96 + 1.15(2.144) = 18.43$$

We see, for example, that the 12.5th percentile of the standard normal distribution is $z(0.125) = -1.15$, and hence, the corresponding estimated 12.5th percentile of the distribution of milk production must be 1.15 standard deviations ($s = 2.144$) below the mean ($\overline{X} = 15.96$). With this method of constructing the class limits,

FIGURE 16.3   **Partitioning a normal distribution into eight intervals of equal probability — Milk production example**

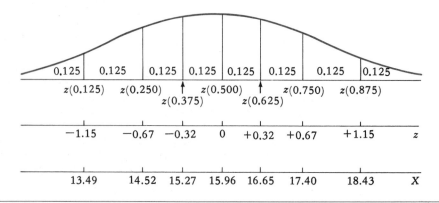

five sample observations are expected to lie in each class — five observations in the class under 13.49, five observations in the class 13.49–14.51, and so on — as shown in Table 16.3b.

The 40 sample observations are now classified into these eight classes, as shown in Table 16.3b, and the testing proceeds in standard fashion. If H_0 holds (that is, milk production is normally distributed), X^2 follows an approximate χ^2 distribution with $k - m - 1 = 8 - 2 - 1 = 5$ degrees of freedom. Remember that $m = 2$ parameters (namely, μ and σ) had to be estimated here. For $\alpha = 0.01$, we require $\chi^2(0.99; 5) = 15.09$. Hence, the decision rule is as follows:

If $X^2 \leq 15.09$, conclude H_0.
If $X^2 > 15.09$, conclude H_1.

Step 4. From Table 16.3b, we see that $X^2 = 8.40 \leq 15.09$. Hence, we conclude H_0 — that the fit of a normal distribution to the sample data is adequate. □

Comments

1. As we have seen in earlier chapters, many statistical tests that assume a normal population are quite robust and, hence, are satisfactory even in the absence of exact normality. Similarly, in the development of models, exact normality is often not essential. In these cases, use of a low α risk with the chi-square test may be reasonable. The risk of a Type II error (concluding that the population is normal when it is not) will then be high if the departure from normality is small, but the normal model is still a reasonable one. However, the risk of a Type II error will be low when the departure from normality is great and the normal model is not reasonable.

2. When H_0 is rejected because of sharp departures between observed and expected frequencies, an examination of the residuals $f_i - F_i$ often yields clues to the true form of the probability distribution.

3. When sample data from a continuous probability distribution are already classified in a frequency distribution (for instance, in published data), one may not be able to construct equal-probability classes for the goodness-of-fit test. The chi-square procedure can still be used with the frequency classes already provided, pooling them where needed to achieve F_i of adequate size. The resulting loss in the discriminating ability of the test will, in most cases, be small.

4. The parameter estimates on which theorem (16.7) is based employ the observed frequencies $f_1, f_2, \ldots, f_k$. As a practical matter, however, it is usually more convenient to estimate the parameters from the original data, as we did when we obtained $\overline{X}$ and s in the milk production example from the original data in Table 16.3a. It can be shown that use of the original data to estimate the parameters leads to an actual α risk level that may be somewhat larger than the specified one. However, when the number of classes utilized in the test is reasonably large and the expected frequencies F_i satisfy the conditions stated earlier, the actual α risk level will be close enough to the specified one to be satisfactory.

16.5 OPTIONAL TOPIC — CONFIDENCE BAND AND TEST FOR CUMULATIVE PROBABILITY FUNCTION

In Chapter 5, we noted that a continuous random variable X can be described by its cumulative probability function $F(x)$, where $F(x) = P(X \leq x)$. In this section, we discuss the Kolmogorov-Smirnov (abbreviated K–S) estimation procedure, which provides a confidence band for the cumulative probability function $F(x)$. This procedure also can be used as a goodness-of-fit test.

Assumptions

The assumptions for the K–S estimation procedure are as follows:

1. The population is continuous.
2. The sample is a simple random one.

K–S Statistic

Cumulative Sample Function. The basis of the K–S estimation procedure is the cumulative sample function, which we denote by $S(x)$.

(16.9)

> The *cumulative sample function,* denoted by $S(x)$, specifies for each value of x the proportion of sample values less than or equal to x.

From this definition, we see that a graph of $S(x)$ is simply a step function ogive, such as the one illustrated in Figure 3.3b.

EXAMPLE

In the hospital stays example, the researcher is studying length of hospital stay for certain psychiatric patients. Figure 16.4a contains the sample results for this study in array form. The step function ogive $S(x)$ for these data is plotted in Figure 16.4b, based on the following cumulative data:

x:	16	17	20	27	31	34	37	55
$S(x)$:	0.3	0.4	0.5	0.6	0.7	0.8	0.9	1.0

The step function $S(x)$ has steps of $1/n = 0.1$ except at $x = 16$, where the step is 0.3 because there are three observations there. ☐

$D(n)$ Statistic. The K–S estimation procedure utilizes a statistic, denoted by $D(n)$, that is based on the differences between the cumulative sample function $S(x)$ and the cumulative probability function $F(x)$.

FIGURE 16.4   Hospital stays data and the cumulative sample function for 10 psychiatric cases—Hospital stays example

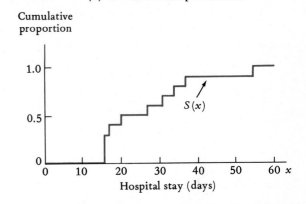

(a) **Array of hospital stays**

Case	Stay (days)	Case	Stay (days)
TAE	16	RJV	27
MSC	16	AAM	31
AGB	16	JJT	34
KFG	17	WKR	37
JCM	20	RAM	55

(b) **Cumulative sample function**

$$(16.10) \qquad D(n) = \max_x |S(x) - F(x)|$$

In other words, $D(n)$ equals the largest absolute deviation of $S(x)$ from $F(x)$ when all values of x are considered.

The statistic $D(n)$ is shown as a function of n because its sampling distribution depends on the sample size. Surprisingly, however, it does not depend on the specific form of $F(x)$. Thus, the sampling distribution of $D(n)$ can be tabulated for different values of n, irrespective of $F(x)$. Table C.11 gives percentiles of the $D(n)$ distribution for selected percentages and values of n, up to $n = 50$. For $n > 50$, the table shows formulas for computing the percentiles. Letting $D(a; n)$ denote the $100a$ percentile of the $D(n)$ distribution, we note, for instance, that $D(0.90; 10) = 0.37$. Thus the probability is 0.90 that the largest absolute deviation of $S(x)$ from $F(x)$ is 0.37 or less when $n = 10$, no matter what is the cumulative probability function $F(x)$.

Confidence Band for $F(x)$

The statistic $D(n)$ provides a simple way of constructing a confidence band from $S(x)$ for the cumulative probability function $F(x)$. Since $D(n)$ is the maximum *absolute* deviation between $S(x)$ and $F(x)$ and is less than $D(1 - \alpha; n)$ with probability $1 - \alpha$, it follows that $S(x)$ has a probability of $1 - \alpha$ of lying within distance $D(1 - \alpha; n)$ of $F(x)$, that is:

$$P[S(x) - D(1 - \alpha; n) \le F(x) \le S(x) + D(1 - \alpha; n)] = 1 - \alpha$$

This probability statement provides us with a confidence band for $F(x)$.

(16.11)

The confidence band for the cumulative probability function $F(x)$ with confidence coefficient $1 - \alpha$ is of the form $L(x) \leq F(x) \leq U(x)$, where:

$$L(x) = S(x) - D(1 - \alpha; n) \qquad U(x) = S(x) + D(1 - \alpha; n)$$

EXAMPLE

For the hospital stays example, a 90 percent confidence band for $F(x)$ is desired. Figure 16.5a shows the construction of the confidence band. We require from Table C.11, for $n = 10$ and $1 - \alpha = 0.90$, $D(0.90; 10) = 0.37$. Thus, $L(x)$ and $U(x)$ are obtained by constructing a pair of step functions that are identical to $S(x)$ but at vertical distances of 0.37 below and above it, respectively. We obtain the following:

x:	16	17	20	27	31	34	37	55
$S(x)$:	0.30	0.40	0.50	0.60	0.70	0.80	0.90	1.0
$L(x)$:	0	0.03	0.13	0.23	0.33	0.43	0.53	0.63
$U(x)$:	0.67	0.77	0.87	0.97	1.0	1.0	1.0	1.0

Note that there is no need to extend the confidence band above 1 or below 0 since $F(x)$ cannot be outside these limits, by definition. In conclusion, we can state with 90 percent confidence that $F(x)$ lies *entirely* within the $L(x)$ and $U(x)$ band in Figure 16.5a.   ☐

Comment

When the underlying population is discrete, the confidence band in (16.11) can be applied to the cumulative probability function, but the actual confidence coefficient will then be greater than $1 - \alpha$.

Goodness-of-Fit Test for $F(x)$

The K–S procedure can also be used to test whether the cumulative probability function $F(x)$ has a specified form, to be denoted by $F_0(x)$. The test alternatives may then be stated as follows.

(16.12)

H_0: $F(x) = F_0(x)$ for all x

H_1: $F(x) \neq F_0(x)$ for some x

Test Based on Confidence Band. The confidence band for $F(x)$ in (16.11) affords a convenient test of the alternatives in (16.12). We simply plot $F_0(x)$ together with the confidence band. If $F_0(x)$ falls entirely within the confidence band, then we conclude H_0; otherwise, we conclude H_1.

FIGURE 16.5   K–S procedures for the hospital stays example

(a) Confidence band for $F(x)$ (b) Goodness-of-fit test

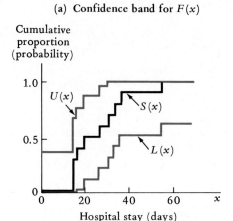
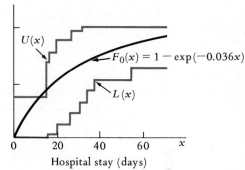

(c) Computer output for K–S goodness-of-fit test

```
MAXIMUM ABSOLUTE DEVIATION OCCURS AT            X   = 16.0000-

CUMULATIVE SAMPLE FUNCTION AT MAXIMUM         S(X) =    .0000

CUMULATIVE PROBABILITY FUNCTION UNDER H0      F(X) =    .4379

MAXIMUM ABSOLUTE DEVIATION                      D  =    .4379
```

Note: Minus sign on $X = 16.0000-$ indicates maximum
occurs just below 16.0000.

EXAMPLE

In the hospital stays example, the researcher wishes to test whether or not hospital stays have an exponential distribution with $\lambda = 0.036$.

Step 1.   We know from (7.16) that the alternatives here are as follows:

H_0: $F(x) = 1 - \exp(-0.036x)$ for all x

H_1: $F(x) \neq 1 - \exp(-0.036x)$ for some x

Thus, $F_0(x) = 1 - \exp(-0.036x)$ here.
Step 2.   The α risk is to be controlled at 0.10.

Step 3.   Figure 16.5b repeats the 90 percent confidence band for $F(x)$ from Figure 16.5a and also shows the plot of $F_0(x)$. We can use Table C.7 for plotting $F_0(x)$, as illustrated by the following examples.

1. At $x = 16$, $\lambda x = 0.036(16) = 0.58$, and $F_0(16) = 0.44$. Alternatively, we can calculate directly $F_0(16) = 1 - \exp[-0.036(16)] = 1 - 0.56 = 0.44$.

2. At $x = 20$, $\lambda x = 0.036(20) = 0.72$, and $F_0(20) = 0.51$. Alternatively, we can calculate directly $F_0(20) = 1 - \exp[-0.036(20)] = 1 - 0.49 = 0.51$.

Step 4.   We note from Figure 16.5b that $F_0(x)$ does not fall entirely within the confidence band. Hence, we conclude H_1—that the probability distribution is not exponential with $\lambda = 0.036$. Thus, either the distribution is not exponential or it is exponential but $\lambda \neq 0.036$.

The researcher noted that the sample mean is $\overline{X} = 26.9$ days, whereas the mean of the exponential distribution hypothesized in H_0 is, by (7.14), $1/\lambda = 1/0.036 = 27.8$ days. Since these two mean values are so close, he concluded that the specified λ value must be applicable here, and therefore, the probability distribution is likely not exponential. The major lack of fit is in the neighborhood of $X = 16$, where three sample observations occur. Upon further investigation, the researcher found that two of these observations were actually for somewhat shorter hospital stays but were incorrectly recorded at the usual minimum stay of 16 days. With the corrected data, the researcher found that an exponential distribution with $\lambda = 0.036$ is an adequate model.   □

Test Based on D^*.   The alternatives in (16.12) can also be tested directly by calculating the maximum absolute deviation of $S(x)$ from $F_0(x)$ for all x. We shall denote this maximum by D^*.

(16.13)

$$D^* = \max_x |S(x) - F_0(x)|$$

where:

$F_0(x)$ is the cumulative probability function specified in H_0

α risk is controlled at $F(x) = F_0(x)$

It can be shown that D^* must occur at or just below one of the steps in $S(x)$, that is, at or just below one of the sample points x_i. Thus, in principle, one can calculate D^* by (1) obtaining the absolute values $|S(x) - F_0(x)|$ at and just below each step in $S(x)$ and (2) noting the maximum among these absolute deviations. Unless n is small, these calculations are best handled by computer.

The appropriate type of decision rule here is one-sided, large values of D^* being inconsistent with H_0.

(16.14)

> If $D^* \leq D(1 - \alpha; n)$, conclude H_0.
> If $D^* > D(1 - \alpha; n)$, conclude H_1.

EXAMPLE

For the hospital stays example, Figure 16.5c shows computer output giving the test statistic D^* and information about where this maximum absolute deviation occurs. We see that $D^* = 0.44$ and that this maximum absolute deviation occurs just below $x = 16$, where $S(x) = 0$ and $F_0(x) = 0.44$. For $\alpha = 0.10$, we require $D(1 - \alpha; n) = D(0.90; 10) = 0.37$. Since $D^* = 0.44 > 0.37$, we conclude H_1—that the probability distribution is not exponential with $\lambda = 0.036$. $\square$

Comments

1. For the K–S test procedure to be exact, $F_0(x)$ in (16.12) must be fully specified, including its parameter value(s). The test procedure may also be used when the parameter value(s) are estimated from the sample. In this case, however, the test procedure will entail an actual α risk less than the specified one, and the power of the test, that is, its ability to detect departures of $F(x)$ from the specified $F_0(x)$, will be smaller.

2. In situations where both the chi-square test and the K–S test might be applied, the latter is often, but not always, more powerful in detecting departures of $F(x)$ from $F_0(x)$, where both tests have the same specified α risk.

3. The K–S test procedure may be employed when $F_0(x)$ is discrete, but the actual α risk level for the test will then be less than the one specified.

PROBLEMS

*16.1 For each of the following applications of the chi-square goodness-of-fit test, state (1) H_1, (2) the number of degrees of freedom associated with X^2 under H_0, (3) the value of the action limit in the decision rule.

  a. H_0 specifies that the distribution is Poisson. There are 9 classes. The α risk is to be controlled at 0.01.

  b. H_0 specifies that the distribution is uniform with $a = 2$ and $b = 10$. There are 6 classes. The α risk is to be controlled at 0.10.

  c. H_0 specifies that the distribution is normal. There are 12 classes. The α risk is to be controlled at 0.05.

16.2 For each of the following applications of the chi-square goodness-of-fit test, state (1) H_1, (2) the number of degrees of freedom associated with X^2 under H_0, (3) the value of the action limit in the decision rule.

  a. H_0 specifies that the distribution is Poisson with $\lambda = 3$. There are 8 classes. The α risk is to be controlled at 0.10.

  b. H_0 specifies that the distribution is binomial with $n = 4$. There are 5 classes. The α risk is to be controlled at 0.01.

  c. H_0 specifies that the distribution is $N(0, 1)$. There are 6 classes. The α risk is to be controlled at 0.05.

16.3   Consider a chi-square goodness-of-fit test where H_0 is, in fact, correct. Explain why the test statistic X^2 will tend to be smaller when the parameters of the probability distribution in H_0 are estimated from the sample data than when the correct parameter values are used in the test procedure. With reference to theorem (16.7), explain how the test procedure takes this tendency into account.

*16.4   **Fire Safety.**   A city report described the results of an intensive promotional campaign concerned with public fire safety. Among other facts, the report provided the following frequency distribution of the number of fire calls answered by one fire station between 12 midnight and 8 A.M. each day during a 60-day period following the campaign.

Number of Calls:	0	1	2	3 or more
Number of Days:	28	20	7	5

Extensive experience for this station before the promotional campaign showed that the mean number of calls during this 8-hour period each day had been 1.5. Treating the 60 days of post-promotional data as a random sample, use the chi-square procedure to test whether or not the number of daily calls currently is Poisson distributed with $\lambda = 1.5$. Control the α risk at 0.05. State the alternatives, the decision rule, the value of the test statistic, and the conclusion.

16.5   Refer to **Fire Safety** Problem 16.4. The sample mean number of calls for the 60 days of data is $\bar{X} = 0.9$.
   a. Use the chi-square procedure to test whether or not the number of calls follows a Poisson distribution (with no specification of the value of λ). Control the α risk at 0.05. State the alternatives, the decision rule, the value of the test statistic, and the conclusion.
   b. The test in Problem 16.4 leads to the conclusion that the population is not Poisson with $\lambda = 1.5$. Is the conclusion in **a** here inconsistent with the conclusion reached in Problem 16.4? Explain.

16.6   In planning the best size for the waiting room in a medical clinic, a consultant recorded the number of persons accompanying each patient in a random sample of 300 patients. The sample results follow.

Number of Persons:	0	1	2	3	4
Frequency:	163	104	24	8	1

The consultant wishes to determine whether or not the number of accompanying persons follows a Poisson distribution. Use the chi-square procedure to conduct the test, controlling the α risk at 0.10. Pool classes as required. State the alternatives, the decision rule, the value of the test statistic, and the conclusion.

16.7   Refer to **Airline Reservations** Problem 10.12. The airline researcher conjectures that the number of no-shows per flight follows a Poisson distribution.
   a. Test the conjecture by using the chi-square procedure. Control the α risk at 0.10. Pool classes as required. State the alternatives, the decision rule, the value of the test statistic, and the conclusion.
   b. What is the reason for pooling classes in conducting the test?

16.8   A researcher observed the final digit in the daily volume of stock transactions during the past 50 days. The frequency distribution follows.

Final Digit:	0	1	2	3	4	5	6	7	8	9
Frequency:	5	2	4	4	7	5	6	5	7	5

Using the chi-square procedure, test whether the final digits were generated from a discrete uniform probability distribution for integer outcomes $0, 1, \ldots, 9$. Assume that the data constitute a random sample. Control the α risk at 0.05. State the alternatives, the decision rule, the value of the test statistic, and the conclusion.

16.9 **Product Testing.** In a product-testing experiment, each of 150 subjects independently examined three patches of fabric. For each patch, the subject was asked whether the fiber is synthetic or natural. The numbers of correct identifications by subjects were as follows:

Number of Correct Identifications by Subject:	0	1	2	3
Number of Subjects:	9	40	36	65

If the subjects simply make a separate guess for each patch, the number of correct identifications by a subject will follow a binomial probability distribution with $n = 3$ and $p = 0.5$.
   a. Verify that if all subjects guess, the expected numbers of subjects giving 0, 1, 2, and 3 correct identifications are 18.75, 56.25, 56.25, and 18.75, respectively.
   b. Test, using the chi-square procedure, whether or not the subjects in this experiment were guessing. Control the α risk at 0.01. State the alternatives, the decision rule, the value of the test statistic, and the conclusion.

*16.10 Refer to **Random Number Generator** Problem 15.32. Use the chi-square procedure to test whether or not the random number generator is generating numbers from a standard normal distribution. Control the α risk at 0.01, and use ten equal-probability classes. State the alternatives, the decision rule, the value of the test statistic, and the conclusion.

16.11 Refer to **Urban Wages** Problem 2.10. Assume that the 60 wage earners are a random sample from a large population. The sample mean and standard deviation are $\bar{X} = 405.13$ and $s = 21.973$. Test by means of the chi-square procedure whether or not the weekly wages in the population are normally distributed. Control the α risk at 0.05. Use ten equal-probability classes. State the alternatives, the decision rule, the value of the test statistic, and the conclusion.

16.12 Refer to **Hospital Costs** Problem 2.11. Assume that the 50 hospitals are a random sample from a large population.
   a. Test by means of the chi-square procedure whether or not the average daily costs in the population are normally distributed. Control the α risk at 0.01. Use ten equal-probability classes. State the alternatives, the decision rule, the value of the test statistic, and the conclusion.
   b. Examine the residuals $f_i - F_i$ of your test in **a**. What do they suggest about the departure, if any, of the distribution pattern of hospital costs from normality?

16.13 **Hydraulic-Hose Ruptures.** An industrial machine has a 1.5-meter hydraulic hose that ruptures occasionally. The manufacturer has recorded the locations of these ruptures for a random sample of 25 ruptured hoses. These locations (measured in meters from the pump end of the hose) are as follows:

1.32	1.19	1.21	1.36	0.64	1.46	0.80	1.37	0.38
1.07	0.13	1.16	0.33	1.42	1.27	0.08	0.75	
1.22	1.37	1.14	1.43	0.97	1.12	0.27	1.46	

   a. Use the chi-square procedure with five equal-probability classes to test whether or not rupture location is uniformly distributed along the length of the hose. Control the α risk at 0.10. State the alternatives, the decision rule, the value of the test statistic, and the conclusion.

   b. Examine the residuals $f_i - F_i$ of your test in **a**. In which section of the hose is the rupture frequency most out of line with the expectations under H_0? What is the implication of your finding for the engineering design of this hose?

16.14 Refer to the data array of time intervals between oil spills before pollution controls in Figure 3.8. Assume that the 40 observations are a random sample. Test by means of the chi-square procedure whether or not the population is exponential with parameter $\lambda = 0.05$. Control the α risk at 0.05. Use eight equal-probability classes. State the alternatives, the decision rule, the value of the test statistic, and the conclusion.

16.15 Refer to **Response Times** Problem 2.18. The sample mean and standard deviation of the original (that is, unclassified) response times are $\overline{X} = 13.7$ and $s = 5.33$. Assume that the 50 observations are a random sample. Use the chi-square procedure to test whether or not response times are normally distributed. Base the test on the class intervals and observed frequencies given in Problem 2.18. Control the α risk at 0.01. State the alternatives, the decision rule, the value of the test statistic, and the conclusion.

*16.16 Rates of return (in percent) realized on ten oil stocks last year were:

    3   18   21   6   25   12   -1   35   9   15

   a. Construct the cumulative sample function for these data, and plot it on a graph as a step function ogive.
   b. Treat the ten rates of return as a random sample from a continuous probability distribution, and obtain the 90 percent confidence band for the cumulative probability function. Plot the confidence band on your graph in **a**.
   c. For the confidence band plotted in **b**, interpret the values of $L(15)$ and $U(15)$.

16.17 Refer to the melting points data for the April shipment in Figure 3.5a.
   a. Construct the cumulative sample function for these data, and plot it on a graph as a step function ogive.
   b. Treat the 15 melting points as a random sample from a continuous population. Obtain the 95 percent confidence band for the cumulative probability function. Plot the confidence band on your graph in **a**.
   c. From the confidence band in **b**, can you conclude that fewer than 90 percent of filaments in the population have melting points of 330°C or less? Explain.

16.18 **Price Test.**   Fifteen consumers were selected randomly to test a new home garden tool. Each was then asked what would be a reasonable price for this tool. The answers (in dollars) were as follows:

    18   23   36   33   21   26   26   16   31   30   41   27   29   40   38

   Treat the 15 observations as a random sample from a continuous probability distribution.
   a. Construct a graph showing the 95 percent confidence band for the cumulative probability function.
   b. From your confidence band in **a**, what is the maximum proportion of consumers who consider a price of $20 or less to be reasonable?

*16.19 The service lives (in thousands of operating hours) of a random sample of ten aircraft digital-display units are as follows (arrayed by magnitude):

    0.4   2.6   4.4   4.9   10.6   11.3   11.8   12.6   23.0   40.8

   It is desired to determine, from the K–S confidence band for the cumulative probability function, whether or not the service lives of these units are exponentially distributed with $\lambda = 0.1$ failure per thousand hours, controlling the α risk at 0.10.

    a. Construct a graph showing the 90 percent confidence band for the cumulative probability function.

    b. State the test alternatives here. Plot $F_0(x)$ on your graph in **a**. What is the appropriate conclusion? Why?

16.20 The length of time before power is restored by the electric company after a power failure is hypothesized to follow an exponential distribution with a mean of 0.8 hour. Data for the last ten power failures (in hours) are as follows (arrayed by magnitude):

0.16	0.23	0.25	0.39	0.40	0.45	0.53	0.71	1.05	1.17

Assume that the ten observations are a random sample. Use the K–S procedure based on the confidence band for the cumulative probability function to test the hypothesis, controlling the α risk at 0.10. State the alternatives and the conclusion.

16.21 Refer to **Price Test** Problem 16.18. Before collecting the data, the market researcher hypothesized that the prices follow a uniform distribution with $a = \$15$ and $b = \$45$. Use the K–S confidence band for the cumulative probability function to test the researcher's hypothesis, controlling the α risk at 0.05. State the alternatives and the conclusion.

*16.22 A financial analyst wishes to test whether or not the rates of return (X, in percent) on common stocks last year were normally distributed with mean 8.0 percent and standard deviation 10.5 percent. He obtained the rates of return for a random sample of 50 stocks and, using a computer routine, found that the K–S maximum absolute deviation is $D^* = 0.269$ at a point just below $x = 9.3$ percent.

    a. What is the value of $F_0(x)$ at the point where the maximum absolute deviation D^* occurs? What does D^* represent here?

    b. Conduct the desired K–S test based on D^*, controlling the α risk at 0.01. State the alternatives, the decision rule, the value of the test statistic, and the conclusion.

16.23 An operations specialist wishes to test whether or not the repair time (X, in hours) for a standard type of pump used in a chemical plant follows an exponential distribution with mean 12 hours. A computer analysis of a random sample of 225 repair times for pumps of this type determined that the K–S maximum absolute deviation is $D^* = 0.053$ at $x = 24$ hours.

    a. What is the value of $F_0(15)$? From the value of D^*, what is the smallest possible value of $S(15)$? What is the largest possible value of $S(15)$?

    b. Conduct the K–S test based on D^*, controlling the α risk at 0.05. State the alternatives, the decision rule, the value of the test statistic, and the conclusion.

EXERCISES

16.24 Refer to **Product Testing** Problem 16.9. Here, p denotes the probability that a subject will correctly identify a patch of fabric.

    a. Estimate the value of p from the sample data.

    b. Suppose that the analyst hypothesizes simply that X is binomial with $n = 3$ (without specifying the value of p). Conduct the appropriate chi-square test, controlling the α risk at 0.01. State the alternatives, the decision rule, the value of the test statistic, and the conclusion.

    c. Does the alternative H_0 being tested in **b** include the possibility that parameter p is different for different subjects? Comment.

16.25 Show that test statistic X^2 in (16.6) can be written as the following weighted sum of the residuals $f_i - F_i$:

$$X^2 = \sum_{i=1}^{k} \frac{f_i}{F_i} (f_i - F_i)$$

16.26 Consider the test statistic X^2 in (16.6). Assume that H_0 fully specifies the population distribution and that, consequently, no parameters need to be estimated from the sample data.

    a. Show that, for any class i, $E\{(f_i - F_i)^2/F_i\} = 1 - (F_i/n)$ when H_0 holds. [*Hint:* f_i is a binomial random variable with mean F_i and variance $F_i(n - F_i)/n$ in this case.]

    b. Use the result in **a** to show that $E\{X^2\} = k - 1$ when H_0 holds.

16.27 The percentile $D(1 - \alpha; n)$ approaches 0 as n increases for any α. What does this fact imply about the probability of a Type II error with the K–S goodness-of-fit procedure when n is large?

16.28 Show that, for a continuous random variable, $S(x)$ is a consistent and unbiased estimator of $F(x)$ at any specified value of x. [*Hint:* Let $F(x) = p$ and $S(x) = \bar{p}$, and employ the Chebyshev inequality (5.10).]

STUDIES

16.29 Refer to **Product Testing** Problem 16.9. A statistician has suggested an alternative model— namely, that a proportion P of the subjects know the identification for certain and the remainder guess.

    a. Use the chi-square procedure to test the aptness of this model when $P = 0.3$, controlling the α risk at 0.01. State the alternatives, the decision rule, the value of the test statistic, and the conclusion.

    b. How might one determine an estimate of P here? (*Hint:* Utilize the X^2 statistic.)

16.30 A newspaper's sports page contained the following frequency distribution of times between consecutive goals for the city's soccer team in a recent season.

Time Between Goals (minutes)	Number of Cases	Time Between Goals (minutes)	Number of Cases
0–under 10	105	30–under 40	21
10–under 20	72	40–under 50	14
20–under 30	28	50 and over	20

For the 260 cases, $\bar{X} = 20.0$.

    a. Treating the 260 cases as a random sample, use the chi-square procedure to test whether or not the times between goals are exponentially distributed. Control the α risk at 0.05. Since the original data are not available for setting up equal-probability classes, the classes given in the newspaper article must be used. State the alternatives, the decision rule, the value of the test statistic, and the conclusion.

    b. A sportswriter, commenting on the data, said they showed that goal scoring has a degree of contagion: Once the team scores a goal, there is a high probability that another goal will follow shortly. Does the residual $f_i - F_i$ for the class 0–under 10 support this conclusion? Discuss.

16.31 Refer to the after-controls data set in Table 3.1a. Consider these data to constitute a random sample, and ignore the rounded nature of the data.

a. Use the chi-square procedure to test whether or not the data were generated from an exponential probability distribution with mean 28. Use eight equal-probability classes and $\alpha = 0.05$. State the alternatives, the decision rule, the value of the test statistic, and the conclusion.

b. The mean of the data set is $\overline{X} = 27.6$. Would the goodness of fit be much improved if you had tested only whether or not the probability distribution is exponential? Explain.

c. Plot the residuals $f_i - F_i$ obtained in your test in **a**. What does the plot show about the lack of fit of the exponential distribution?

16.32 A financial analyst has undertaken a computer simulation study of a complex and risky capital investment proposal. As part of the study, he has obtained, in 20 simulation runs, the following rates of return (in percent), which may be considered as a simple random sample from the true rate-of-return distribution for this investment proposal.

−4.7	22.7	12.3	5.1	20.4	16.7	11.1
−1.5	4.3	2.6	−11.3	1.0	13.4	23.0
12.1	10.3	−2.1	−3.4	1.1	10.1	

a. Construct a graph showing the 90 percent confidence band for the cumulative probability function $F(x)$. Interpret this band.

b. From your confidence band, obtain an interval estimate for $F(0)$, the probability that the investment proposal will produce a negative or zero rate of return.

c. How many rates of return would have to be simulated in order to have probability 0.99 that $S(x)$ will lie within a vertical distance of ± 0.04 of $F(x)$ for all values x?

d. Letting $p = F(0)$, obtain an exact 90 percent confidence interval for p, using (12.13). Why is this interval narrower than the one obtained in **b**?

17 MULTINOMIAL POPULATIONS

In many problems, one is interested in populations containing elements classified into several categories or classes. In this chapter, we study populations of this type and take up some important inference procedures for samples drawn from them.

17.1 NATURE OF MULTINOMIAL POPULATIONS

Populations containing elements classified into several classes or categories are called multinomial populations.

(17.1)
> When each element of a population is assigned to one (and only one) of two or more attribute classes or categories, the population is called a *multinomial population*.

We encountered multinomial populations in Chapter 4. Table 4.1a provides an example. We now consider some other examples.

EXAMPLES

1. All of the eggs laid in a month by a hen flock constitute the population of interest. Each egg is classified according to its size, the attribute categories being extra large, large, medium, and small.

2. All of the households in a city constitute the population of interest. Each household is classified as owning no car, one car, or two or more cars.

3. The process of producing memory chips for computers is the population of interest. A chip is classified as acceptable, having minor defects, or having major defects. Observe that the population here is infinite. ☐

Parameters of Interest

We denote the number of attribute classes into which a population element can be classified by k. Thus, in Example 1, we have $k = 4$, and in Examples 2 and 3, we have $k = 3$. The probability that a population element selected at random comes from the *i*th attribute class is denoted by p_i ($i = 1, 2, \ldots, k$). Of course, these probabilities must sum to 1.

EXAMPLE		

Machine Component. A large lot of used machine components is the population of interest. Each machine component is classified according to its condition, the attribute categories being excellent, good, poor, and scrap. The probabilities p_i for the lot are as follows:

i	Condition of Component	p_i
1	Excellent	$p_1 = 0.4$
2	Good	$p_2 = 0.3$
3	Poor	$p_3 = 0.2$
4	Scrap	$p_4 = \underline{0.1}$
	Total	1.0

The probabilities p_i are the parameters of interest in multinomial populations but are usually unknown. Sampling is then employed to make inferences about these parameters.

Comment When only $k = 2$ categories are used for classifying a population element, a random outcome from the multinomial population is equivalent to a Bernoulli trial. For example:

i	Sex of Employee	p_i
1	Male	$p_1 = p$
2	Female	$p_2 = 1 - p$
	Total	1.0

17.2 MULTINOMIAL PROBABILITY DISTRIBUTIONS

When a sample of n elements is selected from a multinomial population, we denote the number of sample observations from the *i*th attribute category by f_i. Of course, the sum of the f_i is equal to the sample size n.

EXAMPLE		

In the machine component example, a sample of ten machine components yielded the following results.

i	Condition of Component	f_i
1	Excellent	$f_1 = 3$
2	Good	$f_2 = 3$
3	Poor	$f_3 = 2$
4	Scrap	$f_4 = \underline{2}$
	Total	$n = 10$

When a simple random sample is selected from an infinite multinomial population, the probability of obtaining the sample frequencies $f_1, f_2, \ldots, f_k$, denoted by $P(f_1, f_2, \ldots, f_k)$, is given by the multinomial probability function.

(17.2)

The *multinomial probability function* is:

$$P(f_1, f_2, \ldots, f_k) = \frac{n!}{f_1! f_2! \cdots f_k!} p_1^{f_1} p_2^{f_2} \cdots p_k^{f_k}$$

where:

$$\sum_{i=1}^{k} f_i = n$$

$$\sum_{i=1}^{k} p_i = 1$$

Observe that the multinomial probability function is discrete, with parameters n and $p_1, p_2, \ldots, p_k$. Each different set of parameter values defines a different probability function in the multinomial family.

When the multinomial population is finite, the multinomial probability function (17.2) provides approximate probabilities of the sample outcomes as long as the sampling fraction n/N is not large. Throughout this chapter, we shall assume that finite populations are large enough to be treated as infinite ones.

EXAMPLES

1. We wish to find the probability of the sample outcomes in the machine component example. We have:

$$p_1 = 0.4 \qquad p_2 = 0.3 \qquad p_3 = 0.2 \qquad p_4 = 0.1$$

$$n = 10 \qquad f_1 = 3 \qquad f_2 = 3 \qquad f_3 = 2 \qquad f_4 = 2$$

We obtain:

$$P(3, 3, 2, 2) = \frac{10!}{3!\, 3!\, 2!\, 2!} (0.4)^3 (0.3)^3 (0.2)^2 (0.1)^2 = 0.01742$$

2. For the same parameter values, we wish to find the probability of the sample outcomes $f_1 = 8$, $f_2 = 2$, $f_3 = 0$, $f_4 = 0$:

$$P(8, 2, 0, 0) = \frac{10!}{8!\, 2!\, 0!\, 0!} (0.4)^8 (0.3)^2 (0.2)^0 (0.1)^0 = 0.002654 \qquad \square$$

Comment	The binomial probability function (6.3) is a special case of the multinomial probability function (17.2). This result is to be expected since we saw that a Bernoulli trial is a special case of a random outcome from a multinomial population. Specifically, when $k = 2$, (17.2) reduces to (6.3), with:

$$f_1 = x \qquad f_2 = n - x$$
$$p_1 = p \qquad p_2 = 1 - p$$

17.3 INFERENCES CONCERNING THE PARAMETERS p_i

Various types of inferences can be made concerning the parameters p_i of a multinomial population. We discuss two basic types here — inferences on a given parameter p_i and inferences on all the p_i.

Assumptions

1. A simple random sample of size n has been selected from an infinite or large multinomial population.
2. The sample size n is reasonably large.

Inferences for One p_i

The estimation or testing of any one of the multinomial parameters p_i is carried out by the procedures of Chapter 12 for a population proportion. To see why, consider the machine component example again. Suppose we wish to estimate p_1, the probability that a used component is in excellent condition. To estimate p_1, we do not care what the other categories are and thus can combine them into one class consisting of all components that are not in excellent condition. But then we have the Bernoulli situation where there are two categories (excellent condition, not excellent condition), with probabilities p_1 and $1 - p_1$, respectively, and all of our earlier inference procedures for a population proportion are applicable. We consider now two examples illustrating the use of these inference procedures.

Confidence Interval for p_i.   The first example illustrates the construction of a confidence interval for a multinomial parameter p_i.

EXAMPLE	**Bearing Failures.** A railroad engineer is investigating maintenance problems with rolling stock. He is currently studying a certain type of bearing failure to ascertain whether occurrences of these failures are associated with operational levels only or whether other factors are also involved. Table 17.1 contains a record of 1044 bearing failures on rolling stock over the past several years, classified by quarter of the year of occurrence.

TABLE 17.1   Record of bearing failures in rolling stock, classified by quarter of occurrence — Bearing failures example

Quarter of Year i	Number of Failures f_i
1	249
2	256
3	297
4	242
Total $n = 1044$	

The engineer believes it is reasonable to treat these data as a random sample from an infinite multinomial population of bearing failures. Here, p_1 is the probability that a failure occurs in the first quarter, p_2 is the probability that it occurs during the second quarter, and p_3 and p_4 are similarly defined.

To estimate the probability of a failure occurring in the first quarter, we use the sample proportion:

$$\bar{p}_1 = \frac{f_1}{n} = \frac{249}{1044} = 0.2385$$

A 90 percent confidence interval for p_1 is desired. Since $n = 1044$, we employ the large-sample confidence interval (12.9) for a population proportion. We require:

$$s\{\bar{p}_1\} = \sqrt{\frac{\bar{p}_1(1 - \bar{p}_1)}{n - 1}} = \sqrt{\frac{0.2385(1 - 0.2385)}{1044 - 1}} = 0.0132$$

$$z(1 - \alpha/2) = z(0.95) = 1.645$$

Hence, the 90 percent confidence limits for p_1 are $0.2385 \pm 1.645(0.0132)$, and the confidence interval is:

$$0.217 \le p_1 \le 0.260$$

Thus, the engineer can conclude with 90 percent confidence that the probability of a bearing failure occurring during the first quarter is between 0.217 and 0.260.

☐

Test for p_i.   The second example illustrates a test for a multinomial parameter p_i.

EXAMPLE

In the bearing failures example, the proportion of annual traffic carried in the first quarter is 0.210. The engineer would like to test whether the relative frequency of bearing failure occurrences in the first quarter is the same as the proportion of traffic.

Step 1.   The alternatives here are:

H_0: $p_1 = 0.210$

H_1: $p_1 \neq 0.210$

Step 2.   The α risk is to be controlled at 0.10 when $p_1 = 0.210$.

Step 3.   For $\alpha = 0.10$, we require $z(1 - \alpha/2) = z(0.95) = 1.645$. The decision rule for the test, therefore, is as follows:

If $|z^*| \leq 1.645$, conclude H_0.
If $|z^*| > 1.645$, conclude H_1.

Step 4.   To calculate the standardized test statistic z^* defined in (12.11), we use 0.210 for p_0 and obtain:

$$\sigma\{\bar{p}_1\} = \sqrt{\frac{0.210(0.790)}{1044}} = 0.01261$$

Since $\bar{p}_1 = 0.2385$, we obtain:

$$z^* = \frac{0.2385 - 0.210}{0.01261} = 2.26$$

Because $|z^*| = 2.26 > 1.645$, we conclude H_1—that $p_1 \neq 0.210$. $\square$

Inferences for All p_i

At times, one may wish to make inferences about all of the parameters p_i of a multinomial population.

EXAMPLE

In the bearing failures example, the engineer would like to test whether the occurrences of bearing failures by quarter vary in direct proportion to the volume of railroad traffic. The quarterly proportions of annual traffic are as follows:

Quarter:	1 ($i = 1$)	2 ($i = 2$)	3 ($i = 3$)	4 ($i = 4$)
Proportion of Annual Traffic:	0.21	0.23	0.30	0.26

For example, about 21 percent of annual traffic is handled in the first quarter of each year. $\square$

Alternatives.   In this type of situation, the alternatives take the following form.

(17.3)

H_0: $p_i = p_{i0}$ for all i

H_1: $p_i \neq p_{i0}$ for some i

Here, p_{i0} denotes the value specified for the parameter p_i ($i = 1, 2, \ldots, k$) in H_0.

EXAMPLE

In the bearing failures example, the test alternatives are:

$$H_0: p_1 = p_{10} = 0.21 \qquad p_3 = p_{30} = 0.30$$

$$ p_2 = p_{20} = 0.23 \qquad p_4 = p_{40} = 0.26$$

$$H_1: p_i \neq p_{i0} \qquad \text{for some } i$$

Note that H_1 states in formal terms that H_0 does not hold. This does not require that *all* $p_i \neq p_{i0}$. For instance, alternative H_0 would be incorrect if $p_1 = 0.21$ and $p_2 = 0.23$ but $p_3 = 0.40$ and $p_4 = 0.16$, even though p_1 and p_2 do agree with the specified values. ☐

Test Procedure.   The test for the multinomial alternatives (17.3) is precisely of the same form as the goodness-of-fit test (16.8), and Figure 16.1 describes the test procedure. The correspondence to the earlier test can be seen as follows:

1.   In each case, the sample outcomes are classified into one of k classes.
2.   Alternative H_0 provides the expected frequencies F_i when H_0 holds.
3.   The test then compares the observed frequencies f_i with the expected frequencies F_i, utilizing test statistic X^2 and the χ^2 distribution.

(17.4)

Tests for the multinomial alternatives (17.3) are based on the test statistic:

$$X^2 = \sum_{i=1}^{k} \frac{(f_i - F_i)^2}{F_i}$$

When H_0 in (17.3) holds and the sample size is reasonably large so that theorem (16.7) applies:

$$X^2 \simeq \chi^2(k - 1)$$

Large values of X^2 lead to conclusion H_1.

Note that the degrees of freedom for the χ^2 distribution here are $k - 1$ because $m = 0$; no parameters are estimated since all are specified by H_0.

EXAMPLE

We continue with the bearing failures example.

Step 1.   The alternatives for the test were given earlier.
Step 2.   The α risk is to be controlled at 0.05.

Step 3. The degrees of freedom for the χ^2 distribution are, from (17.4), $k - 1 = 4 - 1 = 3$. Thus, we require $\chi^2(0.95; 3) = 7.81$. Hence, the decision rule is as follows:

If $X^2 \leq 7.81$, conclude H_0.
If $X^2 > 7.81$, conclude H_1.

Step 4. Figure 17.1 contains computer output for this chi-square test. The output repeats the observed frequencies from Table 17.1 and also shows the specified probabilities p_{i0} under H_0. The expected frequencies in the output are obtained as usual by multiplying the probabilities p_{i0} by n, that is, $F_i = np_{i0}$. Note that all of the F_i are large enough to meet the χ^2 approximation rule in (16.7).

The test statistic X^2, calculated in the usual way by using (17.4), is given in the output as $X^2 = 9.121$. Since $X^2 = 9.121 > 7.81$, we conclude H_1—that the occurrence of bearing failures by quarter is not entirely consistent with volume of traffic.

The computer output in Figure 17.1 also gives the upper-tail P-value for the test. Since P-value $= 0.0277 < 0.05 = \alpha$, we see again that the appropriate test conclusion is H_1.

The engineer noted, in the output of Figure 17.1, the large relative squared residuals for the first and fourth quarters and that the residuals are positive for the first two quarters and negative for the last two. These findings proved to be useful clues to him in his search for factors (in addition to operating volume) that affect the occurrence of bearing failures by quarter, such as temperature and company maintenance policy. □

FIGURE 17.1   Computer output for the chi-square test—Bearing failures example

CLASS	OBSERVED FREQUENCY	UNDER H0 PROBABILITY	UNDER H0 EXPECTED FREQUENCY	RESIDUAL	RELATIVE SQUARED RESIDUAL
1	249	.2100	219.24	29.76	4.040
2	256	.2300	240.12	15.88	1.050
3	297	.3000	313.20	-16.20	.838
4	242	.2600	271.44	-29.44	3.193
TOTAL	1044	1.0000	1044.00	0.00	9.121

CHI-SQUARE TEST STATISTIC   9.121
DF   3
UPPER-TAIL P-VALUE   .0277

| Comment | When there are $k = 2$ categories only, the chi-square test for the alternatives (17.3) is equivalent to the two-sided test for a population proportion in (12.11). |

17.4 BIVARIATE MULTINOMIAL POPULATIONS

Often, when considering multinomial populations, we are interested in two or more characteristics of the population elements and wish to study the relationships between the variables. In this section, we are concerned with the analysis of bivariate multinomial populations.

(17.5)   A multinomial population having attribute categories arranged in a bivariate classification system is called a *bivariate multinomial population*.

We encountered bivariate multinomial populations in Chapter 4. Table 4.1b provides an example. We now consider several other examples.

EXAMPLES   1. Persons in the labor force may be categorized according to age and employment status with the following bivariate classification system:

Employment Status	Age Class			
	25 and Under	26–45	46–65	Over 65
Unemployed				
Employed part-time				
Employed full-time				

This bivariate classification will facilitate a study of the relationship between the two variables.

2. **Instrument Failures.** Electronic instrument failures may be categorized by the type of failure that is observed (T_1 or T_2) and the location of the failure in the instrument (L_1, L_2, or L_3) as follows:

Type of Failure	Location of Failure		
	L_1	L_2	L_3
T_1			
T_2			

This bivariate classification system will enable the analyst to study whether or not type and location of failure are related variables. If so, the nature of the relationship is important for maintenance troubleshooting and for some aspects of instrument design.   □

| Comment | Note that each of the bivariate multinomial populations in the two examples can be viewed as an ordinary multinomial population, with every cell in the bivariate classification constituting a class. Thus, the classification in Example 1 involves $k = 3(4) = 12$ classes, and in Example 2, there are $k = 2(3) = 6$ classes. The bivariate arrangement facilitates, of course, the analysis of the relationship between the two variables. |

Parameters of Interest

The probability parameters of a bivariate multinomial population are displayed in a convenient form in Table 17.2a. The table shows a bivariate classification system, with the row variable having r classes and the column variable having c classes, so in all, the multinomial population has $k = rc$ classes or cells. The symbol p_{ij} denotes the joint probability that a population element selected at random falls in the ith class of the row variable and the jth class of the column variable. The symbols $p_{i.}$ and $p_{.j}$ denote the marginal probabilities for the ith row and jth column, respectively. An example of the display in Table 17.2a with probability values is found in Table 4.1b.

TABLE 17.2   Bivariate multinomial probability parameters and sample frequencies

(a)
Bivariate multinomial probability parameters

Class for Row Variable	Class for Column Variable				Total
	$j = 1$	$j = 2$	$\cdots$	$j = c$	
$i = 1$	p_{11}	p_{12}	$\cdots$	p_{1c}	$p_{1.}$
$i = 2$	p_{21}	p_{22}	$\cdots$	p_{2c}	$p_{2.}$
$\vdots$	$\vdots$	$\vdots$		$\vdots$	$\vdots$
$i = r$	p_{r1}	p_{r2}	$\cdots$	p_{rc}	$p_{r.}$
Total	$p_{.1}$	$p_{.2}$	$\cdots$	$p_{.c}$	1.0

(b)
Bivariate multinomial sample frequencies

Class for Row Variable	Class for Column Variable				Total
	$j = 1$	$j = 2$	$\cdots$	$j = c$	
$i = 1$	f_{11}	f_{12}	$\cdots$	f_{1c}	$f_{1.}$
$i = 2$	f_{21}	f_{22}	$\cdots$	f_{2c}	$f_{2.}$
$\vdots$	$\vdots$	$\vdots$		$\vdots$	$\vdots$
$i = r$	f_{r1}	f_{r2}	$\cdots$	f_{rc}	$f_{r.}$
Total	$f_{.1}$	$f_{.2}$	$\cdots$	$f_{.c}$	n

Sampling a Bivariate Multinomial Population

When a sample of size n is selected from a bivariate multinomial population, the sample frequencies will be denoted as shown in Table 17.2b. In the table, f_{ij} denotes the number of sample observations in the ith class of the row variable and the jth class of the column variable. We shall let $f_{i.}$ and $f_{.j}$ denote the totals of the sample frequencies for the ith row and jth column, respectively. An example of the display in Table 17.2b with sample values is found in Table 17.3, which we will discuss shortly.

Test for Statistical Independence (Contingency Table Test)

When one is analyzing bivariate multinomial populations, the first question of interest usually is whether or not the two variables are statistically independent. If they are independent, we know from Chapter 4 that there is no relationship between them. If it turns out that they are not independent and a relationship does exist between the two variables, the next step in the analysis is to study the nature of the relationship. We begin with the first step of the analysis, testing whether or not the two variables are independent.

The test for statistical independence has a structure with which we are now well familiar: We compare the observed frequencies f_{ij} with the frequencies F_{ij} that are expected if the variables are independent, and we calculate the test statistic X^2. If X^2 is small, it indicates that the variables are statistically independent; if X^2 is large, the data support the conclusion that the variables are not independent.

Assumptions

1. A simple random sample of size n has been selected from an infinite or large bivariate multinomial population.
2. The sample size n is reasonably large.

EXAMPLE

For the instrument failures example, historical data on $n = 200$ failures are available. They are displayed in Table 17.3. The analyst desires to test whether or not type and location of failure are statistically independent variables. She assumes

TABLE 17.3   **Two hundred electronic instrument failures classified by type of failure and location of failure — Instrument failures example**

Type of Failure	Location of Failure			Total
	L_1	L_2	L_3	
T_1	50	16	31	97
T_2	61	26	16	103
Total	111	42	47	200

that the 200 observations constitute a random sample from the current production process, that is, from the infinite bivariate multinomial population of failures for the electronic instrument under study. ☐

Alternatives. For two variables to be statistically independent, we know from (4.24) that each joint probability in the bivariate distribution must equal the product of the corresponding row and column marginal probabilities; that is, in the notation of Table 17.2a:

$$p_{ij} = p_{i.}p_{.j} \qquad \text{for all } (i,j)$$

Thus, the appropriate alternatives for a test of statistical independence are as follows.

(17.6)

$$H_0: p_{ij} = p_{i.}p_{.j} \qquad \text{for all } (i,j)$$
$$H_1: p_{ij} \neq p_{i.}p_{.j} \qquad \text{for some } (i,j)$$

Here, H_0 represents statistical independence and H_1 statistical dependence. Note in H_1 that statistical dependence exists even if only some (not all) $p_{ij} \neq p_{i.}p_{.j}$.

Computing Expected Frequencies. Under H_0 of (17.6), $p_{ij} = p_{i.}p_{.j}$ for each cell of the bivariate multinomial population. Hence, an estimate of p_{ij} under the assumption of statistical independence can be obtained first by estimating the marginal probabilities $p_{i.}$ and $p_{.j}$ and then by forming their product. Estimates of $p_{i.}$ and $p_{.j}$, which we denote by $\bar{p}_{i.}$ and $\bar{p}_{.j}$, may be computed from the marginal sample frequencies as follows.

(17.7)

$$\bar{p}_{i.} = \frac{f_{i.}}{n} \qquad \bar{p}_{.j} = \frac{f_{.j}}{n}$$

Point estimates of the joint probabilities if H_0 holds, denoted by $\bar{p}_{ij}$, are then obtained as follows.

(17.8)

$$\bar{p}_{ij} = \bar{p}_{i.}\bar{p}_{.j}$$

EXAMPLE

Refer to Table 17.4, where we reproduce the sample frequencies from Table 17.3 for the instrument failures example. Using these frequencies, we find, for example:

$$\bar{p}_{1.} = \frac{97}{200} = 0.4850 \qquad \bar{p}_{.1} = \frac{111}{200} = 0.5550$$

TABLE 17.4   Chi-square test for the statistical independence of type and location of failure — Instrument failures example.   Observed and expected frequencies are displayed in the table in the format f_{ij} (F_{ij}).

Type of Failure	Location of Failure			Total
	L_1 $(j = 1)$	L_2 $(j = 2)$	L_3 $(j = 3)$	
T_1 $(i = 1)$	50 (53.84)	16 (20.37)	31 (22.80)	97
T_2 $(i = 2)$	61 (57.17)	26 (21.63)	16 (24.21)	103
Total	111	42	47	200

Hence, a point estimate of the joint probability p_{11} under the assumption of statistical independence is:

$$\bar{p}_{11} = \bar{p}_{1.}\bar{p}_{.1} = 0.4850(0.5550) = 0.2692$$

As another example, $\bar{p}_{23} = \bar{p}_{2.}\bar{p}_{.3} = (103/200)(47/200) = 0.1210.$ ☐

Finally, to obtain the expected frequencies F_{ij} if H_0 holds, we multiply the estimated joint probabilities by the sample size n.

(17.9)
$$F_{ij} = n\bar{p}_{ij} = n\bar{p}_{i.}\bar{p}_{.j}$$

EXAMPLE

For the instrument failures example, the estimated expected frequency if H_0 holds for failure type T_1 in location L_1 is:

$$F_{11} = n\bar{p}_{11} = 200(0.2692) = 53.84$$

This and the other expected frequencies F_{ij} are shown in parentheses in Table 17.4 alongside the corresponding observed frequencies f_{ij}.   ☐

Test Statistic and Decision Rule.   As usual, we employ the X^2 test statistic.

(17.10)
$$X^2 = \sum_{i=1}^{r} \sum_{j=1}^{c} \frac{(f_{ij} - F_{ij})^2}{F_{ij}}$$

The X^2 statistic is obtained as always by summing over all classes, but a double summation is required here since each class (i, j) is identified in terms of the categories of both variables. (For a review of the double summation notation, refer to Appendix A, Section A.1.)

To control the α risk, we need to know the sampling distribution of X^2 when H_0 holds.

(17.11)

When the variables of a bivariate multinomial population are statistically independent and the sample size is reasonably large so that theorem (16.7) applies:

$$X^2 \simeq \chi^2[(r-1)(c-1)]$$

where:

X^2 is given by (17.10)

Large values of X^2 lead to conclusion H_1.

EXAMPLE

In the instrument failures example, the analyst wishes to test whether or not type of failure and location of failure are statistically independent.

Step 1.   The alternatives are given in (17.6).
Step 2.   The α risk is to be controlled at 0.05.
Step 3.   Since $r = 2$ and $c = 3$, the degrees of freedom for the χ^2 distribution when H_0 holds are:

$$(r-1)(c-1) = (2-1)(3-1) = 2$$

Hence, we require $\chi^2(0.95; 2) = 5.99$. The appropriate decision rule is therefore as follows:

If $X^2 \leq 5.99$, conclude H_0.
If $X^2 > 5.99$, conclude H_1.

Step 4.   To complete the test, we note first that all of the F_{ij} in Table 17.4 are large enough to meet the χ^2 approximation rule in (16.7). We then compute test statistic (17.10) from the data in Table 17.4 as follows:

$$X^2 = \frac{(50 - 53.84)^2}{53.84} + \frac{(16 - 20.37)^2}{20.37} + \cdots + \frac{(16 - 24.21)^2}{24.21} = 8.08$$

Because $X^2 = 8.08 > 5.99$, we conclude H_1—that type and location of failure are statistically dependent. The next step in the investigation therefore will be to study the nature of the relationship.   □

Comments   1.   A shortcut formula for calculating F_{ij} can be obtained by combining (17.7) and (17.9).

(17.12)

$$F_{ij} = \frac{f_i.f_{.j}}{n}$$

2.   A bivariate sample frequency distribution (as shown in Table 17.2b) is often called a *contingency table*. For this reason, the test procedure described here is sometimes called a *contingency table test*.

3.   The degrees of freedom for the χ^2 distribution for the statistical independence test are obtained as follows: Since there are rc multinomial classes, we have $k = rc$. The number of parameters estimated from the sample is $(r - 1) + (c - 1)$. The reason is that both sets of marginal probabilities must add to 1; hence, only $r - 1$ of the row marginal probabilities and $c - 1$ of the column marginal probabilities are independently estimated from the sample data. Thus, the degrees of freedom of the χ^2 distribution for the test are:

$$k - m - 1 = rc - [(r - 1) + (c - 1)] - 1 = (r - 1)(c - 1)$$

Examination of Nature of Statistical Dependence

When the test for statistical independence leads to the conclusion of dependence, a simple way to examine the nature of the statistical dependence is to consider the conditional probability distributions, as discussed in Chapter 4. Of course, since only sample data are available here, we must work with estimated conditional probabilities.

EXAMPLE

In the instrument failures example, the analyst was particularly interested in the effect of location of failure on the type of failure. Table 17.5 shows the estimated conditional probability distributions of type of failure for each of the three locations. The estimated probabilities are computed from the sample data in Table

TABLE 17.5   **Estimated conditional probability distributions of type of instrument failure for each failure location — Instrument failures example**

Type of Failure	Conditional upon Location of Failure		
	L_1	L_2	L_3
T_1	0.45	0.38	0.66
T_2	0.55	0.62	0.34
Total	1.00	1.00	1.00

17.3. For location L_1, for example, the estimated conditional probabilities are $50/111 = 0.45$ for failure type T_1 and $61/111 = 0.55$ for failure type T_2.

Note how the conditional probability distribution for location L_3 differs from the distributions for locations L_1 and L_2. Failure T_2 is the more likely type of failure in locations L_1 and L_2, while T_1 is the more likely type of failure in location L_3. Design engineers can use such insights to identify potential causes of failure and to recommend design modifications aimed at improved instrument reliability.

17.5 COMPARISONS OF SEVERAL MULTINOMIAL POPULATIONS

Frequently, one wishes to compare several multinomial populations to see whether they are identical.

| EXAMPLES |

1. An educator wishes to compare the performance of three sections of the same course, each taught by a different instructional method, to see whether the methods produce the same or different letter grade distributions.

2. **Flavor Preference.** A market researcher wishes to compare the preferences among a sample of men with those among a sample of women to determine whether the proportions of persons preferring each of four different syrup flavors are the same for men and women.

Parameters of Interest

Table 17.6 illustrates the parameters of interest. There are c multinomial populations (for example, 3 instructional methods, 2 sexes), each with the same r attribute classes (for example, 5 letter grades, 4 syrup flavors). The probability that an element selected at random from population j falls into class i is denoted by p_{ij}. Of course, the p_{ij} sum to 1 for each population j.

TABLE 17.6 c multinomial populations with the same r classes

Class	\multicolumn{4}{c}{Multinomial Population}
	$j = 1$	$j = 2$	$\cdots$	$j = c$
$i = 1$	p_{11}	p_{12}	$\cdots$	p_{1c}
$i = 2$	p_{21}	p_{22}	$\cdots$	p_{2c}
$\vdots$	$\vdots$	$\vdots$		$\vdots$
$i = r$	p_{r1}	p_{r2}	$\cdots$	p_{rc}
Total	1.0	1.0	$\cdots$	1.0

Assumptions

1. Independent random samples of sizes $n_1, n_2, \ldots, n_c$ are selected from c infinite or large multinomial populations.
2. Each of the c sample sizes $n_1, n_2, \ldots, n_c$ is reasonably large.

We denote the total number of observations in the study by n_T.

(17.13)
$$n_T = \sum_{j=1}^{c} n_j$$

Alternatives

If all of the multinomial populations are identical, $p_{11} = p_{12} = \cdots = p_{1c}$. Similarly, p_{2j} will be the same value for each population j; and so on. Thus, a test for the identity of the c populations involves the following alternatives.

(17.14)
$$H_0: p_{i1} = p_{i2} = \cdots = p_{ic} \qquad \text{for all } i$$
$$H_1: \text{not all equalities in } H_0 \text{ hold}$$

Test Procedure (Test of Homogeneity)

Again, we shall employ the test statistic X^2, which compares the observed frequencies with the frequencies expected if H_0 holds. The test statistic X^2 is the same as that in (17.10) for testing statistical independence in a bivariate multinomial population, and the procedure parallels those of earlier tests.

(17.15)
When all multinomial populations are identical and the sample sizes n_j are large so that theorem (16.7) applies:

$$X^2 \simeq \chi^2[(r-1)(c-1)]$$

where:
X^2 is given by (17.10)

Large values of X^2 lead to conclusion H_1.

The logic of obtaining the expected frequencies for testing the identity of multinomial populations is not the same as that for testing statistical independence in a bivariate multinomial population, even though the algebraic results are identical. We shall explain the logic by means of the flavor preference example.

EXAMPLE

In the flavor preference example, the market researcher selected a random sample of $n_1 = 250$ men and independently a random sample of $n_2 = 400$ women. Each person was given four syrup flavors to taste and then was asked for the preferred flavor. The sample results are presented in Table 17.7. We let f_{ij} denote the observed frequency in the ith class for the jth population. Here, $i = 1, 2, 3, 4$ and $j = 1, 2$. Thus, we see from Table 17.7 that $f_{11} = 23$, $f_{12} = 174$, and so on. We also note that $n_T = n_1 + n_2 = 250 + 400 = 650$.

Step 1. The researcher would like to test whether the preference distributions for men and women are the same:

H_0: $p_{i1} = p_{i2}$ for $i = 1, 2, 3, 4$

H_1: not all equalities in H_0 hold

Step 2. The α risk is to be controlled at 0.01.

Step 3. There are $c = 2$ multinomial populations to be compared, each having $r = 4$ flavor preference classes. From (17.15), it follows that the required degrees of freedom are:

$$(r - 1)(c - 1) = (4 - 1)(2 - 1) = 3$$

The action limit for the decision rule therefore is $\chi^2(0.99; 3) = 11.34$. Hence, the appropriate decision rule for the test is as follows:

If $X^2 \leq 11.34$, conclude H_0.
If $X^2 > 11.34$, conclude H_1.

TABLE 17.7 **Chi-square test for the identity of syrup flavor preferences among men and women — Flavor preference example.** Observed and expected frequencies are displayed in the table in the format f_{ij} (F_{ij}).

| Multinomial Class | Preferred Syrup | Sex | | Total |
		$j = 1$ Men	$j = 2$ Women	
$i = 1$	Flavor 1	23 (75.8)	174 (121.2)	197
$i = 2$	Flavor 2	169 (78.1)	34 (124.9)	203
$i = 3$	Flavor 3	48 (76.2)	150 (121.8)	198
$i = 4$	Flavor 4	10 (20.0)	42 (32.0)	52
	Total	$n_1 = 250$	$n_2 = 400$	$n_T = 650$

Step 4. If H_0 holds, all multinomial populations are the same. Let p_1, $p_2, \ldots, p_r$ denote the parameters for the common multinomial distribution if H_0 holds. To estimate the frequencies that we expect when H_0 is true, we must first estimate these common parameters. An estimator for p_i is the proportion of observations falling in the ith class among all c samples combined. We denote this estimator by $\bar{p}_i$ and calculate it as follows.

(17.16)
$$\bar{p}_i = \frac{\sum_{j=1}^{c} f_{ij}}{n_T} = \frac{f_{i.}}{n_T}$$

From Table 17.7, we see that $f_{1.} = 197$ persons of the total of $n_T = 650$ in the study preferred syrup flavor 1 (that is, fell in class $i = 1$). Thus, $\bar{p}_1 = 197/650 = 0.3031$. The parameter estimates for the other classes are obtained similarly.

We shall let F_{ij} denote the frequency expected in class i for the sample from the jth population when the multinomial distributions are, in fact, identical — that is, when H_0 holds. It follows from our foregoing argument that the expected frequency has the following form.

(17.17)
$$F_{ij} = n_j \bar{p}_i$$

For instance, in the sample of $n_1 = 250$ men, we would expect $F_{11} = n_1 \bar{p}_1 = 250(0.3031) = 75.8$ men to fall in the first class if H_0 is true. Similarly, $F_{21} = n_1 \bar{p}_2 = 250(203/650) = 250(0.3123) = 78.1$ men would be expected to fall in the second class. Table 17.7 shows the values of F_{ij} in parentheses immediately beside the corresponding observed frequencies f_{ij}.

Note from Table 17.7 that all F_{ij} are large enough for the χ^2 approximation rule in (16.7) to apply.

Test statistic (17.10) is calculated from the data in Table 17.7 as follows:

$$X^2 = \frac{(23 - 75.8)^2}{75.8} + \frac{(169 - 78.1)^2}{78.1} + \cdots + \frac{(42 - 32.0)^2}{32.0} = 256.8$$

Because $X^2 = 256.8 > 11.34$, we conclude H_1 — that the preference patterns for men and women are different.

The market researcher thereupon investigated the nature of the differences. The estimated probabilities of preference for men and women, based on the data in Table 17.7, are as follows:

Preferred Flavor	Men	Women
1	0.092	0.435
2	0.676	0.085
3	0.192	0.375
4	0.040	0.105
Total	1.000	1.000

Clearly, men show a strong preference for flavor 2 and women have strong preferences divided between flavors 1 and 3. We could use the procedures for comparing two population proportions when the samples are independent (Section 13.4) for analyzing further the differences in the probabilities between men and women for any of the flavors.

Comments

1. The degrees of freedom $(r - 1)(c - 1)$ for the χ^2 distribution when testing the identity of multinomial populations is arrived at as follows: For any one population sampled, there would be $r - 1$ degrees of freedom if H_0 specified the common probabilities p_1, $p_2, \ldots, p_r$ completely. It can be shown that independent χ^2 variables are additive, as are their degrees of freedom. Hence, there would be $c(r - 1)$ degrees of freedom for all c populations if the common probabilities were specified by H_0. But the common probabilities are not specified by H_0 and have to be estimated from the sample. Hence, $m = r - 1$ degrees of freedom are lost, and the number of degrees of freedom is:

    $$c(r - 1) - (r - 1) = (r - 1)(c - 1)$$

2. As we noted earlier, the form of the test statistic X^2 and the decision rule for testing the identity of several multinomial populations are the same as those for testing statistical independence. Moreover, formula (17.17) for F_{ij} is identical to formula (17.12) for the expected frequency in the statistical independence test, with $n_j = f_j$ and $n_T = n$. Thus, even though the theoretical basis is not the same for the two tests, they are computationally identical. For this reason, statistical computer packages usually provide only a single computational routine to do both of these tests.

3. The test for the identity of several multinomial populations is sometimes called a *test of homogeneity*.

4. When there are $c = 2$ populations and each has $r = 2$ categories, the test in (17.15) is equivalent to the two-sided test (13.27) for two population proportions.

17.6   OPTIONAL TOPIC—INFERENCES FOR A RATIO OF MULTINOMIAL PROBABILITIES

Often, we are interested in comparing two probabilities p_i and p_j of a multinomial population in terms of their ratio p_i/p_j.

1.   **Election Poll.**   In an election poll, the following results were obtained:

i	Candidate Preference	p_i	f_i
1	A	p_1	$f_1 = 460$
2	B	p_2	$f_2 = 350$
3	Undecided	p_3	$f_3 = \underline{190}$
	Total	1.0	$n = 1000$

It is desired to estimate the margin of preference for candidate A among the "decided" voters; that is, p_1/p_2 is to be estimated. A point estimate is $460/350 = 1.31$. Thus, among the "decided" voters, we estimate that 131 favor candidate A for every 100 favoring candidate B. The ratio 1.31, when expressed in the form 1.31 to 1, is called the *odds,* as discussed in Section 4.2.

2.   In the bearing failures example, the engineer knows that during the third quarter, 143 percent as much volume of railroad traffic is carried as during the first quarter. He would like to estimate p_3/p_1 for the bearing failures to see whether this ratio is near 1.43.   □

Inferences on a ratio p_i/p_j are readily made by utilizing the following statistical theorem pertaining to the natural logarithm of the ratio of the sample frequencies — that is, $\ln(f_i/f_j)$.

(17.18)

When the sample frequencies f_i and f_j are reasonably large, $\ln(f_i/f_j)$ is approximately normally distributed with a standard deviation estimated by:

$$\sqrt{\frac{1}{f_i} + \frac{1}{f_j}}$$

Confidence Interval

Confidence limits for $\ln(p_i/p_j)$ can be obtained in the usual fashion from theorem (17.18). In turn, confidence limits for p_i/p_j can then be obtained by taking the antilogarithms of the logarithmic confidence limits.

(17.19)

The confidence limits L and U for the ratio p_i/p_j with approximate confidence coefficient $1 - \alpha$, when f_i and f_j are reasonably large, are obtained by following these two steps.

Step 1.   Find the confidence limits for $\ln(p_i/p_j)$:

$$\ln L = \ln\left(\frac{f_i}{f_j}\right) - z\sqrt{\frac{1}{f_i} + \frac{1}{f_j}}$$

$$\ln U = \ln\left(\frac{f_i}{f_j}\right) + z\sqrt{\frac{1}{f_i} + \frac{1}{f_j}}$$

where:

$$z = z(1 - \alpha/2)$$

Step 2.   Take the antilogarithms of $\ln L$ and $\ln U$:

$$L = \exp(\ln L) \qquad U = \exp(\ln U)$$

where:

$\exp(x)$ denotes e^x

(A review of natural logarithms and the base e of natural logarithms is given in Appendix A, Section A.2.)

EXAMPLE

In the election poll example, it is desired to estimate the odds p_1/p_2 with a 95 percent confidence interval.

Step 1.   Given that $f_1 = 460$ and $f_2 = 350$, we require:

$$\frac{f_1}{f_2} = \frac{460}{350} = 1.3143 \qquad \ln(1.3143) = 0.2733$$

$$\sqrt{\frac{1}{f_1} + \frac{1}{f_2}} = \sqrt{\frac{1}{460} + \frac{1}{350}} = 0.0709$$

$$z(1 - \alpha/2) = z(0.975) = 1.960$$

Hence, we obtain:

$$\ln L = 0.2733 - 1.960(0.0709) = 0.1343$$

$$\ln U = 0.2733 + 1.960(0.0709) = 0.4123$$

Step 2.   Taking the antilogarithms, we find:

$$L = \exp(0.1343) = 1.14 \qquad U = \exp(0.4123) = 1.51$$

so that the desired confidence interval is:

$$1.14 \le \frac{p_1}{p_2} \le 1.51$$

Hence, we can conclude with 95 percent confidence that, among the "decided" voters, between 114 to 151 voters favor candidate A for every 100 voters favoring candidate B.   □

PROBLEMS

*17.1 **Machine Component.**   Under preventive maintenance, a machine component is replaced routinely after 2500 operating hours. At the time of preventive replacement, the component may be in excellent ($i = 1$), good ($i = 2$), or poor ($i = 3$) condition. The probabilities for the three outcome states are 0.1, 0.2, and 0.7, respectively.

    a. Do the attribute classes here satisfy the conditions for a system of classification as defined in (2.2)? Explain.

    b. Identify the parameters of the multinomial distribution here. Must these parameters necessarily sum to 1? Comment.

    c. For a random sample of 5 components about to be replaced, what is represented by $f_1 = 2$, $f_2 = 1$, $f_3 = 2$?

    d. Obtain $P(0, 2, 3)$. What does this probability represent?

    e. For a random sample of 5 components about to be replaced, what is the probability that three are in poor condition and the other two are not in poor condition? (*Hint:* Redefine the attribute classes as poor and not poor, and obtain the appropriate probabilities.)

17.2 **Fragile Instrument.**   A fragile instrument, when shipped by air freight, will require upon arrival either no realignment ($i = 1$), minor realignment ($i = 2$), major realignment ($i = 3$), or replacement of parts ($i = 4$). The respective probabilities are 0.6, 0.2, 0.1, and 0.1.

    a. For four instruments shipped independently, obtain the probability that two will require no realignment, one will require minor realignment, and one will require replacement of parts.

    b. Obtain $P(1, 2, 0, 1)$. What does this probability represent?

    c. Construct the probability distribution of all possible outcomes for two instruments shipped independently. What is the most probable outcome?

17.3 The probability that an account will be paid within 30 days is 0.7, between 31 and 60 days is 0.2, and after 60 days is 0.1. The timings of payment for different accounts are statistically independent.

    a. What is the probability that, in a random sample of three accounts, two will be paid within 30 days and one will be paid after 60 days? All three will be paid after 60 days?

    b. Construct the probability distribution of all possible payment timing outcomes in a random sample of three accounts. Which outcome is most probable?

    c. What is the probability that, in a random sample of four accounts, one will be paid within 30 days and three will not be paid within 30 days? (*Hint:* Redefine the attribute classes and obtain the appropriate probabilities.)

*17.4 **Cream Preference.**   In a random sample of 180 women who use a certain type of skin cream regularly, each was asked to state which of the three brands on the market (A, B, C) she preferred. The preference responses were as follows:

Brand Most Preferred:	A ($i = 1$)	B ($i = 2$)	C ($i = 3$)
Number of Women:	75	54	51

    a. Construct a 99 percent confidence interval for p_1, the probability that a regular skin cream user prefers brand A. Interpret your interval estimate.

    b. Before the study was made, a product manager hypothesized that the market share of brand B among regular users is 20 percent. Test this hypothesis, controlling the α risk at 0.01 when $p_2 = 0.20$. State the alternatives, the decision rule, the value of the test statistic, and the conclusion.

17.5 **Customer Complaints.** The manager of the customer services department of a large chain discount store tabulated the 250 most recent complaints as follows:

i	Nature of Complaint	Number of Complaints
1	Quality of merchandise	69
2	Price	48
3	Service	122
4	Other	11
	Total	250

Assume that the 250 complaints constitute a random sample.
 a. Construct a 95 percent confidence interval for p_2, the probability that a complaint deals with price. Interpret your confidence interval.
 b. In the past, 24 percent of the complaints have involved service. Test whether or not p_3 equals 0.24 currently, controlling the α risk at 0.01 when $p_3 = 0.24$. State the alternatives, the decision rule, the value of the test statistic, and the conclusion.

17.6 Refer to **Fragile Instrument** Problem 17.2. The 200 most recent shipments of this instrument by air have utilized a newly developed type of packaging. The shipment outcomes with this packaging were as follows:

i	Outcome	Number of Shipments
1	No realignment	114
2	Minor realignment	53
3	Major realignment	24
4	Replacement of parts	9
	Total	200

Assume that the 200 shipments can be treated as a random sample.
 a. Construct a 90 percent confidence interval for p_1, the probability that no realignment is required.
 b. Convert the confidence interval in **a** into a 90 percent confidence interval for $1 - p_1$. Interpret this confidence interval.
 c. Test the alternatives H_0: $p_4 \geq 0.10$ versus H_1: $p_4 < 0.10$. Control the α risk at 0.05 when $p_4 = 0.10$. State the decision rule, the value of the test statistic, and the conclusion.

17.7 **Hot Line.** A free-lance author writing for a microcomputer journal used a computerized device to call an assistance hot line maintained by a software firm. In 100 calls placed at randomly selected times when the hot line was open (weekdays between 10 A.M. and 4 P.M.), the device was connected immediately with a hot-line staff member ($i = 1$) 16 times, received a taped message requesting that the caller stay on the line ($i = 2$) 47 times, and received a busy signal ($i = 3$) 37 times.
 a. Obtain a 99 percent confidence interval for p_3, the probability that a call made to the hot line will result in a busy signal.
 b. The software firm claims that 25 percent or more of the calls to the hot line result in immediate connection with a hot-line staff member. Test the alternatives H_0: $p_1 \geq 0.25$ versus H_1: $p_1 < 0.25$, controlling the α risk at 0.01 when $p_1 = 0.25$. State the decision rule, the value of the test statistic, and the conclusion.

*17.8   Refer to **Cream Preference** Problem 17.4. Current market shares for the three brands of skin cream, reflecting purchases by all users (regular users as well as nonregular users), are as follows:

Brand:	A $(i = 1)$	B $(i = 2)$	C $(i = 3)$
Market Share (percent):	35	45	20

The study director wants to determine whether or not the probabilities of regular users preferring brands A, B, and C are the same as the respective current market shares based on purchases by all users.

a. State the alternatives for the test. Does alternative H_1 necessarily imply that none of the three brands have the same market share among regular users as among all users? Explain.

b. Perform the appropriate test, controlling the α risk at 0.01. State the decision rule, the value of the test statistic, and the conclusion.

c. Examine the residuals $f_i - F_i$ for the test, and analyze any major differences between the preference distributions for regular users and all users.

17.9   Refer to **Customer Complaints** Problem 17.5. In the entire chain, the distribution of complaints is known from extensive experience to be as follows:

i	Nature of Complaint	Proportion of Complaints
1	Quality of merchandise	0.25
2	Price	0.19
3	Service	0.24
4	Other	0.32
	Total	1.00

a. Test whether or not the store's complaint pattern is the same as that for the entire chain, controlling the α risk at 0.05. State the alternatives, the decision rule, the value of the test statistic, and the conclusion.

b. Examine the residuals $f_i - F_i$ for the test, and analyze how the complaint pattern for the store differs from that for the chain.

17.10   Refer to **Fragile Instrument** Problems 17.2 and 17.6. The probabilities cited in Problem 17.2 are based on extensive experience with the original type of packaging for the instrument.

a. Test whether or not the probabilities for the new type of packaging are the same as those for the original type of packaging. Control the α risk at 0.10. State the alternatives, the decision rule, the value of the test statistic, and the conclusion.

b. Does your conclusion in **a** imply that the two types of packaging have different probabilities of an instrument requiring no realignment? Explain.

c. Examine the residuals $f_i - F_i$ for your test in **a**, and comment on any differences in the outcome distributions for the two types of packaging.

17.11   Refer to **Hot Line** Problem 17.7. The software firm had engaged a consultant who studied the hot-line operation. The consultant concluded that the current hot-line system could not yield better performance than the following response probabilities: $p_1 = 0.25$, $p_2 = 0.45$, $p_3 = 0.30$.

a. Test whether the current system is operating at the performance levels specified by the consultant. Control the α risk at 0.025. State the alternatives, the decision rule, the value of the test statistic, and the conclusion.

b. Subsequently, the software firm installed a new computerized hot-line system. In a random sample of 200 calls under the new system, the observed frequencies were $f_1 = 82, f_2 = 90,$

$f_3 = 28$. Test whether the new system is operating at the performance levels specified by the consultant. Control the α risk at 0.025. State the alternatives, the decision rule, the value of the test statistic, and the conclusion.

c. Obtain the residuals $f_i - F_i$ for the tests in **a** and **b**, and compare them to see what changes in the hot-line performance levels appear to have taken place. Describe your findings briefly.

*17.12 An investment analyst is studying the relation between stock price movements in two consecutive weeks in March. A random sample of 100 stocks was selected, and the price movements of each stock during the two weeks were cross-classified, as follows:

Movement in First Week	Movement in Second Week		
	Increase	No Change	Decrease
Increase	28	6	1
No change	6	32	4
Decrease	2	6	15

a. Test whether or not price movements in the two weeks are statistically independent, controlling the α risk at 0.10. State the alternatives, the decision rule, the value of the test statistic, and the conclusion.

b. Examine the residuals $f_i - F_i$ for the test. What do they suggest about how the price movements depart from independence?

c. For each first-week movement category, obtain the estimated conditional probability distribution of price movement in the second week. Describe the nature of the relationship between the price movements in the two weeks.

17.13 Two hundred members of a test panel were shown a preview of a film of a new musical to be released shortly. Each panel member was asked to indicate whether the amount of dancing in the film is too much, about right, or too little. The results, cross-classified by opinion and sex of viewer, were as follows:

Opinion	Sex of Viewer	
	Male	Female
Too much	18	22
About right	63	47
Too little	29	21

Assume that the 200 panel members are a random sample of the target audience for the film.

a. Test whether or not opinion and sex of viewer are statistically independent, controlling the α risk at 0.05. State the alternatives, the decision rule, the value of the test statistic, and the conclusion.

b. Given your conclusion in **a**, should the next step be to examine the residuals $f_i - F_i$ for the test? Explain.

17.14 A pharmaceutical firm has surveyed a random sample of 120 persons suffering from Parkinson's disease. Among the facts obtained was the following bivariate frequency distribution of disease duration and degree of self-reliance.

Self-reliance	Disease Duration (years)			
	Under 5	5–9	10–14	15 or more
Considerable	36	25	19	10
Little	4	7	9	10

a. Test whether or not disease duration and degree of self-reliance are statistically independent, controlling the α risk at 0.01. State the alternatives, the decision rule, the value of the test statistic, and the conclusion.

b. For each disease duration category, obtain the estimated conditional probability distribution of the degree of self-reliance. What do these distributions show about the nature of the relationship between disease duration and self-reliance?

c. Computer output gives a P-value of 0.006 for the contingency table test in **a**. Is this P-value consistent with your conclusion in **a**? Explain.

17.15 A savings bank surveyed a random sample of 400 heads of households to determine how many knew the current interest rate paid on six-month savings certificates. The following results were obtained when the survey data were cross-classified by occupation and response.

Occupation	Response	
	Did Know	Did Not Know
Wage earner, clerical worker	110	100
Manager, professional	60	20
Other	40	70

The bank management would like to know the probability that a household head knows the current interest rate, whether or not this knowledge is related to occupation, and, if so, the nature of the relationship.

a. Estimate, by means of a 90 percent confidence interval, the probability that a household head knows the current interest rate. Interpret your interval estimate.

b. Test whether or not occupation and knowledge of the current interest rate are statistically independent, controlling the α risk at 0.01. State the alternatives, the decision rule, the value of the test statistic, and the conclusion. What are the implications of your conclusion for the bank?

c. For each occupation, obtain the estimated conditional probability distribution of knowledge of the current interest rate. What do these distributions show about the nature of the relationship between occupation and knowledge of the current interest rate?

*17.16 All freshmen enrolled in special-studies programs at Bushnell University must take the course Civilization 100. A student can receive a grade of A, B, C, D, or F in the course. The grade distributions for this course during the last two years for students in the two most popular special-studies programs at Bushnell are as follows:

Grade	Policy Studies	American Studies
A	50	29
B	68	57
C	55	40
D	19	22
F	9	6
Total	201	154

Assume that the students in the two programs constitute independent random samples.
  a. Test whether or not the probability distributions of grades are the same for students in the two programs. Control the α risk at 0.05. State the alternatives, the decision rule, the value of the test statistic, and the conclusion.
  b. Given your conclusion in **a**, would it now be meaningful to study the nature of the differences in the grade distributions for students in the two programs? Explain.

17.17 Refer to **Cream Preference** Problem 17.4. The study described there was preceded by another study conducted two years earlier. This earlier study used an independent random sample of 150 regular users and yielded the following preference responses:

Brand Most Preferred:	A ($i = 1$)	B ($i = 2$)	C ($i = 3$)
Number of Women:	42	55	53

  a. Test whether or not the preference patterns in the two periods are identical, controlling the α risk at 0.10. State the alternatives, the decision rule, the value of the test statistic, and the conclusion.
  b. Use (13.24) to construct a 90 percent confidence interval for estimating the change that has occurred in the probability that a regular user prefers brand A.
  c. Computer output gives a P-value of 0.035 for the test of homogeneity in **a**. Is this P-value consistent with your conclusion in **a**? Explain.

17.18 Random samples of machine parts selected independently from each of three large production runs revealed the following data on the number of parts in acceptable condition.

Quality of Part		Run 1	Run 2	Run 3
Acceptable		80	89	56
Unacceptable		10	31	4
	Total	90	120	60

  a. Test whether or not the probability of an acceptable part is the same for the three runs. Control the α risk at 0.05. State the alternatives, the decision rule, the value of the test statistic, and the conclusion.
  b. Runs 1 and 3 were made in the Global City plant; run 2 was made in the Midville plant. Test whether or not the probability of a part being in acceptable condition is the same for runs 1 and 3, using $\alpha = 0.05$. State the alternatives, the decision rule, the value of the test statistic, and the conclusion. What do the results of this test, together with the result in **a**, suggest about run 2? Discuss.

17.19 A market research organization conducted three surveys at about the same time. All pertained to the population of families in the United States, though the three samples were selected independently. The first survey — the income survey — sought information on the relation between clothing expenditures and family income. The second survey — the brand preference survey — sought information on brand preferences among various packaged foods. The third survey — the family-planning survey — sought information on families' procreative intentions. The respective sizes of the three surveys were 1000, 1500, and 500 families. The numbers of families refusing to participate in the surveys were 150, 140, and 130, respectively. Assume that each sample is a simple random one and that all were carried out in the same manner and with the same care and supervision.
  a. Test whether or not the subject matter of the survey affects the refusal rate, controlling the α risk at 0.01. State the alternatives, the decision rule, the value of the test statistic, and the conclusion.

     b. Test whether or not the refusal rates for the income and brand preference surveys differ, controlling the α risk at 0.01. State the alternatives, the decision rule, the value of the test statistic, and the conclusion.

     c. Given the conclusion in **a**, was the test in **b** superfluous in the sense that the conclusion obtained in **a** logically implies the conclusion obtained in **b**? Comment.

17.20 Refer to **Environmental Amendment** Problems 13.30 and 13.34.

     a. Conduct a chi-square test to decide whether the distribution of opinion (in favor, not in favor) is the same for males and females. Control the α risk at 0.10. State the alternatives, the decision rule, the value of the test statistic, and the conclusion.

     b. Explain why the test alternatives here are the same as those in Problem 13.34**a**.

     c. Confirm that the value of the test statistic in **a** is equal to the square of the value of the test statistic in Problem 13.34**a**.

*__17.21__ Refer to **Cream Preference** Problem 17.4. Construct a 95 percent confidence interval for the ratio of the probability that a regular user prefers brand A to the probability that she prefers brand B. Does your interval indicate that a greater proportion of regular users prefer brand A than prefer brand B? Explain.

17.22 Refer to **Customer Complaints** Problem 17.5. The manager of the customer services department is concerned about how much more frequently complaints are received about service than about quality of mechandise. Construct an appropriate 99 percent confidence interval, and interpret your interval estimate for the manager.

17.23 Each person in a random sample of 144 summer residents at a major recreational lake was interviewed about his or her reaction to a proposal to dig a canal connecting the lake to a chain of other lakes. The results were as follows:

Response:	Favor	Oppose	Undecided
Number of Residents:	26	87	31

     a. Construct a 95 percent confidence interval for the ratio of the proportion of residents who oppose the proposal to the proportion who favor the proposal.

     b. Convert your confidence interval in **a** to one for the proportion of decided residents who oppose the proposal. Does this confidence interval indicate that a majority of residents who are decided about the issue oppose the proposal? Explain. [*Hint:* If $y = a/b$, what does $y/(1 + y)$ equal?]

EXERCISES

17.24 Refer to **Machine Component** Problem 17.1. In a random sample of five components about to be replaced, two are in excellent condition and three are not in excellent condition. What is the conditional probability that the latter three components consist of one component in good condition and two components in poor condition?

17.25 Derive the multinomial probability function (17.2), using (A.28) in Appendix A.

17.26 Refer to multinomial probability function (17.2). Explain why $E\{f_i\} = np_i$ and $\sigma^2\{f_i\} = np_i(1 - p_i)$ for any sample frequency f_i, $i = 1, 2, \ldots, k$.

17.27 Refer to multinomial probability function (17.2). Given that $\sigma\{f_i, f_j\} = -np_ip_j$ for $i \neq j$, derive $E\{f_if_j\}$. [*Hint:* Use (5.26a).]

17.28 Refer to **Divorces and Annulments** Problem 12.29.

　　a. Conduct the appropriate test by using a chi-square test procedure. State the alternatives, the decision rule, the value of the test statistic, and the conclusion.

　　b. Show mathematically that the test statistic in **a** is necessarily the square of the test statistic used in Problem 12.29**a**.

17.29 Refer to **Flight Simulation** Problem 2.30. Before collecting the data, an analyst conjectured that phase of flight and cause of error would have the following bivariate multinomial probability distribution:

Phase of Flight	Cause of Error		Total
	M	O	
T	0.250	0.125	0.375
C	0.125	0.125	0.250
L	0.125	0.250	0.375
Total	0.500	0.500	1.000

Test the conjecture, controlling the α risk at 0.05. Assume that the 45 observations constitute a random sample. State the alternatives, the decision rule, the value of the test statistic, and the conclusion.

STUDIES

17.30 In a study of 1000 randomly selected recent automobile accidents, each accident was classified according to whether it occurred at an intersection (A_1) or elsewhere (A_2), whether it occurred during the day (B_1) or at night (B_2), and whether it involved personal injury (C_1) or no personal injury (C_2). The resulting cross-classification follows.

	B_1		B_2	
	C_1	C_2	C_1	C_2
A_1	52	203	96	34
A_2	47	389	105	74

　　a. Test whether or not the three variables are independent, controlling the α risk at 0.01. What is your conclusion? [*Hint:* (4.24) extends directly to three variables.]

　　b. To analyze the nature of the relationship, obtain the marginal bivariate distributions for variables A and B, variables A and C, and variables B and C. For each, test whether the two variables are independent, controlling the α risk at 0.01 in each case. Are any insights into the nature of the relationships between the three variables provided by these tests? Discuss.

　　c. Consider the conditional bivariate distribution of variables A and B, given C_1. Test whether variables A and B are independent here, controlling the α risk at 0.01. Do the same for the conditional bivariate distribution of variables A and B, given C_2. Discuss your findings.

17.31 Refer to the **Power Cells** data set. Assume that the observations for each of the three ambient temperatures constitute independent random samples.

　　a. Test whether the mode-of-failure distributions are the same at the three ambient temperatures (10°C, 20°C, 30°C). Control the α risk at 0.05. What is your conclusion?

b. Determine whether the test conclusion would be affected if the temperature categories for 10° and 20°C were combined (that is, pooled). Continue to control the α risk at 0.05. Explain why the test statistic here happens to have the same value as the one for the test in **a**.

c. Obtain an exact 90 percent confidence interval for the probability that a cell will fail by shorting (mode 0) when operating at an ambient temperature of 30°C, using (12.13). Do the test results in **a** and **b** assure us that this probability differs from the corresponding probabilities for cells operating at 10° and 20°C? Explain.

17.32 (Computer needed). Refer to the **Investment Fund** data set.

a. Obtain the 44 quarter-to-quarter changes for the equity unit values and also for the fixed-income unit values of the fund. (Note that the unit values for January 1 of year 1 are given in the description of the data set.) Cross-classify the quarter-to-quarter changes according to whether an increase or decrease occurred, as follows:

Fixed-Income	Equity Unit Value	
Unit Value	Decrease	Increase
Decrease		
Increase		

b. Assume that the 44 pairs of quarter-to-quarter changes in **a** constitute a random sample from a bivariate multinomial population. Test whether the change status of the equity unit values is statistically independent of the change status of the fixed-income unit values. Control the α risk at 0.01. What is your conclusion?

c. Obtain the P-value of the test in **b**. Is the test conclusion in **b** consistent with the P-value?

UNIT FIVE

Linear Statistical Models

18 SIMPLE LINEAR REGRESSION

$\overline{\overline{\overline{}}}$

Regression analysis enables us to ascertain and utilize a relation between a variable of interest, called the *dependent* or *response* variable, and one or more *independent, explanatory,* or *predictor* variables.

<table>
<tr>
<td>EXAMPLES</td>
<td>

1. In a recent regression study, the response variable was family expenditures for food last year, and the independent variables were family size and family income last year.

2. In another regression study, the response variable was number of metric tons of lubricating oil consumed regionally during a given period, and the independent variables included the number of firms in the region using this oil and a regional index of industrial activity.
</td>
</tr>
</table>

Regression analysis is often used to predict the response variable from knowledge of the independent variables. In the lubricating oil study, for instance, it was desired to predict the consumption of lubricating oil in a new marketing region from knowledge of the independent variables for this region. At other times, regression analysis is utilized primarily for examining the nature of the relationship between the independent variables and the response variable. For example, the food expenditures study was designed to determine whether food expenditures account for a declining proportion of family income as family income rises.

In this chapter, we take up some basic concepts of regression analysis. In subsequent chapters, we consider important extensions and applications.

18.1 RELATION BETWEEN TWO VARIABLES

Functional Relation Between Two Variables

We begin with the concept of a relation between two variables. It is useful to distinguish between a functional and a statistical relation. We first consider a functional relation.

(18.1)

> A *functional relation* between two variables X and Y is exact; the value of Y is uniquely determined when the value of X is specified.

EXAMPLES

1. The rental fee (Y, in dollars) for an electric motor is related to the number of hours rented (X) as follows:

    $$Y = 1.50 + 2.00X$$

    Here, 1.50 is the fixed service charge and 2.00 is the hourly charge. Thus, for each number of hours rented, there is a unique rental fee. Figure 18.1a shows the line of relationship $Y = 1.50 + 2.00X$, as well as the observations for three recent rentals — for 1, 2, and 4 hours, respectively. The plotted observations are (1, 3.50), (2, 5.50), and (4, 9.50). Since the value of Y is uniquely determined from X, all observations fall on the line of relationship.

2. The area of a square sheet of metal (Y, in square centimeters) is related to the length of its sides (X, in centimeters) by the functional relation $Y = X^2$. Figure 18.1b shows the curve of relationship, as well as the observations for four sheets whose sides are 10, 25, 30, and 35 centimeters, respectively.   □

Statistical Relation Between Two Variables

In most empirical studies where interest centers on a dependent variable Y, the value of Y is not uniquely determined when the level of the independent variable is specified.

FIGURE 18.1   Examples of functional relations.   All observations fall on the curve of relationship.

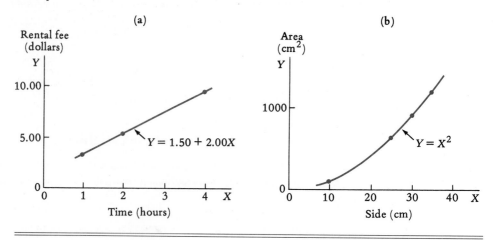

Thus, in studying the relation between family expenditures for food and family income, we are likely to find that families with the same income level differ in their food expenditures. A main reason is that other factors besides family income also play a role, such as ethnic background, family size, and living style.

Relations where Y is not uniquely determined from knowledge of X are called statistical relations.

(18.2)

> A *statistical relation* between two variables X and Y is not exact; the value of Y is not uniquely determined when the value of X is specified.

Some statistical relations are linear.

EXAMPLE

Bag Shipments. Figure 18.2a shows observations on weight (Y) and number of bags (X) in 15 recent shipments of a food product in hundred-weight bags from a company's branch plant overseas to its main plant in this country. One of the shipments contained $X = 40$ bags and weighed $Y = 3963$ pounds, so the point for this shipment is plotted at $(40, 3963)$. The other observations are plotted similarly.

Figure 18.2a shows clearly that weight tends to increase with number of bags. To describe this tendency, we have plotted a line through the concentration of the points, as shown in Figure 18.2b. The relation is not a perfect one, however, and the observations are scattered about the line of relationship. Hence, the relation is a statistical one. Because of the scattering of the points about the line of relationship, Figure 18.2a is called a *scatter diagram* or *scatter plot*. The statistical relation here is *linear* in that it follows a straight line. ☐

Other statistical relations are curvilinear.

FIGURE 18.2 **Scatter plot of a linear statistical relation — Bag shipments example.** The observations are scattered around the line of statistical relationship.

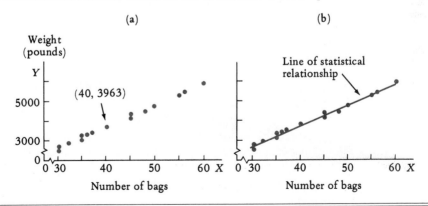

FIGURE 18.3   **Scatter plot of a curvilinear statistical relation — Selling space example**

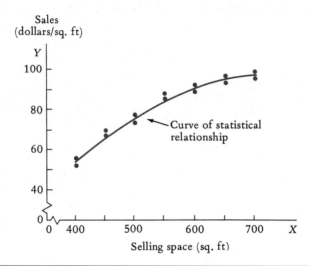

EXAMPLE

Selling Space.   Figure 18.3 shows the observations from an experiment involving 14 supermarkets in a chain. Variable Y is dollar sales per square foot of selling space in a specialty foods section, and X is number of square feet of selling space in the section. The scattering of points indicates that the relation is a statistical one. In contrast to Figure 18.2b, the relation here is *curvilinear*. The growth in sales per square foot diminishes with increasing selling space.   ☐

18.2   SIMPLE LINEAR REGRESSION MODEL

Development of Model

The two examples of statistical relations just presented portray the two main features of statistical relations:

1.   A tendency of the dependent variable Y to vary systematically with the independent variable X, as described by a line or curve of statistical relationship.
2.   A scattering of observations around the line or curve of statistical relationship, partly because factors in addition to the independent variable X affect the dependent variable Y, and partly because of inherent variability in Y.

Regression models incorporate these features of a statistical relation by assuming the following:

1. For each level X of the independent variable, there is a probability distribution of Y.
2. The means of these probability distributions vary in a systematic fashion with X.

EXAMPLE

Consider again the linear statistical relation in the bag shipments example in Figure 18.2b. We portray the corresponding regression model in Figure 18.4. Shown there are the probability distributions of shipment weight Y when the number of bags is $X = 40$ and $X = 50$. The regression model assumes that when, say, $X = 40$ bags are shipped, the observed shipment weight Y will be a random selection from the probability distribution for that level of X.

Note that the means of the probability distributions of Y have a systematic relation to the level of X. This systematic relation is called the *regression function*. In Figure 18.4, the regression function happens to be linear. The fact that the probability distributions of Y are centered around the regression function leads to sample observations that are scattered in a pattern describing the statistical relation, such as the data for the bag shipments example in Figure 18.2a. ☐

Model of an Observation

The features of a regression model that we have just described can also be expressed by assuming that each observation Y is the sum of two components. Let us denote the ith observations on the independent and dependent variables in a regression study by

---

FIGURE 18.4   Illustration of a regression model — Bag shipments example.   The means of the probability distributions are located on the regression function.

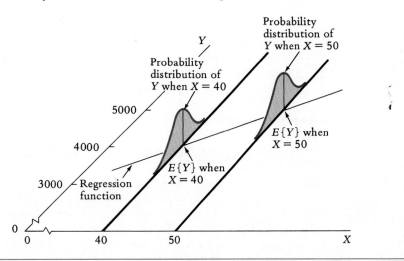

(X_i, Y_i). A regression model then assumes that each observation on the dependent variable, Y_i, is the sum of the following:

1. A component reflecting the line or curve of statistical relationship.
2. A random component reflecting the scatter or deviations about the line or curve of statistical relationship.

For example, if the statistical relation is linear as in Figure 18.2b, the regression model for observation Y_i takes the form:

$$Y_i = \underbrace{\beta_0 + \beta_1 X_i}_{\substack{\text{Line of} \\ \text{statistical} \\ \text{relationship} \\ \text{component}}} + \underbrace{\varepsilon_i}_{\substack{\text{Random} \\ \text{scatter} \\ \text{component}}}$$

The component reflecting the statistical relation is called the *regression component,* and the component reflecting the random scatter is called the *error term.* Since each observation Y_i contains a random component ε_i (Greek epsilon), Y_i is a random variable.

The regression component need not be linear, as just illustrated, but can be curvilinear. Also, the regression component may involve more than one independent variable. Furthermore, various assumptions can be made about the random error terms ε_i in the regression model.

Simple Linear Regression Model

We now consider a basic regression model where the statistical relation is linear and the following assumptions are made about the error terms ε_i:

1. The error terms ε_i are normally distributed.
2. The expected value of ε_i is $E\{\varepsilon_i\} = 0$. (This assumption is necessary so that the random scatter balances around the line or curve of statistical relationship.)
3. The variance of ε_i is constant at all levels of X and is denoted by $\sigma^2\{\varepsilon_i\} = \sigma^2$.
4. The error terms ε_i for different observations are statistically independent.

An equivalent statement of these assumptions is that the error terms ε_i are independent $N(0, \sigma^2)$.

We incorporate these conditions into a model for the observations Y_i as follows.

(18.3)

$$Y_i = \beta_0 + \beta_1 X_i + \varepsilon_i \qquad i = 1, 2, \ldots, n$$

where:

Y_i is the response in the ith case

> X_i is the value of the independent variable in the ith case, assumed to be a known constant
> β_0 and β_1 are parameters
> ε_i are independent $N(0, \sigma^2)$

Model (18.3) is called the *simple linear regression model*. It has three parameters — β_0 (Greek beta), β_1, and σ^2. We shall now consider some important features of this regression model.

Regression Function

Since the error term ε_i is a random variable, so is the response Y_i. We wish to find the expected value of Y_i. Remember that β_0 and β_1 in (18.3) are parameters (hence, constants), and that the value X_i of the independent variable is assumed to be a known constant. Hence, we obtain, by (5.11):

$$E\{Y_i\} = E\{\beta_0 + \beta_1 X_i + \varepsilon_i\} = \beta_0 + \beta_1 X_i + E\{\varepsilon_i\}$$

Since $E\{\varepsilon_i\} = 0$, we find the following result.

(18.4)

$$E\{Y_i\} = \beta_0 + \beta_1 X_i$$

Thus, when the independent variable has the value X_i, the expected value of Y_i is $E\{Y_i\} = \beta_0 + \beta_1 X_i$.

When (18.4) is considered for any value of X, the relationship between X and $E\{Y\}$ is called the regression function.

(18.5)

> The *regression function* relates $E\{Y\}$, the mean of Y, to the value of the independent variable X. For model (18.3), the regression function is:
>
> $$E\{Y\} = \beta_0 + \beta_1 X$$

The regression function thus is the model counterpart to the intuitive curve of statistical relationship. The graph of the regression function is called the *regression line* or *curve*. The parameters β_0 and β_1 are called *regression parameters*. Parameter β_0 is the intercept of the regression line, and β_1 is the slope of the line.

EXAMPLE

UDS. United Data Systems, Incorporated (UDS), a data-processing and equipment firm, leases tape drives to customers for extended periods. These drives must be tested periodically. UDS performs this testing as part of the leasing arrangement. An operations analyst for UDS is considering the relation between number of minutes required on a customer service call to test tape drives (Y) and number of tape drives tested (X). She is employing regression model (18.3) since she considers it to be a reasonable model for this application.

Ordinarily, the regression parameters β_0 and β_1 are unknown and must be estimated from sample data. Suppose, however, that the parameters are $\beta_0 = 15$ and $\beta_1 = 45$. The regression function then is:

$$E\{Y\} = 15 + 45X$$

Figure 18.5 illustrates the meaning of the regression parameters. The intercept, $\beta_0 = 15$, indicates that the value of the regression function at $X = 0$ is 15. The slope, $\beta_1 = 45$, signifies that for each additional tape drive tested, the expected time required to service the tape drives increases by 45 minutes. □

Probability Distributions of Y

We have seen that each response Y_i is a random variable with expected value $E\{Y_i\} = \beta_0 + \beta_1 X_i$. Let us next find the variance of Y_i. Since $\beta_0 + \beta_1 X_i$ is a constant, we have, by (5.12):

$$\sigma^2\{Y_i\} = \sigma^2\{\beta_0 + \beta_1 X_i + \varepsilon_i\} = \sigma^2\{\varepsilon_i\}$$

But model (18.3) assumes that $\sigma^2\{\varepsilon_i\} = \sigma^2$. Hence, we have the following result.

FIGURE 18.5   Illustration of a linear regression function — UDS example.   Parameter β_0 is the intercept of the regression line, and β_1 is the slope.

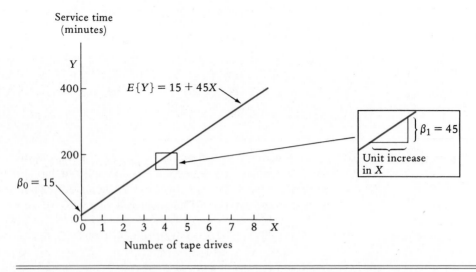

(18.6)

$$\sigma^2\{Y_i\} = \sigma^2$$

Thus, the Y_i have the same variability regardless of the value of X_i.

We further know from theorem (7.8) that each Y_i is normally distributed, because Y_i is a linear function of the normal random variable ε_i. Hence, regression model (18.3) implies the following for any given value X_i of the independent variable:

1. Y_i is normally distributed.
2. $E\{Y_i\} = \beta_0 + \beta_1 X_i$.
3. $\sigma^2\{Y_i\} = \sigma^2$.

Finally, since the ε_i are assumed to be independent for the various observations, so are the Y_i. Hence, an equivalent formulation of regression model (18.3) is the following.

(18.7)

$$Y_i \text{ are independent } N(\beta_0 + \beta_1 X_i, \sigma^2) \qquad i = 1, 2, \ldots, n$$

In words, the Y_i are independent normal random variables, with mean $\beta_0 + \beta_1 X_i$ and variance σ^2.

EXAMPLE

Figure 18.6 shows a graphic representation of the simple linear regression model for the UDS example, assuming that $\beta_0 = 15$, $\beta_1 = 45$, and $\sigma = 8$. Two particular distributions of Y are displayed, at $X = 3$ and $X = 4$, respectively. Note that the means of these distributions are located on the regression line. Also note that both distributions have the same amount of variability and that both are normal.

Suppose that for a customer with $X_i = 4$ tape drives, $Y_i = 185$ minutes are required for the service call. This case is plotted in Figure 18.6. Since the expected value of Y_i is $E\{Y_i\} = 15 + 45(4) = 195$, the error term is $\varepsilon_i = Y_i - E\{Y_i\} = 185 - 195 = -10$. Thus, we see again that the error term ε_i is simply the deviation of Y_i from its mean $E\{Y_i\}$ and hence reflects the scatter around the regression line.

□

Comments

1. The regression function is also called the *response function*, and the term *response curve* is also used for the regression curve.
2. The normality assumption for ε_i is appropriate in many regression applications. A major reason is that there are often many factors influencing Y that are not included among the independent variables of the regression model. Insofar as the effects of these factors are additive and tend to vary with a degree of mutual independence, the composite error term ε_i often will tend to comply with the central limit theorem, and its distribution will be nearly normal when the number of "missing" factors is large.

3.   The designations "response variable" or "dependent variable" on the one hand and "independent variable" on the other carry no connotation that changes in X *cause* changes in Y. No particular cause-effect pattern between the variables is implied necessarily by a regression model. We discuss this point in more detail in Chapter 19.

FIGURE 18.6   Illustration of the simple linear regression model (18.3) for $\beta_0 = 15$, $\beta_1 = 45$, $\sigma = 8$ — UDS example

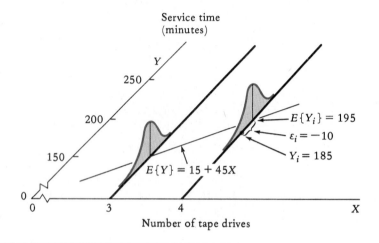

18.3   POINT ESTIMATION OF β_0 AND β_1

Ordinarily, the regression parameters β_0 and β_1 are unknown and must be estimated from sample data. In this section, we take up point estimation of β_0 and β_1. Point estimates of these parameters are often required in their own right and are also needed when interval estimates of these parameters are desired.

When n sample cases are available, we shall denote the first by (X_1, Y_1), the second by (X_2, Y_2), and the ith by (X_i, Y_i), where $i = 1, 2, \ldots, n$.

EXAMPLE

The analyst employed by UDS collected data on number of tape drives and service times for 12 recent customer calls. The data are shown in Table 18.1. With our notation, the observations for the first case are denoted by $(X_1, Y_1) = (4, 197)$, for the second case by $(X_2, Y_2) = (6, 272)$, and so on.

The data are shown as a scatter plot in Figure 18.7. The assumption of a linear response function would appear to be reasonable here.   □

TABLE 18.1   Data for the UDS example

Case i	Number of Tape Drives X_i	Service Time (minutes) Y_i
1	4	197
2	6	272
3	2	100
4	5	228
5	7	327
6	6	279
7	3	148
8	8	377
9	5	238
10	3	142
11	1	66
12	5	239

FIGURE 18.7   Scatter plot — UDS example

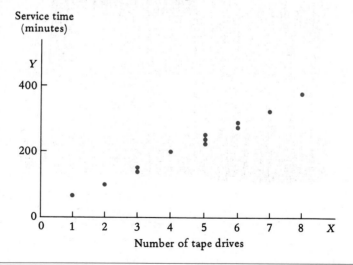

Least Squares Estimators

Point estimates of β_0 and β_1 are usually obtained by a method of estimation called the *method of least squares*. We explain the essential nature of this method by an example.

EXAMPLE

Figure 18.8a shows a scatter plot for four sample cases. We wish to find the straight line that is the "best" fit for these data. Figure 18.8b shows the horizontal straight line $\hat{Y} = 1.8 + 0X$, which obviously is a poor fit. The symbol $\hat{Y}$ (read "Y hat") here denotes an ordinate of the fitted line. The distances, or deviations, of the Y observations from this horizontal line, shown by the vertical rules in Figure 18.8b, are relatively large. For instance, the first deviation is $1 - 1.8 = -0.8$.

FIGURE 18.8   **Finding the least squares regression line for a scatter plot**

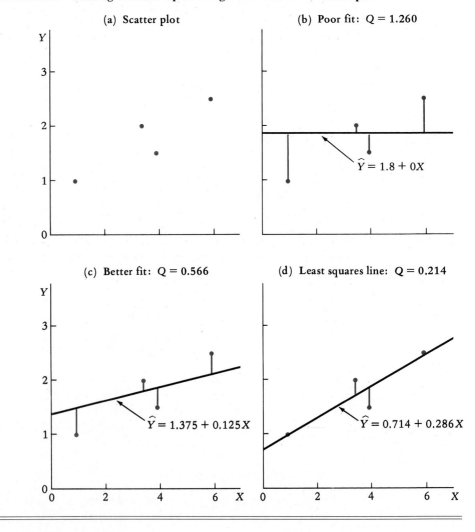

The method of least squares considers as a measure of fit of a straight line the sum of the squared deviations of the Y observations from the given straight line. We shall denote such a sum by Q. For the horizontal straight line in Figure 18.8b, the value of Q is:

$$Q = (1 - 1.8)^2 + (2 - 1.8)^2 + (1.5 - 1.8)^2 + (2.5 - 1.8)^2 = 1.260$$

A better-fitting straight line is the one shown in Figure 18.8c, for which the sum of the squared deviations is $Q = 0.566$. A still better fit is given by the line in Figure 18.8d, for which $Q = 0.214$. In fact, it can be shown that this latter line, $\hat{Y} = 0.714 + 0.286X$, has a smaller sum of squared deviations than any other straight line. The method of least squares considers the best-fitting straight line to be the one for which the criterion Q is a minimum. Thus, $\hat{Y} = 0.714 + 0.286X$ is the least squares regression line, 0.714 is the least squares estimate of β_0 (the intercept of the regression line), and 0.286 is the least squares estimate of β_1 (the slope of the line).   □

The *least squares line,* as we have seen, is that line for which the sum of the squared vertical distances of the Y observations from the line is a minimum. We shall denote the *least squares estimators* of β_0 and β_1 by b_0 and b_1, respectively. The least squares criterion requires that b_0 and b_1 minimize Q, the sum of the squared deviations.

(18.8)

$$Q = \sum_{i=1}^{n} [Y_i - (b_0 + b_1 X_i)]^2$$

It can be shown mathematically that the least squares estimators for the simple linear regression model (18.3) are obtained by solving the following two simultaneous equations.

(18.9)

$$\sum Y_i = nb_0 + b_1 \sum X_i$$
$$\sum X_i Y_i = b_0 \sum X_i + b_1 \sum X_i^2$$

All summations are taken over the sample cases $i = 1, 2, \ldots, n$. We do not show the summation index here and throughout the regression chapters because the nature of the summation is clear.

The equations in (18.9) are called *normal equations.* When they are solved for b_0 and b_1, we obtain the least squares estimators of β_0 and β_1.

(18.10)

For the simple linear regression model (18.3), the least squares estimators of β_1 and β_0 are, respectively:

continues

(18.10a)
$$b_1 = \frac{\sum X_i Y_i - \dfrac{(\sum X_i)(\sum Y_i)}{n}}{\sum X_i^2 - \dfrac{(\sum X_i)^2}{n}}$$

(18.10b)
$$b_0 = \frac{1}{n}(\sum Y_i - b_1 \sum X_i)$$

Thus, the least squares estimates b_0 and b_1 are obtained by calculating the quantities $\sum X_i$, $\sum Y_i$, $\sum X_i Y_i$, and $\sum X_i^2$ from the sample data and substituting into the expressions in (18.10).

Algebraically equivalent formulas for b_1 and b_0 are given next.

(18.11)
$$b_1 = \frac{\sum (X_i - \overline{X})(Y_i - \overline{Y})}{\sum (X_i - \overline{X})^2}$$

(18.12)
$$b_0 = \overline{Y} - b_1 \overline{X}$$

EXAMPLE

The quantities required in (18.10) for the UDS example are calculated from the sample data in Table 18.2. We obtain $\sum X_i = 55$, $\sum Y_i = 2613$, $\sum X_i Y_i = 14{,}060$,

TABLE 18.2   Basic calculations to obtain b_0 and b_1 — UDS example

Case i	Number of Tape Drives X_i	Service Time (minutes) Y_i	$X_i Y_i$	X_i^2
1	4	197	788	16
2	6	272	1,632	36
3	2	100	200	4
4	5	228	1,140	25
5	7	327	2,289	49
6	6	279	1,674	36
7	3	148	444	9
8	8	377	3,016	64
9	5	238	1,190	25
10	3	142	426	9
11	1	66	66	1
12	5	239	1,195	25
Total	$\sum X_i = 55$	$\sum Y_i = 2613$	$\sum X_i Y_i = 14{,}060$	$\sum X_i^2 = 299$

$\Sigma X_i^2 = 299$, and $n = 12$. We calculate then by (18.10):

$$b_1 = \frac{14{,}060 - \dfrac{55(2613)}{12}}{299 - \dfrac{(55)^2}{12}} = 44.41385$$

$$b_0 = \frac{1}{12}[2613 - 44.41385(55)] = 14.18652$$

Thus, the point estimates of β_0 and β_1 are, respectively, $b_0 = 14.187$ and $b_1 = 44.414$.   □

Comments

1. The least squares estimators b_0 and b_1 for regression model (18.3) are unbiased. In addition, we note from (18.10) that both estimators depend linearly on the n values of Y_j. Among all unbiased estimators of β_0 and β_1 that depend linearly on the Y_i, the least squares estimators are the most efficient. These desirable properties of the least squares estimators do not depend on the normality assumption of model (18.3).

2. (Calculus needed.) The least squares estimators can be derived by calculus. For the given observations (X_i, Y_i), the quantity Q to be minimized in (18.8) is a function of b_0 and b_1. To find the values of b_0 and b_1 that minimize Q, we first take partial derivatives of Q with respect to b_0 and b_1:

$$\frac{\partial Q}{\partial b_0} = \Sigma \frac{\partial}{\partial b_0}(Y_i - b_0 - b_1 X_i)^2 = -2\Sigma(Y_i - b_0 - b_1 X_i)$$

$$\frac{\partial Q}{\partial b_1} = \Sigma \frac{\partial}{\partial b_1}(Y_i - b_0 - b_1 X_i)^2 = -2\Sigma X_i(Y_i - b_0 - b_1 X_i)$$

We then set these partial derivatives equal to zero, as follows:

$$-2\Sigma(Y_i - b_0 - b_1 X_i) = 0$$

$$-2\Sigma X_i(Y_i - b_0 - b_1 X_i) = 0$$

After simplifying these equations and rearranging terms, we obtain the normal equations (18.9).

3. When the distribution of the error terms ε_i is specified in the regression model, as in model (18.3), estimators of β_0 and β_1 can also be developed by the method of maximum likelihood described in Chapter 10. For the normal error regression model (18.3), the same normal equations are obtained with this method as with the method of least squares. Thus, the least squares estimators b_0 and b_1 for model (18.3) have the combined properties of least squares estimators and maximum likelihood estimators — they are unbiased, consistent, and most efficient among all unbiased estimators.

18.4   POINT ESTIMATION OF MEAN RESPONSE

Estimated Regression Function

The regression function $E\{Y\} = \beta_0 + \beta_1 X$ is estimated as follows.

(18.13)
$$\hat{Y} = b_0 + b_1 X$$

The equation in (18.13) is called the *estimated regression function*.

EXAMPLE

Earlier we found for the UDS example that $b_0 = 14.18652$ and $b_1 = 44.41385$. Thus, the estimated regression function is:

$$\hat{Y} = 14.18652 + 44.41385X$$

This estimated regression function is plotted in Figure 18.9, together with the original observations. Note that the fit of the estimated regression function to the data appears to be good.   □

Mean Response

As mentioned earlier, the value of the dependent variable Y is often called a response. The mean response refers to the expected value of Y for a given value of X. Thus, in

FIGURE 18.9   Observations and the estimated regression function — UDS example

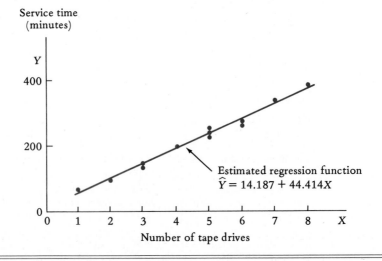

the UDS example the mean response when $X = 4$ refers to the mean of the probability distribution of service times (Y) for the case where four tape drives are serviced.

(18.14)

> The *mean response* when $X = X_h$ is denoted by $E\{Y_h\}$ and for regression model (18.3) is:
>
> $$E\{Y_h\} = \beta_0 + \beta_1 X_h$$

Note that X_h denotes any specified level of X, in contrast to X_i, which refers to the value of the X variable for the ith sample case.

Regression analysis is often utilized to estimate mean responses. For instance, management of UDS may be interested in the expected service time, $E\{Y_h\}$, when there are $X_h = 4$ tape drives in order to estimate the direct labor cost of the service calls. An economist may be interested in the expected amount of savings, $E\{Y_h\}$, of families with incomes of $X_h = \$22,000$ to use in a savings model.

Point Estimator of $E\{Y_h\}$

The point estimator of $E\{Y_h\}$ is given by $\hat{Y}_h$, where $\hat{Y}_h$ is the value of the estimated regression function when $X = X_h$.

(18.15)

> $$\hat{Y}_h = b_0 + b_1 X_h$$

Since b_0 and b_1 are unbiased estimators of β_0 and β_1, respectively, it follows that $\hat{Y}_h$ is an unbiased estimator of $E\{Y_h\}$; that is, $E\{\hat{Y}_h\} = \beta_0 + \beta_1 X_h = E\{Y_h\}$.

EXAMPLES

1.   For the UDS example, we wish to find the point estimate of the mean time required on calls to test $X_h = 4$ tape drives. We substitute into the estimated regression function with $X_h = 4$ and obtain:

$$\hat{Y}_h = 14.18652 + 44.41385(4) = 191.84$$

or 192 minutes.

This result signifies that in a large number of calls to test four tape drives, made under the same general conditions prevailing for the 12 calls in the sample, the mean required time on a call is in the neighborhood of 192 minutes.

2.   For $X_h = 7$ tape drives, the point estimate of the mean service time is:

$$\hat{Y}_h = 14.18652 + 44.41385(7) = 325.08 \text{ minutes}$$

Residuals

For conducting inferences in regression analysis, we need to estimate the magnitude of the random variation in the Y_i. The first step is to measure the scatter of the Y observations around the estimated regression line. We measure this scatter by comparing Y_i, the *observed* value of Y for the ith case, and $\hat{Y}_i$, the value of the estimated mean response when $X = X_i$.

(18.16)

$$\hat{Y}_i = b_0 + b_1 X_i$$

The value $\hat{Y}_i$ is often called the *fitted* value of Y_i. The difference between the observed value Y_i and the fitted value $\hat{Y}_i$ is called the residual for the ith observation.

(18.17)

The *residual* for the ith observation, denoted by e_i, is:

$$e_i = Y_i - \hat{Y}_i$$

EXAMPLE

The residuals for the UDS example are calculated in Table 18.3. For the first case, $X_1 = 4$ and $Y_1 = 197$. The corresponding fitted value and residual are:

$$\hat{Y}_1 = 14.18652 + 44.41385(4) = 191.84$$

$$e_1 = 197 - 191.84 = 5.16$$

The observed Y values, corresponding fitted values, and residuals for the UDS example are portrayed in Figure 18.10.   ☐

TABLE 18.3   Calculation of residuals — UDS example

Case i	Number of Tape Drives X_i	Observed Value Y_i	Fitted Value $\hat{Y}_i$	Residual $e_i = Y_i - \hat{Y}_i$
1	4	197	191.84	5.16
2	6	272	280.67	−8.67
3	2	100	103.01	−3.01
4	5	228	236.26	−8.26
5	7	327	325.08	1.92
6	6	279	280.67	−1.67
7	3	148	147.43	0.57
8	8	377	369.50	7.50
9	5	238	236.26	1.74
10	3	142	147.43	−5.43
11	1	66	58.60	7.40
12	5	239	236.26	2.74
Total	55	2613	2613.0	0.0
Mean	4.583	217.75	217.75	0

FIGURE 18.10   Least squares regression line and residuals — UDS example (observed values and residuals not plotted to scale)

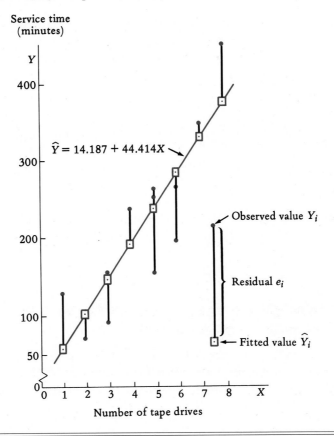

Distinction Between e_i and ε_i. The distinction between the residual e_i and the error term value ε_i is an important one. The former measures the deviation of Y_i from the corresponding fitted value $\hat{Y}_i = b_0 + b_1 X_i$, whereas the latter denotes the deviation of Y_i from the true mean $E\{Y_i\} = \beta_0 + \beta_1 X_i$. Since ε_i usually is unknown, it is estimated by e_i.

The residuals e_i are needed not only for estimating the magnitude of the random variation in the Y_i for making statistical inferences but also for assessing the appropriateness of the regression model employed. We shall discuss this latter use in the next chapter.

Properties of Least Squares Residuals. The least squares residuals (18.17) for regression model (18.3) have a number of important properties:

1. The residuals sum to zero.

(18.18)
$$\sum e_i = 0$$

2. The sum of the squared residuals, $\sum e_i^2$, is a minimum. This property follows because the method of least squares minimizes Q, defined in (18.8).
3. The sum of the weighted residuals is zero when each residual is weighted by the level of the independent variable.

(18.19)
$$\sum X_i e_i = 0$$

4. The sum of the weighted residuals is zero when each residual is weighted by the fitted value.

(18.20)
$$\sum \hat{Y}_i e_i = 0$$

18.5 ANALYSIS OF VARIANCE APPROACH TO SIMPLE LINEAR REGRESSION

Analysis of variance (ANOVA) is a highly useful and flexible mode of analysis for regression models. Here, we introduce some basic ANOVA concepts and procedures and use them to obtain a point estimator of σ^2 and to measure the degree of linear association between X and Y in the observed sample data.

Partitioning of Total Sum of Squares

Total Sum of Squares. Consider again the UDS example. If we wish to predict service time without knowing the number of tape drives of the customer, the uncertainty associated with a prediction is related to the variability of the Y observations around their mean as measured by the following deviations.

(18.21)
$$Y_i - \bar{Y}$$

Here, $\bar{Y}$ is the mean of the Y observations. These deviations for the UDS example are shown in Figure 18.11a. Clearly, if all Y_i are equal, there is no variability in the data and all deviations $Y_i - \bar{Y}$ equal zero. The greater the variability in the data, the larger will be the deviations $Y_i - \bar{Y}$, and the greater is the uncertainty associated with a prediction of Y without utilizing knowledge of X.

FIGURE 18.11 **Partitioning of total sum of squares — UDS example (Y values not plotted to scale)**

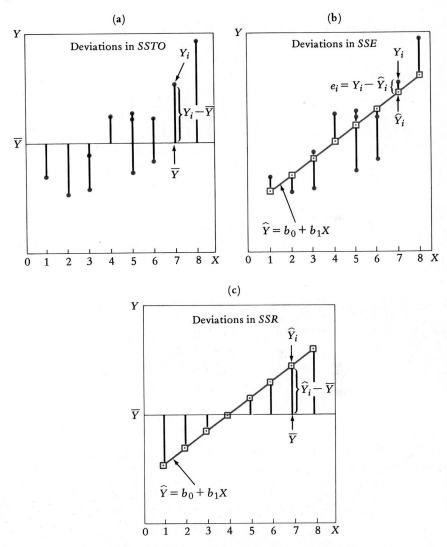

Conventionally, the measure of the variability of the Y observations is expressed in terms of the sum of squares of the deviations $Y_i - \overline{Y}$ and is denoted by $SSTO$.

(18.22)
$$SSTO = \Sigma (Y_i - \overline{Y})^2$$

Here, *SSTO* stands for the *total sum of squares*. If there is no variability in the Y observations, so $Y_i \equiv \overline{Y}$, *SSTO* $= 0$. The greater is the variability in the Y_i, the larger is *SSTO*.

We have actually encountered the total sum of squares earlier but not by this name. Note that the numerator of the sample variance in (8.12b) is *SSTO* (except that we used X instead of Y to designate the variable).

For the UDS sample data, $\overline{Y} = 217.75$ and we obtain for *SSTO* (see the data in Table 18.3):

$$SSTO = (197 - 217.75)^2 + (272 - 217.75)^2 + \cdots + (239 - 217.75)^2$$
$$= 92{,}884.25$$

Error Sum of Squares.   If we use knowledge of the number of tape drives X to predict the service time Y, the uncertainty associated with a prediction is related to the variability of the Y_i around the fitted regression line as measured by the following deviations.

(18.23)
$$Y_i - \hat{Y}_i$$

If all the Y observations fall on the fitted regression line, all deviations $Y_i - \hat{Y}_i$ will equal zero. The larger the deviations $Y_i - \hat{Y}_i$, the greater is the uncertainty associated with a prediction utilizing knowledge of the number of tape drives. The deviations $Y_i - \hat{Y}_i$ for the UDS example are shown in Figure 18.11b.

Again, the conventional measure of the variability around the fitted regression line is the sum of the squared deviations, which is denoted by *SSE*.

(18.24)
$$SSE = \sum (Y_i - \hat{Y}_i)^2$$

Here, *SSE* stands for the *error sum of squares*. If all Y observations fall on the fitted regression line, *SSE* $= 0$. The greater are the deviations $Y_i - \hat{Y}_i$, the larger is *SSE*.

Note that $Y_i - \hat{Y}_i = e_i$ by (18.17); hence, *SSE* is simply the sum of the squared residuals.

(18.24a)
$$SSE = \sum e_i^2$$

We calculate *SSE* for the UDS example by using the residuals in Table 18.3:

$$SSE = (5.16)^2 + (-8.67)^2 + \cdots + (2.74)^2 = 336.88$$

Regression Sum of Squares.   We see in the UDS example that the variability of the Y_i when X is utilized (*SSE* $= 336.88$) is much smaller than when X is not utilized

($SSTO$ = 92,884.25). This reduction in the variability of the Y_i is associated with the utilization of knowledge of the number of tape drives for predicting service time. The reduction is, indeed, another sum of squares, denoted by SSR.

(18.25)

$$SSR = SSTO - SSE$$

Here, SSR stands for the *regression sum of squares*.

For the UDS example, the reduction in the variability of the Y_i is:

$$SSR = 92,884.25 - 336.88 = 92,547.37$$

It can be shown that SSR is a sum of squares involving the deviations:

(18.26)

$$\hat{Y}_i - \overline{Y}$$

Since the mean of the n fitted values $\hat{Y}_i$ equals $\overline{Y}$ (see Table 18.3 for a confirmation), each deviation $\hat{Y}_i - \overline{Y}$ represents the difference between the fitted value and the mean of the fitted values. The deviations $\hat{Y}_i - \overline{Y}$ for the UDS example are shown in Figure 18.11c.

The regression sum of squares then is as follows.

(18.27)

$$SSR = \Sigma(\hat{Y}_i - \overline{Y})^2$$

The regression sum of squares SSR may be viewed as a measure of the effect of the regression relation in reducing the variability of the Y_i. If $SSR = 0$, the regression relation does not reduce the variability at all. The larger is SSR compared with $SSTO$, the greater is the reduction in the variability of the Y_i by utilization of knowledge of X through the regression relation.

For the UDS example, the direct calculation of SSR by (18.27) proceeds as follows, using the data in Table 18.3:

$$SSR = (191.84 - 217.75)^2 + (280.67 - 217.75)^2$$
$$+ \cdots + (236.26 - 217.75)^2$$
$$= 92,547.37$$

This result necessarily agrees with that obtained previously by subtraction.

Formal Statement of Partitioning. Since $SSR = SSTO - SSE$ by (18.25), it follows that $SSTO = SSR + SSE$. We can therefore view the process we have just completed as a partitioning of the total sum of squares $SSTO$ into two additive components: (1) SSR — the portion of the total variability of the Y_i eliminated when the independent variable X is considered, and (2) *SSE* — the portion of the total variability of the Y_i that remains as residual or error variation when X is considered.

(18.28) For simple linear regression, the decomposition of the total sum of squares into two additive components is:

(18.28a)
$$SSTO \quad = \quad SSR \quad + \quad SSE$$

(18.28b)
$$\Sigma (Y_i - \overline{Y})^2 = \Sigma (\hat{Y}_i - \overline{Y})^2 + \Sigma (Y_i - \hat{Y}_i)^2$$

EXAMPLE

In assembling the respective sums of squares for the UDS example, we obtain:

$$92{,}884.25 = 92{,}547.37 + 336.88$$

$$SSTO = SSR + SSE$$

Computational Formulas. The preceding formulas for the three sums of squares are the basic definitional ones. The following formulas are algebraically equivalent and tend to be more convenient for manual computation.

(18.29)
$$SSTO = \Sigma Y_i^2 - \frac{(\Sigma Y_i)^2}{n}$$

(18.30)
$$SSR = \frac{\left(\Sigma X_i Y_i - \dfrac{\Sigma X_i \Sigma Y_i}{n}\right)^2}{\Sigma X_i^2 - \dfrac{(\Sigma X_i)^2}{n}}$$

(18.31)
$$SSE = SSTO - SSR$$

Comment Formula (18.28b) can be derived by writing:

$$\Sigma (Y_i - \overline{Y})^2 = \Sigma [(Y_i - \hat{Y}_i) + (\hat{Y}_i - \overline{Y})]^2$$

By expanding the right side, we obtain:

$$\Sigma (Y_i - \overline{Y})^2 = \Sigma (Y_i - \hat{Y}_i)^2 + \Sigma (\hat{Y}_i - \overline{Y})^2 + 2\Sigma (Y_i - \hat{Y}_i)(\hat{Y}_i - \overline{Y})$$

It can be shown that the last term equals zero.

Partitioning of Degrees of Freedom

A sum of squares has an associated number of degrees of freedom. Recall that the sample variance s^2 in (8.12b) has a denominator $n - 1$. This number is the degrees of freedom associated with the numerator sum of squares in s^2.

Corresponding to the partitioning of the total sum of squares $SSTO$ into the components SSR and SSE, there is a partitioning of the degrees of freedom. $SSTO$ has $n - 1$ degrees of freedom associated with it. To see why, note that there are n deviations $Y_i - \overline{Y}$ that enter into $SSTO$. However, there is one constraint on these deviations, namely, $\Sigma(Y_i - \overline{Y}) = 0$, so there are $n - 1$ degrees of freedom in the n deviations.

SSE has $n - 2$ degrees of freedom. There are n residuals $e_i = Y_i - \hat{Y}_i$. However, two degrees of freedom are lost because of two constraints on the e_i associated with estimating the parameters β_0 and β_1 by the two normal equations.

Finally, SSR has one degree of freedom. There are two parameters in the regression function, but the deviations $\hat{Y}_i - \overline{Y}$ are subject to the constraint $\Sigma(\hat{Y}_i - \overline{Y}) = 0$. We thus see that the degrees of freedom (df) are additive.

(18.32)
$$\underbrace{n - 1}_{\substack{df \text{ for} \\ SSTO}} = \underbrace{1}_{\substack{df \text{ for} \\ SSR}} + \underbrace{n - 2}_{\substack{df \text{ for} \\ SSE}}$$

EXAMPLE

For the UDS example, where $n = 12$, $SSTO$ has associated with it $12 - 1 = 11$ degrees of freedom, SSR has one degree of freedom, and SSE has $12 - 2 = 10$ degrees of freedom. Hence, $11 = 1 + 10$. □

Mean Squares

A sum of squares divided by its associated degrees of freedom is called a *mean square*. The sample variance s^2 is an example of a mean square. Two mean squares of importance in regression analysis are the *regression mean square*, denoted by MSR, and the *error mean square*, denoted by MSE.

(18.33)
$$MSR = \frac{SSR}{1} = SSR$$

(18.34)
$$MSE = \frac{SSE}{n - 2}$$

| EXAMPLE | For the UDS example, we have $SSR = 92{,}547.37$, $SSE = 336.88$, and $n = 12$. Hence: |

$$MSR = \frac{92{,}547.37}{1} = 92{,}547.37 \qquad MSE = \frac{336.88}{10} = 33.688 \qquad \square$$

| Comment | Mean squares are not additive; MSR and MSE do not sum to $SSTO/(n-1)$. For the UDS example, we have: |

$$\frac{SSTO}{n-1} = \frac{92{,}884.25}{11} = 8444.023$$

whereas:

$$MSR + MSE = 92{,}547.37 + 33.688 = 92{,}581.06$$

ANOVA Table

It is useful to collect the sums of squares (SS), degrees of freedom (df), and mean squares (MS) in an ANOVA table. Table 18.4a shows a basic ANOVA table for regres-

TABLE 18.4   ANOVA table for simple linear regression and results for the UDS example

(a)
General format

Source of Variation	SS	df	MS
Regression	$SSR = \Sigma(\hat{Y}_i - \bar{Y})^2$	1	$MSR = \dfrac{SSR}{1}$
Error	$SSE = \Sigma(Y_i - \hat{Y}_i)^2$	$n-2$	$MSE = \dfrac{SSE}{n-2}$
Total	$SSTO = \Sigma(Y_i - \bar{Y})^2$	$n-1$	

(b)
UDS example

Source of Variation	SS	df	MS
Regression	92,547.37	1	92,547.37
Error	336.88	10	33.688
Total	92,884.25	11	

sion analysis, containing the definitional formulas as entries. The corresponding numerical results for the UDS example are displayed in Table 18.4b.

Point Estimation of σ^2

For making statistical inferences in regression applications, we require a point estimate of σ^2, the variance of the error terms ε_i. Just as we use the mean square:

$$s^2 = \frac{\Sigma (X_i - \overline{X})^2}{n - 1}$$

to estimate the variance in a single population where X_i is the ith observation and $\overline{X}$ the estimated mean, we use a mean square to estimate σ^2 in the regression model.

In the regression case, the ith observation is Y_i and its estimated mean is $\hat{Y}_i$. The relevant deviations therefore are the residuals $e_i = Y_i - \hat{Y}_i$, and the relevant mean square for estimating σ^2 is the error mean square $MSE = \Sigma e_i^2/(n - 2)$.

(18.35)

> The variance σ^2 of the error terms ε_i in regression model (18.3) is estimated by MSE as defined in (18.34).

It can be shown that MSE is an unbiased estimator of σ^2; that is, $E\{MSE\} = \sigma^2$.

EXAMPLE

Earlier we found for the UDS example that $MSE = 33.688$ (Table 18.4b). Thus, the point estimate of σ^2 is 33.688. The corresponding point estimate of σ is $\sqrt{MSE} = \sqrt{33.688} = 5.804$.

Consider again the probability distribution of service times (Y) when the customer has $X_h = 4$ tape drives. We found before that $\hat{Y}_h = 191.84$ when $X_h = 4$. Hence, the distribution of service times is normal [assuming that model (18.3) is appropriate], with approximate mean 191.84 minutes and approximate standard deviation 5.8 minutes. In the same way, we can describe the probability distribution of Y for any other number of tape drives serviced. □

18.6 COEFFICIENTS OF SIMPLE DETERMINATION AND CORRELATION

We now turn to two measures that are widely used to describe the degree of linear relationship between the variables X and Y as reflected in the simple linear regression model.

Coefficient of Simple Determination

As we saw, $SSTO$ reflects the variability of the Y_i when knowledge of X is not utilized. In contrast, SSE reflects the remaining variability of the Y_i when knowledge of X is utilized by the simple linear regression model. Finally, $SSR = SSTO - SSE$ reflects the reduction in the total variability of Y_i associated with use of X to predict Y. A relative measure of this reduction in variability is obtained by expressing SSR as a proportion of $SSTO$. The proportion is denoted by r^2 and is called the coefficient of simple determination.

(18.36)

The *coefficient of simple determination* r^2 is defined:

$$r^2 = \frac{SSR}{SSTO} = \frac{SSTO - SSE}{SSTO} = 1 - \frac{SSE}{SSTO}$$

EXAMPLE

For the UDS example, we have $SSTO = 92{,}884.25$ and $SSR = 92{,}547.37$. Thus:

$$r^2 = \frac{92{,}547.37}{92{,}884.25} = 0.99637$$

This value means that the variability in service times is reduced 99.64 percent when the number of tape drives is considered. This large proportionate reduction in the variability of the Y_i by the introduction of X through the simple linear regression model suggests that the regression relation may be highly useful. For example, direct labor costs might be anticipated usefully on the basis of the number of tape drives serviced.   ☐

Limiting Values of r^2.   One limiting value of r^2 occurs when the statistical relation is perfect for the sample data, that is, when all Y observations fall on the fitted regression line so that all residuals e_i equal zero. Then, $SSE = 0$ and $SSR = SSTO - SSE = SSTO$. Hence, $r^2 = SSR/SSTO = 1$. This limiting case is illustrated in Figure 18.12a by the exact fit to the observations of the linear regression function.

The other limiting value of r^2 occurs when the regression relation is of no use in reducing the variability of the Y_i. Then, the remaining variability SSE is the same as $SSTO$. Hence, $SSR = SSTO - SSE = 0$ then, and $r^2 = 0$. Incidentally, the formal condition for SSE to equal $SSTO$ is that the fitted regression line be horizontal, that is, that $b_1 = 0$. This condition is illustrated in Figure 18.12b. Observe that the sample data exhibit no linear statistical relationship between the two variables X and Y.

In general, we have the following bounds for r^2.

(18.37)

$0 \le r^2 \le 1$

Here, the lower bound implies that there is no linear statistical relation exhibited by the observations, and the upper bound implies a perfect linear relation. In practice, r^2 usually falls somewhere between 0 and 1; the closer r^2 is to 1, the greater is the degree of linear statistical relation in the observations. The last two panels in Figure 18.12 illustrate two scatter plots with r^2 values that lie between 0 and 1. Note how the scatter

FIGURE 18.12   Scatter plots showing several degrees of linear statistical relationship

(a) $r^2 = 1; r = +1$

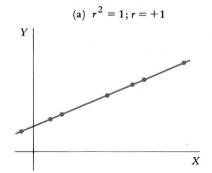

(b) $r^2 = 0; r = 0 \ (b_1 = 0)$

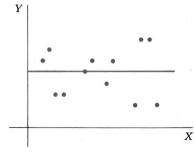

(c) $r^2 = 0.476; r = +0.690$

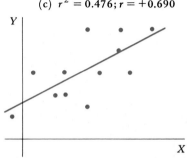

(d) $r^2 = 0.901; r = -0.949$

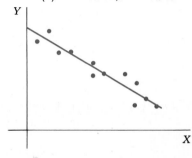

plot in Figure 18.12d, for which $r^2 = 0.901$, is fairly tightly clustered about the regression line, whereas there is quite a bit of scattering in Figure 18.12c, for which $r^2 = 0.476$.

Coefficient of Simple Correlation

The square root of r^2 is another descriptive measure of the degree of linear statistical relation between X and Y. It is called the coefficient of simple correlation.

(18.38)

The *coefficient of simple correlation r* is defined:

$$r = \pm\sqrt{r^2}$$

where:
the sign is that of b_1
r^2 is given by (18.36)

The sign attached to r simply shows whether the slope of the regression line is positive or negative.

Since $0 \leq r^2 \leq 1$, the bounds on r are as follows.

(18.39)

$$-1 \leq r \leq 1$$

The closer the absolute value of r is to 1, the greater is the degree of linear statistical relation in the sample observations.

The coefficient r does not have as clear-cut an operational interpretation as r^2. Even so, there has been a tendency to use r in preference to r^2 in much applied work. Note that r may give the impression of a closer relation between X and Y than does r^2, since the absolute value of r is closer to 1 than is r^2 (except when r^2 equals 0 or 1).

EXAMPLES

1. For the UDS example, $r^2 = 0.99637$ and b_1 is positive. Hence, $r = +\sqrt{0.99637} = +0.99818$.

2. For the scatter plot in Figure 18.12d, $r^2 = 0.901$ and b_1 is negative. Hence, $r = -\sqrt{0.901} = -0.949$.   □

Comment

The coefficient of simple correlation in (18.38) can be calculated directly from the following formula.

(18.40)

$$r = \frac{\sum (X_i - \bar{X})(Y_i - \bar{Y})}{\sqrt{\sum (X_i - \bar{X})^2 \sum (Y_i - \bar{Y})^2}}$$

18.7   COMPUTER OUTPUT

Practically all statistical software packages and many types of calculators have routines for simple regression analysis. Figure 18.13 gives typical regression computer output for the UDS example. For convenience, we have numbered the blocks of information in the output.

Block 1.   The sample mean and the sample standard deviation for the observations on variables X and Y are given.

Block 2.   The point estimates b_0 and b_1 are given, corresponding to the labels CONSTANT and X, respectively. Other information needed for making inferences on the regression parameters β_0 and β_1 is also provided. This information will be explained in the next chapter.

FIGURE 18.13   Computer output for simple linear regression—UDS example

1. VARIABLE   MEAN   STANDARD DEVIATION

    X   4.58333   2.06522
    Y   217.75000   91.89136

2. VARIABLE   REG. COEFF.   STD. DEV.   T STAT.   P-VALUE

    CONSTANT   14.18650   4.22980   3.35   .00732
    X   44.41385   .84737   52.41   .00000

3.   ANALYSIS OF VARIANCE TABLE

    SOURCE   SUM OF SQUARES   DF   MEAN SQUARE

    REGRESSION   92547.3690   1   92547.36901
    RESIDUAL   336.8810   10   33.68810
    TOTAL   92884.2500   11

        F STAT. = 2747.18   UPPER-TAIL P-VALUE = .00000

4. RESIDUAL STD. DEV. = 5.80415   NUMBER OF OBS. = 12
    SIMPLE R SQUARED   = .99637   SIMPLE R   =   .99818

5. CASE NO.   OBSERVED   ESTIMATED   RESIDUAL

    1   197.000   191.842   5.15808
    2   272.000   280.670   -8.66963
    3   100.000   103.014   -3.01421
    4   228.000   236.256   -8.25577
    5   327.000   325.083   1.91652
    6   279.000   280.670   -1.66963
    7   148.000   147.428   .57194
    8   377.000   369.497   7.50266
    9   238.000   236.256   1.74423
    10   142.000   147.428   -5.42806
    11   66.000   58.600   7.39964
    12   239.000   236.256   2.74423

Block 3.   The entries for the ANOVA table are displayed. The format is that of Table 18.4a, with the error source of variation labeled RESIDUAL. The items F STAT. and UPPER-TAIL P-VALUE immediately below the table will be discussed in the next chapter.

Block 4.   The values of $\sqrt{MSE}$, r^2, r, and n are given in this block.

Block 5.   The OBSERVED value Y_i, the ESTIMATED (that is, fitted) value $\hat{Y}_i$, and the RESIDUAL e_i are given by case number.

The calculational results in Figure 18.13 are the same as those presented earlier, except for some minor rounding differences.

PROBLEMS

18.1 A study panel for a professional association suggested the following income guidelines for members in private practice (X is years of experience and Y is target annual income in thousands of dollars):

X:	5	10	15	20	25	30
Y:	39.5	49.0	58.5	68.0	77.5	87.0

a. Examine the relation between experience and income graphically.
b. Do the guidelines represent a statistical or functional relation? Explain.

18.2 For each of the following pairs of variables, explain whether a functional or statistical relation would most likely hold:
a. X = number of beds in hospital; Y = hospital's annual operating cost.
b. X = altitude; Y = atmospheric pressure.
c. X = company's promotional expenditures; Y = company's sales revenues.

18.3 **Vending Machines.**   A vending machine company studied the relation between maintenance cost and dollar sales for six machines. The purpose was to identify machines whose costs are "out of line" with their sales volumes. The data (in dollars) follow.

Machine:	1	2	3	4	5	6
Cost:	95	110	80	100	125	90
Sales:	900	1250	550	850	1500	800

a. Which variable is the dependent variable here? Why?
b. Construct a scatter plot of the data. Does the maintenance cost of any machine seem "out of line"? Comment.

18.4 **Spouses' Ages.**   The ages (in years) of the spouses of six married couples follow.

Couple i:	1	2	3	4	5	6
Wife X_i:	35	24	51	25	53	42
Husband Y_i:	38	24	49	31	55	44

a. Construct a scatter plot of the data.
b. Does it appear that the relation between X and Y here is statistical or functional? Linear or curvilinear? Comment.

18.5 Refer to Figure 18.6. For the case $X_i = 3$, $Y_i = 146$, calculate ε_i. What is the implication of the sign of ε_i?

18.6 Suppose the regression function relating annual agricultural exports (Y) to agricultural imports (X) for a certain country is $E\{Y\} = 320 + 0.4X$, where both Y and X are expressed in millions of dollars.
a. What is the expected level of agricultural exports when agricultural imports are $380 million?

b. For year i, $X_i = 600$ and $\varepsilon_i = 35$. What is the level of agricultural exports for that year? What symbol is used to denote this quantity?

18.7 In an expenditures study, an economist related family expenditures for clothing (Y) to family size (X) by using regression model (18.3). Critics of the study commented that numerous factors other than family size affect clothing expenditures. What are some of these other factors, and how are they accommodated by model (18.3)?

*18.8 Regression model (18.3) applies in a situation with $\beta_0 = -6$, $\beta_1 = 1.3$, and $\sigma^2 = 16$.

    a. What is the probability that ε_i lies between -5 and $+5$? Explain why you can answer without knowing the value of X_i.

    b. What is the value of Y_i if $X_i = 15$ and $\varepsilon_i = 4$?

    c. For $X_i = 6$, obtain (1) $E\{Y_i\}$, (2) the probability that Y_i will exceed 1.

18.9 Regression model (18.3) applies in a situation with $\beta_0 = 15$, $\beta_1 = 10$, and $\sigma^2 = 25$.

    a. For $X_i = 8$, obtain (1) $E\{Y_i\}$, (2) the probability that ε_i lies between -10 and $+10$, (3) the probability that Y_i lies between 85 and 105.

    b. Plot the regression line and the case $(X_i, Y_i) = (8, 92)$ on the same graph. What component of regression model (18.3) is represented by the vertical deviation of the point $(8, 92)$ from the regression line? What is the magnitude of this deviation?

*18.10 Refer to **Vending Machines** Problem 18.3.

    a. Calculate the sum of squared deviations Q for the horizontal line $\hat{Y} = 100 + 0X$, where Y and X denote cost and sales, respectively.

    b. Obtain the least squares line for a regression of cost on sales. Calculate the sum of squared deviations Q for the least squares line.

    c. On the scatter plot of the data, show the horizontal line in **a** and the least squares line in **b**. Which line provides the better visual fit to the scatter plot? Is this result consistent with the values of Q obtained in **a** and **b**? Explain.

18.11 Refer to **Spouses' Ages** Problem 18.4.

    a. Calculate the sum of squared deviations Q for the line $\hat{Y} = 2 + X$.

    b. Obtain the least squares line. Calculate the sum of squared deviations Q for the least squares line.

    c. Compare the values of Q obtained in **a** and **b**. Are the results consistent with theoretical expectations? Explain.

*18.12 **Fuel Efficiency.** Data on automobile weight (in hundreds of pounds) and expressway mileage (in miles per gallon) gathered in a study of the fuel efficiency of eight automobiles follow.

Automobile i:	1	2	3	4	5	6	7	8
Weight X_i:	21	24	23	21	22	18	20	26
Mileage Y_i:	35	27	31	38	36	40	37	28

Assume that regression model (18.3) is appropriate.

    a. Construct a scatter plot of the data.

    b. Obtain the estimated regression function, and plot it on your graph in **a**. Does a linear regression function appear to be a good fit here? Comment.

18.13 **Marketing Experiment.** The data that follow were obtained in an experiment to estimate the relation between the number of showings of a TV commercial in a sales territory and territory sales of a high-technology electric blanket (in thousands of units). Each territory received a single newspaper advertisement and zero to four showings of the TV commercial. The territories were assigned at random to the numbers of showings. Other factors, such as TV channel employed and timing of commercials, were controlled carefully so that their effects would be the same in the different territories.

Territory i:	1	2	3	4	5	6	7	8	9	10
Showings X_i:	3	1	4	0	2	4	0	3	1	2
Sales Y_i:	2.66	1.29	3.02	1.09	2.01	3.64	0.55	3.21	1.85	2.50

Assume that regression model (18.3) is appropriate.
  a. Construct a scatter plot of the data.
  b. Obtain the estimated regression function, and plot it on your graph in **a**. Does a linear regression function appear to be a good description of the statistical relation between X and Y here? Comment.

18.14 **Reconditioning Cost.** Data for nine large incinerators on the cost of the most recent reconditioning (in $000) and operating hours since the preceding reconditioning (in thousands) follow.

Incinerator i:	1	2	3	4	5	6	7	8	9
Hours X_i:	2.2	1.8	2.9	2.5	1.6	2.9	2.7	3.1	1.9
Cost Y_i:	5.0	4.3	6.2	5.1	3.6	5.8	5.9	6.1	4.1

Assume that regression model (18.3) is appropriate.
  a. Construct a scatter plot of the data.
  b. Obtain the estimated regression function, and plot it on your graph in **a**. Does a linear regression function appear to be a good description of the statistical relation between X and Y here? Comment.

18.15 **Bid Preparation.** Data for the past eight weeks on the number of bids prepared by Atlas Structural Corporation and the number of man-hours required for bid preparation follow.

Week i:	1	2	3	4	5	6	7	8
Bids X_i:	6	3	9	6	4	8	2	6
Man-Hours Y_i:	170	78	232	155	107	212	56	162

Assume that regression model (18.3) is appropriate.
  a. Construct a scatter plot of the data.
  b. Obtain the estimated regression function, and plot it on your graph in **a**. Does a linear regression function appear to be a good fit here? Comment.

*18.16 Refer to **Fuel Efficiency** Problem 18.12.
  a. Interpret b_1.
  b. Obtain the estimated mean response when an automobile's weight is 19 hundred pounds. Interpret this number.
  c. Is it reasonable to estimate expressway mileage for a weight of 19 hundred pounds when none of the eight automobiles in the regression study had this weight? Discuss.

18.17 Refer to **Marketing Experiment** Problem 18.13.
  a. Interpret b_0 and b_1.
  b. Obtain a point estimate of mean sales in a territory receiving three showings of the TV commercial.

18.18 Refer to **Reconditioning Cost** Problem 18.14.
  a. What is meant by *mean response* here?

b. Obtain a point estimate of the change in the mean response when the number of operating hours increases by 1.0 thousand.

c. Obtain a point estimate of the mean reconditioning cost when the number of operating hours is 2.0 thousand.

d. Are your point estimates in **b** and **c** subject to sampling errors? If so, do the point estimates convey any information about the magnitudes of these errors?

18.19 Refer to **Bid Preparation** Problem 18.15.

a. The numbers of man-hours required were not the same in the three weeks in which six bids were prepared. Does this represent a departure from regression model (18.3)? Explain.

b. Obtain a point estimate of the change in the mean response when one additional bid is prepared.

c. Estimate the mean response when five bids are prepared.

*18.20 Refer to **Fuel Efficiency** Problem 18.12.

a. Obtain the residuals.

b. Verify that (1) $\sum e_i = 0$, (2) $\sum X_i e_i = 0$, (3) $\sum e_i^2 = 25.283$, (4) $\sum (Y_i - \bar{Y})^2 = 160$.

18.21 Refer to **Marketing Experiment** Problem 18.13.

a. Obtain the residuals.

b. Verify that (18.18), (18.19), and (18.20) hold here.

c. Refer to the residuals for territories 7 and 8. Do these residuals differ solely because the TV commercial was not shown in territory 7 but was shown three times in territory 8? Explain.

18.22 Refer to **Reconditioning Cost** Problem 18.14.

a. Obtain the residuals. For which incinerator does the cost deviate the most from the least squares line in absolute value?

b. Verify that (18.18), (18.19), and (18.20) hold here.

18.23 Refer to **Bid Preparation** Problem 18.15.

a. Obtain the residuals. Verify that the residuals sum to zero.

b. Since e_i is an estimate of ε_i when regression model (18.3) is appropriate, does it necessarily follow that $\sum_{i=1}^{n} \varepsilon_i = 0$? Explain.

*18.24 Refer to **Fuel Efficiency** Problem 18.12.

a. Set up the ANOVA table.

b. Which parameter of regression model (18.3) is estimated by (1) MSE, (2) $\sqrt{MSE}$?

18.25 Refer to **Marketing Experiment** Problem 18.13.

a. Calculate $SSTO$, SSE, and SSR. Do your results satisfy identity (18.28) as expected?

b. Set up the ANOVA table.

c. Which parameter of regression model (18.3) is estimated by $\sqrt{MSE}$? In what units is $\sqrt{MSE}$ expressed?

18.26 Refer to **Reconditioning Cost** Problem 18.14.

a. Set up the ANOVA table.

b. Obtain a point estimate of σ. What information does this estimate provide about the magnitude of the residuals e_i for this regression study?

18.27 Refer to **Bid Preparation** Problem 18.15.

a. Set up the ANOVA table.

b. What are the estimated mean and standard deviation of the number of hours required to prepare (1) six bids during a week, (2) eight bids during a week?

c. How would the entries in your ANOVA table in **a** change if (1) each of the Y_i values were increased by 10 man-hours, (2) each Y_i value were converted from units of man-hours to units of man-weeks (using a basis of 40 hours per week)?

*18.28 Refer to **Fuel Efficiency** Problems 18.12 and 18.24.

a. Is *SSR* a large proportion of *SSTO* here? What does this fact imply about the extent to which variation in expressway mileage among the eight automobiles is reduced when weight of automobile is considered?

b. Calculate r^2 and r. What sign did you attach to r and why?

18.29 Refer to **Marketing Experiment** Problems 18.13 and 18.25. Calculate r^2 and r. Interpret the former measure in the context of this regression application.

18.30 Refer to **Reconditioning Cost** Problems 18.14 and 18.26.

a. Does knowledge of the operating hours since the preceding reconditioning appear to be helpful in reducing the variability of reconditioning costs? Calculate the coefficient of simple determination and comment.

b. Calculate the coefficient of simple correlation. In what units is it expressed?

18.31 Refer to **Bid Preparation** Problems 18.15 and 18.27.

a. Calculate r^2 and r. Which measure has the clearer operational interpretation? Which measure is closer to one in absolute value?

b. What would be the values of r^2 and r in **a** if each Y_i value were converted from units of man-hours to units of man-weeks (using a basis of 40 hours per week)?

18.32 A discussant stated: "Up to a point, most of the typists we hire improve in speed and accuracy as they gain experience. We have developed two tests to predict performance ratings after improvement levels off. The coefficient of correlation between our speed test score and speed performance rating is $+0.85$, while that between the accuracy test score and accuracy performance rating is $+0.82$."

a. Would the predictive usefulness of the tests be reduced if the signs of the coefficients were negative instead of positive? Discuss.

b. Can we conclude from the correlation coefficients that *SSE* is smaller for speed than for accuracy? Explain.

18.33 In a study of the effect of market share on profitability, an economist regressed rate of return on investment (Y) on market share (X) for 18 firms in an industry. She obtained a coefficient of simple correlation of $r = +0.24$.

a. What does the sign of r signify here?

b. Calculate r^2. Does the market share of a firm appear to be helpful in explaining variation in the rate of return on investment for firms in this industry? Explain.

EXERCISES

18.34 Demonstrate the algebraic equivalence of the following pairs of formulas: (1) (18.10b) and (18.12), (2) (18.10a) and (18.11).

18.35 Demonstrate that the estimated regression function (18.13) passes through the point $(\bar{X}, \bar{Y})$, where $\bar{X}$ and $\bar{Y}$ denote the means of the X_i and Y_i values, respectively.

18.36 (Calculus needed.) Confirm that the solution to the normal equations (18.9) does indeed lead to a minimum for Q defined in (18.8).

18.37 (*Calculus needed.*) When β_0 in regression model (18.3) equals zero, the regression function (18.5) becomes $E\{Y\} = \beta_1 X$. This case is called *regression through the origin*. Obtain the least squares estimator of β_1 for this case.

18.38 Refer to Table 18.1 for the UDS example. Suppose a thirteenth case were added to the sample, namely, $X = 2$, $Y = 103.0142$.

    a. Verify that the plotted point for this case will fall directly on the least squares line for the original 12 cases.

    b. If the least squares line were calculated for all 13 cases, would this line differ from the line for the original 12 cases? Explain. (*Hint:* Would the least squares line for the 12 cases still satisfy the least squares criterion as applied to the 13 cases?)

    c. How would the ANOVA table for the 13 cases differ from Table 18.4b? Be specific.

18.39 Prove the following properties of the least squares residuals for regression model (18.3): (1) $\sum e_i = 0$, (2) $\sum X_i e_i = 0$. [*Hint:* Use normal equations (18.9).]

18.40 Prove the following properties of the least squares residuals for regression model (18.3): (1) $\sum \hat{Y}_i e_i = 0$, (2) $\sum Y_i e_i = SSE$.

18.41 Demonstrate that (18.27) and (18.30) are algebraically equivalent formulas.

18.42 In fitting regression model (18.3) to some data, we obtain $b_0 = 2.0$, $b_1 = 0$. Explain which of the following will be true and which false: (1) $SSTO = 0$, (2) $r^2 = 0$, (3) $r = 0$, (4) $SSR = SSE$, (5) $SSE = SSTO$, (6) $SSTO = SSR + SSE$.

18.43 Show that $SSR = b_1^2 \sum_{i=1}^{n} (X_i - \bar{X})^2$.

STUDIES

18.44 Data on work loads for ten distribution centers maintained in overseas locations by a multinational firm follow (X denotes thousands of work units performed during the reporting period and Y denotes thousands of man-hours required).

Center i:	1	2	3	4	5	6	7	8	9	10
X_i:	5.0	3.5	10.0	5.0	6.5	6.0	7.1	2.5	3.0	4.2
Y_i:	27.5	20.1	50.5	26.3	33.5	32.4	36.8	15.5	18.3	22.0

An analyst calculated the number of man-hours required per unit of work (Y/X) for each center. He suggested that center 3 appears to be particularly efficient in that its ratio of man-hours to work units is lowest.

    a. Calculate the ratio Y/X for each center, and present the ratios in the form of a stem-and-leaf display. Does the ratio for center 3 appear to be particularly out of line with respect to those for other centers?

    b. Regress man-hours required on work units performed, using regression model (18.3). Make a scatter plot of the data, and plot your estimated regression function on the same graph.

    c. Contrast your findings with those of the analyst, paying particular attention to the matter of efficiency.

18.45 (Computer needed.) Refer to the **Financial Characteristics** data set.

    a. Regress net sales in year 2 on net assets in year 2 for firms in the electronics industry (firms 249–269 inclusive), using regression model (18.3). State the estimated regression function.

    b. Calculate r and $\sqrt{MSE}$. Which of these two measures provides more direct information about the variability of net sales of firms with given asset size? Explain.

    c. Obtain the residuals. Do any firms seem out of line in terms of the relation of their net sales to net assets? Comment on factors that might explain such "outliers."

    d. Redo the regression analysis in **a**, omitting firm 263 from the set of observations. Did this one observation have a substantial effect on the values of b_0 and b_1? On $\sqrt{MSE}$? In general, why do outlying observations have a substantial influence on b_0 and b_1 with the method of least squares?

19 INFERENCES IN SIMPLE LINEAR REGRESSION

In the previous chapter, we discussed fundamental concepts of simple linear regression. Here, we consider interval estimation of the parameters of the regression model and of the mean response, as well as a number of other topics of importance in applied work.

The simple linear regression model (18.3) will continue to be employed. For convenience, we repeat this model.

(19.1)

$$Y_i = \beta_0 + \beta_1 X_i + \varepsilon_i \qquad i = 1, 2, \ldots, n$$

where:

Y_i is the response in the ith case

X_i is the value of the independent variable in the ith case, assumed to be a known constant

β_0 and β_1 are parameters

ε_i are independent $N(0, \sigma^2)$

19.1 CONFIDENCE INTERVAL FOR $E\{Y_h\}$

As noted in Chapter 18, regression applications frequently entail estimation of the mean of the distribution of Y at a specified level of X.

EXAMPLES

1. A loan officer needs an estimate of the mean repayment period for home improvement loans of the most popular size, namely $5000.

2. An admissions officer at a university wishes to estimate the mean grade point average of freshmen students who receive the minimal acceptable score, 450, on an entrance examination. ☐

Point Estimator of $E\{Y_h\}$

Recall that the point estimator of $E\{Y_h\}$, the mean of Y when $X = X_h$, was given in (18.15) and is denoted by $\hat{Y}_h$.

(19.2)

$$\hat{Y}_h = b_0 + b_1 X_h$$

Sampling Distribution of $\hat{Y}_h$

The estimate $\hat{Y}_h$ will vary in repeated samples since b_0 and b_1 will vary from sample to sample. The sampling distribution of $\hat{Y}_h$ describes this variability in $\hat{Y}_h$ from sample to sample. Specifically, the sampling distribution of $\hat{Y}_h$ refers to the different possible values of $\hat{Y}_h$ that would be obtained if repeated samples were selected, each with the same values $X_1, X_2, \ldots, X_n$ for the independent variable.

Properties.   Statistical theory provides us with information about the properties of the sampling distribution of $\hat{Y}_h$.

(19.3)

For the simple linear regression model (19.1), the sampling distribution of $\hat{Y}_h$ has the following properties:

$$\text{Mean:} \quad E\{\hat{Y}_h\} = \beta_0 + \beta_1 X_h = E\{Y_h\}$$

$$\text{Variance:} \quad \sigma^2\{\hat{Y}_h\} = \sigma^2 \left[\frac{1}{n} + \frac{(X_h - \overline{X})^2}{\Sigma (X_i - \overline{X})^2} \right]$$

Functional form: Normal distribution

We noted in Chapter 18 that $\hat{Y}_h$ is an unbiased estimator of $E\{Y_h\}$.

Estimated Variance of $\hat{Y}_h$.   Usually, $\sigma^2\{\hat{Y}_h\}$, the variance of the sampling distribution of $\hat{Y}_h$, is unknown because σ^2, the variance of the error terms ε_i, is unknown. However, we know from Chapter 18 that MSE is an unbiased estimator of σ^2. Hence, we can replace σ^2 by MSE to obtain the estimated variance of $\hat{Y}_h$, denoted by $s^2\{\hat{Y}_h\}$.

(19.4)

$$s^2\{\hat{Y}_h\} = MSE \left[\frac{1}{n} + \frac{(X_h - \overline{X})^2}{\Sigma (X_i - \overline{X})^2} \right]$$

EXAMPLES

1.   We continue with the UDS example from Chapter 18. We wish to estimate $\sigma^2\{\hat{Y}_h\}$ for $X_h = 4$ tape drives. From the computer output in Figure 18.13, we have $n = 12$, $\overline{X} = 4.583$, and $MSE = 33.688$. The quantity $\Sigma(X_i - \overline{X})^2$ can be calculated from the data in Table 18.3:

$$\Sigma (X_i - \overline{X})^2 = (4 - 4.583)^2 + (6 - 4.583)^2 + \cdots + (5 - 4.583)^2$$
$$= 46.91667$$

The estimated variance of $\hat{Y}_h$ for $X_h = 4$ then is:

$$s^2\{\hat{Y}_h\} = 33.688\left[\frac{1}{12} + \frac{(4 - 4.583)^2}{46.91667}\right] = 3.05139$$

and hence the estimated standard deviation of $\hat{Y}_h$ is:

$$s\{\hat{Y}_h\} = \sqrt{3.05139} = 1.747$$

2. We also wish to find for the UDS example the estimated variance of $\hat{Y}_h$ when $X_h = 7$. We obtain:

$$s^2\{\hat{Y}_h\} = 33.688\left[\frac{1}{12} + \frac{(7 - 4.583)^2}{46.91667}\right] = 7.00204$$

The estimated standard deviation therefore is:

$$s\{\hat{Y}_h\} = \sqrt{7.00204} = 2.646$$ $\square$

The magnitude of the estimated variance $s^2\{\hat{Y}_h\}$ in (19.4) is affected by a number of factors:

1. *MSE.* The larger the variability of the residuals e_i, the larger $s^2\{\hat{Y}_h\}$ tends to be.
2. *Deviation of X_h from $\overline{X}$.* The further the specified level X_h is from $\overline{X}$ in either direction, the larger the quantity $(X_h - \overline{X})^2$ and the greater tends to be $s^2\{\hat{Y}_h\}$. For a given sample, $s^2\{\hat{Y}_h\}$ is smallest at $X_h = \overline{X}$. This is illustrated in the UDS example where we found $s^2\{\hat{Y}_h\} = 7.00204$ when $X_h = 7$ and $s^2\{\hat{Y}_h\} = 3.05139$ when $X_h = 4$. Since $\overline{X} = 4.583$, $X_h = 7$ is further from the mean than $X_h = 4$.
3. *Variability of X_i.* The greater is the variability of the X_i around their mean $\overline{X}$, the larger is $\Sigma (X_i - \overline{X})^2$ and the smaller tends to be $s^2\{\hat{Y}_h\}$.
4. *Sample size n.* The greater is n, the smaller is $1/n$ and the smaller tends to be $s^2\{\hat{Y}_h\}$. In addition, the denominator quantity $\Sigma (X_i - \overline{X})^2$ often is larger for increasing n, so again the smaller tends to be $s^2\{\hat{Y}_h\}$.

These properties can be useful when designing a regression study. For example, suppose the major purpose of the regression study is to estimate the mean $E\{Y_h\}$ at $X = X_h$, and the levels of X in the sample may be chosen freely. It would then be desirable to select levels of X with mean $\overline{X} = X_h$.

Comment   The quantity $\Sigma (X_i - \overline{X})^2$ in (19.4) can usually be obtained from regression computer output. Let s_X denote the standard deviation of the X_i values. From (3.15), we then have that $\Sigma (X_i - \overline{X})^2 = (n - 1)s_X^2$. Thus, $\Sigma (X_i - \overline{X})^2$ can be calculated from knowledge of n and s_X, and these quantities are generally provided in the computer output. In the UDS example, for instance, $n = 12$ and block 1 of the computer output in Figure 18.13 gives $s_X = 2.06522$. Hence, $\Sigma (X_i - \bar{X})^2 = (12 - 1)(2.06522)^2 = 46.916$, as shown earlier.

Development of Confidence Interval

A confidence interval for $E\{Y_h\}$ can be easily constructed on the basis of the following theorem.

(19.5)

> For the simple linear regression model (19.1):
>
> $$\frac{\hat{Y}_h - E\{Y_h\}}{s\{\hat{Y}_h\}} = t(n - 2)$$

The degrees of freedom for the t random variable are the $n - 2$ degrees of freedom associated with *MSE* (Table 18.4a). Recall from (19.4) that *MSE* is used to obtain $s^2\{\hat{Y}_h\}$.

Given theorem (19.5), a confidence interval for $E\{Y_h\}$ is obtained in the usual fashion, utilizing the t distribution.

(19.6)

> The confidence limits for $E\{Y_h\}$ with confidence coefficient $1 - \alpha$ for the simple linear regression model (19.1) are:
>
> $$\hat{Y}_h \pm ts\{\hat{Y}_h\}$$
>
> *where:*
> $t = t(1 - \alpha/2; n - 2)$
> $s\{\hat{Y}_h\}$ is given by (19.4)

EXAMPLES

1. In the UDS example, the service manager desires a 95 percent confidence interval for the mean time required for service calls involving $X_h = 4$ tape drives. We calculated, in Chapter 18, that $\hat{Y}_h = 191.84$ when $X_h = 4$. Earlier in this chapter, we obtained $s\{\hat{Y}_h\} = 1.747$ when $X_h = 4$. Finally, for $n = 12$ and $1 - \alpha = 0.95$, we require $t(0.975; 10) = 2.228$. Hence, the confidence limits are $191.84 \pm 2.228(1.747)$, and the desired confidence interval is:

   $$188 \le E\{Y_h\} \le 196$$

   Thus, with 95 percent confidence, the service manager can conclude that the mean service time for four tape drives is between 188 and 196 minutes.

2. In the UDS example, the service manager also wishes to obtain a 90 percent interval estimate for $E\{Y_h\}$ when $X_h = 7$. In Chapter 18, we found for $X_h = 7$ that $\hat{Y}_h = $

325.08, and in this chapter, we found $s\{\hat{Y}_h\} = 2.646$. We require $t(0.95; 10) = 1.812$. Hence, the 90 percent confidence limits are $325.08 \pm 1.812(2.646)$, and the confidence interval is:

$$320 \leq E\{Y_h\} \leq 330 \qquad \qquad \square$$

Comments

1. The confidence coefficient for confidence interval (19.6) is interpreted in terms of repeated samples in which the set of X levels is the same from sample to sample. Thus, for the UDS example, the 95 percent confidence coefficient attached to the interval estimate of $E\{Y_h\}$ when $X_h = 4$ signifies that if many independent samples were taken where the X levels are the same as in the 12 calls in the observed sample and a 95 percent confidence interval for $E\{Y_h\}$ were constructed from each sample, about 95 percent of these intervals would contain the true value of $E\{Y_h\}$.

2. The confidence limits in (19.6) are robust. The actual confidence coefficient remains close to $1 - \alpha$ even when the error terms of the regression model are not exactly normally distributed, provided the departure from normality is not too marked.

19.2   PREDICTION OF NEW RESPONSE $Y_{h(\text{new})}$

In regression applications, one frequently wishes to predict a new response for a given level of the independent variable, based on a previous sample.

(19.7)

> A *new response*, denoted by $Y_{h(\text{new})}$, is the value of Y to be observed in a *new* observation when the level of the independent variable is $X = X_h$.

It is important to distinguish between a new response $Y_{h(\text{new})}$ and a mean response $E\{Y_h\}$.

EXAMPLES

1. In the UDS example, technicians will test six tape drives on a forthcoming call. The firm wishes to predict the number of minutes required on this particular call. A new response $Y_{h(\text{new})}$ is involved here. The required service time will constitute, in effect, a new observation from the distribution of Y when $X_h = 6$. The prediction is not equivalent to estimating $E\{Y_h\}$, the mean of the distribution of Y at $X_h = 6$, since the new observation in all likelihood will deviate from the mean.

2. Anne Smith has just scored 115 on the company's aptitude test. A prediction of Anne Smith's job performance rating based on her test score of 115 entails a prediction of a new response $Y_{h(\text{new})}$ when $X_h = 115$. This is in contrast to an estimate of the mean performance rating $E\{Y_h\}$ for all persons who score 115 on the aptitude test.

3. An analyst for a food-processing company has investigated the relation between weight loss in wheels of Swiss cheese during storage (Y) and storage temperature

(X). He now wishes to estimate the mean weight loss when the storage temperature is 7°C. Here, an estimate of the mean response $E\{Y_h\}$ is required and not a prediction for a new wheel of cheese.   □

Regression Parameters Known

To illustrate the basic concepts of prediction intervals in a regression setting, we assume first that the parameters β_0, β_1, and σ^2 of the simple linear regression model are known. We then extend the procedure to cover the usual case where the parameters are not known and must be estimated.

EXAMPLE

Music Store.   An operations analyst is studying the mail-order operations of a large music store. The simple linear regression model (19.1) is known to be applicable, where Y is the distance in feet traveled by a clerk in filling the mail order and X is the number of different record albums in the order. The parameters of the regression model are known to be $\beta_0 = 20$, $\beta_1 = 15$, and $\sigma = 6.5$.

An order for $X_h = 3$ albums is to be filled next. Figure 19.1 shows the probability distribution of Y when $X_h = 3$. The distribution is normal, with mean $E\{Y_h\} = 20 + 15(3) = 65$ feet and standard deviation $\sigma = 6.5$ feet.

We wish to predict the distance to be traveled in filling this next order; that is, we wish to predict a new response $Y_{h(new)}$ when $X_h = 3$. We know that the mean of the probability distribution is $E\{Y_h\} = 65$ feet and that the new response will deviate from this mean. We therefore need to set up a prediction interval that recognizes the random deviation of $Y_{h(new)}$ from $E\{Y_h\}$.

Suppose we use an interval of ± 1.96 standard deviations about the mean. The prediction limits then are:

$$65 \pm 1.96(6.5)$$

**FIGURE 19.1   Prediction of a new response $Y_{h(new)}$ when the parameters are known —
Music store example**

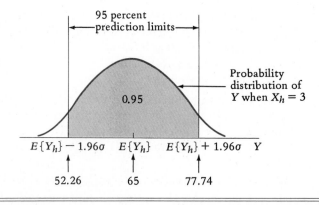

and the prediction interval is:

$$52.26 \leq Y_{h(new)} \leq 77.74$$

Since 95 percent of the area of a normal distribution is within ± 1.96 standard deviations about the mean, our procedure will give correct predictions 95 percent of the time. $\square$

Regression Parameters Unknown

In the usual case when the regression parameters are unknown, β_0, β_1, and σ^2 are estimated by b_0, b_1, and MSE, respectively. To predict a new response $Y_{h(new)}$ at $X = X_h$, we need to utilize $\hat{Y}_h = b_0 + b_1 X_h$ as the estimate of the mean of the probability distribution and MSE as the estimate of the variance of the probability distribution.

EXAMPLE

In the UDS example, the point estimate of the mean service time $E\{Y_h\}$ on the next call with $X_h = 4$ tape drives is:

$$\hat{Y}_h = b_0 + b_1 X_h = 14.18652 + 44.41385(4) = 191.84 \text{ minutes}$$

The estimate of the variance of this probability distribution is $MSE = 33.688$. $\square$

Development of Prediction Interval. There are two assumptions made in the development of the prediction interval:

1. The error term for the new response is independent of the error terms for the sample observations that are used for estimating $E\{Y_h\}$ and σ^2.
2. The regression model utilized for the sample is also appropriate for the new response.

The relevant variance in constructing the prediction interval, denoted here by $\sigma^2\{Y_{h(new)}\}$, is the following.

(19.8)
$$\sigma^2\{Y_{h(new)}\} = \sigma^2 + \sigma^2\{\hat{Y}_h\}$$

Note that there are two components of the variance in (19.8) relating to the following two sources of prediction error:

1. The inherent variability in the probability distribution of Y, represented by σ^2.
2. The uncertainty about the mean response at $X = X_h$, represented by $\sigma^2\{\hat{Y}_h\}$, the variance of $\hat{Y}_h$.

The variance in (19.8) is estimated by the following statistic, denoted by $s^2\{Y_{h(new)}\}$.

(19.9)

$$s^2\{Y_{h(\text{new})}\} = MSE + s^2\{\hat{Y}_h\}$$

where:

$s^2\{\hat{Y}_h\}$ is given by (19.4)

An equivalent expression for the estimated variance in (19.9) is given next.

(19.9a)

$$s^2\{Y_{h(\text{new})}\} = MSE\left[1 + \frac{1}{n} + \frac{(X_h - \overline{X})^2}{\sum (X_i - \overline{X})^2}\right]$$

The prediction interval for a new observation $Y_{h(\text{new})}$ is based on the following theorem.

(19.10)

For the simple linear regression model (19.1):

$$\frac{Y_{h(\text{new})} - \hat{Y}_h}{s\{Y_{h(\text{new})}\}} = t(n - 2)$$

The prediction interval for a new observation $Y_{h(\text{new})}$ when $X = X_h$ then takes the usual form.

(19.11)

The prediction limits for $Y_{h(\text{new})}$ with confidence coefficient $1 - \alpha$ for the simple linear regression model (19.1) are:

$$\hat{Y}_h \pm ts\{Y_{h(\text{new})}\}$$

where:

$t = t(1 - \alpha/2; n - 2)$

$s\{Y_{h(\text{new})}\}$ is given by (19.9)

EXAMPLES

1. In the UDS example, the customer for the next service call has $X_h = 4$ tape drives. We wish to obtain a 95 percent prediction interval for the service time $Y_{h(\text{new})}$. From earlier work, we have for $X_h = 4$:

$$n = 12 \quad \overline{X} = 4.583 \quad MSE = 33.688 \quad \hat{Y}_h = 191.84$$

$$s^2\{\hat{Y}_h\} = 3.05139$$

We begin by obtaining $s^2\{Y_{h(\text{new})}\}$, using (19.9):

$$s^2\{Y_{h(\text{new})}\} \doteq 33.688 + 3.05139 = 36.73939$$

The estimated standard deviation therefore is:

$$s\{Y_{h(\text{new})}\} = \sqrt{36.73939} = 6.0613$$

We also require $t(0.975; 10) = 2.228$. The prediction limits therefore are $191.84 \pm 2.228(6.0613)$, and the desired prediction interval is:

$$178 \leq Y_{h(\text{new})} \leq 205$$

Thus, in the next call in which four tape drives are to be tested, we predict with 95 percent confidence that the time required will be between 178 and 205 minutes.

2. Suppose in the UDS example the customer for the next service call has $X_h = 7$ tape drives. We wish to obtain 99 percent prediction limits for the service time $Y_{h(\text{new})}$. From earlier work, we have for $X_h = 7$:

$$n = 12 \qquad \overline{X} = 4.583 \qquad MSE = 33.688 \qquad \hat{Y}_h = 325.08$$
$$s^2\{\hat{Y}_h\} = 7.00204$$

Hence, $s^2\{Y_{h(\text{new})}\} = 33.688 + 7.00204 = 40.69004$ and $s\{Y_{h(\text{new})}\} = \sqrt{40.69004} = 6.3789$. We require $t(0.995; 10) = 3.169$. The 99 percent prediction limits therefore are $325.08 \pm 3.169(6.3789)$, or 305 and 345 minutes, respectively. □

Comments

1. Prediction limits are useful for control purposes. Suppose in the UDS example that 230 minutes were required for the new call with four tape drives. Since the 95 percent prediction interval ranges from 178 to 205 minutes, management may wish to ascertain why the service time was so high.

2. Note that the 95 percent prediction interval for $Y_{h(\text{new})}$ calculated in Example 1 is wider than the corresponding confidence interval for $E\{Y_h\}$. The reason is that $s^2\{Y_{h(\text{new})}\}$ in (19.9) contains not only $s^2\{\hat{Y}_h\}$ but also an additional component, MSE, which reflects the variability in the probability distribution of Y.

3. The same factors affecting the magnitude of $s^2\{\hat{Y}_h\}$ also affect $s^2\{Y_{h(\text{new})}\}$. For instance, the further X_h is from $\overline{X}$, the larger is $s^2\{Y_{h(\text{new})}\}$ for any given sample, and the wider will be the prediction interval.

4. The relevant variance for predicting a new observation Y, at a given level X_h of the independent variable, can be shown to be based on the deviation of the new observation Y from $\hat{Y}_h$, the estimated mean when $X = X_h$. Because of the independence of the new observation and the earlier sample, we have, by (5.14b):

$$\sigma^2\{Y - \hat{Y}_h\} = \sigma^2\{Y\} + \sigma^2\{\hat{Y}_h\} = \sigma^2 + \sigma^2\{\hat{Y}_h\}$$

In (19.8), this variance is denoted by $\sigma^2\{Y_{h(\text{new})}\}$.

5. Unlike the confidence limits for $E\{Y_h\}$ in (19.6), the prediction limits (19.11) are not robust against departures from normality. Some remedial procedures that may be helpful when the error terms are not normally distributed are discussed in Section 19.7.

19.3   INFERENCES CONCERNING β_1

We turn now to inferences on β_1, the slope of the regression function for regression model (19.1). In some regression studies, the primary objective is to estimate this slope. Recall that β_1 indicates the change in the mean of the distribution of Y when X increases by one unit. Thus, in the UDS example, the service manager may wish to ascertain how much the mean service time increases for each additional tape drive tested on a call.

There are also occasions when a test on β_1 is appropriate. For example, an operations analyst for a large mail-order house wishes to ascertain whether a relation exists between weight of incoming mail (X) and dollar value of mail orders (Y). If a relation does exist, the analyst would like to utilize it to predict dollar value of daily mail-order sales from the weight of the day's mail.

A test whether a relation between X and Y exists, given that regression model (19.1) is appropriate, takes the following form.

(19.12)

$$H_0: \beta_1 = 0$$

$$H_1: \beta_1 \neq 0$$

The reason is that if $\beta_1 = 0$, $E\{Y\} = \beta_0 + \beta_1 X = \beta_0$ for all levels of X. Then, all distributions of Y have the same mean. Since regression model (19.1) requires all distributions of Y to be normal with equal variability, the additional condition of equal means implies that all distributions of Y are identical, regardless of the level of X. Hence, there is no relation between X and Y when $\beta_1 = 0$ for regression model (19.1).

Sampling Distribution of b_1

We require knowledge of the sampling distribution of b_1 for constructing a confidence interval for β_1, as well as for conducting tests about this parameter.

(19.13)

For the simple linear regression model (19.1), the sampling distribution of b_1 has the following properties:

Mean: $E\{b_1\} = \beta_1$

Variance: $\sigma^2\{b_1\} = \dfrac{\sigma^2}{\sum (X_i - \overline{X})^2}$

Functional form: Normal distribution

We noted in Chapter 18 that the least squares estimator b_1 is unbiased.

Estimated Variance of b_1. The variance $\sigma^2\{b_1\}$ usually is unknown and is estimated by replacing the error term variance σ^2 by *MSE*. This estimated variance is denoted by $s^2\{b_1\}$.

(19.14)
$$s^2\{b_1\} = \frac{MSE}{\Sigma(X_i - \bar{X})^2}$$

EXAMPLE

In the UDS example, we know from earlier work that $MSE = 33.688$ and $\Sigma(X_i - \bar{X})^2 = 46.91667$. Hence:

$$s^2\{b_1\} = \frac{33.688}{46.91667} = 0.71804$$

and the estimated standard deviation of the sampling distribution of b_1 is:

$$s\{b_1\} = \sqrt{0.71804} = 0.84737$$

Confidence Interval for β_1

A confidence interval for β_1 can be constructed readily from the following theorem.

(19.15)
For the simple linear regression model (19.1):

$$\frac{b_1 - \beta_1}{s\{b_1\}} = t(n - 2)$$

Again, the degrees of freedom for the t random variable are the $n - 2$ degrees of freedom associated with *MSE*.

On the basis of theorem (19.15), the confidence interval for β_1 takes the usual form.

(19.16)
The confidence limits for β_1 with confidence coefficient $1 - \alpha$ for the simple linear regression model (19.1) are:

$$b_1 \pm ts\{b_1\}$$

where:

$t = t(1 - \alpha/2; n - 2)$
$s\{b_1\}$ is given by (19.14)

EXAMPLE

We wish to obtain a 95 percent confidence interval for β_1 for the UDS example. We have $b_1 = 44.41385$ (from Chapter 18) and $s\{b_1\} = 0.84737$, and we require $t(0.975; 10) = 2.228$. The confidence limits therefore are $44.41385 \pm 2.228(0.84737)$, and the confidence interval is:

$$42.5 \leq \beta_1 \leq 46.3$$

Thus, we estimate, with 95 percent confidence, that the mean service time increases by somewhere between 42.5 and 46.3 minutes for each additional tape drive tested on a call. □

Statistical Tests for β_1

As a result of theorem (19.15), tests on β_1 are constructed in an analogous fashion to those for a population mean μ as outlined in Section 11.5—see Table 11.2 in particular. A test of the alternatives (19.12) uses the following standardized test statistic.

(19.17)

Tests concerning β_1 for the simple linear regression model (19.1) when the alternatives are:

$$H_0: \beta_1 = 0$$

$$H_1: \beta_1 \neq 0$$

are based on the standardized test statistic:

$$t^* = \frac{b_1}{s\{b_1\}}$$

where:

$s\{b_1\}$ is given by (19.14)

α risk is controlled at $\beta_1 = 0$

When $\beta_1 = 0$, t^* follows the t distribution with $n - 2$ degrees of freedom. Large absolute values of t^* lead to conclusion H_1.

EXAMPLE

In the UDS example, we wish to test whether a relation between X and Y exists.

Step 1.   The alternatives here are:

$$H_0: \beta_1 = 0$$

$$H_1: \beta_1 \neq 0$$

Step 2.   The α risk is to be controlled at 0.05.
Step 3.   Since $\alpha = 0.05$ and $n - 2 = 10$, we require $t(0.975; 10) = 2.228$. The decision rule is as follows:

If $|t^*| \leq 2.228$, conclude H_0.
If $|t^*| > 2.228$, conclude H_1.

Step 4. We have $b_1 = 44.41385$ and $s\{b_1\} = 0.84737$. Thus, the standardized test statistic in (19.17) equals:

$$t^* = \frac{44.41385}{0.84737} = 52.41$$

Since $|t^*| = 52.41 > 2.228$, we conclude H_1—that $\beta_1 \neq 0$. This conclusion implies for regression model (19.1) that X and Y are related. □

Computer Output

The main computational results required for inferences on β_1 are generally available from regression computer output.

EXAMPLE

Refer to the computer output for the UDS example in Figure 18.13. Block 2 of the output gives information on β_1 in the following form:

VARIABLE	REG. COEFF.	STD. DEV.	T STAT.	P-VALUE
X	b_1	$s\{b_1\}$	t^*	Two-sided P-value

The values $b_1 = 44.41385$, $s\{b_1\} = 0.84737$, and $t^* = 52.41$ in the output are exactly those calculated earlier from the appropriate formulas. The P-VALUE shown in the output is the two-sided P-value for the test alternatives in (19.17). In the UDS example, the two-sided P-value for the test of H_0: $\beta_1 = 0$ versus H_1: $\beta_1 \neq 0$, expressed to five decimal places of accuracy, is 0.00000. Since this P-value is less than the specified α risk for the test ($\alpha = 0.05$), the appropriate conclusion is H_1 ($\beta_1 \neq 0$), precisely the same as the one based on t^*. □

Comments

1. Both the test and confidence interval for β_1 described in this section are robust inference procedures. The actual α risk and actual confidence coefficient remain close to their specified values even when the error terms of the regression model are not exactly normally distributed, provided the departure from normality is not too marked.

2. Occasionally, one wishes to test whether or not β_1 is some value other than 0. For example, an analyst wishes to test whether or not the slope of the regression line relating total production cost (Y) to lot size (X) exceeds \$15, a standard set by the cost accounting department. The value of β_1 where the α risk is to be controlled is denoted by β_{10}. In this example, $\beta_{10} = 15$ and the alternatives are:

H_0: $\beta_1 \leq 15$

H_1: $\beta_1 > 15$

The test statistic for this test is a generalization of test statistic (19.17).

(19.18)

$$t^* = \frac{b_1 - \beta_{10}}{s\{b_1\}}$$

where:

$s\{b_1\}$ is given by (19.14)

α risk is controlled at $\beta_1 = \beta_{10}$

When $\beta_1 = \beta_{10}$, t^* follows the t distribution with $n - 2$ degrees of freedom.

19.4   INFERENCES CONCERNING β_0

Occasionally, inferences on the intercept β_0 are of interest. For example, sometimes the intercept β_0 can be viewed as a "fixed" component in the relation between X and Y, as when Y is the cost of a production run and X is the number of units in the run. Here, β_0 might represent the fixed setup cost for the production run and β_1 the variable cost per unit.

Inferences concerning β_0 are made in analogous fashion to those for β_1, relying on the following key results.

(19.19)   For the simple linear regression model (19.1):

(19.19a) $E\{b_0\} = \beta_0$

(19.19b) $\sigma^2\{b_0\} = \sigma^2\left[\frac{1}{n} + \frac{\overline{X}^2}{\sum (X_i - \overline{X})^2}\right]$

(19.19c) $s^2\{b_0\} = MSE\left[\frac{1}{n} + \frac{\overline{X}^2}{\sum (X_i - \overline{X})^2}\right]$

(19.19d) $\dfrac{b_0 - \beta_0}{s\{b_0\}} = t(n - 2)$

EXAMPLE

We wish to construct a 95 percent confidence interval for β_0 for the UDS example, where $b_0 = 14.18652$. We begin by calculating the estimated variance of b_0. Since $MSE = 33.688$, $\overline{X} = 4.583$, and $\sum (X_i - \overline{X})^2 = 46.91667$, we obtain, by (19.19c):

$$s^2\{b_0\} = 33.688\left[\frac{1}{12} + \frac{(4.583)^2}{46.91667}\right] = 17.88895$$

$$s\{b_0\} = \sqrt{17.88895} = 4.22953$$

In view of (19.19d), the confidence limits for β_0 are of the usual form:

$$b_0 \pm t s\{b_0\}$$

where $t = t(1 - \alpha/2; n - 2)$. We require $t(0.975; 10) = 2.228$. Hence, the confidence limits are $14.18652 \pm 2.228(4.22953)$, and the desired confidence interval is $4.8 \le \beta_0 \le 23.6$. □

Comments
1. The confidence interval that we have just obtained is equivalent to a confidence interval for $E\{Y_h\}$ when $X_h = 0$. This follows because when $X_h = 0$, we have $E\{Y_h\} = \beta_0 + \beta_1 X_h = \beta_0$.
2. If it is known that $\beta_0 = 0$, regression model (19.1) simplifies to one where the regression line goes through the origin. Such a model is considered in Chapter 25 for time series data.
3. The main computational results for inferences on β_0 are often available from regression computer output. In the output for the UDS example in Figure 18.13, the information is given in block 2 in the following form:

VARIABLE	REG. COEFF.	STD. DEV.	T STAT.	P-VALUE
CONSTANT	b_0	$s\{b_0\}$	t^*	Two-sided P-value

The two-sided P-value and t^* refer to a test of $H_0: \beta_0 = 0$ versus $H_1: \beta_0 \ne 0$. For this test, $t^* = b_0 / s\{b_0\}$.

19.5   COVARIANCE OF b_0 AND b_1

Although b_0 and b_1 calculated from a given sample are estimates of different parameters, they usually contain sampling errors that are interrelated. Figure 19.2 illustrates this point. It shows the true linear regression function for a particular application, together with two estimated regression functions obtained from two different samples. The estimated regression function from sample 1 has a steeper slope and lower intercept than the true regression function. Hence, here b_1 overestimates β_1 while b_0 underestimates β_0. The estimated regression function from sample 2 shows the reverse pattern, with b_1 underestimating β_1 and b_0 overestimating β_0. Note that, in both cases, b_0 and b_1 for a given sample err in opposite directions, so the covariations, as defined in (5.23), are negative. The tendency for b_0 and b_1 to err in opposite directions is not surprising. Intuitively, we sense that in most cases where the slope of an estimated regression line is too high, the intercept will be too low, and vice versa.

The covariance of two random variables was defined in (5.24). It can be shown that the covariance of b_0 and b_1, denoted by $\sigma\{b_0, b_1\}$, is as follows for regression model (19.1).

(19.20)
$$\sigma\{b_0, b_1\} = -\bar{X}\sigma^2\{b_1\}$$

FIGURE 19.2   **Joint error tendencies in b_0 and b_1 when $\sigma\{b_0, b_1\}$ is negative.**   The estimates b_0 and b_1 tend to err in opposite directions.

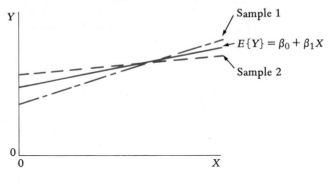

Consequently, when $\overline{X}$ is positive, as in the UDS example, $\sigma\{b_0, b_1\}$ is negative, and b_0 and b_1 tend to err in opposite directions in the same sample. Less commonly, when $\overline{X}$ equals zero, b_0 and b_1 have zero covariance; and when $\overline{X}$ is negative, b_0 and b_1 have positive covariance.

The estimated covariance of b_0 and b_1 is denoted by $s\{b_0, b_1\}$ and is obtained by replacing $\sigma^2\{b_1\}$ by $s^2\{b_1\}$ in (19.20).

(19.21)
$$s\{b_0, b_1\} = -\overline{X}s^2\{b_1\}$$

EXAMPLE

For the UDS example, where $\overline{X} = 4.583$ and $s^2\{b_1\} = 0.71804$, we calculate:

$$s\{b_0, b_1\} = -4.583(0.71804) = -3.29078 \qquad \square$$

19.6   *F* TEST OF $\beta_1 = 0$

In Section 19.3, we discussed a statistical test for ascertaining whether or not $\beta_1 = 0$. This test, which for regression model (19.1) is equivalent to ascertaining whether or not a relation exists between X and Y, is often utilized as a first step in the analysis of a statistical relation between two variables.

We now take up an equivalent test via the analysis of variance (ANOVA) approach. While this approach does not provide us with anything new here, it will be most useful when we consider multiple regression where there are two or more independent variables.

Expected Mean Squares

We stated earlier that the error mean square *MSE*, given in (18.34), is an unbiased estimator of σ^2, the variance of the error terms ε_i.

(19.22)
$$E\{MSE\} = \sigma^2$$

It can also be shown that the expected value of *MSR*, the regression mean square given in (18.33), is as follows for regression model (19.1).

(19.23)
$$E\{MSR\} = \sigma^2 + \beta_1^2 \Sigma (X_i - \overline{X})^2$$

Thus, when $\beta_1 = 0$, $E\{MSR\} = \sigma^2$, so both *MSR* and *MSE* have the same expected value then. This means that when $\beta_1 = 0$, *MSR* and *MSE* tend to be of approximately the same order of magnitude. On the other hand, when $\beta_1 \neq 0$, the term $\beta_1^2 \Sigma (X_i - \overline{X})^2$ will be positive and $E\{MSR\} > E\{MSE\}$. Hence, if $\beta_1 \neq 0$, *MSR* will tend to be larger than *MSE*.

Development of Statistical Test

The ANOVA test statistic is denoted by F^* and is defined as follows.

(19.24)
$$F^* = \frac{MSR}{MSE}$$

If F^* is near 1 (*MSR* and *MSE* are approximately equal), this would suggest by the earlier reasoning that $\beta_1 = 0$. On the other hand, if F^* is substantially greater than 1 (*MSR* is much larger than *MSE*), this would suggest that $\beta_1 \neq 0$. Thus, an upper-tail test is appropriate.

Figure 19.3 shows the appropriate decision rule for the test, with large values of F^* leading to conclusion H_1: $\beta_1 \neq 0$. The sampling distribution of F^* when $\beta_1 = 0$, the case for which the α risk is to be controlled, is given by the following theorem.

(19.25)   For the simple linear regression model (19.1), when $\beta_1 = 0$:

$$F^* = F(1, n - 2)$$

where:

F^* is given by (19.24)

FIGURE 19.3   **General form of the statistical decision rule for an F test of $\beta_1 = 0$ for simple linear regression model (19.1)**

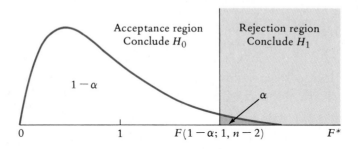

Hence, the distribution in Figure 19.3 is an F distribution, where the degrees of freedom are those associated with *MSR* and *MSE*. (Readers who need to review the F distribution should turn to Appendix B, Section B.3, before proceeding.) The action limit of the decision rule to make the right-tail area in the rejection region equal to α is $F(1 - \alpha; 1, n - 2)$; that is, it is the $100(1 - \alpha)$ percentile of the F distribution with 1 degree of freedom for the numerator and $n - 2$ for the denominator.

We now present the entire decision procedure.

(19.26)

When the alternatives are:

H_0: $\beta_1 = 0$

H_1: $\beta_1 \neq 0$

and the simple linear regression model (19.1) is applicable, the appropriate decision rule to control the α risk is as follows:

If $F^* \leq F(1 - \alpha; 1, n - 2)$, conclude H_0.
If $F^* > F(1 - \alpha; 1, n - 2)$, conclude H_1.

where:
F^* is given by (19.24)

EXAMPLE

For the UDS example, we wish to test whether or not $\beta_1 = 0$ by the ANOVA approach.

Step 1.   The alternatives here are:

H_0: $\beta_1 = 0$

H_1: $\beta_1 \neq 0$

Step 2.   The α risk is to be controlled at 0.05.

Step 3.   For $\alpha = 0.05$ and $n - 2 = 12 - 2 = 10$, we require $F(0.95; 1, 10)$ $= 4.96$. The decision rule therefore is as follows:

If $F^* \leq 4.96$, conclude H_0.
If $F^* > 4.96$, conclude H_1.

Step 4.   We found in Table 18.4b that $MSR = 92{,}547.37$ and $MSE = 33.688$. Hence, the test statistic is:

$$F^* = \frac{MSR}{MSE} = \frac{92{,}547.37}{33.688} = 2747.2$$

Since $F^* = 2747.2 > 4.96$, we conclude H_1—that $\beta_1 \neq 0$. Necessarily, this result is the same as that obtained by the t test. $\square$

Comments   1.   The ANOVA F test is equivalent to the t test. It can be shown that the F^* statistic (19.24) is the square of the t^* statistic (19.17). Recall that we had $t^* = 52.41$, and $(52.41)^2 = 2747 = F^*$. Further, the action limit for the t test was 2.228, and $(2.228)^2 = 4.96$, the action limit for the ANOVA test. This equivalence is based on the second relation in (B.16) of Appendix B.

2.   The F test can be used only for testing $\beta_1 = 0$ versus $\beta_1 \neq 0$, and not for testing one-sided alternatives.

3.   Regression computer output usually shows the value of F^* in conjunction with the ANOVA table. Block 3 of Figure 18.13 for the UDS example, for instance, shows F^* (identified as F STAT.) as well as the corresponding upper-tail P-value for the test. Here, the P-value is 0.00000, expressed to five decimal places of accuracy. The fact that this P-value is less than $\alpha = 0.05$ implies that H_1 ($\beta_1 \neq 0$) is the appropriate conclusion, as we noted in the test based on F^*.

19.7   EVALUATION OF APTNESS OF MODEL

When a regression model is applied in practice, one cannot usually be sure in advance that the model employed is appropriate for the situation at hand. Consequently, the model considered needs to be checked for aptness or suitability, using the observed data for guidance. Indeed, frequently several models need to be investigated before one is finally selected.

Residual Analysis

A basic technique for investigating the aptness of a regression model is based on analyzing the residuals defined in (18.17). If the model is apt, the observed residuals $e_i = Y_i - \hat{Y}_i$ should reflect the properties ascribed to the error terms ε_i. For instance, if the model assumes that the ε_i are normal random variables with constant variance, the observed e_i should show a pattern consistent with these properties.

We now consider how analysis of residuals is used to study the aptness of the simple linear regression model (19.1). Specifically, we shall examine the following departures from regression model (19.1):

1. The regression function is not linear.
2. The distributions of Y do not have constant variances at all levels of X; or equivalently, the ε_i do not have constant variances.
3. The distributions of Y are not normal; or equivalently, the ε_i are not normal.
4. The error terms ε_i are not independent.

Nonlinearity of Regression Function. Whether the regression function is linear or curvilinear can be studied from either a scatter plot or a residual plot. In a *residual plot*, the residuals e_i are plotted against the corresponding fitted values $\hat{Y}_i$. Recall that we utilized residual plots in Chapter 2 to compare observed values with anticipated values.

EXAMPLES

1. **Discount Card Use.** Figure 19.4a contains a scatter plot for a study of discount card use by students at ten community colleges. Here, X is the number of merchants in the discount card plan at a college and Y is the average discount card expenditure per student at the college during the period covered by the study. A linear regression function has been fitted to the data by the method of least squares. The observed values Y_i, the fitted values $\hat{Y}_i$, and the residuals e_i are as follows:

i:	1	2	3	4	5	6	7	8	9	10
Y_i:	75.0	112.0	38.0	120.0	105.0	52.0	116.0	118.0	105.0	110.0
$\hat{Y}_i$:	70.5	94.3	54.6	134.0	86.4	62.6	110.2	126.1	110.2	102.2
e_i:	4.5	17.7	−16.6	−14.0	18.6	−10.6	5.8	−8.1	−5.2	7.8

If the true regression function is linear, the observations should scatter at random around the fitted straight line. This does not appear to be the case in Figure 19.4a; the points systematically fall below, then above, then again below the fitted line as X increases. It would seem that a linear regression function is not in accord with the observations and that a curvilinear function is required instead.

The residual plot of the residuals e_i against the fitted values $\hat{Y}_i$ for the discount card use data is shown in Figure 19.4b. If the specified regression function is apt, the residuals will tend to scatter at random around the zero line when plotted against $\hat{Y}_i$. Note how this fails to occur in Figure 19.4b. Instead, we see a pattern of deviations which mirrors that in Figure 19.4a. We reach the same conclusion from either graph.

2. Figure 19.5 shows the residual plot for the UDS example. Here, no pattern of systematic departure of the points around the zero line is evident. Thus, the residual plot in Figure 19.5 suggests that the linear regression function specified in regression model (19.1) is apt in the UDS case. □

FIGURE 19.4 Scatter and residual plots illustrating a nonlinear regression relation — Discount card use example

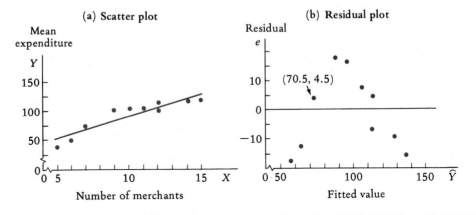

(a) Scatter plot   (b) Residual plot

FIGURE 19.5 Residual plot illustrating the aptness of a linear regression model — UDS example

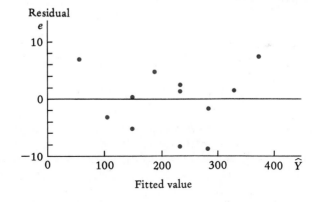

Nonconstancy of Error Variances. A residual plot also provides information as to whether or not the error terms ε_i have constant variance.

| EXAMPLES | 1. Figure 19.6 shows a residual plot for a study of the relation between typing speed (Y) and hours of training (X) for beginning typists using a new keyboard design. Note that the pattern of spread in the residuals becomes greater as $\hat{Y}$ increases. |

FIGURE 19.6   **Residual plot illustrating nonconstant error variance**

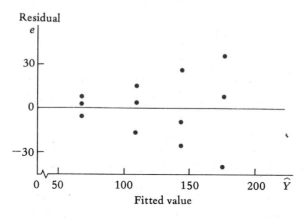

This suggests that the distributions of Y have increasing variances as $E\{Y\}$ becomes larger.

2. The scatter plot in Figure 19.5 for the UDS example illustrates a pattern of residuals when the error variances are constant. Here, the e_i tend to fall at random in a horizontal band when plotted against $\hat{Y}_i$. □

Lack of Normality.   A variety of checks for normality of the error terms, or equivalently, of the distributions of Y, are available. One of these employs *standardized residuals*, denoted by e'_i.

(19.27)
$$e'_i = \frac{e_i}{\sqrt{MSE}}$$

Thus, a standardized residual e'_i is the residual divided by the estimated standard deviation. If the distributions of Y are normal with constant variance and the linear regression function is appropriate, the standardized residuals will tend to follow the standard normal distribution when n is large. Hence, about half of the standardized residuals should then be positive and half negative, about 68 percent should fall between -1 and $+1$, and so on. If n is not large, the standardized residuals should be compared with the $t(n-2)$ distribution.

EXAMPLE

For the UDS example, we have $\sqrt{MSE} = 5.804$. The residual for the first observation is $e_1 = 5.16$ (Table 18.3). Hence, the standardized residual is $e'_1 =$

5.16/5.804 = 0.89. The standardized residuals for the 12 observations are as follows:

i:	1	2	3	4	5	6
e_i':	0.89	−1.49	−0.52	−1.42	0.33	−0.29
i:	7	8	9	10	11	12
e_i':	0.10	1.29	0.30	−0.94	1.27	0.47

We shall analyze the normality of the error terms by comparing the observed frequencies of the residuals with those expected if normality holds. For this purpose, we shall consider the 5th, 25th, 50th, 75th, and 95th percentiles of the appropriate t distribution.

For the t distribution with $n − 2 = 10$ degrees of freedom, we find that the 50th, 75th, and 95th percentiles are $t(0.50; 10) = 0$, $t(0.75; 10) = 0.700$, and $t(0.95; 10) = 1.812$. Hence, we know that $t(0.05; 10) = −1.812$ and $t(0.25; 10) = −0.700$. We now construct the distributions of the observed frequencies (f) and the expected frequencies (F) for the 12 standardized residuals. These distributions are shown in Table 19.1. The expected frequency for the interval 1.812–under ∞, for example, is $12(1 − 0.95) = 0.6$; the expected frequency for the interval 0.700–under 1.812 is $12(0.95 − 0.75) = 2.4$; and so on. Although $n = 12$ observations are too few to draw a firm conclusion, the comparison of the observed and expected frequencies in Table 19.1 provides no evidence of any serious departure from normality. □

Comments

1. When n is large, the formal tests for normality discussed in Chapter 16 can be employed, and a histogram of the residuals can provide a useful visual check.
2. Many regression computer packages print standardized residuals as a routine part of the output. Some packages, however, define standardized residuals somewhat differently from (19.27), using the estimated standard deviation of the residual in the denominator rather than $\sqrt{MSE}$.

Lack of Independence. In regression applications, the observations are often obtained in a time sequence, as when the value of an index of business activity (X) is related to the dollar value of new orders received by a manufacturing corporation (Y) for each of the past 48 months. In such data, there is a strong possibility that the error terms ε_i are not independent. For instance, if the error term is positive (negative) in a given month, it is often likely with such data that the error term for the following month is also positive (negative). Such error terms are said to be *autocorrelated,* in contrast to regression model (19.1) where the ε_i are assumed to be statistically independent.

A plot of the residuals against the time order of the observations helps to assess whether or not autocorrelation is present. We discuss this plot in Chapter 25. We also discuss there a test for independence of the error terms based on the residuals.

TABLE 19.1   **Comparison of observed and expected frequencies for standardized residuals based on the $t(n - 2)$ distribution — UDS example**

	Frequency	
Standardized Residual e'	Observed f	Expected F
$-\infty$ –under -1.812	0	0.6
-1.812–under -0.700	3	2.4
-0.700–under 0.0	2	3.0
0.0 –under 0.700	4	3.0
0.700–under 1.812	3	2.4
1.812–under ∞	0	0.6
Total	12	12.0

Remedial Actions

When residual analysis discloses that regression model (19.1) is not apt for the data under study, remedial action may be required. We now illustrate two types of remedial action — transformations of variables and use of a different regression model.

Transformations of Variables.   Transformations of variables can be employed frequently to make the data for a regression problem conform to the linear regression model (19.1). When the regression relation is not linear, for instance, it is often possible to linearize it by transforming the independent variable. In other cases, lack of linearity and nonconstant error variance are both found to be present. In those cases, a transformation of the dependent variable may be helpful. In still other cases, both the dependent and the independent variables may need to be transformed.

Transformations that are frequently utilized include the logarithmic, square root, and reciprocal transformations.

> EXAMPLE

Figure 19.7a presents a scatter plot of observations for 11 athletes on an index of physical fitness (X) and the performance time by the athlete in a track event (Y). The regression relation is clearly not linear. Here, a reciprocal transformation on the independent variable is helpful. When the independent variable is transformed to $1/X$, the reciprocal of the fitness index, the relation becomes linear, as shown in Figure 19.7b.   ☐

A simple transformation of a complex regression function may sometimes yield a linear regression relation.

> EXAMPLE

The following complex regression relation is often encountered in studies of the growth of a phenomenon over time, such as sales (Y) as a function of time (X).

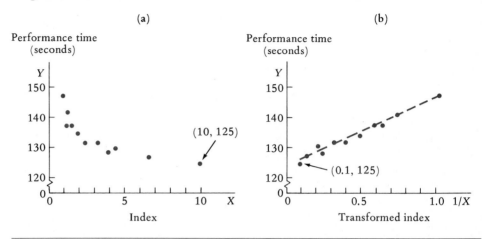

FIGURE 19.7   **Use of a reciprocal transformation on the independent variable to linearize a regression relation**

(19.28)

$$E\{Y\} = \gamma_0 \gamma_1^X$$

When we take the logarithms of both sides, we obtain:

$$\log E\{Y\} = \log \gamma_0 + X \log \gamma_1$$

If we let $E\{Y\}' = \log E\{Y\}$, $\beta_0 = \log \gamma_0$, and $\beta_1 = \log \gamma_1$, we can write the transformed regression function as a linear regression function:

$$E\{Y\}' = \beta_0 + \beta_1 X$$

Thus, a linear regression of $\log Y$ on X may be appropriate here.   ☐

When the distribution of the error terms is far from normal, a mathematical transformation of the Y variable may bring the distribution of the error terms closer to normality. Thus, when the distribution of the ε_i is sharply right-skewed, $\log Y$ might be used as the dependent variable instead of Y. The use of a logarithmic transformation to achieve approximate normality is illustrated in Section 13.5—in particular, see Figure 13.6 and the accompanying discussion. Moderate departures from normality are not serious for estimating the regression function, and even substantial departures usually have little effect on the specified level of confidence or the α risk when the sample size is large.

Use of Different Regression Model.   When the simple linear regression model (19.1) is not apt and transformations of variables are not helpful, one must adopt a different regression model. Frequently, the new model is a *multiple regression model*. These models are taken up in the next chapter. A regression model that is helpful when the error terms are autocorrelated is an *autoregressive model,* discussed in Chapter 25.

19.8 PRACTICAL CONSIDERATIONS IN USING REGRESSION ANALYSIS

We now discuss a number of important considerations in using a regression model.

Scope of Model

Range of Observations.   Caution must be exercised in applying a regression model outside the range of the observations that have been used to estimate the model's parameters and to verify the aptness of the model's assumptions.

> EXAMPLE
>
> In the UDS example, we fitted regression model (19.1) to observations where the number of tape drives (X) was between 1 and 8. The fit appeared to be satisfactory for these observations, but in the absence of other information, we do not know whether the model is apt outside the interval $1 \leq X \leq 8$. Thus, it may be dangerous to use the fitted model to estimate the mean service time when $X_h = 25$ because the linear model might not be a good fit when extended to 25 tape drives. A curvilinear model might be needed instead.   □

Similarly, an estimate of β_0 in regression model (19.1) should only be used as an indication of the "fixed" component of the relationship when there are observations near $X = 0$ or when theoretical considerations indicate that the regression function is linear for the range including $X = 0$.

Situations frequently arise in practice where inferences are required outside the range of past observations. For example, it may be desired to use the economic indicator per capita disposable personal income (X) to predict the level of business activity (Y) for a firm. In an expanding economy, the level of X required for this prediction will frequently be outside the range of the past data. In such cases, inferences obtained with the regression model must be applied with considerable judgment.

Continuation of Causal Conditions.   Whenever a regression model is utilized to make an inference for the future, the validity of the inference requires that future causal conditions be the same as during the period covered by the observed data. Thus, a prediction from the regression model in the UDS example of the service time on a customer call six months later assumes that the causal conditions affecting the servicing of tape drives have not changed.

When causal conditions have changed, the fitted model may no longer be appropriate. For example, the response of butter consumption to changes in the price of butter was different after the availability of margarine than before.

Causality

As noted earlier, the presence of a regression relation between two variables X and Y does not imply a cause-and-effect relation between them. In some instances, a change in X does force a change in Y, as, for example, when an increase in drug dosage X causes a decrease in blood pressure Y. In other instances, both X and Y change in response to changes in one or more other variables without a direct causal linkage between them.

EXAMPLE

Reading ability (Y) was regressed on shoe size (X) for a sample of elementary school children and a positive regression relation was observed. This relation does not imply, of course, that larger shoes cause children to read better. The age of the child is an intervening variable in this case — older children tend to wear bigger shoes and to read better.

The omission of an intervening variable can sometimes also hide a relation between two variables.

EXAMPLE

Figure 19.8a shows a scatter plot of years of education (X) and salary (Y) for 16 corporate middle managers aged 30 to 35 in a multinational firm. The regression line is horizontal, indicating no regression relation between the two variables. In Figures 19.8b and 19.8c, the observations are divided into two groups — sales managers and others. The separate scatter plots show a persistent increase in aver-

---

FIGURE 19.8   Effect of an intervening variable on a regression relation.   The relation between salary and education is hidden when the scatter plots for sales managers and others are combined.

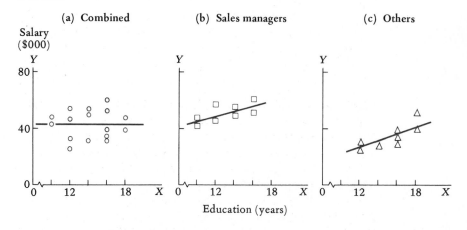

---

age salary with more years of education for each group. The fact that managers not in sales activities tend to be paid less than sales managers for any given number of years of education disguises the relation between years of education and salary when all managers are combined. □

Finally, there are instances where causation acts in the opposite direction to the regression relation in the sense that changes in Y force changes in X.

EXAMPLE

A thermometer is being calibrated by obtaining readings at various known temperatures and regressing the actual temperature (Y) on the thermometer reading (X). Clearly, the thermometer readings X are affected by the actual temperatures Y, and not vice versa. □

Independent Variable Must Be Predicted

Sometimes, we must predict the value to be taken by the independent variable X in obtaining a required prediction for Y. This happens quite often in business forecasting. For example, suppose X_h is the value projected for an economic indicator variable in the next quarter and a prediction interval is desired for company sales $(Y_{h(new)})$ in that period. The sales prediction then is a conditional one, dependent on the correctness of the projected X_h.

Independent Variable Is Random

Sometimes, in regression applications, it is more reasonable to consider both X and Y as random, rather than to take X as fixed in accord with regression model (19.1).

EXAMPLE

An analyst in an electric company is studying the regression relation between daily residential electricity consumption in a service area (Y) and mean daily outdoor temperature in the area (X). Since the outdoor temperature cannot be controlled, X might be considered as random in this application. □

When X is random, the distribution of Y at a given level of X is considered to be a conditional distribution with a conditional mean and a conditional variance. It can be shown that all of the results presented for regression model (19.1) where the X_i are fixed still apply when the X_i are random if the following conditions hold: (1) The conditional distributions of Y are normal, with conditional mean $\beta_0 + \beta_1 X$ and conditional variance σ^2; and (2) the X_i are independent random variables whose probability distribution does not depend on the parameters β_0, β_1, and σ^2.

When the independent variable is random, the interpretations of confidence coefficients and risks of errors differ from before. The interpretations now refer to repeated sampling where *both* the X and Y values change from one sample to the next. Thus, in the electricity consumption example, the confidence coefficient for the interval estimate for β_1 would refer to the proportion of correct interval estimates obtained if repeated random samples of n pairs of (X_i, Y_i) values were taken and the confidence interval for β_1 calculated for each sample. There is no implication that the X levels remain the same from sample to sample.

Comment	When X and Y are both random variables, a population coefficient of simple correlation $\rho\{X, Y\}$ is defined as in (5.27). The coefficient of simple correlation r defined in (18.40) for the n sample values (X_i, Y_i) is then not only a descriptive measure of the degree of linear relationship between X and Y in the sample but also serves as an estimator of $\rho\{X, Y\}$. Under the conditions stated above for the inferential results to apply when X is random, the population coefficient of correlation $\rho\{X, Y\}$ is zero when the slope β_1 of the regression function is zero. Hence, a test for $\beta_1 = 0$ is then equivalent to a test of whether X and Y are uncorrelated random variables.

PROBLEMS

*19.1 Refer to **Fuel Efficiency** Problem 18.12. Obtain a 95 percent confidence interval for $E\{Y_h\}$ when $X_h = 23$. Interpret your confidence interval.

19.2 Refer to **Marketing Experiment** Problem 18.13.
   a. Obtain a 95 percent confidence interval for $E\{Y_h\}$ when (1) $X_h = 0$, (2) $X_h = 2$, (3) $X_h = 4$. Interpret the last confidence interval.
   b. Why are confidence intervals (1) and (3) in **a** wider than interval (2)? Why do confidence intervals (1) and (3) have the same width?

19.3 Refer to **Reconditioning Cost** Problem 18.14. Construct a 99 percent confidence interval for the mean reconditioning cost of incinerators with 2.5 thousand operating hours since the preceding reconditioning. Interpret your interval estimate.

19.4 Refer to **Bid Preparation** Problem 18.15.
   a. Construct a 90 percent confidence interval for the mean number of man-hours required for bid preparation when eight bids are prepared during the week. Interpret your confidence interval. How is the confidence coefficient interpreted here?
   b. An observer comments that the number of man-hours required to prepare a given number of bids cannot be exactly normally distributed as required by regression model (18.3) because the number of man-hours is never a negative number and is always recorded as a whole number. Explain why the confidence coefficient for the interval estimate in **a** will still be approximately 90 percent in spite of these departures from exact normality.

19.5 An analyst, when regressing contributions to the pension plan (Y) on years of job seniority (X) for employees in the company, obtained computer output showing the 95 percent confidence interval for $E\{Y_i\}$ for each X_i value in the data set. The analyst expected to find that about 95 percent of the Y_i values would be contained in the corresponding confidence intervals for $E\{Y_i\}$ but discovered that this occurred in a substantially smaller percentage of the cases. Explain the analyst's error in thinking.

19.6 For each of the following questions, explain whether a confidence interval for $E\{Y_h\}$ or a prediction interval for $Y_{h(new)}$ is appropriate.
   a. What is the average statistics grade of students who receive 530 on the graduate admissions test?

b. What will the Cramer family spend on restaurant meals next year if the family's income is $34,000?

c. What do companies with 500 full-time employees spend each year, on the average, on group life insurance premiums?

19.7 For each of the following questions, explain whether a confidence interval for $E\{Y_h\}$ or a prediction interval for $Y_{h(new)}$ is appropriate.

a. What will be the unemployment rate in this metropolitan area next quarter, given that the index of business activity will be 178.6?

b. What is the average score on the mathematics portion of this admissions test for applicants who score 600 on the verbal portion?

c. How many hours of pain relief will Ms. Jones obtain from this medication when the dosage administered is 10 percent above the standard level?

*19.8 Refer to **Fuel Efficiency** Problems 18.12 and 19.1. Another automobile, not included in the regression study, has a weight of 23 hundred pounds.

a. Construct a 95 percent prediction interval for the expressway mileage of this automobile. Interpret your prediction interval.

b. Why is the prediction interval in **a** wider than the confidence interval in Problem 19.1?

19.9 Refer to **Marketing Experiment** Problem 18.13. Four showings of the TV commercial under the same experimental conditions are scheduled for an eleventh sales territory that is similar to the territories used in the experiment.

a. Construct a 95 percent prediction interval for the sales in this eleventh territory. Interpret your prediction interval.

b. If regression model (18.3) were not appropriate in this application because the error terms ε_i follow a highly skewed distribution, would the confidence coefficient of the prediction interval in **a** still be near 95 percent? Explain.

19.10 Refer to **Reconditioning Cost** Problem 18.14.

a. The reconditioning cost for an incinerator with 2.5 thousand operating hours is to be predicted. Obtain a 99 percent prediction interval. Interpret your prediction interval. Is it precise enough to be useful if a prediction with ±10 percent precision is required? Discuss.

b. If the reconditioning cost for an incinerator with 5.2 thousand operating hours were to be predicted, would the width of the prediction interval be much wider than that in **a**? Would there be other problems in this case?

19.11 Refer to **Bid Preparation** Problems 18.15 and 19.4.

a. Construct a 90 percent prediction interval for $Y_{h(new)}$ when $X_h = 8$. Interpret your prediction interval.

b. Why is the prediction interval in **a** wider than the confidence interval in Problem 19.4a?

c. Which interval — the prediction interval or the confidence interval — would be more useful for evaluating the actual number of man-hours required for bid preparation next week when eight bids are to be prepared? Explain.

*19.12 Refer to **Fuel Efficiency** Problem 18.12. Obtain a 95 percent confidence interval for β_1. Interpret the confidence interval.

19.13 Refer to **Marketing Experiment** Problem 18.13.

a. Construct a 95 percent confidence interval for β_1.

b. The research director of the experiment has calculated that the company must expect to sell over 200 more blankets in a territory in order to cover the cost of an additional showing of the TV commercial in the territory. Calculate the lower 95 percent confidence interval for β_1 to determine whether this requirement is being met. What do you find?

19.14 Refer to **Reconditioning Cost** Problem 18.14. Obtain a 99 percent confidence interval for the change in the mean response when the number of operating hours increases by (1) 1.0 thousand, (2) 1.5 thousand.

19.15 Refer to **Bid Preparation** Problem 18.15. Obtain a 99 percent confidence interval for β_1. Does the confidence interval indicate that the means of the probability distributions of Y at different levels of X are not the same? Explain.

*19.16 Refer to **Fuel Efficiency** Problem 18.12.
  a. Test whether or not $\beta_1 = 0$. Use test statistic t^*, and control the α risk at 0.05 when $\beta_1 = 0$. State the alternatives, the decision rule, the value of the test statistic, and the conclusion. What is the implication of your conclusion?
  b. What is the P-value of the test? Is your test conclusion in **a** consistent with the P-value?

19.17 Refer to **Marketing Experiment** Problem 18.13.
  a. Test whether or not $\beta_1 = 0$. Use test statistic t^*, and control the α risk at 0.05 when $\beta_1 = 0$. State the alternatives, the decision rule, the value of the test statistic, and the conclusion.
  b. What is the P-value of the test?

19.18 Refer to **Reconditioning Cost** Problem 18.14.
  a. Test whether or not $\beta_1 = 0$. Use test statistic t^*, and control the α risk at 0.01 when $\beta_1 = 0$. State the alternatives, the decision rule, the value of the test statistic, and the conclusion.
  b. What is the P-value of the test?
  c. An engineering cost study shows that the expected reconditioning cost should increase by $2.0 thousand for each additional 1 thousand operating hours. Test whether or not $\beta_1 = 2.0$, controlling the α risk at 0.01 when $\beta_1 = 2.0$. State the alternatives, the decision rule, the value of the test statistic, and the conclusion.
  d. Could each of the tests in **a** and **c** be conducted by using a 99 percent confidence interval for β_1? Explain.

19.19 Refer to **Bid Preparation** Problem 18.15.
  a. Test whether or not $\beta_1 = 0$. Use test statistic t^*, and control the α risk at 0.01 when $\beta_1 = 0$. State the alternatives, the decision rule, the value of the test statistic, and the conclusion. What does your conclusion imply about a relationship between X and Y here? Discuss.
  b. What is the P-value of the test?

*19.20 Refer to **Fuel Efficiency** Problem 18.12. Test whether or not $\beta_0 = 0$, controlling the α risk at 0.01 when $\beta_0 = 0$. State the alternatives, the decision rule, the value of the test statistic, and the conclusion.

19.21 Refer to **Marketing Experiment** Problem 18.13. Construct a 95 percent confidence interval for β_0. Interpret your confidence interval.

19.22 Refer to **Reconditioning Cost** Problem 18.14. Test whether or not $\beta_0 = 0$, controlling the α risk at 0.05 when $\beta_0 = 0$. State the alternatives, the decision rule, the value of the test statistic, and the conclusion.

19.23 Refer to **Bid Preparation** Problem 18.15.
  a. Construct a 95 percent confidence interval for β_0.
  b. Does it appear from the confidence interval in **a** that β_0 might be 0? Explain.
  c. Does β_0 have an operational interpretation here? Comment.

*19.24 Refer to **Fuel Efficiency** Problem 18.12. Estimate the covariance of b_0 and b_1. What does this estimate show about the relation between sampling errors in b_0 and b_1 here?

19.25 Refer to **Marketing Experiment** Problem 18.13. Estimate the covariance of b_0 and b_1. If b_1 happens to underestimate β_1 here, what is the likely direction of the sampling error in b_0?

19.26 Refer to **Reconditioning Cost** Problem 18.14. Estimate the covariance of b_0 and b_1. What does this estimate show about the relation between sampling errors in b_0 and b_1 here?

19.27 Refer to **Bid Preparation** Problem 18.15. Estimate the covariance of b_0 and b_1. If b_1 happens to underestimate β_1 here, what is the likely direction of the sampling error in b_0?

19.28 A computer simulation study utilized regression model (19.1) with $\beta_0 = 20.0$ and $\beta_1 = 10.0$ in several trials involving ten cases each, where the X values were held the same in each trial.
  a. In the first trial, the estimated regression function was $\hat{Y} = 18.0 + 11.5X$. What is the covariation of b_0 and b_1 for this trial?
  b. Might the sign of the covariation of b_0 and b_1 in the second trial be different from that in the first trial? Explain.
  c. If $\overline{X}$ is positive in the simulation study, will the covariance of b_0 and b_1 necessarily be negative? Explain.

*19.29 Refer to **Fuel Efficiency** Problems 18.12 and 18.24.
  a. Conduct an F test of whether or not $\beta_1 = 0$, controlling the α risk at 0.05 when $\beta_1 = 0$. State the alternatives, the decision rule, the value of the test statistic, and the conclusion.
  b. Show, numerically, that the value of test statistic F^* in **a** is the square of the test statistic t^* calculated from (19.17).

19.30 Refer to **Marketing Experiment** Problems 18.13 and 18.25.
  a. Conduct an F test of whether or not $\beta_1 = 0$, controlling the α risk at 0.05 when $\beta_1 = 0$. State the alternatives, the decision rule, the value of the test statistic, and the conclusion.
  b. Show, numerically, the equivalence of the test statistic F^* in **a** and the test statistic t^* in (19.17).

19.31 Refer to **Reconditioning Cost** Problems 18.14 and 18.26.
  a. Conduct an F test of whether or not the expected reconditioning cost varies with operating hours. Control the α risk at 0.01. State the alternatives, the decision rule, the value of the test statistic, and the conclusion.
  b. Show, numerically, the equivalence of the test statistic F^* in **a** and the test statistic t^* in (19.17).

19.32 Refer to **Bid Preparation** Problems 18.15 and 18.27. Conduct an F test of whether or not $\beta_1 = 0$, controlling the α risk at 0.01 when $\beta_1 = 0$. State the alternatives, the decision rule, the value of the test statistic, and the conclusion.

*19.33 Refer to **Fuel Efficiency** Problem 18.12.
  a. Obtain the residuals and the fitted values.
  b. Plot the residuals against the fitted values. What does your plot indicate about the aptness of regression model (18.3) with regard to (1) linearity of the regression function, (2) constancy of the error variance? Would a larger number of cases in this regression study increase your confidence in the conclusions? Comment.
  c. Obtain the standardized residuals. Construct a table like Table 19.1, based on the 25th, 50th, and 75th percentiles of the relevant t distribution. Do you see any strong evidence of lack of normality in the error terms? Explain.

19.34 Refer to **Marketing Experiment** Problem 18.13.
  a. Obtain the residuals and the fitted values.

b. Plot the residuals against the fitted values. Does your plot show any systematic pattern that may be inconsistent with regression model (18.3)? Comment.

19.35 Refer to **Reconditioning Cost** Problem 18.14.
a. Obtain the residuals and the fitted values.
b. Plot the residuals against the fitted values. What does your plot indicate about the aptness of regression model (18.3) with regard to (1) linearity of the regression function, (2) constancy of the error variance? Would a larger number of cases in this regression study increase your confidence in the conclusions? Comment.
c. Obtain the standardized residuals. Construct a table like Table 19.1, based on the 25th, 50th, and 75th percentiles of the relevant t distribution. Do you see any strong evidence of lack of normality in the error terms? Explain.

19.36 Refer to **Bid Preparation** Problem 18.15.
a. Obtain the residuals and the fitted values.
b. Plot the residuals against the fitted values. What does your plot indicate about the aptness of regression model (18.3) with regard to (1) linearity of the regression function, (2) constancy of the error variance?
c. The bids were prepared in the time order given. Plot the residuals against time order. What does your plot indicate about the aptness of regression model (18.3)?
d. Obtain the standardized residuals. Construct a table like Table 19.1, based on the 25th, 50th, and 75th percentiles of the relevant t distribution. Do you see any strong evidence of lack of normality in the error terms? Explain.

19.37 A business psychologist created 16 tasks of varying difficulty, ranging from a difficulty rating of 0 (trivial) to one of 100 (extremely complex). She then organized 16 decision-making groups of roughly equal composition and ability, and she randomly assigned one group to each task. The performance of the group in completing its task was rated on a scale from 0 (very low) to 100 (very high). As part of the data analysis, the psychologist regressed group performance rating (Y) on task difficulty (X), using regression model (19.1), and obtained the following fitted values and residuals:

$\hat{Y}_i$	e_i	$\hat{Y}_i$	e_i	$\hat{Y}_i$	e_i	$\hat{Y}_i$	e_i
67.4	3.6	74.5	0.5	77.1	−9.1	71.4	9.6
71.7	13.3	75.0	−10.0	70.0	13.0	67.6	7.4
64.3	−12.3	69.1	10.9	72.4	0.6	65.2	−17.2
65.5	−0.5	66.7	−3.7	72.2	6.8	75.7	−12.7

a. Plot the residuals against the fitted values. Does a linear regression function appear to be appropriate here? Discuss.
b. Can the plot in **a** be readily examined for constancy of the error variance here? For normality of the error terms? Explain.

19.38 **Gear Wear.** A small gear that connects two large gears is made of relatively soft metal to minimize wear on the large gears. An analyst wished to study the relation between operating hours to date of the small gear (X) and remaining operating hours before its replacement is required (Y). He obtained the following data (in hundreds of operating hours) for a random sample of ten gears:

i:	1	2	3	4	5	6	7	8	9	10
X_i:	38.7	5.0	9.5	15.4	30.2	17.1	40.1	14.1	25.0	42.4
Y_i:	7.5	33.7	19.6	15.0	10.3	12.9	7.3	15.0	11.6	7.0

a. Construct a scatter plot of the data. Does it appear that regression model (19.1) is appropriate here? Explain.

b. After further investigation, the analyst decided to employ the reciprocal transformation $X' = 1/X$. Construct a scatter plot of Y versus X'. Does this transformation appear to be successful in linearizing the relation between the variables here? Discuss.

19.39 The relation between extrusion speed of plastic tubing (X) and the number of blemishes in 500 feet of tubing (Y) was investigated for a newly acquired machine by a manufacturing engineer. Twelve trials were run with the machine at extrusion speeds ranging from 0.2 foot per second to 1.2 feet per second. In each run, 500 feet of tubing were produced. The results were as follows:

i:	1	2	3	4	5	6	7	8	9	10	11	12
X_i:	0.2	0.2	0.4	0.4	0.6	0.6	0.8	0.8	1.0	1.0	1.2	1.2
Y_i:	3	0	6	3	12	7	19	15	24	29	39	34

a. Construct a scatter plot of the data. Does it appear that regression model (19.1) is appropriate for studying the relation between the variables here? Explain.

b. After further analysis, the engineer decided to employ the square root transformation $Y' = \sqrt{Y}$. Construct a scatter plot of Y' versus X. Does this transformation appear to be successful in linearizing the relation between the variables here? Discuss.

19.40 Refer to **Fuel Efficiency** Problem 18.12. Would it be appropriate to use the regression results in this study to estimate the mean expressway mileage of a new type of lightweight automobile weighing 10 hundred pounds? Explain.

19.41 Refer to **Marketing Experiment** Problem 18.13. A marketing executive has proposed that the regression results for the experiment be used to predict blanket sales in a territory subjected to "saturation" TV exposure. His plan calls for ten showings of the commercial in a territory that is similar to those in the experiment. Should the executive place much confidence in the prediction from the regression study? Explain.

19.42 A scientist is studying the relation between the volume occupied by a gas (Y) and temperature (X) while pressure is kept constant. The results of eight trials follow.

Trial i:	1	2	3	4	5	6	7	8
Temperature (°K) X_i:	200	250	300	350	200	250	300	350
Volume (milliliters) Y_i:	251	315	374	440	241	302	362	423

The scientist intends to regress gas volume on gas temperature, using regression model (19.1). Trials 1–4 were done by one laboratory technician and trials 5–8 by a second technician. Construct a scatter plot of the data, using a different plotting symbol for trials 1–4 and trials 5–8, respectively. Should the scientist reconsider his intended regression analysis? Explain.

19.43 An analyst finds that a statistical relation holds between thickness of silver film deposited in a process and frequency shift of a crystal used as a sensor. She proposes to investigate whether regression model (19.1) can be used to estimate film thickness (taken as the dependent variable) from knowledge of the magnitude of the frequency shift (taken as the independent variable). An observer states that regression model (19.1) cannot be used in this manner because frequency shift actually depends on film thickness, not vice versa. Comment.

EXERCISES

19.44 Refer to formula (19.8).

    a. Do both components of $\sigma^2\{Y_{h(\text{new})}\}$ become smaller as the sample size increases? Explain.

    b. What does your answer in **a** imply about our ability to make $\sigma^2\{Y_{h(\text{new})}\}$ small by increasing n?

19.45 Derive formulas (19.19a) and (19.19b) from theorem (19.3) by letting $X_h = 0$.

19.46 A student asks why the F test for deciding between $\beta_1 = 0$ and $\beta_1 \neq 0$ is one-sided, even though the latter alternative implies that either $\beta_1 < 0$ or $\beta_1 > 0$. Explain with specific reference to formulas (19.22) and (19.23).

19.47 In an application of regression model (19.1), $\sigma^2 = 100$ and $\Sigma (X_i - \bar{X})^2 = 50$. Obtain $E\{MSE\}$ and $E\{MSR\}$ when (1) $\beta_1 = 0$, (2) $\beta_1 = 10$, (3) $\beta_1 = -10$, (4) $\beta_1 = 50$. What do your results imply about the type of decision rule appropriate for the F test of a regression relation?

19.48 Prove (19.23), using the fact that $MSR = b_1^2 \Sigma (X_i - \bar{X})^2$. [*Hint:* Obtain $E\{b_1^2\}$ by using (5.7a).]

19.49 Show that the F^* test statistic (19.24) is equal to the square of the t^* test statistic in (19.17), using the fact that $MSR = b_1^2 \Sigma (X_i - \bar{X})^2$.

19.50 In a finance study, bond yield (Y) was regressed on bond maturity (X) for 400 issues, using regression model (19.1). About half of the issues were corporate bond issues, and the other half were municipal bond issues. The frequency distribution of the standardized residuals obtained from the regression fit follows.

e'_i	Number of Issues	e'_i	Number of Issues
Under -1.5	14	0 –under 0.5	38
-1.5–under -1.0	70	0.5–under 1.0	80
-1.0–under -0.5	90	1.0–under 1.5	59
-0.5–under 0	29	1.5 or more	20

    a. What probability distribution should the standardized residuals approximately follow under regression model (19.1)? Does this model seem to hold in this application? Explain.

    b. If the standardized residuals are approximately normally distributed, what is the approximate probability that a standardized residual has a value below -1.5 or above 1.5? Are the tails of the distribution of the 400 standardized residuals beyond ± 1.5 consistent with those of a normal distribution?

    c. Would the frequency pattern of the residuals differ from that of the standardized residuals? Explain.

STUDIES

19.51 Refer to **Gear Wear** Problem 19.38. Assume that regression model (19.1) with the transformed independent variable $X' = 1/X$ is appropriate.

    a. Obtain the estimated regression function based on the transformed independent variable $X' = 1/X$. Also, obtain the residuals.

  b. Construct a 95 percent prediction interval for the remaining operating hours of a gear hav-
     ing 25.0 hundred operating hours to date.
  c. If the distributions of Y were normal with constant variance σ^2 before the transformation
     $X' = 1/X$ was applied, will they continue to be so after the transformation? Explain.
  d. The analyst was asked why he did not use the logarithmic transformation $Y' = \log Y$ to lin-
     earize the relation between X and Y here. Regress Y' on X, and state the estimated regres-
     sion function in the original Y units. Also, obtain the residuals in the original Y units.
  e. In separate graphs, plot the residuals obtained in **a** and **d** against their respective fitted val-
     ues in the original units. Does the logarithmic transformation of Y appear to be better than
     the analyst's reciprocal transformation of X? Explain.

19.52 (Computer needed.) Refer to the **Financial Characteristics** data set. Consider the firms in the
  paper industry. Assume regression model (19.1) applies.
  a. Regress net income for year 2 on net assets for year 2.
  b. Construct (1) a 95 percent confidence interval for β_1, (2) a 95 percent confidence interval
     for $E\{Y_h\}$ where X_h is the mean net assets for year 2 for all 24 paper firms, (3) a 95 percent
     prediction interval for $Y_{h(new)}$ where again X_h is the mean net assets for all 24 firms. Interpret
     each interval. Is the prediction interval precise enough to be useful? Comment.
  c. Obtain the standardized residuals. Plot the standardized residuals against the fitted values,
     and check the aptness of regression model (19.1) with regard to (1) linearity of the regres-
     sion function, (2) constancy of the error variance.

MULTIPLE REGRESSION

In many situations, two or more independent variables must be included in a regression model to provide an adequate description of the process under study or to yield sufficiently precise inferences.

EXAMPLES

1. A regression model to control the diameter of plastic pellets produced by an extrusion process uses as independent variables the initial temperature of the process, the die temperature, and the extrusion rate.

2. A regression model for predicting the demands for a firm's product in its 25 sales territories uses as independent variables two socioeconomic variables (mean household income, average years of schooling of head of household), two demographic variables (average family size, percentage of population over 65 years of age), and two environmental variables (mean daily temperature, index of atmospheric pollution). □

Regression models containing two or more independent variables are called *multiple regression* models. In this chapter, we extend the procedures for simple linear regression to multiple regression and also consider some special topics of importance when multiple regression models are used.

20.1 MULTIPLE REGRESSION MODELS

Model in Two Independent Variables

The simple linear regression model (19.1) can easily be extended to include two independent variables, X_1 and X_2.

(20.1)

$$Y_i = \beta_0 + \beta_1 X_{i1} + \beta_2 X_{i2} + \varepsilon_i \qquad i = 1, 2, \ldots, n$$

where:

Y_i is the response in the ith case

continues

X_{i1} and X_{i2} are the values of the independent variables in the ith case,
   assumed to be known constants
β_0, β_1, and β_2 are parameters
ε_i are independent $N(0, \sigma^2)$

As for the simple linear regression model (19.1), we are assuming here that the error terms ε_i are statistically independent normal random variables, with mean zero and constant variance σ^2.

Regression Function.   The regression or response function for model (20.1) is as follows.

(20.2)
$$E\{Y\} = \beta_0 + \beta_1 X_1 + \beta_2 X_2$$

This function is frequently called the regression or response *surface*. The surface for response function (20.2) is a plane, as illustrated in Figure 20.1. The parameters of a multiple regression model are interpreted analogously to those in the simple linear case. Thus, in response function (20.2):

1. β_0 is the Y intercept of the plane; it is the mean of the distribution of Y when $X_1 = 0$ and $X_2 = 0$.

FIGURE 20.1   Example of a response plane. β_0 and β_1 are positive and β_2 is negative here. Point $E\{Y_i\}$ on the plane is the mean of the probability distribution corresponding to observation Y_i at coordinates (X_{i1}, X_{i2}). The error term is $\varepsilon_i = Y_i - E\{Y_i\}$.

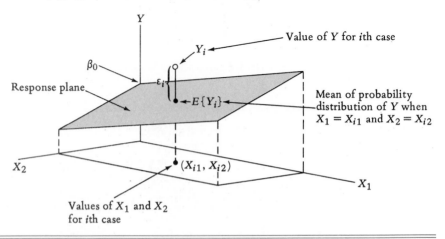

---

2. β_1 indicates the change in $E\{Y\}$ when X_1 increases by one unit while X_2 remains constant.

3. β_2 indicates the change in $E\{Y\}$ when X_2 increases by one unit while X_1 remains constant.

EXAMPLE

Figure 20.1 shows a response plane for regression model (20.1) where β_0 and β_1 are positive and β_2 is negative. Just as the regression line for the simple linear regression model (19.1) gives the mean of the distribution of Y for any level of X, the response plane in Figure 20.1 gives the mean of the distribution of Y for any combination of (X_1, X_2) values. Shown in Figure 20.1 is a response Y_i when $X_1 = X_{i1}$ and $X_2 = X_{i2}$. The mean of this probability distribution, $E\{Y_i\}$, is a point on the regression plane, as indicated in the figure. The error term ε_i for this observation is represented by the vertical distance between Y_i and $E\{Y_i\}$, since $\varepsilon_i = Y_i - E\{Y_i\}$ as for simple linear regression. □

Model in $p - 1$ Independent Variables

The regression model (20.1) in two independent variables can be readily extended to include $p - 1$ independent variables $X_1, X_2, \ldots, X_{p-1}$.

(20.3)

$$Y_i = \beta_0 + \beta_1 X_{i1} + \beta_2 X_{i2} + \cdots + \beta_{p-1} X_{i,p-1} + \varepsilon_i \qquad i = 1, 2, \ldots, n$$

where:

Y_i is the response in the ith case

$X_{i1}, X_{i2}, \ldots, X_{i,p-1}$ are the values of the independent variables in the ith case, assumed to be known constants

$\beta_0, \beta_1, \ldots, \beta_{p-1}$ are parameters

ε_i are independent $N(0, \sigma^2)$

Note that when $p - 1 = 1$, regression model (20.3) becomes the simple linear regression model (19.1); and when $p - 1 = 2$, it becomes regression model (20.1), the multiple regression model with two independent variables.

The parameters in regression model (20.3) are interpreted in the standard manner:

1. β_0, the Y intercept, indicates the mean of the distribution of Y when $X_1 = X_2 = \cdots = X_{p-1} = 0$.

2. β_k ($k = 1, 2, \ldots, p - 1$) indicates the change in $E\{Y\}$ when X_k increases by one unit while the other independent variables remain constant.

3. σ^2 is the common variance of the distributions of Y.

20.2 BASIC CALCULATIONAL RESULTS FOR MULTIPLE REGRESSION MODELS

Most calculational procedures for multiple regression are direct extensions of those already discussed for simple linear regression. We demonstrate these procedures with an illustration involving two independent variables.

EXAMPLE

Promotional Expenditures. An experiment was designed by a market researcher to study the effects of two types of promotional expenditures on sales of a line of food products sold in supermarkets. Sixteen localities were selected for the test. They were similar in market potential and were representative of the target market for this line of products. Different combinations of media advertising expenditures (X_1) and point-of-sale expenditures (X_2) were specified for the study, and the localities were assigned at random to one of these (X_1, X_2) combinations. Table 20.1 shows the two expenditures levels, together with the dollar sales (Y), for the test period for each of the 16 localities.

Figure 20.2 contains computer output for a regression analysis of the results in Table 20.1, using regression model (20.1). We have numbered the blocks of information in Figure 20.2 for ready identification. The format is similar to that of

TABLE 20.1   Data for the promotional expenditures example

Locality i	Media Expenditures ($ thousand) X_{i1}	Point-of-Sale Expenditures ($ thousand) X_{i2}	Sales Volume ($ ten thousand) Y_i
1	2	2	8.74
2	2	3	10.53
3	2	4	10.99
4	2	5	11.97
5	3	2	12.74
6	3	3	12.83
7	3	4	14.69
8	3	5	15.30
9	4	2	16.11
10	4	3	16.31
11	4	4	16.46
12	4	5	17.69
13	5	2	19.65
14	5	3	18.86
15	5	4	19.93
16	5	5	20.51

FIGURE 20.2   Computer output for multiple regression — Promotional expenditures example

1.   VARIABLE   MEAN   STANDARD DEVIATION

     X1   3.50000   1.15470
     X2   3.50000   1.15470
     Y   15.20688   3.62630

2.   SIMPLE CORRELATION MATRIX

                  X1   X2   Y
     X1   1.000   .000   .965
     X2   1.000   .225
     Y   1.000

3.   VARIABLE   REG. COEFF.   STD. DEV.   T STAT.   P-VALUE

     CONSTANT   2.13438   .61036   3.50   .00394
     X1   3.02925   .12028   25.18   .00000
     X2   .70575   .12028   5.87   .00006

4.   ANALYSIS OF VARIANCE TABLE

     SOURCE   SUM OF SQUARES   DF   MEAN SQUARE

     REGRESSION   193.4888   2   96.74439
     RESIDUAL   3.7616   13   .28935
     TOTAL   197.2503   15

        F STAT. = 334.35   UPPER-TAIL P-VALUE = .00000

5.   RESIDUAL STD. DEV. = .53791   NUMBER OF OBS. = 16
     MULTIPLE R SQUARED = .98093   MULTIPLE R   =   .99042

6.   VARIANCE-COVARIANCE MATRIX OF REG. COEFF.

                  B0   B1   B2
     B0   .37254   -.05064   -.05064
     B1   .01447   .00000
     B2   .01447

7.   VARIABLE   SPECIFIED LEVEL

     X1   5.0
     X2   2.0

     Y ESTIMATE = 18.6921   STD. DEV. MEAN RESPONSE = .2884
                                 STD. DEV. NEW RESPONSE = .6104

Continues

FIGURE 20.2   Continued

8.	CASE NO.	OBSERVED	ESTIMATED	RESIDUAL
	1	8.740	9.604	-.86438
	2	10.530	10.310	.21988
	3	10.990	11.016	-.02588
	4	11.970	11.722	.24838
	5	12.740	12.634	.10638
	6	12.830	13.339	-.50938
	7	14.690	14.045	.64488
	8	15.300	14.751	.54913
	9	16.110	15.663	.44713
	10	16.310	16.369	-.05863
	11	16.460	17.074	-.61438
	12	17.690	17.780	-.09013
	13	19.650	18.692	.95788
	14	18.860	19.398	-.53788
	15	19.930	20.104	-.17363
	16	20.510	20.809	-.29938

Figure 18.13 for simple linear regression. For instance, block 1 gives the means and standard deviations for X_1, X_2, and Y, respectively. □

Correlation Matrix

The interpretation of multiple regression results often requires information about the coefficients of correlation between pairs of the variables in the study. Let us denote the coefficient of simple correlation between Y and X_1, as defined in (18.38), by r_{Y1}. Similarly, r_{Y2} denotes the coefficient of simple correlation between Y and X_2, and r_{12} the coefficient of simple correlation between X_1 and X_2. This latter coefficient relates to the degree of linear association between the independent variables X_1 and X_2. The simple correlation coefficients in a multiple regression problem are frequently presented in the form of a *correlation matrix*.

EXAMPLE

The simple correlation coefficients for the promotional expenditures example are given in block 2 of Figure 20.2 in a matrix called SIMPLE CORRELATION MATRIX. The format and the values of the coefficients for the study are as follows:

$$
\begin{array}{c}
\begin{array}{ccc} X_1 & X_2 & Y \end{array} \\
\begin{array}{c} X_1 \\ X_2 \\ Y \end{array}
\begin{bmatrix}
1 & r_{12} & r_{Y1} \\
  & 1 & r_{Y2} \\
  &   & 1
\end{bmatrix}
=
\begin{bmatrix}
1 & 0 & 0.965 \\
  & 1 & 0.225 \\
  &   & 1
\end{bmatrix}
\end{array}
$$

The market researcher noted with interest that the coefficient of simple correlation between sales (Y) and media expenditures (X_1) is substantially greater than that between sales (Y) and point-of-sale expenditures (X_2), namely that $r_{Y1} = 0.965$ while $r_{Y2} = 0.225$. Also, he knew in advance that the coefficient of simple correlation between X_1 and X_2 would be $r_{12} = 0$, since he had designed this feature into the experiment by setting up the (X_1, X_2) combinations in the manner shown in Table 20.1. We will discuss the reason for this design later. □

Estimated Regression Function

For regression model (20.1), the estimated regression function is as follows.

(20.4)
$$\hat{Y} = b_0 + b_1 X_1 + b_2 X_2$$

Here, b_0, b_1, and b_2 are the least squares estimators of the corresponding parameters.

The method of least squares leads to a system of normal equations for obtaining b_0, b_1, and b_2, as follows.

(20.5)
$$\Sigma Y_i = nb_0 + b_1 \Sigma X_{i1} + b_2 \Sigma X_{i2}$$
$$\Sigma X_{i1} Y_i = b_0 \Sigma X_{i1} + b_1 \Sigma X_{i1}^2 + b_2 \Sigma X_{i1} X_{i2}$$
$$\Sigma X_{i2} Y_i = b_0 \Sigma X_{i2} + b_1 \Sigma X_{i1} X_{i2} + b_2 \Sigma X_{i2}^2$$

The least squares estimates are obtained by solving this system of equations simultaneously. Computer programs are generally utilized to find the solution, as is done here.

EXAMPLE

Block 3 in Figure 20.2 presents information on the estimated regression coefficients for the promotional expenditures example. The format is an extension of that in Figure 18.13 for simple linear regression:

VARIABLE	REG. COEFF.	STD. DEV.	T STAT.	P-VALUE
CONSTANT	b_0	$s\{b_0\}$	$t^* = b_0/s\{b_0\}$	Two-sided P-value for t^*
X_1	b_1	$s\{b_1\}$	$t^* = b_1/s\{b_1\}$	Two-sided P-value for t^*
X_2	b_2	$s\{b_2\}$	$t^* = b_2/s\{b_2\}$	Two-sided P-value for t^*

The estimated regression function is seen to be:

$$\hat{Y} = 2.13438 + 3.02925X_1 + 0.70575X_2$$ □

Analysis of Variance

Sums of Squares. The sums of squares for the analysis of variance are defined identically in simple and multiple regression. For convenience, we repeat the definitional formulas for the total sum of squares $SSTO$, the error sum of squares SSE, and the regression sum of squares SSR.

(20.6)

$$SSTO = \Sigma (Y_i - \overline{Y})^2$$

(20.7)

$$SSE = \Sigma e_i^2 = \Sigma (Y_i - \hat{Y}_i)^2$$

(20.8)

$$SSR = \Sigma (\hat{Y}_i - \overline{Y})^2$$

Degrees of Freedom. As usual, the total sum of squares $SSTO$ has $n - 1$ degrees of freedom associated with it.

The error sum of squares SSE has $n - p$ degrees of freedom associated with it. There are n residuals e_i, but p constraints arise on these residuals because p parameters—$\beta_0, \beta_1, \ldots, \beta_{p-1}$—have to be estimated for obtaining the fitted values $\hat{Y}_i$.

The regression sum of squares SSR has $p - 1$ degrees of freedom associated with it. There are p parameters in the regression function, but the deviations $\hat{Y}_i - \overline{Y}$ are subject to one constraint, $\Sigma (\hat{Y}_i - \overline{Y}) = 0$.

Note that when $p - 1 = 1$, the simple linear regression case, the degrees of freedom for SSE are $n - p = n - 2$ and those for SSR are $p - 1 = 1$. These, of course, are the degrees of freedom stated in Table 18.4a for simple linear regression.

The degrees of freedom are additive:

$$n - 1 = (p - 1) + (n - p)$$

Table 20.2 shows the general form of the ANOVA table for multiple regression. The mean squares MSR and MSE are defined as for simple linear regression, that is, as a sum of squares divided by degrees of freedom. Block 4 in Figure 20.2 contains the ANOVA table for the promotional expenditures example.

Point Estimation of σ^2. As in simple linear regression, the error mean square MSE is an unbiased point estimator of σ^2. The ANOVA table in block 4 shows that $MSE = 0.28935$ for the promotional expenditures example. The corresponding point

TABLE 20.2 **General format of an ANOVA table for multiple regression**

Source of Variation	SS	df	MS
Regression	$SSR = \Sigma (\hat{Y}_i - \overline{Y})^2$	$p - 1$	$MSR = \dfrac{SSR}{p - 1}$
Error	$SSE = \Sigma (Y_i - \hat{Y}_i)^2$	$n - p$	$MSE = \dfrac{SSE}{n - p}$
Total	$SSTO = \Sigma (Y_i - \overline{Y})^2$	$n - 1$	

estimate of σ is $\sqrt{MSE} = \sqrt{0.28935} = 0.53791$, labeled in block 5 of Figure 20.2 as RESIDUAL STD. DEV.

Coefficient of Multiple Determination

The coefficient of multiple determination, denoted by R^2, is defined analogously to r^2, the coefficient of simple determination.

(20.9)

The *coefficient of multiple determination R^2* is defined:

$$R^2 = \frac{SSR}{SSTO} = 1 - \frac{SSE}{SSTO}$$

Thus, R^2 measures the proportionate reduction in $SSTO$ associated with the use of the independent variables $X_1, X_2, \ldots, X_{p-1}$ in the multiple regression model. When $p - 1 = 1$, R^2 reduces to r^2. Also, like r^2, R^2 ranges from 0 to 1.

(20.10)

$$0 \le R^2 \le 1$$

We obtain $R^2 = 0$ when $b_1 = b_2 = \cdots = b_{p-1} = 0$, and $R^2 = 1$ when all the observed Y_i fall directly on the estimated regression surface.

EXAMPLE

For the promotional expenditures example, we find from the ANOVA table in Figure 20.2 that $SSR = 193.4888$ and $SSTO = 197.2503$. Hence:

$$R^2 = \frac{193.4888}{197.2503} = 0.98093$$

Thus, the variability of locality sales is reduced by 98 percent when media and point-of-sale promotional expenditures are considered in the regression model.

□

Comment

It can be shown that the coefficient of multiple determination R^2 is identical to the coefficient of simple determination obtained when the observations Y_i are regressed on the fitted values $\hat{Y}_i$ from the multiple regression model.

Coefficient of Multiple Correlation.   The positive square root of R^2 is called the coefficient of multiple correlation.

(20.11)

The *coefficient of multiple correlation R* is defined:

$$R = +\sqrt{R^2}$$

where:

R^2 is given by (20.9)

EXAMPLE

For the promotional expenditures example, we have:

$$R = \sqrt{R^2} = \sqrt{0.98093} = 0.99042$$

The values of R^2 and R for the promotional expenditures example are included in block 5 in Figure 20.2, labeled MULTIPLE R SQUARED and MULTIPLE R, respectively.

☐

Adjusted R^2.   Sometimes, the coefficient of multiple determination defined in (20.9) is modified to take into account the number of independent variables in the model. The reason is that this coefficient can generally be made larger if additional independent variables are added to the model. To see this, note that *SSE* tends to become smaller (it cannot increase) with each additional independent variable, while *SSTO* remains fixed. A measure that recognizes the number of independent variables in the model is called the *adjusted coefficient of multiple determination* and is denoted by R_a^2:

(20.12)

$$R_a^2 = 1 - \left(\frac{n-1}{n-p}\right)\left(\frac{SSE}{SSTO}\right) = 1 - \frac{MSE}{\left(\frac{SSTO}{n-1}\right)}$$

When an independent variable is added to the model, p increases by one and $n - p$ decreases correspondingly. Therefore R_a^2 can become smaller if the decrease in $n - p$ is not offset by a sufficient decrease in *SSE*.

For the promotional expenditures example, we have:

$$R_a^2 = 1 - \frac{15}{13}\left(\frac{3.7616}{197.2503}\right) = 0.97800$$

Here, the adjustment has only a small effect; note that R^2 and R_a^2 are almost equal.

Coefficient of Partial Determination

The coefficient of multiple determination R^2 is a measure of the combined effect of all independent variables $X_1, X_2, \ldots, X_{p-1}$ in the regression model in reducing the total variability *SSTO*. Often, an additional measure, called the coefficient of partial determination, is needed when one is considering whether or not to add another variable to the regression model.

EXAMPLE

In the promotional expenditures example in Table 20.1, a regression of sales volume Y on point-of-sale expenditures X_2 alone yields the estimated regression function $\hat{Y} = 12.73675 + 0.70575X_2$ (computations not shown). The error sum of squares for this simple regression model, which we shall denote by $SSE(X_2)$, is $SSE(X_2) = 187.2886$. The question of concern is whether it is worth adding the other independent variable X_1, media expenditures, to the regression model. We see from block 4 of Figure 20.2 that the error sum of squares when both X_1 and X_2 are in the regression model, to be denoted by $SSE(X_1, X_2)$, is $SSE(X_1, X_2) = 3.7616$. Thus, the residual variability in sales volume when X_2 is already in the regression model, 187.2886, is reduced to 3.7616 by adding X_1 to the regression model. The proportionate reduction is 98 percent:

$$\frac{SSE(X_2) - SSE(X_1, X_2)}{SSE(X_2)} = \frac{187.2886 - 3.7616}{187.2886} = 0.9799$$

This measure indicates the marginal effect of X_1 in reducing the variability in Y when X_2 is already in the model, and it is called a coefficient of partial determination.

$\square$

(20.13)

For regression model (20.1), with two independent variables, the *coefficient of partial determination* between Y and X_1, given that X_2 is already in the model, is denoted by $r^2_{Y1.2}$ and is defined:

$$r^2_{Y1.2} = \frac{SSE(X_2) - SSE(X_1, X_2)}{SSE(X_2)} = 1 - \frac{SSE(X_1, X_2)}{SSE(X_2)}$$

Comments

1. The coefficient of partial determination between Y and X_2, given that X_1 is already in the regression model, is denoted by $r^2_{Y2.1}$ and is defined in an analogous way to (20.13) by reversing the roles of X_1 and X_2.

2. Coefficients of partial determination may also be defined in cases where there are more than two independent variables. For instance, when X_4 enters a regression model already containing the variables X_1, X_2, and X_3, the coefficient of partial determination is given by:

$$r^2_{Y4.123} = \frac{SSE(X_1, X_2, X_3) - SSE(X_1, X_2, X_3, X_4)}{SSE(X_1, X_2, X_3)}$$

Here, $r^2_{Y4.123}$ measures the relative reduction in the error sum of squares by adding X_4 into the model when X_1, X_2, and X_3 are already in the regression model.

Coefficient of Partial Correlation.   The square root of the coefficient of partial determination is called the *coefficient of partial correlation*. It is given the same sign as the corresponding regression coefficient in the fitted regression function which contains all the independent variables. For the promotional expenditures example, $r_{Y1.2} = +\sqrt{0.9799} = +0.9899$ because b_1 in the fitted regression function is positive ($b_1 = +3.02925$ in block 3 of Figure 20.2).

20.3 RESIDUAL ANALYSIS

Usually, the first step in the analysis of a multiple regression model is an examination of the aptness of the model. Residuals are studied for this purpose in the same manner described earlier for the simple linear regression model. Thus, residuals are analyzed for randomness, normality, constancy of error variance, and appropriateness of the regression function. Plots of the residuals e_i against the fitted values $\hat{Y}_i$ and against time order are useful. Plots against the independent variables $X_1, X_2, \ldots, X_{p-1}$ one at a time are also helpful in identifying whether the effects of any of the independent variables differ from those postulated in the model. Finally, plots of the residuals e_i against selected new variables not already in the model help to discover if one or more important variables have been omitted from the model.

EXAMPLE

The observed values Y_i, the fitted values $\hat{Y}_i$ (labeled ESTIMATED), and the residuals e_i for the promotional expenditures example are presented in block 8 of Figure 20.2. Figure 20.3 shows a computer plot in which the residuals e_i are plotted against the fitted values $\hat{Y}_i$. The random scatter of the points about the 0 line indicates that there is no relationship between the residuals and the fitted values, suggesting that multiple regression model (20.1) is an apt one for this example. The market researcher, after studying this and other residual plots, concluded that none of the plots challenged the appropriateness of multiple regression model (20.1).

20.4 *F* TEST FOR REGRESSION RELATION

Frequently, the next step in the analysis of a multiple regression model is to test whether or not there is a relation between the dependent variable Y and the independent variables $X_1, X_2, \ldots, X_{p-1}$. The expectation, of course, is that such a relation exists, because the model is developed in the first place to exploit this relation.

The F test for a multiple regression relation is a direct extension of the F test for H_0: $\beta_1 = 0$ versus H_1: $\beta_1 \neq 0$ in simple linear regression. It is a test whether or not $\beta_1 = \beta_2 = \cdots = \beta_{p-1} = 0$, that is, a test whether or not there is a relation between $E\{Y\}$ and the independent variables $X_1, X_2, \ldots, X_{p-1}$. To see what is involved here, let us return to the case of two independent variables and regression model (20.1). If β_1 and β_2 both equal zero, the response function reduces to $E\{Y\} = \beta_0$. Thus, the means of the distributions of Y are the same for all (X_1, X_2) combinations, and there is no

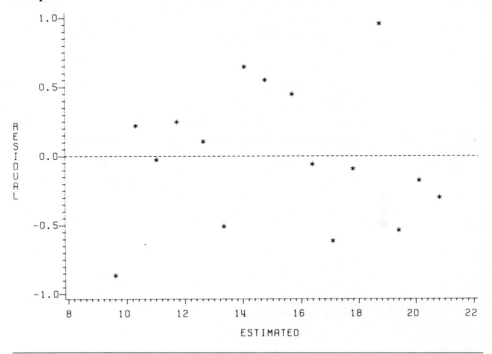

FIGURE 20.3 Computer residual plot of e_i against $\hat{Y}_i$ — Promotional expenditures example

regression relation between Y and the independent variables X_1 and X_2. On the other hand, if β_1 and β_2 are not both equal to zero (at least one of the two is nonzero), the means of Y are not constant for all (X_1, X_2) combinations, and a regression relation does exist between Y and the independent variables.

These ideas extend directly to multiple regression model (20.3) with $p - 1$ independent variables. Thus, a test of whether or not Y is related to $X_1, X_2, \ldots, X_{p-1}$ involves the following alternatives:

$$H_0: \beta_1 = \beta_2 = \cdots = \beta_{p-1} = 0$$

$$H_1: \text{not all } \beta_k = 0 \qquad k = 1, 2, \ldots, p - 1$$

The test procedure is similar to that for the F test of $\beta_1 = 0$ versus $\beta_1 \neq 0$ in simple linear regression. The test statistic is the same.

(20.14)
$$F^* = \frac{MSR}{MSE}$$

The decision rule to control the α risk takes a corresponding form.

(20.15)

When the alternatives are:

H_0: $\beta_1 = \beta_2 = \cdots = \beta_{p-1} = 0$

H_1: not all $\beta_k = 0$ $k = 1, 2, \ldots, p - 1$

and the multiple regression model (20.3) is applicable, the appropriate decision rule to control the α risk is as follows:

If $F^* \leq F(1 - \alpha; p - 1, n - p)$, conclude H_0.
If $F^* > F(1 - \alpha; p - 1, n - p)$, conclude H_1.

where:
F^* is given by (20.14)

EXAMPLE

In the promotional expenditures study, the market researcher, after being satisfied from the residual analysis that regression model (20.1) is apt, wished next to test whether or not a relation between sales and the two types of promotional expenditures does indeed exist. If so, as he expected, he would then analyze the multiple regression model in some detail.

Step 1.   The test alternatives are:

H_0: $\beta_1 = \beta_2 = 0$

H_1: not both $\beta_1 = 0$ and $\beta_2 = 0$

Step 2.   The researcher specified that the α risk be controlled at 0.05.
Step 3.   For $\alpha = 0.05$, $n = 16$, and $p - 1 = 2$, we require $F(0.95; 2, 13) = 3.81$. Hence, the decision rule for the test is as follows:

If $F^* \leq 3.81$, conclude H_0.
If $F^* > 3.81$, conclude H_1.

Step 4.   From the computer output in Figure 20.2 (block 4), we see that $MSR = 96.74439$ and $MSE = 0.28935$. Thus, $F^* = 96.7444/0.28935 = 334 > 3.81$. Hence, we conclude H_1—that there is a regression relation between the two types of promotional expenditures and sales.

Note, incidentally, that the value of F^* for this test is given in the computer output in Figure 20.2 immediately below the ANOVA table. Also, the upper-tail P-value appears immediately below the ANOVA table and is seen to be 0.00000, to five decimal places of accuracy. Since this P-value is less than $\alpha = 0.05$, H_1 is the appropriate conclusion, as noted already.   □

20.5   INFERENCES CONCERNING INDIVIDUAL REGRESSION COEFFICIENTS

Once it has been established that a regression relation exists between the dependent variable and the set of independent variables, estimation and testing of individual regression coefficients are often of interest.

Estimates and tests of individual regression coefficients β_k in the multiple regression model (20.3) are conducted in the same manner as in simple linear regression. The only difference is that the t multiple now involves $n - p$ degrees of freedom, because MSE in multiple regression has $n - p$ degrees of freedom associated with it. The confidence interval for any β_k takes the usual form.

(20.16)

The confidence limits for β_k with confidence coefficient $1 - \alpha$ for multiple regression model (20.3) are:

$$b_k \pm ts\{b_k\}$$

where:

$t = t(1 - \alpha/2; n - p)$
$s\{b_k\}$ is the estimated standard deviation of b_k

The estimated standard deviation of b_k, $s\{b_k\}$, is given in the computer output of almost all regression packages.

A test of whether or not $\beta_k = 0$ is conducted in the manner explained for simple regression.

(20.17)

Tests concerning β_k for multiple regression model (20.3) when the alternatives are:

$H_0: \beta_k = 0$

$H_1: \beta_k \neq 0$

are based on the standardized test statistic:

$$t^* = \frac{b_k}{s\{b_k\}}$$

where:

$s\{b_k\}$ is the estimated standard deviation of b_k
α risk is controlled at $\beta_k = 0$

When $\beta_k = 0$, t^* follows the t distribution with $n - p$ degrees of freedom. Large absolute values of t^* lead to conclusion H_1.

EXAMPLES

1. In the promotional expenditures example, the market researcher wished to estimate β_1 with a 95 percent confidence interval. From the computer output in Figure 20.2 (block 3), we obtain $b_1 = 3.02925$ and $s\{b_1\} = 0.12028$. Since $n - p = 16 - 3 = 13$, we require $t(0.975; 13) = 2.160$. The confidence limits by (20.16) are $3.02925 \pm 2.160(0.12028)$, and the desired confidence interval is:

$$2.77 \le \beta_1 \le 3.29$$

The confidence interval indicates that an increase of $1000 in media expenditures (with point-of-sale expenditures held fixed) is associated with an increase in expected sales between $27,700 and $32,900. Recall from Table 20.1 that the sales volume variable is expressed in units of ten thousand dollars.

2. In the promotional expenditures example, the market researcher also wished to test $H_0: \beta_2 = 0$ versus $H_1: \beta_2 \ne 0$, using an α risk of 0.05. Conclusion H_0 would imply that X_2 could be dropped from the regression model which already contains X_1. Again, we require $t(0.975; 13) = 2.160$. Consulting the computer output in Figure 20.2 (block 3), we find $t^* = b_2/s\{b_2\} = 0.70575/0.12028 = 5.87$. Since $|t^*| = 5.87 > 2.160$, the researcher should conclude H_1 — that $\beta_2 \ne 0$. This conclusion implies that there is a relation between sales volume (Y) and point-of-sale expenditures (X_2) when media expenditures (X_1) are already in the regression model. Thus, X_2 should not be dropped from the regression model.

   The two-sided P-value for t^* in block 3 of Figure 20.2 could have been used to test whether $\beta_2 = 0$. Since $0.00006 < 0.05 = \alpha$, H_1 is the appropriate conclusion, as we have seen already.

Comment

If one concludes that $\beta_k = 0$ for some variable X_k in a multiple regression model, it does not necessarily follow that X_k is not related to Y. It simply means that, when the other independent variables are already in the regression model, the marginal contribution of X_k in further reducing the error sum of squares is negligible. We shall consider this point further in our discussion of multicollinearity.

20.6   CONFIDENCE INTERVAL FOR $E\{Y_h\}$

As with simple linear regression, a key objective in many applications of multiple regression is to make an inference on a mean response, $E\{Y_h\}$. Our discussion of such inferences in simple linear regression extends readily to multiple regression.

Let the levels of the independent variables for which the mean response is to be estimated be denoted by $X_{h1}, X_{h2}, \ldots, X_{h,p-1}$. Then, $E\{Y_h\}$ denotes the mean response when $X_1 = X_{h1}$, $X_2 = X_{h2}, \ldots, X_{p-1} = X_{h,p-1}$. The point estimator of $E\{Y_h\}$ is as follows.

(20.18)

$$\hat{Y}_h = b_0 + b_1 X_{h1} + b_2 X_{h2} + \cdots + b_{p-1} X_{h,p-1}$$

The confidence interval for $E\{Y_h\}$ takes the usual form.

(20.19)

The confidence limits for $E\{Y_h\}$ with confidence coefficient $1 - \alpha$ for multiple regression model (20.3) are:

$$\hat{Y}_h \pm ts\{\hat{Y}_h\}$$

where:

$$t = t(1 - \alpha/2; n - p)$$
$\hat{Y}_h$ is given by (20.18)
$s\{\hat{Y}_h\}$ is the estimated standard deviation of $\hat{Y}_h$

The estimated standard deviation of $\hat{Y}_h$, $s\{\hat{Y}_h\}$, is a complex expression involving the variances and covariances of the regression coefficients and the specified levels of the independent variables. For the simple case of only two independent variables, for instance, $s^2\{\hat{Y}_h\}$ is defined as follows.

(20.20)

$$s^2\{\hat{Y}_h\} = s^2\{b_0\} + X_{h1}^2 s^2\{b_1\} + X_{h2}^2 s^2\{b_2\} + 2X_{h1}s\{b_0, b_1\}$$
$$+ 2X_{h2}s\{b_0, b_2\} + 2X_{h1}X_{h2}s\{b_1, b_2\}$$

Computer programs are almost always used to calculate $s\{\hat{Y}_h\}$.

EXAMPLE

In the promotional expenditures example, the market researcher wished to estimate expected sales when $X_{h1} = 5$ and $X_{h2} = 2$. A 95 percent confidence coefficient is to be employed.

We know from Figure 20.2 (block 3) that $b_0 = 2.13438$, $b_1 = 3.02925$, and $b_2 = 0.70575$. Hence, the point estimate of mean sales in localities where $X_{h1} = 5$ and $X_{h2} = 2$ is:

$$\hat{Y}_h = 2.13438 + 3.02925(5) + 0.70575(2) = 18.6921$$

To obtain $s\{\hat{Y}_h\}$, we refer to the computer printout in Figure 20.2. Many computer programs, like the one here, permit the user to specify the levels of the independent variables of interest and then will calculate $\hat{Y}_h$ and $s\{\hat{Y}_h\}$. We see from Figure 20.2, block 7, that the standard deviation of the mean response when $X_{h1} = 5$ and $X_{h2} = 2$, labeled STD. DEV. MEAN RESPONSE, is $s\{\hat{Y}_h\} = 0.2884$. The estimate $\hat{Y}_h$ is, of course, the same as calculated previously, labeled in the printout Y ESTIMATE.

We require $t(0.975; 13) = 2.160$. The confidence limits then are $18.6921 \pm 2.160(0.2884)$, and the confidence interval is:

$$18.07 \leq E\{Y_h\} \leq 19.32$$

We therefore estimate, with 95 percent confidence, that mean sales are between $181 thousand and $193 thousand when media and point-of-sale expenditures are $5 thousand and $2 thousand, respectively.   □

Comment

The estimated standard deviation of $\hat{Y}_h$ in the example could have been calculated directly by using (20.20). To do so, we need the estimated variances and covariances of the regression coefficients. These are often provided in computer printouts. For instance, the printout in Figure 20.2 provides them in block 6 in the form of a matrix called VARIANCE-COVARIANCE MATRIX OF REG. COEFF. The format and results for the promotional expenditures example are:

$$\begin{bmatrix} s^2\{b_0\} & s\{b_0, b_1\} & s\{b_0, b_2\} \\ & s^2\{b_1\} & s\{b_1, b_2\} \\ & & s^2\{b_2\} \end{bmatrix} = \begin{bmatrix} 0.37254 & -0.05064 & -0.05064 \\ & 0.01447 & 0 \\ & & 0.01447 \end{bmatrix}$$

20.7   PREDICTION INTERVAL FOR $Y_{h(new)}$

Frequently, we wish to employ a multiple regression model to predict a new response $Y_{h(new)}$ when the independent variables are at specified levels $X_{h1}, X_{h2}, \ldots, X_{h,p-1}$. The prediction interval takes the usual form.

(20.21)

The prediction limits for $Y_{h(new)}$ with confidence coefficient $1 - \alpha$ for multiple regression model (20.3) are:

$$\hat{Y}_h \pm ts\{Y_{h(new)}\}$$

where:

$t = t(1 - \alpha/2; n - p)$
$\hat{Y}_h$ is given by (20.18)
$s^2\{Y_{h(new)}\} = MSE + s^2\{\hat{Y}_h\}$

EXAMPLE

In the promotional expenditures example, the market researcher wished to predict sales in a locality for the next period when $X_{h1} = 5$ and $X_{h2} = 2$, using a 95 percent prediction interval. From the preceding example, we know that, for this case, $\hat{Y}_h = 18.6921$ and $s^2\{\hat{Y}_h\} = (0.2884)^2 = 0.08317$. We also know that $MSE = 0.28935$. Hence, we calculate, by (20.21):

$$s^2\{Y_{h(\text{new})}\} = 0.28935 + 0.08317 = 0.3725$$

$$s\{Y_{h(\text{new})}\} = \sqrt{0.3725} = 0.610$$

This statistic could also have been obtained directly from block 7 in Figure 20.2. Finally, we need $t(0.975; 13) = 2.160$. The prediction limits then are $18.6921 \pm 2.160(0.610)$, and the prediction interval is:

$$17.37 \leq Y_{h(\text{new})} \leq 20.01$$

Hence, we predict, with 95 percent confidence, that sales will be between $174 thousand and $200 thousand in the locality when media and point-of-sale expenditures are $5 thousand and $2 thousand, respectively. □

20.8 INDICATOR VARIABLES

In our discussion of regression analysis up to this point, we have utilized *quantitative* independent variables, such as number of tape drives or dollar expenditures. Frequently, however, a qualitative independent variable is of interest. Examples of qualitative variables are job location of employee (plant, office, field), sex of respondent (male, female), nature of financial disclosure (none, footnote only, line item discussion), and season of year (winter, spring, summer, autumn).

In this section, we show how qualitative independent variables can be included in the regression model.

Representation of Qualitative Variables

We begin by defining what we mean by a 0, 1 indicator variable.

(20.22)

> A *0, 1 indicator variable* is a variable that assumes only the two possible values 0 or 1.

A 0, 1 indicator variable is also sometimes called a *binary* or *dummy variable*.

A qualitative variable is represented in a regression model by one or more 0, 1 indicator variables. A qualitative variable that has k mutually exclusive and exhaustive classes requires the use of $k - 1$ indicator variables in the regression model to represent it.

EXAMPLES

1. The independent variable sex of respondent (male, female) requires $k - 1 = 2 - 1 = 1$ indicator variable to represent it in the regression model. We denote this indicator variable by X_1 and define it as follows:

$$X_1 = \begin{cases} 1 & \text{if female} \\ 0 & \text{otherwise} \end{cases}$$

Hence, the numerical values of X_1 are associated with the two classes of the variable sex as follows:

Class	X_1
Male (reference class)	0
Female	1

Thus, $X_1 = 0$ indicates that the respondent is male and $X_1 = 1$ that the respondent is female. (The reason why the first class is called the reference class will be explained shortly.)

2. The independent variable location of restaurant with three location classes (highway, shopping mall, street) requires $k - 1 = 3 - 1 = 2$ indicator variables to represent it in the regression model. We denote these indicator variables by X_2 and X_3 and define them as follows:

$$X_2 = \begin{matrix} 1 & \text{if shopping mall location} \\ 0 & \text{otherwise} \end{matrix}$$

$$X_3 = \begin{matrix} 1 & \text{if street location} \\ 0 & \text{otherwise} \end{matrix}$$

Hence, the numerical values of X_2 and X_3 are associated with the three location classes as follows:

Class	X_2	X_3
Highway (reference class)	0	0
Shopping mall	1	0
Street	0	1

The two examples show why the term *indicator* is used to describe the numerical variables. Each indicator variable is associated with one class of the qualitative variable, in the sense that it takes the value 1 for that class and is 0 for all other classes of the qualitative variable. As a result, there is one class of the qualitative variable for which every one of the indicator variables has been set to 0. This class is called the *reference class* of the system of indicator variables. We have arbitrarily selected the first class (male, highway) as the reference class in each of the two examples.

Comment   Indicator variables with numerical values other than 0 and 1 can also be used to represent a qualitative variable. In some applications, for example, the values 1 and -1 are used. In this text, we shall use only 0, 1 indicator variables.

Regression Analysis with Independent Indicator Variables

No new calculational problems are encountered when a qualitative independent variable is represented in a regression model by its associated set of indicator variables.

The only new element involves the interpretation of the regression coefficients for the indicator variables. An example will make this clear.

| EXAMPLE | **Restaurant Sales.**   A study for a chain of fast-food restaurants examined the relationship between restaurant sales during a recent period (Y, in thousands of dollars) and number of households in the restaurant's trading area (X_1, in thousands) and location of restaurant (highway, shopping mall, street). The coding of the two indicator variables representing the location variable is that of Example 2 on page 674. |

The data for the study appear in Table 20.3. We employ regression model (20.3) with three independent variables.

$$(20.23) \qquad Y_i = \beta_0 + \beta_1 X_{i1} + \beta_2 X_{i2} + \beta_3 X_{i3} + \varepsilon_i$$

The response function is as follows.

$$(20.24) \qquad E\{Y\} = \beta_0 + \beta_1 X_1 + \beta_2 X_2 + \beta_3 X_3$$

TABLE 20.3   Data for the restaurant sales example

Restaurant i	Number of Households X_{i1} (thousand)	Location Qualitative Class	X_{i2}	X_{i3}	Sales Volume Y_i ($ thousand)
1	155	Highway	0	0	135.27
2	93	Highway	0	0	72.74
3	128	Highway	0	0	114.95
4	114	Highway	0	0	102.93
5	158	Highway	0	0	131.77
6	183	Highway	0	0	160.91
7	178	Mall	1	0	179.86
8	215	Mall	1	0	220.14
9	172	Mall	1	0	179.64
10	197	Mall	1	0	185.92
11	207	Mall	1	0	207.82
12	95	Mall	1	0	113.51
13	224	Street	0	1	203.98
14	199	Street	0	1	174.48
15	240	Street	0	1	220.43
16	100	Street	0	1	93.19

For highway restaurants, for which $X_2 = 0$ and $X_3 = 0$, the response function reduces to the following form.

(20.24a)
$$E\{Y\} = \beta_0 + \beta_1 X_1 + \beta_2(0) + \beta_3(0) = \beta_0 + \beta_1 X_1 \qquad \text{highway location}$$

For mall restaurants, for which $X_2 = 1$ and $X_3 = 0$, the response function becomes the following.

(20.24b)
$$E\{Y\} = \beta_0 + \beta_1 X_1 + \beta_2(1) + \beta_3(0) = (\beta_0 + \beta_2) + \beta_1 X_1$$
$$\text{mall location}$$

Finally, for street restaurants, for which $X_2 = 0$ and $X_3 = 1$, the response function becomes the following.

(20.24c)
$$E\{Y\} = \beta_0 + \beta_1 X_1 + \beta_2(0) + \beta_3(1) = (\beta_0 + \beta_3) + \beta_1 X_1$$
$$\text{street location}$$

The response functions for the three locations are portrayed in Figure 20.4. Note that each response function is linear, with the same slope β_1. The intercepts for the highway, mall, and street response functions are β_0, $\beta_0 + \beta_2$, and $\beta_0 + \beta_3$, respectively. Because of the parallel response functions, it follows that, *for any given number of households* (X_1), the mean sales volume $E\{Y\}$ of a mall restaurant differs from that of a highway restaurant by β_2 and the mean sales volume of a street restaurant differs from that of a highway restaurant by β_3. Moreover, the mean sales volume $E\{Y\}$ of a mall restaurant differs from that of a street restaurant by $\beta_2 - \beta_3$ for any given number of households (X_1). In Figure 20.4, β_0, β_1, β_2, and β_3 are portrayed as being positive, with β_2 greater than β_3. Also, note in Figure 20.4 how β_2 and β_3 reflect the differential effects of mall and street location, each with respect to the reference class highway location.

A standard multiple regression program was used to fit model (20.23) to the data in Table 20.3. Figure 20.5 presents the output. The fitted model is:

$$\hat{Y} = -1.81683 + 0.87782X_1 + 27.29787X_2 + 7.39208X_3$$

The regression coefficient $b_2 = 27.29787$ indicates that, for any given number of households, the mean sales volume of a mall restaurant is estimated to be \$27.3 thousand more than that of a highway restaurant. The corresponding differential effect for a street restaurant is estimated from b_3 to be \$7.4 thousand. Finally, it follows that the mean sales volume of a mall restaurant is estimated to be \$19.9 thousand more than that of a street restaurant ($b_2 - b_3 = 27.29787 - 7.39208 = 19.9058$) for any given number of households.

FIGURE 20.4 Meaning of indicator variable parameters — Restaurant sales example

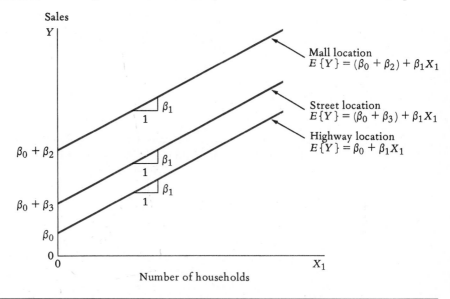

FIGURE 20.5 Computer output for the restaurant sales example

VARIABLE	REG. COEFF.	STD. DEV.	T STAT.	P-VALUE
CONSTANT	-1.81683	5.45267	-.333	.74473
X1	.87782	.03546	24.752	.00000
X2	27.29787	3.62048	7.540	.00001
X3	7.39208	4.17704	1.770	.10216

ANALYSIS OF VARIANCE TABLE

SOURCE	SUM OF SQUARES	DF	MEAN SQUARE
REGRESSION	33438.8567	3	11146.2856
RESIDUAL	403.6035	12	33.6336
TOTAL	33842.4602	15	

F STAT. = 331.40   UPPER-TAIL P-VALUE = .00000

A 90 percent confidence interval for β_2 was desired. It is obtained in the usual fashion from the following:

$$n = 16 \qquad n - p = 12 \qquad t(0.95; 12) = 1.782 \qquad b_2 = 27.29787$$

$$s\{b_2\} = 3.62048$$

Hence, the confidence limits are $27.29787 \pm 1.782(3.62048)$ and the confidence interval is:

$$20.8 \leq \beta_2 \leq 33.7$$

Thus, with 90 percent confidence, we estimate that, for any given number of households in the trading area, the expected sales in a mall location are between \$20.8 thousand and \$33.7 thousand greater than those in a highway location. □

Comment   Regression models containing more than one qualitative independent variable are developed by including a set of indicator variables for each qualitative variable. For example, in a regression of the number of absences of a plant worker (Y) on age of the worker (X_1), sex of the worker (male, female), and work shift (first, second, third), the following regression model might be utilized:

$$Y_i = \beta_0 + \beta_1 X_{i1} + \beta_2 X_{i2} + \beta_3 X_{i3} + \beta_4 X_{i4} + \varepsilon_i$$

where:

X_1 = age of worker

X_2 = 1 if female worker; 0 otherwise

X_3 = 1 if second-shift worker; 0 otherwise

X_4 = 1 if third-shift worker; 0 otherwise

20.9 MODELING CURVILINEAR RELATIONSHIPS

The multiple regression model (20.3):

$$Y_i = \beta_0 + \beta_1 X_{i1} + \cdots + \beta_{p-1} X_{i,p-1} + \varepsilon_i$$

is called the *general linear regression model* because it encompasses many special cases. It is called a *linear model* because it is *linear in the parameters* $\beta_0, \beta_1, \ldots, \beta_{p-1}$. Thus, no parameter appears as an exponent or is multiplied by another parameter. The general linear regression model is not, however, restricted to linear relationships; it encompasses nonlinear relationships, as we shall now see.

Curvilinear relationships between the dependent variable Y and an independent variable X are frequently encountered.

1. Sales (Y) may increase with advertising expenditures (X) but at a decreasing rate as saturation is approached.

2. Crop yield (Y) may increase with the amount of rain (X) up to a point and then decline as the crop becomes adversely affected by excess moisture. □

Quadratic Regression

One of the more widely encountered curvilinear regression relationships is a *quadratic* relationship. The following is a quadratic regression model with one independent variable.

(20.25)

$$Y_i = \beta_0 + \beta_1 X_i + \beta_2 X_i^2 + \varepsilon_i$$

Here, the regression function is $E\{Y\} = \beta_0 + \beta_1 X + \beta_2 X^2$. The regression curve shown in Figure 18.3 for the selling space example is quadratic. Parameter β_1 is called the *linear effect coefficient* and β_2 the *curvature effect coefficient*.

To see that model (20.25) is a special case of the general linear regression model (20.3), let $X_{i1} = X_i$ and $X_{i2} = X_i^2$. We can then write model (20.25) in the standard form:

$$Y_i = \beta_0 + \beta_1 X_{i1} + \beta_2 X_{i2} + \varepsilon_i$$

We now illustrate the analysis of a regression model containing a quadratic effect and show how a standard multiple regression computer program can handle any such model.

EXAMPLE

Job Proficiency. An analyst was asked to investigate whether job proficiency of applicants seeking employment as assemblers could be predicted with satisfactory confidence and precision from the applicants' scores on two tests administered during the interview. Table 20.4 presents a portion of the results for a random sample of 25 applicants who were hired for purposes of the experiment irrespective of their test scores. Here, Y denotes job proficiency score after a learning period, and X_1 and X_2 are the respective scores on manual dexterity and depth perception tests.

From earlier studies of the assembling operation, the analyst knew that manual dexterity and depth perception have linear effects on job proficiency. However, past evidence was inconclusive as to whether or not depth perception also has a curvature effect. Hence, the following model was employed for the study:

$$Y_i = \beta_0 + \beta_1 X_{i1} + \beta_2 X_{i2} + \beta_3 X_{i2}^2 + \varepsilon_i$$

with the intent of examining the data to determine whether or not the curvature effect $\beta_3 X_{i2}^2$ is really needed in the model.

TABLE 20.4   Partial data for the job proficiency example, with a quadratic term added

Applicant i	Manual Dexterity Score X_{i1}	Depth Perception Score X_{i2}	$X_{i3} = X_{i2}^2$	Job Proficiency Score Y_i
1	60	82	6,724	1102
2	135	163	26,569	2333
3	101	61	3,721	1384
⋮	⋮	⋮	⋮	⋮
23	103	102	10,404	1655
24	156	170	28,900	2588
25	77	122	14,884	1513

In entering the data for computation, the analyst employed the transformation $X_{i3} = X_{i2}^2$ to create a third X variable, as shown in Table 20.4. Figure 20.6 presents the computer output obtained by the analyst. The analyst now wished to test for the curvature effect for depth perception:

$H_0: \beta_3 = 0$

$H_1: \beta_3 \neq 0$

Here, H_0 implies that a curvature effect is not present and hence can be dropped from the model, while H_1 implies a curvature effect is present and should be

FIGURE 20.6   Computer output for the job proficiency example

```
VARIABLE          REG. COEFF.        STD. DEV.        T STAT.        P-VALUE

CONSTANT           26.91673          228.39985          .118          .90731
X1                  9.66796            1.43283         6.747          .00000
X2                  9.97013            4.29145         2.323          .03029
X3                  -.03122             .01703        -1.834          .08093


                  ANALYSIS OF VARIANCE TABLE

SOURCE            SUM OF SQUARES        DF        MEAN SQUARE

REGRESSION       3043679.89170          3        1014559.96390
RESIDUAL          407538.66830         21          19406.60325
TOTAL            3451218.56000         24


 F STAT. = 52.279          UPPER-TAIL P-VALUE = .00000
```

retained in the model. The analyst specified that the α risk is to be controlled at 0.10. For $\alpha = 0.10$ and $n - p = 21$, we require $t(0.95; 21) = 1.721$. As shown in Figure 20.6, the standardized test statistic for variable X_3 is $t^* = b_3/s\{b_3\} = -1.834$. Since $|t^*| = 1.834 > 1.721$, the analyst concluded H_1—that a curvature effect is present and should be retained in the model.

The analyst then examined the quadratic regression model for aptness, using various residual plots. Upon being satisfied by the examination, she went on to study how well proficiency scores can be predicted from the two aptitude test scores. □

Comments

1. When the relationship between Y and the independent variables is curvilinear, the response surface is curved rather than a plane. Figure 20.7 shows a computer plot of the estimated quadratic response surface for the job proficiency example. Note how the surface is curved in the direction of the depth perception score axis (X_2) because of the presence in the regression function of a curvature effect for this independent variable.

2. Quadratic regression is a special case of *polynomial regression*. In polynomial regression models, one or more independent variables are present in the model in powers of two or higher, such as X^2, X^3, and so forth. For example, the following is a cubic regression model for one independent variable.

(20.26)
$$Y_i = \beta_0 + \beta_1 X_i + \beta_2 X_i^2 + \beta_3 X_i^3 + \varepsilon_i$$

We see that this model is a special case of the general linear regression model (20.3) by letting $X_{i1} = X_i$, $X_{i2} = X_i^2$, and $X_{i3} = X_i^3$.

Regression with Transformed Variables

Another way to incorporate curvilinear effects in a regression model is to transform one or more of the independent variables, the dependent variable, or both. In Section 19.7, we discussed the use of transformed variables to linearize the regression relation. Equivalently, a linear regression relation in transformed variables can be viewed as a curvilinear relation in the original variables.

EXAMPLE

An engineer studied the cooling of ingots cast in a smelter. The curvilinear regression model utilized relates the rate of heat loss (L) to the temperature differential of the ingot relative to its surroundings (D) and to the ingot's surface area (S), with each variable expressed in logarithms, in the following form:

$$\log L_i = \beta_0 + \beta_1 \log D_i + \beta_2 \log S_i + \varepsilon_i$$

To see that this model is a special case of the general linear regression model (20.3), let $Y_i = \log L_i$, $X_{i1} = \log D_i$, and $X_{i2} = \log S_i$. When this model is viewed in terms of the original variables L, D, and S, the regression relation is curvilinear. □

FIGURE 20.7   Computer plot of the estimated quadratic response surface — Job proficiency example

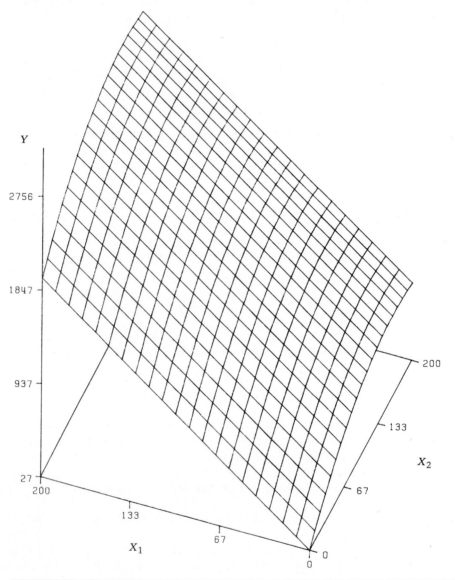

20.10   MODELING INTERACTION EFFECTS

In some regression relationships, the effects of the independent variables on the dependent variable are not additive but rather reinforce or interfere with one another.

EXAMPLES

1. In the promotional expenditures example, the market researcher was interested in whether the two types of promotional expenditures interfere with one another when both are at high levels so that the responsiveness of sales volume to increases in one type of promotional expenditure is reduced when the other type of promotional expenditure is high.

2. A chemical engineer studied the relation between the yield rate of a chemical reaction and the reaction temperature. She knew that the presence of a catalyst in the reaction magnifies the effect of reaction temperature on the yield rate and wished to estimate this reinforcing influence of the catalyst.

Incorporating Interaction Effects in Regression Model

The interference and reinforcement effects described in the two examples are referred to as *interaction effects*. The standard way of incorporating interaction effects in a regression model is by adding *cross-product* or *interaction* terms, such as the term $X_{i1}X_{i2}$ in the following model.

(20.27)
$$Y_i = \beta_0 + \beta_1 X_{i1} + \beta_2 X_{i2} + \beta_3 X_{i1} X_{i2} + \varepsilon_i$$

With this type of model, the effect of X_1 on Y depends on the level of X_2, and similarly, the effect of X_2 on Y depends on the level of X_1. Hence, the effects of X_1 and X_2 on Y are no longer additive.

The fact that the effect of X_1 on Y depends on X_2 in regression model (20.27) can be seen by writing the response function for this model in the following form.

(20.28)
$$E\{Y\} = \beta_0 + \beta_1 X_1 + \beta_2 X_2 + \beta_3 X_1 X_2$$
$$= (\beta_0 + \beta_2 X_2) + (\beta_1 + \beta_3 X_2) X_1$$

For a fixed value of X_2, (20.28) represents a simple linear regression function that relates Y to X_1. Note, however, that both the intercept $(\beta_0 + \beta_2 X_2)$ and the slope $(\beta_1 + \beta_3 X_2)$ of this function depend on the fixed level of X_2. In the same way, the effect of X_2 on Y depends on the level of X_1. Whether the interaction effect is interfering or reinforcing depends on the sign of β_3.

Regression model (20.27) is a special case of the general linear regression model. To see this, let $X_{i3} = X_{i1} X_{i2}$. We can then write model (20.27) in the form of the general linear regression model (20.3).

TABLE 20.5   **Partial data for the promotional expenditures example, with an interaction variable added**

Locality i	Media Expenditures ($ thousand) X_{i1}	Point-of-Sale Expenditures ($ thousand) X_{i2}	Expenditures Interaction $X_{i3} = X_{i1}X_{i2}$	Sales Volume ($ ten thousand) Y_i
1	2	2	4	8.74
2	2	3	6	10.53
⋮	⋮	⋮	⋮	⋮
16	5	5	25	20.51

To test for interaction effects in regression model (20.27), we need only test whether the regression coefficient β_3 of the interaction term equals zero, using the standard test procedure in (20.17).

EXAMPLE

In the promotional expenditures example, the market researcher employed regression model (20.27) to study whether or not interaction effects are present. The data for computation consisted of those in Table 20.1 plus a third X variable, $X_{i3} = X_{i1}X_{i2}$, as illustrated in Table 20.5. Figure 20.8 presents the computer output for this regression analysis. The estimated response function is as follows.

FIGURE 20.8   **Computer output for the promotional expenditures example with an interaction variable added.**   The interaction effect is represented by variable $X_3 = X_1X_2$.

```
VARIABLE        REG. COEFF.      STD. DEV.        T STAT.       P-VALUE

CONSTANT          -.82400        1.18296           -.697         .49935
X1               3.87450          .32196         12.034         .00000
X2               1.55100          .32196          4.817         .00042
X3                -.24150         .08763         -2.756         .01741


                ANALYSIS OF VARIANCE TABLE

SOURCE          SUM OF SQUARES        DF        MEAN SQUARE

REGRESSION        194.94683           3         64.98228
RESIDUAL            2.30352          12           .19196
TOTAL             197.25034          15


  F STAT. = 338.52          UPPER-TAIL P-VALUE = .00000
```

(20.29)

$$\hat{Y} = -0.8240 + 3.8745X_1 + 1.5510X_2 - 0.2415X_1X_2$$

The positive coefficients b_1 and b_2 together with the negative coefficient for the interaction term, $b_3 = -0.2415$, suggest that the two types of promotional expenditures may be interfering with one another, as the researcher suspected. To test whether this effect is indeed present, the researcher used the test statistic in (20.17). The test alternatives are:

$H_0: \beta_3 = 0$ (no interaction present)

$H_1: \beta_3 \neq 0$ (interaction present)

The α risk of the test is to be controlled at 0.05. For $n - p = 12$, we require $t(0.975; 12) = 2.179$. Referring to Figure 20.8, we find $t^* = b_3/s\{b_3\} = -2.756$ for the interaction term $X_3 = X_1X_2$. Since $|t^*| = 2.756 > 2.179$, we conclude H_1—that $\beta_3 \neq 0$ so that the two types of promotional expenditures do interact.

Once the researcher concluded that interaction effects are present, he studied the nature of the interaction effects by plotting the estimated response function relating Y to X_1 for two different levels of X_2. The levels of X_2 selected were $X_2 = 2$, when point-of-sale expenditures are at their lowest level in the experiment, and $X_2 = 5$, when these expenditures are at their highest level. The two estimated response functions are obtained by substituting $X_2 = 2$ and $X_2 = 5$, respectively, in (20.29) and then simplifying the resulting expressions as follows:

$X_2 = 2$: $\hat{Y} = -0.8240 + 3.8745X_1 + 1.5510(2) - 0.2415X_1(2)$

            $= 2.2780 + 3.3915X_1$

$X_2 = 5$: $\hat{Y} = -0.8240 + 3.8745X_1 + 1.5510(5) - 0.2415X_1(5)$

            $= 6.9310 + 2.6670X_1$

These estimated response functions are plotted in Figure 20.9. Evidence of an interaction effect is reflected in the graph by the fact that the two estimated response functions are not parallel. The difference in slopes indicates that an addition to media expenditures (that is, an increase in X_1) is associated with a smaller increase in estimated mean sales when point-of-sale expenditures are high ($X_2 = 5$) than when they are low ($X_2 = 2$). It is in this sense that the two types of promotional expenditures interfere with one another in their relationship with expected sales. □

Interaction Effects Involving Indicator Variables

The variables used to form an interaction or cross-product term can also be indicator variables. An example will illustrate this point.

FIGURE 20.9   **Estimated response functions for the promotional expenditures example with an interaction variable added.**   Nonparallel lines suggest an interaction between the two types of promotional expenditures.

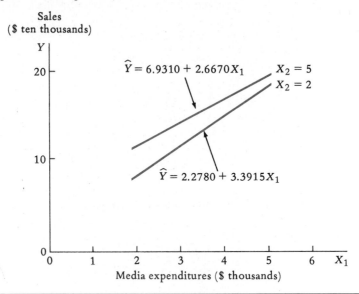

Chemical Reaction.   A chemical engineer utilized regression model (20.27) with interaction effects for relating the yield rate of a chemical reaction (Y, in kilograms per hour) to the reaction temperature (X_1, in degrees Celsius, °C) and whether or not a catalyst is present in the reaction ($X_2 = 1$ if catalyst present, $X_2 = 0$ otherwise). The temperature range in the experiment was 180° to 250°C. The experiment involved $n = 32$ observations. The estimated response function was as follows (the experimental data are not given here):

$$\hat{Y} = 112.0 + 0.103X_1 - 138.3X_2 + 0.607X_1X_2$$
$$\quad\quad\quad (0.00845)\quad (20.67)\quad (0.114)$$

The estimated standard deviations of the regression coefficients are shown in parentheses under their respective coefficients. For the interaction term, we see that $b_3 = 0.607$ and $s\{b_3\} = 0.114$, so that $t^* = 0.607/0.114 = 5.32$. To test whether or not $\beta_3 = 0$ with $\alpha = 0.05$, we require $t(0.975; 28) = 2.048$. Since $|t^*| = 5.32 > 2.048$, we conclude that $\beta_3 \neq 0$, that is, that interaction effects are present.

   To study the nature of the interaction effects, the engineer plotted the estimated response functions corresponding to the catalyst being present ($X_2 = 1$) and not present ($X_2 = 0$):

$$\text{Catalyst present:} \quad \hat{Y} = 112.0 + 0.103X_1 - 138.3(1) + 0.607X_1(1)$$
$$= -26.3 + 0.710X_1$$
$$\text{Catalyst not present:} \quad \hat{Y} = 112.0 + 0.103X_1 - 138.3(0) + 0.607X_1(0)$$
$$= 112.0 + 0.103X_1$$

These estimated response functions are shown in Figure 20.10. Note that the estimated mean yield rate increases with temperature whether the catalyst is present or not but that the amount of increase (that is, the slope) is much greater when the catalyst is present. □

OPTIONAL TOPIC—MULTICOLLINEARITY

In some regression applications, a major objective is to measure the separate effects of the independent variables on the dependent variable. Generally, the regression coefficients are utilized for this purpose because the regression coefficient β_k in the general linear model (20.3) indicates the change in the mean of the distribution of Y when X_k changes by one unit and the other independent variables remain constant.

FIGURE 20.10   Estimated response functions with interaction effects present—Chemical reaction example

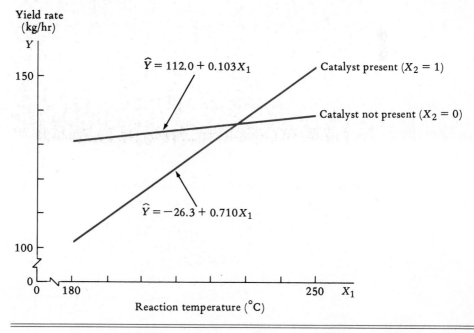

Unfortunately, the separate effects of the different independent variables cannot usually be measured satisfactorily when the sample observations of the independent variables are highly correlated among themselves. Multicollinearity is then said to exist.

(20.30)   *Multicollinearity* is present in a regression analysis when the sample observations of the independent variables, or linear combinations of them, are highly correlated.

Effects of Multicollinearity

When multicollinearity prevails, two major, related problems are encountered in assessing the separate effects of the different independent variables:

1. The estimated regression coefficient b_k for the independent variable X_k may vary substantially, depending on which other independent variables are included in the model. Thus, the value obtained for b_k in any particular fitted model does not indicate the effect of X_k on $E\{Y\}$ in any absolute sense.
2. The estimated regression coefficients tend to have extremely large sampling errors, indicating that they vary widely in repeated samples. As a result, they will give very imprecise information about the regression parameters.

Fortunately, the difficulties just cited for the regression coefficients generally do not carry over to inferences on Y. As long as inferences are made within the region of sample observations on the independent variables, multicollinearity usually causes no special problems in estimating a mean response or predicting a new observation.

We shall illustrate the effects of multicollinearity by an example.

EXAMPLE

Machine Shop.   A machine shop receives rough shafts and smoothes them to specifications. When a shaft is smoothed, it is first carefully mounted on a machine; then, excess metal is machined off. The time required to process a shipment of shafts is affected by the time required to mount the shafts for smoothing and by the time required to machine the shafts once they have been mounted. The total mounting time depends on the number of shafts in the shipment, while the total machining time depends on the weight of the shipment (since the aggregate amount of metal to be machined off varies with the total weight).

Thus, two physically distinct variables affect the processing time for a shipment, namely, the number of shafts in the shipment and the total weight of the shipment. Table 20.6 shows, for the 15 most recent shipments, the number of shafts in the shipment (X_1), the total weight of the shipment (X_2), and the processing time for the shipment (Y). The independent variables are highly correlated, as the scatter plot in Figure 20.11 shows. The coefficient of simple correlation between X_1 and X_2 is above 0.99.

TABLE 20.6 **Data for the machine shop example**

Shipment i	Number of Shafts X_{i1}	Total Weight (kilograms) X_{i2}	Processing Time (minutes) Y_i
1	55	5563	1738
2	20	2041	491
3	35	3594	999
4	45	4523	1370
5	40	4082	1150
6	25	2534	684
7	55	5556	1650
8	30	3044	876
9	60	6095	1910
10	45	4561	1380
11	35	3562	995
12	25	2546	660
13	45	4576	1390
14	35	3529	1025
15	30	3056	821

FIGURE 20.11 **Region of sample observations for the independent variables in the machine shop example.** The (X_1, X_2) observations fall in a narrow band here, the coefficient of correlation between them exceeding 0.99.

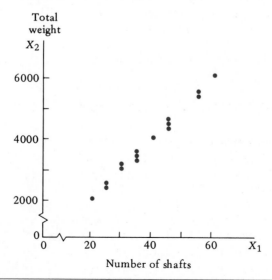

We have fitted three regression models to the data on the 15 shipments, as follows:

Fitted Model	Independent Variables in Model
1	X_1
2	X_1, X_2
3	X_2

Key results are shown in Table 20.7. They illustrate the three points that we made earlier about the effects of multicollinearity:

1.   The point estimate b_1 changes substantially when X_2 is added to the model. When the fitted model contains X_1 alone, we obtain $b_1 = 34.9$; but when the model contains both X_1 and X_2, we obtain $b_1 = 57.8$. Similarly, the value of b_2 changes when X_1 is added. In fact, the sign of the coefficient changes.

Thus, b_k does not indicate the effect of X_k on $E\{Y\}$ in any absolute sense when multicollinearity is present.

2.   Table 20.7 also shows that the estimated standard deviation of b_k tends to become larger when other variables highly correlated with X_k are added to the fitted model. Thus, when X_1 is the only independent variable in the model, we have $s\{b_1\} = 0.62818$; but when both X_1 and X_2 are included, we have $s\{b_1\} = 39.51970$. Similarly, $s\{b_2\}$ increases from 0.00667 to 0.39187 when X_1 is added to the model.

3.   Multicollinearity causes no special problems when inferences on Y are made within the region of sample observations on the independent variables, as shown by the estimation of the mean response when $X_{h1} = 40$ and $X_{h2} = 4060$. Note from Figure 20.11 that this pair of values falls within the region of sample observations on the independent variables. As Table 20.7 shows, the estimated means $\hat{Y}_h$ and the estimated standard deviations $s\{\hat{Y}_h\}$ are practically the same for

TABLE 20.7   Results for three fitted regression models for the machine shop example. Independent variables are highly correlated (multicollinearity present).

Statistic	Fitted Model 1: X_1	Fitted Model 2: X_1, X_2	Fitted Model 3: X_2
b_1	34.85684	57.75047	—
$s\{b_1\}$	0.62818	39.51970	—
b_2	—	−0.22704	0.34553
$s\{b_2\}$	—	0.39187	0.00667
	$X_{h1} = 40, X_{h2} = 4060$		
$\hat{Y}_h$	1189.0758	1188.7534	1189.5460
$s\{\hat{Y}_h\}$	7.3437	7.5593	7.8622

the three fitted models. This shows, incidentally, that in the presence of high correlation between X_1 and X_2, either variable alone contains much of the information provided by the other variable. □

Comments	1.	Note that $s\{\hat{Y}_h\}$ for the fitted model containing both X_1 and X_2 is of about the same magnitude as when only one of the independent variables is utilized despite the much larger sampling errors for b_1 and b_2. The reason is that these large sampling errors tend to be offsetting here.
	2.	The outcome of a test of whether or not $\beta_k = 0$ in the presence of multicollinearity depends very much on which other independent variables are included in the regression model. For instance, note in Table 20.7 for the fitted model containing both X_1 and X_2 that one would conclude that $\beta_1 = 0$ or that $\beta_2 = 0$, for any usual level of the α risk. Yet $R^2 = 0.996$ here. As noted previously, the test for $\beta_k = 0$ is a marginal test and depends on the marginal effect of X_k, given that the other variables are already in the model. When X_1 and X_2 are highly correlated, as here, the marginal effect of either variable may be very small when the other variable is already in the model, yet there may be a strong relation between X_1 and Y and between X_2 and Y.

Uncorrelated Independent Variables

When the independent variables are uncorrelated, the values of the regression coefficients and of their estimated standard deviations do not depend on which other variables are included in the regression model.

EXAMPLE

In the promotional expenditures example, the two types of promotional expenditures were set up so that X_1 and X_2 are uncorrelated. The basic data were presented in Table 20.1. Table 20.8 shows pertinent results for three regression models fitted to these data, as follows:

Fitted Model	Independent Variables in Model
1	X_1
2	X_1, X_2
3	X_2

To simplify the example, we do not consider any interaction effects in these models.

Table 20.8 illustrates two important points:

1. The values of b_1 and b_2 are unaffected by whether or not the other independent variable is included in the model. This is true in general for any number of independent variables that are mutually uncorrelated.

2. Neither $s\{b_1\}$ nor $s\{b_2\}$ becomes enlarged when the other variable is added to the model. □

TABLE 20.8   **Results for three fitted regression models for the promotional expenditures example.**   Independent variables are uncorrelated (no multicollinearity present).

Statistic	Fitted Model 1: X_1	Fitted Model 2: X_1, X_2	Fitted Model 3: X_2
b_1	3.02925	3.02925	—
$s\{b_1\}$	0.22139	0.12028	—
b_2	—	0.70575	0.70575
$s\{b_2\}$	—	0.12028	0.81786

The stability of the regression coefficients when the independent variables are mutually uncorrelated makes it desirable to use experimental control to obtain uncorrelated independent variables. Unfortunately, experimental control is often not feasible, and one may then have to employ correlated independent variables.

Remedial Measures

Remedial measures for the difficulties caused by multicollinearity can be employed at times. Sometimes, one can obtain additional observations that break the pattern of multicollinearity. In other cases, one can estimate some of the regression coefficients from other data. Thus, in the machine shop example, it may be feasible to estimate β_1 by a time-and-motion study. On still other occasions, one may be able to transform some of the independent variables or to drop some from the model in order to lessen the degree of multicollinearity. In all cases, however, one must be cautious in the interpretation of regression coefficients in the presence of multicollinearity.

20.12   OPTIONAL TOPIC — SELECTING THE REGRESSION MODEL

The regression models used in the examples so far have been specified in advance, and there has been little discussion of whether an alternative model with other independent variables or a different functional relationship might have been better. Unfortunately, theoretical knowledge and conceptual understanding of the problem at hand is not sufficient in many applications to provide an exact specification of the most appropriate regression model in advance. Instead, there exist frequently a multitude of potential independent variables that may be helpful in the regression model. Typically, in business, economic, and behavioral applications, these independent variables are highly intercorrelated. As we noted in Section 20.11, it is then very difficult to isolate the effect of an independent variable on the dependent variable. This makes the task of identifying the independent variables that should be included in the model very difficult.

Generally, we would like to include in the regression model only a small subset of all the potential independent variables—a subset that captures most of the explanatory or predictive information available in the whole set of potential independent variables. Since there may be a number of such subsets that are useful, the objective frequently is to develop a few reasonable alternative subsets and to study these intensively before deciding on the final regression model to be employed.

All-Possible-Regressions Method

A number of methods have been developed for selecting a subset of independent variables to be included in the regression model. Here, we describe one method, called the all-possible-regressions selection method.

(20.31)

> The *all-possible-regressions method* of selecting a regression model consists of the following three steps:
>
> *Step 1.* Initially develop a pool of potentially useful independent variables.
> *Step 2.* Fit a regression model for each possible combination of independent variables in the pool developed in step 1.
> *Step 3.* Order the regression models in step 2 by some criterion, and evaluate in detail a small number of the models that are rated as best according to the criterion.

The criterion we shall employ here is that of minimizing the error mean square MSE for the regression model. This criterion considers regression models to be good when the variation of the observations around the fitted regression function is small. It can be shown that the criterion of minimizing MSE is equivalent to maximizing R_a^2, the adjusted coefficient of multiple determination, as defined in (20.12).

Once the criterion of minimizing MSE has led to the identification of several regression models that are good according to this criterion, expert judgment taking into account a variety of other considerations is then used to identify the final regression model to be employed.

EXAMPLE

Property Selling Price. A real estate expert is interested in developing a regression model that relates the selling price of suburban residential properties to characteristics of the properties. Her interest lies in a new, large residential property development on the outskirts of a major city for which she has data on 30 properties that were sold recently. She developed a long list of possible explanatory variables including lot size, property taxes, assessed property value, various house characteristics (such as floor area, number of rooms, style of construction), and traffic access. After carefully sifting through this long list, she decided to include five potential independent variables in the pool:

Variable	Description
X_1	Property taxes (annual taxes, in dollars)
X_2	House size (floor area, in square feet)
X_3	Lot size (in acres)
X_4	Lot size squared ($X_4 = X_3^2$)
X_5	Attractiveness index (a composite measure of ratings for different features of the property)

The reason for including variable X_4 is that there may be a curvilinear effect of lot size on selling price. The data for these independent variables and the dependent variable selling price are given in Table 20.9 for the 30 properties.

Using a computer package that produces regression results for all combinations of independent variables from a prespecified list, she obtained the value of *MSE* for each combination of the five independent variables in the pool. A total of $2^5 = 32$ regression models were analyzed, including the model with an intercept only. The *MSE* values for these 32 models are displayed in Figure 20.12. For each number of independent variables included in the model, the *MSE* values are arrayed in descending order. Figure 20.13 contains a plot of the *MSE* values against

FIGURE 20.12 All-possible-regressions method results for the property selling price example. Values of *MSE* for all possible regression models involving subsets of five independent variables are presented.

VARIABLES IN MODEL	MSE	VARIABLES IN MODEL	MSE
INTERCEPT ONLY	3195	X3 X4 X5	3006
		X1 X3 X4	2277
X4	3262	X1 X4 X5	1906
X3	3260	X1 X2 X4	1843
X5	2877	X1 X2 X3	1812
X1	2669	X2 X3 X4	1792
X2	2013	X1 X3 X5	1791
		X1 X2 X5	1158
X3 X4	3381	X2 X3 X5	1044
X3 X5	2928	X2 X4 X5	1037
X4 X5	2915		
X1 X5	2297	X1 X2 X3 X4	1861
X1 X4	2231	X1 X3 X4 X5	1833
X1 X3	2193	X2 X3 X4 X5	1078
X1 X2	1786	X1 X2 X4 X5	1049
X2 X4	1785	X1 X2 X3 X5	1033
X2 X3	1746		
X2 X5	1408	X1 X2 X3 X4 X5	1071

TABLE 20.9 Data for the property selling price example

Property i	Property Taxes (dollars) X_{i1}	House Size (square feet) X_{i2}	Lot Size (acres) X_{i3}	Lot Size Squared X_{i4}	Attractiveness Index X_{i5}	Selling Price ($000) Y_i
1	7337	3000	3.6	12.96	64	550
2	4204	2300	1.2	1.44	69	461
3	5574	3300	1.3	1.69	72	501
4	5924	2100	3.2	10.24	71	455
5	5182	3900	1.1	1.21	40	503
6	5932	3100	2.0	4.00	74	529
7	5966	3600	1.6	2.56	69	478
8	5574	2900	2.5	6.25	85	562
9	4927	2000	2.6	6.76	70	417
10	5025	3500	1.3	1.69	74	566
11	6210	3100	2.3	5.29	79	494
12	5425	3200	1.5	2.25	75	515
13	4178	2800	1.3	1.69	62	490
14	7048	3300	3.3	10.89	62	537
15	7540	3000	3.9	15.21	70	527
16	5807	3400	2.4	5.76	81	577
17	4875	2800	1.7	2.89	77	490
18	5540	2000	3.4	11.56	67	486
19	5980	2400	2.9	8.41	68	450
20	7324	3600	2.9	8.41	84	674
21	4582	2400	1.9	3.61	75	454
22	6759	3000	2.8	7.84	63	523
23	6444	2200	3.6	12.96	78	469
24	6307	3600	2.4	5.76	73	628
25	4285	2900	1.1	1.21	85	570
26	7722	3000	4.4	19.36	69	564
27	4917	3100	1.8	3.24	54	444
28	5068	2200	2.1	4.41	75	494
29	6500	2500	3.9	15.21	61	479
30	4612	2900	1.1	1.21	74	477

the number of independent variables ($p - 1$) in the model. The independent variables in the best model for each given number of independent variables are identified in Figure 20.13. These minimum *MSE* points are connected by straight lines. Note, for example, that the best regression model with three independent variables ($p - 1 = 3$) contains variables X_2, X_4, and X_5. Figure 20.12 shows that $MSE = 1037$ for this model.

The expert then reviewed the regression output for a few models with *MSE* values close to the minimum. From her assessment of the specific variables in the

FIGURE 20.13   Plot of *MSE* against the number of independent variables for all possible regression models — Property selling price example

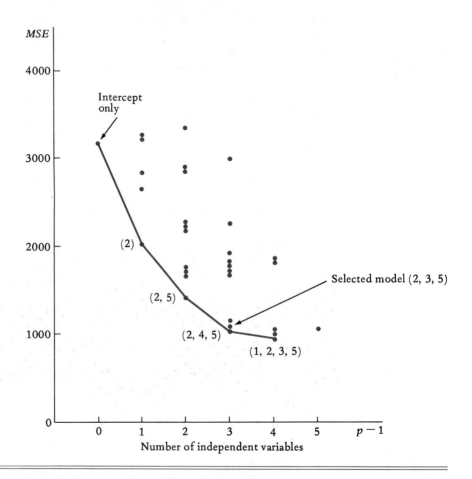

model, the signs, magnitudes, and standard errors of their regression coefficients, and other checks on the aptness of the model, the expert chose the model with variables X_2, X_3, and X_5 for use in subsequent studies. The point corresponding to this model is shown by an arrow in Figure 20.13. Although this model does not give the smallest value of *MSE*, it is a close competitor and contains fewer independent variables than several of the other leading candidates. The three variables in the final model are house size (X_2), lot size (X_3), and attractiveness index (X_5). □

20.13 OPTIONAL TOPIC—*F* TEST FOR SEVERAL REGRESSION COEFFICIENTS

At times, one wishes to determine whether or not a subset of two or more independent variables in a multiple regression model can be dropped from the model, given the presence of the other independent variables in the model. In effect, one wishes to test whether or not all of the regression coefficients in the subset of independent variables equal zero.

An *F* test is available for this purpose. This test involves choosing between two alternative regression models, one that contains all of the independent variables (called the *full model*) and one in which a subset of the independent variables has been dropped (called the *reduced model*).

EXAMPLE   **Arthritis Incidence.**   A market analyst for a pharmaceutical manufacturer studied, for 13 districts, the relation between the incidence of arthritis in the district (*Y*, in cases per thousand population) and the independent variables:

X_1: percent of district population over 65 years of age

X_2: number of physicians in district per thousand population

X_3: mean disposable income of persons over 65 years of age in district (in \$ thousand)

The analyst believed that regression model (20.3) is appropriate. He knew from previous studies that the incidence of arthritis is related to the percent of the population over 65 but was unsure whether number of physicians and mean disposable income are also related to the dependent variable. The analyst therefore wished to choose between:

Full model: $E\{Y\} = \beta_0 + \beta_1 X_1 + \beta_2 X_2 + \beta_3 X_3$

Reduced model: $E\{Y\} = \beta_0 + \beta_1 X_1$

The corresponding test alternatives in terms of the regression coefficients in the full model are:

H_0: $\beta_2 = \beta_3 = 0$

H_1: not both β_2 and β_3 equal zero   ☐

When one is considering whether several independent variables can be simultaneously dropped from the regression model, as in the arthritis incidence example, it is inappropriate to make the decision on the basis of separate applications of test (20.17) for individual regression coefficients when the independent variables are correlated among themselves, as is often the case.

Test Procedure

To test properly whether a subset of independent variables can be dropped from the model, one must consider whether adding this subset of variables to the regression model leads to a significant reduction of the error sum of squares. This involves fitting both the full and the reduced regression models, as follows:

1. Fit the full regression model to the data. Let $SSE(F)$ and df_F denote the associated error sum of squares and its degrees of freedom, respectively.
2. Fit the reduced regression model to the data. Let $SSE(R)$ and df_R denote the associated error sum of squares and its degrees of freedom, respectively.
3. Calculate the following test statistic F^*, which is a function of the amount of reduction in the error sum of squares.

(20.32)
$$F^* = \left(\frac{df_F}{df_R - df_F} \right) \left[\frac{SSE(R) - SSE(F)}{SSE(F)} \right]$$

Large values of F^* are consistent with the full model because the reduction in the error sum of squares is large in that case. Small values of the test statistic are consistent with the reduced model. The decision rule for the test therefore takes the following form.

(20.33)

When the alternatives are:

H_0: all β_k for a subset of variables equal zero
H_1: not all β_k for the subset equal zero

and multiple regression model (20.3) is applicable, the appropriate decision rule to control the α risk is as follows:

If $F^* \leq F(1 - \alpha; df_R - df_F, df_F)$, conclude H_0.
If $F^* > F(1 - \alpha; df_R - df_F, df_F)$, conclude H_1.

where:

F^* is given by (20.32)

| EXAMPLE |

The data set for the arthritis incidence example is given in Table 20.10. The test alternatives here are:

H_0: $\beta_2 = \beta_3 = 0$

H_1: not both β_2 and β_3 equal zero

TABLE 20.10   Data for the arthritis incidence example

District i	Percent of Population over 65 X_{i1}	Number of Physicians (per 1000) X_{i2}	Mean Disposable Income for over 65 ($000) X_{i3}	Arthritis Incidence (cases per 1000) Y_i
1	10.4	1.36	16.5	8.70
2	14.8	1.49	13.8	10.44
3	12.5	1.50	13.2	9.82
4	18.0	1.91	20.6	12.38
5	12.3	1.62	15.8	8.25
6	15.6	1.25	22.6	11.04
7	10.1	1.09	19.4	7.89
8	12.5	1.99	16.1	10.08
9	12.8	1.04	15.9	9.35
10	14.9	1.72	16.2	11.57
11	11.2	1.13	13.2	7.29
12	15.6	1.34	20.7	11.49
13	11.3	1.98	14.0	10.22

Thus, the two models to be fitted are:

Full model: $E\{Y\} = \beta_0 + \beta_1 X_1 + \beta_2 X_2 + \beta_3 X_3$

Reduced model: $E\{Y\} = \beta_0 + \beta_1 X_1$

The α risk is to be controlled at 0.01.

Computer output from the regression analysis of the full model is given in Figure 20.14a. Corresponding output for the reduced model is given in Figure 20.14b. Referring to the ANOVA table in each case, we find the following:

$$SSE(F) = 4.51567 \qquad df_F = 9$$

$$SSE(R) = 6.96186 \qquad df_R = 11$$

Thus, adding variables X_2 and X_3 to the regression model that already contains X_1 reduces the error sum of squares by:

$$SSE(R) - SSE(F) = 6.96186 - 4.51567 = 2.44619$$

For $\alpha = 0.01$, we require $F(0.99; 2, 9) = 8.02$. The decision rule is as follows:

If $F^* \leq 8.02$, conclude H_0.
If $F^* > 8.02$, conclude H_1.

FIGURE 20.14   Computer output for the full and reduced regression models — Arthritis incidence example

(a) Computer output for full model

$$E\{Y\} = \beta_0 + \beta_1 X_1 + \beta_2 X_2 + \beta_3 X_3$$

VARIABLE	REG. COEFF.	STD. DEV.	T STAT.	P-VALUE
CONSTANT	.43914	1.57976	.278	.78731
X1	.46963	.11035	4.256	.00212
X2	1.49976	.67926	2.208	.05463
X3	.05921	.08163	.725	.48664

ANALYSIS OF VARIANCE TABLE

SOURCE	SUM OF SQUARES	DF	MEAN SQUARE
REGRESSION	23.99884	3	7.99961
RESIDUAL	4.51567	9	.50174
TOTAL	28.51451	12	

F STAT. = 15.944   UPPER-TAIL P-VALUE = .00060

(b) Computer output for reduced model

$$E\{Y\} = \beta_0 + \beta_1 X_1$$

VARIABLE	REG. COEFF.	STD. DEV.	T STAT.	P-VALUE
CONSTANT	2.38250	1.30464	1.826	.09506
X1	.56714	.09719	5.836	.00011

ANALYSIS OF VARIANCE TABLE

SOURCE	SUM OF SQUARES	DF	MEAN SQUARE
REGRESSION	21.55264	1	21.55264
RESIDUAL	6.96186	11	.63290
TOTAL	28.51451	12	

F STAT. = 34.054   UPPER-TAIL P-VALUE = .00011

Test statistic (20.32) here is:

$$F^* = \left(\frac{9}{11-9}\right)\left(\frac{2.44619}{4.51567}\right) = 2.44$$

Since $F^* = 2.44 \le 8.02$, we conclude H_0—that number of physicians (X_2) and mean disposable income (X_3) taken together can be dropped from the regression model when the model already contains the variable X_1, the percent of the population over 65. $\square$

Comments
1. The test of whether *all* $\beta_k = 0$ ($k = 1, 2, \ldots, p - 1$) in (20.15) is a special case of the test for a reduced model in (20.33). The equivalence is seen by noting that in this case $SSE(R) = SSTO$, $SSE(F) = SSE$, $df_R = n - 1$, and $df_F = n - p$.
2. A test of whether an *individual* β_k equals zero also is a special case of the test for a reduced model in (20.33). In this case, t^* as defined in (20.17) and F^* as defined in (20.32) are related by $F^* = (t^*)^2$.

PROBLEMS

*20.1 Suppose that a firm's earnings (Y, in millions of dollars) are related to its sales revenues (X_1, in millions of dollars) and its total long-term debt (X_2, in millions of dollars) according to regression model (20.1), and that the parameter values are $\beta_0 = -5$, $\beta_1 = 0.15$, $\beta_2 = -0.08$, $\sigma^2 = 4$.
   a. State the regression function.
   b. Interpret β_1 and σ here.
   c. For $X_{i1} = 100$ and $X_{i2} = 50$, determine (1) the value of $E\{Y_i\}$, (2) the value of Y_i when $\varepsilon_i = -8$, (3) the probability that Y_i is less than 3.

20.2 Consider regression model (20.1) and suppose that $E\{Y\} = -10 + 2X_1 - 3X_2$ and $\sigma^2 = 1$.
   a. Graph the relationship between $E\{Y\}$ and X_2 when $X_1 = 6$. How is β_2 interpreted here?
   b. Describe the probability distribution of Y when $X_1 = 6$ and $X_2 = 3$.

20.3 Consider regression model (20.1) and suppose that $E\{Y\} = 15 - X_1 + 4X_2$ and $\sigma^2 = 4$.
   a. Graph the relationship between $E\{Y\}$ and X_2 when $X_1 = 10$. How is β_2 interpreted here?
   b. Describe the probability distribution of Y when $X_1 = 10$ and $X_2 = 5$.

*20.4 **Fermentation Process.** A chemical plant makes an organic product by means of a fermentation process. The amount of product obtained in a single production run is determined by a distillation procedure that takes several days to complete. To study the factors affecting the amount of product, an analyst has fitted regression model (20.3) from data for the last 23 runs. The dependent variable is amount of product (Y, in liters) and the three independent variables are the peak temperature reached in the run (X_1, in degrees Celsius, °C), the number of hours before the fermentation ceases (X_2), and the level of trace impurities in the key raw material used in the run (X_3, in parts per million). The correlation matrix and key results of fitting the regression model follow.

$$
\begin{array}{c}
\quad\quad X_1 \quad\quad X_2 \quad\quad X_3 \quad\quad Y \\
\begin{array}{c} X_1 \\ X_2 \\ X_3 \\ Y \end{array}
\begin{bmatrix}
1.0000 & 0.4666 & 0.4977 & -0.7737 \\
 & 1.0000 & 0.7945 & -0.5874 \\
 & & 1.0000 & -0.6023 \\
 & & & 1.0000
\end{bmatrix}
\end{array}
$$

k:	0	1	2	3	
					$n = 23$
b_k:	162.876	−1.2103	−0.66591	−8.6130	$SSR = 4133.633$
$s\{b_k\}$:	25.776	0.3015	0.82100	12.2413	$SSE = 2011.584$

a. Which independent variable is most highly correlated with Y?

b. Give the values of (1) r_{Y2}, (2) r_{13}, (3) r_{23}. Interpret r_{23}.

20.5 **Freezer Power Use.** An experiment was conducted for two weeks to study the effect of freezer temperature (X_1) and freezer storage density (X_2) on freezer power use (Y, in kilowatthours) for a certain type of commercial freezer. The independent variables are measured in terms of deviations from the levels normally used. Thus, in the first case, the temperature setting was 10°C below the normal setting and the storage density was 10 percent points less than the normal density. The study results follow.

i:	1	2	3	4	5	6	7	8	9
X_{i1}:	−10	0	+10	−10	0	+10	−10	0	+10
X_{i2}:	−10	−10	−10	0	0	0	+10	+10	+10
Y_i:	228	204	142	247	190	149	248	219	180

Regression model (20.1) was employed, and the following correlation matrix and key regression results were obtained:

$$
\begin{array}{c}
\quad\quad X_1 \quad\quad X_2 \quad\quad Y \\
\begin{array}{c} X_1 \\ X_2 \\ Y \end{array}
\begin{bmatrix}
1.0000 & 0 & -0.9340 \\
 & 1.0000 & 0.2706 \\
 & & 1.0000
\end{bmatrix}
\end{array}
$$

k:	0	1	2	
				$n = 9$
b_k:	200.778	−4.2000	1.2167	$SSR = 11,472.167$
$s\{b_k\}$:	3.500	0.4286	0.4286	$SSE = 661.389$

a. Which independent variable is more highly correlated with Y?

b. Give the value of r_{12}. Interpret this measure.

c. Does the sign of r_{Y1} indicate that power use tends to increase with lower freezer temperature? Explain.

20.6 **Hotel Equipment.** A student intern in a hotel equipment and supply firm studied, for eight of the firm's sales territories, the relation of the number of hotel rooms in the territory (X_1, in thousands) and an index of business activity for the territory (X_2) to the firm's annual sales in the territory (Y, in $000). The data follow.

i:	1	2	3	4	5	6	7	8
X_{i1}:	2.70	3.13	1.64	1.25	1.38	2.17	3.77	3.46
X_{i2}:	128	82	96	107	88	91	96	112
Y_i:	88.7	77.4	71.8	73.2	71.0	75.9	85.8	87.9

The intern assumed that regression model (20.1) is appropriate and obtained the following correlation matrix and key regression results:

$$
\begin{array}{c}
 \quad\quad X_1 \quad\quad X_2 \quad\quad Y \\
\begin{array}{c} X_1 \\ X_2 \\ Y \end{array}
\left[\begin{array}{ccc}
1.0000 & 0.1163 & 0.8274 \\
 & 1.0000 & 0.6386 \\
 & & 1.0000
\end{array}\right]
\end{array}
$$

k:	0	1	2	
b_k:	37.6331	5.7996	0.27193	$SSR = 374.605$
$s\{b_k\}$:	3.0286	0.4492	0.02925	$SSE = 6.5738$

$n = 8$

a. Give the values of (1) r_{12}, (2) r_{Y1}, (3) r_{Y2}. Interpret r_{Y2}.
b. Which independent variable is more highly correlated with Y? What percentage reduction in the variability of Y would result if this independent variable were used alone to predict Y?

*20.7 Refer to **Fermentation Process** Problem 20.4.
   a. State the estimated regression function. Interpret b_1.
   b. Obtain a point estimate of $E\{Y_h\}$ when $X_{h1} = 38$, $X_{h2} = 55$, and $X_{h3} = 2.2$.

20.8 Refer to **Freezer Power Use** Problem 20.5.
   a. State the estimated regression function.
   b. Obtain point estimates of the change in expected power use (1) when freezer temperature is reduced 1°C and density is held fixed, (2) when freezer temperature is reduced 10°C and density is held fixed.
   c. Obtain a point estimate of the expected power use when freezer temperature and storage density are both at their normal levels.

20.9 Refer to **Hotel Equipment** Problem 20.6.
   a. State the estimated regression function. Interpret b_1. In what units is b_1 expressed?
   b. Obtain a point estimate of mean sales in territories where the number of rooms is 1.25 thousand and the index of business activity is 107.

*20.10 Refer to **Fermentation Process** Problem 20.4.
   a. Set up the ANOVA table. What parameter in regression model (20.3) is estimated by MSE?
   b. Calculate R^2 and R from the ANOVA table. Interpret R^2.
   c. Calculate R_a^2 and R_a. What does the latter value take into account that R does not? Is the difference between R_a and R substantial here?

20.11 Refer to **Freezer Power Use** Problem 20.5.
   a. Set up the ANOVA table.
   b. Obtain $\sqrt{MSE}$. The estimated mean power use when freezer temperature and storage density are both at their normal levels is 200.778. Estimate the coefficient of variation of the probability distribution of power use in this case.
   c. Calculate R^2 and R. Interpret R^2.
   d. Calculate R_a^2 and R_a. Must R_a^2 necessarily be smaller than R^2, as it is here? Why?

20.12 Refer to **Hotel Equipment** Problem 20.6.
  a. Set up the ANOVA table.
  b. Obtain a point estimate of σ. In what units is this estimate expressed?
  c. Calculate R^2 and R. Why would one expect R^2 to be larger than either r_{Y1}^2 or r_{Y2}^2, as is the case here?
  d. Calculate R_a^2 and R_a.
  e. The intern added another business indicator (X_3) to the model and obtained $R_a^2 = 0.971$ when Y is regressed on X_1, X_2, and X_3. Compare the values of R_a^2 with and without the added variable X_3. Has SSE decreased relatively more or less than $n - p$ with the addition of X_3 to the regression model?

*20.13 Refer to **Fermentation Process** Problem 20.4. Note that $SSE(X_1, X_2, X_3) = 2011.584$. When Y is regressed on X_1 and X_2 only, the error sum of squares is $SSE(X_1, X_2) = 2063.998$.
  a. What does the difference $SSE(X_1, X_2) - SSE(X_1, X_2, X_3)$ measure here?
  b. Obtain $r_{Y3.12}^2$ and interpret its meaning here. Also obtain $r_{Y3.12}$.

20.14 Refer to **Freezer Power Use** Problem 20.5. Note that $SSE(X_1, X_2) = 661.389$. When Y is regressed on X_1 only, the error sum of squares is $SSE(X_1) = 1549.556$. Likewise, when Y is regressed on X_2 only, $SSE(X_2) = 11{,}245.389$. Obtain (1) $r_{Y2.1}^2$, (2) $r_{Y1.2}^2$. Interpret these measures.

20.15 Refer to **Hotel Equipment** Problem 20.6. Note that $SSE(X_1, X_2) = 6.5738$. The coefficient of simple correlation between Y and X_1 is $r_{Y1} = 0.8274$.
  a. Obtain $SSE(X_1)$.
  b. Obtain $r_{Y2.1}^2$ and interpret its meaning. Also obtain $r_{Y2.1}$.

*20.16 Refer to **Fermentation Process** Problem 20.4. The fitted values and residuals are as follows:

i:	1	2	3	4	5	6	7	8
$\hat{Y}_i$:	48.6	68.9	63.6	68.4	84.8	42.6	59.8	55.8
e_i:	-0.6	-11.9	2.4	1.6	4.2	-6.6	-13.8	-1.8

i:	9	10	11	12	13	14	15	16
$\hat{Y}_i$:	33.7	76.4	75.1	54.9	61.3	68.0	43.8	69.4
e_i:	-7.7	0.6	13.9	12.1	-14.3	-17.0	13.2	-3.4

i:	17	18	19	20	21	22	23
$\hat{Y}_i$:	64.1	80.8	72.2	41.7	73.3	46.0	57.7
e_i:	14.9	7.2	-12.2	7.3	3.7	6.0	2.3

  a. Plot the residuals against the fitted values. Analyze the plot for aptness of the regression function and constancy of the error variance. State your findings.
  b. Obtain the standardized residuals. Construct a table like Table 19.1, based on the 10th, 50th, and 90th percentiles of the relevant t distribution. Do the error terms appear to be normally distributed? Explain. (*Hint:* The relevant degrees of freedom for the t distribution are the degrees of freedom associated with MSE.)

20.17 Refer to **Freezer Power Use** Problem 20.5.
  a. Obtain the fitted values and the residuals. Do the residuals sum to zero as they should?
  b. Plot the residuals against the fitted values. Analyze the plot for aptness of the regression function and constancy of the error variance. State your findings.
  c. Obtain the standardized residuals. Construct a table like Table 19.1, based on the 10th, 50th, and 90th percentiles of the relevant t distribution. Do the error terms appear to be nor-

mally distributed? Explain. (*Hint:* The relevant degrees of freedom for the t distribution are the degrees of freedom associated with *MSE*.)

20.18 Refer to **Hotel Equipment** Problem 20.6.
   a. Obtain the fitted values and the residuals, and prepare a residual plot. Analyze the plot, and state your findings.
   b. Plot the residuals against (1) X_1, (2) X_2. Does the effect of either independent variable on Y appear to differ from that postulated in the regression model? Explain.

*20.19 Refer to **Fermentation Process** Problem 20.4.
   a. Test whether or not a regression relation holds, controlling the α risk at 0.01. State the alternatives, the decision rule, the value of the test statistic, and the conclusion.
   b. Does conclusion H_1 imply that a regression relation holds between Y and each of the independent variables? Does conclusion H_0 imply that a regression relation does not hold between Y and any of the independent variables? Explain.

20.20 Refer to **Freezer Power Use** Problem 20.5. Test whether or not a regression relation holds, controlling the α risk at 0.05. State the alternatives, the decision rule, the value of the test statistic, and the conclusion. Does your conclusion necessarily imply that Y is related to X_1? Explain.

20.21 Refer to **Hotel Equipment** Problem 20.6. The ANOVA block of the computer output for this study gives the upper-tail P-value for F^* as 0.00004. Use the P-value to test whether or not a regression relation holds, controlling the α risk at 0.01. State the alternatives, the decision rule, and the conclusion.

*20.22 Refer to **Fermentation Process** Problem 20.4.
   a. Construct a 95 percent confidence interval for β_1. Interpret your interval estimate.
   b. Test whether or not $\beta_3 = 0$, controlling the α risk at 0.05 when $\beta_3 = 0$. State the alternatives, the decision rule, the value of the test statistic, and the conclusion. Does your conclusion necessarily imply that there is no regression relation between X_3 and Y? Explain.

20.23 Refer to **Freezer Power Use** Problem 20.5.
   a. Construct a 95 percent confidence interval for β_1. Interpret your interval estimate.
   b. Test whether or not $\beta_2 = 0$, controlling the α risk at 0.05 when $\beta_2 = 0$. State the alternatives, the decision rule, the value of the test statistic, and the conclusion. What is the implication of your conclusion?
   c. Given your conclusion in **b**, would you expect $r_{Y2.1}$ to differ substantially from 0 here? Explain.

20.24 Refer to **Hotel Equipment** Problem 20.6.
   a. Construct a 99 percent confidence interval for β_2. Use the confidence interval to estimate the change in expected annual sales in territories whose index of business activity increases by 10 points, everything else remaining constant.
   b. Test whether or not $\beta_1 = 0$, controlling the α risk at 0.05 when $\beta_1 = 0$. State the alternatives, the decision rule, the value of the test statistic, and the conclusion. What is the P-value of the test?

*20.25 Refer to **Fermentation Process** Problem 20.4. Computer calculations give $s\{\hat{Y}_h\} = 4.932$ when $X_{h1} = 38$, $X_{h2} = 55$, and $X_{h3} = 2.2$.
   a. Construct a 95 percent confidence interval for $E\{Y_h\}$ when $X_{h1} = 38$, $X_{h2} = 55$, and $X_{h3} = 2.2$. Interpret your confidence interval.
   b. Construct a 95 percent prediction interval for the amount of product that will be obtained from a new run for which the independent variables are at the levels specified in **a**. Interpret your prediction interval.

20.26 Refer to **Freezer Power Use** Problem 20.5. The values of $s\{b_0, b_1\}$, $s\{b_0, b_2\}$, and $s\{b_1, b_2\}$ are all zero here.

    a. State the estimated variance-covariance matrix of the regression coefficients. Are the sampling errors in the estimated regression coefficients related here? Explain.

    b. Use (20.20) to construct a 95 percent confidence interval for the mean power use when $X_{h1} = -10$ and $X_{h2} = -10$. Interpret your confidence interval.

    c. Construct a 95 percent prediction interval for the power use of a particular freezer when $X_{h1} = -10$ and $X_{h2} = -10$. Explain why it is reasonable that this prediction interval is wider than the confidence interval in **b**.

    d. Explain why β_0 is a meaningful parameter in this application, and obtain a 95 percent confidence interval for it.

20.27 Refer to **Hotel Equipment** Problem 20.6. Computer calculations give $s\{\hat{Y}_h\} = 0.71846$ when $X_{h1} = 1.25$ and $X_{h2} = 107$.

    a. Construct a 90 percent confidence interval for $E\{Y_h\}$ when $X_{h1} = 1.25$ and $X_{h2} = 107$. Interpret your confidence interval.

    b. Construct a 90 percent prediction interval for sales in a new territory where $X_{h1} = 1.25$ and $X_{h2} = 107$. Interpret your prediction interval.

    c. Would you expect the prediction interval in **b** to be much narrower if the regression function had been estimated from a very large number of territories rather than from only eight? Discuss.

20.28 Refer to regression model (20.23) for the restaurant sales example in the text. An observer has stated that when $\beta_2 > 0$ and $\beta_3 = 0$, location does have an impact on sales if the restaurant is in a shopping mall or on a highway, but it has no impact if the restaurant is on a city street. Comment.

*20.29 **Bond Issues.** A financial analyst is studying the relation between bond yield (Y, in percent), bond maturity (X_1, in years), and whether the bond is from a corporate or municipal issue (X_2). Here, $X_2 = 1$ denotes a corporate bond issue and $X_2 = 0$ a municipal bond issue. The analyst has performed a regression analysis based on model (20.1) for a random sample of 300 bond issues. Some regression results are $b_0 = 8.223$, $b_1 = 0.1421$, $b_2 = 0.3381$, $s\{b_1\} = 0.006270$, $s\{b_2\} = 0.04816$.

    a. Plot, on the same graph, $\hat{Y}$ against X_1 for a corporate issue ($X_2 = 1$) and also for a municipal issue ($X_2 = 0$). Let X_1 range between 0 and 15 years in your plot. Interpret b_1 and b_2 here.

    b. Construct a 95 percent confidence interval for the difference in the mean yields of corporate and municipal bonds of the same maturity. Should one conclude from this interval estimate that there is a difference in mean yields? Explain.

20.30 Prior to general distribution of a successful hardcover novel in paperback form, an experiment was conducted in nine test markets having equal sales potential. The experiment sought to assess the effect of three different price discount levels for the paperback (50, 75, and 95 cents off the printed cover price) and the effect of three different cover designs (abstract, photograph, drawing) on sales of the paperback. Each of the nine combinations of price discount and cover design was assigned at random to one of the test markets. Regression model (20.3) was employed, where Y denotes thousands of copies sold in a test market during the experiment, X_1 denotes the price discount, and X_2 and X_3 are indicator variables coded as follows:

Class	X_2	X_3
Abstract (reference)	0	0
Photograph	1	0
Drawing	0	1

The data and some key regression results follow.

i:	1	2	3	4	5	6	7	8	9
X_{i1}:	50	50	50	75	75	75	95	95	95
X_{i2}:	0	0	1	0	0	1	0	0	1
X_{i3}:	0	1	0	0	1	0	0	1	0
Y_i:	14.89	16.34	15.02	20.21	21.55	19.31	23.41	25.42	22.13

k:	0	1	2	3	
					$n = 9$
b_k:	6.03685	0.18363	-0.68333	1.60000	$SSR = 111.091$
$s\{b_k\}$:	0.75311	0.00942	0.42468	0.42468	$SSE = 1.35266$

a. Plot, on the same graph, $\hat{Y}$ against X_1 for each cover design. Let X_1 range between 50 and 100 in your plot. What differential effect is estimated by (1) b_2, (2) b_3, (3) $b_3 - b_2$?

b. Construct a 99 percent confidence interval for (1) β_1, (2) β_3. Interpret each confidence interval.

c. Construct a 99 percent confidence interval for the mean sales in test markets when the cover design is a drawing and the price discount is 75 cents. Computer output shows that $s\{\hat{Y}_h\} = 0.30071$ for this case.

20.31 An industrial psychologist used regression model (20.3) to study the relationship between an employee's work involvement (Y, measured by a composite behavorial rating on a 100-point scale) and the employee's age (X_1, in years), marital status ($X_2 = 1$, married; $X_2 = 0$, not married), and sex ($X_3 = 1$, male; $X_3 = 0$, female). Data were collected for 64 employees of the same firm, and the following regression results were obtained:

k:	0	1	2	3
b_k:	51.620	0.1514	4.9936	2.2510
$s\{b_k\}$:	11.299	0.2589	1.1627	1.6442

a. Plot, on the same graph, $\hat{Y}$ against X_1 for each combination of marital status and sex. Let X_1 range between 20 and 65 years in your plot. Give point estimates of the difference in the mean work involvement rating between (1) married men and single men of the same age, (2) married men and married women of the same age.

b. Construct a 90 percent confidence interval for (1) β_1, (2) β_3. Does either interval span 0? If so, what is the implication?

*20.32 The following data and fitted values were obtained when the number of defective pipes in a shipment (Y) was regressed on the total number of pipes in the shipment (X, in hundreds), using the simple linear regression model (18.3) for 12 recent shipments.

i:	1	2	3	4	5	6	7	8	9	10	11	12
X_i:	5	10	4	10	7	8	8	5	10	5	12	6
Y_i:	30	51	26	52	40	43	45	31	52	30	59	36
$\hat{Y}_i$:	30.8	51.7	26.7	51.7	39.2	43.3	43.3	30.8	51.7	30.8	60.0	35.0

An analyst reviewing this regression application was concerned whether the regression function actually is quadratic. Some results of fitting the quadratic model (20.25) to the data follow.

k:	0	1	2	
				$n = 12$
b_k:	4.084	5.8478	-0.10736	$SSR = 1273.920$
$s\{b_k\}$:	2.194	0.6036	0.03830	$SSE = 4.3300$

a. Obtain the residuals for the simple linear regression model, and plot them against the fitted values. Does the plot suggest that the regression function is curvilinear? Explain.

b. State the estimated quadratic regression function. Show the data setup for computer input for the first three shipments, using the format of Table 20.4.

c. Test whether or not the curvature effect can be dropped from the quadratic regression model, controlling the α risk at 0.05. State the alternatives, the decision rule, the value of the test statistic, and the conclusion. What is the implication of your conclusion?

20.33 A sports statistician studied the relation between the time (Y, in seconds) for a particular competitive swimming event and the swimmer's age (X, in years) for 20 swimmers with ages ranging from 8 to 18. She employed quadratic regression model (20.25) and obtained the following results: $\hat{Y} = 147.3 - 11.11X + 0.2730X^2$, $s\{b_2\} = 0.1157$.

a. Plot the estimated regression function. Would it be reasonable to use this regression function when the swimmer's age is 40? Comment.

b. Construct a 99 percent confidence interval for the curvature effect coefficient. Interpret your interval estimate.

c. Test whether or not the curvature effect can be dropped from the quadratic regression model, controlling the α risk at 0.01. State the alternatives, the decision rule, the value of the test statistic, and the conclusion. What is the P-value of the test?

20.34 Shown below are the number of telephones in service in a telephone company system (N, in millions) and the calls handled annually by the system (C, in millions) for each of the past ten years. A company engineer theorized that C and N are related by the curvilinear relation:

$$\log C = \beta_0 + \beta_1 \log N$$

i:	1	2	3	4	5	6	7	8	9	10
N_i:	1.01	1.16	1.24	1.32	1.50	1.68	1.74	1.88	1.92	1.96
C_i:	1512	1689	1778	1884	2112	2292	2380	2527	2586	2613

a. Letting $Y = \log C$ and $X = \log N$, fit the simple linear regression model (18.3) to the data; use common logarithms.

b. Construct a scatter plot of the original data for C and N. Plot the estimated regression function expressed in the original units on the same graph. Is the curvature of the estimated regression function pronounced? Comment.

*20.35 The personnel department of a company studied salaries of employees who did not complete high school. In the study, employee's salary (Y, in $000) was regressed on number of years of work experience (X_1) and number of years of high school completed (X_2). The interaction regression model (20.27) was employed, and the estimated regression function was $\hat{Y} = 18.90 + 1.10X_1 + 3.05X_2 - 0.20X_1X_2$.

a. Plot $\hat{Y}$ against X_1 when $X_2 = 0$. On the same graph, plot $\hat{Y}$ against X_1 when $X_2 = 3$. Describe how the effect of X_1 on $\hat{Y}$ depends on the level of X_2 for this interaction model.

b. The two-sided P-value for the test statistic $b_3/s\{b_3\}$ is 0.0271. Test whether or not the interaction term should be kept in the regression model, controlling the α risk at 0.05. State the alternatives, the decision rule, and the conclusion.

c. Given the signs of the regression coefficients and your conclusion in b, does it appear that X_1 and X_2 are reinforcing or interfering in their relationship with mean salary? Explain.

20.36 Refer to **Freezer Power Use** Problem 20.5. When interaction model (20.27) is fitted to these data, one obtains $b_3 = 0.04500$ and $s\{b_3\} = 0.05387$. Test whether or not the interaction term can be dropped from the regression model, controlling the α risk at 0.05. State the alternatives, the decision rule, the value of the test statistic, and the conclusion. What is the *P*-value of the test?

20.37 Refer to **Hotel Equipment** Problem 20.6. The intern also fitted interaction model (20.27) to the data and obtained $b_3 = 0.05981$ and $s\{b_3\} = 0.04765$. Test whether or not the interaction term in the regression model can be dropped, controlling the α risk at 0.05. State the alternatives, the decision rule, the value of the test statistic, and the conclusion. What is the *P*-value of the test?

20.38 Refer to **Reconditioning Cost** Problem 18.14. Two incinerator types (A, B) were involved in the study. Incinerators 1, 2, 3, and 7 are type A and the others are type B. Let operating hours be denoted by X_1, and let X_2 be an indicator variable so that $X_2 = 1$ if the incinerator is type A and $X_2 = 0$ if it is type B. Interaction regression model (20.27) was fitted to the data, and the following results were obtained:

k:	0	1	2	3	
					$n = 9$
b_k:	0.91561	1.67683	0.25061	0.06641	$SSR = 7.2336$
$s\{b_k\}$:	0.03791	0.01536	0.06743	0.02755	$SSE = 0.0019357$

a. Test whether or not $\beta_3 = 0$, controlling the α risk at 0.10 when $\beta_3 = 0$. State the alternatives, the decision rule, the value of the test statistic, and the conclusion. What is the implication of your conclusion?

b. State the estimated response functions for the two types of incinerators. Plot both functions on the same graph. Describe how the effect of operating hours on mean reconditioning cost differs for the two types of incinerators.

20.39 In an analysis of the relation between sales (Y) and target population (X_1) and advertising expenditures (X_2) for ten sales territories of a large corporation, the following correlation matrix was obtained:

$$
\begin{array}{c}
\\
X_1 \\
X_2 \\
Y
\end{array}
\begin{array}{ccc}
X_1 & X_2 & Y \\
\left[\begin{array}{ccc}
1.000 & 0.999 & 0.992 \\
 & 1.000 & 0.992 \\
 & & 1.000
\end{array}\right]
\end{array}
$$

a. What difficulties arise here in attempting to measure the separate effects of population and advertising expenditures on sales based on a multiple regression of Y on X_1 and X_2?

b. Will the difficulties cited in **a** necessarily carry over to the making of inferences on Y? Explain.

c. Would simple regressions of Y on X_1 alone or of Y on X_2 alone show a high degree of linear association? Explain.

20.40 Refer to **Freezer Power Use** Problem 20.5. Results for three fitted regression models follow.

Statistic	Model 1: X_1	Model 2: X_1, X_2	Model 3: X_2
b_1	−4.20000	−4.20000	—
$s\{b_1\}$	0.60741	0.42862	—
b_2	—	1.21667	1.21667
$s\{b_2\}$	—	0.42862	1.63630

a. Construct a scatter plot of X_2 against X_1, as in Figure 20.11. Is multicollinearity present in these data? Is the same conclusion drawn from the fact that $r_{12} = 0$ here? Explain.

b. Is the value of b_1 here dependent on whether or not X_2 is in the fitted model? Is the value of b_2 dependent on whether or not X_1 is in the fitted model? Are these findings consistent with your conclusion in **a**? Comment.

20.41 Data on the hardening of eight boat propellor shafts follow, where X_1 is the number of hours the shaft was in the hardening process, X_2 is the process temperature in degrees Celsius (°C), and Y is a measure of hardness of the shaft.

i:	1	2	3	4	5	6	7	8
X_{i1}:	11.0	13.0	15.0	14.5	12.0	12.5	13.5	14.0
X_{i2}:	122	160	180	152	112	144	131	137
Y_i:	382	463	476	460	374	422	444	439

Results for three fitted regression models are as follows:

Statistic	Model 1: X_1	Model 2: X_1, X_2	Model 3: X_2
b_1	24.48120	13.47826	—
$s\{b_1\}$	5.62024	6.07759	—
b_2	—	0.91999	1.52585
$s\{b_2\}$	—	0.37259	0.32575
	$X_{h1} = 13.0, X_{h2} = 160$		
$\hat{Y}_h$	427.910	446.303	459.584
$s\{\hat{Y}_h\}$	7.095	9.094	8.800

a. Construct a scatter plot of X_2 against X_1, as in Figure 20.11. Is multicollinearity present in these data? Is the same conclusion drawn from the fact that $r_{12} = 0.7332$ here?

b. Describe how the regression results demonstrate the problems encountered in assessing the separate effects of the independent variables in a multiple regression model when multicollinearity is present.

c. Do the difficulties cited in **b** carry over to inferences on Y when such inferences are within the region of sample observations on the independent variables? Make reference to the inference results when $X_{h1} = 13.0$ and $X_{h2} = 160$ in your comments.

*20.42 Refer to **Fermentation Process** Problem 20.4. The values of *MSE* for all possible regression models involving subsets of the three independent variables follow.

Variables in Model	*MSE*	Variables in Model	*MSE*
Intercept only	279.33	X_2, X_3	185.91
X_2	191.65	X_1, X_3	104.06
X_3	186.47	X_1, X_2	103.20
X_1	117.47	X_1, X_2, X_3	105.87

a. Plot *MSE* against the number of independent variables for all possible regression models, as in Figure 20.13.

b. According to the *MSE* criterion, which three models are best among all the models and should be considered for further analysis? Would these same models be identified if the criterion were to minimize *SSE*? Explain.

c. An observer suggests that, given its simplicity and predictive power, the model with X_1 only ought to be chosen to predict the amount of product from a run. Should one reach such a conclusion without further analysis? Discuss.

20.43 Refer to the **Power Cells** data set. The *MSE* values for all possible regression models involving subsets of the four independent variables follow, where Y is the number of discharge-charge cycles before failure, X_1 is the charge rate, X_2 is the discharge rate, X_3 is the depth of discharge, and X_4 is the ambient temperature.

Variables in Model	MSE	Variables in Model	MSE
Intercept only	18,000	X_2, X_4	11,275
X_2	18,312	X_3, X_4	10,219
X_3	17,276	X_1, X_4	8,768
X_1	15,854	X_1, X_2, X_3	15,340
X_4	11,092	X_2, X_3, X_4	10,388
X_2, X_3	17,581	X_1, X_2, X_4	8,908
X_1, X_2	16,130	X_1, X_3, X_4	7,831
X_1, X_3	15,074	X_1, X_2, X_3, X_4	7,955

a. Plot *MSE* against the number of independent variables for all possible regression models, as in Figure 20.13.

b. According to the *MSE* criterion, which three models are best among all the models and should be considered for further analysis?

c. What further analysis should be conducted before deciding on the regression model to be employed?

*20.44 Refer to **Fermentation Process** Problem 20.4. When X_1 only is included in the regression model, $SSE = 2466.782$.

a. Test whether or not both X_2 and X_3 should be dropped from the regression model containing X_1, X_2, and X_3. Control the α risk at 0.05. State the alternatives, the full and reduced regression functions, the decision rule, the value of the test statistic, and the conclusion.

b. If H_1 were concluded in **a**, would this necessarily imply that both X_2 and X_3 should be retained in the model? Explain.

20.45 Refer to the restaurant sales example in the text (p. 675). The data are given in Table 20.3, and the regression results for the fitted model containing X_1, X_2, and X_3 are given in Figure 20.5. If sales volume (Y) is regressed on number of households (X_1) alone, $SSE = 2566.029$.

a. Test whether or not X_2 and X_3 should be dropped from the regression model containing X_1, X_2, and X_3. Control the α risk at 0.01. State the alternatives, the full and reduced regression functions, the decision rule, the value of the test statistic, and the conclusion.

b. What is the implication of your conclusion in **a** about the effect of location of restaurants?

EXERCISES

20.46 (Calculus needed.) Derive the normal equations (20.5) for regression model (20.1).

20.47 Show that the coefficient of multiple determination R^2 is identical to the coefficient of simple determination obtained when the observations Y_i are regressed on the fitted values $\hat{Y}_i$ from the multiple regression model.

20.48 **a.** Show that $r_{Y1.2}^2 = (R^2 - r_{Y2}^2)/(1 - r_{Y2}^2)$, where R^2 is the coefficient of multiple determination for the regression of Y on X_1 and X_2.

  b. Prove that F^* in (20.14) equals $(n - p)R^2/(p - 1)(1 - R^2)$ for regression model (20.3).

20.49 Refer to **Bond Issues** Problem 20.29. What would have been the values of b_0, b_1, and b_2 if X_2 had been coded so that $X_2 = 1$ for a municipal issue and $X_2 = 0$ for a corporate issue? Would the fitted values $\hat{Y}_i$ be affected by this change in the coding scheme? Explain. (*Hint:* The original indicator variable X_2 and the new one X_2' are related by $X_2' = 1 - X_2$.)

20.50 Consider the regression function $E\{Y\} = \beta_0 X_1^{\beta_1} X_2^{\beta_2}$, where β_0, β_1, and β_2 denote regression coefficients and X_1 and X_2 denote independent variables. Show how the transformation $\log E\{Y\}$ converts this function into a linear regression function, and explain the relation of the regression coefficients in the resulting linear regression function to the original regression coefficients.

20.51 A business researcher has a theory that a firm's rate of return on investment (Y) tends to increase with its accumulated research and development expenditures (X_1) according to a linear statistical relation. She also anticipates that this linear relation between Y and X_1 may differ from one industry to another. To test the theory, she gathered data on Y and X_1 for 15 firms in each of two industries (coded $X_2 = 0$ and $X_2 = 1$, respectively). She then regressed Y on X_1, X_2, and $X_3 = X_1 X_2$, using interaction regression model (20.27). Finally, she used procedure (20.33) to test the hypothesis $H_0: \beta_2 = \beta_3 = 0$. How should the researcher interpret the test result if (1) H_0 is concluded, (2) H_1 is concluded?

STUDIES

20.52 (Computer needed.) The manager of a university computer system wishes to employ regression model (20.1) in a study of the relation between the average response time of the system (Y, in seconds) and the number of university terminal users (X_1) and commercial terminal users (X_2) signed on the system. The data for 24 cases follow.

X_1	X_2	Y	X_1	X_2	Y	X_1	X_2	Y	X_1	X_2	Y
10	0	0.08	48	4	0.52	23	3	0.13	69	12	1.06
36	8	0.59	21	1	0.13	66	7	0.81	58	10	0.74
75	5	0.77	66	10	0.88	10	2	0.15	26	2	0.23
16	4	0.21	30	3	0.28	70	10	1.00	14	4	0.18
35	5	0.45	55	9	0.70	44	0	0.28	62	9	0.72
50	8	0.56	49	6	0.63	42	4	0.43	63	3	0.55

  a. Obtain (1) the estimated regression function; (2) $s\{b_1\}$, $s\{b_2\}$; (3) the fitted values and residuals; (4) the correlation matrix.

  b. Obtain 95 percent confidence intervals to estimate the change in the mean response time of the system when (1) an additional university terminal user signs on the system and the number of commercial users is held constant, (2) an additional commercial terminal user signs on the system and the number of university users is held constant.

  c. Is multicollinearity a problem in this regression study? Comment, referring to appropriate results in **a** to support your answer.

  d. Obtain the residuals and plot them against the fitted values to check the aptness of regression model (20.1). What are your findings? Also, plot the residuals against X_2. What does this plot show about the aptness of the regression model?

20.53 (Computer needed.) Data on the region, number of beds (X_1), and number of admissions (Y) last year for each of 24 small acute-care hospitals follow.

Hospital i:	1	2	3	4	5	6	7	8
Region:	A	A	A	A	A	A	B	B
Beds X_{i1}:	19	120	49	100	33	22	96	48
Admissions Y_i:	460	3374	2244	3606	950	703	2958	1487

Hospital i:	9	10	11	12	13	14	15	16
Region:	B	B	B	B	B	B	B	C
Beds X_{i1}:	148	101	66	138	25	193	44	76
Admissions Y_i:	4700	3308	2696	4845	1159	5692	1576	2648

Hospital i:	17	18	19	20	21	22	23	24
Region:	C	C	C	C	C	C	C	C
Beds X_{i1}:	75	84	13	40	69	125	13	32
Admissions Y_i:	2757	2881	402	1600	1646	4825	370	987

Use X_2 and X_3 to define the regions, as follows:

Region	X_2	X_3
A	1	0
B	0	1
C	0	0

Assume regression model (20.3) is appropriate.

a. Obtain (1) the estimated regression function; (2) $s\{b_k\}$, $k = 1, 2, 3$; (3) the ANOVA table; (4) the fitted values and residuals; (5) the correlation matrix.

b. Interpret the meanings of b_1, b_2, and b_3. Give a point estimate of the mean number of admissions for 100-bed hospitals in region A.

c. Calculate $\sqrt{MSE}$. What does this number measure here?

d. Construct a 95 percent prediction interval for the number of admissions to a particular 100-bed hospital in region A. Computer output shows that $s\{\hat{Y}_h\} = 170.927$ in this case.

e. Test whether or not a regression relation holds, controlling the α risk at 0.01. State the alternatives, the decision rule, the value of the test statistic, and the conclusion. Does your conclusion necessarily imply that mean admissions differ among the three regions for hospitals with a given number of beds? Comment.

f. Test whether or not the two indicator variables for regions should be dropped from the regression model, controlling the α risk at 0.01. State the alternatives, the decision rule, the value of the test statistic, and the conclusion. [*Hint:* Use the value of r_{Y1} from **a** to obtain $SSE(R)$.]

20.54 (Computer needed.) A student, after studying the text discussion for the machine shop example in Table 20.6, stated that the multicollinearity in the regression of Y on X_1 and X_2 would largely disappear if the total weight in excess of 100 pounds per shaft for each shipment were used in place of the total weight of shafts in the shipment, that is, if the variable $X'_2 = X_2 - 100X_1$ were to replace X_2 in the regression model.

a. Calculate the coefficient of correlation between X_1 and X'_2. Do you agree with the student's statement? Explain.

b. For the new regression of Y on X_1 and X_2', obtain the regression coefficients denoted by b_1' and b_2', and also obtain their estimated standard deviations $s\{b_1'\}$ and $s\{b_2'\}$. How do these quantities compare with the corresponding values in Table 20.7 when Y is regressed on X_1 and X_2? Comment.

c. Obtain R^2 for the regression of Y on X_1 and X_2'. Is it the same, larger, or smaller than R^2 for the regression of Y on X_1 and X_2? (Recall from the text that, in the latter case, $R^2 = 0.996$.) Discuss.

21 ANALYSIS OF VARIANCE

In this chapter, we consider analysis of variance (ANOVA) models, which are useful for studying the statistical relation between a dependent variable and one or more independent variables. Although these models may be viewed as special cases of the general linear regression model, they allow a study of statistical relations from a different perspective and are widely used. The nature of the ANOVA approach was partly explained in the preceding chapters on regression; we now undertake a fuller study of this approach, with particular emphasis on ANOVA models.

21.1 BASIC CONCEPTS

We illustrate the nature of ANOVA models by two examples.

EXAMPLE

Soft Drink. A firm developing a new citrus-flavored soft drink conducted an experiment to study consumer preferences for the color of the drink. Four colors were under consideration: colorless, pink, orange, and lime green. Twenty test localities were selected that were similar in sales potential and representative of the target market for this product. Each color was then randomly assigned to five of these localities for test marketing. The dependent variable was number of cases sold during the test period per 1000 population, and the independent variable was color. Other factors, such as price, flavor, degree of carbonation, sweetness, and calorie content, were held fixed in all the localities.

An ANOVA model was employed to study the comparative effects of color on sales. Figure 21.1a portrays this model. For each color, there is a probability distribution of sales per 1000 population. Each of the distributions is normal, with the same variability, but the means of the distributions differ. In Figure 21.1a, the illustration shows that the green color has the largest mean and thus tends to lead to highest sales. Since all probability distributions are normal with constant variance, the means μ_1, μ_2, μ_3, and μ_4 of the four probability distributions convey the information about the effects of color on sales. Thus, Figure 21.1a illustrates the situation where sales of orange and colorless drinks tend to be substantially smaller than those of pink and green drinks. □

FIGURE 21.1   Representations of the single-factor ANOVA model

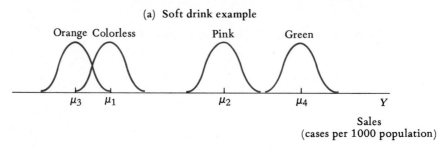

(a) Soft drink example

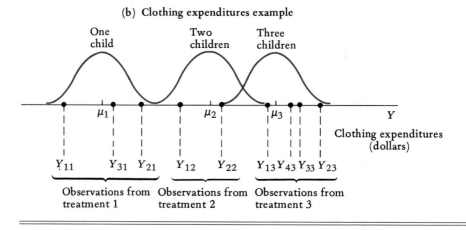

(b) Clothing expenditures example

In the soft drink example, the independent variable is qualitative (color of drink). ANOVA models may also be used when the independent variable is quantitative, as shown by the next example.

EXAMPLE

Clothing Expenditures.   Families with one, two, or three children were selected in a middle-income school district to study expenditures on children's clothing during the past year. Here, the dependent or response variable is annual family expenditures for children's clothing (in dollars). The independent variable is number of children, and three levels of the independent variable are included in the study, namely one, two, and three children.

Figure 21.1b portrays the ANOVA model for this case. Note that there is a probability distribution of clothing expenditures for each level of the independent variable, that is, for each number of children, and that these probability distributions are normal with the same variance. Thus, μ_1, μ_2, and μ_3 convey information about the effect of number of children on annual children's clothing expenditures. As might be expected, the positions of the means μ_1, μ_2, and μ_3 in Figure 21.1b indicate that a family's expenditures for children's clothing increase with number of children. $\square$

Factors and Factor Levels

In the special terminology of ANOVA, an independent variable is called a *factor*. Thus, in the soft drink example, the factor is color of drink; and in the clothing expenditures example, it is number of children. A *factor level* or *treatment* is a particular outcome of the independent variable, such as lime green color of the soft drink or two children in the family.

We distinguish between *single-factor* and *multifactor* ANOVA just as we did between simple and multiple regression. In multifactor ANOVA, there are two or more independent variables. A study to assess the effects of number of children (1, 2, 3) and family income (low, medium, high) on children's clothing expenditures would be a multifactor ANOVA study.

Finally, we distinguish between *experimental* and *observational* factors just as we did between experimental and observational (nonexperimental) studies in Chapter 1. Experimental factors are under the control of the investigator, and the levels are assigned at random to the experimental units. In the soft drink example, color of the drink is an experimental factor because the colors were randomly assigned to test localities. In contrast, observational factors pertain to existing characteristics of the units under study, and the levels of the observational factors cannot be assigned at random. In the clothing expenditures example, the number of children in a family is an observational factor since it is a characteristic of the family that is not randomly assigned by the investigator.

As explained in Chapter 1, experimental control with random assignments of factor levels helps avoid biases that can otherwise arise. On the other hand, experimental control is frequently not possible in social science and management applications, so ANOVA studies in these fields often are nonexperimental and use observational factors.

21.2 ANALYSIS OF VARIANCE MODEL

Development of Model

The ANOVA model, which we illustrated graphically for two examples, has the following features:

1. There are r factor levels or treatments under study.
2. For each treatment, the probability distribution of Y is normal.
3. All probability distributions of Y have constant variance, denoted by σ^2.
4. The mean of the probability distribution of Y for the jth treatment $(j = 1, 2, \ldots, r)$ is denoted by μ_j. The means μ_j may differ, reflecting the varying effects of the treatments.

In an ANOVA study, n_j observations are selected for the jth treatment. The ith observation ($i = 1, 2, \ldots, n_j$) for the jth treatment is denoted by Y_{ij}. This observation is assumed to represent a random selection from the probability distribution of Y for the jth treatment. The sampling process is illustrated in Figure 21.1b for the clothing expenditures example. Note that the sample sizes n_j need not be the same for all treatments.

As with regression models, we state the ANOVA model formally by indicating the components that make up an observation Y_{ij}.

(21.1)

$$Y_{ij} = \mu_j + \varepsilon_{ij} \qquad i = 1, 2, \ldots, n_j; j = 1, 2, \ldots, r$$

where:

Y_{ij} is the response in the ith observation for the jth treatment

μ_j are parameters

ε_{ij} are independent $N(0, \sigma^2)$

Model (21.1) is called a *single-factor ANOVA model*.

Important Features of Model

1. Model (21.1) assumes that the observed value Y_{ij} is the sum of two components — a constant term μ_j and a random deviation or error term ε_{ij}. Thus, Y_{ij} is a random variable.
2. Since the error terms ε_{ij} are independent and normally distributed, so are the Y_{ij}.
3. Since $E\{\varepsilon_{ij}\} = 0$, it follows that $E\{Y_{ij}\} = \mu_j$. The parameter μ_j thus is the mean response for the jth treatment.
4. The parameter σ^2 is the common variance of the distributions of Y because $\sigma^2\{Y_{ij}\} = \sigma^2\{\varepsilon_{ij}\} = \sigma^2$ by (5.12a).
5. In view of these features, model (21.1) can be expressed alternatively as follows.

(21.2)

$$Y_{ij} \text{ are independent } N(\mu_j, \sigma^2) \qquad i = 1, 2, \ldots, n_j; j = 1, 2, \ldots, r$$

6. Model (21.1) is a linear statistical model because it is linear in the parameters $\mu_1, \mu_2, \ldots, \mu_r$. Thus, no parameter μ_j appears as an exponent or is multiplied by another parameter.

Steps in Analysis

Since the probability distributions of Y corresponding to the different treatments are each normal with the same variance, differences in treatment effects are associated with differences between the treatment means μ_j. For this reason, ANOVA studies concentrate on making inferences about the μ_j and usually proceed in two stages:

1. Determine whether or not the μ_j are equal. If they are equal, we say there is no factor effect, and no further analysis is required.
2. If the μ_j are not equal, study the nature of the treatment effects.

We now consider each of these stages of analysis in turn.

21.3 TEST FOR EQUALITY OF TREATMENT MEANS

EXAMPLE

Table 21.1 contains the sales data for the soft drink example. Each flavor was assigned to five localities for test marketing, and the number of cases sold per 1000 population during the study period was recorded for each locality. Thus, we see from Table 21.1 that, in the first locality that was assigned the colorless drink, 26.5 cases per 1000 population were sold during the study period, or $Y_{11} = 26.5$. Similarly, $Y_{12} = 31.2$.   □

Point Estimation of μ_j

It can be shown that the least squares estimator of the treatment mean μ_j is the sample mean of the observations for the jth treatment, denoted by $\overline{Y}_j$.

TABLE 21.1   Sales data for the soft drink example (cases per 1000 population)

Observation i	1 Colorless	2 Pink	3 Orange	4 Green	Total
1	26.5	31.2	27.9	30.8	
2	28.7	28.3	25.1	29.6	
3	25.1	30.8	28.5	32.4	
4	29.1	27.9	24.2	31.7	
5	27.2	29.6	26.5	32.8	
Total	136.6	147.8	132.2	157.3	573.9
Mean	$\overline{Y}_1 = 27.32$	$\overline{Y}_2 = 29.56$	$\overline{Y}_3 = 26.44$	$\overline{Y}_4 = 31.46$	

Number of observations $n_1 = 5$ $n_2 = 5$ $n_3 = 5$ $n_4 = 5$

Total number of observations $n_T = 5 + 5 + 5 + 5 = 20$

Overall mean $\overline{\overline{Y}} = \dfrac{573.9}{20} = 28.695$

$$(21.3) \qquad \overline{Y}_j = \frac{\sum\limits_{i=1}^{n_j} Y_{ij}}{n_j}$$

The sample mean $\overline{Y}_j$ is an unbiased estimator of μ_j.

EXAMPLE

We see from Table 21.1 for the soft drink example that the point estimate of μ_1, the mean sales for the colorless drink, is $\overline{Y}_1 = 27.32$ cases per 1000 population. Note from Table 21.1 that mean sales were highest for the green color. ☐

Partitioning of Total Sum of Squares

Total Sum of Squares. As the term *analysis of variance* suggests and as was illustrated by the use of ANOVA in regression analysis, ANOVA procedures utilize partitions of the total sum of squares. We denote the overall mean for all the sample data by $\overline{\overline{Y}}$.

$$(21.4) \qquad \overline{\overline{Y}} = \frac{\sum\sum Y_{ij}}{n_T}$$

Here, n_T represents the total number of sample observations in the entire study.

$$(21.5) \qquad n_T = \sum n_j$$

Again, we omit the indexes of summation when the nature of the summation is clear.

Recall from regression analysis that the *total sum of squares, SSTO*, is based on the deviations of the observations around their mean. In our current notation, these deviations are as follows.

$$(21.6) \qquad Y_{ij} - \overline{\overline{Y}}$$

SSTO is then the sum of the squares of these deviations.

$$(21.7) \qquad SSTO = \sum\sum (Y_{ij} - \overline{\overline{Y}})^2$$

EXAMPLE

In the soft drink example, the total sample size is $n_T = 20$ (Table 21.1), and the overall mean is:

$$\overline{\overline{Y}} = \frac{26.5 + 28.7 + \cdots + 32.8}{20} = \frac{573.9}{20} = 28.695$$

Hence, we find *SSTO* as follows:

$$SSTO = (26.5 - 28.695)^2 + (28.7 - 28.695)^2 + \cdots + (32.8 - 28.695)^2$$
$$= 115.92950 \qquad \square$$

Error Sum of Squares. The *error sum of squares, SSE*, measures the variability of the observations Y_{ij} when information about the treatments is utilized. Hence, the deviations are taken around the sample treatment means $\overline{Y}_j$, as follows.

(21.8)

$$Y_{ij} - \overline{Y}_j$$

The error sum of squares is then simply the sum of squares of these deviations, first summed within a treatment and then summed over all treatments.

(21.9)

$$SSE = \sum_j \left[\sum_i (Y_{ij} - \overline{Y}_j)^2 \right]$$

Note that if for each treatment all observations within the treatment are the same, we have:

$$\sum_i (Y_{ij} - \overline{Y}_j)^2 = 0$$

for all treatments, and $SSE = 0$. At the other extreme, if all treatment means $\overline{Y}_j$ are equal, so that $\overline{Y}_j \equiv \overline{\overline{Y}}$, a comparison of (21.7) and (21.9) shows that $SSE = SSTO$ then. Thus, the closer the treatment means $\overline{Y}_j$ are to each other, the closer will SSE be to $SSTO$.

EXAMPLE

For the soft drink example, we obtain *SSE* by first calculating the sums of squares within each treatment. Thus, for treatment 1, we have:

$$\sum (Y_{i1} - \overline{Y}_1)^2 = (26.5 - 27.32)^2 + (28.7 - 27.32)^2 + \cdots$$
$$+ (27.2 - 27.32)^2$$
$$= 10.68800$$

The corresponding calculations for the other three treatments are:

$$(31.2 - 29.56)^2 + \cdots + (29.6 - 29.56)^2 = 8.57200$$
$$(27.9 - 26.44)^2 + \cdots + (26.5 - 26.44)^2 = 13.19200$$
$$(30.8 - 31.46)^2 + \cdots + (32.8 - 31.46)^2 = 6.63200$$

Then, *SSE* is the sum of these within-treatments sums of squares:

$$SSE = 10.68800 + 8.57200 + 13.19200 + 6.63200 = 39.08400 \qquad \square$$

Treatment Sum of Squares.   The difference between *SSTO* and *SSE* is called the *treatment sum of squares* and is denoted by *SSTR*.

(21.10)
$$SSTR = SSTO - SSE$$

As we noted earlier, if the $\overline{Y}_j$ are close to each other, *SSE* will be close to *SSTO* and *SSTR* will then be small. If, however, the sample treatment means $\overline{Y}_j$ are not close to each other, *SSTR* will tend to be larger.

The treatment sum of squares is actually a sum of squares made up of deviations of the sample treatment means $\overline{Y}_j$ around the overall mean $\overline{\overline{Y}}$.

(21.11)
$$\overline{Y}_j - \overline{\overline{Y}}$$

The treatment sum of squares, *SSTR*, is expressed in terms of these deviations as follows.

(21.12)
$$SSTR = \Sigma\, n_j (\overline{Y}_j - \overline{\overline{Y}})^2$$

Thus, *SSTR* is the sum of the squared deviations of $\overline{Y}_j$ around the overall mean $\overline{\overline{Y}}$, weighted by the number of observations n_j. Formula (21.12) makes it very clear that *SSTR* reflects the variability of the sample treatment means $\overline{Y}_j$. The further apart they are, the larger will be *SSTR*. If all $\overline{Y}_j$ are equal, then *SSTR* = 0.

EXAMPLE

In the soft drink example, we obtain, by (21.10):

$$SSTR = SSTO - SSE = 115.92950 - 39.08400 = 76.84550$$

Here, *SSE* is not close to *SSTO*, so *SSTR* is comparatively large.

The direct calculation of *SSTR* by (21.12) is as follows:

$$\begin{aligned}
SSTR = {}& 5(27.32 - 28.695)^2 + 5(29.56 - 28.695)^2 \\
& + 5(26.44 - 28.695)^2 + 5(31.46 - 28.695)^2 \\
= {}& 76.84550
\end{aligned}$$

This result must be the same as that obtained by subtraction. $\qquad \square$

Formal Statement of Partitioning.   As in regression analysis, we have now obtained a partitioning of the total sum of squares into two components.

(21.13) For single-factor ANOVA, the decomposition of the total sum of squares into two additive components is:

(21.13a)
$$SSTO = SSTR + SSE$$

(21.13b)
$$\sum\sum(Y_{ij} - \bar{\bar{Y}})^2 = \sum n_j(\bar{Y}_j - \bar{\bar{Y}})^2 + \sum\sum(Y_{ij} - \bar{Y}_j)^2$$

EXAMPLE

We obtain the following decomposition for the soft drink example:

$$115.92950 = 76.84550 + 39.08400$$

$$SSTO = SSTR + SSE$$

☐

Calculational Formulas. We have used the basic definitional formulas for $SSTO$, $SSTR$, and SSE to bring out the meaning of these different sums of squares. The following formulas are algebraically equivalent and tend to be more convenient for hand computation.

(21.14)
$$SSTO = \sum\sum Y_{ij}^2 - \frac{\left(\sum\sum Y_{ij}\right)^2}{n_T}$$

(21.15)
$$SSTR = \sum_j \frac{\left(\sum_i Y_{ij}\right)^2}{n_j} - \frac{\left(\sum\sum Y_{ij}\right)^2}{n_T}$$

(21.16)
$$SSE = SSTO - SSTR$$

Comments

1. $SSTR$ and SSE are sometimes called, respectively, the *between-treatments* and *within-treatments* sums of squares.

2. Formula (21.13b) can be derived by writing:

$$\sum\sum(Y_{ij} - \bar{\bar{Y}})^2 = \sum\sum[(Y_{ij} - \bar{Y}_j) + (\bar{Y}_j - \bar{\bar{Y}})]^2$$

Upon expanding the right side, we obtain:

$$\sum\sum(Y_{ij} - \bar{\bar{Y}})^2 = \sum\sum(Y_{ij} - \bar{Y}_j)^2 + \sum\sum(\bar{Y}_j - \bar{\bar{Y}})^2 + 2\sum\sum(Y_{ij} - \bar{Y}_j)(\bar{Y}_j - \bar{\bar{Y}})$$

The second sum on the right can be written $\sum n_j(\bar{Y}_j - \bar{\bar{Y}})^2$ since the term $(\bar{Y}_j - \bar{\bar{Y}})^2$ is constant when summing over i and is picked up n_j times ($i = 1, 2, \ldots, n_j$). The last sum on the right can be shown to equal zero.

Partitioning of Degrees of Freedom

As in regression analysis, each sum of squares for the single-factor ANOVA model has associated with it a number of degrees of freedom. $SSTO$ has $n_T - 1$ degrees of freedom associated with it. There are n_T deviations $Y_{ij} - \overline{\overline{Y}}$, but one constraint exists, namely, $\Sigma \Sigma (Y_{ij} - \overline{\overline{Y}}) = 0$.

$SSTR$ has $r - 1$ degrees of freedom associated with it. There are r deviations $\overline{Y}_j - \overline{\overline{Y}}$, but one constraint exists, namely, $\Sigma n_j(\overline{Y}_j - \overline{\overline{Y}}) = 0$.

Finally, SSE has $n_T - r$ degrees of freedom associated with it. Each of the r levels of the factor contributes a component sum of squares $\sum_i (Y_{ij} - \overline{Y}_j)^2$ to SSE. The component sum of squares for the jth treatment is equivalent in form to a total sum of squares and therefore has $n_j - 1$ degrees of freedom. The degrees of freedom associated with SSE are then the sum of the degrees of freedom for each of the r components:

$$\sum_{j=1}^{r} (n_j - 1) = n_T - r$$

We see again that the degrees of freedom are additive.

(21.17)
$$\underbrace{n_T - 1}_{\substack{df \text{ for} \\ SSTO}} = \underbrace{r - 1}_{\substack{df \text{ for} \\ SSTR}} + \underbrace{n_T - r}_{\substack{df \text{ for} \\ SSE}}$$

EXAMPLE

For the soft drink example, where $n_T = 20$ and $r = 4$, the degrees of freedom associated with the sums of squares are as follows:

$$SSTR: \quad 4 - 1 = 3$$
$$\underline{SSE: \quad 20 - 4 = 16}$$
$$SSTO: 20 - 1 = 19$$

☐

Mean Squares

Our interest is in the *treatment mean square,* denoted by *MSTR,* and in the *error mean square,* denoted by *MSE.* These are obtained in the usual way by dividing the respective sums of squares by their associated degrees of freedom.

(21.18)
$$MSTR = \frac{SSTR}{r - 1}$$

(21.19)
$$MSE = \frac{SSE}{n_T - r}$$

EXAMPLE

For the soft drink example, we found earlier that $SSTR = 76.84550$ and $SSE = 39.08400$. Also, $r - 1 = 3$ and $n_T - r = 16$. Thus, we calculate:

$$MSTR = \frac{76.84550}{3} = 25.61517$$

$$MSE = \frac{39.08400}{16} = 2.44275$$

ANOVA Table

The ANOVA table for single-factor analysis of variance is shown in Table 21.2. Table 21.2a displays the general format of the table, and Table 21.2b shows the results for the soft drink example.

F Test for Equality of Treatment Means

We are now in a position to conduct an F test for equality of treatment means. The alternative conclusions for the test are the following.

TABLE 21.2   ANOVA table for a single-factor study and the results for the soft drink example

(a)
General format

Source of Variation	SS	df	MS
Treatments	$SSTR = \sum n_j(\bar{Y}_j - \bar{\bar{Y}})^2$	$r - 1$	$MSTR = \dfrac{SSTR}{r - 1}$
Error	$SSE = \sum\sum(Y_{ij} - \bar{Y}_j)^2$	$n_T - r$	$MSE = \dfrac{SSE}{n_T - r}$
Total	$SSTO = \sum\sum(Y_{ij} - \bar{\bar{Y}})^2$	$n_T - 1$	

(b)
Soft drink example

Source of Variation	SS	df	MS
Drink color	76.84550	3	25.61517
Error	39.08400	16	2.44275
Total	115.92950	19	

(21.20)
$$H_0: \mu_1 = \mu_2 = \cdots = \mu_r$$
$$H_1: \text{not all } \mu_j \text{ are equal}$$

Here, H_0 implies that all distributions of Y have the same mean, and hence for model (21.1) are identical, while H_1 implies they do not all have the same mean. Note that Figure 21.1 portrays two situations where H_1 is true — one for the soft drink example and another for the clothing expenditures example.

The test statistic is analogous to that for regression analysis.

(21.21)
$$F^* = \frac{MSTR}{MSE}$$

Large values of F^* lead to conclusion H_1 because $MSTR$ tends to be larger than MSE when H_1 holds, whereas the two mean squares tend to be of the same magnitude when H_0 holds. This can be seen from the expected values of the mean squares for model (21.1), which are as follows.

(21.22)
$$E\{MSE\} = \sigma^2$$

(21.23)
$$E\{MSTR\} = \sigma^2 + \frac{\sum n_j(\mu_j - \mu)^2}{r - 1}$$

where:

$$\mu = \frac{\sum n_j \mu_j}{n_T}$$

Note that MSE is always an unbiased estimator of the error variance σ^2. Also, note that $E\{MSTR\} = \sigma^2$ when all μ_j are equal because then $\mu_j \equiv \mu$, where μ is a weighted average of the μ_j. On the other hand, when the μ_j are not all equal, the second term on the right in (21.23) is positive and $E\{MSTR\} > E\{MSE\} = \sigma^2$.

The test is illustrated in Figure 21.2, where the appropriate upper-tail decision rule is shown. It is known that F^* follows the F distribution when H_0 holds.

(21.24)
For the single-factor ANOVA model (21.1), when $\mu_1 = \mu_2 = \cdots = \mu_r$:

$$F^* = F(r - 1, n_T - r)$$

where:
F^* is given by (21.21)

FIGURE 21.2   **General form of the statistical decision rule for an F test of equality of treatment means in a single-factor ANOVA study**

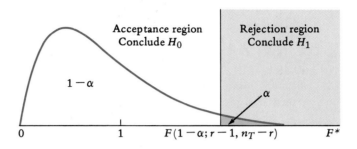

Figure 21.2 shows this sampling distribution of F^* when H_0 holds, the case for which the α risk is to be controlled. The figure also shows that the action limit of the decision rule is $F(1 - \alpha; r - 1, n_T - r)$, since the risk of concluding H_1 when H_0 actually is true, represented by the shaded right-tail area, is to be controlled at α. The entire decision procedure may be summarized as follows.

(21.25)

When the alternatives are:

H_0: $\mu_1 = \mu_2 = \cdots = \mu_r$

H_1: not all μ_j are equal

and the single-factor ANOVA model (21.1) is applicable, the appropriate decision rule to control the α risk is as follows:

If $F^* \leq F(1 - \alpha; r - 1, n_T - r)$, conclude H_0.
If $F^* > F(1 - \alpha; r - 1, n_T - r)$, conclude H_1.

where:
F^* is given by (21.21)

EXAMPLE

In the soft drink example, we wish to test whether or not color of the drink affects mean sales.

Step 1.   The alternatives here are:

H_0: $\mu_1 = \mu_2 = \mu_3 = \mu_4$

H_1: not all μ_j are equal

Step 2. It is desired to control the α risk at 0.05.

Step 3. Since $\alpha = 0.05$, $r - 1 = 3$, and $n_T - r = 16$, we require $F(0.95; 3, 16) = 3.24$. The decision rule is as follows:

If $F^* \leq 3.24$, conclude H_0.
If $F^* > 3.24$, conclude H_1.

Step 4. From Table 21.2b, we calculate $F^* = 25.61517/2.44275 = 10.49 > 3.24$. Hence, we conclude H_1—that mean sales are not the same for the different colors. The next step in the analysis is to study the nature of the color effects in some detail. □

Comment   In Chapter 13, we discussed the comparison of two population means when the populations are normal with equal variance and the samples are independent. The test whether or not $\mu_1 = \mu_2$ based on test statistic (13.11a) is a special case of the ANOVA F test.

21.4   ANALYSIS OF TREATMENT EFFECTS

We now take up some procedures for investigating treatment effects when it is concluded that the treatment means μ_j are not all equal. Important avenues of investigation include interval estimation of the mean response for a given treatment and comparison of mean responses for different treatments. We consider each of these in turn.

Interval Estimation of μ_j

We noted earlier that the least squares estimator of the mean response μ_j is the sample mean $\overline{Y}_j$. The estimated variance of $\overline{Y}_j$, denoted by $s^2\{\overline{Y}_j\}$, is as follows.

(21.26)
$$s^2\{\overline{Y}_j\} = \frac{MSE}{n_j}$$

A two-sided confidence interval for μ_j takes the usual form.

(21.27)
The confidence limits for μ_j with confidence coefficient $1 - \alpha$ for the single-factor ANOVA model (21.1) are:

$$\overline{Y}_j \pm ts\{\overline{Y}_j\}$$

where:
$t = t(1 - \alpha/2; n_T - r)$
$s\{\overline{Y}_j\}$ is given by (21.26)

The t multiple involves $n_T - r$ degrees of freedom because these are the degrees of freedom associated with MSE.

EXAMPLE

In the soft drink example, it is desired to estimate mean sales for the colorless version of the drink (treatment 1) with a 90 percent confidence interval. We know from Tables 21.1 and 21.2 that $\bar{Y}_1 = 27.32$, $MSE = 2.44275$, $n_T - r = 16$, and $n_1 = 5$. We calculate, by (21.26):

$$s^2\{\bar{Y}_1\} = \frac{2.44275}{5} = 0.48855$$

Hence, $s\{\bar{Y}_1\} = \sqrt{0.48855} = 0.69896$. We require $t(0.95; 16) = 1.746$. The confidence limits therefore are $27.32 \pm 1.746(0.69896)$, and the confidence interval is:

$$26.1 \leq \mu_1 \leq 28.5$$

Thus, it can be reported with 90 percent confidence that mean sales for the colorless version are between 26.1 and 28.5 cases per 1000 population. □

Comparison of Two Treatment Means

Frequently, we wish to compare the mean responses for two treatments. Let the two treatments of interest be denoted by j and j'. The difference of interest then is as follows.

(21.28)
$$\mu_j - \mu_{j'}$$

Such a difference between a pair of treatment means is called a *pairwise comparison*. An unbiased point estimator of $\mu_j - \mu_{j'}$ is as follows.

(21.29)
$$\bar{Y}_j - \bar{Y}_{j'}$$

Because all error terms are independent, $\bar{Y}_j$ and $\bar{Y}_{j'}$ are independent, and by (5.14b), the variance of $\bar{Y}_j - \bar{Y}_{j'}$ is simply the sum of the variances of $\bar{Y}_j$ and $\bar{Y}_{j'}$. Thus, the estimated variance of $\bar{Y}_j - \bar{Y}_{j'}$, which is denoted by $s^2\{\bar{Y}_j - \bar{Y}_{j'}\}$, is as follows.

(21.30)
$$s^2\{\bar{Y}_j - \bar{Y}_{j'}\} = \frac{MSE}{n_j} + \frac{MSE}{n_{j'}}$$

A two-sided confidence interval for $\mu_j - \mu_{j'}$ is of the usual form.

(21.31)

> The confidence limits for $\mu_j - \mu_{j'}$ with confidence coefficient $1 - \alpha$ for the single-factor ANOVA model (21.1) are:
>
> $$(\overline{Y}_j - \overline{Y}_{j'}) \pm ts\{\overline{Y}_j - \overline{Y}_{j'}\}$$
>
> *where:*
> $$t = t(1 - \alpha/2; n_T - r)$$
> $s\{\overline{Y}_j - \overline{Y}_{j'}\}$ is given by (21.30)

EXAMPLE

In the soft drink example, the firm's advertising agency had initially recommended, on the basis of interviews with a panel of consumers, that either lime green or pink be used in launching the product nationally. Hence, a major purpose of the experiment was to compare the effects of these two colors in an actual market setting.

The pairwise comparison of interest is $\mu_4 - \mu_2$, and it is to be estimated with a 90 percent confidence interval. The point estimator is $\overline{Y}_4 - \overline{Y}_2$. From Table 21.1, we see that $\overline{Y}_4 - \overline{Y}_2 = 31.46 - 29.56 = 1.90$. We obtain $t(0.95; 16) = 1.746$, and require $s\{\overline{Y}_4 - \overline{Y}_2\}$. Since $MSE = 2.44275$, $n_4 = 5$, and $n_2 = 5$, we calculate, by (21.30):

$$s^2\{\overline{Y}_4 - \overline{Y}_2\} = \frac{2.44275}{5} + \frac{2.44275}{5} = 0.97710$$

and hence $s\{\overline{Y}_4 - \overline{Y}_2\} = \sqrt{0.97710} = 0.98848$. The confidence limits therefore are $1.90 \pm 1.746(0.98848)$, and the desired confidence interval is:

$$0.2 \leq \mu_4 - \mu_2 \leq 3.6$$

Thus, it is estimated, with 90 percent confidence, that mean sales with green color are somewhere between 0.2 and 3.6 cases per 1000 population greater than those with pink color. □

Simultaneous Comparisons

In studies with three or more treatments, interest often centers on several pairwise comparisons. In the soft drink example, for instance, the sales manager might wish to compare mean sales for both green and pink colors with those for the colorless mixture. The set of comparisons then would be as follows:

Pairwise Comparison	Colors	Treatment Mean Difference
1	Green − colorless	$\mu_4 - \mu_1$
2	Pink − colorless	$\mu_2 - \mu_1$

Since the results of both comparisons will affect management's final marketing decision, the sales manager would like to have known assurance that both confidence intervals in the set will be correct simultaneously. This situation calls for *simultaneous confidence intervals,* where the *joint confidence coefficient* indicates the probability that the procedure will lead to all confidence intervals in the set being simultaneously correct. There are a variety of procedures available for obtaining simultaneous confidence intervals. We shall discuss a simple procedure that is highly useful when the number of confidence intervals in the set is not too large.

Simultaneous Confidence Intervals. If a $1 - \alpha$ confidence interval is constructed for each of m pairwise comparisons, using the usual procedure of the preceding section, the level of confidence that *all m* confidence intervals are simultaneously correct will be less than $1 - \alpha$. The reason is that each confidence statement involves a risk α of being incorrect, and these risks compound when several such statements are made. Hence, if one desires a joint confidence coefficient $1 - \alpha$ that all comparisons in the set are simultaneously correct, each individual confidence interval will need to be constructed with a confidence coefficient that involves a risk smaller than α.

The following theorem states how much smaller than α the risk of an incorrect confidence interval must be so that the overall risk for all m confidence intervals in the set does not exceed α.

(21.32)

> If each of m confidence intervals is constructed with confidence coefficient $1 - (\alpha/m)$, the joint confidence coefficient for the set of m confidence intervals is at least $1 - \alpha$.

In other words, if the confidence interval for each pairwise comparison is constructed with confidence coefficient $1 - (\alpha/m)$, the probability that all m confidence intervals are simultaneously correct will be at least $1 - \alpha$.

Theorem (21.32) thus provides a simple procedure for controlling the joint confidence level at $1 - \alpha$ when m pairwise comparisons are to be made.

(21.33)

> The confidence limits for m simultaneous pairwise comparisons $\mu_j - \mu_{j'}$ with joint confidence coefficient of at least $1 - \alpha$ for the single-factor ANOVA model (21.1) are:
>
> $$(\overline{Y}_j - \overline{Y}_{j'}) \pm t s\{\overline{Y}_j - \overline{Y}_{j'}\}$$
>
> *where:*
>
> $t = t(1 - \alpha/2m; n_T - r)$
> $s\{\overline{Y}_j - \overline{Y}_{j'}\}$ is given by (21.30)

Note that the percentile of t that must be used is $100(1 - \alpha/2m)$. With risk $\alpha/2m$ placed in each tail, the combined risk in both tails is α/m and the confidence coefficient is $1 - (\alpha/m)$ for each confidence interval.

EXAMPLE

For the two pairwise comparisons in the soft drink example given earlier, it is desired to control the joint confidence coefficient at 90 percent. Hence, for $\alpha = 0.10$ and $m = 2$, each pairwise comparison requires $t(0.975; 16) = 2.120$ because $1 - (\alpha/2m) = 0.975$ and $n_T - r = 16$. The estimated standard deviation $s\{\bar{Y}_4 - \bar{Y}_2\}$ was calculated earlier from (21.30) to be 0.98848; it is applicable for all pairwise comparisons in this study because each treatment has the same number of observations. The remaining calculational details are as follows:

Colors	$\mu_j - \mu_{j'}$	$\bar{Y}_j - \bar{Y}_{j'}$	$\pm ts\{\bar{Y}_j - \bar{Y}_{j'}\}$
Green − colorless	$\mu_4 - \mu_1$	$31.46 - 27.32 = 4.14$	$\pm 2.120(0.98848) = \pm 2.096$
Pink − colorless	$\mu_2 - \mu_1$	$29.56 - 27.32 = 2.24$	$\pm 2.120(0.98848) = \pm 2.096$

Hence, the two confidence intervals are:

$$2.04 \leq \mu_4 - \mu_1 \leq 6.24$$

and:

$$0.14 \leq \mu_2 - \mu_1 \leq 4.34$$

The joint confidence coefficient for the two intervals is at least 0.90. Hence, with confidence of at least 90 percent, it can be concluded that both the green and pink drinks have greater mean sales than the colorless drink. □

21.5   COMPUTER OUTPUT

Computer program packages for ANOVA are widely available, and a number of pocket calculators also can perform ANOVA calculations. Figure 21.3 shows output for the soft drink example from a computer package.

Block 1 contains the data input. The observations here are shown in columns, with each column representing a treatment.

Block 2 shows, for each treatment, the sample size n_j, the sample mean $\bar{Y}_j$, and the sample standard deviation. It also shows the corresponding statistics for the entire study.

Block 3 contains the ANOVA table.

21.6   RESIDUAL ANALYSIS

Residuals

The aptness of ANOVA model (21.1) needs to be evaluated, just as evaluation of the aptness of regression models is required. Residual plots again are most helpful. The residual e_{ij} for the ANOVA model (21.1) is defined as follows.

FIGURE 21.3 Computer output for single-factor ANOVA — Soft drink example

```
SINGLE FACTOR ANOVA

1. DATA
                    26.5      31.2      27.9      30.8
                    28.7      28.3      25.1      29.6
                    25.1      30.8      28.5      32.4
                    29.1      27.9      24.2      31.7
                    27.2      29.6      26.5      32.8


2. TREATMENT        N         MEAN          STD. DEV.

      1             5         27.32          1.6346
      2             5         29.56          1.4639
      3             5         26.44          1.816
      4             5         31.46          1.2876
      TOTAL         20        28.695         2.4701


3. ANOVA TABLE

      SOURCE          SS            DF        MS

      TREATMENTS      76.8455        3        25.61517
      ERROR           39.0840       16         2.44275
      TOTAL          115.9295       19
```

(21.34)

$$e_{ij} = Y_{ij} - \overline{Y}_j$$

Thus, e_{ij} represents the deviation of the observation Y_{ij} from the sample mean for the jth treatment. Table 21.3 contains the residuals for the soft drink example, calculated from the data in Table 21.1. For example, $e_{11} = Y_{11} - \overline{Y}_1 = 26.5 - 27.32 = -0.82$.

TABLE 21.3 Residuals for the soft drink example

Observation i	Treatment j			
	1	2	3	4
1	−0.82	1.64	1.46	−0.66
2	1.38	−1.26	−1.34	−1.86
3	−2.22	1.24	2.06	0.94
4	1.78	−1.66	−2.24	0.24
5	−0.12	0.04	0.06	1.34
Total	0.00	0.00	0.00	0.00

Comment   Note from (21.9) and (21.34) that $SSE = \Sigma\,\Sigma\,e_{ij}^2$. Also, note from Table 21.3 that the residuals sum to zero for each treatment.

(21.35)
$$\sum_i e_{ij} = 0 \qquad j = 1, 2, \ldots, r$$

Since there are r treatments, the n_T residuals are subject to r constraints of the form (21.35). This is consistent with our earlier discussion where we noted that SSE is associated with $n_T - r$ degrees of freedom.

Residual Plots

As in regression, the residuals e_{ij} should reflect the properties ascribed to the error terms ε_{ij} if the model is apt. ANOVA model (21.1), as we know, requires the error terms ε_{ij} to be independent and normally distributed with constant variance σ^2. The construction and use of residual plots for ANOVA proceeds in a manner similar to that for regression. We will illustrate the use of residual plots for two departures from the ANOVA model (21.1) by means of new examples, since the residual plots for the soft drink example do not suggest any departures from the ANOVA model.

Lack of Independence.   In ANOVA studies, the observations frequently are taken in a time sequence. The residuals should then be plotted against time order and checked for independence over time.

EXAMPLE

Figure 21.4 contains residual plots for a single-factor ANOVA study based on model (21.1), in which respondents were asked to estimate the price of a rug. The three treatments were different pile lengths of the rug. The residuals are plotted against observation number, and the plots show tracklike patterns instead of random scatter. Subsequent investigation found that many respondents had been able to overhear the price estimate of the preceding respondent for the same rug.   ☐

Lack of independence in the error terms ε_{ij} can have serious effects on inferences in ANOVA, causing the true risks of errors and confidence coefficients to differ materially from the stated ones. Thus, it is important to prevent this difficulty whenever possible.

Nonconstancy of Error Variances.   A convenient graphic check for constancy of the error variance from treatment to treatment can be made by comparing the scatter in the residual plots for the several treatments.

EXAMPLE

Figure 21.5 contains the residuals taken from a study of productivity of salespeople under four different compensation plans. The plots suggest that the error

FIGURE 21.4   Residual plots suggesting lack of independence

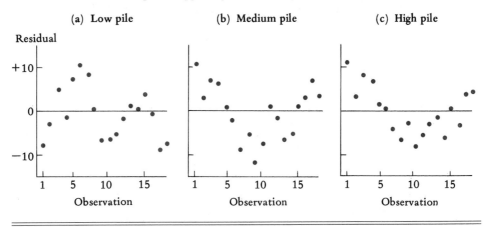

FIGURE 21.5   Residual plots suggesting unequal treatment variances

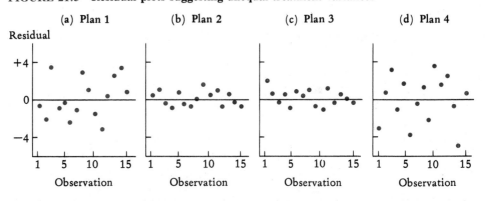

variances are unequal, those for plans 1 and 4 being greater than those for plans 2 and 3. ☐

Unequal error variances can also arise as a result of time-related effects, as when random measurement errors become smaller in the course of an experiment because the laboratory technicians become increasingly proficient with the instruments.

Inferences for model (21.1) involving the F test are not seriously affected by unequal error variances if the sample sizes n_j are approximately equal. However, pairwise comparisons can be seriously affected so that the actual and specified confidence coefficients may differ markedly. Frequently, it is possible to find a mathematical transformation of Y that will produce approximately equal error variances.

21.7 OPTIONAL TOPIC—CORRESPONDENCE BETWEEN ANOVA AND REGRESSION

As was noted earlier, ANOVA models may be viewed as special cases of the general linear regression model. We now illustrate the regression formulation of the single-factor ANOVA model.

Development of Regression Model

In Chapter 20, we saw how indicator variables can be used in a regression model to represent a qualitative variable. Since the factor of interest in a single-factor ANOVA study (for example, color of soft drink) is equivalent to a qualitative variable, with each factor level (for instance, green) corresponding to one outcome of the variable, the factor levels are represented in a regression function by indicator variables.

EXAMPLE

For the soft drink example, the factor is color of soft drink, and the factor levels are the four colors: colorless, pink, orange, and green. The ANOVA model for the soft drink example is:

$$Y_{ij} = \mu_j + \varepsilon_{ij} \qquad j = 1, 2, 3, 4$$

To develop the equivalent regression model, we require three indicator variables, which we shall define as follows.

(21.36)

$$X_1 = \begin{matrix} 1 & \text{if color is pink} \\ 0 & \text{otherwise} \end{matrix}$$

$$X_2 = \begin{matrix} 1 & \text{if color is orange} \\ 0 & \text{otherwise} \end{matrix}$$

$$X_3 = \begin{matrix} 1 & \text{if color is green} \\ 0 & \text{otherwise} \end{matrix}$$

Note that the factor level colorless has been chosen as the reference class in this system of indicator variables, that is, the class for which all three indicator variables are set equal to zero.

The equivalent regression model then is as follows.

(21.37)

$$Y_i = \beta_0 + \beta_1 X_{i1} + \beta_2 X_{i2} + \beta_3 X_{i3} + \varepsilon_i \qquad i = 1, 2, \ldots, n_T$$

Here, the index i now ranges over all $n_T = 20$ cases in the experiment, and the indicator variables are as defined in (21.36).

Table 21.4 contains the soft drink data with the indicator variables coded in accord with (21.36). For example, for each case for reference treatment 1 (colorless), the associated values of the independent variables are $X_1 = 0$, $X_2 = 0$, and $X_3 = 0$. Similarly, for each case for treatment 2 (pink), the values of the independent variables are $X_1 = 1$, $X_2 = 0$, and $X_3 = 0$.   □

Correspondences of Parameters

We now develop the correspondences between the ANOVA parameters μ_j and the regression parameters β_k for the soft drink example. From (21.37), we know that the response function for the regression model is as follows.

$$(21.38) \qquad E\{Y\} = \beta_0 + \beta_1 X_1 + \beta_2 X_2 + \beta_3 X_3$$

For a case from reference treatment 1 (colorless), for which $X_1 = X_2 = X_3 = 0$, the response function simplifies to:

$$E\{Y\} = \beta_0 + \beta_1(0) + \beta_2(0) + \beta_3(0) = \beta_0$$

TABLE 21.4   Data for a regression analysis in the soft drink example

Case i	Color	Indicator Variables X_{i1}	X_{i2}	X_{i3}	Y_i
1	Colorless	0	0	0	26.5
2	Colorless	0	0	0	28.7
3	Colorless	0	0	0	25.1
4	Colorless	0	0	0	29.1
5	Colorless	0	0	0	27.2
6	Pink	1	0	0	31.2
7	Pink	1	0	0	28.3
8	Pink	1	0	0	30.8
9	Pink	1	0	0	27.9
10	Pink	1	0	0	29.6
11	Orange	0	1	0	27.9
12	Orange	0	1	0	25.1
13	Orange	0	1	0	28.5
14	Orange	0	1	0	24.2
15	Orange	0	1	0	26.5
16	Green	0	0	1	30.8
17	Green	0	0	1	29.6
18	Green	0	0	1	32.4
19	Green	0	0	1	31.7
20	Green	0	0	1	32.8

But according to the ANOVA model, the mean for treatment 1 is μ_1, so we have $\beta_0 = \mu_1$.

For a case from treatment 2 (pink), for which $X_1 = 1$, $X_2 = 0$, $X_3 = 0$, the response function (21.38) reduces to:

$$E\{Y\} = \beta_0 + \beta_1(1) + \beta_2(0) + \beta_3(0) = \beta_0 + \beta_1$$

Since the mean for treatment 2 in the ANOVA model is μ_2, it follows that $\beta_0 + \beta_1 = \mu_2$. Moreover, we have just shown that $\beta_0 = \mu_1$. Hence, $\beta_1 = \mu_2 - \mu_1$.

Continuing in this fashion, we establish the following correspondences between the parameters of the single-factor ANOVA model and the equivalent regression model.

(21.39a)

$$\beta_0 = \mu_1 \qquad\qquad \beta_2 = \mu_3 - \mu_1$$

$$\beta_1 = \mu_2 - \mu_1 \qquad \beta_3 = \mu_4 - \mu_1$$

(21.39b)

$$\mu_1 = \beta_0 \qquad\qquad \mu_3 = \beta_0 + \beta_2$$

$$\mu_2 = \beta_0 + \beta_1 \qquad \mu_4 = \beta_0 + \beta_3$$

Note in (21.39a) that β_1, β_2, and β_3 represent the *differential effects* of treatments 2, 3, and 4, respectively, compared with reference treatment 1.

Inferences with Regression Model

Inferences about the treatment means in the ANOVA model can be carried out in terms of inferences about the regression coefficients of the equivalent regression model. We demonstrate two of these equivalent procedures for the soft drink example.

EXAMPLE 1

Estimating Treatment Means.   Figure 21.6 shows selected computer output from fitting regression model (21.37) to the soft drink data in Table 21.4. Observe that b_0 is identical to the sample mean for treatment 1 as given in Table 21.1; that is, $\bar{Y}_1 = b_0 = 27.32$. Likewise, observe that:

$$\bar{Y}_2 = b_0 + b_1 = 27.32 + 2.24 = 29.56$$

$$\bar{Y}_3 = b_0 + b_2 = 27.32 - 0.88 = 26.44$$

$$\bar{Y}_4 = b_0 + b_3 = 27.32 + 4.14 = 31.46$$

These correspondences follow from the relations between the μ_j $(j = 1, 2, 3, 4)$ and β_k $(k = 0, 1, 2, 3)$ given in (21.39b). $\qquad\qquad\Box$

EXAMPLE 2

Testing the Equality of Treatment Means.   The ANOVA test for equality of means in the soft drink example involves the alternatives:

$$H_0: \mu_1 = \mu_2 = \mu_3 = \mu_4$$

$$H_1: \text{not all } \mu_j \text{ are equal}$$

FIGURE 21.6   Computer output for a regression analysis in the soft drink example

VARIABLE	REG. COEFF.	STD. DEV.	T STAT.	P-VALUE
CONSTANT	27.32000	.69896	39.086	.00000
X1	2.24000	.98848	2.266	.03767
X2	-.88000	.98848	-.890	.38652
X3	4.14000	.98848	4.188	.00070

ANALYSIS OF VARIANCE TABLE

SOURCE	SUM OF SQUARES	DF	MEAN SQUARE
REGRESSION	76.84550	3	25.61517
RESIDUAL	39.08400	16	2.44275
TOTAL	115.92950	19	

F STAT. = 10.486   UPPER-TAIL P-VALUE = .00047

We see from (21.39a) that, when $\mu_1 = \mu_2 = \mu_3 = \mu_4$, the regression coefficients β_1, β_2, and β_3 equal zero. Hence, the equivalent test in the regression framework is:

$$H_0: \beta_1 = \beta_2 = \beta_3 = 0$$

$$H_1: \text{not all } \beta_k = 0 \quad k = 1, 2, 3$$

These are precisely the alternatives for testing the existence of a regression relation. The appropriate test statistic, as given in (20.14), is $F^* = MSR/MSE$, and the test procedure is given in (20.15).

Referring to the regression output in Figure 21.6, we see that the entries in the regression ANOVA table are identical to those in Table 21.2b for ANOVA model (21.1), with *SSR* corresponding to *SSTR*. Thus, the F test for the equality of treatment means is identical to the F test for the existence of a regression relation when indicator variables are used in the regression model.   □

Comment   When the factor under study is qualitative, such as color of soft drink, the ANOVA model and the regression model with indicator variables are equivalent models, and there is no choice of any other type of regression model. When the factor is quantitative, however, the analyst can choose whether to use (1) an ANOVA model or the equivalent regression model with indicator variables, or (2) a regression model with the factor treated as a continuous variable. Consider a study of the effect of size of crew (4, 12, 20 persons) on volume of output. If interest centers solely on the three crew sizes under study, the factor can be treated as a qualitative one by means of the ANOVA model or the equivalent regression model with indicator variables, requiring no specification of a functional relation between crew size and volume of output. On the other hand, if the factor is treated as a quantitative variable in an ordinary regression model, a specification of the nature of the relation between crew size and output volume (for instance, linear, quadratic) must be made. Making this specification is more restrictive than treating the independent variable as qualitative, but it does have the advantage of permitting inferences about crew sizes other than those in the study (for example, crews of 5 or 17 persons).

PROBLEMS

21.1 In each of the following single-factor ANOVA studies, identify (1) the factor, stating whether it is experimental or observational; (2) the factor levels; (3) the dependent variable.

   a. An automobile service chain operates three kinds of service centers: tire centers, brake system centers, exhaust system centers. An analyst wishes to investigate quarterly profit rates for the different kinds of centers.

   b. Three interviewers have been hired for a household survey. Households to be surveyed are randomly assigned to the interviewers in order to compare the effects of the interviewers on responses about household income.

21.2 In each of the following single-factor ANOVA studies, identify (1) the factor, stating whether it is experimental or observational; (2) the factor levels; (3) the dependent variable.

   a. In a study of heat resistance of a certain type of wood paint, identical painted wood panels are exposed to one of three different temperatures in laboratory test rooms. Temperature levels are randomly assigned to the panels, and resistance is measured by the time until heat blistering occurs.

   b. Financial reports of companies from five different extractive industries are examined to compare the effects of a depletion tax allowance on reported annual earnings in these industries.

21.3 **Television Commercials.** Each of 15 persons in a consumer panel was shown one of three versions of a television commercial and was asked to rate its appeal on a scale from 1 (poor) to 12 (excellent). Five persons were assigned at random to each of the three versions. Assume that ANOVA model (21.1) is applicable and that the parameter values are $\mu_1 = 8.5$, $\mu_2 = 5.3$, $\mu_3 = 7.6$, $\sigma = 1.0$.

   a. Portray the model for this case graphically, as in Figure 21.1a.

   b. Explain the meaning of the following symbols: (1) μ_1, (2) Y_{32}, (3) ε_{41}, (4) σ.

   c. Will Y_{23} necessarily be smaller than Y_{41}? Are Y_{23} and Y_{41} statistically independent? Explain.

21.4 In a study of bonus pay earned under four bonus plans, five salespeople were assigned at random to each of these plans. Assume that ANOVA model (21.1) is applicable and that the parameter values (in dollars) are $\mu_1 = 370$, $\mu_2 = 850$, $\mu_3 = 450$, $\mu_4 = 590$, $\sigma = 40$.

   a. Portray the model for this case graphically, as in Figure 21.1a.

   b. Explain the meaning of the following symbols: (1) μ_3, (2) Y_{14}, (3) ε_{32}, (4) σ.

   c. If $Y_{52} = 800$, what is ε_{52}?

*21.5 For each of the following cases, state the degrees of freedom associated with (1) *SSTR*, (2) *SSE*, (3) *SSTO*.
a. Four treatments, seven observations for each treatment.
b. Four treatments, 20 observations for each treatment.
c. Five treatments, three observations for each of treatments 1, 2, 3, and 4, and four observations for treatment 5.

21.6 For each of the following cases, state the degrees of freedom associated with (1) *SSTR*, (2) *SSE*, (3) *SSTO*.
a. Three treatments, four observations for each treatment.
b. Five treatments, 16 observations for each treatment.
c. Four treatments, three observations for each of treatments 1 and 2, and eight observations for each of treatments 3 and 4.

*21.7 **Instructional Modes.** Fifteen students enrolled in a mathematics course were randomly divided into three groups of 5 students each. Each group was then randomly assigned to one of three instructional modes, augmenting the traditional course materials: (1) programmed text, (2) videotapes, (3) conversational computer programs. At the end of the course, each student was given the same achievement test. The test scores follow.

Observation i	Treatment j		
	1 Programmed Text	2 Videotapes	3 Computer Programs
1	86	90	78
2	82	79	70
3	94	88	65
4	77	87	74
5	86	96	63

Assume that ANOVA model (21.1) is applicable.
a. State the values of (1) n_T, (2) n_1, (3) Y_{32}, (4) $\bar{Y}_2$, (5) $\bar{\bar{Y}}$.
b. For which instructional mode is the sample mean $\bar{Y}_j$ largest? Is the treatment mean μ_j for this instructional mode necessarily the largest of the three treatment means? Explain.
c. Obtain the ANOVA table.

21.8 **Portable Radios.** A product-rating organization tested battery life for four comparable makes of portable radios. Six radios of each make were purchased off the shelf from local retail stores, and identical fully charged batteries were inserted on a random basis. The observations on battery playing hours at high volume follow.

Observation i	Treatment j			
	1	2	3	4
1	5.5	4.7	6.1	4.5
2	5.0	3.9	5.7	5.1
3	5.2	4.3	5.0	4.3
4	5.3	4.5	5.3	4.1
5	4.8	4.1	5.2	4.5
6	4.8	4.3	6.3	5.1

Assume that ANOVA model (21.1) is applicable.
  a. State the values of (1) n_T, (2) n_3, (3) Y_{24}, (4) $\bar{Y}_3$, (5) $\bar{\bar{Y}}$.
  b. Compute *SSTO*, *SSE*, and *SSTR*. Are your results consistent with identity (21.13a)?
  c. Obtain the ANOVA table.

21.9 **Low-Energy Cookware.**   A kitchen utensils manufacturer selected 18 similar stores to try out three different promotional displays for a new low-energy cooking pot. The display that generates the highest sales in this study is to be used in the manufacturer's national promotion program. Each display was assigned at random to six stores. Sales (in dollars) for the stores during a two-week observation period follow (note that the treatments are placed in rows).

| Display | Store i | | | | | |
j	1	2	3	4	5	6
1	2161	1769	2748	1782	2830	3183
2	2379	1913	1119	1208	1962	1689
3	1479	1024	1598	963	1913	2251

Assume that ANOVA model (21.1) is applicable.
  a. Calculate the three sample treatment means. Which display generated the highest mean sales per store in the experiment? Does this display necessarily have the largest treatment mean μ_j? Explain.
  b. Obtain the ANOVA table.

21.10 **Cake Ingredient.**   A food scientist studied the effect of the amount of a dairy ingredient on the volume of a baked cake (measured in milliliters per 100 grams). Twenty-one cakes were baked, using identical recipes and procedures except for the amount of the ingredient. Seven of the 21 cakes were each baked by using one of three progressively larger amounts of the ingredient (denoted by 1, 2, and 3). The cake volumes follow (note that the treatments are placed in rows).

| Ingredient Amount | Cake i | | | | | | |
j	1	2	3	4	5	6	7
1	351	369	381	386	370	358	398
2	390	394	406	407	415	375	374
3	398	409	415	399	434	427	414

Assume that ANOVA model (21.1) is applicable.
  a. Calculate the three sample treatment means. Do the values of these means suggest that cake volume tends to increase with the amount of the ingredient? Can one be certain of this tendency from these data? Explain.
  b. Obtain the ANOVA table.

*21.11 Refer to **Instructional Modes** Problem 21.7.
  a. State the alternatives for the test of equality of treatment means here. Specify the distribution of F^* if H_0 is correct.
  b. Test for the equality of the treatment means, controlling the α risk at 0.05. State the decision rule, the value of the test statistic, and the conclusion.
  c. Given your conclusion in **b**, should the next step involve an examination of the differences among the individual treatment means? Comment.

21.12 Refer to **Portable Radios** Problem 21.8.

    a. Test for the equality of the treatment means, controlling the α risk at 0.01. State the alternatives, the decision rule, the value of the test statistic, and the conclusion.

    b. Does your conclusion in **a** imply that radio makes 2 and 4 have different mean playing hours at high volume? Comment.

    c. If playing volume had not been controlled in the tests, would the interpretation of the results be affected? Discuss.

    d. Is it likely that the differences between makes are actually due to differences in the test batteries? Explain.

21.13 Refer to **Low-Energy Cookware** Problem 21.9.

    a. Test whether or not the displays are equally effective in generating sales, controlling the α risk at 0.05. State the alternatives, the decision rule, the value of the test statistic, and the conclusion.

    b. Obtain $\sqrt{MSE}$. What parameter does this statistic estimate? Interpret this parameter in the context of this experiment.

21.14 Refer to **Cake Ingredient** Problem 21.10. Test for the equality of the treatment means, controlling the α risk at 0.01. State the alternatives, the decision rule, the value of the test statistic, and the conclusion.

*21.15 Refer to **Instructional Modes** Problem 21.7.

    a. Obtain a 95 percent confidence interval for μ_1. Interpret your interval estimate.

    b. Obtain a 95 percent confidence interval for the pairwise comparison $\mu_2 - \mu_1$. In light of the test for the equality of treatment means in Problem 21.11, what does your confidence interval suggest about the mean achievement score for the computer programs instructional mode? Explain.

21.16 Refer to **Portable Radios** Problem 21.8.

    a. Obtain a 99 percent confidence interval for the mean playing hours at high volume for radio make 3.

    b. Obtain a 99 percent confidence interval for the difference in mean playing hours at high volume between radio makes 3 and 1. Interpret your confidence interval.

    c. If another radio of make 3 were to be tested under the same conditions, would your confidence interval in **a** provide appropriate limits for the number of playing hours for this radio? Explain.

21.17 Refer to **Low-Energy Cookware** Problem 21.9.

    a. The marketing manager had surmised in advance of seeing the experiment's sales figures that display 1 would be the most effective. Estimate the treatment mean for this display, using a 95 percent confidence interval. Interpret your interval estimate.

    b. Obtain a 95 percent confidence interval for the difference in mean sales per store between displays 1 and 2. Does this confidence interval partly bear out the marketing manager's surmise described in **a**? Comment.

21.18 Refer to **Cake Ingredient** Problem 21.10.

    a. Obtain a 95 percent confidence interval for (1) μ_1, (2) μ_2, (3) $\mu_2 - \mu_1$.

    b. The confidence intervals for μ_1 and for μ_2 are overlapping, yet the confidence interval for $\mu_2 - \mu_1$ does not include 0. Are the intervals inconsistent with one another? Comment.

*21.19 Refer to **Instructional Modes** Problem 21.7. Construct simultaneous confidence intervals for the pairwise comparisons $\mu_2 - \mu_1$, $\mu_2 - \mu_3$, and $\mu_1 - \mu_3$ such that the joint confidence coefficient is at least 94 percent. Do your confidence intervals indicate that the computer programs mode is clearly inferior to the other two modes? Explain. What other information is provided by the confidence intervals?

21.20 Refer to **Portable Radios** Problem 21.8. It is desired to construct simultaneous confidence intervals for the difference in mean playing hours for the two least expensive makes of radio (makes 2 and 4) and for the difference between the two makes with best tone (makes 1 and 4). Obtain these confidence intervals with a joint confidence coefficient of at least 90 percent. Interpret your confidence intervals.

21.21 Refer to **Low-Energy Cookware** Problem 21.9. Construct simultaneous confidence intervals for the differences in mean sales per store between displays 1 and 2 and between displays 1 and 3. Use a joint confidence coefficient of at least 0.96. Do your intervals indicate that display 1 is clearly superior to both of the others? Comment.

21.22 Refer to **Cake Ingredient** Problem 21.10. Construct simultaneous confidence intervals for $\mu_2 - \mu_1$, $\mu_3 - \mu_1$, and $\mu_3 - \mu_2$, using a joint confidence coefficient of at least 0.97. Do the confidence intervals indicate that mean cake volume increases with the amount of the ingredient? Comment.

*21.23 Refer to **Instructional Modes** Problem 21.7.
   a. Obtain the residuals. Do they sum to zero for each treatment? Is this necessarily the case?
   b. Construct a residual plot, as in Figure 21.5, and analyze it for constancy of the treatment error variances. State your findings.

21.24 Refer to **Portable Radios** Problem 21.8.
   a. Obtain the residuals.
   b. Construct a residual plot, as in Figure 21.5, and analyze it for constancy of the treatment error variances. State your findings.

21.25 Refer to **Low-Energy Cookware** Problem 21.9. Obtain the residuals and make a residual plot, as in Figure 21.5. Analyze this plot for unequal treatment error variances. State your findings.

21.26 Refer to **Cake Ingredient** Problem 21.10.
   a. Obtain the residuals.
   b. For each treatment, the cakes were baked in the order indicated (from 1 to 7). Construct a residual plot as in Figure 21.4, and analyze it for lack of independence. State your findings.
   c. The sample standard deviations of the seven cake volumes for each treatment are 16.3, 16.0, and 13.4, respectively. Does the assumption of constant standard deviation for the three treatments appear to be fairly reasonable here? Comment.

*21.27 Refer to **Instructional Modes** Problem 21.7. Let two indicator variables be defined as follows:

$$X_1 = 1 \text{ if instructional mode is videotapes, 0 otherwise}$$

$$X_2 = 1 \text{ if instructional mode is computer programs, 0 otherwise}$$

   a. State the regression model equivalent to ANOVA model (21.1) for this study.
   b. Arrange the data for computer input in the format of Table 21.4.
   c. The following results were obtained in fitting the regression model in **a**: $b_0 = 85.0$, $s\{b_0\} = 2.769$, $b_1 = 3.0$, $s\{b_1\} = 3.916$. Construct a 95 percent confidence interval for (1) β_0, (2) β_1. What treatment characteristics are estimated by these confidence intervals?

21.28 Refer to **Portable Radios** Problems 21.8 and 21.12. Let three indicator variables be defined as follows:

$$X_1 = 1 \text{ if radio is make 2, 0 otherwise}$$

$$X_2 = 1 \text{ if radio is make 3, 0 otherwise}$$

$$X_3 = 1 \text{ if radio is make 4, 0 otherwise}$$

a. State the regression model equivalent to ANOVA model (21.1) for this experiment.

b. For the regression model in **a**, $SSR = 5.8800$ and $SSE = 3.0200$. Test for equality of the treatment means via the regression approach, controlling the α risk at 0.01. State the alternatives, the decision rule, the value of the test statistic, and the conclusion. Show the correspondence to the ANOVA approach in Problem 21.12a.

c. For the regression model in **a**, $b_2 = 0.50$ and $s\{b_2\} = 0.2244$. Construct a 99 percent confidence interval for β_2. What differential effect is estimated by this interval?

21.29 Refer to **Low-Energy Cookware** Problem 21.9. Let two indicator variables be defined as follows:

$$X_1 = 1 \text{ if store is assigned display 2, 0 otherwise}$$

$$X_2 = 1 \text{ if store is assigned display 3, 0 otherwise}$$

Selected results for the regression of Y on X_1 and X_2 follow.

k:	0	1	2	
b_k:	____	−700.5	____	SSR = ____
$s\{b_k\}$:	____	304.1	____	SSE = 4,160,290

a. Use the sample treatment means and the ANOVA results from Problem 21.9 to obtain the missing regression results.

b. Test for the equality of treatment means via the regression approach, controlling the α risk at 0.05. State the alternatives, the decision rule, the value of the test statistic, and the conclusion.

c. Construct a 95 percent confidence interval for β_1. What differential effect is estimated by this interval?

21.30 Refer to **Cake Ingredient** Problem 21.10. Let two indicator variables be defined as follows:

$$X_1 = 1 \text{ if cake has amount 2, 0 otherwise}$$

$$X_2 = 1 \text{ if cake has amount 3, 0 otherwise}$$

Selected results for the regression of Y on X_1 and X_2 follow.

k:	0	1	2	
b_k:	373.29	21.143	____	MSR = ____
$s\{b_k\}$:	____	____	____	MSE = 233.143

a. Use the sample treatment means and the ANOVA results from Problem 21.10 to obtain the missing regression results.

b. Test for the equality of treatment means via the regression approach, controlling the α risk at 0.01. State the alternatives, the decision rule, the value of the test statistic, and the conclusion.

c. Construct a 95 percent confidence interval for (1) β_0, (2) β_1. What treatment characteristics are estimated by these confidence intervals?

EXERCISES

21.31 Refer to **Television Commercials** Problem 21.3. Obtain the following probabilities: (1) $P(Y_{23} < 8.0)$, (2) $P(Y_{41} - Y_{23} > 0)$.

21.32 If $\bar{Y}_j = \bar{\bar{Y}}$ for all r treatments, which of the following will be true? Which false? Why? (1) $SSTR = 0$, (2) $SSTO = SSTR + SSE$, (3) $SSE = 0$.

21.33 In an experiment, three treatments were investigated and five subjects were utilized for each treatment. Assume that ANOVA model (21.1) is applicable and that $\sigma^2 = 9$.

  a. Obtain $E\{MSTR\}$ and $E\{MSE\}$ when (1) $\mu_1 = \mu_2 = \mu_3 = 10$; (2) $\mu_1 = 6$, $\mu_2 = 10$, $\mu_3 = 14$; (3) $\mu_1 = 0$, $\mu_2 = 10$, $\mu_3 = 20$.

  b. What do your results in **a** imply about the type of decision rule appropriate for the F test of the equality of treatment means? Explain.

  c. Redo **a** assuming that ten subjects were utilized for each treatment. What do your results imply about the power of the test when the treatment sample sizes are increased? Discuss.

21.34 Refer to Comment 2 on p. 723. Show that $\sum \sum (Y_{ij} - \bar{Y}_j)(\bar{Y}_j - \bar{\bar{Y}}) = 0$.

21.35 Use the properties of ANOVA model (21.1) to prove (21.22).

$$\left[\textit{Hint: } E\left\{ \sum_{i=1}^{n_j} (Y_{ij} - \bar{Y}_j)^2/(n_j - 1) \right\} = \sigma^2. \right]$$

21.36 Use the properties of ANOVA model (21.1) to prove that $MSE/\sigma^2 = \chi^2(n_T - r)/(n_T - r)$. [*Hint:* Use (13.28) and Exercise B.15b in Appendix B.]

21.37 Two confidence intervals are being constructed from the same sample data, so they are not statistically independent. Let α_1 and α_2 denote the probabilities for the two intervals, respectively, that the procedure will lead to an *incorrect* interval. Use addition theorem (4.16) to prove that the probability both intervals will be correct is at least $1 - \alpha_1 - \alpha_2$. [*Hint:* $P(E_1 \cap E_2) = 1 - P(E_1^* \cup E_2^*)$.]

STUDIES

21.38 Refer to the **Financial Characteristics** data set. Consider all firms in the electronic computer equipment industry with net assets in year 2 of $100 million or less (that is, exclude firms 217, 219, 220, 221, 224, and 227). Divide these firms into two groups — those that made a profit in year 1 and those that did not. View the net income amounts in year 2 for the two groups as independent samples from normal populations with equal variances.

  a. Investigate whether mean net income in year 2 differs between firms that made a profit in year 1 and those that did not. In your investigation, perform an ANOVA test for the equality of population means, controlling the α risk at 0.05. Also, estimate the difference between the two population means by a 95 percent confidence interval.

  b. For the test in **a**, obtain test statistic t^* in (13.11a), and show numerically the equivalence of the value of this test statistic to the test statistic F^* in **a**.

  c. Test whether or not the two populations of net incomes in year 2 have equal variances, controlling the α risk at 0.02 when $\sigma_1^2 = \sigma_2^2$. [*Hint:* $F(0.99; 18, 11) = 4.15$, $F(0.99; 11, 18) = 3.43$.]

  d. Using the assumption that the populations of net incomes are normal, obtain a point estimate of the probability that a profitable firm in year 1 is unprofitable in year 2. Compare this estimate with a direct estimate of this probability based on the sample proportion. Does the direct estimate depend on any assumptions about the population distribution?

21.39 (Computer needed.) Refer to the **Power Cells** data set. It is desired to study the statistical relation between the number of cycles before failure (CYCLES) and the charge rate (CHARGE) and depth of discharge (DEPTH) for cells operating at an ambient temperature of 30°C. The study data consist of the 18 cells numbered 3, 6, 9, and so on. The six treatments under investigation are as follows:

Treatment j:	1	2	3	4	5	6
Charge Rate:	0.4	0.4	1.0	1.0	1.6	1.6
Depth of Discharge:	40	80	40	80	40	80

Assume that ANOVA model (21.1) is appropriate.

a. Identify the three observations associated with each of the six treatments. Obtain the ANOVA table, and test for the equality of treatment means, controlling the α risk at 0.10. Interpret your conclusion.

b. Regress the number of cycles before failure (Y) on the charge rate (X_1) and the depth of discharge (X_2), employing multiple regression model (20.1). State the estimated regression function. Test whether or not a regression relation holds, controlling the α risk at 0.10. Does the conclusion here have the same interpretation as the one in **a**? Explain.

c. Which model, the ANOVA model in **a** or the multiple regression model in **b**, provides a better fit to the response data here, as measured by the magnitude of *MSE*?

d. Which combination of the charge rate (0.4, 1.0, 1.6) and the depth of discharge (40, 80) appears to yield the largest mean number of cycles before failure at an ambient temperature of 30°C? Construct a 95 percent confidence interval for this mean parameter, using (1) the ANOVA model in **a**, (2) the multiple regression model in **b** (computer calculations show that $s\{\hat{Y}_h\} = 44.86$ when $X_{h1} = 0.4$ and $X_{h2} = 40$). Are the two interval estimates similar? Comment.

21.40 A maintenance report for five aircraft of the same type in an airline's fleet contains data on the time intervals (in operating hours) between successive breakdowns in each aircraft's air-conditioning equipment. The numbers of operating hours between breakdowns, shown in chronological order for each aircraft, follow.

	Aircraft				
Breakdown	1	2	3	4	5
1	17	100	54	6	13
2	31	98	15	36	270
3	179	87	33	281	91
4	45	230	68	54	38
5	27	81	132	254	603
6	198	98	67		118
7		140	60		450
8		488	229		
9		48	209		
10		89	102		

Assume that ANOVA model (21.1) is applicable when the logarithms of the time intervals are employed.

a. Test whether the aircraft differ in terms of the mean log-time between successive break-downs, using the log-time data. Control the α risk at 0.01. Use logarithms to base 10. [*Hint:* $F(0.99; 4, 33) = 3.95$.]

b. Given your conclusion in **a**, would it be appropriate to combine all 38 sample observations and treat them as a random sample from the same population? Discuss.

c. Obtain the residuals for the log-time data, and construct a residual plot, as in Figure 21.4. Examine the plot for constancy of the error variance and for independence of the error terms for each aircraft. State your findings.

d. Calculate the standardized residuals $e_{ij}/\sqrt{MSE}$, and combine them into one group. Construct a table like Table 19.1, based on the 25th, 50th, and 75th percentiles of the standard normal distribution (since $n_T - r$ is large here). Do you see any strong evidence of lack of normality? Which of your findings in **c** justifies combining the residuals into one group?

UNIT SIX

Bayesian Decision Making

22

BAYESIAN DECISION MAKING I: NO SAMPLE INFORMATION

Decision making has been studied intensively by researchers in the social and behavioral sciences. Basic principles and methods of decision making have been developed that are widely applied in administration, economics, and other traditional fields as well as in special fields, such as genetic counseling and environmental conservation. In this chapter and the next, we extend our discussion of decision making begun in earlier chapters. To understand the nature of the extensions, one must recognize three pertinent points about statistical decision making presented in earlier chapters:

1. Decision rules were constructed and evaluated in terms of risks of errors at different possible values of a population parameter, such as the population mean or the population proportion.
2. Monetary and other consequences of a decision were considered in formulating the alternatives (for example, one-sided versus two-sided alternatives) and in specifying the α and β risks, but these consequences were not measured explicitly and were not incorporated into the calculations.
3. The only information used in reaching a conclusion about the population parameter was the information in the sample itself.

We now broaden our approach to decision making by incorporating monetary and other consequences of a decision into our calculations, and by using any information about the population that is available prior to sampling. In this chapter, we discuss decision making when no sample information is available, and in the next chapter, we take up decision making when sample information is available.

First in this chapter, we consider the common structure of decision problems. Then, we take up decision making under certainty and decision making under uncertainty, in both of which the consequences of decisions are explicitly taken into account. Finally, we consider Bayesian decision making, where not only the consequences of decisions are explicitly taken into account but prior information about the population is also utilized.

22.1 DECISION PROBLEMS

Decision problems, whether in business, government, or any other sphere, have a common structure. To illustrate this common structure, we consider two examples.

1. **Engine Production.** Firm A has developed a new engine that will give greater fuel economy than conventional engines. Tooling up for mass production can be initiated now or can be delayed for another year in the hope that further research and development will yield additional major refinements. The engineering group assesses the probability of major additional refinements in the next year to be 0.6. However, a rival—firm B—has also developed a new engine and has just begun to tool up for mass production. This engine is competitive with the present version of firm A's new engine. Thus, firm A will be at a long-term competitive disadvantage if it delays mass production of its new engine and does not discover major additional refinements in the next year. However, if additional major refinements are discovered within the next year, firm A will more than recoup the losses from the delay in getting into mass production. The executive committee of firm A must now decide whether to initiate immediate tooling up for mass production or undertake further research and development for another year.

2. **Tanker.** A small oil tanker has just discharged its cargo and is about to begin another round trip. The tanker's operator must decide whether to air-blow the ship's wing tanks to clear the gas residue or to fill the tanks with water ballast. The former procedure costs $200, while pumping the water ballast costs $2000. Air blowing will suffice unless there is a collision affecting a wing tank. If this occurs, a wing tank that is merely air-blown will explode, with damage to the ship and possible risk to the crew—about a $10 million estimated loss altogether— whereas a wing tank filled with water ballast cannot explode. The tanker operator is not able to assess the probability of a wing tank collision, except that he knows the possibility cannot be ignored since two narrow and busy channels must be negotiated on the return trip for another cargo of oil. ☐

Structural Elements of Decision Problems

The two examples just presented illustrate the common structural elements of decision problems.

(22.1)

The structural elements of decision problems are as follows:

1. *Decision maker.* The decision maker is either a single individual (the tanker operator) or a group of persons acting as a single individual (the executive committee).
2. *Acts or strategies A_j.* These are the alternative courses of action open to the decision maker (tool up now, delay; air-blow, use water ballast). They

reflect the *controllable variables* in the decision problem—that is, the variables under the decision maker's control.

3. *Outcome states S_i.*   These are the different situations that may prevail and affect the consequences of the acts (further refinements in engine will be discovered, no further refinements in engine will be discovered; wing tank collision, no wing tank collision). The outcome states reflect the *uncontrollable variables* in the decision problem—that is, the variables not under the decision maker's control.

4. *Consequences or payoffs C_{ij}.*   These are the different gains, rewards, and so on, that may be experienced by the decision maker, measured in monetary or other units. The term C_{ij} denotes the gain or payoff experienced if the jth course of action is chosen and the ith outcome state prevails (if the operator uses air blowing and a wing tank collision occurs).

5. *Outcome state probabilities $P(S_i)$.*   These are probabilities that the decision maker may assign to the outcome states. Outcome state probabilities are not present in all decision problems (they are present in the engine production example but not in the tanker example).

6. *Criterion.*   This is a basis for identifying the act that is "best" among the courses of action available to the decision maker (the examples did not illustrate any criterion).

Payoff Tables and Decision Trees

Two convenient devices for displaying the acts, outcome states, consequences, and probabilities in a decision problem are the *payoff table* and the *decision tree*. They present the same information; hence, either one can be used depending on whether a tabular or graphic display is more effective in the given situation.

Payoff Table.   Table 22.1a shows the format and notation we shall employ for payoff tables, when there are r outcome states and c acts. Table 22.1b presents the entries for the tanker example, where $r = c = 2$. Since the tanker operator was unable to assign probabilities $P(S_i)$ to the different possible outcome states S_i, "n.a." (not applicable) is entered in place of the $P(S_i)$. Ordinarily, we shall omit this column when probabilities are not used. The entries in Table 22.1b are negative because they entail losses to the operator. Thus, $C_{11} = -200$, since the cost of air blowing is $200 and no other losses are encountered when no wing tank collision occurs.

Note how effectively Table 22.1b consolidates the information in the tanker decision problem. We see quickly that water ballast entails a loss of $2000 regardless of whether or not there is a collision. Air blowing, on the other hand, entails a loss of only $200 if there is no wing tank collision but a loss of $10 million if such a collision occurs.

Decision Tree.   Figure 22.1a shows the format of a decision tree for a decision problem with two acts and two outcome states, and Figure 22.1b illustrates the decision tree for the tanker example. A decision tree consists of a sequence of *nodes* and

TABLE 22.1   General format of a payoff table and payoff table for the tanker example

(a)
General format

Outcome State	Probability	Act			
		A_1	A_2	$\cdots$	A_c
S_1	$P(S_1)$	C_{11}	C_{12}	$\cdots$	C_{1c}
S_2	$P(S_2)$	C_{21}	C_{22}	$\cdots$	C_{2c}
$\vdots$	$\vdots$	$\vdots$	$\vdots$		$\vdots$
S_r	$P(S_r)$	C_{r1}	C_{r2}	$\cdots$	C_{rc}

(b)
Tanker example
(payoffs in dollars)

Outcome State	Probability	Act	
		Air-Blow	Water Ballast
No wing tank collision	n.a.	-200	-2000
Wing tank collision	n.a.	-10 million	-2000

branches. We enter from the left. The branches emanating from the first node represent the alternative courses of action A_j. Since this node entails a controllable variable, it is called a *decision node*. The next set of branches represents the possible outcome states S_i for each course of action. The nodes from which these branches emanate involve uncontrollable variables and are called *chance nodes*. The outcome state probabilities $P(S_i)$ — when applicable — are shown on the branches emanating from each chance node. Finally, the payoff for each sequence of A_j and S_i is shown at the right. We follow the convention of indicating decision nodes by squares, and chance nodes by circles.

Comments

1. Two or more courses of action must be available to the decision maker before a decision problem exists. If only one course of action is available, there is no decision problem.

2. The identification of feasible acts and relevant outcome states, and the measurement of payoffs, can be highly demanding in actual decision problems. For instance, in our tanker example, it is very difficult to assess the loss resulting from a wing tank explosion. In addition, it is entirely possible that some desirable courses of action are overlooked when the available acts are formulated. One may need to examine the causal basis of a problem in great detail to identify all available acts for solving it.

3. Note that the losses considered in the tanker example are those directly produced by the course of action. Losses that would run on unaffected by the action or that occurred in the past are excluded. Thus, in the tanker example, there is no need to consider the crew's wages, fuel costs, and other costs of the next trip, which will occur regardless of whether water ballast is used or the wing tanks are air-blown.

4.   In the tanker example, the payoffs are expressed as losses and hence are negative. In many decision problems, the payoffs are expressed as gains, such as profits, and are therefore positive.

FIGURE 22.1   **General format of a decision tree and decision tree for the tanker example**

(a) **General Format**

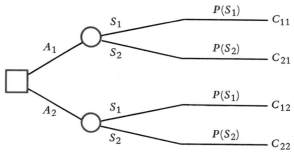

(b) **Tanker Example**

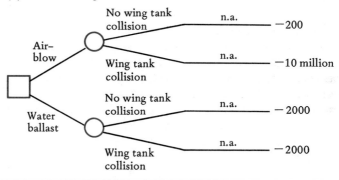

22.2   **DECISION MAKING UNDER CERTAINTY**

The first class of decision problems where the consequences of acts are explicitly considered is one where each act has a unique consequence. Decision problems of this type involve decision making under certainty.

(22.2)   In *decision making under certainty,* each act A_j has a unique consequence that is known in advance.

A key feature of decision making under certainty is that there are no uncontrollable variables present.

A municipality had to choose an investment firm to handle its forthcoming bond issue. Three highly reputable firms made written bids guaranteeing the net interest rate the municipality would pay. The payoff table, indicating the guaranteed interest rates, is as follows:

	Firm A	Firm B	Firm C
Guaranteed interest rate	9.92%	9.80%	9.88%

Note there is only a single outcome state here, or equivalently, there are no uncontrollable variables present.

The municipality decided to employ the criterion: Minimize the total interest to be paid. Hence, firm B was selected. ☐

Comment

Decision problems involving certainty sometimes are referred to as *deterministic*. These decision problems are not necessarily simple. Sophisticated mathematical methods may be required to find the act that is best according to the criterion. But the characteristic feature of all such problems, whether simple or complex, is that for each act there is a unique payoff that is known in advance.

22.3 DECISION MAKING UNDER UNCERTAINTY

We turn next to decision problems where uncontrollable variables are present. Here, there are several outcome states, and the payoff for act A_j depends on the outcome state S_i that prevails. Since the outcome state S_i is not known with certainty, two different approaches may be taken:

1. The decision maker is unwilling or unable to assign probabilities to the outcome states S_i and regards them simply as uncertain.
2. The decision maker assigns probabilities $P(S_i)$ to the outcome states.

We consider the first approach in this section. It is called decision making under uncertainty.

(22.3)

> In *decision making under uncertainty,* there are two or more outcome states S_i that affect the payoff, and there is no assignment of probabilities $P(S_i)$ to these states.

The uncertainty approach tends to be appealing when there is no relevant past experience or other basis for assessing the probabilities $P(S_i)$. We now describe an example where this approach was utilized.

EXAMPLE

Camera Firm. A firm planning to produce a special camera and film for making photographs with a distinct three-dimensional effect is being sued for patent infringements in separate lawsuits by a camera manufacturer and a film manufacturer. The alternative courses of action being considered by management are shown in Table 22.2. Note that they involve different degrees of tooling up to produce the camera and film. For example, act A_1 entails tooling up for both camera and film, while A_2 entails tooling up for the camera but delaying tooling up for the film until the outcomes of the lawsuits are known. The firm has decided to oppose both suits and obtain definitive legal decisions rather than seek any compromise settlements. Hence, the outcome states are the four possible combinations of legal decisions in the two suits, as shown in Table 22.2. The firm's legal counsel is unwilling to assess probabilities for the outcome states because legal issues without precedent are involved. Consequently, management has decided to regard the outcome states simply as uncertain. Thus, the problem is one of decision making under uncertainty.

The payoffs C_{ij} are shown in the payoff table in Table 22.2. Each payoff represents the estimated present value of future net income in millions of dollars. These payoffs are affected by two factors: (1) If the firm loses a given lawsuit, it can still manufacture the item but at a higher production cost because of modifications to avoid patent infringement. (2) Competing firms will develop their own versions of the camera and film, so it is advantageous for the firm to place its product on the market quickly if the legal cases are won.

TABLE 22.2 **Payoff table for the camera firm example ($ million)**

	Act			
Outcome State	A_1 Tool Up for Both Camera and Film	A_2 Tool Up for Camera Only	A_3 Tool Up for Film Only	A_4 Do Not Tool Up for Either
S_1: Win both suits	1450	1400	1100	1100
S_2: Win camera suit only	1300	1425	1000	1000
S_3: Win film suit only	700	825	850	850
S_4: Lose both suits	400	775	800	810
Minimum payoff	400	775		810

Admissible Acts

Sometimes, not all available acts need be considered. An act may be so poor that it should be dropped immediately. To determine whether such poor acts are present, we compare the payoffs for each pair of acts. In the camera firm example, suppose we compare acts A_3 and A_4 in Table 22.2. We note that the payoffs are equally good under outcome states S_1, S_2, and S_3. However, under S_4, the payoff for A_3 ($C_{43} = 800$) is less than that for A_4 ($C_{44} = 810$). Hence, there is no need to consider A_3 any further because the payoff with A_3 cannot be higher than with A_4 but can be lower. We say that A_4 dominates A_3 here, and hence A_3 is inadmissible.

(22.4)

> Act A_j *dominates* act $A_{j'}$ if A_j offers as good a payoff as $A_{j'}$ in every outcome state and a better payoff in at least one outcome state, that is, if $C_{ij} \geq C_{ij'}$ for all outcome states i, and $C_{ij} > C_{ij'}$ for at least one outcome state.
> Act A_j is *admissible* if it is not dominated by any other available act, and it is *inadmissible* if it is dominated by another act.

Note from Table 22.2 that acts A_1, A_2, and A_4 are not dominated by any other act and hence are each admissible. Thus, in the camera firm example, we shall only need to consider A_1, A_2, and A_4 as candidates for best act.

Criteria for Selecting Best Act

To choose the "best" act, we require a criterion. We now discuss two criteria that have been proposed by decision theorists for decision making under uncertainty.

Maximin Criterion. A decision maker who tends to be cautious might emphasize the guaranteed payoff for each act — that is, the level below which the actual payoff cannot go — and select the act that offers the best guarantee. To put this another way, the decision maker would select the act that maximizes the minimum possible payoff.

(22.5)

> The *maximin criterion* considers the minimum payoff that can be obtained with each act A_j — namely, $\min_i C_{ij}$ — and selects that act for which the minimum payoff is maximized:
>
> $$\max_j \left(\min_i C_{ij} \right)$$

EXAMPLE

In the camera firm example, the minimum payoffs for the three admissible acts A_1, A_2, and A_4 are the respective column minimums shown in Table 22.2. For example, the minimum payoff for act A_1 is 400. We see from Table 22.2 that act A_4 maximizes the minimum possible payoff and would be chosen under the maximin criterion. Of course, the actual payoff with A_4 can be higher than the minimum 810 (if the outcome state is S_1, S_2, or S_3), but it cannot be lower. ☐

Comment

In some decision problems involving uncertainty, the outcome states are under the control of an adversary (for example, a business competitor) whose objective is to hold the decision maker to as poor a payoff as possible. The maximin criterion may have particular appeal to the decision maker here since it guarantees the best of the worst possible payoffs that can be forced on the decision maker by the adversary.

Minimax Regret Criterion. A second criterion for decision making under uncertainty appeals to a decision maker who reasons as follows: "I must commit myself now to a course of action before I know the outcome state. But once the outcome state is known, the 'Monday morning quarterbacks' will compare my result against the best that could have been obtained in light of the actual outcome state that prevailed. I would like to protect myself against falling too far short!"

A decision maker taking this position is concerned about regrets or opportunity losses, which we denote by O_{ij} for the ith outcome state and jth act.

(22.6)

The *regret* or *opportunity loss* O_{ij} is the difference between the best payoff attainable under the ith outcome state — namely, $\max_j C_{ij}$ — and the payoff attained with the jth act under that state:

$$O_{ij} = \left(\max_j C_{ij}\right) - C_{ij}$$

The minimax regret criterion considers the maximum regret for each act and selects the act with the smallest maximum regret as best.

(22.7)

The *minimax regret criterion* considers the maximum regret that can be obtained with each act A_j — namely, $\max_i O_{ij}$ — and selects that act for which the maximum regret is minimized:

$$\min_j \left(\max_i O_{ij}\right)$$

EXAMPLE

The regrets for the camera firm example are obtained in Table 22.3. The payoffs C_{ij} for the admissible acts are repeated in Table 22.3a. The maximum payoffs attainable under the four outcome states are the respective row maximums. Thus, $C_{11} = 1450$ is the maximum payoff attainable under S_1.

TABLE 22.3   **Payoffs and regrets for the camera firm example ($ million)**

(a)
Payoffs (C_{ij})

Outcome State	Act A_1	A_2	A_4	Maximum Payoff
S_1	1450	1400	1100	1450
S_2	1300	1425	1000	1425
S_3	700	825	850	850
S_4	400	775	810	810

(b)
Regrets or opportunity losses (O_{ij})

Outcome State	Act A_1	A_2	A_4
S_1	0	50	350
S_2	125	0	425
S_3	150	25	0
S_4	410	35	0
Maximum regret	410	50	425

The regrets O_{ij} are shown in Table 22.3b. Note that $O_{11} = 1450 - 1450 = 0$, because A_1 is the best act when S_1 prevails. Similarly, $O_{12} = 1450 - 1400 = 50$, which indicates that the payoff under S_1 with act A_2 is $50 million worse than with the best act for this outcome state. The other regrets are obtained in the same manner.

To obtain the maximum regret for each act, we find the largest regret within each column. These are shown in Table 22.3b; for example, the maximum regret for act A_1 is 410. The smallest maximum regret is for act A_2. The decision maker, in selecting A_2, is assured that the payoff actually achieved when the outcome state materializes will fall short of the best attainable under that outcome state by no more than $50 million.   ☐

Choice of Criterion.   We have considered two criteria for selecting a course of action in decision making under uncertainty. Still other criteria might have been used. Unfortunately, different criteria frequently select different acts in the same problem. Thus, in the camera firm example, the maximin criterion selected A_4 while the minimax regret criterion selected A_2. Since the theory of decision making under uncertainty does not identify any criterion as objectively superior to the others, the choice of a criterion remains subjective.

22.4 BAYESIAN DECISION MAKING WITH PRIOR INFORMATION ONLY

We now consider the second of the two approaches listed earlier for decision problems in which uncontrollable variables are present — namely, the approach where the decision maker assigns probabilities $P(S_i)$ to the outcome states. This approach is called Bayesian decision making.

(22.8)

> In *Bayesian decision making,* there are two or more outcome states S_i that affect the payoff, and probabilities $P(S_i)$ are assigned to these outcome states.

Bayesian decision making plays an important role in both the theory and practice of decision making, and we shall study it in the remainder of this chapter and in the following one. In this chapter, we take up Bayesian decision making using only prior information about the outcome states in the form of probabilities $P(S_i)$ but no sample information. In the next chapter, we consider Bayesian decision making when both prior information about the outcome states and sample information are available. Bayesian decision making with prior information only is also called *decision making under risk.*

EXAMPLE

Island Tour. Mr. Adams operates a tourist service on an island noted for its historic sites. Tourists on incoming packaged tours to a nearby mainland area may make optional side trips by air to this island. Mr. Adams handles the travel and lodging reservations for these side trips. He must make reservations for lodging and travel before he knows how many tourists will make the side trip. He gains $50 for each filled reservation up to four reservations but loses a $30 advance deposit for each unfilled reservation. Mr. Adams neither gains nor loses for any tourists in excess of the number of reservations made. Finally, if five or more persons in an incoming tour take the side trip, the travel agency handling the tour makes the side trip arrangements, reimbursing Mr. Adams for any deposits made and giving him a fixed fee of $150 for servicing the group's visit.

The alternative courses of action available to Mr. Adams for a tour are the number of reservations to be made, as shown in Table 22.4. The outcome states are the actual numbers of tourists making the side trip, and these are also shown in Table 22.4. We illustrate the entries in the payoff table by considering the payoffs for act A_3 (two reservations). If no tourists make the side trip, Mr. Adams forfeits two $30 deposits for a total loss of $60. If one tourist makes the side trip, he gains $50 for that one and loses $30 on the one unfilled reservation, for a net gain of $20. If two tourists make the side trip, he gains $50 for each, for a total gain of $100. If three or four tourists make the side trip, Mr. Adams' gain still is $100 since he neither gains nor loses on tourists in excess of the number of reservations.

TABLE 22.4   **Payoff table for the island tour example (dollars)**

Number of Tourists Making Side Trip	Probability $P(S_i)$	Number of Reservations				
		A_1 None	A_2 One	A_3 Two	A_4 Three	A_5 Four
S_1: None	0.25	0	−30	−60	−90	−120
S_2: One	0.10	0	50	20	−10	−40
S_3: Two	0.20	0	50	100	70	40
S_4: Three	0.20	0	50	100	150	120
S_5: Four	0.15	0	50	100	150	200
S_6: Five or more	0.10	150	150	150	150	150

Finally, if five or more tourists make the side trip, the gain is the fixed fee of $150 paid to him by the travel agency.

No more than four reservations need be considered as available acts in the decision problem. Any act entailing more than four reservations is dominated by act A_5 (four reservations).

Mr. Adams was able to assign probabilities to each of the outcome states from past experience. For example, in about 25 percent of past tours, no person elected the side trip; hence, he assessed the probability that no person will elect the side trip on the next tour as $P(S_1) = 0.25$. The outcome state probabilities are also shown in Table 22.4.

Comment   An objective interpretation can be given to Mr. Adams' outcome state probabilities since they are based on observed relative frequencies for past tours conducted under conditions similar to those prevailing currently. In many cases, such relative frequencies are not available; for instance, there may be no relevant past experience. In these cases, subjective or personal probability assessments may be employed. We discussed such assessments in Chapter 4. Bayesian decision making proceeds in the same fashion whether the probabilities are objective or subjective.

Expected Payoff Criterion

When probabilities $P(S_i)$ are assigned to the outcome states, the payoff for an act is considered to be a random variable with an associated probability distribution. Thus, we note from Table 22.4 that if Mr. Adams selects act A_5 (four reservations), the probability distribution of payoffs is as follows:

Payoff C_{i5}	Probability $P(S_i)$
-120	0.25
-40	0.10
40	0.20
120	0.20
200	0.15
150	0.10
Total	1.00

For brevity, we use the term *lottery* to denote a probability distribution of payoffs. When a decision maker selects a given act in Bayesian decision making, he or she chooses, in effect, a given lottery in preference to the other lotteries available.

Which of the available lotteries should Mr. Adams choose? Since his handling of the side trips is a recurrent activity, Mr. Adams may well prefer the lottery that yields the largest expected payoff per tour. This is equivalent to maximizing his total gain over the long run. The act with the largest expected payoff is said to satisfy the expected payoff criterion.

(22.9)   The *expected payoff criterion* selects that act for which the expected payoff is maximized.

We calculate the expected payoff in the manner of any other expected value. In adapting formula (5.5) to our present notation, we obtain the following:

(22.10)
$$EP(A_j) = \sum_i C_{ij} P(S_i)$$

Here, $EP(A_j)$ denotes the expected payoff for act A_j. The act with the largest expected payoff is called the Bayes act.

(22.11)   The act selected by the expected payoff criterion is called the *Bayes act*. The expected payoff for the Bayes act is denoted by *BEP*, where:

$$BEP = \max_j EP(A_j)$$

EXAMPLE

To find the Bayes act for the island tour example, we need to obtain the expected payoff for each act. We illustrate the calculations for act A_5:

$$EP(A_5) = \sum_{i=1}^{6} C_{i5} P(S_i) = (-120)(0.25) + (-40)(0.10) + \cdots + 150(0.10)$$

$$= 43$$

Thus, if Mr. Adams makes four reservations, his expected gain per tour is $43. The other expected payoffs are obtained in similar fashion. We find:

Act:	A_1	A_2	A_3	A_4	A_5
$EP(A_j)$:	15	40	57	58	43

We see that the expected payoff is largest for A_4, so act A_4 (three reservations) is the Bayes act and the expected payoff for the Bayes act is $BEP = EP(A_4) = 58 per tour. $\square$

Comments   1.   Note that $EP(A_j)$ in the island tour example becomes larger and then declines as the number of reservations increases. The act of making three reservations strikes the best balance here between losses arising from unused reservations and gains from reservations that are utilized.

2.   In the island tour example, as in most applications of Bayesian decision making, the outcome state probabilities do not depend on the acts. Hence, the same set of probabilities $P(S_i)$ is used to calculate the expected payoff for each act. In some decision problems, however, the outcome state probabilities are affected by the acts. For example, a firm's bid price for a construction contract will affect its probability of winning the contract. In this type of decision problem, different sets of outcome state probabilities must be used to calculate the expected payoffs for the different acts.

Decision Tree Representation.   Bayesian decision problems can be represented by decision trees, though this representation becomes impractical for large problems without use of specialized computer software.

EXAMPLE   Figure 22.2 shows the decision tree for the island tour example. We have truncated most of the branches for convenience. The act branches are shown emanating from the decision node. Let us follow the branch for act A_4. This branch leads to a chance node with six outcome state branches. The probabilities $P(S_i)$ and payoffs C_{i4} are shown on the tree, and the expected payoff $EP(A_4) = 58$ is recorded in the chance node. The other portions of the tree are completed in a similar manner. The decision maker, standing as it were at the decision node, now selects the act branch that leads to the highest expected payoff, namely, branch A_4. The other act branches are marked by blocking signs to indicate they are not optimal. $\square$

Value of Information

Frequently, a decision maker has the option of obtaining information about the outcome state before making a decision. Of course, this information normally entails a

FIGURE 22.2 **Decision tree for the island tour example**

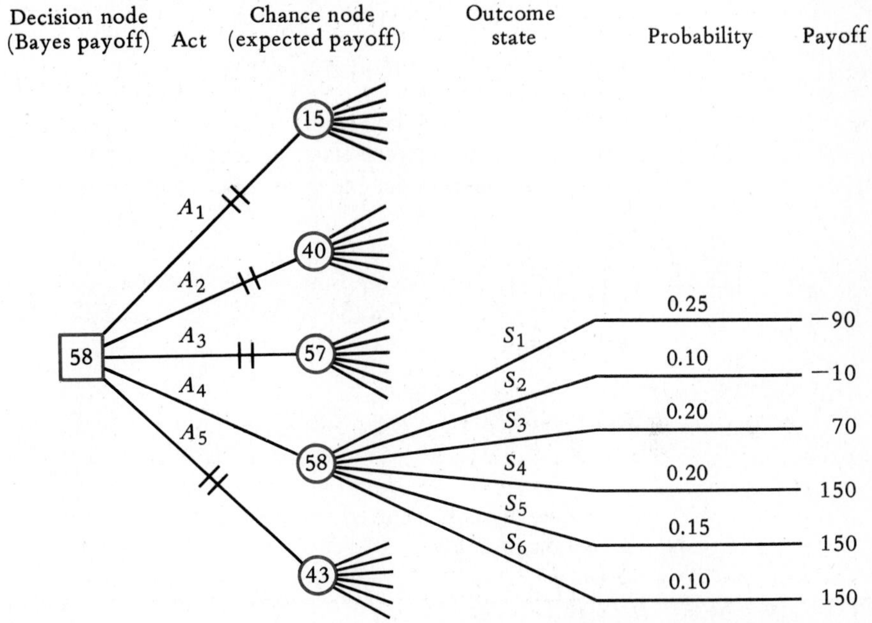

cost. We now consider how much a decision maker should be willing to pay to learn with certainty which outcome state prevails. We continue with the island tour example.

Expected Payoff with Perfect Information. Suppose it were possible for Mr. Adams to ascertain in advance how many tourists will make the side trip (for example, by radiotelephone to the mainland). In that case, Mr. Adams would know the outcome state for each tour—that is, he would have *perfect information* about the outcome state. He would then be in the position of choosing the best act for the given outcome state. Thus, we see from Table 22.4 that if four tourists will make the side trip, he should select act A_5 (make four reservations), which will give him a payoff of $200. The best act and its payoff for each outcome state, and the probability of each outcome state, are as follows:

Outcome State:	S_1	S_2	S_3	S_4	S_5	S_6
Best act:	A_1	A_2	A_3	A_4	A_5	Any
$\max_j C_{ij}$:	0	50	100	150	200	150
$P(S_i)$:	0.25	0.10	0.20	0.20	0.15	0.10

Note that we have another probability distribution of payoffs here, this time for the case where there is perfect information about the outcome state and the best act is chosen for the outcome state. We can now obtain the expected payoff with perfect information, denoted by *EPPI*:

$$EPPI = 0(0.25) + 50(0.10) + 100(0.20) + 150(0.20) + 200(0.15)$$
$$+ 150(0.10) = 100$$

Thus, if Mr. Adams knows the outcome state in advance for each tour and selects the best act for that outcome state, his expected payoff is *EPPI* = $100 per tour.

We now define the expected payoff with perfect information formally.

(22.12)

The *expected payoff with perfect information,* denoted by *EPPI*, is the expected payoff when the act is chosen based on information that indicates the exact outcome state, and is defined:

$$EPPI = \sum_i \left(\max_j C_{ij}\right) P(S_i)$$

Note that any cost incurred in obtaining the perfect information is not considered in the definition of *EPPI*.

Expected Value of Perfect Information.   In the island tour example, we found earlier that the expected payoff of the Bayes act, which utilizes only the probabilistic information about the outcome states, is *BEP* = 58. Since the expected payoff with perfect information is *EPPI* = 100, the difference *EPPI* − *BEP* = 100 − 58 = 42 is the incremental gain due to perfect information about the outcome state. This incremental gain is called the expected value of perfect information. Assuming that the expected payoff criterion is relevant, Mr. Adams should be willing to pay up to $42 per tour for perfect information about the number of tourists who will make the side trip.

We now define the expected value of perfect information formally.

(22.13)

The *expected value of perfect information,* denoted by *EVPI*, is the difference between the expected payoff with perfect information (*EPPI*) and the expected payoff of the Bayes act (*BEP*):

$$EVPI = EPPI - BEP$$

Comments   1.   Assuming that the expected payoff criterion is relevant, *EVPI* is the maximum amount that should be spent for perfect information about the outcome state. As we shall demonstrate in the next chapter, sample information is imperfect and so is worth less than perfect information.

2.   Regrets or opportunity losses can be used in Bayesian decision making. The Bayes act is then the one that minimizes the expected regret. Two results can be shown to hold:
   (a) The same act that maximizes expected payoff minimizes expected regret.
   (b) The magnitude of the minimum expected regret is identical to *EVPI*.

Sensitivity Analysis

Frequently, the data on outcome state probabilities and payoffs in decision problems are unreliable to some degree. Consequently, it is advisable that one study whether the choice of the best act is sensitive to variations in the data. This analysis is called *sensitivity analysis*. Sensitivity analysis in large-scale decision problems is conducted by mathematical methods or computer simulation. In small decision problems, one can simply insert new values for the probabilities or payoffs in the problem to see whether the optimal act remains the same, and if not, whether the expected payoffs of the different optimal acts vary greatly.

EXAMPLE

In the island tour example, if the probabilities:

Outcome State:	S_1	S_2	S_3	S_4	S_5	S_6
$P(S_i)$:	0.20	0.20	0.20	0.15	0.15	0.10

had been used instead of those given in Table 22.4, the expected payoffs of the different acts would have been:

Act:	A_1	A_2	A_3	A_4	A_5
$EP(A_j)$:	15	44	57	54	39

Note that the Bayes act would change from A_4 to A_3, although the expected payoff with act A_4 would only be \$3 less than that of the Bayes act for the second set of outcome state probabilities.   □

22.5 MULTISTAGE DECISION PROBLEMS

The decision problems discussed so far in this chapter are called *single-stage* problems since a single decision is involved. Other decision problems entail a sequence of two or more decisions and are called *multistage* decision problems. Decision trees are useful for studying multistage problems.

EXAMPLE

Computer Manufacturer.   A manufacturer of minicomputers has determined that the firm's product line must be broadened to include a line of office machines utilizing the latest technology. Investigation shows that two alternative courses of action are worth considering:

A_1: undertake research and development to achieve production of the desired line

A_2: acquire a company already producing a satisfactory line and undertake manufacturing integration

These alternative acts are represented in Figure 22.3 by the branches emanating from the first-stage decision node. (Ignore for now the blocking marks on some branches and the numerical entries in the nodes.) Note that if the research and development (R & D) strategy is followed (A_1), the possible outcomes are "unsuccessful" and "successful" and the respective probabilities are assessed at 0.4 and 0.6. The payoff is the estimated net income position (in million dollars) of the company five years hence.

In turning to the path for acquisition and manufacturing integration (A_2), we note that the attempt to integrate manufacturing could be successful (with probability 0.7) or unsuccessful (with probability 0.3). If success is achieved, the overall management structure will be consolidated. The alternative courses of action in this second-stage decision are:

B_1: consolidation on a product basis

B_2: consolidation on a geographic basis

Thus, a second-stage decision node is indicated in the tree at this point. The branches emanating from this decision node lead to respective chance nodes and the possible outcomes failure and success, with probabilities as shown. Finally, the payoffs are given for each of the branches. □

Backward Induction

Bayesian decision making for multistage problems can be based on the principle of *backward induction*. This principle requires that the optimal acts at the last decision nodes be identified first, that the optimal acts at the preceding set of decision nodes then be identified by using the expected payoffs of the previously selected optimal acts, and so forth. In terms of a decision tree, the principle of backward induction requires us to work from right to left, using the expected payoffs of the optimal acts to the right in succeeding calculations.

EXAMPLE

We illustrate the principle of backward induction for the computer manufacturer example represented in Figure 22.3. The steps are as follows.

1. There is only a single decision node at the second stage, which arises if first-stage act A_2 is successful. We therefore need to ascertain whether B_1 or B_2 is the Bayes act at this stage. The expected payoffs for the two acts are as follows:

$$EP(B_1) = 70(0.2) + 90(0.8) = 86$$

$$EP(B_2) = 72(0.5) + 86(0.5) = 79$$

FIGURE 22.3   **Decision tree for a two-stage decision problem — Computer manufacturer example**

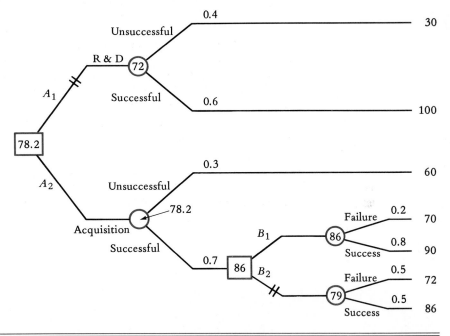

These expected payoffs are recorded for convenience in the respective chance nodes. Act B_1 maximizes the expected payoff in the second-stage decision. The expected payoff of the Bayes act, 86, is then entered in the second-stage decision node, and the B_2 path is blocked.

2. The expected payoff in a decision node consolidates the data on probabilities and payoffs on all paths emanating rightward from that node. Thus, in Figure 22.3, act A_2 in the first-stage decision is now considered to lead to two chance outcomes offering, respectively, a payoff of 60 with probability 0.3 and a payoff of 86 with probability 0.7.

3. We are now in position to calculate the expected payoffs for the acts emanating from the first-stage decision node. We obtain:

$$EP(A_1) = 30(0.4) + 100(0.6) = 72$$

$$EP(A_2) = 60(0.3) + 86(0.7) = 78.2$$

Thus, A_2 maximizes the expected payoff and is selected under the expected payoff criterion. The A_1 path is blocked off.

4. The optimal strategy can now be seen by *forward induction,* that is, by tracing through the unblocked branches from left to right. We see from Figure 22.3 that the firm should acquire a company already producing a satisfactory line and undertake manufacturing integration (A_2). If this is successful, managerial consolidation on a product basis should be carried out (B_1). The possible payoffs for the optimal strategy, with attendant probabilities, are as follows:

Payoff	Probability
60	0.30
70	0.7(0.2) = 0.14
90	0.7(0.8) = 0.56
Total	1.00

The expected payoff for the optimal strategy by forward induction is 60(0.30) + 70(0.14) + 90(0.56) = 78.2, the same as calculated before by backward induction. ☐

Comments

1. In multistage decision problems, second- or later-stage decisions are not necessarily acted on. If events intervene and new conditions arise, later-stage decisions must be determined anew in the light of the then-existing situation.
2. The probabilities in the branches emanating from any chance node always sum to 1. Thus, for any chance node after the first, these probabilities are conditional ones, the conditioning events being the relevant outcomes earlier in the tree.
3. Multistage decision problems, like single-stage ones, can be handled in terms of uncertainty—that is, without assigning probabilities $P(S_i)$ to the branches emanating from the chance nodes. For instance, if the computer manufacturer example in Figure 22.3 were approached in terms of uncertainty and the maximin criterion employed, the acquisition strategy would be implemented; the minimum possible payoff for this strategy is 60 while that for the R & D strategy is 30.

22.6　OPTIONAL TOPIC—UTILITY

Problem with Monetary Payoffs

Our discussion of Bayesian decision making so far has utilized the expected payoff criterion where the payoffs are measured in monetary units. However, payoffs in monetary units may not fully reflect the decision maker's attitude toward risk in a given situation. Hence, the comparative expected payoffs may not be in accord with the decision maker's actual preferences. Consider the owner of a small construction company who must decide whether to be the general contractor on a large project (A_1) or simply a subcontractor (A_2). The payoff table is as follows:

Outcome State	Probability	Act	
		A_1	A_2
S_1: No delays	0.8	$80,000	$30,000
S_2: Delays	0.2	−$40,000	−$5,000

The contractor knows that a loss of $40,000 with act A_1 if delays occur would place him in extreme financial difficulties, whereas a loss of $5000 with act A_2 could be absorbed. In other words, a $40,000 loss would be much more than eight times as severe for him as a $5000 loss. Although the expected payoffs are $EP(A_1) = \$56,000$ and $EP(A_2) = \$23,000$ (calculations not shown), the contractor finds A_2 preferable.

Basic Concepts of Utility Theory

Expected Utility Axioms.   Utility theory has been developed to take into account the decision maker's attitude toward risk. Recall that in Bayesian decision making, each act entails a probability distribution of payoffs, which is called a lottery. Utility theory makes some basic assumptions about the consistency of the decision maker's preferences among lotteries. One of these is the *transitivity* assumption or axiom, which states that if lottery A_1 is preferred to lottery A_2 and lottery A_2 is preferred to lottery A_3, then the decision maker will prefer A_1 to A_3. The transitivity axiom, together with several other axioms, make up a set of axioms called the *expected utility axioms*. If a decision maker's preferences are consistent with this set of axioms, then it can be shown that a utility function $u(C)$ exists that properly reflects the decision maker's attitude toward risk. Here, C is the monetary consequence, and $u(C)$ is the *utility number* associated with the money amount C. We call the function $u(C)$ the decision maker's *utility function for money*. Roughly interpreted, $u(C)$ is the subjective worth to the decision maker of a dollar payoff of amount C.

Expected Utility Criterion.   Utility theory shows that when utility numbers rather than monetary payoffs are used in Bayesian decision making, the preferred act will be the one with the largest expected value of the outcomes expressed in utility numbers. Consider a Bayesian decision problem where C_{ij} is the monetary payoff when the jth act is selected and the ith outcome state prevails. The expected payoff for the jth act, according to (22.10), is:

$$EP(A_j) = \sum_i C_{ij} P(S_i)$$

If we now express each monetary payoff C_{ij} in terms of its corresponding utility number $u(C_{ij})$, we can calculate, for the jth act, the expected value of the outcomes in utility numbers. We denote this expected value by $EU(A_j)$.

(22.14)

$$EU(A_j) = \sum_i u(C_{ij})P(S_i)$$

Here, $EU(A_j)$ stands for the expected utility of act A_j.

As mentioned before, utility theory shows that the decision maker's preferred act is the one with the greatest expected utility.

(22.15)

The *expected utility criterion* selects that act for which the expected utility, defined in (22.14), is maximized.

The act selected by the expected utility criterion is called the *Bayes act*.

Assessing the Utility Function

The major problem in utilizing utility theory is the assessment of the utility function for the decision maker. Once the utility function $u(C)$ has been ascertained, the finding of the Bayes act proceeds as before, except that utility numbers rather than monetary payoffs are used. We shall use an illustration to explain how the utility function for a decision maker may be assessed.

EXAMPLE

Investment Client.   An investment counselor is offering a speculative venture to a client. If the venture is successful, the client will gain $10 thousand; but if it is unsuccessful, she will lose $6 thousand. Alternatively, the client can decline the venture, in which case her payoff is $0 — she neither gains nor loses. Thus, there are three possible dollar outcomes, namely, $10 thousand, $0, and −$6 thousand. We wish to assign utility numbers to these dollar amounts that will reflect the client's attitude toward risk.   ☐

Specifying the Utility Scale.   Since utility is synonymous with subjective worth, the more desirable an outcome, the larger is the utility number assigned to it. Thus, utility functions for money assign larger utility numbers $u(C)$ as the monetary amount C increases.

There is no unique scale for utility, and two points on the utility function can be defined arbitrarily. For convenience, we shall usually follow the convention of assigning to the smallest monetary outcome in the decision problem the utility number 0 and to the largest monetary outcome the utility number 1. Thus, for the investment client example, we shall let $u(-6) = 0$ and $u(10) = 1$, where the monetary outcomes are expressed in thousands of dollars.

We have all encountered other scales where two points can be assigned arbitrary values. The temperature scale is one such scale. The Celsius scale assigns 0 to freezing temperature and 100 to boiling, while the Fahrenheit scale assigns 32 and 212 to these temperatures, respectively.

Reference Lottery Method. A variety of methods are available for assessing utility functions. We shall explain one of these, the *reference lottery method*.

For the investment client example, we have defined $u(-6) = 0$ and $u(10) = 1$. We still require the utility of \$0, $u(0)$, as well as some other utility numbers so that we can assess the shape of the decision maker's utility function.

To find $u(0)$, we give the decision maker a choice between \$0 for certain and the following reference lottery (made up of equal probabilities for the two extreme outcomes):

Outcome ($000)	Probability
10	0.5
−6	0.5

The decision maker prefers \$0 for certain to this reference lottery. Thus, $u(0)$ exceeds the expected utility of the reference lottery. Since $u(10) = 1$ and $u(-6) = 0$, the expected utility of the reference lottery is:

$$u(10)\,(0.5) + u(-6)\,(0.5) = 1(0.5) + 0(0.5) = 0.5$$

Since we wish to find the indifference point between \$0 for certain and a reference lottery, we need to increase the probability of gaining \$10 thousand in the reference lottery.

After several steps, we reach the reference lottery:

Outcome ($000)	Probability
10	0.7
−6	0.3

Now, the client indicates that she is indifferent between \$0 for certain and this reference lottery. Hence, $u(0)$ is equal to the expected utility of this reference lottery:

$$u(0) = u(10)\,(0.7) + u(-6)\,(0.3) = 1(0.7) + 0(0.3) = 0.7$$

To ascertain some other points for the utility function, we wish next to find $u(2)$. We utilize the same approach, giving the client a choice each time between a \$2 thousand gain for certain and a reference lottery. We eventually reach the indifference point for the following reference lottery:

Outcome ($000)	Probability
10	0.8
−6	0.2

Hence, we have:

$$u(2) = 1(0.8) + 0(0.2) = 0.8$$

By continuing in this manner, we obtain the following results for the investment client:

C:	-6	-3	-1	0	$+2$	$+5$	$+8$	$+10$
$u(C)$:	0	0.45	0.65	0.70	0.80	0.90	0.98	1.00

These points are plotted in Figure 22.4. When enough points are obtained, a curve can be drawn through them as shown in the figure. This curve is the graph of the client's utility function for money. ☐

Comments

1. Often, a mathematical function is fitted to the utility assessment data, such as a *quadratic utility function*:

(22.16)
$$u(C) = b_0 + b_1 C + b_2 C^2$$

or an *exponential utility function*:

(22.17)
$$u(C) = b_0 + b_1 \exp(b_2 C)$$

where b_0, b_1, and b_2 are constants.

2. Since persons may give inconsistent responses when utility functions are assessed, it is good practice to check the responses for internal consistency with respect to the expected utility rule. In the investment client example, for instance, the client might be given an additional choice between $2 thousand for certain and a reference lottery offering a 50–50 chance at $0 and $5 thousand. Recall from earlier questioning that $u(2) = 0.8$ and note that the expected utility of the reference lottery is also 0.8, that is, $0.9(0.5) + 0.7(0.5)$. If the client is not indifferent here, additional questioning and possible revisions of earlier responses will be required.

Application of Utility Function

Once the decision maker's utility function has been assessed, the finding of the Bayes act is routine, with the utility numbers simply replacing the monetary outcomes.

EXAMPLES

1. For the investment client example, the speculative venture lottery is as follows:

Outcome ($000)	Probability
10	0.5
−6	0.5

We have $u(10) = 1$ and $u(-6) = 0$. If A_1 denotes pursuing the speculative venture, we therefore have $EU(A_1) = 1(0.5) + 0(0.5) = 0.5$. Let A_2 denote the act

---

FIGURE 22.4 **Utility function for money—Investment client example**

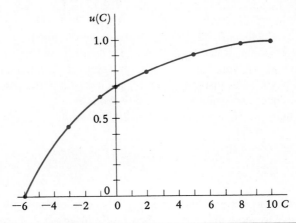

---

not to pursue the speculative venture. Then, $EU(A_2) = u(0) = 0.7$. Hence, act A_2 has the higher expected utility and is the preferred act.

2. Management of a firm has the choice of conducting mineral exploration alone (A_1), as a joint venture with another firm (A_2), or not conducting the exploration (A_3). The payoff table is as follows:

Outcome State	Probability	A_1	A_2	A_3
Success	0.4	17	8.5	0
Failure	0.6	−10	−5.0	0

The utility function for the firm, within the range of payoffs for this problem, is $u(C) = 0.2 + 0.03C + 0.001C^2$.

We obtain the needed utility numbers by substituting into the utility function. For example:

$$u(17) = 0.2 + 0.03(17) + 0.001(17)^2 = 1.0$$

We obtain:

$$u(-10) = 0 \quad u(-5) = 0.075 \quad u(0) = 0.2 \quad u(8.5) = 0.53$$
$$u(17) = 1.0$$

We now calculate the expected utility for each act:

$$EU(A_1) = 1.0(0.4) + 0(0.6) = 0.4$$
$$EU(A_2) = 0.53(0.4) + 0.075(0.6) = 0.257$$
$$EU(A_3) = u(0) = 0.2$$

Hence, act A_1 is the preferred act.

Types of Utility Functions for Money

Researchers have investigated utility functions for money in some detail. Three proto-type utility functions are graphed in Figure 22.5. The *concave utility function* in Fig-ure 22.5a, labeled *risk-averse,* is the type exhibited by the investment client in the earlier example. It can be shown that a decision maker with a concave utility function will prefer a given dollar payoff for certain to a lottery offering an equal expected dollar payoff. For instance, such a person would prefer a sure gain of $5 thousand to a 50–50 chance of obtaining $0 or $10 thousand.

Figure 22.5b shows a *linear utility function.*

(22.18)
$$u(C) = b_0 + b_1 C \qquad b_1 > 0$$

A decision maker with this type of utility function for money is said to be *risk-neutral.* It is easy to show that the expected utility of an act A_j for a risk-neutral person can be expressed in terms of the expected payoff of the act as follows.

(22.19)
$$EU(A_j) = b_0 + b_1[EP(A_j)] \qquad b_1 > 0$$

Hence, the act that maximizes expected payoff also maximizes expected utility. Thus, there is no need to use utility numbers for risk-neutral persons; the expected payoff criterion will correctly indicate the preferred act.

The *convex utility function* in Figure 22.5c pertains to decision makers said to be *risk-seeking.* The attitude toward risk here is the opposite of risk aversion. For instance,

---

FIGURE 22.5   Examples of utility functions for money showing different attitudes toward risk

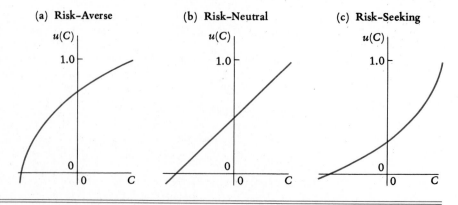

a decision maker with a convex utility function for money will prefer a lottery offering a 50–50 chance at $0 or $10 thousand to $5 thousand for certain. Thus, an individual who buys a ticket in a state lottery is exhibiting risk-seeking behavior because the cost of the ticket exceeds the expected amount of winning. The quadratic utility function in the mineral exploration example is convex, indicating that the management of the firm is risk-seeking.

Comments

1. Our discussion of utility focused on utility functions for money. Utility functions can also be developed for other types of outcomes, such as rate of return on investment, health status, and location of next job.

2. Utility is subjective. We cannot say that an individual is correct or incorrect in his or her attitude toward risk. Moreover, a decision maker's utility function may change over time as new circumstances arise. Finally, because utility is a subjective measure and arbitrarily scaled, it is meaningless to compare the utility numbers for different individuals.

3. In many practical situations, the decision maker's utility function for money is approximately linear within the range of the monetary outcomes encountered. Hence, as we have seen, in these situations utility assessments are not required since the act with the largest expected payoff (in monetary terms) will be the decision maker's preferred act.

PROBLEMS

22.1 A food company is launching a new product. The company's marketing group must choose one of three alternative package designs for the product. It is anticipated that sales of the new product will depend in part on the extent to which consumers find the chosen design to be both functional and attractive. The marketing group is somewhat uncertain, however, about consumer reactions to each design. Identify each of the structural elements of a decision problem in (22.1) for this decision problem.

22.2 An executive of a multinational corporation told a group of graduate students: "In inflation-prone countries, we employ a flexible credit policy adjusted to the short-run outlook for inflation in the local currency. If inflation will be slight, we allow retailers a six-week credit limit, since experience shows this limit will generate the highest remitted profits in dollar terms. If inflation will be moderate, we cut this limit to four weeks, since a four-week limit then will result in the highest remitted dollar profits, and so on down to a limit of one week if inflation is severe. In one of the countries, the government has told us we must announce a credit policy that will be fixed for the next two years. We no longer can shift our limit back and forth as tactical considerations dictate. In the next few days, I must decide what our two-year credit policy will be in that country." Identify each of the structural elements of a decision problem in (22.1) for this decision problem.

*22.3 **Substitute Teachers.** Substitute teachers for a school can be hired on a contract or noncontract basis. Contract substitutes receive $30 each school day for the whole year simply for being available to teach and also receive an additional $150 on any day they actually teach. When the number of contract substitutes is inadequate on any school day, noncontract substitutes must be called in. Noncontract substitutes receive $220 for each day they teach but nothing when they do not teach. The school administrator must now decide how many substitutes should be given contracts for the coming school year. Records show that the following probability distribution describes the number of substitutes required on any school day:

S_i:	0	1	2
$P(S_i)$:	0.4	0.5	0.1

Construct the payoff table for the administrator's decision problem, showing the acts, outcome states, and payoffs for any given school day. Display the costs as negative payoffs.

22.4 **Venture Capital.** A venture investor is about to invest $100 thousand for a two-year period. Two ventures (V_1, V_2) are under consideration. The investor must choose among the following alternatives: (1) Invest the entire amount in V_1, (2) invest the entire amount in V_2, (3) invest half the amount in each of V_1 and V_2, (4) do not invest in either V_1 or V_2. The amount invested in V_1 will be doubled with probability 0.8 or lost completely with probability 0.2. The amount invested in V_2 will be tripled with probability 0.5 or lost completely with probability 0.5. The investment outcomes for V_1 and V_2 are statistically independent. If the investor chooses to invest in neither venture, the $100 thousand amount can be invested in riskless securities that will be worth $130 thousand at the end of the two-year period.

   a. Construct the payoff table for the investor's decision problem, showing the acts, outcome states, and payoffs at the end of the two-year period.

   b. Determine the probability associated with each of the outcome states identified in **a**.

22.5 **Odd-Even Game.** Consider playing the following game with a friend. Your friend writes either "0¢" or "$1.01" on a slip of paper. Concurrently, you write either "50¢" or "51¢" on your slip of paper. The two amounts are then added. If the sum is even, you pay your friend that amount; if the sum is odd, your friend pays you that amount.

   a. Construct your payoff table for a single trial of this game.

   b. If you knew the number your friend will write on the next play, would you have a decision problem on this play? If so, would it involve certainty or uncertainty?

22.6 Refer to **Substitute Teachers** Problem 22.3. Construct the decision tree corresponding to the payoff table. Show the probabilities associated with the branches of each chance node.

22.7 Refer to **Venture Capital** Problem 22.4. Construct the decision tree corresponding to the payoff table in **a**. Show the probabilities associated with the branches of each chance node.

22.8 Refer to **Odd-Even Game** Problem 22.5.

   a. Construct the decision tree corresponding to the payoff table. Does your decision tree contain the same information as your payoff table? Explain.

   b. Suppose you have already played this game five times and are 50¢ behind. Should this loss be included in the payoffs in your decision tree for the sixth play? Explain.

*22.9 Three workers (W_1, W_2, W_3) are to be assigned to carry out three tasks (T_1, T_2, T_3), one worker doing one task. Any worker can do any task, but they require different amounts of time (in hours), as shown by the following table:

Worker	Task		
	T_1	T_2	T_3
W_1	43	35	30
W_2	28	32	29
W_3	36	37	33

a. Explain why this assignment problem is one of decision making under certainty.

b. Construct a payoff table for this decision problem where the consequence is the total time required to do the three tasks. Which assignment yields the smallest total?

c. Construct another payoff table where the consequence is the maximum of the three times required for the tasks. Which assignment yields the smallest maximum?

d. How many acts would there be for this type of decision problem if ten workers were to be assigned to ten tasks?

22.10 Loose tea is to be packaged in a box having a total volume of 1000 cubic centimeters. It is desired to choose the height (h), width (w), and depth (d) for the box so that the surface area will be as small as possible in order to minimize boxing material requirements.

a. Explain why the choice of the dimensions of the box is a decision problem involving certainty.

b. Which of the following alternative sets of dimensions is best in this decision problem?

	Dimensions (in centimeters)		
Alternative	h	w	d
1	5	20	10
2	10	10	10
3	5	25	8

*22.11 Homecoming Brochure. Paul McQue helps finance his education at Bushnell University by undertaking small projects. McQue's next project requires a decision concerning the number of copies of a brochure to be printed for homecoming weekend. The printing order must be a multiple of 500. The size of the homecoming crowd is uncertain because it depends on weather conditions. The payoff table follows (payoffs are net profits in dollars).

	Printing Order				
	A_1	A_2	A_3	A_4	A_5
Size of Crowd	1000	1500	2000	2500	3000
S_1: Above average	200	300	400	500	450
S_2: Average	200	300	250	200	150
S_3: Below average	200	150	100	50	0

a. Which acts are admissible? Does it appear that McQue has missed an opportunity by not including printing sizes above 3000 among his possible courses of action? Comment.

b. Which act satisfies the maximin criterion? What is the guaranteed payoff with this act? Can you tell what will be the actual payoff here with the maximin act? Can you always tell?

c. Convert the payoffs to regrets, and ascertain which act satisfies the minimax regret criterion. Is this act the same one selected by the maximin criterion in **b**? Is this necessarily the case?

22.12 Cooking Utensils. Union Appliance, Ltd., is a market leader in cast-iron cooking utensils. Last year, a firm specializing in advanced metal technology introduced a line of competitive lightweight utensils containing a new design. These utensils are made of a new alloy that exactly duplicates the properties of cast iron for cooking. The competitive line has cut sharply into sales of Union's line. Union can incorporate the new design into its own line but not the lighter weight. Changing its lines to the new design will be expensive for Union, so management does not wish to undertake it unless the new design is the major factor in consumer acceptance of the competitive line. The payoff table follows (payoffs are discounted profits in millions of dollars).

	Act	
Importance of New Design	A_1 Incorporate New Design	A_2 Do Not Incorporate New Design
S_1: Major factor	3.5	0.8
S_2: Not major factor	1.9	3.9

a. Which act satisfies the maximin criterion? What minimum payoff does this act guarantee? Could the actual payoff exceed the guaranteed minimum? Explain.

b. Convert the payoffs to opportunity losses, and ascertain which act satisfies the minimax regret criterion. Is this act the same one selected by the maximin criterion in **a**? Is this always the case?

22.13 **Computer Graphics Manufacturer.**   A small manufacturer of computer graphics equipment has accumulated substantial cash reserves. The executive committee will now decide whether to expand vertically by buying another computer graphics company, or expand horizontally by buying a company that manufactures medical monitoring equipment. The outcome states correspond to different possible degrees of elimination of financially weaker companies in the computer graphics field in the next several years. The payoff table follows (payoffs are present values of future earnings in millions of dollars).

	Act	
Degree of Elimination	A_1 Vertical	A_2 Horizontal
S_1: Moderate	8	15
S_2: Severe	16	10

a. Which act satisfies the maximin criterion?

b. Convert the payoffs to regrets, and ascertain which act satisfies the minimax regret criterion.

22.14 A marine salvage company about to raise a sunken cargo vessel must now decide how to proceed. The payoff table describing the decision problem follows. The payoffs represent the salvage value less salvage costs (in millions of dollars). The acts are different salvage strategies. The outcome states reflect uncertain weather and ocean conditions and the position and condition of the sunken vessel.

Outcome State	Act		
	A_1	A_2	A_3
S_1	1.3	1.0	0.2
S_2	1.8	1.8	2.0
S_3	1.9	2.0	2.9
S_4	2.3	2.8	3.6

a. Which act satisfies the maximin criterion?

b. Convert the payoffs to regrets, and ascertain which act satisfies the minimax regret criterion.

c. If it were known that outcome state S_1 could not occur, would the company's best act be clear? Explain.

*22.15 Refer to **Substitute Teachers** Problem 22.3. The payoff table for the school administrator's decision problem follows.

		Contract Substitutes to Be Hired		
Substitutes Required for School Day	Probability $P(S_i)$	A_1 0	A_2 1	A_3 2
S_1: 0	0.4	0	-30	-60
S_2: 1	0.5	-220	-180	-210
S_3: 2	0.1	-440	-400	-360

a. Find the Bayes act. What is its expected payoff?

b. What will be the expected earnings per school day of a contract substitute if the administrator adopts the Bayes act in **a**?

22.16 Refer to **Homecoming Brochure** Problem 22.11. Suppose McQue is able to assign probabilities to each outcome state, and that these are $P(S_1) = 0.3$, $P(S_2) = 0.5$, $P(S_3) = 0.2$.

a. Which act is the Bayes act? What is the expected payoff with this act?

b. Under some other set of probabilities, could A_1 be the Bayes act here? Could A_5 ever be the Bayes act? Explain.

22.17 Refer to **Cooking Utensils** Problem 22.12. Management has assessed the probabilities of the outcome states as follows: $P(S_1) = 0.6$, $P(S_2) = 0.4$.

a. Which act will give Union the larger expected payoff? What is the expected payoff with the Bayes act?

b. Give the probability distribution of payoffs that Union faces if it chooses the Bayes act.

22.18 Refer to **Computer Graphics Manufacturer** Problem 22.13. The executive committee assesses the outcome state probabilities to be $P(S_1) = 0.4$, $P(S_2) = 0.6$.

a. Show the probability distributions of payoffs associated with acts A_1 and A_2, respectively.

b. Find the Bayes act. What is its expected payoff?

*22.19 Refer to **Substitute Teachers** Problems 22.3 and 22.15. Obtain *EPPI* and *EVPI*. Interpret their meanings here.

22.20 Refer to **Homecoming Brochure** Problems 22.11 and 22.16.

a. Obtain *EPPI* and *EVPI*. Interpret their meanings here.

b. By paying a $65 premium for overtime work, McQue can delay the printing order until just prior to the weekend, when the size of the crowd is known. Do you recommend that he follow this option? Explain.

22.21 Refer to **Cooking Utensils** Problems 22.12 and 22.17. A consumer research study will be undertaken to obtain information on which outcome state prevails.

a. What is the maximum amount management should be willing to spend on a study providing perfect information? Explain.

b. Since the proposed study involves a sample survey, information obtained from it about the prevailing outcome state will be subject to sampling error and perhaps also other errors. What effect does this have on the amount management should be willing to spend on the study?

22.22 Refer to **Computer Graphics Manufacturer** Problems 22.13 and 22.18. Obtain *EPPI* and *EVPI*. Interpret their meanings here.

22.23 For a decision problem, it is known that *EPPI* = 160 and *EVPI* = 30. What is the expected payoff of the Bayes act? What is the expected regret or opportunity loss of the Bayes act?

*22.24 Refer to Figure 22.3, containing the decision tree for the computer manufacturer example. Suppose that if management follows the R & D route and the outcome is unsuccessful, there is still an opportunity to broaden the product line by undertaking negotiations with another firm on either a merger or a license arrangement. (It would be extremely disadvantageous to attempt to negotiate in both areas, so one or the other must be selected for negotiation.) The acts, outcomes, payoffs, and probabilities of success for the negotiations are:

> Merger: success, 65; failure, 20; probability of success = 0.70
>
> License: success, 60; failure, 30; probability of success = 0.60

    a. Draw a decision tree to reflect the modified situation.
    b. Using backward induction, determine the optimal strategy. What is the expected payoff for the optimal strategy?
    c. Confirm the expected payoff for the optimal strategy by forward induction.

22.25 **Consumer By-product.** A firm making industrial products developed a marketable consumer item as a by-product of its research and development. A decision is to be made on whether to sell this item through the company's own sales and distribution channels (A_1) or to use independent sales agents (A_2). A second-stage decision problem will arise if the firm selects A_1 and is relatively unsuccessful, since management must then decide between switching to sales agents immediately (B_1) or first attempting to dispose of the item by selling the key patent at favorable terms to another firm and switching only if the attempt is unsuccessful (B_2). The decision tree for this problem is shown in Figure 22.6 (payoffs are present values of estimated future earnings in millions of dollars).

    a. Using backward induction, determine the optimal strategy. What is the expected payoff for the optimal strategy?
    b. Confirm the expected payoff for the optimal strategy by forward induction.
    c. Suppose probabilities could not be assigned to the outcome states in the decision tree. Which strategy would be selected with the maximin criterion?

FIGURE 22.6   Decision tree for Problem 22.25

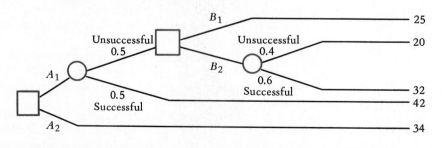

22.26 A manufacturing plant uses a sophisticated robot welder that has just broken down. The plant manager must now decide whether the plant repair staff should attempt to repair it (A_1) or to bring in outside repair specialists immediately (A_2). The manager does not know if the repair problem is a simple or complex one but assesses the probabilities of these outcomes as 0.4 and 0.6, respectively. If the plant staff attempts to repair the robot welder and discovers that the repair problem is a simple one, they will succeed in repairing the robot. On the other hand, if the repair problem turns out to be a complex one, the manager must decide either to have the staff attempt the repair (B_1) or to bring in the specialists at that stage (B_2). The probability is 0.3 that the staff will succeed in a complex repair problem. If the staff fails in a complex repair problem, the specialists would need to be brought in. Figure 22.7 shows the decision tree for this problem. The payoffs are costs (in thousands of dollars), which reflect the facts that (1) the specialists are more expensive than the plant staff, (2) a complex repair problem is more costly than a simple one, and (3) delays in repair are expensive.

a. Using backward induction, determine the optimal strategy. Remember that the consequences shown in Figure 22.7 are cost amounts.

b. What is the expected cost for the optimal strategy identified in a?

c. Give the probability distribution of costs associated with the optimal strategy in a.

FIGURE 22.7   **Decision tree for Problem 22.26**

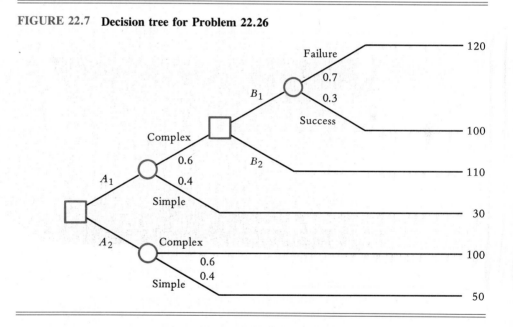

22.27 Refer to **Computer Graphics Manufacturer** Problems 22.13 and 22.18. A third act available to the manufacturer (A_3) is to delay the choice between A_1 and A_2 for several years until it is known which outcome state (S_1, S_2) will prevail and then to make the choice. Assume that such a delay would reduce the payoff by $3 million whichever outcome state prevails and whichever act is chosen.

a. Draw a decision tree to represent this two-stage decision problem.

b. Using backward induction, determine the optimal strategy. What is the expected payoff for the optimal strategy?

*22.28 A decision maker's utility function for money is $u(C) = C^2/90,000$, where $0 \le C \le 300$ and C is in thousands of dollars.

a. Plot $u(C)$. Is the decision maker risk-averse, risk-neutral, or risk-seeking?

b. Which will the decision maker prefer: (1) a lottery offering an even chance at $0 or $200 thousand, or (2) $80 thousand for certain?

22.29 A decision maker's utility function for money is $u(C) = \sqrt{C/300}$, where $0 \le C \le 300$ and C is in thousands of dollars.

a. Plot $u(C)$. Is the decision maker risk-averse, risk-neutral, or risk-seeking?

b. Which will the decision maker prefer: (1) a lottery offering an even chance at $0 or $100 thousand, or (2) $40 thousand for certain?

22.30 A decision maker's utility function for money is $u(C) = 1 - \exp(-0.05C)$, where $0 \le C < \infty$ and C is in thousands of dollars.

a. Plot $u(C)$. Is the decision maker risk-averse, risk-neutral, or risk-seeking?

b. Which will the decision maker prefer: (1) a lottery offering an even chance at $0 or $400 thousand, or (2) $160 thousand for certain?

22.31 Refer to Figure 22.4, containing the utility function for the investment client example. Suppose this client is offered a lottery in which there are 2 chances in 3 of losing $3 thousand and 1 chance in 3 of winning $8 thousand, and the client indicates indifference between playing this lottery and not playing it. Is this response consistent with the client's utility function in Figure 22.4? Explain.

*22.32 Refer to **Venture Capital** Problem 22.4. Assume that the investor's utility function for money is the one given in Problem 22.29.

a. Which of the four acts is optimal according to the expected utility criterion?

b. For each act, determine the amount of money offered with certainty that would be equally attractive to the investor. Does the optimal act identified in **a** correspond to the largest certain amount? Is this result to be expected? Explain.

c. Is the optimal act according to the expected utility criterion identified in **a** the same as the optimal act according to the expected payoff criterion? Is this result to be expected? Explain.

22.33 Refer to **Cooking Utensils** Problems 22.12 and 22.17. The utility function $u(C) = \sqrt{(C - 0.8)/3.1}$, where C is in millions of dollars, reflects management's utility values for payoffs in this decision problem.

a. Which act is optimal according to the expected utility criterion?

b. What monetary amount C, offered with certainty, would give management the same utility level as the optimal act identified in **a**?

22.34 Refer to **Computer Graphics Manufacturer** Problems 22.13 and 22.18. The utility function $u(C) = 2 - (16/C)$, where C is in millions of dollars, reflects the executive committee's utility values for payoffs in this decision problem. Which act is optimal according to the expected utility criterion?

EXERCISES

22.35 Show that the act selected by the maximin criterion is necessarily admissible.

22.36 **a.** Show that the act that maximizes expected payoff minimizes expected regret.
    b. Show that the expected regret for the Bayes act equals *EVPI*.

22.37 Refer to **Cooking Utensils** Problems 22.12 and 22.17. For what value of $P(S_1)$ do the two acts have equal expected payoffs? How is this value useful if management is unsure about the exact outcome state probabilities but does believe that $P(S_1)$ is somewhere between 0.15 and 0.35? Between 0.50 and 0.80?

22.38 Refer to **Computer Graphics Manufacturer** Problems 22.13 and 22.18. The executive committee is somewhat tentative about its assessed value for $P(S_1)$. How far off could this assessment be before it would affect the optimal act?

22.39 Can the optimal strategy in a multistage decision problem be ascertained by either forward or backward induction? If so, what are the advantages of the backward induction approach? Explain.

22.40 Refer to Figure 22.3, containing the decision tree for the computer manufacturer example. The manufacturer conjectures that the payoff of 86 found on the bottom branch of the tree might be an understatement. How much larger could this payoff be before it would alter the manufacturer's optimal strategy, assuming that the other payoffs and the probabilities remained unchanged?

STUDIES

22.41 **Lottery Ticket.** Someone has given you a ticket in a state lottery. You now find that your ticket number is one of ten in a final runoff. Three numbers will be drawn from the ten at random without replacement. If your number is drawn first, second, or third, you will win $250,000, $100,000, or $50,000, respectively. If your number remains undrawn, you will win $1000. A consortium has offered you $25,000 for your ticket before the runoff. Assume that this offer is the only one you will receive.

    a. Construct the decision tree associated with your decision problem. Also, present this information in a payoff table. State the probability distribution of your winnings if you keep the lottery ticket.

    b. Which act (accepting the consortium's offer, keeping the ticket) is optimal under (1) the maximin criterion, (2) the minimax regret criterion, (3) the expected payoff criterion?

    c. Calculate the expected value of perfect information. Interpret its meaning here.

22.42 Refer to **Odd-Even Game** Problem 22.5. Suppose many trials of the game are to be played and you decide to employ a randomized strategy whereby with probability 0.5 you will write "50¢" on your slip of paper and with probability 0.5 you will write "51¢," the random choices being made independently from one trial to the next. Suppose your friend uses the same probabilities to choose between "0¢" and "$1.01," the random choices being made independently of yours. What is your expected payoff? Is the game played in this fashion a fair one in the sense that neither you nor your friend can expect to gain over many trials of the game?

22.43 Refer to **Lottery Ticket** Problem 22.41. Use the reference lottery method to assess your own utility function for the payoffs in this decision problem. Test your utility function for consistency and revise it if necessary, and then apply it to ascertain your optimal act. Is the result obtained from your derived utility function consistent with your actual preference? If not, what are the implications of this?

22.44 Refer to **Consumer By-product** Problem 22.25 and the corresponding decision tree in Figure 22.6. Suppose management's utility function is $u(C) = 1 - \exp(-0.05C)$, where $0 \leq C < \infty$ and C is in millions of dollars.

    a. What is management's optimal strategy in terms of maximizing expected utility? Is this the same strategy as the one that maximizes expected payoff? Discuss.

    b. Suppose that the payoff associated with act A_2 (shown as 34 in Figure 22.6) were, in fact, somewhat uncertain. For what range of values for this payoff would A_1 be the optimal first-stage decision? For what range would A_2 be the optimal first-stage decision?

23 BAYESIAN DECISION MAKING II: SAMPLE INFORMATION

In this chapter, we discuss the use of Bayesian decision making when (1) prior information about the outcome states in the form of probabilities $P(S_i)$ is available, and (2) it is possible to obtain information about the prevailing outcome state by sampling. This type of decision situation is encountered frequently.

EXAMPLES

1. A decision must be made whether or not to accept a shipment from a supplier. The payoff is affected by the proportion of defective items in the shipment. The outcome states here are the different possible proportions defective. From past experience with many shipments from the same supplier, probabilities can be assigned to the outcome states. In addition, a sample can be selected from the particular shipment under consideration to obtain information on the proportion defective in this shipment.

2. A decision must be made whether or not to market a product currently sold by competing firms. The payoff is affected by the number of retailers who now carry this product. The outcome states here are the different possible number of retailers who now carry the product. No firm information is available on this, but management is willing to assign subjective probabilities to the different outcome states. In addition, it is possible to sample the retailers in the industry to obtain information about the number currently carrying the product. □

When it is possible to obtain sample information about the prevailing outcome state to supplement prior information in the form of outcome state probabilities, two major questions arise:

1. For given sample size n, how does one select the best act based on both the prior probabilities and the sample information?
2. If the sample size is not predetermined, how does one ascertain the best sample size?

We shall discuss each of these questions in turn, employing a detailed case illustration. *Throughout this chapter, we assume that the expected payoff criterion is the appropriate guide for finding the best act.*

In the context of decision making with sample information, the outcome state probabilities $P(S_i)$ are called *prior probabilities* because the sample information subsequently provides further information about the prevailing outcome state.

Comment	If the decision maker's utility function for money is not linear (see Section 22.6), utility numbers will need to replace the monetary outcomes in the calculations for identifying the best act.

23.1   POLLUTION DETECTOR EXAMPLE

A firm manufactures an air pollution detector. A key component is a large membrane produced periodically by the firm in batches of several hundred and inventoried to meet production and replacement needs. A chemical purchased from a supplier and used in producing the membrane must be pure, or else some membranes will be inert and have to be discarded. From experience, it is known that if there are slight trace impurities, about 10 percent of membranes will be inert. With higher levels of impurity, the proportion of inert membranes will be higher.

To keep the problem relatively simple, we shall assume that the proportion of inert membranes as a result of the impurities in the chemical will be one of only five values — these are the outcome states for the decision problem:

Outcome State	Proportion of Inert Membranes
S_1	0
S_2	0.1
S_3	0.2
S_4	0.3
S_5	0.4

The chemical tends to deteriorate in storage, so the firm receives a fresh shipment from the supplier just before making a batch of membranes. The decision problem then arising involves the following alternative acts:

A_1: use the shipment as is

A_2: put the shipment through a purification process

The purification process costs $1700 but removes all impurities.

The outcome states S_i, prior probabilities $P(S_i)$, and payoffs C_{ij} for this decision problem are shown in Table 23.1. The prior probabilities are based on experience with previous shipments. The payoffs represent the anticipated dollar gain to the firm in producing a batch of the membranes. For act A_1, the payoff varies with the purity of the shipment. For A_2, the payoff is fixed at $500 — the gain when the chemical is pure ($2200) less the cost of purification ($1700).

TABLE 23.1 Outcome states, prior probabilities, and payoffs for the pollution detector example (payoffs in dollars)

Proportion of Inert Membranes	Prior Probability $P(S_i)$	Act	
		A_1 Do Not Purify	A_2 Purify
S_1: 0	0.2	2200	500
S_2: 0.1	0.2	1600	500
S_3: 0.2	0.1	1000	500
S_4: 0.3	0.2	400	500
S_5: 0.4	0.3	−200	500

Analysis Without Sample Information

If a decision had to be made without sample information, we would follow the procedures presented in Chapter 22. For review, we shall apply these to the pollution detector example. The expected payoffs for the two acts are, by (22.10):

$$EP(A_1) = 2200(0.2) + 1600(0.2) + 1000(0.1) + 400(0.2)$$
$$+ (-200)(0.3) = 880$$

$$EP(A_2) = 500(0.2) + 500(0.2) + 500(0.1) + 500(0.2) + 500(0.3) = 500$$

Hence, if no information about the purity of a particular shipment is available, the best act is A_1—to use the shipment as is. Act A_1 therefore is the Bayes act, and the Bayes expected payoff is $BEP(0) = 880$. We now use $BEP(0)$ to denote the Bayes expected payoff, indicating that a sample of size $n = 0$ is utilized.

If perfect information about the purity of shipments were available, the best act would be A_1 for outcome states S_1, S_2, and S_3, and A_2 for outcome states S_4 and S_5 (see Table 23.1). Hence, the expected payoff with perfect information is, by (22.12):

$$EPPI = 2200(0.2) + 1600(0.2) + 1000(0.1) + 500(0.2) + 500(0.3)$$

$$= 1110$$

Finally, the expected value of perfect information is, by (22.13):

$$EVPI = EPPI - BEP(0) = 1110 - 880 = 230$$

Thus, the decision maker should pay no more than $230 per shipment for perfect information about the purity of the shipment.

The results of the Bayesian analysis without sample information are summarized in Table 23.2.

TABLE 23.2   **Bayesian decision analysis without sample information in the pollution detector example**

Expected Payoffs	Expected Payoff with Perfect Information
$EP(A_1) = 880$ $EP(A_2) = 500$	$EPPI = 1110$
Bayes Act	Expected Value of Perfect Information
A_1 $BEP(0) = 880$	$EVPI = 230$

Sample Information

Management need not make a decision here without sample information about the purity of a particular shipment of chemical, because an engineer has developed a process for making test membranes by hand. While a test membrane is not sensitive enough to be used in the monitoring equipment, it can be checked for inertness. Thus, it is possible for the company to obtain sample information on the outcome state for an incoming shipment by making a number of test membranes and ascertaining how many of these are inert.

23.2   OPTIMAL DECISION RULE FOR GIVEN SAMPLE SIZE

We now consider the first of the two questions posed earlier: For given sample size n, how does one select the best act based on both the prior probabilities and the sample information? In practice, this case arises when a sample has already been taken and the results are now available, or when the sample size is predetermined by policy or by constraints involving time, money, or availability of personnel.

In the pollution detector example, management has decided to make $n = 6$ test membranes for each shipment of chemical before choosing its course of action. We let X denote the number of test membranes in the sample that are inert.

Decision Rule

In Bayesian decision making with sample information, we use the same type of statistical decision rule as discussed in earlier chapters for statistical decision making. Thus, since A_1 yields a higher payoff than A_2 when the chemical is relatively pure and a lower payoff when the chemical is relatively impure, small values of X (number of inert membranes in sample) will lead to act A_1 while large values will lead to act A_2. Hence, the appropriate form of the decision rule is the following (where K denotes the action limit).

(23.1)

For a sample of size n:

If $X \leq K$, select A_1 (do not purify).

If $X > K$, select A_2 (purify).

This decision rule is denoted by (n, K).

In contrast to our earlier approach to constructing statistical decision rules, in Bayesian decision analysis the action limit K is determined by explicitly utilizing information about the outcome state probabilities and payoffs.

Set of Decision Rules Under Consideration

The choice of the optimal action limit K when $n = 6$ test membranes are utilized must be made from among eight possibilities, as shown in Table 23.3. For compactness, we express decision rule (23.1) by (n, K). Note that decision rule $(6, 6)$ always leads to act A_1 regardless of the sample outcome. The reason is that this rule leads to act A_1 whenever $X \leq 6$, and in a sample of size $n = 6$, $X \leq 6$ always. Similarly, rule $(6, -1)$ always leads to act A_2 since X can never be less than zero.

Expected Payoff for Decision Rule

We now seek the decision rule among those in Table 23.3 that has the greatest expected payoff. Consider rule $(6, 0)$. This rule directs that six test membranes be made and that A_1 be selected if none of the six is inert; otherwise, A_2 is to be selected. The

TABLE 23.3   **Set of decision rules under consideration for the pollution detector example: $n = 6$**

Decision Rule	Select A_1 If:	Select A_2 If:
$(6, 0)$	$X = 0$	$X > 0$
$(6, 1)$	$X \leq 1$	$X > 1$
$(6, 2)$	$X \leq 2$	$X > 2$
$(6, 3)$	$X \leq 3$	$X > 3$
$(6, 4)$	$X \leq 4$	$X > 4$
$(6, 5)$	$X \leq 5$	$X > 5$
$(6, 6)$	Always	Never
$(6, -1)$	Never	Always

TABLE 23.4   Calculation of the expected payoff for the decision rule (6, 0) — Pollution detector example

(1)	(2)	(3)	(4)	(5)	(6) Conditional Expected Payoff (2) × (4) + (3) × (5)	(7) Prior Probability $P(S_i)$	(8)
Proportion of Inert Membranes	Payoffs C_{i1}	C_{i2}	Act Probabilities $P(A_1 \mid S_i)$	$P(A_2 \mid S_i)$			(6) × (7)
S_1: 0	2200	500	1.0000	0	2200.00	0.2	440.000
S_2: 0.1	1600	500	0.5314	0.4686	1084.54	0.2	216.908
S_3: 0.2	1000	500	0.2621	0.7379	631.05	0.1	63.105
S_4: 0.3	400	500	0.1176	0.8824	488.24	0.2	97.648
S_5: 0.4	−200	500	0.0467	0.9533	467.31	0.3	140.193
							$EP(6, 0) = 957.85$

procedure for calculating the expected payoff for this rule is shown in Table 23.4. The basic steps are as follows:

1. Determine the probabilities of the sample outcome leading to acts A_1 and A_2 for a given outcome state.
2. Obtain the expected payoff for that outcome state.
3. Determine the expected payoff over all outcome states.

In columns 1 through 3 of Table 23.4, we repeat from Table 23.1 the outcome states and payoffs for the pollution detector example.

Act Probabilities.   The next two columns of Table 23.4 show the respective probabilities of selecting A_1 and A_2 for rule (6, 0), conditional on the outcome state S_i. We designate these probabilities as act probabilities.

(23.2)

> An *act probability*, denoted by $P(A_j \mid S_i)$, is the conditional probability that act A_j will be selected under the given decision rule when the outcome state is S_i.

The act probabilities $P(A_j \mid S_i)$ are analogous to the probabilities of selecting H_0 and H_1 in a statistical test, discussed in earlier chapters. These latter probabilities are conditional on the value of a parameter, such as the population mean μ.

Sampling Distribution of X.   To ascertain the act probabilities for the pollution detector example, we must obtain the sampling distribution of X, the number of inert membranes in the sample of six. It is reasonable to assume here that X is a binomial random variable, whose probability function is given by (6.3). The parameters of this probability function are $n = 6$ and the proportion of inert membranes for the batch, denoted by p, which is given for each outcome state.

Calculating Act Probabilities. Consider first outcome state S_1. Here, the proportion of inert membranes for the batch is $p = 0$; thus, the probability of an inert membrane is $p = 0$. Hence, no test membrane will be inert under this outcome state, and rule $(6, 0)$ will always lead to A_1. The respective act probabilities therefore are $P(A_1|S_1) = 1.0$ and $P(A_2|S_1) = 0$. These are entered in columns 4 and 5, respectively, for outcome state S_1.

When the outcome state is S_2, the proportion of inert membranes for the batch is $p = 0.1$. Since decision rule $(6, 0)$ leads to A_1 if $X = 0$, we need to find $P(X = 0)$ when $n = 6$ and $p = 0.1$. We use Table C.5 and find $P(0) = 0.5314$. Hence, $P(A_1|S_2) = 0.5314$, and by complementation theorem (4.19), we obtain $P(A_2|S_2) = 1 - 0.5314 = 0.4686$. These act probabilities are entered in Table 23.4 for outcome state S_2.

For outcome state S_3, when $p = 0.2$, we find from Table C.5 that $P(0) = 0.2621$. Hence, $P(A_1|S_3) = 0.2621$ and $P(A_2|S_3) = 1 - 0.2621 = 0.7379$.

The act probabilities for the other outcome states are obtained in similar fashion.

Conditional Expected Payoffs. We can now obtain the expected payoff for rule $(6, 0)$ for each outcome state. Such an expected payoff is called a conditional expected payoff, because it is conditional on a particular outcome state.

(23.3)

The *conditional expected payoff* for decision rule (n, K) and outcome state S_i, denoted by $EP(n, K|S_i)$, is the expected payoff for this rule conditional on outcome state S_i:

$$EP(n, K|S_i) = \sum_j C_{ij} P(A_j|S_i)$$

For outcome state S_1, we see from Table 23.4 that $C_{11} = 2200$, $C_{12} = 500$, $P(A_1|S_1) = 1.0$, and $P(A_2|S_1) = 0$. Hence, the conditional expected payoff for rule $(6, 0)$ when the outcome state is S_1 is:

$$EP(6, 0|S_1) = 2200(1.0) + 500(0) = 2200$$

We interpret this to mean that if rule $(6, 0)$ is used repeatedly with many incoming shipments for which the proportion of inert membranes is $p = 0$, the average payoff is $2200 per shipment.

For outcome state S_2, the conditional expected payoff for rule $(6, 0)$ is:

$$EP(6, 0|S_2) = 1600(0.5314) + 500(0.4686) = 1084.54$$

Thus, if rule $(6, 0)$ is used repeatedly with many incoming shipments for which the proportion of inert membranes is $p = 0.1$, the average payoff is $1084.54 per shipment.

The conditional expected payoffs under S_3, S_4, and S_5 are calculated and interpreted in a similar manner. Column 6 of Table 23.4 shows all of the conditional expected payoffs for rule $(6, 0)$.

Expected Payoff. We can now calculate the expected payoff for decision rule $(6, 0)$ over all outcome states. The prior probabilities $P(S_i)$ assigned on the basis of past experience to the outcome states S_i are repeated in column 7 of Table 23.4. For instance, $P(S_1) = 0.2$. Thus, the probability of conditional expected payoff $EP(6, 0 | S_1) = 2200$ is 0.2 since S_1 will be the outcome in 20 percent of the shipments. Similarly, the probability of conditional expected payoff $EP(6, 0 | S_2) = 1084.54$ is $P(S_2) = 0.2$. In taking the expected value of the conditional expected payoffs, we obtain the expected payoff for rule $(6, 0)$.

(23.4)

> The *expected payoff* for decision rule (n, K), denoted by $EP(n, K)$, is the expected value of the conditional expected payoffs under the prior probabilities $P(S_i)$:
>
> $$EP(n, K) = \sum_i EP(n, K | S_i)P(S_i)$$

For the pollution detector example, we calculate (as shown in column 8 of Table 23.4):

$$EP(6, 0) = 2200(0.2) + 1084.54(0.2) + 631.05(0.1) + 488.24(0.2)$$
$$+ 467.31(0.3) = 957.85$$

We interpret this to mean that if decision rule $(6, 0)$ is used repeatedly with many incoming shipments, the average payoff per shipment in the long run will be $957.85.

Optimal Decision Rule

Now that we have obtained the expected payoff for decision rule $(6, 0)$, we need to find the expected payoffs for the other decision rules in Table 23.3.

In Table 23.5, we show the calculations for decision rule $(6, 1)$. We illustrate how the act probabilities for outcome state S_2 are obtained. Rule $(6, 1)$ states that act A_1 should be selected when $X \leq 1$; otherwise, act A_2 should be selected. Since outcome state S_2 pertains to a proportion of inert membranes of $p = 0.1$, we need to find $P(X \leq 1)$ when $n = 6$ and $p = 0.1$. From Table C.5, we find $P(X \leq 1) = P(0) + P(1) = 0.5314 + 0.3543 = 0.8857$. Hence, $P(A_1 | S_2) = 0.8857$ and $P(A_2 | S_2) = 1 - 0.8857 = 0.1143$.

The expected payoffs for all of the decision rules in Table 23.3 are as follows (further calculations are not shown):

K:	0	1	2	3	4	5	6	-1
$EP(6, K)$:	958	1010	972	918	889	881	880	500

TABLE 23.5 Calculation of the expected payoff for the decision rule (6, 1)—Pollution detector example

(1) Proportion of Inert Membranes	(2) Payoffs C_{i1}	(3) C_{i2}	(4) Act Probabilities $P(A_1\|S_i)$	(5) $P(A_2\|S_i)$	(6) Conditional Expected Payoff $(2) \times (4)$ $+ (3) \times (5)$	(7) Prior Probability $P(S_i)$	(8) $(6) \times (7)$
S_1: 0	2200	500	1.0000	0	2200.00	0.2	440.000
S_2: 0.1	1600	500	0.8857	0.1143	1474.27	0.2	294.854
S_3: 0.2	1000	500	0.6553	0.3447	827.65	0.1	82.765
S_4: 0.3	400	500	0.4201	0.5799	457.99	0.2	91.598
S_5: 0.4	-200	500	0.2333	0.7667	336.69	0.3	101.007
						$EP(6, 1) =$	1010.22

Bayes Decision Rule. The expected payoff criterion leads to the selection of the decision rule with the largest expected payoff. In the pollution detector example, this rule is (6, 1). Rule (6, 1) is called the Bayes decision rule for sample size $n = 6$.

(23.5)

The *Bayes decision rule* for a sample of size n is the one that maximizes the expected payoff among all possible decision rules for that sample size.

The expected payoff for the Bayes decision rule for sample size n is denoted by $BEP(n)$, where:

$$BEP(n) = \max_K EP(n, K)$$

Thus, in the pollution detector example, the expected payoff for the Bayes rule (6, 1) is $BEP(6) = EP(6, 1) = 1010$.

Bayes Acts. The acts to be selected according to the Bayes decision rule for each sample outcome are called the *Bayes acts*. In the pollution detector example, where the Bayes rule for $n = 6$ is (6, 1), the Bayes acts are as follows for the different sample outcomes X:

X:	0	1	2	3	4	5	6
Bayes Act:	A_1	A_1	A_2	A_2	A_2	A_2	A_2

In the most recent shipment sampled, $X = 2$ test membranes were found to be inert. Since the Bayes act for $X = 2$ is A_2, the shipment should be purified.

Summary of Procedure

We now summarize the procedure for finding the Bayes decision rule when the sample size is given.

(23.6)

> The steps for finding the Bayes decision rule for a given sample size n are as follows:
>
> 1. Identify all possible decision rules (n, K) for the given sample size n.
> 2. For each possible decision rule:
>   a. Calculate the act probabilities $P(A_j | S_i)$.
>   b. Obtain the conditional expected payoff $EP(n, K | S_i)$ for each outcome state S_i.
>   c. Use the prior probabilities $P(S_i)$ to obtain the expected payoff $EP(n, K)$ for the decision rule (n, K).
> 3. Select the decision rule for which the expected payoff $EP(n, K)$ is largest.

In actual applications, Bayesian decision rules are often calculated by using computer packages.

Comments

1. The action limit $K = 1$ is optimal in the pollution detector example for the given sample size, payoffs, and prior probabilities. If any of these elements change, some other action limit might be optimal.
2. The expected payoffs for rules $(6, 6)$ and $(6, -1)$, which do not utilize the sample information, are those for decision making with prior information only, as may be seen from Table 23.2. Decision rule $(6, 6)$ has expected payoff $EP(6, 6) = EP(A_1) = 880$, while rule $(6, -1)$ has expected payoff $EP(6, -1) = EP(A_2) = 500$.
3. If a decision maker is unable or unwilling to assign prior probabilities $P(S_i)$, he or she can follow the uncertainty approach described in the preceding chapter. The chosen criterion for decision making under uncertainty is then simply applied to the conditional expected payoffs $EP(n, K | S_i)$.

DETERMINATION OF OPTIMAL SAMPLE SIZE

We now come to the second question raised at the beginning of this chapter: If the sample size is not predetermined, how does one ascertain the best sample size? In practice, this case arises when no firm decision on sample size has been made as yet.

The procedure for finding the optimal sample size is conceptually very simple, though extensive computations may be required. We determine, for each possible sample size, the expected payoff of the Bayes rule net of sampling costs and select that sample size for which this net expected payoff is a maximum. Use of available computer packages can be very helpful. We now consider this procedure in more detail.

Cost of Sampling

We denote the cost of a sample of size n by $s(n)$. We shall assume here that the total sample cost is a linear function of sample size.

(23.7)

$$s(n) = \begin{array}{ll} a + bn & n > 0 \\ 0 & n = 0 \end{array}$$

where:

$a \geq 0$
$b > 0$

Here, a is the overhead cost of setting up the sample, and b is the variable cost of including one additional observation in the sample. Note that the sample cost is zero according to (23.7) when no sample is employed. This simple cost model is often an adequate representation.

Upper Bound on Optimal Sample Size

If many possible sample sizes had to be examined in order to find the optimal n, the computational task would be very large. Fortunately, it is possible to obtain an upper bound for the optimal sample size, beyond which no sample size can be optimal. Recall that the expected value of perfect information, *EVPI*, represents the maximum amount that a decision maker should spend for perfect information about the outcome state. Hence, the sample cost must not exceed *EVPI*; that is:

$$s(n) \leq EVPI$$

For cost model (23.7), the inequality becomes:

$$a + bn \leq EVPI$$

or:

$$n \leq \frac{EVPI - a}{b}$$

(23.8)

For cost model (23.7), an *upper bound* on the optimal sample size, denoted by $\max(n)$, is:

$$\max(n) = \begin{array}{ll} \dfrac{EVPI - a}{b} & \text{if } a < EVPI \\ 0 & \text{if } a \geq EVPI \end{array}$$

Optimal Sample Size

We can now state more formally the procedure for finding the optimal sample size n.

(23.9)

> The steps for finding the optimal sample size n are as follows:
>
> 1. For each sample size in the interval $0 \leq n \leq \max(n)$, ascertain the Bayes decision rule.
> 2. For each Bayes rule, determine the expected payoff $BEP(n)$ and the *net expected payoff*, denoted by $BNEP(n)$, where:
>
> $BNEP(n) = BEP(n) - s(n)$
>
> 3. Select that sample size and associated Bayes decision rule that maximizes the net expected payoff $BNEP(n)$.

In the pollution detector example, suppose that the sample size had not yet been determined. An analyst has been asked to obtain the optimal sample size. She has determined that cost model (23.7) is applicable and that there are no overhead costs, that is, $a = 0$. The variable cost per test membrane is $b = \$7.50$, so the cost function is:

$$s(n) = 7.5n$$

Since $EVPI = 230$ (see Table 23.2), we obtain, by (23.8):

$$\max(n) = \frac{230}{7.5} = 30.7$$

The optimal sample size therefore cannot exceed 30 test membranes.

Consequently, the analyst has to obtain the Bayes decision rule for each sample size from 0 to 30 and then determine the net expected payoff for each. Recall that we found earlier for $n = 6$ that the Bayes rule is $(6, 1)$ and its expected payoff is $BEP(6) = 1010$. Since the cost of six test membranes is $7.50(6) = 45$, the net expected payoff for the Bayes rule is:

$$BNEP(6) = BEP(6) - s(6) = 1010 - 45 = 965$$

The analyst used a computer package to perform the needed calculations. Figure 23.1 shows the computer printout she obtained. For each $n > 0$, the printout shows the action limit K for the Bayes rule for that sample size and the net expected payoff $BNEP(n)$. Note that the net expected payoff for the Bayes decision rule for $n = 6$ is shown to be \$965.237, the same as our result except for rounding.

The analyst could readily see from the printout in Figure 23.1 that the optimal sample size is $n = 6$ and that the Bayes rule for this sample size is $(6, 1)$.

FIGURE 23.1 Computer output showing Bayes decision rules and their expected payoffs, net of sampling costs, for sample sizes 0 to 30 — Pollution detector example

| BAYES RULE | | BAYES RULE |
N	K	NEP
0	–	880.000
1	0	930.500
2	0	949.800
3	0	951.260
4	1	946.628
5	1	960.117
6	1	965.237
7	1	963.502
8	1	956.604
9	2	959.824
10	2	960.661
11	2	957.439
12	2	950.946
13	3	948.537
14	3	946.992
15	3	942.743
16	3	936.259
17	4	931.306
18	4	928.187
19	4	923.158
20	4	916.534
21	5	910.427
22	5	906.189
23	5	900.552
24	5	893.740
25	6	887.155
26	6	882.091
27	6	875.977
28	6	868.974
29	7	862.244
30	7	856.560

OPT N = 6   NEP = 965.237

Comments

1. The printout in Figure 23.1 for $n = 0$ provides the expected payoff for the Bayes act with prior information only, that is, $BEP(0) = 880$ (see Table 23.2).

2. The output in Figure 23.1 can also be used to ascertain the optimal action limit when the sample size is fixed in advance. For instance, if only $n = 3$ test membranes can be made in the time available, it is seen that the Bayes decision rule is $(3, 0)$ — that is, act A_1 should be selected if there are no inert membranes in the sample of three; otherwise, A_2 should be selected. The net expected payoff of this rule is \$951.26, not very much less than the net expected payoff for the optimal sample size $n = 6$.

3. If the upper bound (23.8) on the optimal sample size is zero, the optimal approach is to make the decision using prior information only.

4. The optimal sample size and decision rule depend on the interrelations between the pay-offs, prior probabilities, and sample cost function. Computer-assisted calculations allow

the decision maker to check whether the optimal sample size and decision rule are sensitive to variations in the entries of the payoff table. For example, suppose management is considering modifying the prior probabilities for the pollution detector example as follows because of a suspected recent decline in the quality of shipments:

| Outcome | Prior Probabilities | |
State	Original	Modified
S_1	0.2	0.1
S_2	0.2	0.2
S_3	0.1	0.2
S_4	0.2	0.2
S_5	0.3	0.3

A computer run employing the modified prior probabilities readily shows that the Bayes decision rule still is $(6, 1)$ but with a net expected payoff of $828. Thus, modification of the prior probabilities is not a pressing matter here since the optimal decision rule is unaffected. The difference in the net expected payoffs, $965 - 828 = \$137$, measures the expected cost per shipment to the firm if the suspected quality decline is real.

Value of Sample Information

We now present two measures that deal with the value of sample information in Bayesian decision making, and we shall illustrate them in terms of the pollution detector example.

Expected Value of Sample Information.   The first measure indicates the gain with use of sample information, neglecting the cost of sampling, thus producing a bound on how much should be paid for sample information.

(23.10)

> For a sample of size n, the *expected value of sample information*, denoted by $EVSI(n)$, is the difference between the expected payoff of the Bayes decision rule for that n and the expected payoff for the Bayes act with prior information only:
>
> $$EVSI(n) = BEP(n) - BEP(0)$$

In the pollution detector example, we calculate from results for $n = 6$ obtained earlier:

$$EVSI(6) = 1010 - 880 = 130$$

In a choice between making a decision using prior information only or taking a sample of size $n = 6$, the decision maker should not pay more than $130 for the sample information.

Expected Net Gain from Sampling.   The second measure reflects the net gain with use of sample information, taking into account the cost of this sample information.

(23.11)

> For a sample of size n, the *expected net gain from sampling*, denoted by $ENGS(n)$, is the difference between the net expected payoff for the Bayes rule for that sample size and the expected payoff of the Bayes act with prior information only:
>
> $$ENGS(n) = BNEP(n) - BEP(0)$$

For the pollution detector example for $n = 6$, we have (see Figure 23.1):

$$ENGS(6) = 965 - 880 = 85$$

Note that this is equivalent to $EVSI(6) - s(6) = 130 - 45 = 85$. In using the optimal sample size, we automatically maximize $ENGS(n)$.

23.4 DETERMINING BAYES ACT BY REVISION OF PRIOR PROBABILITIES

Once a sample has been selected and the result observed, it is possible to ascertain directly the Bayes act for that given sample result without developing the decision rule covering all possible sample results. The procedure employs Bayes theorem. (A discussion of Bayes theorem is presented in Section 4.6.)

We shall first restate Bayes theorem (4.25) in terms of our present notation, using the pollution detector example for illustration. The variable of interest here is the outcome state S_i, denoting the proportion of inert membranes. The prior probabilities of the outcome states are denoted by $P(S_i)$. The additional information variable is the number of inert membranes among the n test membranes, denoted by X_j. We shall use the notation X_0 to represent no inert membrane in the sample, X_1 to denote one inert membrane in the sample, and, in general, X_j to denote j inert membranes in the sample. With this notation, Bayes theorem is as follows.

(23.12)

> $$P(S_i \mid X_j) = \frac{P(S_i)P(X_j \mid S_i)}{\sum\limits_{i} P(S_i)P(X_j \mid S_i)}$$

The posterior probability $P(S_i \mid X_j)$ is the probability that outcome state S_i is the prevailing one, given the sample outcome X_j.

To find the Bayes act for a given sample result X_j by means of Bayes theorem, we employ a three-step procedure.

(23.13)

> The steps for finding the Bayes act for a given sample result by utilizing Bayes theorem are as follows:

continues

1. For the prior probabilities $P(S_i)$ and the observed sample result X_j, obtain the posterior probabilities $P(S_i|X_j)$, using Bayes theorem (23.12).
2. Calculate, for each act, the expected payoff under the posterior probabilities $P(S_i|X_j)$.
3. Select the act for which the expected payoff is largest.

We now illustrate this procedure for the pollution detector example where $n = 6$ test membranes were examined and two were found inert; that is, the sample outcome was X_2.

Obtaining Posterior Probabilities

The steps for revising the prior probabilities by Bayes theorem (23.12) are shown in Table 23.6. Columns 1 and 2 repeat the outcome states and prior probabilities given initially in Table 23.1. Column 3 shows the conditional probabilities $P(X_2|S_i)$ of obtaining the observed sample result X_2 for the different possible outcome states S_i. For instance, if the state is S_2—that is, the proportion of inert membranes is $p = 0.1$—we find from Table C.5 for $n = 6$ and $p = 0.1$ that $P(2) = 0.0984$. Thus, $P(X_2|S_2) = 0.0984$. For state S_3, the proportion of inert membranes is $p = 0.2$, so for $n = 6$ we find $P(2) = 0.2458$; hence, $P(X_2|S_3) = 0.2458$. The other conditional probabilities are found in a similar manner.

The joint probabilities $P(S_i \cap X_2)$ are calculated in column 4 by multiplying $P(X_2|S_i)$ by $P(S_i)$ for each outcome state. The sum of this column is the denominator of Bayes theorem (23.12). The posterior probabilities $P(S_i|X_2)$ are obtained in column 5 by dividing each entry in column 4 by the sum of column 4. We see, for instance, that $P(S_3|X_2) = 0.02458/0.20238 = 0.1215$. Thus, given that two membranes were inert among the six test membranes, the conditional probability that the shipment

TABLE 23.6   Revision of prior probabilities for the pollution detector example.   Here $n = 6$ membranes were tested and X_2 (that is, two membranes) were found inert.

(1) Proportion of Inert Membranes	(2) Prior Probability $P(S_i)$	(3) $P(X_2|S_i)$	(4) Joint Probability $P(S_i \cap X_2)$ (2) × (3)	(5) Posterior Probability $P(S_i|X_2)$ (4) ÷ 0.20238
S_1: 0	0.2	0	0	0
S_2: 0.1	0.2	0.0984	0.01968	0.0972
S_3: 0.2	0.1	0.2458	0.02458	0.1215
S_4: 0.3	0.2	0.3241	0.06482	0.3203
S_5: 0.4	0.3	0.3110	0.09330	0.4610
Total	1.0		0.20238	1.0000

will produce proportion $p = 0.2$ of inert membranes (that is, that the outcome state is S_3) is 0.1215. The other posterior probabilities are similarly interpreted.

It is interesting to note how the prior probabilities have been revised in light of the sample information of two inert membranes among the six test membranes. The prior probabilities $P(S_1)$ and $P(S_2)$ have been revised downward substantially, and the prior probabilities $P(S_4)$ and $P(S_5)$ have been revised upward substantially. Thus, finding two inert test membranes makes it much more likely that the proportion p of inert membranes in the shipment will be high.

Determining Expected Payoffs

Proceeding now to step 2 of (23.13), we need to find the expected payoff for each act under the posterior probabilities. We utilize the payoffs in Table 23.1 and the posterior probabilities in Table 23.6. The expected payoffs are calculated in the usual way and are denoted by $EP(A_1|X_2)$ and $EP(A_2|X_2)$ to indicate that they are conditional on the given sample outcome. We obtain:

$$EP(A_1|X_2) = 2200(0) + 1600(0.0972) + 1000(0.1215) + 400(0.3203)$$
$$+ (-200)(0.4610)$$
$$= 312.94$$
$$EP(A_2|X_2) = 500(0) + 500(0.0972) + 500(0.1215) + 500(0.3203)$$
$$+ 500(0.4610)$$
$$= 500$$

Determining Bayes Act

It is evident from the preceding calculations that A_2 is the Bayes act here since it has the larger expected payoff. Thus, when two test membranes among six are inert, the company should purify the shipment. Of course, this result is in accord with the Bayes decision rule (6, 1) obtained earlier.

Extensive and Normal Forms of Bayesian Analysis

The procedure described in this section for ascertaining the Bayes act for a given sample result can be extended to determine the Bayes decision rule for a given sample size and to determine the optimal sample size. It is called the *extensive form* of Bayesian decision analysis. It contrasts with the form employed in the preceding sections of this chapter, called the *normal form*. As we just saw, the extensive form introduces the prior probabilities in the first stage of the analysis and revises them via Bayes theorem. The normal form, on the other hand, introduces the prior probabilities as weights for the conditional expected payoffs in the final stage of the calculations and does not employ Bayes theorem.

Each form of Bayesian analysis has its respective advantages. As noted earlier, if we wish to ascertain the Bayes act for a specific sample result, the extensive form will identify this act directly. The normal form, in contrast, develops the complete Bayes decision rule for that sample size. Another advantage of the extensive form is its adaptability to shortcut mathematical procedures.

On the other hand, the normal form of Bayesian analysis has the advantage that the prior probabilities are introduced last. The decision maker then has the option of treating the outcome states S_i as uncertain or of assigning prior probabilities $P(S_i)$ to them. Also, many users find the normal form to be more suitable for sensitivity analysis. In part, this is because the more objective elements of the decision problem (payoffs and sample cost function) are introduced first. Finally, many statisticians and management scientists find the normal form easier to explain to managers because it proceeds in straightforward steps. The extensive form, on the other hand, requires Bayes theorem at the onset, and its rationale is grasped somewhat less easily.

PROBLEMS

*23.1   Refer to the pollution detector example in the text. Assume that the payoff table remains as in Table 23.1 except that new prior probabilities have been determined, as follows:

S_i:	S_1	S_2	S_3	S_4	S_5
$P(S_i)$:	0.4	0.2	0.1	0.1	0.2

a. Determine the Bayes act based on prior information alone. What is the expected payoff of the Bayes act?

b. Obtain *EPPI* and *EVPI*. Since perfect information is assumed, why are these quantities expected values?

23.2   **Solar Heater.** A home products firm has developed a new solar energy heater and now must decide whether or not to market it. The success of the product, as measured by its profit contribution, depends on the proportion of households in the target market that will purchase it. Research by the marketing department has produced the following payoff table and prior probabilities (payoffs are in thousands of dollars):

Proportion Purchasing Product	Prior Probability $P(S_i)$	Act	
		A_1 Don't Market Product	A_2 Market Product
S_1: 0.05	0.5	0	−500
S_2: 0.10	0.1	0	0
S_3: 0.15	0.1	0	500
S_4: 0.20	0.3	0	1000

Assume that the expected payoff is to be maximized.

a. Determine the Bayes act if a decision must be made without sample information. What is the expected payoff of the Bayes act? What is the probability of a negative payoff with the Bayes act?

b. Obtain *EVPI*, and interpret your result.

23.3 Imported Fruit. A fruit importer is considering the purchase of a large shipment of grapefruit from a supplier abroad. The value of the shipment to the importer depends on the proportion of the shipment that is of premium grade. If the importer does not purchase the shipment from the supplier abroad, he will buy from a local source. The following payoff table shows the profit contribution (in thousands of dollars) to the importer of each act for different proportions of premium-grade fruit in the shipment. The importer's prior probabilities for the shipment, based on extensive past experience with other shipments from the same supplier, are also shown.

		Act	
	Prior	A_1	A_2
Shipment Proportion	Probability	Buy from	Buy from
of Premium-Grade Fruit	$P(S_i)$	Local Source	Supplier Abroad
S_1: 0.1	0.1	5.0	−2.0
S_2: 0.3	0.6	5.0	4.0
S_3: 0.5	0.3	5.0	10.0

Assume that the expected payoff is to be maximized.
a. Determine the Bayes act based on prior information alone. What is the expected payoff of the Bayes act?
b. Obtain *EPPI* and *EVPI*. Interpret the meanings of these measures here.
c. Suppose that a source of perfect information about the shipment proportion of premium-grade fruit were available. What is the probability, prior to having this information, that acting on the basis of it will lead to a higher payoff than is offered by the Bayes act here?

23.4 Control Boxes. A firm manufactures electronic control boxes for electric motors that are sold to industrial customers in large lots. The firm must choose between two policies for selling the boxes. The first, called the inspection policy, requires that every box be checked for malfunctions prior to shipment. The second policy, called the returns policy, entails shipment without inspection, replacement of malfunctioning boxes returned by customers, and reimbursement of customers for direct expenses and inconvenience. The payoff table showing the costs (expressed as negative payoffs in thousands of dollars) of these two policies for different values of the proportion of malfunctioning boxes in a lot follows. Prior probabilities, based on the firm's extensive past experience with its production process for this type of box, are also given.

		Act	
	Prior	A_1	A_2
Lot Proportion	Probability		
Malfunctioning	$P(S_i)$	Returns Policy	Inspection Policy
S_1: 0.01	0.8	−1.5	−2.5
S_2: 0.05	0.2	−7.5	−4.5

Assume that the expected payoff is to be maximized.
a. From prior information alone, which policy should the firm adopt? What is the expected cost per lot of this policy?
b. What would constitute perfect information about a lot in this decision problem? What is the expected value of perfect information to the firm?

*23.5 Refer to Problem 23.1. For the new prior probabilities, the expected payoffs for the decision rules that need to be considered when six test membranes are made follow.

$(6, K)$:	$(6, 0)$	$(6, 1)$	$(6, 2)$	$(6, 3)$	$(6, 4)$	$(6, 5)$	$(6, 6)$	$(6, -1)$
$EP(6, K)$:	_____	1371	1358	_____	1306	1301	_____	_____

   a. Obtain the missing expected payoffs.

   b. For the new prior probabilities, which is the Bayes decision rule for sample size $n = 6$? What is the value of $BEP(6)$? Interpret the meaning of this value.

   c. If six test membranes are made for a shipment and one of these is inert, which act is best?

23.6   Refer to **Solar Heater** Problem 23.2. Suppose a random sample of $n = 15$ households is to be selected from the target market to obtain information about the proportion of households that will purchase the heater. The decision rule to be employed is of the form: If $X \le K$, select A_1; if $X > K$, select A_2. Here, X is the number of households in the sample that will purchase the heater, and it can be treated as a binomial random variable.

   a. For outcome state S_1 and the decision rule $(n, K) = (15, 2)$, obtain (1) the act probabilities, (2) the conditional expected payoff. Interpret the latter result.

   b. Obtain the expected payoff for decision rule $(15, 2)$.

   c. The expected payoffs for selected decision rules based on sample size 15 follow.

$(15, K)$:	$(15, 0)$	$(15, 1)$	$(15, 2)$	$(15, 3)$	$(15, 4)$	$(15, 15)$	$(15, -1)$
$EP(15, K)$:	201	241	_____	113	52	_____	_____

   Obtain the missing expected payoffs. What appear to be the Bayes decision rule for $n = 15$ and the value of $BEP(15)$? Comment.

23.7   Refer to **Solar Heater** Problems 23.2 and 23.6. The Bayes rule is $(15, 1)$.

   a. What is the probability of a negative payoff with the decision rule $(15, 1)$? How does this probability compare with the corresponding one for the Bayes rule based on no sampling?

   b. What other factors should be considered in deciding whether it is worthwhile to select a sample of 15 households?

23.8   Refer to **Imported Fruit** Problem 23.3. The importer plans to select a simple random sample of $n = 10$ grapefruit from the shipment prior to deciding whether to purchase the shipment from the supplier abroad or to buy from a local source. He will buy from a local source (A_1) if the number of premium-grade grapefruit in the sample (X) is less than or equal to a stipulated number (K), and he will purchase the shipment from the supplier abroad (A_2) if X exceeds K. Assume that X is a binominal random variable here.

   a. Suppose that the shipment proportion of premium-grade fruit is actually 0.5 (outcome state S_3). What is the probability that the importer will purchase the shipment from the supplier abroad if he uses $K = 1$ for the decision rule? What is the conditional expected payoff for outcome state S_3 with $K = 1$?

   b. Obtain the expected payoff for decision rule $(n, K) = (10, 1)$.

   c. The expected payoffs for selected decision rules based on sample size 10 follow.

$(10, K)$:	$(10, 0)$	$(10, 1)$	$(10, 2)$	$(10, 3)$	$(10, 4)$	$(10, 5)$	$(10, 10)$	$(10, -1)$
$EP(10, K)$:	5.46	_____	6.00	_____	5.84	5.54	_____	_____

   Obtain the missing expected payoffs. What appear to be the Bayes decision rule for $n = 10$ and the value of $BEP(10)$? Comment.

23.9   Refer to **Control Boxes** Problem 23.4. It is proposed that a random sample of $n = 12$ control boxes be selected from each lot and checked for malfunctions before deciding on the policy to be adopted for the lot. The decision rule to be employed is of the form: If $X \leq K$, select A_1; if $X > K$, select A_2. Here, X is the number of malfunctioning boxes in the sample, and it can be treated as a binomial random variable.

    a. Obtain the expected payoffs for the following decision rules: (1) (12, 12), (2) (12, −1), (3) (12, 0), (4) (12, 1).

    b. Decision rule (12, 0) is, in fact, the Bayes decision rule. If this rule is adopted here, what policy will be chosen for a lot when $X = 0$? When $X = 2$?

23.10 Refer to **Control Boxes** Problems 23.4 and 23.9. What proportion of all lots will be shipped under the returns policy (A_1) if the Bayes decision rule (12, 0) is adopted? [*Hint:* $P(A_1) = P(A_1|S_1)P(S_1) + P(A_1|S_2) P(S_2)$.]

*23.11 Refer to Problem 23.1. Suppose that the cost function is $s(n) = 10n$. What is the upper bound on the optimal sample size given by (23.8)? Interpret this number.

23.12 Refer to **Solar Heater** Problem 23.2. The setup cost of a survey is $8 thousand and the cost per respondent is $0.05 thousand.

    a. State the cost function for sampling.

    b. Obtain the upper bound on the optimal sample size given by (23.8).

23.13 Refer to **Imported Fruit** Problem 23.3. The cost function for sampling (in $000) is $s(n) = 0.2 + 0.005n$. Obtain the upper bound on the optimal sample size given by (23.8). Interpret this bound.

23.14 Refer to **Control Boxes** Problem 23.4. The firm has calculated that the setup cost for sampling a lot is $0.5 thousand and the variable cost of checking each control box in the sample is $0.002 thousand. Obtain the upper bound on the optimal sample size given by (23.8).

*23.15 Refer to Table 23.1 and Figure 23.1, pertaining to the pollution detector example.

    a. Obtain $BEP(n)$ for $n = 0, 1, 2, 3, 4$. Recall that the cost function for sampling is $s(n) = 7.5n$.

    b. What would be the optimal sample size and decision rule if the sampling cost function were (1) $s(n) = 100 + 7.5n$, (2) $s(n) = 30 + 20n$? Give the net expected payoff for the optimal decision rule in each case.

23.16 Refer to **Solar Heater** Problem 23.2. The cost function for sampling (in $000) is $s(n) = 8 + 0.05n$. The values of $BEP(n)$ for selected values of n follow.

n:	0	20	50	100	300
$BEP(n)$:	100.0	254.4	315.2	340.4	349.7

    a. Find $BNEP(n)$, $EVSI(n)$, and $ENGS(n)$ for each of the values of n considered here. Which of these sample sizes leads to the largest expected payoff net of sampling costs?

    b. What is the meaning of (1) $BNEP(50)$, (2) $ENGS(50)$?

    c. For the sample sizes considered here, how large would the setup cost of the survey have to be before it would not pay to conduct a survey?

23.17 Refer to **Imported Fruit** Problem 23.3. The importer's cost function for sampling (in $000) is $s(n) = 0.2 + 0.005n$. The values of $BEP(n)$ for selected values of n follow.

n:	0	20	30	40	50
$BEP(n)$:	_____	6.177	6.264	6.324	6.367

    a. Obtain $BEP(0)$ and $BNEP(0)$. Why are these two values identical?

 b. Find $BNEP(n)$, $EVSI(n)$, and $ENGS(n)$ for each of the values of n considered here. Which of these sample sizes leads to the largest expected payoff net of sampling costs?

23.18 Refer to **Control Boxes** Problem 23.4. The cost function for sampling (in \$000) is $s(n) = 0.5 + 0.002n$. Computer output for $n > 0$ shows that $BNEP(n)$ attains its maximum value, -2.932, at $n = 59$.

 a. Obtain the values of (1) $BEP(0)$, (2) $BEP(59)$, (3) $EVSI(59)$, (4) $ENGS(59)$. Interpret the meaning of the last value here.

 b. Should the firm do any sampling of a lot before shipping it? Explain.

23.19 A government auditor is to check the accuracy of a company's entertainment expense vouchers by examining a sample of n vouchers. The expected payoffs of the Bayes decision rules for several sample sizes follow. The payoffs represent recaptured tax revenues from ineligible entertainment expenses in thousands of dollars.

n:	0	100	200	500	1000
$BEP(n)$:	0	63	86	98	99

Assume the cost function for sampling is $s(n) = 6 + 0.01n$.

 a. Find $BNEP(n)$ for each of the values of n considered here. Which of these sample sizes leads to the largest expected payoff net of sampling costs?

 b. Find $EVSI(n)$ and $ENGS(n)$ for each of the values of n considered here. Interpret the two measures for sample size 100.

*23.20 Refer to Table 23.1, pertaining to the pollution detector example. Suppose eight test membranes are made from a shipment of chemical and none of them is found to be inert.

 a. Revise the prior probabilities to obtain the posterior probabilities for the outcome states. Interpret the posterior probability for outcome state S_2.

 b. Determine the Bayes act for this sample result. What is the expected payoff of the Bayes act, given the sample result?

23.21 Refer to **Solar Heater** Problem 23.2. Suppose that two respondents in a random sample of 20 indicate they will purchase the heater.

 a. Revise the prior probabilities to obtain the posterior probabilities for the outcome states.

 b. Determine the Bayes act for this sample result. What is the expected payoff of the Bayes act, given the sample result?

 c. Explain how the procedure in **a** and **b** could be extended to determine the Bayes decision rule for the given sample size.

 d. In advance of selecting the sample, what is the probability that the sample will contain exactly two respondents who indicate they will purchase the heater?

23.22 Refer to **Imported Fruit** Problem 23.3. The importer has selected a simple random sample of $n = 10$ grapefruit from the shipment and found that $X = 1$ grapefruit in the sample is of premium grade.

 a. Obtain the importer's posterior probabilities for the shipment outcome states based on this sample result.

 b. Obtain the Bayes act and its expected value, given this sample result.

23.23 Refer to **Imported Fruit** Problems 23.3 and 23.22.

 a. Once the importer knows this sample result, what is the $EVPI$ of the decision problem?

 b. Is the $EVPI$ now larger or smaller than it was prior to knowing the sample result? Must the $EVPI$ necessarily change in this direction when sample information is gathered? Discuss.

23.24 Refer to **Control Boxes** Problem 23.4. A random sample of $n = 12$ control boxes from a lot contains exactly two with malfunctions.
  a. Obtain the posterior probabilities for the lot proportion malfunctioning based on this sample result.
  b. Determine the Bayes act for this sample result. What is its expected payoff, given this sample result?
  c. The Bayes decision rule for a sample of this size is $(12, 0)$. Is the Bayes act identified in **b** consistent with the rule? Explain.

EXERCISES

23.25 a. Prove that the expected opportunity loss associated with decision rule (n, K) equals $EPPI - EP(n, K)$.
  b. Explain, for a given sample size, why maximizing the expected payoff and minimizing the expected opportunity loss yield the same optimal decision rule.

23.26 Consider a decision problem and a sampling cost function of the types discussed in this chapter. Assume that some sampling is optimal. Show that the optimal sample size is the smallest one for which the incremental cost of an additional observation exceeds the incremental expected value of the additional observation.

23.27 Can the upper bound for the optimal sample size in (23.8) be employed when the decision maker's utility function is not linear? Discuss.

23.28 Show that revision of prior probabilities by (23.12) must yield posterior probabilities that are numbers between 0 and 1 and sum to 1.

STUDIES

23.29 A quality control official for a government agency inspects large shipments of shirts that are received periodically from a supplier. A shirt is classified as defective if workmanship, fabric quality, or pattern does not meet specifications. Rejecting a shipment costs $150 because it entails 100 percent inspection and administrative costs of obtaining replacements for defective shirts. The cost of accepting a shipment is $15,000p$, where p is the proportion of shirts in the shipment that are defective. Extensive past experience with the supplier indicates that shipment quality has the following probability distribution:

Proportion Defective:	0	0.01	0.04	0.08
Probability:	0.30	0.50	0.15	0.05

  a. Set up the payoff table for the official's decision problem. Show costs as negative payoffs. Denote the acts of accepting and rejecting a shipment by A_1 and A_2, respectively.
  b. A random sample of ten shirts contains one defective. Assume the number of defective shirts in the sample can be taken as a binomial random variable. Revise the prior probabilities, and determine the Bayes act.

c. Calculate the values of $EP(n, K)$ for $n' = 10$ and $K = -1, 0, 1, 2$. What appears to be the Bayes decision rule for this sample size? Is the Bayes act identified in **b** consistent with this rule?

d. Calculate the values of $EP(n, K)$ for $n = 50$ and $K = -1, 0, 1, 2$. Use a Poisson distribution with $\lambda = np$ to approximate the required binomial probabilities (see p. 376). What appear to be the Bayes decision rule and the value of $BEP(50)$?

23.30 A food company imports nuts in the shell in large lots. Since consumer reaction to purchasing a package of nuts in the shell containing a high proportion of spoiled nuts is quite adverse, each lot is sampled to decide whether (1) to package the nuts in the shell for sale; or (2) to shell all nuts first, remove the spoiled ones manually by sorters, and then package the remaining shelled nuts for sale. The company currently follows a zero-defects inspection plan whereby a random sample of 20 nuts is selected from each lot and shelled. If no nuts in the sample are spoiled, the lot is judged acceptable and the nuts are packaged in the shell; otherwise, the lot is judged unacceptable and the nuts are shelled and sorted before being packaged.

The payoffs and prior probabilities for the quality of a lot (as measured by the proportion spoiled) follow. The payoffs (in dollars) represent profit contributions.

Proportion Spoiled	Prior Probability $P(S_i)$	Act	
		A_1 Acceptable	A_2 Unacceptable
S_1: 0	0.6	700	600
S_2: 0.01	0.1	650	580
S_3: 0.10	0.3	−300	400

Assume that the expected payoff is to be maximized.

a. What is the expected payoff of the company's zero-defects inspection plan?

b. Construct the probability distribution of payoffs for a lot randomly received under the company's zero-defects inspection plan.

c. Does the company's zero-defects inspection plan correspond to the Bayes decision rule for a sample size of $n = 20$? [*Hint:* Consider the decision rules $(20, 0)$, $(20, 1)$, $(20, 2)$, $(20, 20)$, and $(20, -1)$.]

23.31 Refer to **Solar Heater** Problem 23.2. Suppose that the utility function for money of the decision maker in this firm is $u(C) = 1 - \exp[-0.001(C + 500)]$, where $-500 \leq C < \infty$ and C is in thousands of dollars, and that the expected utility criterion is to be employed.

a. Which act is better, based on the prior probabilities alone?

b. Which act is better if two respondents in a random sample of 20 indicate they will purchase the heater?

c. What is the maximum amount that the firm should be willing to pay for perfect information about the proportion of households in the target market that will purchase the heater?

UNIT SEVEN

Time Series Analysis and Index Numbers

TIME SERIES AND FORECASTING I: CLASSICAL METHODS

anagers and social scientists often deal with processes that vary as time passes. Observations on such a process in time sequence are called a time series. Examples of time series are monthly sales of a product for the past 240 months and annual gross national product in the United States since 1900.

Time series are analyzed to better understand, describe, control, and predict the underlying process. The analysis usually involves a study of components of the time series — such as trend, cyclical, and seasonal components — the particular components of interest tending to vary from one problem to another. Thus, an economist studying generating capacity needed by a power company 20 years hence will focus on the trend component of electric sales, which reflects the effects of long-term factors on sales (for example, population growth and technological changes). In contrast, the sales manager of a carpet company who wishes to incorporate seasonal effects into quarterly sales forecasts for next year will be concerned with the seasonal component of the company's sales.

In this chapter, we consider basic aspects of time series analysis and discuss some descriptive methods that are widely used. In the next chapter, we take up some additional methods and models for time series analysis.

24.1 TIME SERIES

We begin with a definition of a time series.

(24.1) A *time series* is a sequence of n observations $Y_1, Y_2, \ldots, Y_t, \ldots, Y_n$ on a process at equally spaced points in time.

The equally spaced points in time might be months (as in a time series of monthly sales), years (as in annual population of Canada), or hours (as in hourly electricity consumption).

Time series arise in two different ways:

1. Readings are taken on a variable at consecutive points in time. Examples are (a) the raw materials inventory of a manufacturer recorded at the close of each fiscal

quarter; (b) the level of atmospheric pollution at a downtown location recorded at 5 P.M. each day.
2.   A variable is aggregated or accumulated over an interval of time. Examples are (a) sales aggregated to obtain monthly sales data; (b) snowfalls accumulated to obtain annual data on total snowfall.

Comment   Time series as defined in (24.1) are *discrete* time series, because the observations pertain to separated points in time. There are also *continuous* time series, where the variable Y is measured continuously over time. An electrocardiogram of heart action is a graph of a continuous time series.

 **SMOOTHING**

Smoothing is a statistical procedure used to dampen or average out fluctuations in a time series to obtain a *smooth component* that reflects the systematic movement of the series. We explain smoothing by utilizing the following example.

EXAMPLE

Department Store.   A department store extended its evening openings to five nights, Monday through Friday. The store was staffed during these extended hours primarily by new, part-time sales clerks. Shortly after the new hours went into effect, customers began to complain increasingly about incidents of rudeness and inefficiency. Management quickly initiated a remedial program. Data on the daily number of complaints on Mondays through Fridays for the two weeks prior to the initiation of the remedial program and for four weeks thereafter are shown in Figure 24.1a.

An operations analyst perceived two characteristics of relevance in the time series data:

1.   A downward shift occurred in the underlying direction of the series coinciding with the initiation of the remedial program.
2.   Marked fluctuations within each 5-day period are evident both before and after the initiation of the remedial program.

To present these points effectively at a management review meeting, the analyst wished to average out the fluctuations to obtain a clearer picture of the underlying movement of the series.

Moving Average

A moving average dampens fluctuations in a time series by taking successive averages of groups of observations.

FIGURE 24.1   **Complaints series and its smooth and fluctuating components—Department store example**

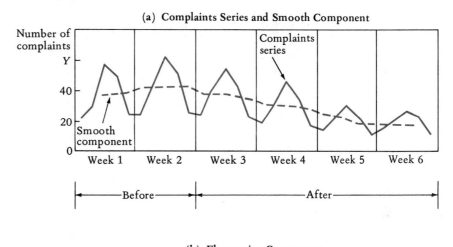

(a) Complaints Series and Smooth Component

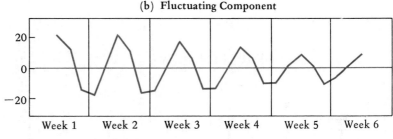

(b) Fluctuating Component

(24.2)

A *moving average* of a time series is obtained by replacing each successive, overlapping sequence of k observations in the series by the mean of that sequence. The first sequence contains observations $Y_1, Y_2, \ldots, Y_k$; the second sequence contains observations $Y_2, Y_3, \ldots, Y_{k+1}$; and so on. Here, k denotes the *term* of the moving average.

The term k of the moving average should be the same as the period of fluctuations in the time series or a multiple of this period.

EXAMPLE

In the department store example, Figure 24.1a shows clearly that there is a 5-day cycle of fluctuations that is repeated week after week. Therefore, the analyst decided to use a 5-term moving average. Note that with this term, each sequence $(Y_1, Y_2, \ldots, Y_5)$, $(Y_2, Y_3, \ldots, Y_6)$, and so on, contains an observation for each of the five days of the week the store is open in the evening. Thus, each average contains a Wednesday, when the volume of complaints reaches a peak, and a Monday and a Friday, when the volume is lower.

Table 24.1 shows the procedure for taking a 5-term moving average. The observations are given in column 1. Column 2 contains the totals of the overlapping sequences of five observations. For example, the total for the first sequence is $22 + 30 + 57 + 51 + 24 = 184$. This total is placed at the middle of its sequence at $t = 3$. Succeeding totals can be obtained readily by dropping the ear-

TABLE 24.1   Calculation of a 5-term moving average — Department store example

Week and Day		t	(1) Number of Complaints Y_t	(2) 5-Term Moving Total	(3) 5-Term Moving Average	(4) Fluctuating Component (1) − (3)
1	M	1	22			
	T	2	30			
	W	3	57	184	36.8	20.2
	T	4	51	186	37.2	13.8
	F	5	24	197	39.4	−15.4
2	M	6	24	203	40.6	−16.6
	T	7	41	204	40.8	0.2
	W	8	63	205	41.0	22.0
	T	9	52	205	41.0	11.0
	F	10	25	205	41.0	−16.0
3	M	11	24	198	39.6	−15.6
	T	12	41	190	38.0	3.0
	W	13	56	186	37.2	18.8
	T	14	44	181	36.2	7.8
	F	15	21	170	34.0	−13.0
4	M	16	19	159	31.8	−12.8
	T	17	30	150	30.0	0.0
	W	18	45	146	29.2	15.8
	T	19	35	142	28.4	6.6
	F	20	17	135	27.0	−10.0
5	M	21	15	120	24.0	−9.0
	T	22	23	108	21.6	1.4
	W	23	30	101	20.2	9.8
	T	24	23	99	19.8	3.2
	F	25	10	96	19.2	−9.2
6	M	26	13	93	18.6	−5.6
	T	27	20	92	18.4	1.6
	W	28	27	92	18.4	8.6
	T	29	22	10			
	F	30	10			

liest observation in the preceding total and adding the next observation. Thus, for the total corresponding to $t = 4$, we have $184 - 22 + 24 = 186$, and the total corresponding to $t = 5$ is $186 - 30 + 41 = 197$. The totals in column 2 are called *moving totals*.

When the totals in column 2 are divided by 5, we obtain the moving averages. Thus, the first moving total, 184 at $t = 3$, is divided by 5 and the average, 36.8, is placed in column 3 at $t = 3$. The averages in column 3 constitute the moving average of the complaints series. Note that there are no moving average values for $t = 1$ and 2 and for $t = 29$ and 30, the first two and last two days of the time series. In general, for a k-term moving average, there will be no moving average values for the first and last $(k - 1)/2$ time periods when k is odd. The moving average of the complaints series, representing the smooth component, is plotted in Figure 24.1a, together with the time series of number of complaints.

Finally, the analyst subtracted the moving averages from the original observations to obtain the series in column 4. This series, plotted in Figure 24.1b, is the *fluctuating component* of the series—in the present case, the pattern of day-to-day fluctuations in complaints.

The analyst presented Figure 24.1 at the management review meeting with the following interpretation:

1. The smooth component of the complaints series, portrayed by the moving average, turned downward shortly after the remedial program was introduced. However, the cause of the downturn cannot be established from the smoothed series alone. The downturn could have resulted from the remedial program, from experience gained by the clerks, or from some combination of these or other factors. Further study is required to establish the cause of the downturn.

2. The smooth component appears to be leveling off in the last week or so. There may be no further improvement unless the situation receives additional attention from management.

3. The fluctuating component exhibits a midweek peak. Further investigation is required to ascertain whether this peak reflects a corresponding peak in the number of transactions handled or whether some other factors are responsible. ☐

Centered Moving Average. When an even number of terms is required in a moving average—for instance, in a 4-term or 12-term moving average—a special problem arises. Consider the following case where a 4-term moving average is taken:

t:	1	2	3	4	5	6	7
Y_t:	2	6	4	8	6	7	2
Moving Total ($k = 4$):		20	24	25	23		
Moving Average ($k = 4$):		5.0	6.0	6.25	5.75		

The 4-term totals and averages, placed at the center of their sequences, fall at $t = 2.5$, $t = 3.5$, and so on. This is unsatisfactory when original observations and moving averages are to be compared. The remedy is to *center* the moving averages as follows:

t:	1	2	3	4	5	6	7
Y_t:	2	6	4	8	6	7	2
Moving Total ($k = 4$):			20	24	25	23	
Moving Average ($k = 4$):			5.0	6.0	6.25	5.75	
Centered Moving Average:				5.5	6.13	6.0	

Note that centering simply involves taking a 2-term moving average of the 4-term moving average. Thus, the first centered average is $(5.0 + 6.0)/2 = 5.5$, corresponding to $t = 3$. Similarly, the second centered average is $(6.0 + 6.25)/2 = 6.13$ at $t = 4$.

With centered moving averages, we do not obtain moving average values for the first and last $k/2$ time periods. Here, $k/2 = 2$, so there are no moving average values for the first two and last two periods.

Comment   A centered moving average may be viewed as a *weighted moving average,* where the observations in a sequence do not receive equal weights. Specifically, the centered moving average in our previous illustration is a weighted 5-term moving average where the observations in a sequence receive weights $(1, 2, 2, 2, 1)$. Thus, the first centered moving average can be obtained as follows:

$$\frac{1}{8}[2(1) + 6(2) + 4(2) + 8(2) + 6(1)] = \frac{44}{8} = 5.5$$

where 8 is the sum of the weights.

Effect of Term of Moving Average

The term k of the moving average can materially affect the moving average series. We illustrate this for the department store example. Since the fluctuating component had a period of five days there, it was natural to use a 5-term moving average to obtain a smoothed series that portrays effectively the systematic movement of the series. If 3-term or 7-term moving averages had been calculated, they would not have been effective in portraying the systematic movement.

EXAMPLE   Figures 24.2a and 24.2b show the 3-term and 7-term moving averages, respectively, for the complaints data. The moving average series are anything but smooth. Since $k = 3$ and $k = 7$ are not multiples of the period of the fluctuations in the series, each moving average contains a mixture of the smooth and fluctuating components. Furthermore, in the 7-term moving average the pattern of fluctuation is the reverse of that in the original series.

Figure 24.2c shows the centered 10-term moving average for the complaints data. Here, the moving average is smooth because 10 is a multiple of 5, the period of the fluctuating component. However, since moving averages are not obtained

---

FIGURE 24.2 3-term, 7-term, and 10-term moving averages — Department store example

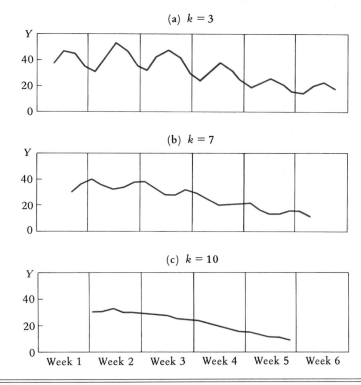

(a) $k = 3$

(b) $k = 7$

(c) $k = 10$

Week 1   Week 2   Week 3   Week 4   Week 5   Week 6

---

for the first and last five time periods when $k = 10$, important information is lost on the possible leveling off of the smooth component in the last week. ☐

This example demonstrates that the term k of a moving average used for smoothing should equal the period of fluctuation in the time series. Unfortunately, many time series in business, economics, and the social sciences oscillate with considerable irregularity. In these cases, the best compromise is to select the smallest value of k that will dampen the fluctuations. Computer programs for time series smoothing are available, and these enable one to experiment with different values of k.

Comment	Since a time series observation Y_t enters k successive moving averages (except at the beginning and end of the series), an unusually small or large observation can produce oscillations in the moving average series even though the original time series contains no periodicity. For this and other reasons, care must be used in interpreting oscillations in a moving average series.

CLASSICAL TIME SERIES MODEL

We have seen from the department store example how a time series can be decomposed into a smooth component, reflecting the systematic movement of the series, and a fluctuating component, reflecting the shorter-term and erratic movements in the time series. For many business and economic time series, however, this decomposition into two components is inadequate because the moving average series tends to combine long-term growth patterns and business cycle movements that need to be studied as separate components.

Multiplicative Model

Economists have developed a model for studying economic and business time series that contains four components. These are called the trend, cyclical, seasonal, and irregular components. In the model that is most widely used, the components are assumed to act in multiplicative fashion.

(24.3)

> The *classical multiplicative time series model* is:
>
> $$Y = T \cdot C \cdot S \cdot I$$
>
> *where:*
>
> T, C, S, and I denote, respectively, the *trend, cyclical, seasonal,* and *irregular* components of the time series

EXAMPLE

Fixture Shipments.   A company manufactures a large outdoor lighting fixture. Quarterly shipments of this fixture between 1977 and 1986 are shown in Figure 24.3a. The four components of this time series according to time series model (24.3) are shown in panels b through e of this same figure. We shall use this figure to illustrate the nature and meaning of the components in the classical time series model (24.3).   □

Trend Component

The trend component of the fixture shipments time series is shown in Figure 24.3b. This component describes the net influence of long-term factors whose effects on the shipments tend to change gradually. Generally, these factors include (1) changes in size, demographic characteristics, and geographic distribution of the population; (2) technological improvements; (3) economic development; and (4) gradual shifts in habits and attitudes. Since these effects tend to operate fairly gradually and in one direction or the other over long periods of time, the trend component usually is modeled by a smooth, continuous curve spanning the entire time series. The curve employed is

FIGURE 24.3 **Time series components according to the classical multiplicative model—Fixture shipments example**

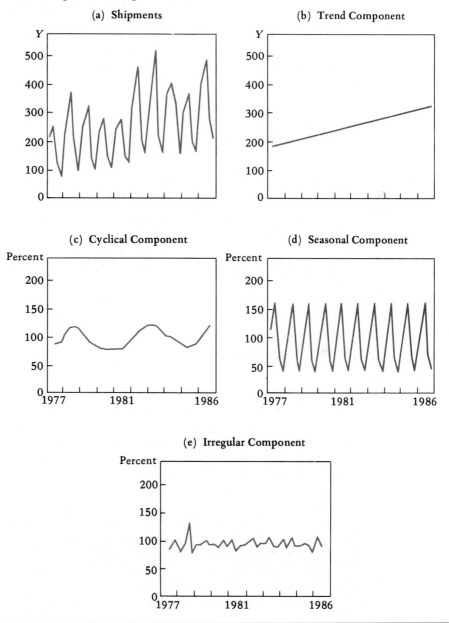

(a) Shipments

(b) Trend Component

(c) Cyclical Component

(d) Seasonal Component

(e) Irregular Component

called the *trend curve*. The trend curve in Figure 24.3b is linear, but in another case, a different type of trend curve might be appropriate.

A major use of trend analysis is for long-term forecasting.

Cyclical Component

The cyclical component of the fixture shipments time series is shown in Figure 24.3c. Generally, business and economic activities do not grow (or decline) at a steady rate. Instead, there are alternating periods of relative expansion and contraction. The cyclical component of a time series describes this characteristic. It measures the net effect of a variety of interrelated factors that tend to shift in direction from time to time and to vary in intensity and impact. As a result, the cyclical component shows cycles of expansion and contraction of uneven duration and amplitude. Factors leading to cyclical movements in business and economic time series include buildups and depletions of inventories, shifts in rates of capital expenditures by businesses, year-to-year variations in harvests, and changes in governmental monetary and fiscal policy. This list could be continued at length. The cyclical component in the shipments of outdoor lighting fixtures in Figure 24.3c reflects changes in sales arising from variations in the rate of development of new shopping centers and outdoor recreational facilities, which in turn are influenced by variations in underlying economic conditions.

Cyclical movements are studied for information on changes in rates of current activity. Such information is useful for assessing current conditions and making short-term forecasts.

Seasonal Component

The seasonal component of the fixture shipments time series is shown in Figure 24.3d. This component describes effects that occur regularly over a period of a year, quarter, month, week, or day. Seasonal effects generally are associated with the calendar or the clock. For example, electricity consumption in the southern United States reaches a peak in summer, year after year, because of air-conditioning demand. Department store sales reach a peak every December, because of the Christmas holidays. Shipments of outdoor lighting fixtures in the example reach a peak in each second quarter, because the height of construction activity is in the summer. Subway traffic reaches a peak each weekday morning and late afternoon, because of passengers going to and from work.

Seasonal effects tend to recur fairly systematically. Consequently, the pattern of movement in the seasonal component tends to be more regular than the cyclical pattern and therefore is more predictable. Note that the seasonal pattern in Figure 24.3d is stable over the entire period. In another series, however, we might find that the seasonal pattern is undergoing gradual modification.

Seasonal movements are measured so that seasonal effects can be taken into account in evaluating past and current activity, and so that they can be incorporated into forecasts of future activity.

Irregular Component

The irregular component describes residual movements that remain after the other components have been taken into account. Irregular movements reflect effects of unique and nonrecurring factors, such as strikes, unusual weather conditions, and international crises.

In some business and economic time series, the cyclical component is itself so irregular that any breakdown into separate cyclical and irregular components would be arbitrary. In such cases, a combined cyclical-irregular component is often developed.

Figure 24.3e contains the irregular component in the fixture shipments series. The unusual peak in the third quarter of 1978 was due to the fact that a major competitor had a strike and could make no deliveries.

Definitions

We now define the time series components in the classical time series model (24.3) more formally.

(24.4)   The components of a business or economic time series according to the classical time series model (24.3) are as follows:

(24.4a)   *Trend* component, which describes the long-term sweep of the series; it is usually modeled by a smooth curve.

(24.4b)   *Cyclical* component, which describes the alternating periods of relative expansion and contraction of more than one year's duration in the series; it consists of cycles that vary in amplitude and duration.

(24.4c)   *Seasonal* component, which describes the pattern of change recurring within periods of a year or less; it consists of a sequence of relatively repetitious cycles.

(24.4d)   *Irregular* component, which describes the effects of all other factors; it tends to have an irregular, saw-toothed pattern.

24.4   LINEAR TREND

Trend Function

As we noted earlier, the trend component is usually described by a smooth curve. We first consider a linear trend, such as the one for the fixture shipments example in Figure 24.3b.

(24.5)

The *linear trend* function is:

$$T_t = b_0 + b_1 X_t$$

where:

T_t is the trend value for period t ($t = 1, 2, \ldots, n$)
X_t is a numerical code denoting period t
b_0 is the intercept of the trend line
b_1 is the slope of the trend line

We shall follow the practice of denoting time periods by $X_1 = 1, X_2 = 2, \ldots, X_t = t, \ldots, X_n = n$. Thus, for annual data from 1960 to 1986 ($n = 27$), 1960 is represented by $X_1 = 1$, 1970 by $X_{11} = 11$, and 1986 by $X_{27} = 27$.

Fitting Trend Line

Ordinarily, b_0 and b_1 are calculated by the method of least squares as described in Chapter 18. Recall that this method does not require an explicit specification of the distribution of the Y observations; hence, it is useful for time series model (24.3), which does not make any such specification. Our discussion of fitting a linear regression model in Chapter 18 applies entirely to fitting a linear trend function. For convenience, we repeat the least squares formulas for b_0 and b_1, using the subscript t now instead of the earlier subscript i for the X and Y observations.

(24.6)

$$b_1 = \frac{\sum X_t Y_t - \dfrac{(\sum X_t)(\sum Y_t)}{n}}{\sum X_t^2 - \dfrac{(\sum X_t)^2}{n}}$$

(24.7)

$$b_0 = \frac{1}{n}\left(\sum Y_t - b_1 \sum X_t\right)$$

Since all summations are over the time series data ($t = 1, 2, \ldots, n$), we do not show the summation index in the formulas.

We now illustrate the calculation of a trend line and its uses by an example.

EXAMPLE

Executive Jet.   Figure 24.4a and Table 24.2 present data obtained by an investment analyst on the number of twin-engine executive jets sold annually between 1977 and 1986 by an aircraft manufacturer. The analyst was preparing a report on sales trend patterns for selected companies and now wished to isolate the trend component for this company. He noted that the annual sales observations tend to

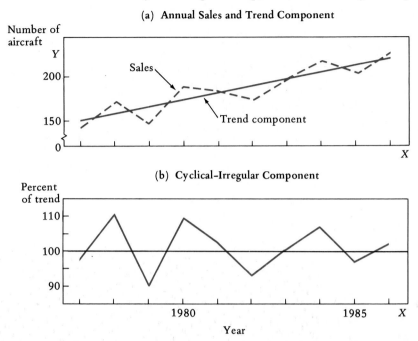

FIGURE 24.4   Linear trend and cyclical-irregular components — Executive jet example

(a) Annual Sales and Trend Component

(b) Cyclical-Irregular Component

TABLE 24.2   Computations in fitting a linear trend — Executive jet example

Year	(1) Year, in Coded Units X_t	(2) Number of Aircraft Sold Y_t	(3) $X_t Y_t$	(4) X_t^2	(5) Trend Value T_t	(6) Percent of Trend
1977	1	147	147	1	152.09	96.7
1978	2	175	350	4	160.25	109.2
1979	3	150	450	9	168.41	89.1
1980	4	191	764	16	176.56	108.2
1981	5	188	940	25	184.72	101.8
1982	6	179	1,074	36	192.88	92.8
1983	7	200	1,400	49	201.04	99.5
1984	8	220	1,760	64	209.19	105.2
1985	9	208	1,872	81	217.35	95.7
1986	10	230	2,300	100	225.51	102.0
Total	55	1888	11,057	385		

fall along a straight line when plotted against time (see Figure 24.4a). This finding, together with an analysis of the factors underlying the long-term sales pattern, led him to conclude that the trend of sales could be modeled appropriately by a straight line. We shall show the calculational steps, although the analyst obtained the linear trend equation directly from a hand-held calculator with a routine for trend fitting.

The preliminary calculations required for obtaining b_0 and b_1 are given in Table 24.2, columns 1 through 4. Note that $n = 10$ here. We obtain:

$$b_1 = \frac{11{,}057 - \dfrac{55(1888)}{10}}{385 - \dfrac{(55)^2}{10}} = 8.15758$$

$$b_0 = \frac{1}{10}[1888 - 8.15758(55)] = 143.93331$$

The fitted trend equation therefore is:

$$T_t = 143.93331 + 8.15758X_t$$

$$X_t = 1 \text{ at } 1977; X \text{ in one-year units}$$

Note that we indicate the coding scheme for X_t when we report the trend equation. The slope $b_1 = 8.2$ indicates that the trend value increases by 8.2 aircraft from one year to the next.

The trend line is plotted in Figure 24.4a. Trend values are calculated in the usual way. For example, the trend value for 1979 ($X_3 = 3$) is:

$$T_3 = 143.93331 + 8.15758(3) = 168.41$$

The trend values are presented in Table 24.2, column 5.   □

Projection of Trend Values

Projection or extrapolation of trend values is performed by simply substituting the X_t value of interest into the trend equation.

EXAMPLE

In the executive jet example, it was desired to project the trend to 1991 ($X_{15} = 15$):

$$T_{15} = 143.93331 + 8.15758(15) = 266.30$$

Thus, if the long-term sweep of jet sales continues as in the past, the trend level of sales in 1991 will be 266 aircraft.   □

Percents of Trend

Additional components of the time series can now be isolated in accord with model (24.3). Since the seasonal component pertains to periodic movements with a duration of one year or less, this component is omitted from the model when annual data are used.

(24.8)

The classical multiplicative time series model (24.3) for *annual* data reduces to:

$$Y = T \cdot C \cdot I$$

Thus if we divide each observation Y_t in an annual time series by its trend value T_t, we obtain the combined cyclical-irregular component, denoted by $C \cdot I$.

(24.9)

$$C \cdot I = \frac{Y}{T} = \frac{T \cdot C \cdot I}{T} \qquad \text{(annual data)}$$

The $C \cdot I$ values are conventionally expressed as percents and are called *percents of trend*.

EXAMPLE

In the executive jet example, for 1977 we have $Y_1 = 147$ and $T_1 = 152.09$. Hence, the percent of trend is $100(147/152.09) = 96.7$. The percents of trend for this example are given in Table 24.2, column 6, and are plotted in Figure 24.4b. Thus, Figure 24.4 shows the decomposition of the executive jet series into the trend and cyclical-irregular components.

Comment

The percents of trend $C \cdot I$ may be smoothed by calculating a moving average; the resulting moving average series is then taken as the cyclical component C. In accord with model (24.8), we can then divide each moving average value C into the corresponding percent of trend $C \cdot I$ to isolate the irregular component I.

24.5 EXPONENTIAL TREND

Nature of Exponential Trend

An exponential trend is one where the trend is changing at a constant *rate* from one period to another. This is in contrast to a linear trend, where the trend is changing by a constant amount. Exponential trends are encountered quite often.

EXAMPLE

Agricultural Imports.   Figures 24.5a and 24.5b present time series computer plots for Canadian imports of agricultural products from the European Economic Community during 1955–1983, expressed in millions of Canadian dollars. The series is plotted against an ordinary arithmetic Y-scale in Figure 24.5a and against a *ratio* or *logarithmic* scale in Figure 24.5b. Whereas equal distances on an arithmetic scale represent equal *amounts* of increase or decrease, equal distances on a ratio or logarithmic scale represent equal *rates* or *percents* of increase or decrease. For instance, the distances from 20 to 40, 40 to 80, and 80 to 160 on the ratio scale in Figure 24.5b are all equal, since in each case the increase is 100 percent. The plotting grid in Figure 24.5b is often called a *semilogarithmic grid,* in contrast to the *arithmetic grid* in Figure 24.5a.

As seen from Figure 24.5a, the long-term pattern in the series bends upward, indicating that the trend component is increasing by larger increments each year. The same series in Figure 24.5b shows a long-term pattern that is approximately a straight line, indicating that the trend component is increasing by a constant rate from year to year. An exponential trend is appropriate for such a time series.   □

FIGURE 24.5   Canadian imports of agricultural products from the European Economic Community, 1955–1983, in millions of Canadian dollars — Agricultural imports example. The time series is plotted on arithmetic and semilogarithmic grids.

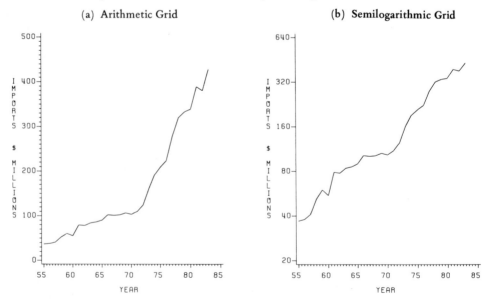

SOURCE: Agriculture Canada.

Fitting Exponential Trend

If the long-term sweep of a series shows an increase (or decrease) from one period to the next at a constant rate, the logarithms of the series will show a linear trend. (A review of logarithms is found in Appendix A, Section A.2.) Hence, we obtain an exponential trend by first taking logarithms of the observations Y_t and then fitting a linear trend to the logarithms of the observations.

(24.10)

The *exponential trend* function, when fitted to the logarithms of the observations, is:

$$T_t' = b_0 + b_1 X_t$$

where:

$$T_t' = \log T_t$$

To obtain the trend value T_t for the original series, we simply take the antilogarithm of T_t'. Thus, there is nothing really new in fitting an exponential trend once we have taken the logarithms of Y_t.

EXAMPLE

Table 24.3a contains the data for the agricultural imports example in Figure 24.5. Table 24.3b shows the basic calculational steps for fitting an exponential trend to the series. These calculations are readily handled by a regression or trend-fitting computer routine.

Column 2 of Table 24.3b contains a portion of the data Y_t from the time series in Table 24.3a. Column 3 contains log Y_t, denoted by Y_t'. The logarithms as usual are to base 10. Columns 4 and 5 contain $X_t Y_t'$ and X_t^2. The calculations of b_0 and b_1 are also shown in the table. The least squares formulas are the same as before, except Y_t' is used instead of Y_t.

The fitted trend equation is:

$$T_t' = 1.53442 + 0.037329 X_t$$

$$X_t = 1 \text{ at } 1955; X \text{ in one-year units}$$

Trend values for the fitted exponential trend curve are obtained in the standard manner. For instance, the logarithmic trend value for 1955 ($X_1 = 1$) is:

$$T_1' = 1.53442 + 0.037329(1) = 1.57175$$

We then take the antilogarithm to obtain the trend value in the original units:

$$T_1 = \text{antilog } 1.57175 = \$37.3 \text{ million}$$

TABLE 24.3   Computations in fitting an exponential trend — Agricultural imports example

(a)
Agricultural imports time series

Year	Imports ($ million)	Year	Imports ($ million)	Year	Imports ($ million)
1955	37	1965	90	1975	208
1956	38	1966	102	1976	222
1957	41	1967	101	1977	277
1958	52	1968	102	1978	319
1959	60	1969	106	1979	332
1960	55	1970	103	1980	338
1961	79	1971	110	1981	388
1962	78	1972	124	1982	379
1963	84	1973	160	1983	427
1964	86	1974	191		

(b)
Basic calculational steps

Year	(1) Year, in Coded Units X_t	(2) Agricultural Imports Y_t	(3) $Y_t' = \log Y_t$	(4) $X_t Y_t'$	(5) X_t^2
1955	1	37	1.56820	1.56820	1
1956	2	38	1.57978	3.15957	4
⋮	⋮	⋮	⋮	⋮	⋮
1983	29	427	2.63043	76.28241	841
Total	435		60.73630	986.82251	8555

$$b_1 = \frac{\sum X_t Y_t' - \dfrac{(\sum X_t)(\sum Y_t')}{n}}{\sum X_t^2 - \dfrac{(\sum X_t)^2}{n}} = \frac{986.82251 - \dfrac{435(60.73630)}{29}}{8555 - \dfrac{(435)^2}{29}} = 0.037329$$

$$b_0 = \frac{1}{n}(\sum Y_t' - b_1 \sum X_t) = \frac{1}{29}[60.73630 - 0.037329(435)] = 1.53442$$

SOURCE: Agriculture Canada.

Rate of Growth

The constant rate of growth implicit in a fitted exponential trend curve, denoted by g, is calculated as follows.

(24.11)

$$g = (\text{antilog } b_1) - 1$$

EXAMPLE

In the agricultural imports example, $b_1 = 0.037329$, so we have:

$$g = (\text{antilog } 0.037329) - 1 = 1.0898 - 1$$

$$= 0.0898 \quad \text{or} \quad 8.98 \text{ percent per year}$$

Comments

1. The exponential trend sometimes is called *log-linear* trend because it is a linear trend for the logarithms of the observations Y_t.

2. Once the trend values in the original units are obtained by taking antilogs of the trend values T'_t, the usual decomposition of the time series can be developed by calculating percents of trend and proceeding from there in accordance with model (24.3) or (24.8), depending upon whether or not a seasonal component is present.

3. The observations Y_t must be positive for log Y_t to be defined, but such is often the case in business and economic series.

24.6 GROWTH CURVE TREND

Many business and economic activities tend to undergo a period of rapid expansion once they are established; but as the size of the activity increases, the rate of expansion slackens and a saturation point is eventually approached. A representative trend pattern for such an activity is shown in computer plots in Figure 24.6, where the trend values are plotted on both arithmetic and semilogarithmic grids. On the arithmetic grid, the pattern forms an elongated S, indicating a slow increase first, then a rapid one, and finally a retardation in absolute growth. The semilogarithmic grid shows a pattern with a declining slope throughout, indicating that the rate of growth is declining during the entire course of the series.

Curves that fit the general pattern of Figure 24.6 are called *growth curves*. Two important growth curves are the *Gompertz* and *logistic trend* functions.

(24.12a)

Gompertz Trend Function:

$$T_t = b_0 b_1^{b_2^{X_t}} \quad \text{where} \quad b_0 > 0, \quad 0 < b_1 < 1, \quad 0 < b_2 < 1$$

(24.12b)

Logistic Trend Function:

$$T_t = \frac{b_0}{1 + b_1 b_2^{X_t}} \quad \text{where} \quad b_0 > 0, \quad b_1 > 0, \quad 0 < b_2 < 1$$

FIGURE 24.6   An example of a growth curve trend function.   The graphs show a Gompertz curve plotted on arithmetic and semilogarithmic grids.

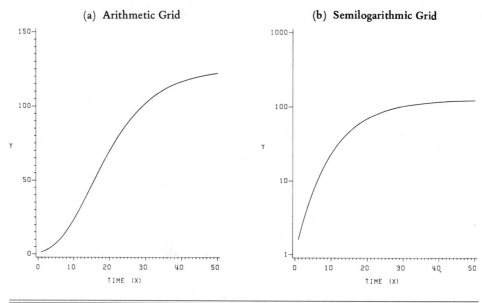

Both of these trend functions are fitted by special methods that can be readily computerized. For each trend function, the lower asymptote of the curve is zero, and the upper asymptote is b_0. The growth rate pattern itself is determined by the constants b_1 and b_2 in each case. In particular, b_2 reflects the degree of retardation from one period to the next, and it is referred to as the *coefficient of retardation*.

EXAMPLE

The growth curve plotted in Figure 24.6 is a Gompertz trend function with $b_0 = 125$, $b_1 = 0.008$, and $b_2 = 0.90$:

$$T_t = 125(0.008)^{0.90^{X_t}}$$

The upper asymptote of this curve is $b_0 = 125$ and the coefficient of retardation is $b_2 = 0.90$.   □

The upper asymptote of a growth curve must be interpreted with caution because it is very sensitive to the observations and may change materially with one or two new observations. Also, there are always serious risks in extrapolating far beyond the range of observations. Finally, while a growth curve may provide a good fit to past data, the assumption of a fixed upper limit may not be reasonable for many business and economic activities in an open and expanding economy.

Comments	1.	A logistic growth curve is also known as a *Pearl-Reed growth curve*.
	2.	The Gompertz and logistic trend functions can also describe a pattern of decline; in that case, $b_2 > 1$ in either (24.12a) or (24.12b). The trend curve then declines from an upper asymptote at b_0 to a lower asymptote at zero.

 ## FORECASTING BY TREND PROJECTIONS

Trend fitting is used often as an aid in forecasting. This use is inherently subjective since the fitted model is extrapolated into a future that never is seen clearly. We now take up some considerations in using trend fitting for forecasting.

Short-Term Versus Long-Term Forecasting

Short-term operating forecasts must be relatively precise. If the cyclical-irregular component in the activity to be forecast has been negligible and is likely to continue so in the future, it may be feasible to project the trend curve a year or two ahead for short-term forecasting. In most cases, however, the cyclical-irregular component is potentially too large to permit a short-term forecast to be based on trend alone. On the other hand, trend projections often are of assistance in longer-term forecasting, where long-term effects are important and shorter-term fluctuations are of secondary consideration.

Time Period

Ideally, the observed data to which the trend curve is fitted should be relevant to the period into which the curve will be projected. If important new conditions affecting trend have arisen in the recent past (such as a new technological breakthrough), data prior to the change of conditions may have to be discarded or given less weight. If important new conditions affecting trend are expected in the future (such as new environmental protection legislation), the trend projection may have to be modified to take into account the effects of the expected new conditions.

Choice of Trend Curve

The choice of a trend curve cannot be made solely on the basis of how good the fit is to the past data. Usually, there will be a number of trend curves that fit the data well. However, when the trend curves are projected beyond the range of the data, the different trend curves may diverge sharply. Figure 24.7 illustrates this point. Exponential and Gompertz curves both fit the sales data closely but diverge increasingly as the curves are projected beyond the range of the data.

Thus, the choice of the trend curve for forecasting must not only be based on a good fit to the past data but must also reflect an expert's assessment of the long-term

FIGURE 24.7   Diverging trend projections from exponential and Gompertz trend curves fitted to the same series.   The trends are plotted on a semilogarithmic grid.

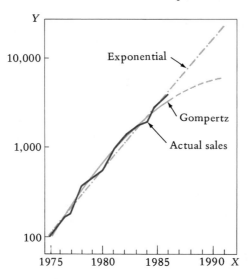

factors determining the future course of the activity. If an exponential trend appears to be apt for the past data and it is expected that the constant rate of growth in the past will continue in the future, an exponential trend equation should be used. If the past rate of growth is expected to taper off, on the other hand, a Gompertz or logistic growth curve might be used instead.

The choice of the trend curve is also affected by the length of the projection. An exponential trend might be reasonable for a projection of 5 years but not for a projection of 20 years.

Computer-Assisted Screening

Computer programs are available for assistance in screening alternative trend functions. In a typical program, a number of different types of trend curves are available for fitting to the time series. For each type of curve, the fitted equation can be obtained together with a descriptive measure, such as r^2, showing the degree of relationship between the observations Y_t and the fitted trend values T_t. Of course, as we noted previously, the degree to which a trend curve fits the past observations should be only one of the factors to be considered in selecting a trend curve for forecasting.

24.8   CYCLICAL ANALYSIS AND SHORT-TERM FORECASTING

Cyclical movements in business and economic data are studied to assess the direction in which an economic sector is moving, to evaluate the short-term outlook for an industry or company, and to prepare intermediate and short-term forecasts used in planning and controlling day-to-day operations. While many time series in engineering and the physical sciences have cyclical components that are quite regular and can be described by periodic functions, cyclical components in business and economic series tend to vary widely in both duration and amplitude from one cycle to the next and hence do not lend themselves to fitting by mathematical periodic functions.

Identification of Cyclical Turning Points

A major objective of cyclical analysis of business and economic time series is to predict cyclical downturns and upturns in the activity. Unfortunately, this is a difficult task. Analysts are often satisfied simply to be able to identify cyclical turning points shortly after they have occurred.

(24.13)　　A *cyclical turning point* is the time point at which the cyclical component of a time series changes direction.

Even the task of identifying a turning point retrospectively is difficult; often, it takes months before a cyclical turn can be recognized.

The reasons for the difficulties in ascertaining current and future cyclical developments are illustrated in Figure 24.8. Presented there is the deseasonalized quarterly series on residential gross private domestic fixed investment (in 1972 dollars) for the United States between 1959 and 1983. We can visually trace the cyclical pattern in this series and, with some arbitrariness, identify the cyclical turning points. Note the variability of the durations and amplitudes of the cycles, typical of most business and eco-

FIGURE 24.8　Example of a leading cyclical indicator series for the United States — Deseasonalized quarterly residential gross private domestic fixed investment in 1972 dollars. The series is expressed as an annual rate in billions of dollars and plotted on a semilogarithmic grid.

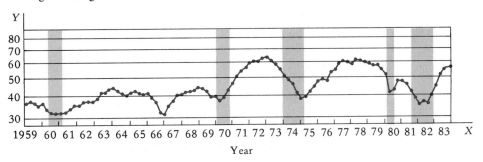

SOURCE: U.S. Department of Commerce.

nomic series. In addition to the variability in the cyclical swings, note the substantial irregular movements present in the series. Both of these factors make it virtually impossible to identify current or imminent cyclical behavior from inspection of the series alone.

We shall comment shortly on the meaning of the clear and shaded intervals in Figure 24.8.

Cyclical Indicators

Cyclical Indicators Approach.   A useful framework for cyclical analysis of business conditions is based on the cyclical indicators approach originally developed by the National Bureau of Economic Research (NBER). In this approach, economic time series are sought that are *leading, coinciding,* and *lagging* in comparison with cyclical movements in aggregate economic activity. A leading series, for instance, is one whose cyclical upturns and downturns tend to occur before the turns in aggregate activity.

The residential fixed investment series in Figure 24.8 is an example of a series classified as a leading series. The clear and shaded intervals shown in Figure 24.8 correspond, respectively, to consecutive periods of cyclical expansion and contraction in *aggregate economic activity,* as identified from an assessment of a whole range of key economic time series. We can see in Figure 24.8 how the cyclical turning points in the residential fixed investment series tend to precede the general business cycle as represented by the unshaded and shaded intervals. Note, however, that the extent of the lead is variable, and in some instances, such as the downturn in the series in 1966, cyclical turning points in the series are not matched by subsequent turns in general economic activity.

Although most series are not sufficiently consistent to serve as reliable indicators, there are exceptions. The Bureau of Economic Analysis (BEA) of the Department of Commerce, working with NBER, has examined many hundreds of series and identified and listed the consistent ones. The classification process is ongoing, and the list is subject to change; it currently contains about 150 series. The BEA has identified a short list of cyclical indicator series, culled from the larger list, that represent key activities in the cyclical process, perform particularly well as indicators, and are collected and published on a timely basis. Table 24.4 shows the short list current in the late 1980s.

Use of Cyclical Indicators.   The leading series are studied primarily to help in anticipating cyclical turning points. The coincident and lagging series are studied mainly for assistance in recognizing a cyclical turn once it has occurred. Unfortunately, the situation is not always clear-cut. At a given time, 8 of the 12 leading indicators on the short list may be moving in one direction while the others are indeterminate or are moving in the opposite direction. Similar mixed pictures can exist in the coincident and lagging series.

Various indexes have been developed to assist in the monitoring of the indicators. Each index summarizes some aspect of the movements in a group of indicators. For example, the BEA composite index of leading indicators is based on the 12 series in Table 24.4, each of which has been standardized. Similarly, there are composite in-

---

TABLE 24.4   Cyclical indicators in BEA composite indexes

(a) Leading Indicators

1. Average weekly hours of production or nonsupervisory workers, manufacturing
2. Average weekly initial claims for unemployment insurance, state programs
3. Manufacturers' new orders, consumer goods and materials industries, in 1982 dollars
4. Index of net business formation
5. Index of stock prices, 500 common stocks
6. Contracts and orders for plant and equipment in 1982 dollars
7. Index of new private housing units authorized by local building permits
8. Vendor performance, percent of companies receiving slower deliveries
9. Change in manufacturing and trade inventories on hand and on order in 1982 dollars
10. Change in sensitive materials prices
11. Money supply M2 in 1982 dollars
12. Change in business and consumer credit outstanding

(b) Coincident Indicators

1. Employees on nonagricultural payrolls
2. Index of industrial production
3. Personal income less transfer payments in 1982 dollars
4. Manufacturing and trade sales in 1982 dollars

(c) Lagging Indicators

1. Index of labor cost per unit of output, manufacturing
2. Ratio of manufacturing and trade inventories to sales in 1982 dollars
3. Average duration of unemployment in weeks
4. Ratio of consumer installment credit outstanding to personal income
5. Commercial and industrial loans outstanding in 1982 dollars
6. Average prime rate charged by banks

SOURCE: U.S. Department of Commerce.

---

dexes of coincident and lagging indicators based on the respective indicator series in Table 24.4.

The cyclical indicators approach, despite some difficulties because of inconsistent behavior among the series, has proven to be highly useful, and a broad apparatus of business cycle analysis has developed around it. A key source that brings together the series used in the cyclical indicators approach is *Business Conditions Digest,* published monthly by the Bureau of Economic Analysis, U.S. Department of Commerce. Figure 24.8 is derived from this publication.

Canadian Experience.   The cyclical indicators concept also has been evaluated for Canada. The Canadian series perform almost identically, in relation to the Canadian cycle, as their U.S. counterparts perform for the U.S. cycle.

CALCULATION OF SEASONAL INDEXES

The seasonal component frequently is the chief source of pronounced short-term fluctuations in business and economic time series, as Figures 24.1 and 24.3 demonstrate. In each of these cases, the seasonal component dominates the short-term movements. In this section, we discuss how to measure the seasonal component in the form of seasonal indexes, and in Section 24.10 we show how seasonal indexes are used in analysis of past and current activity and for short-term forecasting.

Method of Ratio to Moving Average

A method commonly used for measuring the seasonal component is the *method of ratio to moving average*. This method is appropriate for the multiplicative model (24.3). We illustrate it for a quarterly time series, but it can be readily adapted to monthly or weekly series.

EXAMPLE

Taxair.   Figure 24.9 presents the number of revenue passenger miles flown quarterly between 1981 and 1986 by Taxair, Inc., an air taxi service. We note a recurrent seasonal pattern marked by relatively low mileage in the first and fourth quarters and peak mileage in the third quarter. Of course, the fluctuations are not perfectly repetitive, in large part because irregular effects are present in the series.

FIGURE 24.9   **Revenue passenger miles flown quarterly, 1981–1986 — Taxair example**

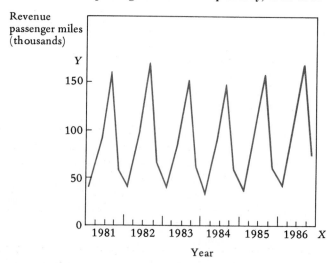

Basic Procedure.   The procedure for measuring the seasonal component by the method of ratio to moving average involves three steps.

(24.14)

The procedure for measuring the seasonal component by the *method of ratio to moving average* is as follows:

1. Smooth the series; the smoothed values reflect $T \cdot C$, the trend-cyclical component.
2. Obtain the seasonal-irregular component, denoted by $S \cdot I$, by dividing the original series by the smoothed values:

$$\frac{Y}{T \cdot C} = \frac{T \cdot C \cdot S \cdot I}{T \cdot C} = S \cdot I$$

3. Isolate the seasonal component S by averaging out the irregular effects in $S \cdot I$.

Smoothing to Obtain $T \cdot C$ Component.   Table 24.5, column 1, contains the quarterly revenue passenger miles for the Taxair example. Since the seasonal pattern repeats every four quarters, a centered 4-quarter moving average is appropriate here, as discussed earlier. We present the centered 4-quarter moving average in column 2.

As an illustration of the computational steps, the first two 4-quarter moving averages are:

$$\frac{43.20 + 90.00 + 162.00 + 64.80}{4} = 90.0000$$

$$\frac{90.00 + 162.00 + 64.80 + 42.35}{4} = 89.7875$$

and the centered average is $(90.0000 + 89.7875)/2 = 89.894$. This is the first entry in column 2, for quarter 3, 1981.

The centered 4-quarter moving average in column 2 represents $T \cdot C$, the trend-cyclical component.

Obtaining $S \cdot I$ Component.   The seasonal-irregular component $S \cdot I$ is isolated in column 3 by dividing each Y observation in column 1 by the corresponding $T \cdot C$ value in column 2 and expressing the result as a percent. For example, $S \cdot I$ for quarter 3, 1981, is $100(162.00/89.894) = 180.21$.

The entries in column 3, representing the seasonal-irregular component $S \cdot I$, are called *specific seasonal relatives* because they measure seasonal and irregular effects specific to each quarter.

Isolation of Stable Seasonal Component S.   The procedure for carrying out the third step—that is, isolating the seasonal component—depends on whether this com-

TABLE 24.5   Computation of quarterly specific seasonal relatives by the method of ratio to moving average — Taxair example

Year and Quarter	(1) Revenue Passenger Miles (thousands) Y	(2) Centered 4-Quarter Moving Average $T \cdot C$	(3) Specific Seasonal Relative $S \cdot I$
1981 1	43.20		
2	90.00		
3	162.00	89.894	180.21
4	64.80	91.050	71.17
1982 1	42.35	93.719	45.19
2	100.10	95.688	104.61
3	173.25	96.326	179.86
4	69.30	95.078	72.89
1983 1	42.96	91.786	46.80
2	89.50	89.660	99.82
3	157.52	88.874	177.24
4	68.02	88.272	77.06
1984 1	37.95	87.582	43.33
2	89.70	86.559	103.63
3	151.80	86.525	175.44
4	65.55	87.450	74.96
1985 1	40.15	89.656	44.78
2	94.90	91.231	104.02
3	164.25	92.186	178.17
4	65.70	93.666	70.14
1986 1	47.64	95.514	49.88
2	99.25	98.034	101.24
3	174.68		
4	75.43		

ponent is stable from year to year or is undergoing gradual modification. A convenient way to study this question is to assemble the specific seasonal relatives $S \cdot I$ by quarter and to plot them in time sequence. This is done in Figure 24.10.

We examine Figure 24.10 to determine whether any systematic modification is taking place in the seasonal component. For instance, if the seasonal pattern were flattening in recent years, the line graphs for the first and fourth quarters would tend upward while that for the third quarter would tend downward. In Figure 24.10, the line graphs do not appear to have any systematic upward or downward tendencies, though they do show substantial irregular effects. Hence, we conclude that the seasonal pattern in the Taxair series is stable from year to year. (In practice, we would supplement our examination of the data with an analysis of the factors causing the seasonal effects.)

FIGURE 24.10 Specific seasonal relatives by quarter — Taxair example. No systematic upward or downward tendencies are evident, which suggests that the seasonal pattern has been stable.

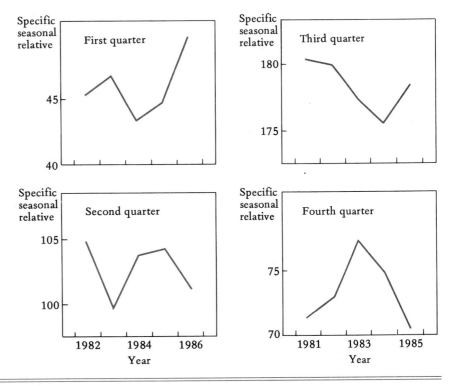

When the seasonal pattern is stable, as here, the seasonal component S is isolated by averaging the specific seasonal relatives for each quarter. The idea is that irregular effects will tend to cancel one another in the average, and as a result, the average will reflect primarily the seasonal effect for that quarter. To avoid the possibility that extreme irregular effects will distort the results, one usually employs an average that is not affected by extreme values, such as a trimmed mean or the median. We shall employ the median here.

The specific seasonal relatives $S \cdot I$ from column 3 of Table 24.5 are assembled by quarter in Table 24.6. We find the median of the specific seasonal relatives for the first quarter to be 45.19. The medians for the four quarters are shown in Table 24.6.

One last step remains to obtain seasonal indexes. Since a seasonal pattern must balance out over a year, the multiplicative model (24.3) requires that the seasonal indexes average out to 100 percent. The method of ratio to moving average does not contain a built-in procedure to constrain the medians to average out to 100 percent; hence, a final adjustment is usually required.

TABLE 24.6   Specific seasonal relatives and seasonal indexes — Taxair example

| | Quarter | | | | |
Year	1	2	3	4	Total
1981			180.21	71.17	
1982	45.19	104.61	179.86	72.89	
1983	46.80	99.82	177.24	77.06	
1984	43.33	103.63	175.44	74.96	
1985	44.78	104.02	178.17	70.14	
1986	49.88	101.24			
Median	45.19	103.63	178.17	72.89	399.88
Seasonal index	45.2	103.7	178.2	72.9	400.0

In the Taxair example, the medians sum to 399.88. They should sum to 400.00 to average out to 100. Hence, we simply multiply each median by the factor $400.00/399.88 = 1.00030$. The adjusted medians are the *seasonal indexes;* they are shown in Table 24.6.

We thus see that for Taxair the seasonal component ranges from a low of 45.2 in the first quarter to a high of 178.2 in the third quarter, indicating large seasonal effects.

Comments

1. At times, the seasonal pattern shifts gradually because of technological developments, changes in customs or habits, or managerial actions such as promotional campaigns timed to dampen seasonal effects. In these cases, we can measure the degree of shift and then project it ahead in calculating seasonal indexes for future periods. One method of doing this is to fit "trend" functions to the specific seasonal relatives for each time unit (for instance, to the specific seasonal relatives in each of the four quarters) and then project each of these "trend" functions.

2. Many statistical computer packages have routines that calculate moving averages, fit trend functions, and calculate seasonal indexes. A widely used computer routine for compiling seasonal indexes is the X–11 Census II Method, originally developed by the U.S. Bureau of the Census and since extended in many ways by other users. This routine employs a refined version of the method of ratio to moving average described here.

24.10   APPLICATIONS OF SEASONAL INDEXES

When one is working with time series data, questions such as the following arise:

1. Is the recent increase in unemployment normal for this time of year?
2. Has a cyclical contraction of retail sales begun that currently is masked by the pre-Christmas buying rush?

3. A sales target of 400,000 units is reasonable for the next 12 months. In breaking down this figure into monthly sales targets, how do we allow for seasonal effects?

Obtaining answers to these questions illustrates two important uses of seasonal indexes:

1. To adjust for seasonal effects in past and current data in order to reveal movements in the other time series components more clearly (questions 1 and 2).
2. To incorporate seasonal effects into forecasts (question 3).

We now take up each of these uses in turn.

Adjusting for Seasonal Effects

For the classical multiplicative time series model (24.3), adjusting for seasonal effects can be represented as follows.

(24.15)
$$\frac{Y}{S} = \frac{T \cdot C \cdot S \cdot I}{S} = T \cdot C \cdot I$$

Thus, the observed value for a period is simply divided by the seasonal index for that period (in decimal form) to obtain the trend-cyclical-irregular component, denoted by $T \cdot C \cdot I$. The $T \cdot C \cdot I$ value for a period is called the *seasonally adjusted value* or *deseasonalized value* of the series for that period.

> EXAMPLE

Department Store Transactions. Table 24.7, column 1, presents the number of transactions of a department store by quarter for the year just ended. The seasonal indexes for the volume of transactions are given in column 2. We first wish to ob-

TABLE 24.7 Calculation of seasonally adjusted quarterly and annual rates—Department store transactions example

Quarter	(1) Transactions (in thousands) Y	(2) Seasonal Index S	(3) Seasonally Adjusted Transactions $T \cdot C \cdot I$	(4) Seasonally Adjusted Annual Rate of Transactions $4(T \cdot C \cdot I)$
1	17.36	70	24.80	99.20
2	21.11	85	24.84	99.36
3	21.68	92	23.57	94.28
4	32.25	153	21.08	84.32

tain the deseasonalized number of transactions for the first quarter; the actual number is 17.36 thousand. The first-quarter seasonal index of 70 indicates that the actual number of transactions in the first quarter is 70 percent as great as if no seasonal effects were present. Thus, to find the $T \cdot C \cdot I$ component (that is, the rate of transactions if seasonal effects were not present), we employ (24.15) as follows:

$$T \cdot C \cdot I = \frac{Y}{S} = \frac{17.36}{0.70} = 24.80 \text{ thousand}$$

Thus, if no seasonal effects had been present, the trend-cyclical-irregular factors for the first quarter would have led to a quarterly rate of 24.80 thousand transactions.

The rate 24.80 thousand transactions is the deseasonalized or seasonally adjusted rate of transactions for the first quarter. The seasonally adjusted transactions for the other quarters are obtained in like manner, by dividing the number of transactions by the seasonal index for that quarter (in decimal form). The results are shown in column 3 of Table 24.7. Note that the seasonally adjusted transactions declined in the third and fourth quarters. Although the actual transactions increased during these quarters, this increase was less than would be expected on a seasonal basis alone. Hence, a decline has occurred in the trend-cyclical-irregular component that is possibly cyclical in nature and warrants attention.   □

Seasonally Adjusted Annual Rates.   Often, seasonally adjusted data are expressed in terms of annual rates for comparison with annual data for past years. The *seasonally adjusted annual rates* for the department store transactions example are given in column 4 of Table 24.7. They are obtained by multiplying the entries in column 3 by 4. The seasonally adjusted annual rate of 99.20 thousand transactions for the first quarter, for instance, indicates that if the trend-cyclical-irregular factors for the first quarter prevailed for an entire year, the annual number of transactions would be 99.20 thousand. We note again from the seasonally adjusted annual rates that the rate of nonseasonal activity declined in the third and fourth quarters.

Deseasonalized Percents of Trend.   If interest is in the cyclical-irregular component $C \cdot I$, rather than in the trend-cyclical-irregular component $T \cdot C \cdot I$, we first obtain percents of trend:

(24.16)
$$\frac{Y}{T} = C \cdot S \cdot I$$

and then deseasonalize these percents of trend:

(24.17)
$$\frac{C \cdot S \cdot I}{S} = C \cdot I$$

TABLE 24.8 Procedure for calculating deseasonalized percents of trend

Year and Quarter	(1) Number of Units Sold Y	(2) Trend Value T	(3) Percent of Trend $C \cdot S \cdot I$ $100[(1) \div (2)]$	(4) Seasonal Index S	(5) Deseasonalized Percent of Trend $C \cdot I$ $100[(3) \div (4)]$
⋮ ⋮	⋮	⋮	⋮	⋮	⋮
1986 1	100,000	90,000	111	97	114
2	150,000	95,000	158	135	117
3	120,000	100,000	120	106	113
4	67,000	105,000	64	62	103
⋮ ⋮	⋮	⋮	⋮	⋮	⋮

If the irregular effects should also be eliminated so that the cyclical effects stand out, we would then smooth the cyclical-irregular component.

The procedure for obtaining deseasonalized percents of trend is illustrated in Table 24.8 for a fragment of a quarterly series.

Incorporation of Seasonal Effects into Forecasts

In short-term forecasting, it is usually essential to incorporate the seasonal effects into the forecast. The basic procedure is to start with an annual forecast. If quarterly forecasts are desired and no strong trend-cyclical effects are expected during the year, the annual forecast is divided by 4 to obtain equal quarterly rates in which no seasonal effects are present. These quarterly rates are then multiplied by the seasonal indexes (in decimal form).

EXAMPLE

A sales manager has received an annual sales forecast of 200 thousand units for a product. She desires sales forecasts for each of the four quarters and expects that quarter-to-quarter changes in the trend-cyclical component will be negligible. The procedure is illustrated in Table 24.9. First, we break down the annual forecast into the quarterly rates without seasonal effects, shown in column 1. The quarterly seasonal indexes for sales of this product are given in column 2. For the multiplicative model (24.3), the seasonal component now is incorporated by multiplying each entry in column 1 by the corresponding seasonal index (in decimal form). The results, entered in column 3, are the desired quarterly sales forecasts. ☐

Comment   If strong trend-cyclical effects are expected during the year, they need to be incorporated into the quarterly (or monthly) forecasts before the seasonal effects are introduced.

TABLE 24.9   **Procedure for incorporating seasonal effects into forecasts**

Quarter	(1) Forecast Without Seasonal Effects	(2) Seasonal Index	(3) Forecast With Seasonal Effects $(1) \times \dfrac{(2)}{100}$
1	50	123	61.5
2	50	108	54.0
3	50	79	39.5
4	50	90	45.0
Total	200		200.0

PROBLEMS

*24.1 **Pulp and Paper Production.**   Data on total production of pulp and paper (in millions of metric tons) in one overseas division of a multinational company during 1967–1986 follow. A 3-term moving average is also presented.

Year:	1967	1968	1969	1970	1971	1972	1973	1974	1975	1976
Production:	3.3	3.7	3.9	3.5	2.9	3.1	3.5	3.6	3.2	3.5
Moving Average:		3.63	3.70	3.43	3.17	3.17	3.40	3.43	3.43	3.57

Year:	1977	1978	1979	1980	1981	1982	1983	1984	1985	1986
Production:	4.0	4.6	4.1	4.8	4.7	4.3	4.5	4.0	3.6	3.8
Moving Average:	4.03	4.23	4.50	4.53	4.60	4.50	____	____	____	

a. Calculate the remaining moving average values.

b. Plot the original series and the moving average series on one graph. What information is provided by the moving average series that is not obtained readily from the original series?

24.2 **Bus Traffic.**   Data on daily city bus traffic (in thousands of passengers) for a recent four-week winter period follow. A 7-term moving average is also presented.

	Week 1		Week 2	
Day	Number of Passengers	Moving Average	Number of Passengers	Moving Average
M	211		163	136.3
T	182		143	136.4
W	199		131	142.1
T	170	158.0	173	144.7
F	206	151.1	207	151.7
S	109	145.6	149	158.0
S	29	135.9	47	166.9

| | Week 3 | | Week 4 | |
Day	Number of Passengers	Moving Average	Number of Passengers	Moving Average
M	212	166.9	200	165.1
T	187	166.4	193	
W	193	166.1	196	____
T	173	165.3	175	____
F	204	163.6	202	____
S	147	164.4	144	
S	41	164.9	46	

a. Calculate the remaining moving average values.
b. Plot the original series and the moving average series on one graph. What information is provided by the moving average series? Is a 7-term moving average a good choice here? Explain.

24.3 **Data Transmission.** The numbers of erroneous data bits transmitted each hour during the past 24 hours over a communications link follow.

Hour	Number	Hour	Number	Hour	Number
1	5	9	1	17	0
2	24	10	2	18	1
3	36	11	4	19	8
4	2	12	20	20	2
5	1	13	4	21	27
6	3	14	1	22	0
7	19	15	24	23	2
8	55	16	30	24	22

a. Calculate a 5-term moving average for the series.
b. Why can 5-term moving averages not be computed for the first two and last two observations of the original series?
c. Plot the original series and the moving average series on one graph. Does it appear that the 5-term moving average series reflects the smooth component of the original series? Would a centered 10-term moving average yield a smoother series?

24.4 Suppose that quarterly production of a processed feed were stable at 9, 2, 6, and 7 million pounds, respectively, in each year during a three-year period.
   a. Calculate for the series (1) a centered 4-term moving average, (2) a 3-term moving average, (3) a 5-term moving average.
   b. Why is the centered 4-term moving average series the smoothest one?

24.5 Refer to **Pulp and Paper Production** Problem 24.1. Calculate the first two values of a centered 6-term moving average for the production series. Would a 3-term moving average or a centered 6-term moving average smooth the series more? Explain.

24.6 Refer to **Data Transmission** Problem 24.3. Calculate the first two values of a centered 10-term moving average. How many observations enter the first centered moving average value? What are the weights of the observations in the centered moving average?

*24.7 Refer to **Pulp and Paper Production** Problem 24.1.
   a. Fit a linear trend function to the series. Set $X_t = 1$ at 1967.

b. Plot the trend line and the original series on one graph. Does the linear trend appear to provide a good description of the trend of the series? Comment.

c. What is the year-to-year change in the trend line obtained in **a**? Does the intercept of the trend line have any meaning here?

d. Obtain the projected trend level for production in 1993. Assuming that the linear trend fitted in **a** is appropriate through 1993 and that the best estimate of the cyclical component in 1993 is 115 percent, obtain a forecast of actual production in 1993.

24.8 **Industrial Customers.** Data on the number of industrial customers (in thousands, at midyear) of a gas company during the period 1969–1986 follow.

Year:	1969	1970	1971	1972	1973	1974	1975	1976	1977
Customers:	3.92	4.12	4.13	4.31	4.23	4.22	4.19	4.45	4.59

Year:	1978	1979	1980	1981	1982	1983	1984	1985	1986
Customers:	4.51	4.50	4.67	4.64	4.67	4.63	4.60	4.79	4.86

a. Does this time series contain a seasonal pattern? Explain.

b. Fit a linear trend function to this series. Set $X_t = 1$ at 1969.

c. Plot the trend line and the original series on one graph. Does the trend line provide a good description of the trend of the series? Comment.

d. Project the trend to midyear 1993. Is this projection a forecast of the number of industrial customers at midyear 1993? Explain.

24.9 **Consulting Contracts.** Data on the total value of contracts (in $000) earned by a management consulting firm in each of the past seven years follow.

Year:	1	2	3	4	5	6	7
Total Value:	603	715	809	920	1008	1124	1203

a. Fit a linear trend function to this series. Set $X_t = 1$ at year 1.

b. What is the year-to-year change in the trend level of the series?

c. Obtain the projected trend level for the total value of contracts in year 9, assuming that the fitted linear trend will be appropriate through year 9.

24.10 **Egg Prices.** A series representing the average annual closing price to producers of one dozen medium-sized eggs (in cents) for the most recent seven years follows.

Year:	1	2	3	4	5	6	7
Price:	83	81	78	80	77	78	78

a. Fit a linear trend function to the series. Set $X_t = 1$ at year 1.

b. By how many cents does the fitted trend level decline between years 1 and 7 for the trend function in **a**?

c. What is the projected trend level of the average closing price for year 8, assuming that the fitted linear trend will be appropriate through year 8?

*24.11 Refer to **Pulp and Paper Production** Problems 24.1 and 24.7.

a. Calculate the percents of trend.

b. Obtain a 5-term moving average of the percents of trend. Explain why the moving average series reflects the cyclical component of the original series.

c. Divide each moving average value into the corresponding percent of trend value to isolate the irregular component.

d. Plot the original series, trend, cyclical, and irregular components in separate graphs, as in Figure 24.3. Has the fluctuation of the series about the trend curve been due predominantly to cyclical or irregular forces? Does the cyclical component appear to be sufficiently regular in period and amplitude that you could confidently project it forward?

24.12 Refer to **Industrial Customers** Problem 24.8. Calculate the percents of trend and plot them. Which components of time series model (24.3) are represented by these percents of trend?

24.13 Refer to **Consulting Contracts** Problem 24.9.
   a. Calculate the percents of trend.
   b. What is the meaning of the percent of trend for year 4?
   c. If the value of the irregular component for year 4 is 101.6, what is the value of the cyclical component for that year?

24.14 Refer to **Egg Prices** Problem 24.10. Calculate the percents of trend and plot them. Are any cyclical movements apparent in the series?

24.15 Refer to Table 24.2 in the text, pertaining to the executive jet example.
   a. Calculate a 5-term moving average series of the percents of trend in column 6 of the table.
   b. Divide the percents of trend in column 6 by the corresponding moving averages of the percents of trend obtained in **a**.
   c. Which component of time series model (24.3) is represented by the series in **a**? By the series in **b**? Interpret the meaning of the value for 1980 in the series in **b**.

*24.16 Sailboats. Data on the number of sailboats sold annually by a manufacturer during 1972–1986 follow.

Year:	1972	1973	1974	1975	1976	1977	1978	1979
Number:	2089	2138	2317	2305	2473	2548	2530	2651

Year:	1980	1981	1982	1983	1984	1985	1986
Number:	2750	2906	2996	3234	3218	3366	3434

   a. Fit an exponential trend function to the series. Set $X_t = 1$ at 1972.
   b. Plot the original series and the trend curve (expressed in the original units) on a graph with arithmetic grid. Does the trend curve provide a good description of the trend in the series? Explain.
   c. Obtain the annual rate of growth in the trend function.
   d. Project the trend curve to 1995 and also to 2050, and express the projected trend values in the original units. Do you believe that the trend projection for 2050 is reasonable? Discuss.

24.17 Life Insurance. Data on the amounts of life insurance (in billions of Canadian dollars) underwritten in Canada by U.S. companies during 1975–1984 follow.

Year:	1975	1976	1977	1978	1979	1980	1981	1982	1983	1984
Amount:	43.7	49.8	51.7	56.6	64.4	70.2	77.9	83.4	87.3	94.7

SOURCE: *Canadian Life Insurance Facts,* published by Canadian Life and Health Insurance Association Inc.

   a. Fit an exponential trend function to the series. Set $X_t = 1$ at 1975. Plot the original series and the trend function (expressed in the original units) on a graph with arithmetic grid.

b. Obtain the annual rate of growth in the trend function.

c. Use the trend function to project the trend level for the year 2000, and express the projected trend level in the original units.

d. The linear trend function for this series is $T_t = 36.3133 + 5.75576X_t$ ($X_t = 1$ at 1975). Plot the linear trend function on the graph in **a**. Do the two trend functions provide equally good fits to the trend of the series? Do they give similar projected trend levels for the year 2000? Explain.

24.18 Production Index.   The U.S. industrial production index (1977 = 100) during 1977–1984 follows.

Year:	1977	1978	1979	1980	1981	1982	1983	1984
Index:	100.0	106.5	110.7	108.6	111.0	103.1	109.2	121.8

SOURCE: *Federal Reserve Bulletin.*

a. Fit an exponential trend function to the series. Set $X_t = 1$ at 1977.

b. Plot the original series and the trend function (expressed in the original units) on a graph with arithmetic grid.

c. Obtain the annual rate of growth in the trend function.

d. What is the projected trend level of the index in the original units for 1995? Is this projection a forecast of the 1995 index? Explain.

*24.19 Refer to **Sailboats** Problem 24.16. Calculate the percents of trend and plot them. Which components of time series model (24.3) are represented by these percents of trend?

24.20 Refer to **Life Insurance** Problem 24.17. Calculate the percents of trend and plot them. Which components of time series model (24.3) are represented by these percents of trend?

24.21 Refer to **Production Index** Problem 24.18. Calculate the percents of trend and plot them. Does the index series appear to have a major cyclical component? Comment.

*24.22 A business researcher obtained the following Gompertz trend function when fitting a trend curve for the annual number of residential dwellings (in thousands) that converted from oil to natural gas for heating in a community:

$$T_t = 300(0.1)^{0.7X_t} \qquad X_t = 1 \text{ at year } 1$$

a. Compute the trend values for years 2, 3, 9, and 10.

b. What percent change in the trend level occurs between years 2 and 3? Between years 9 and 10? Is the fact that the latter percent change is smaller than the former consistent with the Gompertz trend pattern in Figure 24.6b? Explain.

c. What is the upper asymptote of the Gompertz trend curve here?

24.23 A demographer fitted a Gompertz trend function to annual population data for a country and obtained $b_0 = 50$, $b_1 = 0.19$, $b_2 = 0.99$ (T_t is in millions of persons and $X_t = 1$ at 1900).

a. Identify the upper asymptote of the Gompertz trend curve. What is the coefficient of retardation here? How is each of these values interpreted here?

b. Compute the trend values for 1900, 1950, and 2000. Using these points and your answers in **a**, sketch the Gompertz trend curve on graph paper with arithmetic grid.

24.24 In a seminar on product life cycles, the instructor presented the following logistic trend function for a typical product:

$$T_t = \frac{400}{1 + 100(0.30)^{X_t}}$$

Here, T_t represents the trend level of cumulative lifetime sales (in \$ million) of the product by year X_t after its introduction to the market.

  a. What is the value of the upper asymptote of this logistic trend function? Interpret its meaning here.

  b. Compute the trend values for $X_t = 1, 5, 10,$ and 20. The instructor told the class that this product would likely be withdrawn from the market by year $X_t = 10$ because of its poor prospects for sales after that year. Do your trend values support the instructor's claim? Explain.

24.25 A speaker stated: "We have developed a computer program to make short-term and long-term sales forecasts for our products. For each product, the computer will access annual sales data for the past 12 years and will fit five different trend functions — linear, exponential, and three others. Coefficients of correlation between sales and trend will be calculated for each trend function. The trend function with the highest coefficient will be projected one year and ten years ahead for the short-term and long-term sales forecasts for this product." Discuss the problems involved in this approach to sales forecasting.

24.26 A coffee shop patron, on reading in the newspaper that the state's unemployment rate went from 5.0 percent to 6.0 percent in the previous 12 months, remarked: "If this trend continues at the same rate, the whole labor force of the state will be unemployed in 16 years." Should the patron really be concerned with trend or with a different time series component? Discuss.

*24.27** Refer to **Pulp and Paper Production** Problems 24.1 and 24.11. Assume that the 5-term moving average series in Problem 24.11b represents the cyclical component of the original series. Identify the turning points of the cyclical component. From the information given, can you confidently predict the year in which the next turning point will occur? Discuss.

24.28 Data on the cyclical components of number of building permits issued for private housing units (series A) and value of home mortgage loans made by major lending institutions (series B) during a recent four-year period follow.

Year	Quarter	Series A	Series B	Year	Quarter	Series A	Series B
1	1	115	105	3	1	99	96
	2	106	110		2	105	99
	3	101	103		3	103	102
	4	99	100		4	95	108
2	1	90	95	4	1	92	101
	2	86	91		2	96	97
	3	91	90		3	105	93
	4	95	88		4	114	97

  a. Plot the cyclical component of series B on a graph. Identify its turning points.

  b. Plot the cyclical component of series A on the same graph. Does the cyclical behavior of series A lead, lag, or coincide with that of series B? Explain. Is this behavior consistent?

24.29 Explain why it is reasonable that average duration of unemployment is a lagging cyclical indicator while contracts and orders for plant and equipment is a leading cyclical indicator according to the BEA list in Table 24.4.

*24.30** **Electricity Sales.**   Data on quarterly sales revenues (in \$ million) of a southern electric company during 1981–1986 follow.

Quarter	1981	1982	1983	1984	1985	1986
1	172	169	182	169	179	170
2	227	218	218	245	235	241
3	310	309	313	299	292	307
4	222	209	224	221	213	217

a. Plot the time series. Does a seasonal component appear to be present? Explain.
b. Calculate specific seasonal relatives for the series.
c. Assume that the seasonal pattern is stable, and obtain the quarterly seasonal indexes for the series. Describe the seasonal pattern in the series.

24.31 **University Enrollment.** The enrollments at a university operating on a trimester system during 1979–1986 follow.

Trimester	1979	1980	1981	1982	1983	1984	1985	1986
I	10,284	10,724	11,052	11,301	12,240	12,958	12,751	13,253
II	9,445	9,828	10,636	10,757	10,946	11,617	12,177	12,880
III	6,307	6,015	6,922	6,883	7,300	7,153	8,170	7,617

a. Plot the time series. Does the series have a noticeable seasonal component? A noticeable trend-cyclical component? Comment.
b. Calculate specific seasonal relatives for the series. Was there any need to center the moving average in your calculations? Explain.
c. Assume that the seasonal pattern is stable, and obtain the seasonal indexes. Describe the pattern of seasonal variation.

24.32 **Interest Rates.** Quarterly interest rates paid on savings accounts at a major bank, calculated as a quarterly average of the daily rate (expressed in percent), were as follows during the past five years:

Quarter	Year 1	Year 2	Year 3	Year 4	Year 5
1	6.32	7.66	8.92	10.56	7.96
2	7.18	8.65	9.57	9.39	8.68
3	7.56	8.73	9.80	9.03	8.79
4	7.47	9.25	10.69	8.16	9.17

a. Plot the time series. Does the series have a noticeable seasonal component? A noticeable trend-cyclical component? Comment.
b. Calculate specific seasonal relatives for the series.
c. Obtain the quarterly seasonal indexes for the series, assuming that the seasonal pattern is stable.
d. The bank's chief economist expected the series to exhibit very little seasonality except possibly for slightly lower rates in the first quarter of the year. Are your results consistent with the economist's expectation? Explain.

24.33 Refer to **Bus Traffic** Problem 24.2. Let *season* refer to the recurring variation in daily bus traffic within a week.
a. Calculate the specific seasonal relatives for the series. (Note that the 7-term moving average series is given in Problem 24.2.)
b. Obtain the daily seasonal indexes for the series, assuming that the seasonal pattern is stable from week to week. On which day of the week is bus traffic heaviest? Interpret the index value for that day.

24.34 Refer to **Electricity Sales** Problem 24.30. Plot the specific seasonal relatives in a graph, as in Figure 24.10. Is the assumption of a stable seasonal pattern reasonable here? Comment.

24.35 Refer to **University Enrollment** Problem 24.31.
  a. Plot the specific seasonal relatives in a graph, as in Figure 24.10. Is the assumption of a stable seasonal pattern reasonable here? Comment.
  b. Why are there seven specific seasonal relatives for the first and third trimesters but eight for the second trimester?

24.36 Refer to **Interest Rates** Problem 24.32. Plot the specific seasonal relatives in a graph, as in Figure 24.10. Is the assumption of a stable seasonal pattern reasonable here? Comment.

24.37 Refer to **Bus Traffic** Problems 24.2 and 24.33. Plot the specific seasonal relatives in a graph, as in Figure 24.10. Is the assumption of a stable seasonal pattern reasonable here? Comment.

24.38 Suppose that automobile manufacturers shift their annual retooling shutdown for new models to earlier in the year. Will such a shift affect the seasonal component of the quarterly series for automobile production? Comment.

*24.39 Refer to **Electricity Sales** Problem 24.30.
  a. Obtain the deseasonalized quarterly sales revenues for 1986.
  b. Express the deseasonalized sales revenues for the fourth quarter of 1986 as a deseasonalized annual rate. Compare this annual rate with the total sales revenues for 1986. Interpret the results.

24.40 Refer to **University Enrollment** Problem 24.31. Obtain the deseasonalized trimester enrollment levels for 1986. Interpret the results.

24.41 Data on quarterly sales (in $ million) of Nettles, Ltd., during 1986, together with the seasonal indexes for these sales, follow.

Quarter:	1	2	3	4
Sales:	3.40	3.01	4.21	5.33
Seasonal Index:	83	75	102	140

  a. Obtain the deseasonalized sales volume for each quarter of 1986. Should management be pleased with the relatively large sales volume experienced in the fourth quarter? Comment.
  b. Suppose that only seasonal changes in sales were expected between the fourth quarter of 1986 and the first quarter of 1987. What would be the expected sales volume for the first quarter of 1987, assuming that the seasonal pattern is stable? If actual sales in the first quarter of 1987 were $3.06 million, what would this suggest? Discuss.

24.42 Refer to Table 24.6 in the text, pertaining to the Taxair example.
  a. Deseasonalize the specific seasonal relatives to obtain the irregular component of the series.
  b. Taxair had a price promotion on its service during much of the first quarter of 1986. Does the promotion appear to have had an impact on the revenue passenger miles flown for that quarter? Comment.

*24.43 Refer to **Electricity Sales** Problem 24.30.
  a. In 1986, company executives forecast that 1987 annual sales revenues would be $930 million and that trend-cyclical movements during the year would be negligible. Convert this forecast into quarterly forecasts for 1987, assuming that the seasonal pattern is stable.
  b. Actual sales revenues in the first quarter of 1987 were $168 million. Obtain the deseasonalized level of sales revenues and the seasonally adjusted annual rate of sales revenues for this quarter, and interpret each.

24.44 Refer to **Bus Traffic** Problems 24.2 and 24.33. City bus traffic is expected to total 1150 thousand passengers in week 5. Assuming that trend-cyclical movements will be negligible during week 5 and that the seasonal pattern in daily traffic is stable, obtain daily traffic forecasts for week 5.

24.45 The quarterly seasonal indexes for the total loans made by a small credit company are $75, 91, 102, 132$. Late in 1986, a forecast of total loans of $15.7 million for 1987 was made, and it was expected that trend-cyclical movements during 1987 would be very small. Convert the annual forecast into quarterly forecasts for 1987. Why do the quarterly forecasts sum to the annual forecast of $15.7 million here? Explain.

24.46 The monthly seasonal indexes for the number of visitors to a national park follow.

J	F	M	A	M	J	J	A	S	O	N	D
45	41	46	65	103	171	180	178	152	98	53	68

   a. Total visitors for next year are forecast to be 880 thousand. Assuming that trend-cyclical movements will be negligible next year and that the seasonal pattern is stable, convert the annual forecast into monthly forecasts for next year. What percent of all visitors for the year are expected between January and June inclusive?
   b. The actual number of visitors during January of next year was 34.2 thousand. What is the seasonally adjusted annual rate for January? What is the significance of the difference between this annual rate and the annual forecast of 880 thousand visitors?

EXERCISES

24.47 An analyst was studying the trend of a company's monthly production volume. Before fitting any trend curves, however, she divided each month's production by the number of working days in that month.
   a. Why did she take this preliminary step before fitting any trend curves? Is it always necessary to divide production data by the number of working days before studying trend? Explain.
   b. The analyst obtained the equation $T_t = 891 + 1.23X_t$ when fitting a linear trend to the adjusted monthly production data. Project the trend line to month $X_t = 40$. In this month, there will be 22 working days. What is the projected trend level of total production for this month?

24.48 The personnel manager of a large firm fitted a linear trend function to a series representing the annual number of job applicants handled by his department and obtained $T_t = 3085 + 72X_t$, where $X_t = 1$ in 1981 and is in one-year units. The manager wishes to convert this trend function to one of the form $T_t' = b_0' + b_1' X_t'$, where T_t' is the quarterly trend level of number of job applicants, and $X_t' = 1$ in the first quarter of 1981 and is in one-quarter units. Obtain the coefficients b_0' and b_1' of the converted trend function.

24.49 Construct a logarithmic scale with gradation points corresponding to $1, 2, \ldots, 9, 10, 20, \ldots, 90, 100$. Confirm that the distances on your scale between 2 and 6 and between 30 and 90 are the same, as theoretically they should be.

24.50 Refer to **Sailboats** Problem 24.16. Plot the logarithms of the original series and the linear trend function fitted to the log-observations on graph paper with arithmetic grid. Is the trend function a good description of the trend of the log-series?

24.51 Refer to Table 24.3 and Figure 24.5 in the text, pertaining to the agricultural imports example. An observer has suggested that the trend component for this series consists of one linear trend during 1955–1972 and a different linear trend during 1973–1983.
   a. Fit separate linear trend functions to these two periods of the series. Set $X_t = 1$ at the first year of each period.
   b. Plot the original series and the fitted trend lines on one graph. Comment on the combined fit of the two trend lines to the original series.

STUDIES

24.52 (Computer needed.) Refer to the **Investment Fund** data set. Consider the two quarterly unit value series for the equity and fixed-income components for periods 1 to 44 inclusive. Ignore any seasonal component that may be present in either series.
   a. Fit an exponential trend function to each series. Set $X_t = 1$ at period 1. For each series, plot the original observations and the fitted trend function (expressed in the original units) on a graph with arithmetic grid. Evaluate the fit of the exponential trend function to each series. Compare the growth rates of the two series.
   b. Obtain the percents of trend for each series and smooth them, using an 8-term moving average, to isolate the cyclical component of each series. Identify the cyclical turning points of each series. Are any difficulties encountered in making this identification? Do the cyclical components of the two series appear to be in phase? Discuss.

24.53 (Computer needed.) Refer to **Marriages** Problem 15.33.
   a. Decompose the marriages series into its trend, cyclical, seasonal, and irregular components, using the following procedures: (1) Obtain the monthly seasonal indexes by the method of ratio to moving average, using a centered 12-term moving average. (2) Deseasonalize the series. (3) Fit a linear trend function to the deseasonalized series, setting $X_t = 1$ at July of year 1. (4) Obtain the deseasonalized percents of trend and smooth them, using a 5-term moving average.
   b. Plot the original series and its four components on separate graphs, as in Figure 24.3, and briefly describe the nature of the components.
   c. Obtain a forecast of the number of marriages for July of year 5. Assume that the cyclical component will be 100 in that month and that the seasonal pattern is stable.

24.54 The following specific seasonal relatives for quarterly sales of large grade A eggs by a supermarket chain were obtained for 1981–1986:

		Quarter		
Year	1	2	3	4
1981			98.9	97.1
1982	102.9	100.4	97.9	99.5
1983	101.5	99.5	99.2	107.7
1984	100.9	95.1	100.0	105.3
1985	99.5	95.1	100.4	106.7
1986	98.0	92.7		

   a. Describe the shift in the seasonal pattern of egg sales that has occurred during this period. What factors might have produced this shift?

b. Linear "trend" functions fitted to the specific seasonal relatives for the first three quarters are:

First quarter: $S = 104.10 - 1.18X;   X = 1$ in 1982

Second quarter: $S = 102.50 - 1.98X;   X = 1$ in 1982

Third quarter: $S = 97.75 + 0.510X;   X = 1$ in 1981

Obtain the "trend" function for the fourth quarter, letting $X = 1$ in 1981.

c. Calculate the seasonal indexes for 1987 by projecting each "trend" function to 1987 and adjusting the indexes so that they add to 400.0.

24.55 Consult the latest issue of *Business Conditions Digest* and examine the following three cyclical indicator series: (1) contracts and orders for plant and equipment, (2) change in manufacturing and trade inventories on hand and on order, (3) manufacturing and trade sales.

a. For each of these series, indicate whether it is a leading, lagging, or coincident indicator series.

b. Find the most recent value for each of these series. Are these values seasonally adjusted?

c. What is the current cyclical phase — expansion or contraction — of each of the series? How have you established the cyclical phase for each series? Is the phase of each series consistent with its classification as a leading, lagging, or coincident indicator? Comment.

25 TIME SERIES AND FORECASTING II: EXPONENTIAL SMOOTHING AND REGRESSION METHODS

In this chapter, we present additional methods and models that are extensively used in time series analysis and forecasting. We begin with the presentation of a forecasting method called exponential smoothing. Then, we discuss the use of regression models for time series data. Finally, we briefly consider a system of autoregressive and moving average models that integrates principles from several of the methodologies for time series analysis discussed in this chapter and the preceding one.

25.1 EXPONENTIAL SMOOTHING FOR STATIONARY TIME SERIES

Introduction

Many forecasting applications involve the analysis of a large number of time series on a frequent, periodic basis, necessitating an analytical approach that adapts more or less automatically to structural changes in the time series.

EXAMPLES

1. A manufacturer stocks thousands of different types of spare parts, raw materials, components, and finished products. Monthly demand forecasts for these items are required for procurement and production decisions. The forecasts are the key to inventory cost control. Inventory carrying costs and out-of-stock costs can be kept at low levels only if reasonably reliable forecasts of requirements are available.

2. An airline analyst requires forecasts of weekly passenger traffic on each of the airline's several hundred regularly scheduled flights. The forecasts are used in operations management, budget planning, and financial control.

3. An investment firm that manages many investment portfolios requires the daily monitoring of rates of return for a large number of securities.

Many methods of time series analysis and forecasting, including those discussed in the preceding chapter, require much attention by skilled professionals and large amounts of data handling and storage, and would be very expensive to apply to hundreds or thousands of time series on a frequent, periodic basis. Exponential smoothing is a forecasting method that is well suited to this type of situation. It is readily computerized, requires a minimum of professional attention, makes modest demands on data handling and storage, and adapts effectively to structural changes in time series.

The system of exponential smoothing models to be presented is known as the *Holt-Winters exponential smoothing system*. This system has been successfully applied in many diverse settings. We begin by presenting in this section the exponential smoothing model for time series where neither trend nor seasonal components are present. We then extend this basic model to time series with a trend component and to time series with both trend and seasonal components.

Stationary Time Series

The concept of a stationary process was introduced in (6.1). The concept extends directly to a time series.

(25.1)

> A *stationary time series* $Y_1, Y_2, \ldots, Y_t, \ldots, Y_n$ is one in which the consecutive observations Y_t are outcomes of random variables having the same probability distribution.

By definition, therefore, a stationary time series has an irregular or random component but does not have trend, cyclical, or seasonal components. The following are examples of stationary time series.

EXAMPLES

1. The number of traffic lights failing daily in a city.
2. The monthly per capita demand for a food staple, such as rice.
3. The closing monthly balance of a family's checking account.

Smoothing Procedure

To forecast a stationary time series with the exponential smoothing model, we first smooth the time series with a moving average similar to those described in Section 24.2 in order to isolate the systematic or smooth component of the series. We then project this smooth component into the future. The moving average employed by the exponential smoothing model for a stationary time series is a special type of weighted moving average.

The smooth component of a stationary time series may be considered as a succession of estimates of the underlying mean level of the stationary process. We shall denote the estimate in period t of the mean level of the process by A_t, and we shall refer to A_t as the *smoothed estimate* for period t. The exponential smoothing model calculates the current smoothed estimate A_t as a weighted average of the current observation Y_t, with weight a, and the smoothed estimate of the preceding period A_{t-1}, with weight $1 - a$. The weight a is called the *smoothing constant* and generally is a value between 0 and 1. The following formula describes the procedure.

(25.2)

Update the smoothed estimate:

$$A_t = aY_t + (1 - a)A_{t-1} \qquad t = 1, 2, \ldots$$

where:

A_t is the smoothed estimate for the current period t

A_{t-1} is the smoothed estimate for the preceding period $t - 1$

Y_t is the observation for the current period t

a is the smoothing constant, $0 < a < 1$

To apply formula (25.2), we require values for the smoothing constant a and the *starting value* A_0 for the mean level of the series. In the example we now consider, the constant a and the starting value A_0 are prespecified. We shall discuss how these are chosen in practice after the example.

EXAMPLE

Rice Orders. A food store chain restocks inventories of staple items in its stores every week based on orders received by its central warehouse from each of the stores. Column 1 of Table 25.1a shows a time series Y_t of total weekly orders (in cases) received by the warehouse for a certain package of long-grain rice. The analyst has concluded that this time series is stationary, so formula (25.2) may be used to smooth and forecast the series. The weekly smoothed estimates of the current mean level of the series, A_t, are shown in column 2. These smoothed estimates were computed from (25.2) by using $A_0 = 4000$ as a starting value and a smoothing constant of $a = 0.10$. For instance, the first two smoothed estimates were obtained as follows:

$$A_1 = aY_1 + (1 - a)A_0 = 0.10(3169) + 0.90(4000) = 3917$$

$$A_2 = aY_2 + (1 - a)A_1 = 0.10(3682) + 0.90(3917) = 3893$$

Figure 25.1 presents a graph of the original and the smoothed series. Note how the smoothed series remains more or less centered in the original series and "tracks" the original series as it unfolds week by week. ☐

Forecasting Procedure

The exponential smoothing model uses the current smoothed estimate of the process mean level as the forecast for a stationary time series. The reason is that the smoothing procedure largely removes the irregular component from the stationary series. If the time series is anticipated to remain stationary with the same mean level in future periods $t + 1$, $t + 2$, and so on, then the current estimate A_t is the most appropriate fore-

TABLE 25.1   Exponential smoothing without trend or seasonal components — Rice orders example

(a)
Smoothing procedure

Week t	(1) Total Orders (cases) Y_t	(2) Smoothed Estimate A_t	Week t	(1) Total Orders (cases) Y_t	(2) Smoothed Estimate A_t
1	3169	3917	10	2502	3960
2	3682	3893	11	5006	4064
3	2655	3770	12	6885	4346
4	4500	3843	13	4196	4331
5	3682	3827	14	2728	4171
6	3568	3801	15	5262	4280
7	5045	3925	16	3719	4224
8	4733	4006	17	5707	4372
9	5164	4122	18	4580	4393

(b)
Forecasting procedure

Week $t + k$	Forecast F_{t+k}
19	4393
20	4393
21	4393

cast of the series for these future periods. This forecasting procedure may be expressed notationally as follows.

(25.3)

Forecast for period $t + k$:

$$F_{t+k} = A_t \qquad k = 1, 2, \ldots$$

where:

A_t is the smoothed estimate for current period t

F_{t+k} is the forecast for k periods ahead

Note that the exponential smoothing model provides a single number for the forecast for period $t + k$ and not an interval. A point forecast is often required, such as in the rice orders example.

FIGURE 25.1 **Plot of the original and the exponentially smoothed series—Rice orders example**

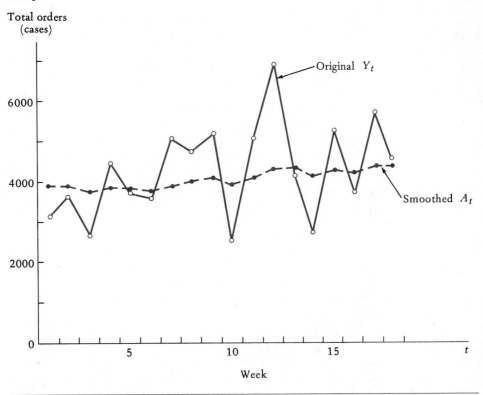

EXAMPLE

For the rice orders example in Table 25.1a, forecasts are desired at the end of week 18 for weeks 19, 20, and 21. The current smoothed estimate, for week 18, is $A_{18} = 4393$. Hence, this value is the forecast of total orders for each of the following three weeks. In notation, we write $F_{19} = F_{20} = F_{21} = A_{18} = 4393$. These forecasts are shown in Table 25.1b. ☐

Properties of Exponential Smoothing Model

The exponential smoothing procedure in (25.2) has a number of important properties.

Weighting of Observations. Exponential smoothing is a special kind of *weighted moving average* procedure. Consider the smoothed estimate for time period t, which in (25.2) is written:

$$A_t = aY_t + (1 - a)A_{t-1}$$

Now, in turn, smoothed estimate A_{t-1} is given by:

$$A_{t-1} = aY_{t-1} + (1 - a)A_{t-2}$$

Substituting this expression for A_{t-1} in the preceding equation gives:

$$A_t = aY_t + (1 - a)[aY_{t-1} + (1 - a)A_{t-2}]$$
$$= aY_t + a(1 - a)Y_{t-1} + (1 - a)^2 A_{t-2}$$

Repeating this process by substituting for A_{t-2}, then for A_{t-3}, and so on, one finds that the current smoothed estimate is the following weighted average of observations for all earlier periods and the starting value A_0.

(25.4)

$$A_t = aY_t + a(1 - a)Y_{t-1} + a(1 - a)^2 Y_{t-2} +$$
$$\cdots + a(1 - a)^{t-1}Y_1 + (1 - a)^t A_0$$

This formula shows that the earlier the period, the less weight is given to that observation. This follows because the constant a is a value between 0 and 1. Hence, the weights in (25.4), of the form $a(1 - a)^r$, decline as r increases. Indeed, the term *exponential smoothing* refers to the fact that the weight $a(1 - a)^r$ declines geometrically the larger is r — that is, the further the period is in the past. For example, if $a = 0.10$, then the current observation Y_t has weight $a = 0.10$, while the observation 8 periods earlier, Y_{t-8}, has weight $0.10(1 - 0.10)^8 = 0.043$.

Thus, observations in periods far in the past have little weight in determining the current smoothed estimate. In this sense, the smoothed estimate is a moving average because updating by means of (25.2) takes into account the most recent observation with weight a and reduces the weights of all earlier observations by a factor of $1 - a$. For instance, if $a = 0.10$, the weights of all earlier observations are reduced by 10 percent.

Adaptive Capability.   Although the procedure in (25.2) is strictly appropriate only for stationary time series, it is capable of adapting to occasional step changes or to cyclical changes in the mean level of the series. This adaptive capability is a consequence of the moving average property of the smoothing process. For instance, if the series shifts to a new mean level, the smoothed estimate will gradually approach the new level as it incorporates observations from the new level and gives less weight to observations from the old level.

Data Storage Requirements.   An advantage of exponential smoothing over alternative computer-based forecasting systems is its economical use of data storage. With exponential smoothing, the only information required to calculate the current smoothed estimate A_t is the current observation Y_t and the preceding smoothed estimate A_{t-1}. Thus, A_{t-1} is the only number that must be stored and carried over to compute A_t in the next period. Although the cost of computer data storage is rapidly declining, the small storage requirement of the exponential smoothing procedure is still an attractive feature for systems with large numbers of time series.

Selection of Starting Value and Smoothing Constant

For a stationary time series, the starting value A_0 may be set equal to the mean of a few observations from the immediate past. If no past observations are available, an informed judgment can be used for the starting value A_0.

The responsiveness and performance of the exponential forecasting system depends in an important way on the smoothing constant a. The larger is the smoothing constant a, the smaller is the weight given to past observations relative to more recent ones. For example, we noted earlier that when $a = 0.1$, Y_{t-8} is given weight $0.1(1 - 0.1)^8 = 0.043$ in the calculation of A_t. When $a = 0.5$, however, then Y_{t-8} is only given weight $0.5(1 - 0.5)^8 = 0.0020$. Thus, if a is too large, the smoothed estimates can be influenced unduly by the irregular components of current and recent observations. On the other hand, if a is too small, the smoothed estimates lose their adaptability and respond too slowly to any changes in the underlying mean level of the series. Hence, the smoothing constant a must be chosen to strike an appropriate balance between these two extremes. One may experiment to see which value of the constant yields the smallest forecasting errors in aggregate for the time series under consideration. Values of a between 0.10 and 0.30 are commonly used in practice. It is always possible to change the value of a should the irregular component of the series change in importance or should the frequency or magnitude of changes in the mean level of the series diverge from past experience.

Comments		
	1.	Exponential smoothing calculations are usually done by computer. Some packages have an option that automatically selects the starting value A_0 and the smoothing constant a based on an input of some initial observations for the series and a specified criterion.
	2.	When an exponential smoothing system involves numerous time series, the series may be categorized into relatively homogeneous groups, with each series in a group having the same smoothing constant.

25.2 EXPONENTIAL SMOOTHING FOR TIME SERIES WITH TREND COMPONENT

We have seen that exponential smoothing for a stationary time series involves one updating step for each new observation, as described in (25.2). When a time series contains a trend component, exponential smoothing involves two updating steps — one for the smoothed estimate and one for a trend estimate. The smoothed estimate A_t for period t still describes the mean level of the series at period t, including now the trend component up to that time period. The *trend estimate* for period t, which we shall denote by B_t, represents the period-to-period change in the mean level of the series at period t and is required for forecasting and succeeding smoothing. The term B_t may be considered as the estimated slope coefficient of the linear trend component at period t.

Updating Procedure

To obtain the smoothed estimate A_t, we compute a weighted average of two separate estimates of the mean level of the series at period t. The first estimate is Y_t, the current observation for period t. The second estimate is $A_{t-1} + B_{t-1}$. This latter estimate of the current mean level of the time series in period t is based on the past observations Y_{t-1}, Y_{t-2}, and so on. It is the sum of (1) the smoothed estimate A_{t-1} of the mean level in the preceding period, and (2) the estimated trend change B_{t-1} as of the preceding period to take account of the anticipated trend change in mean level between periods $t - 1$ and t. The respective weights of the two estimates in the average are a and $1 - a$, where, as before, a is the smoothing constant. The resulting updating formula for the smoothed estimate is as follows.

(25.5a)
$$A_t = aY_t + (1 - a)(A_{t-1} + B_{t-1}) \qquad 0 < a < 1$$

The updating formula for the trend estimate B_t is also based on a weighted average of two estimates. The first estimate is $A_t - A_{t-1}$, which measures the change in the smoothed series between periods $t - 1$ and t and thus reflects the trend component of the most recent observation Y_t. The second estimate is B_{t-1}, the estimate of the trend change as of the preceding period. This estimate is determined solely from the past observations Y_{t-1}, Y_{t-2}, and so on. The respective weights of the two estimates in the average are b and $1 - b$, where the weight b is called the *trend adjustment constant*. This constant, like the smoothing constant, is generally a value between 0 and 1. The resulting updating formula for the trend estimate is as follows.

(25.5b)
$$B_t = b(A_t - A_{t-1}) + (1 - b)B_{t-1} \qquad 0 < b < 1$$

As noted earlier, the updated trend estimate is used for forecasting and for smoothing in the next time period.

In summary, exponential smoothing for a time series with a trend component involves a two-step updating procedure as follows.

(25.5)
Step 1. Update the smoothed estimate:

(25.5a)
$$A_t = aY_t + (1 - a)(A_{t-1} + B_{t-1})$$

Step 2. Update the trend estimate:

(25.5b)
$$B_t = b(A_t - A_{t-1}) + (1 - b)B_{t-1}$$

> *where:*
>
> A_t, A_{t-1}, Y_t, a are defined in (25.2)
>
> B_t is the trend estimate for the current period t
>
> B_{t-1} is the trend estimate for the preceding period $t-1$
>
> b is the trend adjustment constant, $0 < b < 1$

Starting values A_0 and B_0 are required in order to apply the updating formulas in (25.5). These values correspond, respectively, to the mean level and the slope coefficient of the trend component for period $t = 0$. We shall discuss how these starting values and the weighting constants a and b are chosen in practice after the following example. In the example, the weighting constants and starting values are prespecified.

EXAMPLE

Lake Level. The time series for the lake level example in Table 15.4 represents the maximum level of a lake each year for 30 consecutive years (in meters above 190). A test of runs up and down established that the series has a trend. The original observations are reproduced in column 1 of Table 25.2a. The corresponding smoothed estimates and trend estimates, calculated by using the two-step updating procedure in (25.5), appear in columns 2 and 3 of the table. Starting values of $A_0 = 6.70$ and $B_0 = -0.060$ were used to begin the procedure. The smoothing and trend adjustment constants were set at $a = 0.10$ and $b = 0.20$. The following calculations illustrate how the smoothed and trend estimates are obtained for the first two years:

Year 1

Step 1	$A_1 = aY_1 + (1 - a)(A_0 + B_0) = 0.10(6.63) + 0.90(6.70 - 0.060) = 6.639$
Step 2	$B_1 = b(A_1 - A_0) + (1 - b)B_0 = 0.20(6.639 - 6.70) + 0.80(-0.060) = -0.0602$

Year 2

Step 1	$A_2 = aY_2 + (1 - a)(A_1 + B_1) = 0.10(6.59) + 0.90(6.639 - 0.0602) = 6.580$
Step 2	$B_2 = b(A_2 - A_1) + (1 - b)B_1 = 0.20(6.580 - 6.639) + 0.80(-0.0602) = -0.0600$

Figure 25.2 contains a plot of the original and the smoothed series. Observe that the actual lake level (Y_t) is fluctuating about the smoothed estimate A_t and that the smoothed series is tracking the original series quite well. □

Forecasting Procedure

The smoothed series can be extrapolated to give forecasts for future periods by making an allowance for the expected trend change. For a forecast k periods ahead to period $t + k$, the forecast will be based on the smoothed estimate A_t of the current mean level, to which is added the amount kB_t, representing the current anticipated change in the mean level of the series resulting from the trend operating for k periods. The forecasting procedure is therefore as follows.

TABLE 25.2   Exponential smoothing with a trend component — Lake level example (lake level in meters above 190)

(a)
Smoothing procedure

Year t	(1) Lake Level Y_t	(2) Smoothed Estimate A_t	(3) Trend Estimate B_t	Year t	(1) Lake Level Y_t	(2) Smoothed Estimate A_t	(3) Trend Estimate B_t
1	6.63	6.639	−0.0602	16	5.91	5.740	−0.0581
2	6.59	6.580	−0.0600	17	5.81	5.695	−0.0556
3	6.46	6.514	−0.0612	18	5.64	5.639	−0.0555
4	6.49	6.456	−0.0604	19	5.51	5.576	−0.0570
5	6.45	6.402	−0.0594	20	5.31	5.498	−0.0612
6	6.41	6.349	−0.0580	21	5.36	5.430	−0.0627
7	6.38	6.300	−0.0562	22	5.17	5.347	−0.0667
8	6.26	6.245	−0.0559	23	5.07	5.259	−0.0709
9	6.09	6.179	−0.0579	24	4.97	5.167	−0.0753
10	5.99	6.108	−0.0605	25	5.00	5.082	−0.0771
11	5.92	6.035	−0.0631	26	5.01	5.006	−0.0770
12	5.93	5.968	−0.0639	27	4.85	4.921	−0.0786
13	5.83	5.896	−0.0654	28	4.79	4.837	−0.0796
14	5.82	5.830	−0.0656	29	4.73	4.755	−0.0801
15	5.95	5.783	−0.0619	30	4.76	4.683	−0.0784

(b)
Forecasting procedure

Year $t + k$	Forecast F_{t+k}
31	$4.683 + 1(-0.0784) = 4.605$
32	$4.683 + 2(-0.0784) = 4.526$
33	$4.683 + 3(-0.0784) = 4.448$
34	$4.683 + 4(-0.0784) = 4.369$
35	$4.683 + 5(-0.0784) = 4.291$

(25.6)

Forecast for period $t + k$:

$$F_{t+k} = A_t + kB_t$$

where:

A_t is the smoothed estimate for the current period t

B_t is the trend estimate for the current period t

F_{t+k} is the forecast k periods ahead for future period $t + k$

FIGURE 25.2   **Plot of the original and the exponentially smoothed series — Lake level example**

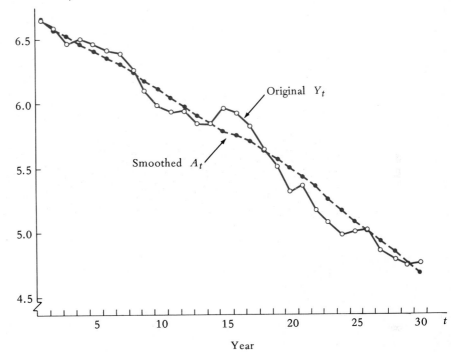

Figure 25.3 illustrates this forecasting procedure.

EXAMPLE

For the lake level example, forecasts in year 30 of the maximum lake level for years 31 through 35 are given in Table 25.2b. The forecasts are calculated from (25.6) using the smoothed estimate and trend estimate for year 30, namely, $A_{30} = 4.683$ and $B_{30} = -0.0784$. For instance, the forecast lake level for year 31 is:

$$F_{31} = 4.683 + 1(-0.0784) = 4.605 \text{ meters above } 190$$

and for year 35 is:

$$F_{35} = 4.683 + 5(-0.0784) = 4.291 \text{ meters above } 190$$

FIGURE 25.3   Illustration of the exponential smoothing forecasting procedure with a trend adjustment

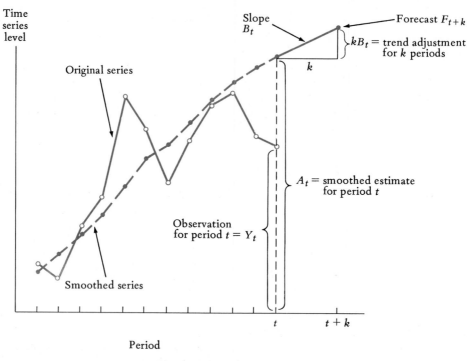

Selection of Starting Values and Weighting Constants

Past observations of the time series may be used to set the starting values of the exponential smoothing procedure by applying the classical time series methods in Chapter 24 or the time series regression methods to be discussed in Sections 25.4 and 25.5. For example, a trend line can be fitted to the past observations and the slope of this line used as the starting value B_0 for the trend estimate. The level of this trend line at period $t = 0$ or the estimated trend-cyclical component at period $t = 0$ can be taken as the starting value A_0 for the smoothed estimate.

The values of the weighting constants a and b need to be chosen with care and need not be the same. In practical applications, typical value ranges for these constants are 0.10 to 0.30. If the constants are too small, the estimates will react sluggishly to underlying changes. On the other hand, if the constants are too large, the estimates will respond excessively to random changes in the time series.

As noted previously, some computer packages for exponential smoothing have an option whereby the starting values and weighting constants are chosen automatically, based on some initial observations for the series and a specified criterion.

25.3   EXPONENTIAL SMOOTHING FOR TIME SERIES WITH TREND AND SEASONAL COMPONENTS

For a time series with both trend and seasonal components, three updates are required for each new observation. As before, the smoothed estimate A_t and the trend estimate B_t for period t must be updated. In addition, an update is required for the *seasonal index estimate* of the series for period t when the time series contains a seasonal component. We shall denote this estimate by C_t; it will be expressed on a base of 1, rather than as a percentage. We also need a symbol for the number of periods in the seasonal cycle and shall use p for this purpose. Thus, for a quarterly series, $p = 4$. For a monthly series, $p = 12$.

Updating Procedure

The updating formula for the smoothed estimate A_t is a weighted average of two estimates of the current mean level.

(25.7a)
$$A_t = a\left(\frac{Y_t}{C_{t-p}}\right) + (1 - a)(A_{t-1} + B_{t-1}) \qquad 0 < a < 1$$

The first estimate in (25.7a), Y_t/C_{t-p}, is simply the current observation Y_t seasonally adjusted. The term C_{t-p} in the denominator is the most recent seasonal index estimate made in period $t - p$, one seasonal cycle prior to current period t. To illustrate this, consider a quarterly series where the current period $t = 11$ is the third quarter. Then, $C_{t-p} = C_{11-4} = C_7$ represents the seasonal index estimate for quarter 7, the third quarter of the preceding year.

The second estimate of the mean level, $A_{t-1} + B_{t-1}$, and the weighting of the two estimates are the same as in updating formula (25.5) for time series with trend only. Since Y_t is seasonally adjusted in updating the smoothed estimate A_t in (25.7a), it follows that the smoothed series is deseasonalized.

The updating formula for the trend estimate B_t is identical to the one in (25.5).

(25.7b)
$$B_t = b(A_t - A_{t-1}) + (1 - b)B_{t-1} \qquad 0 < b < 1$$

The updating of the seasonal index estimate C_t is made by averaging two estimates of the seasonal index.

(25.7c)
$$C_t = c\left(\frac{Y_t}{A_t}\right) + (1 - c)C_{t-p} \qquad 0 < c < 1$$

Here, c is a weighting constant which we shall call the *seasonal index adjustment constant*. The first seasonal index estimate is the current one, Y_t/A_t. This is a seasonal index estimate because A_t is an estimate of the deseasonalized mean level of the series for period t. Hence, the ratio Y_t/A_t is a specific seasonal relative for period t (on a base of 1). The second seasonal index estimate, C_{t-p}, is the seasonal index estimate from p periods earlier, that is, one seasonal cycle earlier.

In summary, exponential smoothing for a time series with trend and seasonal components involves the following three-step updating procedure.

(25.7)   *Step 1.*   Update the smoothed estimate:

(25.7a)
$$A_t = a\left(\frac{Y_t}{C_{t-p}}\right) + (1 - a)(A_{t-1} + B_{t-1})$$

Step 2.   Update the trend estimate:

(25.7b)
$$B_t = b(A_t - A_{t-1}) + (1 - b)B_{t-1}$$

Step 3.   Update the seasonal index estimate:

(25.7c)
$$C_t = c\left(\frac{Y_t}{A_t}\right) + (1 - c)C_{t-p}$$

where:
A_t, A_{t-1}, Y_t, a are defined in (25.2)
B_t, B_{t-1}, b are defined in (25.5)
C_t is the seasonal index estimate for current period t
C_{t-p} is the seasonal index estimate for period $t - p$, one seasonal cycle earlier
c is the seasonal index adjustment constant, $0 < c < 1$

The updated trend and seasonal index estimates are used for forecasting and for smoothing in the next time period. Also, note that only one seasonal index estimate is updated in each period t — namely, the one that corresponds to the seasonal phase of period t. For instance, if t is a third quarter in a quarterly series, then only the seasonal index for the third quarter is updated in that period.

To begin the updating procedure in (25.7), starting values A_0, B_0, and the seasonal indexes $C_{-p+1}, C_{-p+2}, \ldots, C_{-1}$, and C_0 are required. In the following example, these starting values, as well as the weighting constants a, b, and c, are prespecified. After the example, we discuss how these starting values and weighting constants are set in practical applications.

EXAMPLE

Taxair.   The quarterly time series of revenue passenger miles for the Taxair example is presented in Table 24.5. We have applied the exponential smoothing procedure (25.7) to this series in Table 25.3a, utilizing a computer package. The

TABLE 25.3 **Exponential smoothing with trend and seasonal components — Taxair example**

(a)
Smoothing procedure

Year and Quarter	Period t	(1) Miles (thousands) Y_t	(2) Smoothed Estimate A_t	(3) Trend Estimate B_t	(4) Seasonal Index Estimate C_t
1981 1	1	43.20	88.80	1.160	0.465
2	2	90.00	89.54	1.075	1.032
3	3	162.00	90.55	1.063	1.796
4	4	64.80	91.71	1.082	0.703
1982 1	5	42.35	92.63	1.049	0.462
2	6	100.10	94.01	1.116	1.045
3	7	173.25	95.26	1.143	1.805
4	8	69.30	96.62	1.187	0.708
1983 1	9	42.96	97.34	1.092	0.454
2	10	89.50	97.15	0.836	0.996
3	11	157.52	96.92	0.622	1.733
4	12	68.02	97.38	0.592	0.704
1984 1	13	37.95	96.55	0.306	0.429
2	14	89.70	96.18	0.171	0.970
3	15	151.80	95.47	−0.005	1.676
4	16	65.55	95.22	−0.053	0.698
1985 1	17	40.15	95.01	−0.086	0.427
2	18	94.90	95.21	−0.029	0.981
3	19	164.25	95.46	0.028	1.694
4	20	65.70	95.35	0.001	0.694
1986 1	21	47.64	96.98	0.327	0.452
2	22	99.25	97.70	0.404	0.995
3	23	174.68	98.60	0.505	1.725
4	24	75.43	100.06	0.695	0.718

(b)
Forecasting procedure

Year and Quarter	Period $t + k$		Forecast F_{t+k}
1987 1	25	$[100.06 + 1(0.695)](0.452) =$	45.54
2	26	$[100.06 + 2(0.695)](0.995) =$	100.94
3	27	$[100.06 + 3(0.695)](1.725) =$	176.20
4	28	$[100.06 + 4(0.695)](0.718) =$	73.84
1988 1	29	$[100.06 + 5(0.695)](0.452) =$	46.80
2	30	$[100.06 + 6(0.695)](0.995) =$	103.71
3	31	$[100.06 + 7(0.695)](1.725) =$	181.00
4	32	$[100.06 + 8(0.695)](0.718) =$	75.84

weighting constants used in the calculations were $a = 0.1$, $b = 0.2$, and $c = 0.4$. The starting values for the smoothed estimate and the trend estimate were $A_0 = 87.00$ and $B_0 = 1.000$. The starting values for the seasonal index estimates were:

Quarter:	1	2	3	4
Starting Value:	$C_{-3} = 0.450$	$C_{-2} = 1.050$	$C_{-1} = 1.800$	$C_0 = 0.700$

The calculations of the estimates A_t, B_t, and C_t for the first five quarters are illustrated in Table 25.4. □

Forecasting Procedure

A time series with trend and seasonal components can be projected into the future by making allowances for the expected trend change and the seasonal effects. First, we

---

TABLE 25.4   Calculations of smoothed, trend, and seasonal index estimates for periods 1 to 5 of a quarterly time series — Taxair example

Period t	Smoothed Estimate A_t $$a\left(\frac{Y_t}{C_{t-p}}\right) + (1 - a)(A_{t-1} + B_{t-1})$$	Trend Estimate B_t $$b(A_t - A_{t-1}) + (1 - b)B_{t-1}$$	Seasonal Index Estimate C_t $$c\left(\frac{Y_t}{A_t}\right) + (1 - c)C_{t-p}$$
1	$0.1\left(\dfrac{43.20}{0.450}\right) + 0.9(87.00 + 1.000)$ $= 88.80$	$0.2(88.80 - 87.00) + 0.8(1.000)$ $= 1.160$	$0.4\left(\dfrac{43.20}{88.80}\right) + 0.6(0.450)$ $= 0.465$
2	$0.1\left(\dfrac{90.00}{1.050}\right) + 0.9(88.80 + 1.160)$ $= 89.54$	$0.2(89.54 - 88.80) + 0.8(1.160)$ $= 1.075$	$0.4\left(\dfrac{90.00}{89.54}\right) + 0.6(1.050)$ $= 1.032$
3	$0.1\left(\dfrac{162.00}{1.800}\right) + 0.9(89.54 + 1.075)$ $= 90.55$	$0.2(90.55 - 89.54) + 0.8(1.075)$ $= 1.063$	$0.4\left(\dfrac{162.00}{90.55}\right) + 0.6(1.800)$ $= 1.796$
4	$0.1\left(\dfrac{64.80}{0.700}\right) + 0.9(90.55 + 1.063)$ $= 91.71$	$0.2(91.71 - 90.55) + 0.8(1.063)$ $= 1.082$	$0.4\left(\dfrac{64.80}{91.71}\right) + 0.6(0.700)$ $= 0.703$
5	$0.1\left(\dfrac{42.35}{0.465}\right) + 0.9(91.71 + 1.082)$ $= 92.63$	$0.2(92.63 - 91.71) + 0.8(1.082)$ $= 1.049$	$0.4\left(\dfrac{42.35}{92.63}\right) + 0.6(0.465)$ $= 0.462$

project the deseasonalized series k periods ahead, adjusting for the expected trend. This yields the projection $A_t + kB_t$, just as in (25.6) for forecasting a series with only a trend component. We then incorporate the seasonal component by multiplying the deseasonalized projection $A_t + kB_t$ by the appropriate seasonal index estimate C, depending on the period in the seasonal cycle that corresponds to the forecast period $t + k$. Thus, the forecasting procedure is the following.

(25.8)

Forecast for period $t + k$:

$$F_{t+k} = (A_t + kB_t)C$$

where:

A_t is the smoothed estimate for the current period t

B_t is the trend estimate for the current period t

F_{t+k} is the forecast k periods ahead for future period $t + k$

C is the current seasonal index estimate corresponding to the seasonal phase of period $t + k$

EXAMPLE

For the Taxair example, quarterly forecasts at the end of 1986 for 1987 and 1988 are given in Table 25.3b. The smoothed estimate and the trend estimate used in the forecasts are those for the most recent quarter, namely, $A_{24} = 100.06$ and $B_{24} = 0.695$. The seasonal index estimates used in the forecasts are those from the last four quarters, namely:

Quarter	Period	Current Seasonal Index Estimate
1	21	$C_{21} = 0.452$
2	22	$C_{22} = 0.995$
3	23	$C_{23} = 1.725$
4	24	$C_{24} = 0.718$

Selection of Starting Values and Weighting Constants

The starting values A_0 and B_0 and the weighting constants a and b in the updating formulas in (25.7) are set on the basis of the same considerations as for a time series with only a trend component.

The starting values for the seasonal index estimates may be the seasonal indexes computed from past observations of the series by using the method of ratio to moving average described in Section 24.9. The seasonal index adjustment constant c might need to be relatively large compared with the weighting constants a and b because each individual seasonal index estimate is only updated every pth period, that is, only once in each seasonal cycle. Thus, the seasonal index estimates have fewer opportunities to adjust to changes in the underlying seasonal pattern. Larger values of c compensate for this reduced opportunity.

Comment	Seasonal indexes produced by the method of ratio to moving average of Section 24.9 are adjusted to give them an average value of 1 (when expressed in decimal form). There is no mechanism in the exponential smoothing system that forces the p seasonal index estimates to have an average value of 1. As a general rule, however, their average value will be close to 1, as is illustrated by the index values in column 4 of Table 25.3a.

 ## 25.4   REGRESSION MODELS WITH INDEPENDENT ERROR TERMS

The classical time series model discussed in Chapter 24 and the exponential smoothing models discussed in the preceding sections are descriptive models. No attempt is made to define the irregular component in probabilistic terms, and no parameters are specified about which formal statistical inferences can be made. Uncertainties are handled by judgment, not by statistical theory.

Regression time series models, on the other hand, are formal models permitting statistical inferences and predictions. We now consider the use of several regression models for time series data. In this section, we take up a number of regression models in which the error terms are independent. In the next section, we consider regression models where the error terms are correlated.

Regression models for time series data with independent error terms are special cases of multiple regression model (20.3), which we repeat here for convenience.

(25.9)
$$Y_t = \beta_0 + \beta_1 X_{t1} + \beta_2 X_{t2} + \cdots + \beta_{p-1} X_{t,p-1} + \varepsilon_t$$

where:

ε_t are independent $N(0, \sigma^2)$

We now use the subscript t to denote the time period ($t = 1, 2, \ldots, n$). With this type of model, time-related effects are incorporated into the model entirely by means of the independent variables. The random error terms are assumed to be time-independent.

Trend Model

If a time series, such as annual company sales for the past ten years, contains a linear trend and a random component that is independent from one time period to another, regression model (25.9) takes the following form.

(25.10)
$$Y_t = \beta_0 + \beta_1 X_t + \varepsilon_t$$

where:

$X_t = t$

In other words, X_t denotes the coded time period; we have $X_1 = 1$, $X_2 = 2$, and so on.

Additive Trend and Seasonal Components Model

When a time series, such as quarterly shipments of television sets during the past 48 quarters, contains both a trend component and a seasonal component and they act additively, regression model (25.9) again can be used if the random effects are time-independent. For a quarterly time series and a linear trend, the regression model takes the following form.

(25.11)

$$Y_t = \beta_0 + \beta_1 X_{t1} + \beta_2 X_{t2} + \beta_3 X_{t3} + \beta_4 X_{t4} + \varepsilon_t$$

where:

$X_{t1} = t$

$X_{t2} = \begin{matrix} 1 \\ 0 \end{matrix}$ if period t is second quarter
          otherwise

$X_{t3} = \begin{matrix} 1 \\ 0 \end{matrix}$ if period t is third quarter
          otherwise

$X_{t4} = \begin{matrix} 1 \\ 0 \end{matrix}$ if period t is fourth quarter
          otherwise

Here, $\beta_1 X_1$ is the linear trend effect, and X_2, X_3, and X_4 are three indicator variables (discussed in Section 20.8) that denote the four quarters of the year. These indicator variables take on the following values for the four quarters:

Quarter	X_2	X_3	X_4
First	0	0	0
Second	1	0	0
Third	0	1	0
Fourth	0	0	1

The regression coefficient β_2 in model (25.11) represents the differential effect on the dependent variable of the second quarter as compared with the first quarter, and the regression coefficients for the other indicator variables represent corresponding differential effects for the other quarters.

Business Indicators and Other Predictors as Independent Variables

In many cases where the dependent variable is a time series, the independent variables in the regression model include business indicators or other predictor series. Business indicators are time series measuring activities in key sectors of the economy.

EXAMPLES

1. Annual industry production (Y) is related to gross national product (X_1) and the Consumer Price Index (X_2), for the past 16 years.

2. Annual kindergarten enrollment (Y) is related to number of births five years earlier (X), for the past 20 years.

3.   Monthly sales of a firm (Y) are related to the firm's promotional expenditures in the previous month (X), for the past 36 months.

4.   Quarterly shipments of cement (Y) are related to two independent variables—the volume of construction contracts awarded (X) during the preceding quarter and the volume during the quarter before that—for the past 20 quarters.   ☐

Coincident Independent Variables.   Regression model (25.9) for Example 1 takes the following form, assuming the effects of the independent variables are linear and additive.

(25.12)
$$Y_t = \beta_0 + \beta_1 X_{t1} + \beta_2 X_{t2} + \varepsilon_t$$

Here, Y_t is industry production, X_{t1} gross national product, and X_{t2} the Consumer Price Index, each for year t.

The variables Y_t, X_{t1}, and X_{t2} in (25.12) represent *coincident time series;* that is, they refer to the same time period. With coincident independent variables, a forecast of the dependent variable requires forecasts of the independent variables. Suppose in Example 1 the regression model is fitted to data for the years 1971–1986 inclusive. If industry production for 1987 were to have been forecast, forecasts for GNP and the Consumer Price Index for 1987 would have been needed. Forecasting GNP and the Consumer Price Index in order to obtain a forecast of industry production may seem to be a circuitous procedure. If, however, accurate forecasts of the independent variables are available and there is a close relation between them and industry production, an accurate forecast of the dependent variable can result.

Lagged Independent Variables.   Sometimes, it is possible to construct a regression model where some or all of the independent variables are lagged. Example 2 is an illustration. A simple linear regression model would be as follows.

(25.13)
$$Y_t = \beta_0 + \beta_1 X_{t-5} + \varepsilon_t$$

Here, Y_t is kindergarten enrollment in year t and X_{t-5} is the number of births five years earlier. The variable X_{t-5} represents a *lagged time series* relative to Y_t, the lag being five years.

The advantage of a regression model with lagged independent variables is that the independent variables need not be predicted in order to forecast the dependent variable. Suppose that the data for Example 2 span the years 1967–1986 inclusive. If a forecast of kindergarten enrollment for 1990 had been desired in 1987, the independent variable would have been number of births in 1985. This would have been known in 1987, so there would have been no need to forecast the independent variable.

A regression model may contain more than one lagged term of an independent variable. In Example 4, cement shipments in quarter t are related to the volume of construction contracts awarded in quarter $t - 1$ and to the volume in quarter $t - 2$. The regression model here, with linear and additive effects, would be as follows.

(25.14)
$$Y_t = \beta_0 + \beta_1 X_{t-1} + \beta_2 X_{t-2} + \varepsilon_t$$

This model is said to contain a *distributed time lag* because the lag is distributed over more than one period.

More Complex Models

Regression models for time series data often combine several of the features explained previously. Also, they may involve a transformed dependent variable, such as log Y_t. An example of a more complex regression model is the following for a quarterly time series.

(25.15)
$$Y_t' = \beta_0 + \beta_1 X_{t1} + \beta_2 X_{t2} + \beta_3 X_{t3} + \beta_4 X_{t4} + \beta_5 X_{t-1,5} + \varepsilon_t$$

where:
$Y_t' = \log Y_t$
$X_{t1} = t$
X_{t2}, X_{t3}, X_{t4} are indicator variables for quarterly seasonal effects
$X_{t-1,5}$ is a business indicator lagged one quarter

Note that regression model (25.15) involves an exponential trend effect $\beta_1 X_{t1}$ because the dependent variable is log Y_t (Section 24.5).

We illustrate the use of a regression model with independent error terms by considering an annual time series where a linear trend component and a lagged business indicator are employed in the model.

EXAMPLE

Reroofing Contracts.   A consultant, developing forecasting procedures for a firm specializing in commercial and industrial roofing, found that the dollar volume of contracts received annually for reroofing followed a linear trend over the course of the observations. In any year, the volume also reflected the state of general business activity in the firm's service region during the preceding year. The consultant decided tentatively to use the following model to forecast the dollar volume of reroofing contracts one year ahead:

$$Y_t = \beta_0 + \beta_1 X_{t1} + \beta_2 X_{t-1,2} + \varepsilon_t$$

where:
Y_t is the dollar volume of reroofing contracts in year t

$X_{t1} = t$ is the linear trend variable

$X_{t-1,2}$ is the index of general business activity in year $t - 1$

The consultant's data are presented in Table 25.5. The fitted regression model is (computer output not shown):

$$\hat{Y}_t = -188.62891 + 8.77231X_{t1} + 4.35074X_{t-1,2}$$

The consultant was particularly interested in whether or not the error terms could be assumed to be independent. The residuals are shown in Table 25.5, and the residual plot by time order is presented in Figure 25.4. On the basis of this residual plot, the consultant concluded that a regression model with independent error terms is not unreasonable here. A residual plot against the fitted values $\hat{Y}$ was also prepared by the consultant. He concluded from it that the fitted regression function is reasonable and that there is no indication that the error variance is not constant.

A forecast of Y_t for year $t = 12$ is desired, by means of a 90 percent prediction interval. The consultant ascertained that the index of business activity for the current year $t = 11$ is $X_{11,2} = 105.8$. Also, $X_{12,1} = 12$ for next year. Hence:

$$\hat{Y}_{12} = -188.62891 + 8.77231(12) + 4.35074(105.8) = 376.9$$

For $n = 11$ and $1 - \alpha = 0.90$, we require $t(0.95; 8) = 1.860$. The computer output provided the consultant with the estimated standard deviation for the predic-

---

TABLE 25.5   Regression with a linear trend and a lagged business indicator — Reroofing contracts example

Year t	Linear Trend Variable X_{t1}	Index of General Business Activity (lagged one year) $X_{t-1,2}$	Dollar Volume of Reroofing Contracts ($000) Y_t	Residual e_t
1	1	79.7	173.7	6.8
2	2	83.1	187.0	−3.5
3	3	90.9	232.1	−1.1
4	4	99.1	280.0	2.4
5	5	100.0	287.0	−3.3
6	6	97.5	282.8	−5.4
7	7	91.3	273.9	3.9
8	8	86.6	254.7	−3.6
9	9	94.4	300.8	−0.2
10	10	105.5	355.7	−2.4
11	11	108.7	387.2	6.4

FIGURE 25.4   **Residual plot against time order — Reroofing contracts example**

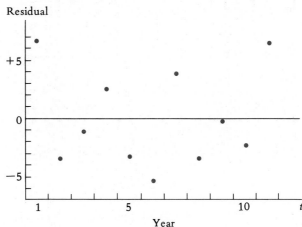

tion interval, as given in (20.21); it is $s\{Y_{h(\text{new})}\} = 5.6551$. The prediction limits, by (20.21), therefore are $376.9 \pm 1.860(5.6551)$, and the prediction interval is:

$$366.4 \leq Y_{h(\text{new})} \leq 387.4$$

Hence, with 90 percent confidence, the consultant predicts that the volume of reroofing contracts in year 12 will be between \$366.4 thousand and \$387.4 thousand.   □

25.5   REGRESSION MODELS WITH AUTOCORRELATED ERROR TERMS

The regression models in the previous section all assume that the error terms ε_t are independent from period to period. In many applications, however, the error terms in the different periods are correlated. When this is the case, the error terms are said to be *autocorrelated* or *serially correlated*.

Autocorrelated Error Terms

Autocorrelated error terms arise for a variety of reasons. One major cause is the omission of one or more key variables from the fitted regression model, such as the omission of population size in a model predicting sales where population size has a time-ordered effect on sales. Another important reason is that major random effects often tend to persist for several periods, as when the effects of a very poor harvest of an agricultural commodity are felt for several years thereafter.

First-Order Autoregressive Error Model.   When the error terms are related over time, a model for the error terms frequently employed is the first-order autoregressive error model.

(25.16)

> The *first-order autoregressive error model* is:
>
> $$\varepsilon_t = \rho\varepsilon_{t-1} + u_t$$
>
> *where:*
>
> ρ (Greek rho) is the autocorrelation parameter, $-1 < \rho < 1$
> u_t are independent $N(0, \sigma^2)$

This model assumes that the error term ε_t for period t contains a component resulting from the error term ε_{t-1} for the preceding period (when $\rho \neq 0$) and a random disturbance term u_t that is independent of earlier time periods.

The parameter ρ is called the *autocorrelation parameter* and, in fact, represents the coefficient of correlation between ε_t and ε_{t-1} as defined in (5.27). When ρ is positive, the autocorrelation is positive; when ρ is negative, the autocorrelation is negative. In most business and economic series with time-dependent error terms, the autocorrelation is positive. For example, large sales in one quarter because of good cyclical conditions are likely to persist into the next quarter.

When $\rho = 0$, the error terms ε_t and ε_{t-1} are uncorrelated and error model (25.16) simplifies to $\varepsilon_t = u_t$. The ε_t are then independent $N(0, \sigma^2)$ — the case assumed for all regression models up to now.

Effects of Autocorrelation.   If the method of least squares is employed in the usual fashion for fitting a regression model when the error terms are autocorrelated according to (25.16), it can be shown that the following consequences arise:

1.   The least squares regression coefficients are still unbiased estimators, but they no longer have the minimum variance property; they tend to be relatively inefficient.
2.   The error mean square *MSE* can seriously underestimate the true variance of the error terms.
3.   Standard procedures for confidence intervals and tests using the t and F distributions are no longer strictly applicable.

There are several methods available to avoid these difficulties caused by autocorrelated error terms. One involves estimation of the autocorrelation parameter ρ. Another method involves a simple transformation of the variables. We shall discuss this latter method after presenting a test for the presence of autocorrelation.

Durbin-Watson Test

A widely used test for examining whether or not the error terms in a regression model are autocorrelated is the Durbin-Watson test. This test is based on the first-order autoregressive error model (25.16). When one is concerned with positive autocorrelation, the usual case for business and economic applications, the test alternatives are as follows.

(25.17)

$$H_0: \rho \leq 0$$

$$H_1: \rho > 0$$

Here, H_0 implies that the error terms are uncorrelated or negatively correlated, while H_1 implies that they are positively correlated.

The Durbin-Watson test statistic d is based on the differences between adjacent residuals, $e_t - e_{t-1}$, and is of the following form.

(25.18)

$$d = \frac{\sum_{t=2}^{n} (e_t - e_{t-1})^2}{\sum_{t=1}^{n} e_t^2}$$

where:

e_t is the regression residual for period t

n is the number of time periods used in fitting the regression model

When the error terms are positively autocorrelated, adjacent residuals will tend to be of similar magnitude so that the numerator of the test statistic d will be small. If the error terms are not correlated or are negatively correlated, e_t and e_{t-1} will tend to differ more and the numerator of the test statistic will be larger. Hence, small values of the test statistic d are consistent with H_1, while large values are consistent with H_0.

The exact action limit for the Durbin-Watson test is difficult to calculate. Hence, the test is used with a lower bound d_L and an upper bound d_U for the action limit. When the test statistic d in (25.18) is less than the lower bound d_L, we conclude H_1, that positive autocorrelation is present. When the test statistic exceeds the upper bound d_U, we conclude H_0, that positive autocorrelation is not present. When the test statistic falls in the interval between d_L and d_U, we do not know for certain which conclusion to reach, although it is usually prudent to conclude that positive autocorrelation may be present.

The decision rule therefore has the following form.

(25.19)

When the alternatives are:

$$H_0: \rho \leq 0$$

$$H_1: \rho > 0$$

continues

and the first-order autoregressive error model (25.16) applies, the appropriate decision rule to control the α risk at $\rho = 0$ is as follows:

If $d > d_U$, conclude H_0.
If $d < d_L$, conclude H_1.
If $d_L \leq d \leq d_U$, the test is inconclusive.

where:

d is given by (25.18)

Appendix Table C.12 contains values of the bounds d_L and d_U for various sample sizes (n), number of independent variables in the regression model ($p - 1$), and two α levels, 0.05 and 0.01. We illustrate the use of the Durbin-Watson test by means of an example.

The Durbin-Watson test statistic d is often included in the output of regression computer routines.

EXAMPLE

Company Sales.   A securities analyst was studying the relation between sales of a company and industry sales. She obtained deseasonalized data for the past 20 quarters. These are presented (in part) in Table 25.7, columns 1 and 2 (p. 886). The analyst expected high positive autocorrelation. Her first step was to fit the simple linear regression model $Y_t = \beta_0 + \beta_1 X_t + \varepsilon_t$ by the method of least squares in order to study the residuals. These are plotted in time order in Figure 25.5a. Note the succession of positive residuals, then negative residuals, and finally positive residuals, with generally gradual changes in the residuals from one quarter to the next. This pattern suggests high positive autocorrelation.

FIGURE 25.5   Residual plots against time order — Company sales example

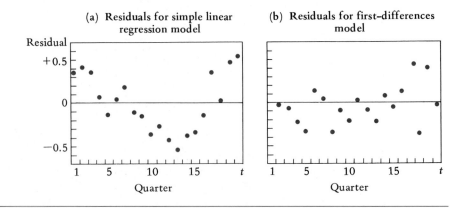

(a) **Residuals for simple linear regression model**

(b) **Residuals for first–differences model**

To formally test for positive autocorrelation, the analyst utilized the Durbin-Watson test. The alternatives are as given in (25.19). A computer run of a regression package provided her with the value of the test statistic, $d = 0.4765$. She wished to control the α risk at 0.05 when $\rho = 0$. Hence, she entered Table C.12 for $\alpha = 0.05$, $p - 1 = 1$ independent variable, and $n = 20$ cases to obtain $d_L = 1.20$ and $d_U = 1.41$. The decision rule according to (25.19) therefore is as follows:

If $d > 1.41$, conclude H_0 ($\rho \le 0$).
If $d < 1.20$, conclude H_1 ($\rho > 0$).
If $1.20 \le d \le 1.41$, the test is inconclusive.

Since $d = 0.4765 < 1.20$, she concluded H_1, that the error terms are positively autocorrelated. $\square$

Comment   The test of runs up and down discussed in Section 15.4 is not strictly applicable to regression residuals but can be used as an approximate test when the number of residuals is large.

Method of First Differences

The method of first differences is a simple method for dealing with the difficulties caused by positively autocorrelated error terms. The basic idea is to use a simple transformation of the variables that will make the error terms approximately uncorrelated so that ordinary regression methods can be used to their full advantage. We shall explain the method of first differences for the simple linear regression model, but the method extends readily to multiple regression applications.

Development of Method.   The simple linear regression model with first-order autoregressive error terms is as follows.

(25.20)

$$Y_t = \beta_0 + \beta_1 X_t + \varepsilon_t$$

where:

$\varepsilon_t = \rho \varepsilon_{t-1} + u_t$

u_t are independent $N(0, \sigma^2)$

As noted before, the autocorrelation parameter in business and economic applications when the error terms are serially correlated is usually positive and frequently very large. Let us see what happens when $\rho = 1$. The error term ε_t in (25.20) then becomes:

(25.21)

$$\varepsilon_t = \varepsilon_{t-1} + u_t$$

and the *first difference* $Y_t - Y_{t-1}$ becomes:

$$
\begin{aligned}
Y_t - Y_{t-1} &= (\beta_0 + \beta_1 X_t + \varepsilon_t) - (\beta_0 + \beta_1 X_{t-1} + \varepsilon_{t-1}) \\
&= (\beta_0 + \beta_1 X_t + \varepsilon_{t-1} + u_t) - (\beta_0 + \beta_1 X_{t-1} + \varepsilon_{t-1}) \qquad \text{by (25.21)}
\end{aligned}
$$

or:

(25.22)

$$Y_t - Y_{t-1} = \beta_1(X_t - X_{t-1}) + u_t$$

We shall use the following notation to represent the first differences in (25.22).

(25.23)

$$Y_t' = Y_t - Y_{t-1} \qquad X_t' = X_t - X_{t-1}$$

Then, we can express (25.22) as follows.

(25.24)

$$Y_t' = \beta_1 X_t' + u_t$$

where:

u_t are independent $N(0, \sigma^2)$

X_t' and Y_t' are given by (25.23)

Note that (25.24) is a simple linear regression model with the intercept $\beta_0 = 0$ and *independent* error terms u_t.

The result in (25.24) suggests that when the error terms are positively correlated and the autocorrelation parameter ρ is large, the ordinary regression procedures become approximately applicable when used in the following way:

1. *First differences* $Y_t - Y_{t-1}$ and $X_t - X_{t-1}$ are used for the dependent and independent variables, respectively.
2. *Regression through the origin* is employed, that is, regression with $\beta = 0$.

Note that the parameter β_1 in the first-differences model (25.24) is the same slope β_1 as in the original regression model (25.20). Thus, we are still estimating the same slope parameter as in our original regression model.

Regression Through Origin.   Procedures for regression through the origin follow those already discussed for the standard regression model. The key formulas for regression through the origin based on model (25.24) are given in Table 25.6. In

TABLE 25.6 Key formulas in simple linear regression through the origin using first-differences data: $Y_t' = \beta_1 X_t' + u_t$

(a) Least squares estimator of β_1

$$b_1 = \frac{\sum X_t' Y_t'}{\sum (X_t')^2}$$

(b) Error mean square

$$MSE = \frac{\sum (Y_t' - b_1 X_t')^2}{n_d - 1}$$

(c) Estimated variance of b_1

$$s^2\{b_1\} = \frac{MSE}{\sum (X_t')^2}$$

(d) Estimated variance of $\hat{Y}_h'$

$$s^2\{\hat{Y}_h'\} = \frac{(X_h')^2 MSE}{\sum (X_t')^2}$$

(e) Estimated variance for predicting new observation

$$s^2\{Y_{h(\text{new})}'\} = MSE\left[1 + \frac{(X_h')^2}{\sum (X_t')^2}\right]$$

where: $Y_t' = Y_t - Y_{t-1}$
  $X_t' = X_t - X_{t-1}$
  n_d is the number of first differences

Table 25.6, n_d denotes the number of first differences for the time series. Confidence intervals and tests are set up in the usual way by means of the t distribution with $n_d - 1$ degrees of freedom, the number of degrees of freedom associated with MSE in regression through the origin.

In contrast to the residuals of a regression model with an intercept term, the residuals of a regression model through the origin usually do not sum to zero when fitted by the method of least squares.

EXAMPLE

In the company sales example of Table 25.7, the analyst decided to use the method of first differences to overcome the difficulties associated with the presence of high positive autocorrelation in the error terms. She fitted the first-differences model (25.24). The first differences for the dependent and independent variables are shown (in part) in columns 3 and 4 of Table 25.7. Note that the first-differences transformation yields one fewer first-difference observation than the number of original observations (that is, $n_d = 19$, while $n = 20$). The necessary calculations with the first differences are shown in columns 5 and 6 of Table 25.7. The estimate of β_1, using the formula in Table 25.6a, is:

$$b_1 = \frac{\sum X_t' Y_t'}{\sum (X_t')^2} = \frac{667.250}{6612.948} = 0.10090$$

and the fitted first-differences model is:

$$\hat{Y}_t' = 0.10090 X_t'$$

TABLE 25.7   **First-differences calculations—Company sales example (sales in millions of dollars)**

	(1)	(2)	(3)	(4)	(5)	(6)
	Company	Industry	First Differences			
Quarter	Sales	Sales				
t	Y_t	X_t	Y_t'	X_t'	$X_t' Y_t'$	$(X_t')^2$
1	77.044	746.512	—	—	—	—
2	78.613	762.345	1.569	15.833	24.842	250.684
3	80.124	778.179	1.511	15.834	23.925	250.716
⋮	⋮	⋮	⋮	⋮	⋮	⋮
20	102.481	1006.882	1.745	17.592	30.698	309.478
				Total	667.250	6612.948

The 19 residuals for this model are plotted in time order in Figure 25.5b. The scatter now appears reasonably consistent with time-independent error terms, and the analyst was satisfied by the first-differences model in this regard. Note in Figure 25.5b that, as stated earlier, the least squares residuals for the regression model through the origin do not balance out around the zero line.

The analyst desired to forecast company sales for quarter $t = 21$. She obtained a reliable industry forecast for that quarter, $X_{21} = 1015.9$. Since industry sales in quarter 20 were $X_{20} = 1006.9$, the projected first difference is $X_{21}' = 1015.9 - 1006.9 = 9.0$. Hence, a point estimate of $E\{Y_{21}'\}$ is:

$$\hat{Y}_{21}' = 0.10090(9.0) = 0.91$$

This estimate indicates that deseasonalized company sales are expected to increase by \$0.91 million between quarters 20 and 21.

The final step in developing the forecast was to take the deseasonalized company sales in quarter 20, $Y_{20} = 102.5$, and adjust them by the expected increase:

Forecast for quarter $21 = 102.5 + 0.91 = 103.4$

The analyst thus predicted that deseasonalized company sales in quarter 21 will be \$103.4 million.   ☐

25.6   OPTIONAL TOPIC—AUTOREGRESSIVE AND MOVING AVERAGE (ARIMA) MODELS

A class of models useful for studying time series in which the observations exhibit various degrees and patterns of autocorrelation is known as the *autoregressive integrated moving average* class, or ARIMA class for short. This class of time series models combines features of autoregression, moving averages, and differencing already discussed.

ARIMA Models

ARIMA models are designed for stationary time series as defined in (25.1). Hence, time series that contain trend or other departures from stationarity must first have the nonstationary component removed. We will discuss later how this is accomplished. Once a time series has been made stationary, one of three different models in the ARIMA class is applied to the stationary time series.

Moving Average Model. The first model, called the moving average model, expresses the observations Y_t of the stationary time series as a function of random disturbances in the current period t and in earlier periods. We shall let μ denote the expected value of Y_t, and $u_t, u_{t-1}, \ldots, u_{t-q}$ denote independent random disturbances in periods $t, t - 1, \ldots, t - q$. Then, the moving average model for Y_t is as follows.

(25.25)

$$Y_t = \mu + \varepsilon_t$$

where:

$$\varepsilon_t = u_t - \theta_1 u_{t-1} - \cdots - \theta_q u_{t-q}$$

u_t are independent

Thus, the error term ε_t is expressed as a weighted moving average of the current disturbance u_t and the earlier disturbances $u_{t-1}, \ldots, u_{t-q}$, where the earlier disturbances have weights $\theta_1, \ldots, \theta_q$ (Greek theta). The number q is called the *order* of the moving average model, and $\theta_1, \ldots, \theta_q$ are called the *moving average parameters*. These parameters are preceded by negative signs in (25.25) by convention but may take on either positive or negative values.

Autoregressive Model. The second model, called the autoregressive model, expresses the observations Y_t of the stationary time series as a function of earlier observations $Y_{t-1}, \ldots, Y_{t-p}$.

(25.26)

$$Y_t = \phi_0 + \phi_1 Y_{t-1} + \cdots + \phi_p Y_{t-p} + \varepsilon_t$$

where:

ε_t are independent

This model is called an autoregressive time series model of *order p*. Here, $\phi_1, \ldots, \phi_p$ (Greek phi) are parameters that determine the dependence of Y_t on each of the p earlier values in the series and are called the *autoregressive parameters*. Note that model (25.26) has the form of a multiple regression model in which Y_t is regressed on $Y_{t-1}, \ldots, Y_{t-p}$ —hence, the term *autoregressive*. The independent variables in model (25.26) are the p lagged values of the stationary time series itself.

Autoregressive Moving Average Model. The third model is a combination of the moving average and autoregressive models and is used for complex business and economic time series where neither of the first two models is adequate. This mixed model is called the autoregressive moving average model.

(25.27)

$$Y_t = \phi_0 + \phi_1 Y_{t-1} + \cdots + \phi_p Y_{t-p} + u_t - \theta_1 u_{t-1} - \cdots - \theta_q u_{t-q}$$

where:

u_t are independent

Model (25.27) is said to be of *order* (p, q) because its autoregressive and moving average components are of order p and q, respectively.

Comment   The ARIMA approach to time series analysis and forecasting is often called the *Box-Jenkins approach* — named after the authors of the original key reference book on the subject.

Use of ARIMA Models

Applications of ARIMA models entail five analytical steps: (1) removing nonstationary components, (2) identifying the particular ARIMA model, (3) estimating the parameters, (4) analyzing the residuals, and (5) evaluating the model's forecasting performance. We consider each step briefly.

Removal of Nonstationary Components.   When a time series contains nonstationary components, such as trend or seasonal, it must be suitably transformed to make it stationary. Differencing is commonly employed for this purpose. For example, if Y_t, $t = 1, 2, \ldots, n$, is a time series with a linear trend component, then the series of first differences:

(25.28)

$$Y_t' = Y_t - Y_{t-1} \qquad t = 2, \ldots, n$$

will not contain a linear trend component. Similarly, if the time series has a stable additive quarterly seasonal component, then the series consisting of consecutive four-quarter differences:

(25.29)

$$Y_t' = Y_t - Y_{t-4} \qquad t = 5, 6, \ldots, n$$

will not contain a seasonal component. Other types of transformations also may help to make a time series stationary.

Model Identification. The selection of an appropriate ARIMA time series model is generally a complicated task. First, a decision must be made as to which of the three models is most appropriate. Then, suitable orders (p, q, or both) must be chosen for the model. Computer-assisted analytical techniques are available to aid with the model identification.

Parameter Estimation. Once a model of a given order has been selected, its parameters must be estimated from the available data for the stationary time series. The estimation method generally used is one akin to least squares, described earlier in connection with fitting regression functions. The estimation is usually accomplished by special computer routines.

Residual Analysis. As with regression models, it is advisable to examine the residuals associated with the fitted time series model as a check on its aptness. The various techniques and considerations described in Section 19.7 in connection with residual analysis for regression models apply to the time series setting with little modification. If residual analysis indicates that the ARIMA model does not fit well, the model needs to be modified.

Forecasting Performance. The acid test of the estimated model is an evaluation of how well it performs in forecasting. The ARIMA method of selecting and fitting time series models has been employed successfully in many cases. However, applications of the method tend to be complicated, even with the assistance of ARIMA computer packages. Also, difficulties have been encountered in some applications where other forecasting methods (such as exponential smoothing) proved to be superior.

PROBLEMS

*25.1 **Paper Mill.** The inventory manager of a paper mill uses exponential smoothing procedure (25.2) with smoothing constant $a = 0.25$ to smooth the weekly demand for a certain type of newsprint, which is a stationary time series. The smoothed estimate from last week is 41.3 metric tons.
   a. Give a forecast of demand for this week.
   b. Demand for this week is 46.5 metric tons. Give (1) the smoothed estimate for this week, (2) the forecast for next week, (3) the forecast for four weeks hence.

25.2 Data on the number of units of a spare part demanded in the past six months follow.

Month t:	1	2	3	4	5	6
Demand Y_t:	638	612	623	651	640	624

Assume that the demand series is stationary.
   a. Use the exponential smoothing procedure (25.2) with smoothing constant $a = 0.2$ and starting value $A_0 = 625$ to obtain the smoothed estimates for this series. What is estimated by the smoothed estimate for month 6?
   b. Use the results in **a** to forecast demand for months 7 and 8.
   c. If the series were expected to shift to a new mean level starting in month 7, would the forecasts in **b** continue to be appropriate? Comment.

25.3 Data on the number of bed sheets washed by a hospital laundry in the past five weeks follow.

Week t:	1	2	3	4	5
Sheets Y_t:	1863	1961	1940	1876	1890

Assume that the number of sheets washed is a stationary time series.
- **a.** Use the exponential smoothing procedure (25.2) with smoothing constant $a = 0.15$ and starting value $A_0 = 1900$ to obtain the smoothed estimates for this series.
- **b.** Use the results in **a** to forecast the number of sheets washed for weeks 6, 7, and 8.
- **c.** If the actual number of sheets washed in week 6 is 1926, what is the error in your forecast? Revise your forecasts for weeks 7 and 8 based on the actual outcome for week 6.

25.4 Refer to **Data Transmission** Problem 24.3. The use of exponential smoothing procedure (25.2) with smoothing constant $a = 0.10$ and starting value $A_0 = 15$ gives the following smoothed estimates for the series:

Hour t:	1	2	3	4	5	6	7	8
A_t:	14.0	15.0	17.1	15.6	14.1	13.0	13.6	17.8
Hour t:	9	10	11	12	13	14	15	16
A_t:	16.1	14.7	13.6	14.2	13.2	12.0	13.2	14.9
Hour t:	17	18	19	20	21	22	23	24
A_t:	13.4	12.2	11.7	10.8	12.4	___	___	___

Assume that the number of erroneous data bits transmitted each hour is a stationary time series.
- **a.** Obtain the smoothed estimates for hours 22, 23, and 24.
- **b.** Plot the original series and the smoothed estimates on the same graph. Does the exponential smoothing procedure appear to have effectively isolated the smooth component of the series? Comment.
- **c.** Forecast the number of erroneous data bits for hours (1) 25, (2) 30, (3) 25 to 48 in total.

25.5 Refer to **Paper Mill** Problem 25.1.
- **a.** In the calculation of the smoothed estimate for this week, what weight is given to the observation for last week? To the observation for three weeks ago? Are these results consistent with the fact that in exponential smoothing procedure (25.2), earlier observations receive less weight in determining the current smoothed estimate?
- **b.** What would be the weights in **a** if the smoothing constant had been $a = 0.10$? Does the smaller smoothing constant give more weight to earlier observations in the calculation of the current smoothed estimate? Explain.

25.6 Consider the following statement: "The best indication of sales of a fashion good next month is given by sales in the past few months. Sales in months further in the past are relatively unimportant." In a forecast of sales of this good one month ahead using exponential smoothing procedure (25.2), should the smoothing constant be close to 0 or to 1? Explain.

25.7 An inventory manager currently uses a procedure based on 6-month moving averages to forecast a multitude of stationary demand series for minor hardware items. Compare this procedure with the exponential smoothing procedure (25.2) with respect to the amount of data that must be carried over from one month to the next to support the forecasting system.

25.8 Consider the following series and smoothed estimates:

t:	1	2	3	4	5	6
Y_t:	200	200	300	300	300	300
A_t:	200	200	____	____	____	____

Obtain the missing smoothed estimates for periods 3 to 6 inclusive, using exponential smoothing procedure (25.2) with smoothing constant (1) 0.1, (2) 0.3. What is your conclusion about how the magnitude of the smoothing constant affects the responsiveness of the smoothed estimate to a permanent shift in the level of a series?

*25.9 Refer to Table 25.2 in the text pertaining to the lake level example. The weighting constants used in this example are $a = 0.10$ and $b = 0.20$.

 a. The lake level for year 31 is 4.84 (meters above 190). Obtain the smoothed and trend estimates for year 31.

 b. Use the estimates for year 31 in **a** to revise the forecasts of the lake level for years 32 to 35 inclusive.

25.10 Refer to **Pulp and Paper Production** Problem 24.1. The smoothed and trend estimates for this series follow. They were obtained by using exponential smoothing procedure (25.5) with starting values $A_0 = 3.50$ and $B_0 = 0.050$ and weighting constants $a = 0.10$ and $b = 0.10$.

Year t:	1967	1968	1969	1970	1971	1972	1973	1974	1975	1976
A_t:	3.53	3.59	3.66	3.69	3.66	3.64	3.66	3.68	3.66	3.67
B_t:	0.048	0.049	0.051	0.049	0.041	0.035	0.033	0.032	0.027	0.025

Year t:	1977	1978	1979	1980	1981	1982	1983	1984	1985	1986
A_t:	3.72	3.84	3.90	4.02	4.13	4.20	4.28	____	____	____
B_t:	0.028	0.037	0.039	0.048	0.054	0.055	0.058	____	____	____

 a. Obtain the smoothed and trend estimates for 1984, 1985, and 1986.

 b. Plot the original series and smoothed estimates on the same graph. Do the smoothed estimates track the original series well? Comment.

 c. Forecast production for 1987 and for 1988. What precisely do these forecasts represent?

25.11 Refer to **Consulting Contracts** Problem 24.9.

 a. Obtain smoothed and trend estimates for this series by means of exponential smoothing procedure (25.5). Use starting values $A_0 = 500$ and $B_0 = 100$ and weighting constants $a = 0.2$ and $b = 0.1$.

 b. Forecast the total value of contracts for year 8 and also for year 9.

25.12 The year-end prices (in dollars per share) of the common stock of Shipland Freighters, Inc., for the past six years are:

Year t:	1	2	3	4	5	6
Price Y_t:	22.50	24.13	25.25	24.87	27.50	30.13

 a. Obtain smoothed and trend estimates of this series by means of exponential smoothing procedure (25.5). Use starting values $A_0 = 20.00$ and $B_0 = 2.00$ and weighting constants $a = 0.2$ and $b = 0.2$.

 b. Forecast the year-end price of this stock for year 7 and also for year 8.

c. If the year-end stock price series had an exponential trend component rather than a linear trend component, would exponential smoothing procedures (25.5) and (25.6) still be appropriate for smoothing and forecasting the series? Explain.

25.13 Refer to **Consulting Contracts** Problems 24.9 and 25.11. What starting values for the smoothed and trend estimates are suggested by the fitted linear trend function obtained in Problem 24.9a if exponential smoothing procedure (25.5) will be applied to the series beginning with the observation for year 8?

25.14 A business researcher wishes to apply exponential smoothing procedure (25.5) to an annual business indicator series with trend. Data for the series are available for years 1 to 4 inclusive. The linear trend function fitted to these data is $T_t = 103.2 + 1.264X_t$, where $X_t = 1$ in year 1. What starting values are suggested by this trend line if exponential smoothing procedure (25.5) will be applied first to the observation for year 5?

*25.15 Refer to **Electricity Sales** Problem 24.30. Exponential smoothing procedure (25.7) has been applied to this quarterly sales revenues series, using weighting constants $a = 0.1$, $b = 0.1$, and $c = 0.2$. The values of the smoothed, trend, and seasonal index estimates and the original observations for several recent quarters follow.

Year:	1985				1986			
Quarter:	1	2	3	4	1	2	3	4
Y_t:	179	235	292	213	170	241	307	217
A_t:	235.6	236.2	235.3	234.6	234.1	235.1	___	___
B_t:	0.582	0.585	0.438	0.330	0.239	0.320	___	___
C_t:	0.753	0.994	1.300	0.939	0.747	1.000	___	___

a. Obtain the smoothed, trend, and seasonal index estimates for quarters 3 and 4 of 1986.
b. Calculate, at the end of 1986, quarterly forecasts for the company's sales revenues for 1987 and 1988.
c. At the end of the fourth quarter of 1986, what was the quarter-to-quarter increase in deseasonalized quarterly sales revenues over the preceding quarter? What was the specific seasonal relative corresponding to the fourth quarter of 1986?

25.16 Exponential smoothing procedure (25.7) has been applied to the quarterly number of travel bookings to Europe made by a travel agency. The weighting constants used in the procedure were $a = 0.1$, $b = 0.2$, and $c = 0.3$. The values of the smoothed, trend, and seasonal index estimates and the original observations for several recent quarters follow.

Year:	1985				1986			
Quarter:	1	2	3	4	1	2	3	4
Y_t:	381	408	344	353	370	391	335	347
A_t:	369.1	366.0	362.5	360.8	358.3	355.2	___	___
B_t:	−3.48	−3.39	−3.43	−3.08	−2.96	−3.00	___	___
C_t:	1.017	1.107	0.952	0.950	1.021	1.105	___	___

a. Obtain the smoothed, trend, and seasonal index estimates for quarters 3 and 4 of 1986.
b. Forecast, at the end of 1986, the number of bookings for each quarter of 1987. What is the forecast total number of bookings for 1987?
c. Does the seasonal pattern of this series appear to have been stable for years 1985 and 1986? Comment.

25.17 Refer to **Bus Traffic** Problem 24.2. Exponential smoothing procedure (25.7) has been used to smooth and project this series, with weighting constants $a = 0.1$, $b = 0.1$, and $c = 0.3$. At the end of Friday of week 4, the smoothed and trend estimates were 163.58 and 0.46, respectively. The current seasonal index estimates at that time were:

Day of Week:	S	S	M	T	W	T	F
Seasonal Index Estimate C:	0.90	0.25	1.26	1.13	1.15	1.08	1.28

  a. Obtain the smoothed, trend, and seasonal index estimates for Saturday and Sunday of week 4.

  b. Forecast, at the end of week 4, the daily traffic for week 5.

  c. According to the set of current seasonal index estimates at the end of Friday of week 4, which day of the week tends to have the most bus traffic? Does this set of current seasonal index estimates sum to 7? Must the set sum to 7? Comment.

25.18 An assistant asked a company's forecasting specialist: "Why does one not set the seasonal index adjustment constant to $c = 0$ when smoothing and forecasting a series that has a stable seasonal component?" Answer the assistant's question. Itemize the factors that should be taken into account in setting the value of the seasonal index adjustment constant c in exponential smoothing procedure (25.7).

25.19 An analyst has data for only one year for a monthly series that currently has a pronounced seasonal component but no significant trend-cyclical component. The data for the year follow.

Month:	J	F	M	A	M	J	J	A	S	O	N	D
Y_t:	44	38	40	45	55	68	81	79	65	60	55	49

The analyst requires starting values for exponential smoothing procedure (25.7) to apply when the observation for the coming January becomes available. Suggest starting values for the smoothed, trend, and seasonal index estimates.

25.20 Refer to Tables 24.5 and 24.6 in the text for the Taxair example. Exponential smoothing procedure (25.7) is to be applied to this series, commencing with the first quarter of 1987. Using the seasonal indexes in Table 24.6 and a linear trend function fitted to the deseasonalized observations of the series for the eight quarters of 1985 and 1986, suggest starting values for the smoothed, trend, and seasonal index estimates.

*25.21 Refer to **Electricity Sales** Problem 24.30. When regression model (25.11) is fitted to the quarterly sales revenues series (with $X_{t1} = 1$ in the first quarter of 1981), the estimated regression coefficients are $b_0 = 172.34$, $b_1 = 0.10536$, $b_2 = 57.06$, $b_3 = 131.29$, $b_4 = 43.85$.

  a. Obtain the fitted value and the residual for the third quarter of 1985.

  b. By how much do you estimate that expected revenues in any third quarter differ from expected revenues in the preceding first quarter because of differential seasonal effects? Because of linear trend?

  c. By how much do you estimate that expected revenues in any third quarter differ from expected revenues in the following fourth quarter because of differential seasonal effects? Because of linear trend?

  d. Obtain a point estimate of expected revenues in the second quarter of 1987.

  e. Give a point estimate of the change in expected revenues between the second quarter of one year and the second quarter of the following year. Is the expected year-to-year change the same for the other quarters? Explain.

*25.22 Refer to **Electricity Sales** Problems 24.30 and 25.21. Additional regression results are $s\{b_0\} = 4.322$, $s\{b_1\} = 0.2464$, $s\{b_2\} = 4.767$, $s\{b_3\} = 4.786$, $s\{b_4\} = 4.818$, $SSR = 53,761.2$, $SSE = 1291.7$.

   a. Construct a 90 percent prediction interval for revenues in the second quarter of 1987. A computer printout shows that $s\{Y_{h(new)}\} = 9.551$. Interpret the prediction interval.

   b. Test whether or not $\beta_4 = 0$, controlling the α risk at 0.05 when $\beta_4 = 0$. State the alternatives, the decision rule, the value of the test statistic, and the conclusion. What is the implication of your conclusion?

25.23 **Bergen Corporation.**   Quarterly sales (in millions of dollars) by the Bergen Corporation for years 1 to 6 follow.

			Year			
Quarter	1	2	3	4	5	6
1	38	37	37	40	41	43
2	34	36	36	38	45	42
3	40	40	38	45	46	50
4	52	55	55	53	54	56

When regression model (25.11) is fitted to this series (with $X_{t1} = 1$ in the first quarter of year 1), the estimated regression coefficients are $b_0 = 35.424$, $b_1 = 0.35536$, $b_2 = -1.189$, $b_3 = 3.123$, $b_4 = 13.767$.

   a. Obtain the fitted value and the residual for the second quarter of year 6.

   b. By how much do you estimate that expected sales in any fourth quarter differ from expected sales in the preceding first quarter because of differential seasonal effects? Because of linear trend?

   c. By how much do you estimate that expected sales in any third quarter differ from expected sales in the preceding second quarter because of differential seasonal effects? Because of linear trend?

   d. Obtain a point estimate of (1) expected sales in the first quarter of year 7, (2) total sales in year 7.

   e. Give the trend function for sales in the second quarter.

25.24 Refer to **Bergen Corporation** Problem 25.23. Additional regression results are $s\{b_0\} = 1.1475$, $s\{b_1\} = 0.065418$, $s\{b_2\} = 1.2657$, $s\{b_3\} = 1.2707$, $s\{b_4\} = 1.2791$, $SSR = 1076.890$, $SSE = 91.068$.

   a. Construct a 95 percent prediction interval for sales in the first quarter of year 7. A computer printout shows that $s\{Y_{h(new)}\} = 2.5359$.

   b. Test whether or not $\beta_3 = 0$, controlling the α risk at 0.05 when $\beta_3 = 0$. State the alternatives, the decision rule, the value of the test statistic, and the conclusion. What is the implication of your conclusion?

   c. Construct a 95 percent confidence interval for β_1. Interpret its meaning.

25.25 **Dishwashers.**   The following regression model was employed to forecast annual sales of dishwashers by a firm:

$$Y_t = \beta_0 + \beta_1 X_{t1} + \beta_2 X_{t-1,2} + \varepsilon_t$$

where:

ε_t are independent $N(0, \sigma^2)$

Here, Y is thousands of units sold, X_1 is a variable for linear trend, and X_2 is thousands of utility connections for new housing. Data for 1977–1986 follow.

Year:	1977	1978	1979	1980	1981	1982	1983	1984	1985	1986
t:	1	2	3	4	5	6	7	8	9	10
X_{t1}:	1	2	3	4	5	6	7	8	9	10
$X_{t-1,2}$:	22.3	28.1	19.6	24.7	26.9	27.3	19.9	20.4	26.7	20.7
Y_t:	6.06	7.55	6.40	7.78	8.53	9.06	8.01	8.42	10.14	9.37

The estimated regression coefficients are $b_0 = 1.38041$, $b_1 = 0.398955$, $b_2 = 0.192618$.
  a. Give a point estimate of the expected year-to-year increase in dishwasher sales when the number of utility connections for new housing is kept constant.
  b. Obtain the fitted value for 1984.
  c. In 1986, there were 24.3 thousand utility connections for new housing. Obtain a point estimate of expected sales of dishwashers by the firm in 1987.

25.26 Refer to **Dishwashers** Problem 25.25. Additional regression results are $s\{b_0\} = 0.11878$, $s\{b_1\} = 0.0052215$, $s\{b_2\} = 0.0046142$, $SSR = 14.4225$, $SSE = 0.015217$.
  a. Construct a 99 percent prediction interval for sales in 1987. A computer printout shows that $s\{Y_{h(new)}\} = 0.05706$. Does it seem likely that 1987 sales will be higher than sales in 1986?
  b. Construct a 99 percent confidence interval for β_1. Interpret your interval estimate.

25.27 **Advertising Impact.** An advertising executive is examining the relationship between monthly sales of a beverage product (Y_t, in thousands of cases) and monthly advertising expenditures for the product (X_t, in thousands of dollars). He conjectures that distributed-lag regression model (25.14) is appropriate. The available data follow.

Month t:	1	2	3	4	5	6	7	8	9	10
Y_t:	—	—	210	208	207	223	199	192	189	195
X_t:	66	74	78	75	80	65	66	57	60	59
Month t:	11	12	13	14	15	16	17	18	19	20
Y_t:	190	233	211	216	228	207	215	227	186	219
X_t:	99	62	80	88	59	77	83	52	91	60

The estimated regression coefficients when regression model (25.14) is fitted to these data are $b_0 = 102.810$, $b_1 = 1.09300$, $b_2 = 0.37348$.
  a. Obtain the fitted value and the residual for month 12.
  b. What is the estimated change in expected monthly sales with an additional \$1 thousand of advertising expenditures in each of the two preceding months?
  c. Obtain a point estimate of expected sales in month 21.

25.28 Refer to **Advertising Impact** Problem 25.27. Additional regression results are $s\{b_0\} = 12.0702$, $s\{b_1\} = 0.099356$, $s\{b_2\} = 0.105508$, $SSR = 3139.237$, $SSE = 389.041$.
  a. Test whether or not a regression relation exists, controlling the α risk at 0.01. State the alternatives, the decision rule, the value of the test statistic, and the conclusion.
  b. Construct a 99 percent confidence interval for β_1. Interpret its meaning.
  c. Obtain $\sqrt{MSE}$. Does the standard deviation of the error terms appear to be relatively large or small for this model? Explain.

*25.29 Refer to **Electricity Sales** Problems 24.30 and 25.21. The following residuals were obtained in the regression analysis:

t:	1	2	3	4	5	6	7	8
e_t:	−0.45	−2.61	6.05	5.39	−3.87	−12.03	4.63	−8.03
t:	9	10	11	12	13	14	15	16
e_t:	8.71	−12.46	8.21	6.54	−4.71	14.12	−6.21	3.12
t:	17	18	19	20	21	22	23	24
e_t:	4.87	3.70	−13.63	−5.30	−4.55	9.28	0.95	−1.72

a. Plot the residuals against time order. Do the errors appear to be serially correlated? Comment.

b. The Durbin-Watson test statistic for the residuals is $d = 2.674$. Test whether or not the errors are positively autocorrelated, controlling the α risk at 0.05 when $\rho = 0$. State the alternatives, the decision rule, and the conclusion.

25.30 Refer to **Bergen Corporation** Problem 25.23. The Durbin-Watson test statistic for the residuals of the fitted regression model is $d = 2.241$. Test whether or not the errors are positively autocorrelated, controlling the α risk at 0.01 when $\rho = 0$. State the alternatives, the decision rule, and the conclusion.

25.31 In a simple regression of annual sales of ovens on number of housing units for the past 20 years, the Durbin-Watson test statistic for the residuals of the fitted regression model is $d = 2.382$. Test whether or not the errors are positively autocorrelated, controlling the α risk at 0.05 when $\rho = 0$. State the alternatives, the decision rule, and the conclusion.

25.32 Refer to **Advertising Impact** Problem 25.27. The following residuals were obtained in the regression analysis:

t:	3	4	5	6	7	8	9	10	11
e_t:	1.66	−7.70	−6.92	4.74	−4.73	−7.22	−0.76	5.32	0.29
t:	12	13	14	15	16	17	18	19	20
e_t:	−0.05	3.45	2.59	−0.87	6.84	5.99	4.71	−4.64	−2.69

a. Plot the residuals against time order. Do the errors appear to be serially correlated? Comment.

b. The Durbin-Watson test statistic for the residuals is $d = 1.548$. Test whether or not the errors are positively autocorrelated, controlling the α risk at 0.05 when $\rho = 0$. State the alternatives, the decision rule, and the conclusion.

*25.33 **University Applicants.**   Data for the past 12 years on number of applicants accepted to a master's program at a state university (X_t), number of applicants who actually entered the program (Y_t), and residuals obtained in fitting regression model (19.1) to the observations follow.

t:	1	2	3	4	5	6
X_t:	330	342	348	355	365	373
Y_t:	140	151	154	161	167	171
e_t:	1.76	1.57	−1.02	−0.55	−3.87	−7.33

t:	7	8	9	10	11	12
X_t:	375	367	376	381	375	385
Y_t:	170	176	184	192	185	192
e_t:	-10.19	3.26	2.87	6.21	4.81	2.48

a. Plot the residuals against time order. Do the error terms appear to be positively autocorrelated? Explain.

b. Fit the first-differences model (25.24), and state the estimated regression function.

c. The number of applicants accepted for year 13 is 380. Predict, by means of a point estimate, the number of applicants entering the program in year 13.

*25.34 Refer to **University Applicants** Problem 25.33.

a. Calculate MSE for the fitted first-differences model. How many degrees of freedom are associated with MSE here?

b. Obtain a 90 percent confidence interval for β_1. Interpret your interval estimate.

25.35 **Empire Data Processing.** Data for the past 11 years on number of payroll-processing accounts serviced by Empire Data Processing (X_t), annual payroll-processing revenues in thousands of dollars received by Empire (Y_t), and residuals obtained in fitting regression model (19.1) to the observations follow.

t:	1	2	3	4	5	6
X_t:	20	25	30	40	35	45
Y_t:	402	526	603	803	756	972
e_t:	26.7	22.5	-28.8	-85.4	-4.1	-44.7

t:	7	8	9	10	11
X_t:	50	50	45	50	55
Y_t:	1172	1172	1072	1165	1258
e_t:	27.0	27.0	55.3	20.0	-15.3

a. Plot the residuals against time order to assess whether the errors are time-dependent. State your findings.

b. Fit the first-differences model (25.24), and state the estimated regression function.

c. Obtain the residuals for the first-differences model in **b**, and plot them in time order. Do the error terms of this model appear to be uncorrelated? Explain.

d. The firm has contracted to service 50 payroll-processing accounts in year 12. Predict, by means of a point estimate, the total payroll-processing revenues to be received in year 12.

25.36 Refer to **Empire Data Processing** Problem 25.35.

a. Calculate MSE for the first-differences model. How many degrees of freedom are associated with MSE here?

b. Obtain a 95 percent confidence interval for β_1. Interpret your interval estimate.

25.37 **Policy Loans.** Data for the past ten years from an insurance company on the value of total assets (X_t, in millions of dollars), the value of policy loans (Y_t, in millions of dollars), and residuals obtained in fitting regression model (19.1) to the observations follow.

t:	1	2	3	4	5
X_t:	241	268	300	339	384
Y_t:	12.1	12.9	13.5	14.4	16.5
e_t:	0.432	0.240	-0.336	-0.869	-0.423

t:	6	7	8	9	10
X_t:	439	489	541	602	669
Y_t:	19.5	21.4	22.9	24.7	27.2
e_t:	0.556	0.619	0.209	−0.233	−0.195

   a. Plot the residuals against time order. Do the errors appear to be positively autocorrelated? Explain.
   b. Fit the first-differences model (25.24), and state the estimated regression function. Interpret b_1.
   c. Total assets of the company in year 11 are anticipated to be $683 million. Predict, by means of a point estimate, the value of policy loans in year 11.

25.38 Refer to **Policy Loans** Problem 25.37.
   a. Calculate $\sqrt{MSE}$. What does this statistic estimate here?
   b. Construct a 99 percent confidence interval for β_1. Interpret your interval estimate.

EXERCISES

25.39 The smoothed estimate obtained when exponential smoothing procedure (25.2) is applied to a time series with linear trend will lag behind the series. Let $Y_t = b_0 + b_1t$ denote a series with linear trend but no irregular component. Use the expansion of A_t in (25.4) to show that the difference $Y_t - A_t$, when A_t is calculated from (25.2), approaches the value $b_1(1 - a)/a$ as t increases.

$$\left[Hint: \sum_{r=0}^{\infty} a(1 - a)^r = 1 \text{ and } \sum_{r=1}^{\infty} ar(1 - a)^r = (1 - a)/a. \right]$$

25.40 Refer to regression model (25.11).
   a. What is the implication if $\beta_1 = 0$?
   b. What is the implication if $\beta_4 = 0$?
   c. What is the implication if $\beta_2 = \beta_3 = \beta_4 = 0$?

25.41 An analyst stated: "Quarterly sales of this product are primarily affected by three factors — the size of the current target market, advertising expenditures, and quarter of the year. The current size of the market has an immediate impact. Advertising expenditures in a given quarter affect sales in the following two quarters. Summer-related factors, which tend to boost sales in the second and third quarters, are absent in the first and fourth quarters. There are still other factors that are not as important, but together, they tend to have a linear trend effect on sales."
   a. Construct a regression model that might be suitable here. What does each regression coefficient denote?
   b. Does your model contain any of the following: (1) a trend component, (2) a seasonal component, (3) a distributed time lag? If so, identify the relevant terms.
   c. Will all of the independent variables in your model have to be predicted in order to forecast sales? Explain.

25.42 The following residuals were obtained in an application of regression model (25.9) with $p - 1 = 3$ independent variables:

t:	1	2	3	4	5	6	7	8	9	10
e_t:	−5.58	−0.60	2.10	6.64	5.28	7.46	7.71	7.57	2.68	2.06

t:	11	12	13	14	15	16	17	18	19	20
e_t:	5.52	2.27	−4.69	−8.21	−5.31	−12.43	−11.17	−10.68	−9.44	−5.98
t:	21	22	23	24	25	26	27	28	29	30
e_t:	−3.55	2.07	5.21	4.18	−0.13	6.73	1.44	4.69	3.24	0.93

   a. Test the error terms for randomness, using the lower-tail test of runs up and down in Section 15.4 as an approximate test. Control the α risk at 0.05 when the error terms are generated by a random process. State the alternatives, the decision rule, the value of the test statistic, and the conclusion.

   b. The Durbin-Watson test statistic for these residuals is $d = 0.363$. Use this statistic to test whether or not the error terms are positively autocorrelated. Control the α risk at 0.05 when $\rho = 0$. State the alternatives, the decision rule, and the conclusion. Is your conclusion the same as the one in **a**? Comment.

25.43 Show that when one is applying the first-differences model (25.24) to a series with linear trend defined by:

$$Y_t = \beta_0 + \beta_1 X_t + \varepsilon_t$$

with $X_t = t$, then $b_1 = (Y_n - Y_1)/(n - 1)$, where n is the number of terms in the original series.

25.44 With the first-differences model (25.24), if X_t is the same in two consecutive time periods, will Y_t also remain unchanged? Explain.

25.45 Refer to moving average model (25.25). Explain why observations Y_t that are more than q periods apart in the series will be uncorrelated, while those that are q or fewer periods apart may be correlated.

25.46 Explain why the first-order autoregressive error model (25.16) is a special case of autoregressive model (25.26). Set up the appropriate correspondence of notation as part of your explanation.

25.47 A student has concluded that applying the four-quarter differences transformation in (25.29) to a time series behaving in accordance to regression model (25.11) with additive trend and seasonal components would yield a stationary series of four-quarter differences. Demonstrate that the student is correct.

STUDIES

25.48 (Computer needed.) Refer to **Marriages** Problem 15.33.

   a. Use exponential smoothing procedure (25.7) to obtain the smoothed, trend, and seasonal index estimates of the series for the period from July of year 3 to June of year 5, as follows:

     (1) Use weighting constants $a = 0.1$, $b = 0.1$, and $c = 0.3$.

     (2) Calculate the monthly seasonal indexes by the method of ratio to moving average for the first 24 months of data (July of year 1 to June of year 3). Use these seasonal indexes as starting values for the seasonal index estimates.

     (3) Deseasonalize the first 24 months of data by using the seasonal indexes just computed. Fit a linear trend function to the deseasonalized series of 24 observations (set $X_t = 1$ at July of year 1). Use the trend line to set the starting values for the smoothed and trend estimates.

   b. Use exponential smoothing procedure (25.8) to forecast the number of marriages for each of the six months from July to December of year 5.

25.49 (Computer needed.) Refer to **University Enrollment** Problem 24.31. Use a computer routine to fit regression model (25.9) to the enrollment data, employing a linear trend variable X_1 (with $X_{t1} = 1$ in the first trimester of 1979) and two indicator variables (X_2 and X_3) for differential seasonal effects for trimesters II and III, respectively.

   a. State the estimated regression function.
   b. Obtain point estimates of the expected enrollment in each trimester of 1987.
   c. Obtain the Durbin-Watson test statistic, and test whether or not the error terms of the regression model are positively autocorrelated. Control the α risk at 0.05 when $\rho = 0$. State the alternatives, the decision rule, the value of the test statistic, and the conclusion.
   d. Examine the aptness of regression model (25.9) by means of plots of the residuals against time order and against the fitted values. Discuss your findings.

25.50 Consider the first-order autoregressive error model $\varepsilon_t = \rho \varepsilon_{t-1} + u_t$ in (25.16). In Appendix Table C.9, the first 15 entries in column 8 constitute a random sequence of u_t observations from the $N(0, 1)$ distribution.

   a. Letting $\varepsilon_0 = 0$, compute ε_t, $t = 1, 2, \ldots, 15$, under the assumption that (1) $\rho = 0.8$, (2) $\rho = -0.8$, (3) $\rho = 0$.
   b. Plot the three sets of ε_t values obtained in **a** in time order in three separate graphs. Contrast the patterns of values in the graphs.

26 PRICE AND QUANTITY INDEXES

Price and quantity indexes are summary measures of relative price and quantity changes over time in a set of items. Thus, the Consumer Price Index summarizes relative price changes in goods and services purchased by urban households; the Producer Price Indexes measure relative changes in prices received in primary markets by producers of commodities in all stages of processing; and the Index of Industrial Production measures relative changes in output in manufacturing, mining, and utilities.

Price and quantity indexes not only are used as summary measures of price and quantity changes but also are employed in statistical analysis. Regression models, for instance, frequently contain price or quantity indexes as independent variables. Furthermore, price indexes are employed to "adjust" time series in dollar units so that the series can be studied without the effects of price changes. In this chapter, we discuss price and quantity indexes in their several roles. First, we take up the construction of price indexes and consider various uses of them. Then, we discuss quantity indexes. Finally, we briefly consider three major published indexes: Consumer Price Index, Producer Price Indexes, and Industrial Production Index.

26.1 PRICE RELATIVES AND LINK RELATIVES

First, we take up the measurement of price changes for a single item, such as gasoline, butter, or socks.

Price Relatives

Price relatives are useful for studying the pattern of relative changes in the price of an item over time.

(26.1)   A *price relatives series* expresses the unit price of an item in each period as a percent of the item's unit price in the base period.

EXAMPLE

Chemical Compound.   The prices of a chemical compound during the period 1983–1986 are shown in Table 26.1, column 1. The corresponding price relatives,

TABLE 26.1   Calculation of price relatives and link relatives series — Chemical compound example

	(1)	(2)	(3)	(4)
		Price Relatives		
	Unit			Link
Year	Price	1983 = 100	1986 = 100	Relative
1983	20	100	40	—
1984	22	110	44	110
1985	36	180	72	164
1986	50	250	100	139

calculated on the base period 1983, are shown in column 2. Thus, the price relative for 1984 is 100(22/20) = 110. This means that the price of the compound in 1984 was 110 percent as great as the price in 1983. The other price relatives are calculated and interpreted in a similar manner.   ☐

Any time period that facilitates the comparisons of interest may be used as the base period. The choice of 1983 as the base period in the chemical compound example facilitates study of the increase in the price of the compound during a period of transition.

Since the price relative for the base period is 100, the base period for a price relatives series is usually identified for reporting purposes in the manner shown in Table 26.1, for example, 1983 = 100.

Interpretation of Price Relatives.   Price relatives must be interpreted with care. We discuss briefly four important considerations:

1.   The absolute magnitudes of price relatives are affected by the choice of the base period, but their proportional magnitudes are not affected by this choice. We illustrate this by presenting in column 3 of Table 26.1 the price relatives calculated on the base period 1986. The price relative for 1983 now is 100(20/50) = 40, not 100 as for the 1983 base period series. However, the proportional magnitudes of the price relatives in columns 2 and 3 are the same. Thus, the price relatives for 1984 and 1983, with base period 1983, have the ratio 110/100 = 1.1. For the price relatives with the 1986 base period, this ratio is the same, that is, 44/40 = 1.1. Similarly, for both series of price relatives, the ratio of the 1985 price relative to the 1984 price relative is 1.64.

2.   A comparison of price relatives for two different series indicates nothing about the absolute prices unless we know the absolute magnitudes in the base period. Thus, information that the price relative for large grade A eggs in Montreal last month

was 120 and that for Toronto was 123 tells us nothing about how the egg prices in the two cities compared unless we know the prices in the base period.

3. In analyzing price relatives, we must distinguish between *percent points of change* and *percent change*. To illustrate the difference, refer to the price relatives with base period 1983 in column 2 of Table 26.1. Between 1985 and 1986, the percent points of change were $250 - 180 = 70$. On the other hand, the percent change was $100[(250 - 180)/180] = 100(70/180) = 38.9$. Thus, percent points of change refers to the absolute change in the price relatives and is dependent on the choice of the base period. Percent change, in contrast, refers to the relative change in the price relatives and, as we noted previously, does not vary with the choice of the base period.

4. A price relative below 100 indicates that the price in the given period is less than the price in the base period.

Link Relatives

Link relatives are useful in studying relative period-to-period price changes instead of changes from a fixed base period. For instance, link relatives enable us to study whether or not the price of an item is increasing at a constant rate, an increasing rate, or a decreasing rate.

(26.2)

> A *link relatives series* expresses the unit price of an item in each period as a percent of the item's unit price in the immediately preceding period.

EXAMPLE

The link relatives for the chemical compound example are given in column 4 of Table 26.1. The link relative for 1984 is $100(22/20) = 110$; thus, the price in 1984 was 110 percent as great as that in 1983. The link relative for 1985 is $100(36/22) = 164$; thus, the price in 1985 was 164 percent as great as that in 1984. Note that the annual percent increase declined at the end of the period (column 4), even though the amount of annual price increase did not decline during the period (column 1).

Comment

When the values in any time series (for example, annual sales, monthly production) are expressed relative to the value in the base period, the resulting ratios expressed as percentages are called *percent relatives*. Price relatives are therefore a special type of percent relative, computed from a price series.

Similarly, link relatives may be computed for time series other than price series.

26.2 PRICE INDEXES BY METHOD OF WEIGHTED AGGREGATES

Need for Price Indexes

In some applications, information on price changes for a single item in the form of price relatives is all that is required. For instance, during the mid 1970s, energy costs began rising rapidly and many rental agreements in office buildings and apartment houses had clauses allowing rents to increase automatically as fuel costs increased. A typical clause would specify that the annual rent will increase by a given dollar amount for each one percent point increase in the price relative for the relevant fuel.

In many other applications, however, knowledge of individual price relatives is insufficient. In New York City, for instance, many apartment buildings have been "rent stabilized." This involves controlling rents by law at levels prevailing in a base period. Rent increases are allowed only to the degree that operating costs have increased because of price increases. In this situation, the prices of many items, such as electricity, water, and gas, are involved. Price relatives for these individual items will not provide suitable information about the aggregate effect of the individual price changes on the cost of operating an apartment building.

Price indexes are summary measures that combine the price changes for a group of items, using weights to give each item its appropriate importance. The Consumer Price Index is such an index, measuring the combined effect of price changes in many goods and services purchased by urban households. It has been estimated that about half of the population of the United States is affected directly by statutes and agreements tied to movements of the Consumer Price Index.

Two basic methods of calculating price indexes are widely used, the method of weighted aggregates and the method of weighted average of relatives. We shall explain each of these in turn, using the following illustration.

EXAMPLE

School Maintenance Supplies.   Officials of a large school district needed to develop a price index for maintenance supplies for the school buildings in the district. This was part of an effort to develop a variety of indexes, including ones for salaries and wages, instructional materials, and capital items. The set of indexes has been useful in budgeting and cost analysis and also in negotiations with legislators and state officials for state aid.

In compiling the price index for school maintenance supplies, the officials could not include all supply items in the index because a large number of items are used. Instead, a sample of supply items was selected to be representative of all items used. The items selected are shown in Table 26.2, column 1. Also shown in this table, in columns 3, 4, and 5, are the unit prices of these items for 1983, 1984, and 1985, respectively. Finally, column 2 of Table 26.2 presents typical quantities consumed for each of the supply items.

TABLE 26.2 Calculation of a price index series by the method of weighted aggregates — School maintenance supplies example

	(1)	(2)	(3)	(4)	(5)	(6)	(7)	(8)
				Unit Prices				
	Schedule of	Quantity	1983	1984	1985			
i	Items	Q_{ia}	P_{i0}	P_{i1}	P_{i2}	$P_{i0}Q_{ia}$	$P_{i1}Q_{ia}$	$P_{i2}Q_{ia}$
1	Window glass	15 sheets	14.70	15.10	15.75	220.50	226.50	236.25
2	Fluorescent tube	130 boxes	28.60	30.01	32.40	3,718.00	3,901.30	4,212.00
3	Floor detergent	290 cans	20.00	22.00	24.20	5,800.00	6,380.00	7,018.00
4	Floor finish	100 cans	45.00	50.00	52.50	4,500.00	5,000.00	5,250.00
5	Latex interior paint	175 cans	9.99	11.57	12.10	1,748.25	2,024.75	2,117.50
6	Mop head	200 pieces	2.84	2.98	3.08	568.00	596.00	616.00
					Total	16,554.75	18,128.55	19,449.75

$$I_{83} = 100\left(\frac{16,554.75}{16,554.75}\right) = 100.0$$

$$I_{84} = 100\left(\frac{18,128.55}{16,554.75}\right) = 109.5$$

$$I_{85} = 100\left(\frac{19,449.75}{16,554.75}\right) = 117.5$$

Method of Weighted Aggregates

The method of weighted aggregates for compiling a price index is conceptually very simple. We shall first explain the notation and terminology to be used.

The *schedule of items* is the list of items included in the price index. Often, the schedule of items consists of a sample of all items of interest, as in the school maintenance supplies example. Sometimes, it is possible to include all items of interest in the schedule.

The price of the ith item in the schedule in any *given period* t is denoted by P_{it}, and the price of this item in the *base period* by P_{i0}. The typical quantity of the ith item that is consumed is denoted by Q_{ia}, where the subscript a stands for an average or typical period. These quantities are used as weights reflecting the importance of the items.

The method of weighted aggregates simply compares the cost of the typical quantities at period t prices with the cost of the same quantities at base period prices. The cost at period t prices is:

$$\text{Cost at period } t \text{ prices} = \sum_i P_{it}Q_{ia}$$

where the summation is over all the items in the schedule. The cost at base period prices is:

$$\text{Cost at period 0 prices} = \sum_i P_{i0} Q_{ia}$$

The price index for period t by the method of weighted aggregates is simply the ratio of these two costs, expressed as a percent.

(26.3)

> The price index for period t, denoted by I_t, by the *method of weighted aggre-gates* is:
>
> $$I_t = 100 \frac{\sum_i P_{it} Q_{ia}}{\sum_i P_{i0} Q_{ia}}$$
>
> *where:*
>
> P_{i0} is the unit price of the ith item in period 0 (base period)
> P_{it} is the unit price of the ith item in period t (given period)
> Q_{ia} is the quantity weight assigned to the ith item

The price index calculated by the method of weighted aggregates has a simple interpretation. It shows how much the typical quantities of the items in the schedule cost at period t prices relative to their cost at base period prices.

EXAMPLE

The price index series for the school maintenance supplies example for 1983, 1984, and 1985 is calculated by the method of weighted aggregates in Table 26.2. The costs of the typical quantities at the prices of each year are shown in columns 6, 7, and 8. For example, the cost of 15 sheets of window glass at 1983 prices is 15(14.70) = 220.50, while at 1984 prices, it is 15(15.10) = 226.50. The aggregate costs of the schedule of items at the prices of the three periods are shown at the bottom of columns 6, 7, and 8, respectively, and the index numbers are calculated at the bottom of the table.

Since the same items and quantities are priced in each period, the relative differences in the index numbers are attributable wholly to changes in the prices of the items. Thus, the price index $I_{84} = 109.5$ indicates that the prices of school maintenance supplies increased, in terms of their aggregate effect, by 9.5 percent between 1983 and 1984. Since the index for 1985 ($I_{85} = 117.5$) is above the index for 1984 ($I_{84} = 109.5$), the prices increased between 1984 and 1985 also. We can calculate the percent change by expressing the percent point change, 117.5 − 109.5 = 8.0, as a percent of the 1984 index and obtain 100(8.0/109.5) = 7.3 percent as the relative price increase between 1984 and 1985. □

Comments

1. Just as with price relatives, the proportional magnitudes of the index numbers in Table 26.2 are not affected by which period is chosen as the base period.

2. When the typical quantity weights Q_{ia} are the quantities consumed in the base period, the weighted aggregates price index is called a *Laspeyres* price index.

(26.4)
$$\text{Laspeyres price index} = 100\,\frac{\sum_i P_{it} Q_{i0}}{\sum_i P_{i0} Q_{i0}}$$

3. When the quantity weights are the quantities consumed in each given period t, the weighted aggregates price index is called a *Paasche* price index.

(26.5)
$$\text{Paasche price index} = 100\,\frac{\sum_i P_{it} Q_{it}}{\sum_i P_{i0} Q_{it}}$$

With a Paasche index series, all comparisons must be made with the base period only. For all other comparisons, the quantity weights are not constant, and differences in index numbers therefore reflect both price and quantity changes. For example, if a Paasche price index series were to be calculated for the school maintenance supplies example with base period 1983, the 1984 price index would utilize quantities consumed in 1984 while the 1985 index would utilize quantities consumed in 1985. Since the quantities consumed in different years generally are not the same, a comparison of the 1984 and 1985 Paasche indexes on a 1983 base would reflect both price and quantity changes. The Paasche price index is rarely used in practice.

4. The period from which the quantity weights are derived for the weighted aggregates price index is called the *weight period*. The base period for the price index and the weight period need not coincide, as we have seen.

Maintenance of Price Index Series

A key issue in the compilation of price index series involves the handling of items to be included in the index and the weights to be assigned to the items. Should the schedule of items and the weights be changed frequently so that they are always up to date, or should they be held fixed over a relatively long sequence of years? If frequent changes are made, changes in the price index series over some years will reflect both changes in prices and changes in the composition of the index. Another disadvantage of frequent revision is that frequent updating of the schedule of items and of the

weights may be very expensive. On the other hand, an index can become badly out-
dated if the schedule of items and the weights are held fixed over too long a period.

The procedure generally followed in practice is to hold the schedule of items and
the weights essentially fixed for five to ten years and then to revise them. The schedule
and weights need not be held exactly fixed, since procedures are available for making
interim adjustments without disturbing the index. For instance, a new item can be substituted for an item that has been taken off the market, and seasonal foods can be
included in the schedule in season.

In addition to periodic updating of the schedule of items and weights, revisions
often also include upgrading of data collection and tabulation procedures and modern-
izing of definitions. A new series of index numbers is then initiated with each revision.
The new series can be "spliced" into the preceding series, if desired, by procedures to
be illustrated shortly.

26.3   PRICE INDEXES BY METHOD OF WEIGHTED AVERAGE OF RELATIVES

A second method of compiling a price index series that is used in practice is the
method of weighted average of relatives. This method is also conceptually quite simple. It utilizes the price relatives for each item in the schedule. The index for period t
is a weighted average of the price relatives for that period for all items in the schedule.

Quantity weights, used for the method of weighted aggregates, cannot be used
here. Recall that a price relative is the ratio of the price of an item in a given period to
its price in the base period, expressed as a percent. In the ratio, the units cancel and
the relative is therefore unit-free. If quantity weights were to be used in the school
maintenance supplies example and multiplied by unit-free relatives, we would obtain a
mixture of units, such as sheets of window glass, boxes of fluorescent tubes, and cans
of floor detergent. Clearly, it would not be appropriate to add these terms.

Usually, *value weights* are used instead. Value weights are dollar values that
reflect the importance of each item in the schedule. For example, the value weights
may be obtained from the typical quantities consumed, Q_{ia}, by multiplying these quantities by typical prices. We shall denote the value weight for the ith item by V_{ia}.

Thus, the method of weighted average of relatives simply calls for a weighted
average of the price relatives $100(P_{it}/P_{i0})$ for the items in the schedule, using value
weights V_{ia}.

(26.6)

The price index for period t, denoted by I_t, by the *method of weighted average of relatives* is:

$$I_t = \frac{\sum_i \left(100 \frac{P_{it}}{P_{i0}}\right) V_{ia}}{\sum_i V_{ia}}$$

> *where:*
>
> P_{i0} is the unit price of the ith item in period 0 (base period)
>
> P_{it} is the unit price of the ith item in period t (given period)
>
> V_{ia} is the value weight assigned to the ith item

EXAMPLE

For the school maintenance supplies example, we shall calculate a price index series by the method of weighted average of relatives with base period 1983. The value weights will be based on the typical quantities and the unit prices in 1983 from Table 26.2, that is:

$$V_{ia} = P_{i0} Q_{ia}$$

Table 26.3 contains the necessary calculations. In column 1 is the schedule of items, and in column 2 are the value weights as obtained from Table 26.2. For example, the value weight for window glass is $V_{1a} = P_{10} Q_{1a} = 14.70(15) = 220.50$. The sum of the value weights is $\Sigma V_{ia} = 16,554.75$.

Columns 3, 4, and 5 contain the price relatives for the three years, based on the prices in Table 26.2. Note that all price relatives are expressed on a 1983 base because the price index is to have 1983 as the base period. For example, the 1984 price relative for window glass is $100(15.10/14.70) = 102.72$.

For each year, we take a weighted average of the price relatives in the appropriate column. The weighting of the price relatives is done in columns 6, 7, and 8, and the indexes are obtained at the bottom of the table. The index of 109.5 for 1984 indicates that prices of school maintenance supplies increased, in terms of their aggregate effect, by 9.5 percent between 1983 and 1984. □

Comments

1. The price indexes in Tables 26.2 and 26.3 are identical. This happened because we used base year prices P_{i0} in obtaining the value weights for the price index series by the method of weighted average of relatives. In that case, the method of weighted average of relatives index (26.6) reduces to the method of weighted aggregates index (26.3), that is:

$$\frac{\sum_i \left(100 \frac{P_{it}}{P_{i0}}\right) V_{ia}}{\sum_i V_{ia}} = \frac{\sum_i \left(100 \frac{P_{it}}{P_{i0}}\right) P_{i0} Q_{ia}}{\sum_i P_{i0} Q_{ia}} = 100 \frac{\sum_i P_{it} Q_{ia}}{\sum_i P_{i0} Q_{ia}}$$

If prices other than base period prices are used for obtaining the value weights V_{ia}, the two methods will not lead to identical results. Generally, the differences in the indexes by the two methods will not be great.

2. Our earlier discussion about the need for periodic updating of the schedule of items and weights applies also to price index series compiled by the method of weighted average of relatives.

TABLE 26.3   Calculation of a price index series by the method of weighted average of relatives—School maintenance supplies example

(1)	(2)	(3)	(4)	(5)	(6)	(7)	(8)
		Price Relatives, 1983 = 100			Price Relative × Value Weight		
		1983	1984	1985	1983	1984	1985
Schedule of Items i	Value Weight $V_{ia} = P_{i0}Q_{ia}$	$100\left(\dfrac{P_{i0}}{P_{i0}}\right)$	$100\left(\dfrac{P_{i1}}{P_{i0}}\right)$	$100\left(\dfrac{P_{i2}}{P_{i0}}\right)$	$\left(100\dfrac{P_{i0}}{P_{i0}}\right)V_{ia}$	$\left(100\dfrac{P_{i1}}{P_{i0}}\right)V_{ia}$	$\left(100\dfrac{P_{i2}}{P_{i0}}\right)V_{ia}$
1 Window glass	220.50	100.00	102.72	107.14	22,050	22,650	23,625
2 Fluorescent tube	3,718.00	100.00	104.93	113.29	371,800	390,130	421,200
3 Floor detergent	5,800.00	100.00	110.00	121.00	580,000	638,000	701,800
4 Floor finish	4,500.00	100.00	111.11	116.67	450,000	500,000	525,000
5 Latex interior paint	1,748.25	100.00	115.82	121.12	174,825	202,475	211,750
6 Mop head	568.00	100.00	104.93	108.45	56,800	59,600	61,600
Total	16,554.75				1,655,475	1,812,855	1,944,975

$$I_{83} = \frac{1,655,475}{16,554.75} = 100.0$$

$$I_{84} = \frac{1,812,855}{16,554.75} = 109.5$$

$$I_{85} = \frac{1,944,975}{16,554.75} = 117.5$$

BASE PERIOD FOR INDEX SERIES

We now take up a number of considerations concerned with the base period of an index series.

Choice of Base Period

Any period within the coverage of the index series can be used as the base period. Of course, when a special-purpose index is designed to measure price changes occurring since a particular period, such as since the lifting of price controls, that period would be taken as the base period. Whenever possible, the base period should involve relatively normal or standard conditions, because many index users assume that the base period represents such conditions. Sometimes, an interval of two or more years is selected as the base period. In this case, the index numbers are calculated so that the indexes for the base period years average to 100. Examples of base periods used for indexes compiled by statistical agencies of the federal governments in the United States and Canada include 1967, 1971, and 1977.

Shifting the Base of an Index Series

Sometimes, it is necessary to shift the base period of an index series that is already compiled — as when two index series on different base periods need to be placed on a common base period to facilitate comparison. Usually, it is not possible to recalculate a published index on the new base period because needed data are not available. However, a shortcut method can be employed that does not entail recalculation of the index.

Consider the following price index series with 1984 as the base period:

Year:	1981	1982	1983	1984	1985	1986
Price Index:	75	88	92	100	110	122

Suppose the base period is to be shifted to 1981. We simply divide each index number by 0.75, the index number for the new base period in decimal form. The index number for 1981 becomes $75/0.75 = 100.0$ and that for 1982 becomes $88/0.75 = 117.3$. The new price index series with base period 1981 is:

Year:	1981	1982	1983	1984	1985	1986
Price Index:	100.0	117.3	122.7	133.3	146.7	162.7

Because the proportional magnitudes of the index numbers are not affected by this shift in the base period, the shifted series conveys the same information about year-to-year relative price changes as the original series.

When the index series is calculated by the method of weighted aggregates employing fixed quantity weights, the shortcut procedure yields results identical to those obtained by recalculating the index series on the new base period. With most other formulas, however, the shortcut procedure only approximates such results.

Splicing

We noted earlier that when we compile index numbers on an ongoing basis, the usual procedure is to revise the index periodically and to initiate a new series with each revision. The new series can be joined or spliced to the older series to yield a single continuous series by a procedure similar to that just described for shifting the base period of an index series. Suppose the following two price index series are to be joined:

Year:	1978	1979	1980	1981	1982	1983	1984	1985	1986
Old Series:	100	103	106	113	130				
Revised Series:					100	120	126	131	134

We simply shift the level of one of the series so that both have a common value for 1982. For instance, to obtain a combined series with 1978 as the base period, we multiply every index in the revised series by 1.30. The spliced series is:

Year:	1978	1979	1980	1981	1982	1983	1984	1985	1986
Spliced Series:	100	103	106	113	130	156	164	170	174

26.5 COLLECTION OF PRICE AND QUANTITY DATA

We now take up some important problems pertaining to the collection of price and quantity data for price indexes.

Collection Methods

All of the data collection methods discussed in Chapter 1 are used in collecting data for price indexes. Price and quantity data are obtained by observation, interview, and self-enumeration. Often sampling is used, as in the Consumer Price Index where probability samples of retail outlets and of items carried by the selected outlets are employed to obtain price data. Sampling is also used for the Consumer Price Index in periodic consumer expenditures surveys to obtain quantity data for weights.

Specification Pricing

In collecting price data for an item over a period of years, one must hold quality and other price-determining characteristics as constant as possible. This typically requires that detailed specifications be developed and adhered to in pricing the item. Here is an example of such a specification:

> Interior latex paint. Professional- or commercial-grade latex interior house paint, matte or flat finish, off white, first-line or quality.

Note that color, finish, and quality are specified because each of these characteristics can affect the price.

Quality Changes

A different problem in collecting and using price data in ongoing index series is the adjustment for quality changes in the items covered by the index. In principle, a price change that is caused by a quality change should not be reflected in the price index. In practice, minor quality changes are ignored. On the other hand, attempts are made in better price indexes to adjust for important quality changes. However, the conceptual and measurement problems are difficult. If the price of a car increases by $200 because of improved emission control equipment, should this increase be treated as a price increase (and reflected in the index) or as a price adjustment necessitated by a quality improvement (and not reflected in the index)? If an oven door is modified by the manufacturer to exclude the glass window while the list price remains unchanged, should this be treated as a price increase? If so, what should be the magnitude of the imputed increase? Extensive research has been undertaken on how these problems connected with quality changes should be handled.

Weights

In some indexes, the schedule includes all or almost all of the items used in the activity. Thus, in a price index for highway construction, where a relatively small number of major items account for most of the total cost, it suffices for practical purposes to limit the schedule to these major items. Here, each weight reflects the importance of the particular item.

Often, however, the schedule must be limited to a sample of items, since thousands of items may be involved in the activity and it is neither feasible nor necessary to include them all. Each item is then selected for the schedule to represent a class of related items. In such cases, the weight normally reflects the importance of the entire class and not just of the item itself. For instance, a particular type of insulation is selected for a price index for residential building construction to represent all types of insulation and vapor barriers. The weight for the particular type of insulation in the schedule represents the importance of the entire class of insulation and vapor barriers.

26.6 USES OF PRICE INDEXES

We noted earlier that price indexes are summary measures of price changes. Indexes such as the Consumer Price Index and the Producer Price Indexes are widely followed as indicators of price changes. In addition, price indexes play important roles in a variety of applications. We now consider several of these.

Measuring Real Earnings

As prices change, the quantities of goods and services that can be purchased by a fixed sum of money change. In economic analysis, it is frequently important to measure changes in *real earnings* (that is, quantities of goods and services that can be purchased). Price indexes are a basic tool in making this measurement.

EXAMPLE

Table 26.4 shows, in column 1, average weekly earnings of private nonagricultural workers in the United States during the period 1979–1983. These actual earnings data are expressed in *current dollars*. For example, 1979 earnings are expressed in terms of the buying power of a wage earner's dollar in 1979, and earnings in 1980 are expressed in terms of the buying power of the dollar in 1980. Column 1 of Table 26.4 shows that the earnings in current dollars increased steadily during the period.

The extent to which prices of goods and services purchased for daily living by workers and their dependents changed during the period is shown in column 2 by the Consumer Price Index for Urban Wage Earners and Clerical Workers. This index is expressed on a 1979 base period in the table. Note that prices in 1980 stood at 113.5 percent of their 1979 level. Thus, families, on the average, had to spend $1.135 in 1980 for every $1 in 1979 in purchasing goods and services for daily living. Consequently, the average money earnings of $235.10 in 1980, expended at the prices of that year, were equivalent in purchasing power to 235.10/1.135 = $207.14 at 1979 prices.

The average money earnings in the other years at 1979 prices are obtained by the same procedure: Each current earnings figure is divided by the decimal value of the price index for that year. Thus, the money earnings of $255.20 in 1981 are equivalent to earnings of 255.20/1.251 = $204.00 at 1979 prices. Column 3 of Table 26.4 contains the average weekly earnings data at 1979 prices. These data are said to be in *constant dollars* or *1979 dollars* since they are expressed in terms of the purchasing power of the dollar in 1979.

Relative changes in constant-dollars earnings data indicate the relative changes in purchasing power associated with the money earnings, that is, the relative changes in *real earnings*. To show these relative changes explicitly, we have expressed the constant-dollars data as link relatives in column 4 of Table 26.4 and as

TABLE 26.4   Weekly earnings at current and 1979 prices

	(1)	(2)	(3)	(4)	(5)
				Relatives for Column 3	
		Consumer	Weekly Earnings		Percent
	Weekly	Price Index	at 1979	Link	Relative
Year	Earnings	(1979 = 100)	Prices	Relative	(1979 = 100)
1979	219.91	100.0	219.91	—	100.0
1980	235.10	113.5	207.14	94.2	94.2
1981	255.20	125.1	204.00	98.5	92.8
1982	266.92	132.6	201.30	98.7	91.5
1983	280.35	136.6	205.23	102.0	93.3

SOURCE: Basic data from *Monthly Labor Review*.

percent relatives in column 5 (1979 = 100). The link relatives show that real earnings declined between 1979 and 1982 and then increased between 1982 and 1983. The percent relatives show that real earnings in 1983 were only at 93.3 percent of their 1979 level.   □

Comment	We noted earlier that proportional magnitudes in an index series are not affected by the choice of the base period but that the absolute magnitudes are affected. The same holds for constant-dollars series. Thus, in the weekly earnings example of Table 26.4, if the Consumer Price Index had been expressed on a 1983 base, the constant-dollars earnings would have differed from column 3 of Table 26.4, but the percent relatives and link relatives based on these 1983 constant-dollars data would have remained exactly the same.

Measuring Quantity Changes

Many important business and economic series on volume of activity are expressed in dollars, such as total retail sales and GNP. Changes in dollar volume reflect quantity changes, price changes, or both. Often, there is interest in the quantity changes alone. For example, if retail sales this year are five percent higher than last year's sales, is this due to price increases only or did the physical volume of goods sold increase? We can measure relative changes in quantities by use of price indexes. The procedure is similar to that for expressing current-dollars earnings as constant-dollars earnings, as we shall now illustrate.

EXAMPLE	Table 26.5, column 1, shows for the period 1983–1986 annual dollar sales by a manufacturer of household appliances. A price index for the selling prices of the appliances sold by the manufacturer is shown in column 2 of Table 26.5. The base

TABLE 26.5   **Sales by a manufacturer of household appliances at current and 1983 prices**

	(1)	(2)	(3)	(4)	(5)
				Relatives for Column 3	
			Sales at		Percent
	Annual	Product	1983		Relative
	Sales	Price Index	Prices	Link	(1983 = 100)
Year	($000)	(1983 = 100)	($000)	Relative	
1983	38,500	100	38,500	—	100.0
1984	43,538	103	42,270	109.8	109.8
1985	49,050	105	46,714	110.5	121.3
1986	54,950	107	51,355	109.9	133.4

year of the index is 1983. In column 3, sales are expressed at constant 1983 prices. For instance, 1984 sales of $43,538 thousand are equivalent to sales of 43,538/1.03 = $42,270 thousand at 1983 prices. Columns 4 and 5, respectively, present link relatives and percent relatives for the constant-dollars sales in column 3. These relatives reflect changes in the quantity of appliances sold, since prices are held constant. The link relatives show that the quantity of appliances sold increased by about 10 percent per year during the period. The percent relatives show, in turn, that by 1986, the quantity sold was about 33 percent higher than in 1983. □

Comment   When we convert current-dollars series into constant dollars, the price index must be relevant to the series. Thus, we would not use a food price index to adjust the series on appliances sales, or a Consumer Price Index to adjust the sales of a steel manufacturer.

Escalator Clauses in Contracts

Escalator clauses keyed to price indexes are widely employed. They are used in long-term contracts for the production of manufactured goods and leasing of office space, where sellers' costs can go up or down markedly because of price changes beyond their control. The most familiar use is for cost-of-living adjustments in labor contracts. In a typical escalator clause, wage rates will increase automatically by some fraction of a percent (for example, two-thirds of a percent) for each one percent increase in an index such as the Consumer Price Index for Urban Wage Earners and Clerical Workers. The clause may or may not provide for wage rate decreases if the index declines.

Escalator clauses are also used to make automatic adjustments in pension and retirement benefits, funding for school programs, poverty threshold specifications, construction contracts, and alimony and child support agreements.

Price Comparisons Between Locations

Our discussion up to this point has involved comparisons of prices over time. Price indexes also are used for comparisons between different locations. For example, the U.S. Department of State maintains indexes of living costs abroad that are used in setting up allowances for personnel serving at overseas posts. An index for an overseas post measures the cost of representative goods and services required by a family for a U.S. "pattern of living" at that post, relative to the cost of goods and services for a comparable pattern of living in Washington, D.C.

The indexes are calculated quarterly. For instance, in March 1984 the index value for Paris was 102. This indicates that the cost of the representative goods and services in Paris in March 1984 for U.S. government employees was about 2 percent greater than the cost of comparable goods and services in Washington, D.C.

QUANTITY INDEXES

A quantity index is a summary measure of relative changes over time in quantities of a set of items — for instance, of the relative changes since 1980 in the quantities of automobiles, trucks, and other vehicles imported by a country. In such index series, the quantities can change from one period to another, but the prices or other weights remain fixed. The index formulas are analogous to those for price indexes.

(26.7)

The quantity index for period t is denoted by I_t. The quantity index by the *method of weighted aggregates* is:

(26.7a)
$$I_t = 100 \frac{\sum_i Q_{it} P_{ia}}{\sum_i Q_{i0} P_{ia}}$$

The quantity index by the *method of weighted average of relatives* is:

(26.7b)
$$I_t = \frac{\sum_i \left(100 \frac{Q_{it}}{Q_{i0}} \right) V_{ia}}{\sum_i V_{ia}}$$

Comment

A special problem in constructing quantity indexes arises in the choice of weights. The usual price or value weights may not be appropriate here. Consider a company that wishes to measure quantity changes in the outputs of different types of electronic circuits assembled from purchased components. Here, price weights or value weights for the different types of circuits would reflect mainly the prices or values of the purchased components and not the quantities of output of the company per se. Weights derived from the typical number of man-hours expended in assembling a unit of each type of circuit or from the value added to a unit of each type of circuit in the assembly would be more appropriate. Value added would be the value of the circuit when shipped from the plant less the cost of the purchased components and other purchased goods and services (for example, fuel, electricity, containers) used in assembling the circuit and packaging it for shipment. The Index of Industrial Production utilizes value-added weights in many of its segments.

 THREE MAJOR INDEXES

In this chapter, we have cited the Consumer Price Index, the Producer Price Indexes, and the Index of Industrial Production. These indexes are major indicators of price and quantity changes in the U.S. economy. Similar indexes are compiled in Canada and other countries for their economies. We now briefly describe the main features of the three U.S. indexes.

Consumer Price Index

Indexes in the Consumer Price Index (CPI) system are published monthly by the Bureau of Labor Statistics, U.S. Department of Labor. Two versions of the CPI are published — the CPI for All Urban Consumers and the CPI for Urban Wage Earners and Clerical Workers. The two index versions are based on schedules, or *market baskets,* of goods and services that are representative of purchases by families in the two respective population groups. Periodic consumer expenditures surveys are taken to determine the purchasing patterns on which the market baskets are based. Since the two population groups have somewhat different purchasing patterns, the respective market baskets differ. Pricing of the items included in the market baskets is done in outlets selected to be representative of places where households in the respective groups shop. The indexes are calculated by a method that gives results equivalent to those of formula (26.3) for the method of weighted aggregates.

For each version of the CPI, the All-Items Index covers all categories of goods and services purchased. Component indexes are also compiled, which cover major categories, such as food and beverages and housing. At still lower levels of aggregation, component indexes are compiled for specific types of commodities and services.

Producer Price Indexes

Indexes in the Producer Price Indexes (PPI) system are published monthly by the Bureau of Labor Statistics, U.S. Department of Labor. Producer Price Indexes measure average changes in prices received in primary markets of the United States by producers of commodities at all stages of processing. The indexes are based on a sample of about 10,000 price quotations for some 2800 major commodities. The quotations are obtained monthly, primarily through mail questionnaires. Price indexes for individual commodities are aggregated into indexes for successively broader commodity groups according to three separate systems of classification: (1) stage of processing (that is, finished goods, semifinished goods, crude materials), (2) type of commodity, and (3) industry sector. Weights are based on value of shipments and are obtained from periodic industrial censuses and special surveys. The indexes are calculated by the method of weighted average of relatives.

Industrial Production Index

The Industrial Production Index (IPI) is a system of quantity indexes published monthly by the Board of Governors of the Federal Reserve System. The indexes measure changes in the physical volume of output in manufacturing, mining, and utilities in the United States. The system contains 235 individual series at various levels of aggregation. Output data are supplied by industry, and the index is calculated by the method of weighted average of relatives. Value-added or man-hour weights are used.

PROBLEMS

*26.1   An aquarium supply company sells two exotic strains of guppy (Monarch and Emperor). Prices (in dollars) for breeding pairs of these strains during 1982–1986 follow.

Year:	1982	1983	1984	1985	1986
Monarch:	6.00	6.25	7.00	7.50	8.25
Emperor:	8.00	8.50	9.00	10.00	10.75

a. Calculate, for each of the two price series, (1) price relatives on a 1982 base, (2) link relatives.

b. What is the percent point change in the price relatives for a Monarch pair between 1984 and 1985? What is the corresponding percent change?

c. For which strain was the relative price increase between 1982 and 1986 greater? For which strain was the relative price increase between 1985 and 1986 greater?

26.2   Prices (in cents) per 20 grams of protein for two food products carried by a supermarket chain during 1983–1986 follow.

Year:	1983	1984	1985	1986
Peanut Butter:	32	33	36	39
Bologna:	122	130	145	146

a. Calculate, for each of the two price series, (1) price relatives on a 1983 base, (2) link relatives.

b. What is the percent point change in the price relatives for peanut butter between 1985 and 1986? What is the corresponding percent change?

c. For which price series was the absolute increase between 1983 and 1986 greater? For which series was the relative increase greater? Is any contradiction involved here? Explain.

d. If 1986 instead of 1983 were used as the base period for the two series of price relatives, would the magnitudes of the price relatives differ from those in a? Would the proportional magnitudes of the price relatives differ from those in a? Explain.

26.3   **Annual Catch.**   The annual catches (in thousands of metric tons) of three species of fish brought in at a port during 1983–1986 follow.

Year:	1983	1984	1985	1986
Flounder:	13.4	12.2	15.0	16.8
Haddock:	2.1	2.0	2.3	2.6
Cod:	4.0	4.3	4.3	4.5

a. Calculate, for each of the three series, (1) percent relatives on a 1983 base, (2) link relatives.

b. For which species was the relative increase in annual catch between 1983 and 1986 greatest? For which was it greatest between 1985 and 1986?

c. Were the calculations in a sufficient to answer b or were additional calculations required? Explain.

*26.4   **Resort Complex.**   A firm manages a resort cottage complex. Many types of items must be replaced periodically in routine maintenance of appliances and plumbing in the complex. A list of items representative of all these items follows, together with quantity weights for the representative items and the unit prices of the representative items in 1984, 1985, and 1986. A price index series for appliances and plumbing supplies is to be compiled.

		Unit Price ($)		
Item	Quantity	1984	1985	1986
Ice-making machine	1	909.00	920.00	940.00
Air conditioner	12	246.00	246.00	250.00
Color television	15	440.00	416.50	395.00
Sink faucet	80	22.16	22.45	23.00

a. Obtain the aggregate cost of the schedule of representative items in each year. Does this cost vary from year to year because of price differences only, quantity differences only, or both price and quantity differences?

b. Calculate the price index series by the method of weighted aggregates. Use 1984 as the base period. Interpret the index number for 1986.

26.5   Refer to **Annual Catch** Problem 26.3. Prices received by fishermen at the port (in $000 per metric ton) follow.

Year:	1983	1984	1985	1986
Flounder:	0.93	1.27	1.25	1.19
Haddock:	0.85	0.87	1.09	1.29
Cod:	0.60	0.73	0.80	0.75

Calculate a price index series by the method of weighted aggregates. Use 1983 as the base period and the annual catches in 1983 for weights. Interpret the index number for 1986.

26.6   **Guard Agency.**   The following data for part-time personnel of the Marlowe Guard Agency show hourly wage rates and numbers of man-hours utilized in 1984, 1985, and 1986 for each job classification:

Job Classification	Wage Rate ($)			Man-Hours Utilized (thousand)		
	1984	1985	1986	1984	1985	1986
Plant protection	7.30	7.45	7.60	50	58	55
Construction site patrol	6.65	6.90	7.18	40	44	42
Escort service	9.70	10.38	10.70	48	47	48

a. Calculate an index series of wage rates for the part-time personnel by the method of weighted aggregates. Employ 1984 as the base period and 1984 man-hours as weights. Interpret the index number for 1986.

b. Would the index series in **a** be different if 1986 were used as the weight period of the index? Explain.

*26.7   Refer to **Resort Complex** Problem 26.4.

a. Calculate a price index series by the method of weighted average of relatives. Use 1984 as the base period and 1984 prices in the value weights.

b. Must your results in **a** be the same as those in Problem 26.4**b**? Explain.

c. If the base period for the index series had been specified as 1985, what price relatives would have been utilized?

26.8 Refer to **Annual Catch** Problems 26.3 and 26.5.
  a. Calculate a price index series by the method of weighted average of relatives. Use 1983 as the base period, and derive the value weights from 1984 prices and annual catches. Interpret the index number for 1986.
  b. Would the price index series in **a** be different if the value weights were derived from 1983 prices and annual catches? Explain.

26.9 Refer to **Guard Agency** Problem 26.6.
  a. Calculate an index series of wage rates by the method of weighted average of relatives. Use 1984 as the base period, and derive the value weights from 1984 wage rates and man-hours. Interpret the index number for 1985.
  b. Must your results in **a** agree with those in Problem 26.6a? Explain.

26.10 Why are value weights used when averaging price relatives while it suffices to use quantity weights when averaging unit prices? Explain.

26.11 In a large city, rents are stabilized at the levels existing in April 1980. Rent increases are allowed only to the degree that operating costs have increased since April 1980 because of price increases. In the compilation of an index series to measure price changes in operating costs for rental units in this city, why would April 1980 be an appropriate base period?

26.12 In the compilation of a price index series for raw materials used principally by a particular industry, why would it be inappropriate to select as a base period for the index a year when the industry was economically depressed and demand for the raw materials was very low?

*26.13 A price index series for products manufactured by the Zarthan Company has been compiled on a 1983 base and is as follows for 1981–1986:

Year:	1981	1982	1983	1984	1985	1986
Index:	96.2	99.1	100.0	101.1	103.7	105.2

Shift the index series to a 1985 base. Are the proportional magnitudes of the index numbers maintained by this shift? Explain.

26.14 A price index series for lumber products exports follows for 1982–1986. The series has been compiled on a 1983 base.

Year:	1982	1983	1984	1985	1986
Index:	108.6	100.0	97.6	104.7	110.8

  a. Shift the index series to a 1985 base.
  b. Will the link relatives for the price index series on the 1985 base be the same as those for the price index series on the 1983 base? Explain.

*26.15 Splice the following two price index series into one continuous series with a 1983 base:

Year:	1981	1982	1983	1984	1985	1986
Old Series (1981 = 100):	100	111	110			
New Series (1985 = 100):			84	92	100	115

26.16 The following two price index series are to be joined into a continuous series:

Year:	1980	1981	1982	1983	1984	1985	1986
Old Series (1980 = 100):	100	109	113	120			
New Series (1983 = 100):				100	109	119	130

   a. Splice the two series into one continuous series with a 1980 base, and calculate the link relatives for this continuous series.

   b. Would the link relatives be the same if a 1986 base had been used for splicing the two series? Explain.

26.17 Lead pencils are one of the items to be included in a price index of school supplies. What are some price-affecting characteristics you would consider in preparing a specification for obtaining price data for this item?

26.18 Canned mushrooms is one of the items to be included in a food price index. What are some price-affecting characteristics you would consider in preparing a specification for obtaining price data for this item?

26.19 Several years ago, some manufacturers of clothes washers planned to eliminate the warm-rinse option because rinsing is a mechanical function and water temperature plays no part in the actual rinsing process. This change would yield an 8 percent saving in energy use of clothes washers. However, many users appeared unwilling to accept the concept of cold rinsing. An increase in customer complaints and service calls could be anticipated if the warm-rinse option were eliminated. There would be no price changes in the washers.

   a. From the point of view of collecting price data for a price index, should this design change be treated as a quality improvement (hence, a price decrease) or as a quality deterioration (hence, a price increase)? Discuss.

   b. What problem in the compilation of ongoing price index series is illustrated here?

26.20 An ammonia product has been selected to represent all cleaning products in a schedule of items used to compile a household maintenance price index. Should household expenditures for this particular product only or for all cleaning products be used as the value weight in calculating the price index based on this schedule of items? Explain.

*26.21 Average weekly earnings (in current dollars) of workers on private construction payrolls in the United States during 1979–1983 follow.

Year:	1979	1980	1981	1982	1983
Weekly Earnings:	342.99	367.78	399.26	426.45	442.97

SOURCE: *Monthly Labor Review.*

   a. Use the Consumer Price Index in Table 26.4 to calculate each year's earnings in constant 1979 dollars. Interpret the constant-dollars earnings for 1983.

   b. Calculate, for the constant-dollars earnings series in **a**, (1) percent relatives on a 1979 base, (2) link relatives.

   c. Make the following comparisons between the results in **b** and those in Table 26.4 for all private nonagricultural workers in the United States: (1) the percent change in real earnings between 1979 and 1983, (2) the percent change in real earnings between 1982 and 1983.

   d. Would the comparisons in **c** be affected if each year's earnings were expressed in constant 1983 dollars? Explain.

26.22 The annual pensions (in thousands of current dollars) during 1981–1986 of plant workers of the Penbrook Corporation who retired at the end of 1980 with a C–8 classification follow. An appropriate price index series, based on expenditures patterns of retirees, is also shown for the same period. The corporation's pension plan requires adjustments for cost-of-living changes every three years.

Year:	1981	1982	1983	1984	1985	1986
Annual Pension ($000):	12.100	12.100	12.100	13.074	13.074	13.074
Price Index (1980 = 100):	111.8	115.2	120.8	124.4	127.2	131.7

a. Calculate the purchasing power of each year's pension at 1980 prices. Interpret the purchasing power value for 1986.

b. Calculate the link relatives and percent relatives on a 1981 base for the constant-dollars pension series in **a**. Describe the changes in real pension earnings during 1981–1986.

26.23 Data on the value of shipments by a manufacturing firm and an index of selling prices for the firm's products during 1981–1986 follow:

Year:	1981	1982	1983	1984	1985	1986
Shipments ($ million):	10.45	10.01	10.30	12.75	14.88	18.45
Price Index (1983 = 100):	124.2	108.1	100.0	112.6	119.8	133.4

a. Obtain the value of shipments at 1983 prices for each year.

b. Calculate, from the constant-dollars shipments series in **a**, (1) percent relatives on a 1981 base, (2) link relatives.

c. Has the increase in the value of the shipments between 1983 and 1986 been due primarily to increases in the selling prices? Have the relative year-to-year changes in the constant-dollars series been fairly stable during the period? Discuss.

26.24 Annual imports and exports of merchandise by a developing country (in millions of national currency units) during 1981–1986, as well as indexes of prices paid for imported merchandise and prices received for exported merchandise, were as follows:

Year:	1981	1982	1983	1984	1985	1986
Imports:	380	383	435	420	525	706
Exports:	397	410	435	485	545	620
Imports Index:	100	106	110	140	233	249
Exports Index:	100	102	106	118	132	150

a. Express the imports and exports series in constant 1981 currency units. Interpret the price-adjusted imports value for 1986.

b. Which constant-currency unit series in **a**—imports or exports—experienced the greater relative change between (1) 1981 and 1986, (2) 1985 and 1986?

c. Was the 56 percent increase in exports between 1981 and 1986 accounted for mainly by higher prices received for exported merchandise or by a larger volume of exported merchandise? Explain.

d. Is the difference between unadjusted imports and exports for any one year a meaningful measure? Is the difference between price-adjusted imports and exports for any one year a meaningful measure? Explain.

26.25 Two cost-of-living escalator clauses are under discussion in contract negotiations. One clause calls for a 10¢ change in the wage rate for each 1 percent change in the Consumer Price Index for Urban Wage Earners and Clerical Workers. The other clause calls for a 10¢ change in the wage rate for each 1 percent point of change in the same index. Are the two clauses equivalent? Explain.

26.26 A speaker, noting that state welfare and social service payments are automatically increased to keep pace with increases in the Consumer Price Index, pointed out that television costs are included in the index. He objected to the inclusion of "luxuries" in an index used to determine increases in welfare payments. Does the inclusion of television in the Consumer Price Index automatically cause the index to increase by more than it would otherwise? Explain.

26.27 The value of the Dallas–Fort Worth Consumer Price Index for All Urban Consumers was 343.9 in October 1985, while the value of the corresponding index for Houston was 337.6 in the same month. Each index is on the base 1967 = 100. Can any conclusions be reached from these two index values about comparative living costs in the two areas in October 1985? Explain.

*26.28 Refer to **Annual Catch** Problems 26.3 and 26.5.
   a. Calculate a quantity index series for the annual catches by the method of weighted aggregates. Use 1983 as the base period and prices received in 1983 for weights.
   b. By what percent has the quantity of annual catches changed from 1983 to 1986 according to the index series?

26.29 Refer to **Guard Agency** Problem 26.6. Calculate a quantity index series for the utilization of part-time personnel by the method of weighted aggregates. Employ 1984 as the base period and 1984 wage rates for weights. Interpret the index number for 1986.

*26.30 Refer to **Annual Catch** Problems 26.3 and 26.5. Calculate a quantity index series for the annual catches by the method of weighted average of relatives. Use 1983 as the base period, and derive the value weights from 1984 quantities and prices. Interpret the index number for 1986.

26.31 Refer to **Guard Agency** Problem 26.6.
   a. Calculate a quantity index series for the utilization of part-time personnel by the method of weighted average of relatives. Use 1984 as the base period, and derive the value weights from 1984 wage rates and man-hours.
   b. By what percent has the quantity of part-time personnel utilized changed from 1984 to 1986 according to the index series?

════════ EXERCISES

26.32 Refer to Table 26.1 in the text.
   a. Obtain the geometric mean of the link relatives for 1984–1986 in column 4. On the basis of this mean value, what has been the annual growth rate of the unit price for the chemical compound during 1983–1986?
   b. Show how the price relative for 1986 (on a 1983 base) can be calculated directly from the geometric mean in **a**.

26.33 Refer to Table 26.2. Calculate the price index series based on 1983–1984 = 100. Does the use of a two-year base period affect the year-to-year relative changes in the price index here? Explain.

26.34 Refer to **Resort Complex** Problem 26.4. You have been asked to examine whether the index should be updated. What are some of the questions you would investigate?

26.35 Explain why shifting the base of an index series by the shortcut method gives the same results as those obtained by recalculating the index series from the original data when the series is based on the method of weighted aggregates but may not give the same results if the series is based on the method of weighted average of relatives.

======   **STUDIES**

26.36 Total annual construction costs of restaurants built by a fast-food chain during 1980–1983 follow, together with a construction cost index for restaurants of this type.

Year:	1980	1981	1982	1983
Cost ($ thousand):	695	1501	3503	938
Construction Cost Index:	100	108	126	135

Today's value of the construction cost index is 180. What is the approximate replacement cost today of the restaurants constructed during the period 1980–1983, that is, what would be the total construction cost if all of these restaurants were built today?

26.37 Explain how you would construct a daily index of stock prices for common stocks traded on a major exchange. In your explanation, describe (1) the types of stocks that would appear in your schedule of items, (2) the quantity or value weights to be used, (3) choice of the base period, (4) whether the method of weighted aggregates or the method of weighted average of relatives would be used in computing the index. Also, comment on any special factors that would need to be considered in compiling and maintaining the index.

26.38 **a.** Consult the *Monthly Labor Review* to obtain the index values of the Consumer Price Index for All Urban Consumers for each of the most recent six months reported. Also, obtain the index values for the following components: (1) food and beverages, (2) housing, (3) apparel and upkeep, (4) transportation.

   b. Calculate the percent changes in the All-Items Index and in each of the four component indexes for the six-month period covered by your data. Which components showed larger relative price changes than the All-Items Index? Which ones showed smaller relative changes?

MATHEMATICAL REVIEW

 ## SUMMATION NOTATION

Single Summation

A number of formulas in this book use the symbol Σ (Greek capital sigma) to designate summation. The expression:

$$\sum_{i=1}^{n} X_i$$

is a shorthand form of the sum of the n values $X_1, X_2, \ldots, X_n$. In other words:

(A.1)
$$\sum_{i=1}^{n} X_i = X_1 + X_2 + \cdots + X_n$$

The subscript i on X is a label or index given to each value of X so that the n values may be distinguished from one another.

Given next are interpretations of several expressions that employ the Σ symbol.

(A.2)
$$\left(\sum_{i=1}^{n} X_i\right)^2 = (X_1 + X_2 + \cdots + X_n)^2$$

(A.3)
$$\sum_{i=1}^{n} X_i^2 = X_1^2 + X_2^2 + \cdots + X_n^2$$

(A.4)
$$\sum_{i=1}^{n} c = c + c + \cdots + c = nc$$

where:

c is a constant

(A.5)
$$\sum_{i=1}^{n} cX_i = cX_1 + cX_2 + \cdots + cX_n = c\left(\sum_{i=1}^{n} X_i\right)$$

$$\text{(A.6)} \qquad \sum_{i=1}^{n} (X_i + Y_i) = (X_1 + Y_1) + (X_2 + Y_2) + \cdots + (X_n + Y_n) = \sum_{i=1}^{n} X_i + \sum_{i=1}^{n} Y_i$$

EXAMPLE

Suppose $n = 3$, $c = 11$, $X_1 = 3$, $X_2 = -1$, and $X_3 = 7$. Then:

$$\sum_{i=1}^{3} X_i = 3 - 1 + 7 = 9$$

$$\left(\sum_{i=1}^{3} X_i \right)^2 = (3 - 1 + 7)^2 = 9^2 = 81$$

$$\sum_{i=1}^{3} X_i^2 = 3^2 + (-1)^2 + 7^2 = 59$$

$$\sum_{i=1}^{3} 11 = 11 + 11 + 11 = 3(11) = 33$$

$$\sum_{i=1}^{3} 11 X_i = 11 \sum_{i=1}^{3} X_i = 11(9) = 99$$

Double Summation

Some formulas use the double summation expression, such as $\sum_{i=1}^{r} \sum_{j=1}^{c} X_{ij}$. The double subscript ij on X is again a label or index given to each value of X so that the values may be distinguished from one another. In this case, the subscript i ranges from 1 to r and the subscript j ranges from 1 to c. The definition of $\sum_{i=1}^{r} \sum_{j=1}^{c} X_{ij}$ is as follows.

$$\text{(A.7)} \qquad \sum_{i=1}^{r} \sum_{j=1}^{c} X_{ij} = \sum_{i=1}^{r} \left(\sum_{j=1}^{c} X_{ij} \right) = \sum_{j=1}^{c} X_{1j} + \sum_{j=1}^{c} X_{2j} + \cdots + \sum_{j=1}^{c} X_{rj}$$

where, for instance:

$$\sum_{j=1}^{c} X_{1j} = X_{11} + X_{12} + \cdots + X_{1c} \qquad \sum_{j=1}^{c} X_{2j} = X_{21} + X_{22} + \cdots + X_{2c}$$

EXAMPLE

Consider the following table of X_{ij} values:

	$j = 1$	$j = 2$	$j = 3$
$i = 1$	4	-3	1
$i = 2$	5	10	-6

The term X_{ij} is the value at the intersection of the ith row and jth column in the table; for example, $X_{11} = 4$, $X_{12} = -3$, $X_{21} = 5$. Note that:

$$\sum_{j=1}^{3} X_{1j} = 4 - 3 + 1 = 2 \qquad \sum_{j=1}^{3} X_{2j} = 5 + 10 - 6 = 9$$

Hence:

$$\sum_{i=1}^{2} \sum_{j=1}^{3} X_{ij} = \sum_{j=1}^{3} X_{1j} + \sum_{j=1}^{3} X_{2j} = (4 - 3 + 1) + (5 + 10 - 6) = 11$$

Observe that the double summation is simply the sum of all six values in the table.

□

Comment   If the ranges of the subscripts do not depend on one another, the order of summation can be reversed.

(A.8)
$$\sum_{i=1}^{r} \sum_{j=1}^{c} X_{ij} = \sum_{j=1}^{c} \sum_{i=1}^{r} X_{ij}$$

For the previous example, we have:

$$\sum_{j=1}^{3} \sum_{i=1}^{2} X_{ij} = \sum_{i=1}^{2} X_{i1} + \sum_{i=1}^{2} X_{i2} + \sum_{i=1}^{2} X_{i3} = (4 + 5) + (-3 + 10) + (1 - 6) = 11$$

Otherwise, the order of summation must be maintained.

Abbreviated Notation

Where the context makes it clear which values are to be summed, it is common to omit the range of the summation index or even the index itself. In this case, we write:

$$\sum_{i} X_i \quad \text{or} \quad \sum X_i \qquad \text{instead of} \qquad \sum_{i=1}^{n} X_i$$

and:

$$\sum_{i} \sum_{j} X_{ij} \quad \text{or} \quad \sum \sum X_{ij} \qquad \text{instead of} \qquad \sum_{i=1}^{r} \sum_{j=1}^{c} X_{ij}$$

Abbreviated summation notation of this type is used frequently in this text.

ALGEBRAIC RULES FOR EXPONENTS AND LOGARITHMS

Exponents

The following is a summary of algebraic rules for exponents. Here a, b, m, and n denote numbers. Each rule is illustrated.

	Rule	Illustration
(A.9)	$a^0 = 1$ if $a \neq 0$	$3^0 = 1$
(A.10)	$a^m a^n = a^{m+n}$	$0.6^2(0.6^3) = 0.6^5 = 0.07776$
(A.11)	$(a^m)^n = a^{mn}$	$(3^2)^4 = 3^8 = 6561$
(A.12)	$a^{-n} = \dfrac{1}{a^n}$	$10^{-2} = \dfrac{1}{10^2} = 0.01$
(A.13)	$\dfrac{a^m}{a^n} = a^{m-n}$	$\dfrac{5^2}{5^4} = 5^{-2} = 0.04$
(A.14)	$(ab)^n = a^n b^n$	$[5(7)]^2 = 5^2(7^2) = 1225$
(A.15)	(i) $a^{1/n} = \sqrt[n]{a}$ if $a > 0$	$8^{1/3} = \sqrt[3]{8} = 2$
	(ii) $a^{1/2} = \sqrt[2]{a}$ is understood to mean $\sqrt{a}$, the square root of a	$9^{1/2} = \sqrt[2]{9} = \sqrt{9} = 3$
(A.16)	$a^{m/n} = \sqrt[n]{a^m}$ if $a > 0$	$5^{3/2} = \sqrt[2]{5^3} = 11.180$

Logarithms

Definition. The symbol $\log_b A$, for any positive number A, denotes the exponent to which b must be raised to equal A. In other words, if $\log_b A = C$, then by definition, $b^C = A$. The number b is called the *base* of the logarithm and may be any positive number except 1. Logarithms with base 10 are called *common logarithms*. Logarithms to base e (where $e = 2.71828\ldots$) are called *natural* or *Naperian logarithms*. The quantity $C = \log_b A$ is called the *logarithm* or *log* of A to base b, and A is called the *antilogarithm* or *antilog* of C to base b. Frequently, natural logarithms are denoted by ln (pronounced "lawn") rather than by $\log_e$. In this text, we use log for common logarithms to base 10 and ln for logarithms to base e. Most uses of logarithms in this text involve common logarithms, as in the following examples.

EXAMPLES

1. $\log 1 = 0$ because $10^0 = 1$
2. $\log 10 = 1$ because $10^1 = 10$
3. $\log 100 = 2$ because $10^2 = 100$
4. $\log 0.1 = -1$ because $10^{-1} = 1/10 = 0.1$
5. $\log 3.162 = 0.5$ because $10^{0.5} = \sqrt{10} = 3.162$
6. antilog $2 = 100$ because $10^2 = 100$
7. antilog $0.5 = 3.162$ because $10^{0.5} = 3.162$

Rules.   The following is a summary of the algebraic rules for logarithms. Here, A and B denote positive numbers, while c denotes any number. Each rule is illustrated.

	Rule	Illustration
(A.17)	$\log AB = \log A + \log B$	$\log[(10)(100)] = \log 10 + \log 100 = 1 + 2 = 3$
(A.18)	$\log \dfrac{A}{B} = \log A - \log B$	$\log\left(\dfrac{3.162}{10}\right) = \log 3.162 - \log 10 = 0.5 - 1 = -0.5$
(A.19)	$\log A^c = c \log A$	$\log(100)^3 = 3 \log 100 = 3(2) = 6$
		$\log \sqrt[3]{10} = \log 10^{1/3} = \dfrac{1}{3} \log 10 = \dfrac{1}{3}(1) = 0.3333$

Calculator Usage

Most calculators have one or more of the following operator keys for computing exponents and logarithms:

Key	Operation	Illustration
10^x	Computes the antilog of x to base 10	antilog $0.5 = 10^{0.5} = 3.162$
$\log x$	Computes the log of x to base 10	$\log 3.162 = 0.5$
e^x	Computes the antilog of x to base e	antiln $2.0 = e^2 = (2.71828)^2 = 7.39$
$\ln x$	Computes the log of x to base e	$\ln 7.39 = 2.0$
y^x	Raises y to the power x	antilog $0.5 = 10^{0.5} = 3.162$
		antiln $2.0 = e^2 = (2.71828)^2 = 7.39$

SET NOTATION, OPERATIONS, AND ALGEBRAIC RULES

Basic Concepts

A *set* is a collection of distinct objects called *elements* of the set. We shall specify a set by listing its elements within a pair of braces. For example, $\{a, c, i, s, t\}$ is the set of letters in the word *statistics,* and $\{0, 1, 2, 3\}$ is the set of number of defective units that can occur in a sample of size three.

If a is an element of set A, then we write $a \in A$. If a is not an element of set A, then we write $a \notin A$. For instance, if $S = \{a, c, i, s, t\}$, then $c \in S$ while $m \notin S$. If all the elements of a set A are also elements of set B, then we say A is a *subset* of B and we write this symbolically as $A \subseteq B$. For instance, if $T = \{a, c, t\}$ and $S = \{a, c, i, s, t\}$, then $T \subseteq S$. Two sets are equal if they contain the same elements.

If I denotes the set of all elements of interest in a particular context, we call I the *universal set* for that context. If A is a subset of I, then A^* denotes the *complement* of A (relative to I) and represents the set of all elements of I that are not elements of A. For instance, if $I = \{1, 2, 3, 4, 5\}$ and $E = \{2, 4\}$, then $E^* = \{1, 3, 5\}$. The set that contains no elements is called the *empty set* and is denoted by $\varnothing$.

Set Operations

The two most common operations involving pairs of sets are set union and set intersection.

Union. For any pair of sets A and B, $A \cup B$ denotes the set of all elements that belong to either A or B or both A and B, and is called their *set union*. For example, if $A = \{1, 11, 21\}$ and $B = \{1, 3, 8, 11\}$, then $A \cup B = \{1, 3, 8, 11, 21\}$.

Intersection. For any pair of sets A and B, $A \cap B$ denotes the set of all elements that belong to both A and B, and is called their *set intersection*. For example, if A and B are the sets in the preceding example, then $A \cap B = \{1, 11\}$.

Algebraic Rules for Set Operations

	Name of Rule	Rule
(A.20)	*Associative Rules*	$A \cup (B \cup C) = (A \cup B) \cup C$
		$A \cap (B \cap C) = (A \cap B) \cap C$
(A.21)	*Commutative Rules*	$A \cup B = B \cup A$
		$A \cap B = B \cap A$

continues

(A.22)	*Distributive Rules*	$A \cap (B \cup C) = (A \cap B) \cup (A \cap C)$
		$A \cup (B \cap C) = (A \cup B) \cap (A \cup C)$
(A.23)	*Idempotency Rules*	$A \cup A = A$
		$A \cap A = A$
(A.24)	*I and $\varnothing$ Rules*	$A \cup \varnothing = A$
		$A \cup I = I$
		$A \cap \varnothing = \varnothing$
		$A \cap I = A$
(A.25)	*Complementation Rules*	$A \cup A^* = I$
		$A \cap A^* = \varnothing$
		$(A \cup B)^* = A^* \cap B^*$
		$(A \cap B)^* = A^* \cup B^*$
(A.26)	*Involution Rules*	$(A^*)^* = A$
		$\varnothing^* = I$
		$I^* = \varnothing$

A.4   PERMUTATIONS AND COMBINATIONS

Permutations

If r objects are selected from a set of n different objects and arranged in a particular order, the arrangement is called a *permutation of n objects taken r at a time*. The number of permutations of n different objects taken r at a time is denoted by $_nP_r$ and is as follows.

(A.27)
$$_nP_r = n(n - 1)(n - 2) \cdots (n - r + 1)$$

The number of permutations of n different objects taken together is as follows.

(A.27a)
$$_nP_n = n! = n(n - 1)(n - 2) \cdots (1)$$

The notation $x!$ (read "x factorial") stands for $x(x - 1)(x - 2) \cdots (1)$. Thus, $5! = 5(4)(3)(2)(1) = 120$. By definition, $0! = 1$.

| EXAMPLES |

1. An experiment requires that two persons be selected from a group of four persons — A, B, C, and D — the first person selected being tested on day 1 and the second person tested on day 2. The number of different permutations of the $n = 4$ persons taken $r = 2$ at a time by (A.27) is $_4P_2 = 4(3) = 12$. The 12 permutations are:

AB	AD	BC	CA	CD	DB
AC	BA	BD	CB	DA	DC

Note that the order of persons matters — AB and BA, for instance, are different permutations because the testing of the two persons follows different sequences.

2. In Example 1, if all $n = 4$ persons are taken for the experiment and scheduled on different days, the total number of different permutations by (A.27a) is $4! = 4(3)(2)(1) = 24$. ☐

Permutations of Like Objects. Consider n objects that are of c different types. Suppose there are r_1 objects of type 1, r_2 of type 2, and so on, where $r_1 + r_2 + \cdots + r_c = n$. The number of distinct permutations (that is, distinguishable ordered arrangements) when all n objects are taken together is as follows.

(A.28)

$$\frac{n!}{r_1! r_2! \cdots r_c!}$$

where:

$$r_1 + r_2 + \cdots + r_c = n$$

| EXAMPLE |

For $n = 5$ bottles coming off a filling line, $r_1 = 2$ have acceptable fills (A) and $r_2 = 3$ have defective fills (D). One possible sequence or permutation for the five bottles as they are processed is ADDDA. The number of distinct permutations by (A.28) here is:

$$\frac{5!}{2! \, 3!} = \frac{120}{2(6)} = 10$$ ☐

| Comment |

The binomial coefficient $\binom{n}{r}$ defined in (6.4) equals the number of distinct permutations that can be formed from n objects, of which r are of one type and $n - r$ are of a second type.

Combinations

A set of r objects selected from n different objects, considered without regard to their order in the set, is called a *combination of n objects taken r at a time*. The number of combinations of n different objects taken r at a time is denoted by $_nC_r$ and is as follows.

(A.29)

$$_nC_r = \frac{n!}{r!\,(n - r)!}$$

Note that $_nC_r = {_nC_{n-r}}$.

EXAMPLE

A litter contains five pups—A, B, C, D, and E—four of which are to be sold to a family. The number of different combinations of these $n = 5$ pups taken $r = 4$ at a time, by (A.29), is $_5C_4 = 5!/4!\,1! = 5$. These five combinations are:

    ABCD   ABCE   ABDE   ACDE   BCDE

Note that the order of the pups is immaterial here—ABCD and DCBA, for instance, are the same combination.

B CHI-SQUARE, *t*, AND *F* DISTRIBUTIONS

In this appendix, we discuss the chi-square, *t*, and *F* distributions, which are all related to the normal distribution and are widely used in statistical analysis. For each distribution, we describe its main characteristics and explain how to use its table of percentiles. In a final section, we consider the theoretical relations between the chi-square, *t*, *F*, and normal distributions.

In Figure B.1, we have set out in summary form, for subsequent reference, several of the important characteristics of the chi-square, *t*, and *F* distributions, as well as of the standard normal distribution.

B.1 CHI-SQUARE DISTRIBUTIONS

Characteristics

The χ^2 (Greek chi-square) distribution is a continuous probability distribution. It has one parameter, ν (Greek nu), which is called its *degrees of freedom* (abbreviated *df*). The parameter ν may be any positive whole number. Each different value of ν corresponds to a different member of the χ^2 family of distributions. The random variable associated with the χ^2 distribution is denoted by $\chi^2(\nu)$ and may take on any positive value, that is, $0 < \chi^2(\nu) < \infty$. Graphs of the χ^2 probability density function for several values of ν are shown in Figure B.1. Figure B.1 also summarizes some of the properties of the χ^2 distribution, which we now discuss.

Shape. From the graphs of the χ^2 distribution in Figure B.1, it is seen that the χ^2 distribution is unimodal and right-skewed, although the skewness becomes smaller as ν increases. In fact, it can be shown that as ν increases, the shape of the χ^2 distribution approaches that of a normal distribution.

Mean and Variance. The mean and the variance of the $\chi^2(\nu)$ random variable are as follows.

FIGURE B.1   **Summary of the characteristics of standard normal, chi-square, *t*, and *F* distributions**

Distribu-tion	Random variable	Sample space	Parameters	Mean	Variance	Skewness	Density functions
Standard normal	Z	$(-\infty, +\infty)$	None	0	1	Symmetrical	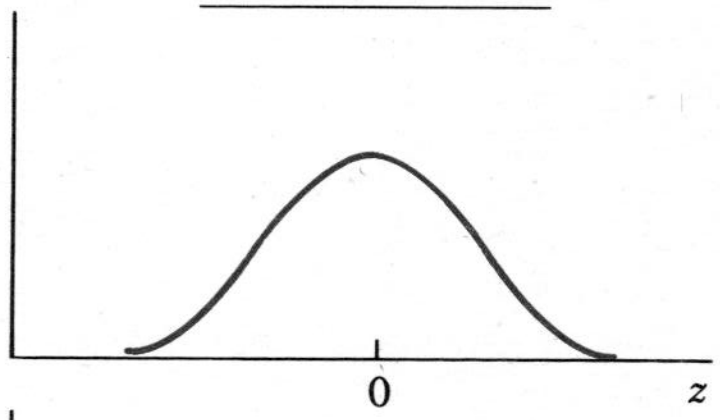
χ^2	$\chi^2(v)$	$(0, \infty)$	v	v	$2v$	Right-skewed	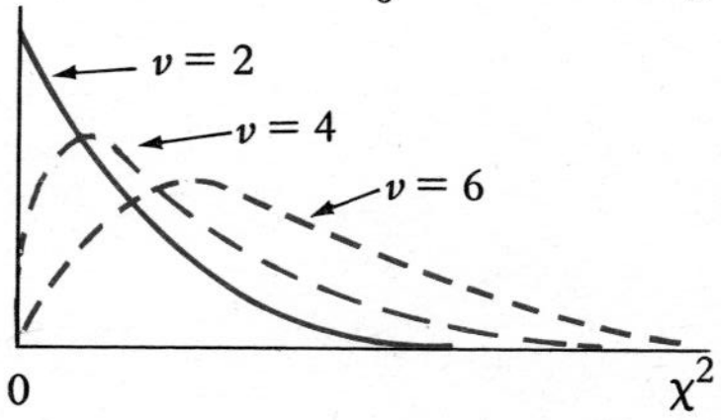
t	$t(v)$	$(-\infty, +\infty)$	v	0	$\dfrac{v}{(v-2)}$ $(v>2)$	Symmetrical	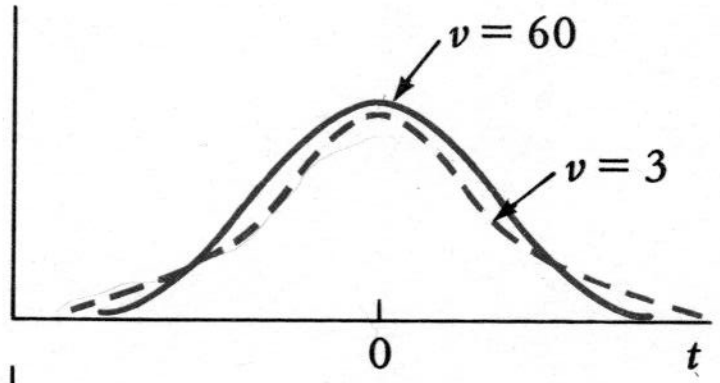
F	$F(v_1, v_2)$	$(0, \infty)$	v_1, v_2	$\dfrac{v_2}{v_2-2}$ $(v_2>2)$	$\dfrac{2v_2^2(v_1+v_2-2)}{v_1(v_2-2)^2(v_2-4)}$ $(v_2>4)$	Right-skewed	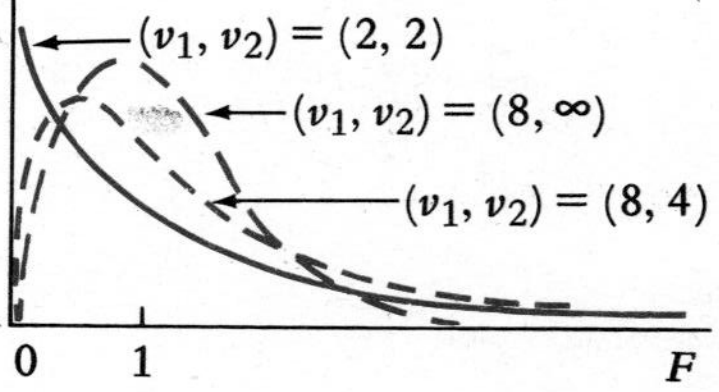

(B.1) $E\{\chi^2(\nu)\} = \nu$

(B.2) $\sigma^2\{\chi^2(\nu)\} = 2\nu$

For example, $\chi^2(5)$ has a mean of 5 and a variance of $2(5) = 10$. Observe that the expected value is equal to the degrees of freedom ν and that the variance equals twice the degrees of freedom. Hence, both the mean and the variance of the χ^2 distribution increase as ν increases.

Determining Percentiles for Chi-square Distributions

Table of Percentiles.   Table C.2 contains selected percentiles of χ^2 distributions for various degrees of freedom from 1 to 100. We shall let $\chi^2(a; \nu)$ denote the $100a$ percentile of the χ^2 distribution with ν degrees of freedom.

(B.3) $P[\chi^2(\nu) \leq \chi^2(a; \nu)] = a$

The illustration at the top of Table C.2 shows this relationship.

EXAMPLES

1. For 5 degrees of freedom, find $\chi^2(0.90; 5)$. From Table C.2, we see that the 90th percentile of the χ^2 distribution with 5 degrees of freedom is $\chi^2(0.90; 5) = 9.24$. Hence, the probability is 0.90 that $\chi^2(5)$ is less than or equal to 9.24.

2. For $df = 60$, find the 5th percentile. From Table C.2, we see that $\chi^2(0.05; 60) = 43.19$. Hence, the probability is 0.05 that $\chi^2(60)$ is less than or equal to 43.19.

Normal Approximation.   It was noted earlier that the shape of the χ^2 distribution approaches that of a normal distribution as ν increases. This allows one to approximate χ^2 probabilities for large values of ν by normal probabilities. A good approximation of a chi-square percentile $\chi^2(a; \nu)$ in terms of a standard normal percentile $z(a)$ is given by the following.

$$\chi^2(a; \nu) \simeq \frac{1}{2}[z(a) + \sqrt{2\nu - 1}]^2$$

EXAMPLES

1. For 100 degrees of freedom, find the 95th percentile by using the normal approximation. We know from Table C.1 that $z(0.95) = 1.645$. Hence, we obtain by (B.4):

$$\chi^2(0.95; 100) \simeq \frac{1}{2}[1.645 + \sqrt{2(100) - 1}]^2 = 124.1$$

The exact percentile as shown in Table C.2 is 124.3, so the approximation is
quite good.

2.   = 250, find the 25th percentile. Since we know from Table C.1 that
= −0.67, we obtain:

$$\chi^2(0.25; 250) \approx \frac{1}{2}[-0.67 + \sqrt{2(250) - 1}\,]^2 = 234.8$$

B.2 *t* DISTRIBUTIONS

Characteristics

The *t* distribution is a continuous probability distribution. Like the χ^2 distribution, it
has only one parameter, ν, which is called its *degrees of freedom* (*df*). The parameter
ν may be any positive whole number. Each different value of ν corresponds to a different member of the family of *t* distributions. The random variable associated with the *t*
distribution is denoted by $t(\nu)$ and may take on any value, that is, $-\infty < t(\nu) < +\infty$.

Graphs of the *t* probability density function for two values of ν are shown in Figure B.1. Figure B.1 also presents some of the properties of the *t* distribution.

Shape. From the graphs of the *t* distribution in Figure B.1, it is seen that the *t* distribution is symmetrical and almost bell-shaped, and its appearance is like the standard
normal distribution when ν is large. In fact, it can be shown mathematically that the
t distribution approaches the standard normal distribution as ν becomes large.

Mean and Variance. The mean and the variance of the $t(\nu)$ random variable are
as follows.

(B.5) $$E\{t(\nu)\} = 0$$

(B.6) $$\sigma^2\{t(\nu)\} = \frac{\nu}{\nu - 2} \qquad \text{when } \nu > 2$$

The variance is undefined for $\nu \leq 2$.

For the $t(6)$ random variable, for instance, the mean and the variance are 0 and
$6/(6 - 2) = 1.50$, respectively. For $t(60)$, the mean still is 0 but the variance now is
1.03.

Note that the *t* distribution has a mean of 0 for all ν, and that the variance
approaches 1 as ν gets larger. These facts are consistent with the earlier statement that
the *t* distribution approaches the standard normal distribution as ν gets larger, since the
latter distribution has mean 0 and variance 1.

Determining Percentiles for *t* Distributions

Table of Percentiles.   Table C.3 contains selected percentiles of *t* distributions for various degrees of freedom ν. We shall let $t(a; \nu)$ denote the $100a$ percentile of the *t* distribution with ν degrees of freedom.

(B.7)
$$P[t(\nu) \leq t(a; \nu)] = a$$

The illustration at the top of Table C.3 shows this relationship.

EXAMPLES

1. For 10 degrees of freedom, find the 95th percentile. From Table C.3, we see that $t(0.95; 10) = 1.812$. Hence, the probability is 0.95 that $t(10)$ is less than or equal to 1.812.

2. For $df = 30$, find the 90th percentile. From Table C.3, we see that $t(0.90; 30) = 1.310$.

3. For $df = 10$, find the 5th percentile. Because of the symmetry of the *t* distribution about its mean 0, the 5th percentile is the negative of the 95th percentile. The same is true for the standard normal distribution, as illustrated in Figure 7.4. Hence, $t(0.05; 10) = -t(0.95; 10) = -1.812$. Note that the symmetry property has made it unnecessary to show percentiles less than the 50th in Table C.3. □

Normal Approximation.   Since the *t* distribution approaches the standard normal distribution as ν increases, for large ν the percentiles of the standard normal distribution may be used to approximate those of the *t* distribution. In fact, the last row in Table C.3, which corresponds to $\nu = \infty$, has the same percentiles as those for the standard normal distribution. For instance, we know from Table C.1 that $z(0.975) = 1.960$. In Table C.3, we see that $t(0.975; \infty) = 1.960$. As Table C.3 shows, once ν is larger than about 30, the standard normal distribution provides approximate percentiles for the *t* distribution that are adequate for most practical purposes. Thus, $t(0.975; 217) \simeq z(0.975) = 1.960$.

B.3   *F* DISTRIBUTIONS

Characteristics

The *F* distribution is a continuous probability distribution. It has two parameters, ν_1 and ν_2, which are positive whole numbers called *degrees of freedom (df)*. To each pair (ν_1, ν_2) corresponds a different member of the family of *F* distributions. The random variable associated with the *F* distribution is denoted by:

$$F(\nu_1, \nu_2)$$

Numerator   Denominator
   df   *df*

where the parameter in the left position denotes the *numerator degrees of freedom* and the parameter in the right position denotes the *denominator degrees of freedom*. The $F(v_1, v_2)$ random variable may take on any positive value, that is, $0 < F(v_1, v_2) < \infty$.

Graphs of the F probability density function for several pairs of (v_1, v_2) values are shown in Figure B.1. Figure B.1 also summarizes some of the properties of the F distribution.

Shape.   From the graphs of the F distribution in Figure B.1, it is seen that the F distribution is unimodal and right-skewed. The mode of the F distribution (that is, the value of F for which the density is largest) approaches 1 as both degrees of freedom get large.

Mean and Variance.   The mean and the variance of the $F(v_1, v_2)$ random variable are as follows.

(B.8)
$$E\{F(v_1, v_2)\} = \frac{v_2}{v_2 - 2} \qquad \text{when } v_2 > 2$$

(B.9)
$$\sigma^2\{F(v_1, v_2)\} = \frac{2v_2^2(v_1 + v_2 - 2)}{v_1(v_2 - 2)^2(v_2 - 4)} \qquad \text{when } v_2 > 4$$

The mean is undefined for $v_2 \leq 2$, and the variance is undefined for $v_2 \leq 4$.

For the $F(3, 5)$ random variable, for instance, the mean and the variance are, respectively:

$$\frac{5}{5 - 2} = 1.67 \qquad \frac{2(5^2)(3 + 5 - 2)}{3(5 - 2)^2(5 - 4)} = 11.11$$

Hence, the standard deviation of $F(3, 5)$ is $\sqrt{11.11} = 3.33$.

Note from (B.8) and (B.9) that the mean approaches 1 as v_2 get large, while the variance approaches 0 as both degrees of freedom become large.

Determining Percentiles for *F* Distributions

Because the F distribution has two parameters, any extensive tabulation of it requires considerable space. Table C.4 contains the 95th and 99th percentiles of F distributions for selected numerator and denominator degrees of freedom. As usual, we shall let $F(a; v_1, v_2)$ denote the $100a$ percentile of the F distribution with v_1 and v_2 degrees of freedom.

(B.10)
$$P[F(v_1, v_2) \leq F(a; v_1, v_2)] = a$$

The illustration at the top of Table C.4 shows this relationship.

Percentiles below 50 percent are not presented in Table C.4 because they can be obtained from the percentiles presented by means of the following relationship.

(B.11)

$$F(a; \nu_1, \nu_2) = \frac{1}{F(1 - a; \nu_2, \nu_1)}$$

Note the reversal of the degrees of freedom in the denominator on the right of (B.11).

EXAMPLES

1. Find the 95th percentile for $F(3, 5)$. From Table C.4, we find that the 95th percentile of the F distribution with 3 degrees of freedom for the numerator and 5 degrees of freedom for the denominator is $F(0.95; 3, 5) = 5.41$. Hence, the probability is 0.95 that $F(3, 5)$ is less than or equal to 5.41.

2. Find the 99th percentile for $F(10, 6)$. From Table C.4, we find $F(0.99; 10, 6) = 7.87$.

3. Find the 5th percentile of $F(3, 5)$. To use the relationship (B.11), we first need to find $F(0.95; 5, 3)$. From Table C.4, we find this to be $F(0.95; 5, 3) = 9.01$. Hence, by (B.11):

$$F(0.05; 3, 5) = \frac{1}{F(0.95; 5, 3)} = \frac{1}{9.01} = 0.111$$

Thus, the probability is 0.05, that $F(3, 5)$ is less than or equal to 0.111.

4. Find $F(0.01; 6, 10)$. We found in Example 2 that $F(0.99; 10, 6) = 7.87$. Hence, $F(0.01; 6, 10) = 1/7.87 = 0.127$. □

B.4 RELATIONS BETWEEN Z, t, χ^2, and F

We first describe the relations of χ^2, t, and F to the standard normal random variable Z, and then we explain how Z, t, and χ^2 may be viewed as special cases of the random variable F.

χ^2 Random Variable

The χ^2 random variable is related directly to the standard normal variable. Let $Z_1, Z_2, \ldots, Z_\nu$ be ν independent standard normal variables. Then a $\chi^2(\nu)$ random variable can be defined in terms of these Z_i variables, as follows.

(B.12)

$$\chi^2(\nu) = Z_1^2 + Z_2^2 + \cdots + Z_\nu^2$$

where:

Z_i are independent $N(0, 1)$

In other words, the sum of ν independent squared standard normal variables is a χ^2 random variable with ν degrees of freedom.

Comment

We can find $E\{\chi^2(\nu)\}$ by applying theorem (5.15a) to definition (B.12). We obtain:

$$E\{\chi^2(\nu)\} = E\{Z_1^2\} + E\{Z_2^2\} + \cdots + E\{Z_\nu^2\}$$

For the standard normal variable Z, $E\{Z_i\} = 0$ and $\sigma^2\{Z_i\} = 1$. Hence, it follows from (5.7a) that $E\{Z_i^2\} = 1$. Consequently:

$$E\{\chi^2(\nu)\} = 1 + 1 + \cdots + 1 = \nu$$

which is the result given in (B.1).

_t_ Random Variable

We have just shown that the χ^2 random variable is related to the standard normal variable. The $t(\nu)$ random variable is related to both the $\chi^2(\nu)$ and Z random variables, as follows.

(B.13)

$$t(\nu) = \frac{Z}{\left[\dfrac{\chi^2(\nu)}{\nu}\right]^{1/2}}$$

where:

Z and $\chi^2(\nu)$ are independent

In other words, a t random variable with ν degrees of freedom is obtained by (1) taking a standard normal random variable Z and a χ^2 random variable with ν degrees of freedom that are independent of one another, and (2) forming the ratio of Z and the positive square root of $\chi^2(\nu)$ divided by its degrees of freedom ν.

Comment

We can see intuitively why the t distribution approaches the standard normal distribution when ν is large by considering the random variable $\chi^2(\nu)/\nu$ in the denominator on the right of (B.13). The expected value of this random variable is:

$$E\left\{\frac{\chi^2(\nu)}{\nu}\right\} = \frac{1}{\nu}E\{\chi^2(\nu)\} = \frac{\nu}{\nu} = 1 \qquad \text{by (5.11b) and (B.1)}$$

while its variance is:

$$\sigma^2\left\{\frac{\chi^2(\nu)}{\nu}\right\} = \frac{1}{\nu^2}\sigma^2\{\chi^2(\nu)\} = \frac{2\nu}{\nu^2} = \frac{2}{\nu} \qquad \text{by (5.12b) and (B.2)}$$

Note that the variance approaches 0 as ν gets large. Hence, $\chi^2(\nu)/\nu$ becomes concentrated within an arbitrarily small neighborhood of the expected value 1 as ν approaches ∞. This suggests that we can think of t as approximately $Z/\sqrt{1} = Z$ when ν is large.

F Random Variable

It was shown in (B.12) that the χ^2 random variable is related to the standard normal variable. The $F(\nu_1, \nu_2)$ random variable in turn is related to two independent χ^2 random variables—$\chi^2(\nu_1)$ and $\chi^2(\nu_2)$—as follows.

(B.14)
$$F(\nu_1, \nu_2) = \frac{\dfrac{\chi^2(\nu_1)}{\nu_1}}{\dfrac{\chi^2(\nu_2)}{\nu_2}}$$

where:

$\chi^2(\nu_1)$ and $\chi^2(\nu_2)$ are independent

In other words, to obtain an F random variable, one divides two independent χ^2 random variables by their respective degrees of freedom and forms the ratio of these ratios.

Comment   The symmetrical relationship between the numerator and denominator in (B.14) makes it easy to see the following.

(B.15)
$$F(\nu_1, \nu_2) = \frac{1}{F(\nu_2, \nu_1)}$$

This fact makes it possible to compute the $100a$ percentiles of $F(\nu_1, \nu_2)$ from knowledge of the $100(1 - a)$ percentiles of $F(\nu_2, \nu_1)$ by the relationship (B.11).

Z, *t*, and χ^2 as Special Cases of *F*

Relations Between Random Variables.   The random variables Z, t, and χ^2 may all be considered special cases of the random variable F.

(B.16)
$$Z^2 = F(1, \infty)$$
$$[t(\nu)]^2 = F(1, \nu)$$
$$\frac{\chi^2(\nu)}{\nu} = F(\nu, \infty)$$

The relationship between Z and F follows because $\nu_2 = \infty$ implies, as we have seen, that $\chi^2(\nu_2)/\nu_2$ is essentially equivalent to 1. Thus:

$$F(1, \infty) = \frac{\dfrac{\chi^2(1)}{1}}{1} = Z^2$$

The other relations follow in the same manner.

Relations Between Percentiles.   The percentiles of the Z, t, and χ^2 distributions are related to the percentiles of the F distribution, in ways illustrated by the following examples.

| EXAMPLES |

1.   Consider $z(0.975) = 1.960$. We have $(1.960)^2 = 3.84 = F(0.95; 1, \infty)$.
2.   Consider $t(0.975; 10) = 2.228$. We have $(2.228)^2 = 4.96 = F(0.95; 1, 10)$.
3.   Consider $\chi^2(0.95; 8) = 15.51$. We have $15.51/8 = 1.94 = F(0.95; 8, \infty)$.   □

Comment   The reason why the corresponding percentiles for the Z and F distributions in Example 1 are 97.5 and 95 percent, respectively, and similarly for the t and F distributions in Example 2, is related to the fact that Z^2 and t^2 equal F. Hence, both negative and positive outcomes of Z and t lead to the same F value.

To show this for Example 1, consider the following probability statement for Z:

$$P[z(0.025) \leq Z \leq z(0.975)] = 0.95$$

But we know that $z(0.025) = -z(0.975)$, so we can write the probability statement as follows:

$$P[-z(0.975) \leq Z \leq z(0.975)] = 0.95$$

In turn, this can be expressed as follows:

$$P\{Z^2 \leq [z(0.975)]^2\} = 0.95$$

because an inequality of the form $-a \leq b \leq a$ is mathematically equivalent to $b^2 \leq a^2$. Since $Z^2 = F(1, \infty)$ by (B.16), it follows that $[z(0.975)]^2$ must equal $F(0.95; 1, \infty)$.

PROBLEMS

*B.1   Consider a $\chi^2(2)$ random variable.
    a. Obtain the 90th and 10th percentiles of the probability distribution.
    b. What is the value of the mean? The standard deviation?
    c. Are the percentiles in **a** equally spaced about the mean? What does this fact imply about the symmetry of the probability distribution?

B.2   Consider a $\chi^2(4)$ random variable.
    a. Obtain the 95th and 5th percentiles of the probability distribution.

     b. What is the value of the mean? The standard deviation?

     c. Are the percentiles in **a** equally spaced about the mean? What does this fact imply about the symmetry of the probability distribution?

*B.3   **a.** Obtain $\chi^2(0.95; 8)$. What does this value represent?

     b. Estimate $\chi^2(0.05; 145)$ by using the normal approximation.

B.4   **a.** Obtain $\chi^2(0.01; 12)$. What does this value represent?

     b. Estimate $\chi^2(0.90; 113)$ by using the normal approximation.

*B.5 Consider a $t(5)$ random variable.

     a. Obtain the 95th and 5th percentiles of the probability distribution.

     b. What is the value of the mean? The standard deviation?

B.6   Consider a $t(7)$ random variable.

     a. Obtain the 90th and 10th percentiles of the probability distribution.

     b. What is the value of the mean? The standard deviation?

*B.7   **a.** Obtain $t(0.90; 15)$ and $t(0.05; 21)$. What does the latter value represent?

     b. Estimate $t(0.01; 250)$ by using the normal approximation.

B.8   **a.** Obtain $t(0.995; 11)$ and $t(0.10; 18)$. What does the latter value represent?

     b. Estimate $t(0.975; 410)$ by using the normal approximation.

B.9   Plot the cumulative probability function of the $t(5)$ random variable and that of the standard normal random variable on the same graph. Compare and contrast the two functions.

B.10 Compare the interquartile range of the $t(3)$ random variable and that of the standard normal random variable. Does the comparison reflect the greater variance of t random variables? Explain.

*B.11 Consider an $F(2, 5)$ random variable.

     a. Obtain the 99th and 1st percentiles of the probability distribution.

     b. What is the value of the mean? The standard deviation?

     c. Are the percentiles in **a** equally spaced about the mean? What does this fact imply about the symmetry of the probability distribution?

B.12 Consider an $F(2, 6)$ random variable.

     a. Obtain the 95th and 5th percentiles of the probability distribution.

     b. What is the value of the mean? The standard deviation?

     c. Are the percentiles in **a** equally spaced about the mean? What does this fact imply about the symmetry of the probability distribution?

*B.13 **a.** Obtain $F(0.95; 2, 40)$. What does this value represent?

     b. Obtain $F(0.01; 8, 3)$.

B.14 **a.** Obtain $F(0.99; 8, 7)$. What does this value represent?

     b. Obtain $F(0.05; 4, 4)$.

EXERCISES

B.15 **a.** Given $\sigma^2\{Z^2\} = 2$ for a standard normal random variable Z, show that $\sigma^2\{\chi^2(\nu)\} = 2\nu$.

     b. Show that $\chi^2(\nu_1 + \nu_2) = \chi^2(\nu_1) + \chi^2(\nu_2)$ when $\chi^2(\nu_1)$ and $\chi^2(\nu_2)$ are statistically independent.

B.16 **a.** Explain why $[t(\nu)]^2 = F(1, \nu)$.

     b. Find $F(0.95; 1, 3)$ from Table C.3.

B.17 **a.** Explain why $\chi^2(\nu)/\nu = F(\nu, \infty)$.

     b. Find $F(0.99; 8, \infty)$ from Table C.2.

TABLES

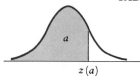

$z(a)$

TABLE C.1 Cumulative probabilities and percentiles of the standard normal distribution

(a) Cumulative probabilities

Entry is area a under the standard normal curve from $-\infty$ to $z(a)$.

z	.00	.01	.02	.03	.04	.05	.06	.07	.08	.09
.0	.5000	.5040	.5080	.5120	.5160	.5199	.5239	.5279	.5319	.5359
.1	.5398	.5438	.5478	.5517	.5557	.5596	.5636	.5675	.5714	.5753
.2	.5793	.5832	.5871	.5910	.5948	.5987	.6026	.6064	.6103	.6141
.3	.6179	.6217	.6255	.6293	.6331	.6368	.6406	.6443	.6480	.6517
.4	.6554	.6591	.6628	.6664	.6700	.6736	.6772	.6808	.6844	.6879
.5	.6915	.6950	.6985	.7019	.7054	.7088	.7123	.7157	.7190	.7224
.6	.7257	.7291	.7324	.7357	.7389	.7422	.7454	.7486	.7517	.7549
.7	.7580	.7611	.7642	.7673	.7704	.7734	.7764	.7794	.7823	.7852
.8	.7881	.7910	.7939	.7967	.7995	.8023	.8051	.8078	.8106	.8133
.9	.8159	.8186	.8212	.8238	.8264	.8289	.8315	.8340	.8365	.8389
1.0	.8413	.8438	.8461	.8485	.8508	.8531	.8554	.8577	.8599	.8621
1.1	.8643	.8665	.8686	.8708	.8729	.8749	.8770	.8790	.8810	.8830
1.2	.8849	.8869	.8888	.8907	.8925	.8944	.8962	.8980	.8997	.9015
1.3	.9032	.9049	.9066	.9082	.9099	.9115	.9131	.9147	.9162	.9177
1.4	.9192	.9207	.9222	.9236	.9251	.9265	.9279	.9292	.9306	.9319
1.5	.9332	.9345	.9357	.9370	.9382	.9394	.9406	.9418	.9429	.9441
1.6	.9452	.9463	.9474	.9484	.9495	.9505	.9515	.9525	.9535	.9545
1.7	.9554	.9564	.9573	.9582	.9591	.9599	.9608	.9616	.9625	.9633
1.8	.9641	.9649	.9656	.9664	.9671	.9678	.9686	.9693	.9699	.9706
1.9	.9713	.9719	.9726	.9732	.9738	.9744	.9750	.9756	.9761	.9767
2.0	.9772	.9778	.9783	.9788	.9793	.9798	.9803	.9808	.9812	.9817
2.1	.9821	.9826	.9830	.9834	.9838	.9842	.9846	.9850	.9854	.9857
2.2	.9861	.9864	.9868	.9871	.9875	.9878	.9881	.9884	.9887	.9890
2.3	.9893	.9896	.9898	.9901	.9904	.9906	.9909	.9911	.9913	.9916
2.4	.9918	.9920	.9922	.9925	.9927	.9929	.9931	.9932	.9934	.9936
2.5	.9938	.9940	.9941	.9943	.9945	.9946	.9948	.9949	.9951	.9952
2.6	.9953	.9955	.9956	.9957	.9959	.9960	.9961	.9962	.9963	.9964
2.7	.9965	.9966	.9967	.9968	.9969	.9970	.9971	.9972	.9973	.9974
2.8	.9974	.9975	.9976	.9977	.9977	.9978	.9979	.9979	.9980	.9981
2.9	.9981	.9982	.9982	.9983	.9984	.9984	.9985	.9985	.9986	.9986
3.0	.9987	.9987	.9987	.9988	.9988	.9989	.9989	.9989	.9990	.9990
3.1	.9990	.9991	.9991	.9991	.9992	.9992	.9992	.9992	.9993	.9993
3.2	.9993	.9993	.9994	.9994	.9994	.9994	.9994	.9995	.9995	.9995
3.3	.9995	.9995	.9995	.9996	.9996	.9996	.9996	.9996	.9996	.9997
3.4	.9997	.9997	.9997	.9997	.9997	.9997	.9997	.9997	.9997	.9998

(b) Selected percentiles

Entry is $z(a)$ where $P[Z \le z(a)] = a$.

a:	.10	.05	.025	.02	.01	.005	.001
$z(a)$:	-1.282	-1.645	-1.960	-2.054	-2.326	-2.576	-3.090

a:	.90	.95	.975	.98	.99	.995	.999
$z(a)$:	1.282	1.645	1.960	2.054	2.326	2.576	3.090

EXAMPLE: $P(Z \le 1.96) = 0.9750$ so $z(0.9750) = 1.96$.
TEXT REFERENCE: Use of this table is discussed on pp. 214–220.

948

TABLE C.2 Percentiles of the chi-square distribution

Entry is $\chi^2(a;v)$ where $P[\chi^2(v) \leq \chi^2(a;v)] = a$.

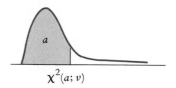

$\chi^2(a;v)$

df					a					
v	.005	.010	.025	.050	.100	.900	.950	.975	.990	.995
1	0.0^4393	0.0^3157	0.0^3982	0.0^2393	0.0158	2.71	3.84	5.02	6.63	7.88
2	0.0100	0.0201	0.0506	0.103	0.211	4.61	5.99	7.38	9.21	10.60
3	0.072	0.115	0.216	0.352	0.584	6.25	7.81	9.35	11.34	12.84
4	0.207	0.297	0.484	0.711	1.064	7.78	9.49	11.14	13.28	14.86
5	0.412	0.554	0.831	1.145	1.61	9.24	11.07	12.83	15.09	16.75
6	0.676	0.872	1.24	1.64	2.20	10.64	12.59	14.45	16.81	18.55
7	0.989	1.24	1.69	2.17	2.83	12.02	14.07	16.01	18.48	20.28
8	1.34	1.65	2.18	2.73	3.49	13.36	15.51	17.53	20.09	21.96
9	1.73	2.09	2.70	3.33	4.17	14.68	16.92	19.02	21.67	23.59
10	2.16	2.56	3.25	3.94	4.87	15.99	18.31	20.48	23.21	25.19
11	2.60	3.05	3.82	4.57	5.58	17.28	19.68	21.92	24.73	26.76
12	3.07	3.57	4.40	5.23	6.30	18.55	21.03	23.34	26.22	28.30
13	3.57	4.11	5.01	5.89	7.04	19.81	22.36	24.74	27.69	29.82
14	4.07	4.66	5.63	6.57	7.79	21.06	23.68	26.12	29.14	31.32
15	4.60	5.23	6.26	7.26	8.55	22.31	25.00	27.49	30.58	32.80
16	5.14	5.81	6.91	7.96	9.31	23.54	26.30	28.85	32.00	34.27
17	5.70	6.41	7.56	8.67	10.09	24.77	27.59	30.19	33.41	35.72
18	6.26	7.01	8.23	9.39	10.86	25.99	28.87	31.53	34.81	37.16
19	6.84	7.63	8.91	10.12	11.65	27.20	30.14	32.85	36.19	38.58
20	7.43	8.26	9.59	10.85	12.44	28.41	31.41	34.17	37.57	40.00
21	8.03	8.90	10.28	11.59	13.24	29.62	32.67	35.48	38.93	41.40
22	8.64	9.54	10.98	12.34	14.04	30.81	33.92	36.78	40.29	42.80
23	9.26	10.20	11.69	13.09	14.85	32.01	35.17	38.08	41.64	44.18
24	9.89	10.86	12.40	13.85	15.66	33.20	36.42	39.36	42.98	45.56
25	10.52	11.52	13.12	14.61	16.47	34.38	37.65	40.65	44.31	46.93
26	11.16	12.20	13.84	15.38	17.29	35.56	38.89	41.92	45.64	48.29
27	11.81	12.88	14.57	16.15	18.11	36.74	40.11	43.19	46.96	49.64
28	12.46	13.56	15.31	16.93	18.94	37.92	41.34	44.46	48.28	50.99
29	13.12	14.26	16.05	17.71	19.77	39.09	42.56	45.72	49.59	52.34
30	13.79	14.95	16.79	18.49	20.60	40.26	43.77	46.98	50.89	53.67
40	20.71	22.16	24.43	26.51	29.05	51.81	55.76	59.34	63.69	66.77
50	27.99	29.71	32.36	34.76	37.69	63.17	67.50	71.42	76.15	79.49
60	35.53	37.48	40.48	43.19	46.46	74.40	79.08	83.30	88.38	91.95
70	43.28	45.44	48.76	51.74	55.33	85.53	90.53	95.02	100.4	104.2
80	51.17	53.54	57.15	60.39	64.28	96.58	101.9	106.6	112.3	116.3
90	59.20	61.75	65.65	69.13	73.29	107.6	113.1	118.1	124.1	128.3
100	67.33	70.06	74.22	77.93	82.36	118.5	124.3	129.6	135.8	140.2

SOURCE: Tabulated values adapted by permission from C. M. Thompson, "Table of Percentage Points of the Chi-Square Distribution," *Biometrika*, Vol. 32 (1941), pp. 188–189.
EXAMPLE: $\chi^2(0.900;4) = 7.78$ so $P[\chi^2(4) \leq 7.78] = 0.900$.
TEXT REFERENCE: Use of this table is discussed on p. 937.

TABLE C.3   Percentiles of the *t* distribution

Entry is $t(a;\nu)$ where $P[t(\nu) \le t(a;\nu)] = a$.

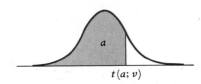

$$t(a;\nu)$$

df	a						
ν	.75	.90	.95	.975	.99	.995	.9995
1	1.000	3.078	6.314	12.706	31.821	63.657	636.619
2	0.816	1.886	2.920	4.303	6.965	9.925	31.599
3	0.765	1.638	2.353	3.182	4.541	5.841	12.924
4	0.741	1.533	2.132	2.776	3.747	4.604	8.610
5	0.727	1.476	2.015	2.571	3.365	4.032	6.869
6	0.718	1.440	1.943	2.447	3.143	3.707	5.959
7	0.711	1.415	1.895	2.365	2.998	3.499	5.408
8	0.706	1.397	1.860	2.306	2.896	3.355	5.041
9	0.703	1.383	1.833	2.262	2.821	3.250	4.781
10	0.700	1.372	1.812	2.228	2.764	3.169	4.587
11	0.697	1.363	1.796	2.201	2.718	3.106	4.437
12	0.695	1.356	1.782	2.179	2.681	3.055	4.318
13	0.694	1.350	1.771	2.160	2.650	3.012	4.221
14	0.692	1.345	1.761	2.145	2.624	2.977	4.140
15	0.691	1.341	1.753	2.131	2.602	2.947	4.073
16	0.690	1.337	1.746	2.120	2.583	2.921	4.015
17	0.689	1.333	1.740	2.110	2.567	2.898	3.965
18	0.688	1.330	1.734	2.101	2.552	2.878	3.922
19	0.688	1.328	1.729	2.093	2.539	2.861	3.883
20	0.687	1.325	1.725	2.086	2.528	2.845	3.850
21	0.686	1.323	1.721	2.080	2.518	2.831	3.819
22	0.686	1.321	1.717	2.074	2.508	2.819	3.792
23	0.685	1.319	1.714	2.069	2.500	2.807	3.768
24	0.685	1.318	1.711	2.064	2.492	2.797	3.745
25	0.684	1.316	1.708	2.060	2.485	2.787	3.725
26	0.684	1.315	1.706	2.056	2.479	2.779	3.707
27	0.684	1.314	1.703	2.052	2.473	2.771	3.690
28	0.683	1.313	1.701	2.048	2.467	2.763	3.674
29	0.683	1.311	1.699	2.045	2.462	2.756	3.659
30	0.683	1.310	1.697	2.042	2.457	2.750	3.646
40	0.681	1.303	1.684	2.021	2.423	2.704	3.551
60	0.679	1.296	1.671	2.000	2.390	2.660	3.460
120	0.677	1.289	1.658	1.980	2.358	2.617	3.373
∞	0.674	1.282	1.645	1.960	2.326	2.576	3.291

EXAMPLE: $t(0.95;10) = 1.812$ so $P[t(10) \le 1.812] = 0.95$.
TEXT REFERENCE: Use of this table is discussed on p. 939.

TABLE C.4 Percentiles of the F distribution

Entry is $F(a; \nu_1, \nu_2)$ where $P[F(\nu_1, \nu_2) \leq F(a; \nu_1, \nu_2)] = a$.

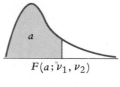

$$F(a; \nu_1, \nu_2)$$

$$a = .95$$

denominator df	numerator df								
	1	**2**	**3**	**4**	**5**	**6**	**7**	**8**	**9**
1	161.4	199.5	215.7	224.6	230.2	234.0	236.8	238.9	240.5
2	18.51	19.00	19.16	19.25	19.30	19.33	19.35	19.37	19.38
3	10.13	9.55	9.28	9.12	9.01	8.94	8.89	8.85	8.81
4	7.71	6.94	6.59	6.39	6.26	6.16	6.09	6.04	6.00
5	6.61	5.79	5.41	5.19	5.05	4.95	4.88	4.82	4.77
6	5.99	5.14	4.76	4.53	4.39	4.28	4.21	4.15	4.10
7	5.59	4.74	4.35	4.12	3.97	3.87	3.79	3.73	3.68
8	5.32	4.46	4.07	3.84	3.69	3.58	3.50	3.44	3.39
9	5.12	4.26	3.86	3.63	3.48	3.37	3.29	3.23	3.18
10	4.96	4.10	3.71	3.48	3.33	3.22	3.14	3.07	3.02
11	4.84	3.98	3.59	3.36	3.20	3.09	3.01	2.95	2.90
12	4.75	3.89	3.49	3.26	3.11	3.00	2.91	2.85	2.80
13	4.67	3.81	3.41	3.18	3.03	2.92	2.83	2.77	2.71
14	4.60	3.74	3.34	3.11	2.96	2.85	2.76	2.70	2.65
15	4.54	3.68	3.29	3.06	2.90	2.79	2.71	2.64	2.59
16	4.49	3.63	3.24	3.01	2.85	2.74	2.66	2.59	2.54
17	4.45	3.59	3.20	2.96	2.81	2.70	2.61	2.55	2.49
18	4.41	3.55	3.16	2.93	2.77	2.66	2.58	2.51	2.46
19	4.38	3.52	3.13	2.90	2.74	2.63	2.54	2.48	2.42
20	4.35	3.49	3.10	2.87	2.71	2.60	2.51	2.45	2.39
21	4.32	3.47	3.07	2.84	2.68	2.57	2.49	2.42	2.37
22	4.30	3.44	3.05	2.82	2.66	2.55	2.46	2.40	2.34
23	4.28	3.42	3.03	2.80	2.64	2.53	2.44	2.37	2.32
24	4.26	3.40	3.01	2.78	2.62	2.51	2.42	2.36	2.30
25	4.24	3.39	2.99	2.76	2.60	2.49	2.40	2.34	2.28
26	4.23	3.37	2.98	2.74	2.59	2.47	2.39	2.32	2.27
27	4.21	3.35	2.96	2.73	2.57	2.46	2.37	2.31	2.25
28	4.20	3.34	2.95	2.71	2.56	2.45	2.36	2.29	2.24
29	4.18	3.33	2.93	2.70	2.55	2.43	2.35	2.28	2.22
30	4.17	3.32	2.92	2.69	2.53	2.42	2.33	2.27	2.21
40	4.08	3.23	2.84	2.61	2.45	2.34	2.25	2.18	2.12
60	4.00	3.15	2.76	2.53	2.37	2.25	2.17	2.10	2.04
120	3.92	3.07	2.68	2.45	2.29	2.17	2.09	2.02	1.96
∞	3.84	3.00	2.60	2.37	2.21	2.10	2.01	1.94	1.88

TABLE C.4   Percentiles of the F distribution (continued)

$a = .95$

numerator df										denominator df
10	12	15	20	24	30	40	60	120	∞	
241.9	243.9	245.9	248.0	249.1	250.1	251.1	252.2	253.3	254.3	1
19.40	19.41	19.43	19.45	19.45	19.46	19.47	19.48	19.49	19.50	2
8.79	8.74	8.70	8.66	8.64	8.62	8.59	8.57	8.55	8.53	3
5.96	5.91	5.86	5.80	5.77	5.75	5.72	5.69	5.66	5.63	4
4.74	4.68	4.62	4.56	4.53	4.50	4.46	4.43	4.40	4.36	5
4.06	4.00	3.94	3.87	3.84	3.81	3.77	3.74	3.70	3.67	6
3.64	3.57	3.51	3.44	3.41	3.38	3.34	3.30	3.27	3.23	7
3.35	3.28	3.22	3.15	3.12	3.08	3.04	3.01	2.97	2.93	8
3.14	3.07	3.01	2.94	2.90	2.86	2.83	2.79	2.75	2.71	9
2.98	2.91	2.85	2.77	2.74	2.70	2.66	2.62	2.58	2.54	10
2.85	2.79	2.72	2.65	2.61	2.57	2.53	2.49	2.45	2.40	11
2.75	2.69	2.62	2.54	2.51	2.47	2.43	2.38	2.34	2.30	12
2.67	2.60	2.53	2.46	2.42	2.38	2.34	2.30	2.25	2.21	13
2.60	2.53	2.46	2.39	2.35	2.31	2.27	2.22	2.18	2.13	14
2.54	2.48	2.40	2.33	2.29	2.25	2.20	2.16	2.11	2.07	15
2.49	2.42	2.35	2.28	2.24	2.19	2.15	2.11	2.06	2.01	16
2.45	2.38	2.31	2.23	2.19	2.15	2.10	2.06	2.01	1.96	17
2.41	2.34	2.27	2.19	2.15	2.11	2.06	2.02	1.97	1.92	18
2.38	2.31	2.23	2.16	2.11	2.07	2.03	1.98	1.93	1.88	19
2.35	2.28	2.20	2.12	2.08	2.04	1.99	1.95	1.90	1.84	20
2.32	2.25	2.18	2.10	2.05	2.01	1.96	1.92	1.87	1.81	21
2.30	2.23	2.15	2.07	2.03	1.98	1.94	1.89	1.84	1.78	22
2.27	2.20	2.13	2.05	2.01	1.96	1.91	1.86	1.81	1.76	23
2.25	2.18	2.11	2.03	1.98	1.94	1.89	1.84	1.79	1.73	24
2.24	2.16	2.09	2.01	1.96	1.92	1.87	1.82	1.77	1.71	25
2.22	2.15	2.07	1.99	1.95	1.90	1.85	1.80	1.75	1.69	26
2.20	2.13	2.06	1.97	1.93	1.88	1.84	1.79	1.73	1.67	27
2.19	2.12	2.04	1.96	1.91	1.87	1.82	1.77	1.71	1.65	28
2.18	2.10	2.03	1.94	1.90	1.85	1.81	1.75	1.70	1.64	29
2.16	2.09	2.01	1.93	1.89	1.84	1.79	1.74	1.68	1.62	30
2.08	2.00	1.92	1.84	1.79	1.74	1.69	1.64	1.58	1.51	40
1.99	1.92	1.84	1.75	1.70	1.65	1.59	1.53	1.47	1.39	60
1.91	1.83	1.75	1.66	1.61	1.55	1.50	1.43	1.35	1.25	120
1.83	1.75	1.67	1.57	1.52	1.46	1.39	1.32	1.22	1.00	∞

TABLE C.4   **Percentiles of the F distribution (continued)**

$$a = .99$$

denominator df	numerator df								
	1	**2**	**3**	**4**	**5**	**6**	**7**	**8**	**9**
1	4052	4999.5	5403	5625	5764	5859	5928	5981	6022
2	98.50	99.00	99.17	99.25	99.30	99.33	99.36	99.37	99.39
3	34.12	30.82	29.46	28.71	28.24	27.91	27.67	27.49	27.35
4	21.20	18.00	16.69	15.98	15.52	15.21	14.98	14.80	14.66
5	16.26	13.27	12.06	11.39	10.97	10.67	10.46	10.29	10.16
6	13.75	10.92	9.78	9.15	8.75	8.47	8.26	8.10	7.98
7	12.25	9.55	8.45	7.85	7.46	7.19	6.99	6.84	6.72
8	11.26	8.65	7.59	7.01	6.63	6.37	6.18	6.03	5.91
9	10.56	8.02	6.99	6.42	6.06	5.80	5.61	5.47	5.35
10	10.04	7.56	6.55	5.99	5.64	5.39	5.20	5.06	4.94
11	9.65	7.21	6.22	5.67	5.32	5.07	4.89	4.74	4.63
12	9.33	6.93	5.95	5.41	5.06	4.82	4.64	4.50	4.39
13	9.07	6.70	5.74	5.21	4.86	4.62	4.44	4.30	4.19
14	8.86	6.51	5.56	5.04	4.69	4.46	4.28	4.14	4.03
15	8.68	6.36	5.42	4.89	4.56	4.32	4.14	4.00	3.89
16	8.53	6.23	5.29	4.77	4.44	4.20	4.03	3.89	3.78
17	8.40	6.11	5.18	4.67	4.34	4.10	3.93	3.79	3.68
18	8.29	6.01	5.09	4.58	4.25	4.01	3.84	3.71	3.60
19	8.18	5.93	5.01	4.50	4.17	3.94	3.77	3.63	3.52
20	8.10	5.85	4.94	4.43	4.10	3.87	3.70	3.56	3.46
21	8.02	5.78	4.87	4.37	4.04	3.81	3.64	3.51	3.40
22	7.95	5.72	4.82	4.31	3.99	3.76	3.59	3.45	3.35
23	7.88	5.66	4.76	4.26	3.94	3.71	3.54	3.41	3.30
24	7.82	5.61	4.72	4.22	3.90	3.67	3.50	3.36	3.26
25	7.77	5.57	4.68	4.18	3.85	3.63	3.46	3.32	3.22
26	7.72	5.53	4.64	4.14	3.82	3.59	3.42	3.29	3.18
27	7.68	5.49	4.60	4.11	3.78	3.56	3.39	3.26	3.15
28	7.64	5.45	4.57	4.07	3.75	3.53	3.36	3.23	3.12
29	7.60	5.42	4.54	4.04	3.73	3.50	3.33	3.20	3.09
30	7.56	5.39	4.51	4.02	3.70	3.47	3.30	3.17	3.07
40	7.31	5.18	4.31	3.83	3.51	3.29	3.12	2.99	2.89
60	7.08	4.98	4.13	3.65	3.34	3.12	2.95	2.82	2.72
120	6.85	4.79	3.95	3.48	3.17	2.96	2.79	2.66	2.56
∞	6.63	4.61	3.78	3.32	3.02	2.80	2.64	2.51	2.41

TABLE C.4   Percentiles of the F distribution (concluded)

$$a = .99$$

numerator df										denominator df
10	**12**	**15**	**20**	**24**	**30**	**40**	**60**	**120**	**∞**	
6056	6106	6157	6209	6235	6261	6287	6313	6339	6366	1
99.40	99.42	99.43	99.45	99.46	99.47	99.47	99.48	99.49	99.50	2
27.23	27.05	26.87	26.69	26.60	26.50	26.41	26.32	26.22	26.13	3
14.55	14.37	14.20	14.02	13.93	13.84	13.75	13.65	13.56	13.46	4
10.05	9.89	9.72	9.55	9.47	9.38	9.29	9.20	9.11	9.02	5
7.87	7.72	7.56	7.40	7.31	7.23	7.14	7.06	6.97	6.88	6
6.62	6.47	6.31	6.16	6.07	5.99	5.91	5.82	5.74	5.65	7
5.81	5.67	5.52	5.36	5.28	5.20	5.12	5.03	4.95	4.86	8
5.26	5.11	4.96	4.81	4.73	4.65	4.57	4.48	4.40	4.31	9
4.85	4.71	4.56	4.41	4.33	4.25	4.17	4.08	4.00	3.91	10
4.54	4.40	4.25	4.10	4.02	3.94	3.86	3.78	3.69	3.60	11
4.30	4.16	4.01	3.86	3.78	3.70	3.62	3.54	3.45	3.36	12
4.10	3.96	3.82	3.66	3.59	3.51	3.43	3.34	3.25	3.17	13
3.94	3.80	3.66	3.51	3.43	3.35	3.27	3.18	3.09	3.00	14
3.80	3.67	3.52	3.37	3.29	3.21	3.13	3.05	2.96	2.87	15
3.69	3.55	3.41	3.26	3.18	3.10	3.02	2.93	2.84	2.75	16
3.59	3.46	3.31	3.16	3.08	3.00	2.92	2.83	2.75	2.65	17
3.51	3.37	3.23	3.08	3.00	2.92	2.84	2.75	2.66	2.57	18
3.43	3.30	3.15	3.00	2.92	2.84	2.76	2.67	2.58	2.49	19
3.37	3.23	3.09	2.94	2.86	2.78	2.69	2.61	2.52	2.42	20
3.31	3.17	3.03	2.88	2.80	2.72	2.64	2.55	2.46	2.36	21
3.26	3.12	2.98	2.83	2.75	2.67	2.58	2.50	2.40	2.31	22
3.21	3.07	2.93	2.78	2.70	2.62	2.54	2.45	2.35	2.26	23
3.17	3.03	2.89	2.74	2.66	2.58	2.49	2.40	2.31	2.21	24
3.13	2.99	2.85	2.70	2.62	2.54	2.45	2.36	2.27	2.17	25
3.09	2.96	2.81	2.66	2.58	2.50	2.42	2.33	2.23	2.13	26
3.06	2.93	2.78	2.63	2.55	2.47	2.38	2.29	2.20	2.10	27
3.03	2.90	2.75	2.60	2.52	2.44	2.35	2.26	2.17	2.06	28
3.00	2.87	2.73	2.57	2.49	2.41	2.33	2.23	2.14	2.03	29
2.98	2.84	2.70	2.55	2.47	2.39	2.30	2.21	2.11	2.01	30
2.80	2.66	2.52	2.37	2.29	2.20	2.11	2.02	1.92	1.80	40
2.63	2.50	2.35	2.20	2.12	2.03	1.94	1.84	1.73	1.60	60
2.47	2.34	2.19	2.03	1.95	1.86	1.76	1.66	1.53	1.38	120
2.32	2.18	2.04	1.88	1.79	1.70	1.59	1.47	1.32	1.00	∞

SOURCE: Tabulated values adapted from Table 5 of Pearson and Hartley, *Biometrika Tables for Statisticians,* Volume 2, 1972, published by the Cambridge University Press for the Biometrika Trustees, with the permission of the authors and publishers.

EXAMPLE: $F(0.99; 8, 24) = 3.36$ so $P[F(8, 24) \leq 3.36] = 0.99$.

TEXT REFERENCE: Use of this table is discussed on pp. 940–941.

TABLE C.5 Binomial probabilities

Entry is probability $P(X = x) = \binom{n}{x}p^x(1 - p)^{n-x}$.

						p							
n	x	.01	.02	.03	.04	.05	.06	.07	.08	.09		
2	0	0.9801	0.9604	0.9409	0.9216	0.9025	0.8836	0.8649	0.8464	0.8281	2	
	1	0.0198	0.0392	0.0582	0.0768	0.0950	0.1128	0.1302	0.1472	0.1638	1	
	2	0.0001	0.0004	0.0009	0.0016	0.0025	0.0036	0.0049	0.0064	0.0081	0	2
3	0	0.9703	0.9412	0.9127	0.8847	0.8574	0.8306	0.8044	0.7787	0.7536	3	
	1	0.0294	0.0576	0.0847	0.1106	0.1354	0.1590	0.1816	0.2031	0.2236	2	
	2	0.0003	0.0012	0.0026	0.0046	0.0071	0.0102	0.0137	0.0177	0.0221	1	
	3	0.0000	0.0000	0.0000	0.0001	0.0001	0.0002	0.0003	0.0005	0.0007	0	3
4	0	0.9606	0.9224	0.8853	0.8493	0.8145	0.7807	0.7481	0.7164	0.6857	4	
	1	0.0388	0.0753	0.1095	0.1416	0.1715	0.1993	0.2252	0.2492	0.2713	3	
	2	0.0006	0.0023	0.0051	0.0088	0.0135	0.0191	0.0254	0.0325	0.0402	2	
	3	0.0000	0.0000	0.0001	0.0002	0.0005	0.0008	0.0013	0.0019	0.0027	1	
	4	0.0000	0.0000	0.0000	0.0000	0.0000	0.0000	0.0000	0.0000	0.0001	0	4
5	0	0.9510	0.9039	0.8587	0.8154	0.7738	0.7339	0.6957	0.6591	0.6240	5	
	1	0.0480	0.0922	0.1328	0.1699	0.2036	0.2342	0.2618	0.2866	0.3086	4	
	2	0.0010	0.0038	0.0082	0.0142	0.0214	0.0299	0.0394	0.0498	0.0610	3	
	3	0.0000	0.0001	0.0003	0.0006	0.0011	0.0019	0.0030	0.0043	0.0060	2	
	4	0.0000	0.0000	0.0000	0.0000	0.0000	0.0001	0.0001	0.0002	0.0003	1	
	5	0.0000	0.0000	0.0000	0.0000	0.0000	0.0000	0.0000	0.0000	0.0000	0	5
6	0	0.9415	0.8858	0.8330	0.7828	0.7351	0.6899	0.6470	0.6064	0.5679	6	
	1	0.0571	0.1085	0.1546	0.1957	0.2321	0.2642	0.2922	0.3164	0.3370	5	
	2	0.0014	0.0055	0.0120	0.0204	0.0305	0.0422	0.0550	0.0688	0.0833	4	
	3	0.0000	0.0002	0.0005	0.0011	0.0021	0.0036	0.0055	0.0080	0.0110	3	
	4	0.0000	0.0000	0.0000	0.0000	0.0001	0.0002	0.0003	0.0005	0.0008	2	
	5	0.0000	0.0000	0.0000	0.0000	0.0000	0.0000	0.0000	0.0000	0.0000	1	
	6	0.0000	0.0000	0.0000	0.0000	0.0000	0.0000	0.0000	0.0000	0.0000	0	6
7	0	0.9321	0.8681	0.8080	0.7514	0.6983	0.6485	0.6017	0.5578	0.5168	7	
	1	0.0659	0.1240	0.1749	0.2192	0.2573	0.2897	0.3170	0.3396	0.3578	6	
	2	0.0020	0.0076	0.0162	0.0274	0.0406	0.0555	0.0716	0.0886	0.1061	5	
	3	0.0000	0.0003	0.0008	0.0019	0.0036	0.0059	0.0090	0.0128	0.0175	4	
	4	0.0000	0.0000	0.0000	0.0001	0.0002	0.0004	0.0007	0.0011	0.0017	3	
	5	0.0000	0.0000	0.0000	0.0000	0.0000	0.0000	0.0000	0.0001	0.0001	2	
	6	0.0000	0.0000	0.0000	0.0000	0.0000	0.0000	0.0000	0.0000	0.0000	1	
	7	0.0000	0.0000	0.0000	0.0000	0.0000	0.0000	0.0000	0.0000	0.0000	0	7
8	0	0.9227	0.8508	0.7837	0.7214	0.6634	0.6096	0.5596	0.5132	0.4703	8	
	1	0.0746	0.1389	0.1939	0.2405	0.2793	0.3113	0.3370	0.3570	0.3721	7	
	2	0.0026	0.0099	0.0210	0.0351	0.0515	0.0695	0.0888	0.1087	0.1288	6	
	3	0.0001	0.0004	0.0013	0.0029	0.0054	0.0089	0.0134	0.0189	0.0255	5	
	4	0.0000	0.0000	0.0001	0.0002	0.0004	0.0007	0.0013	0.0021	0.0031	4	
	5	0.0000	0.0000	0.0000	0.0000	0.0000	0.0000	0.0001	0.0001	0.0002	3	
	6	0.0000	0.0000	0.0000	0.0000	0.0000	0.0000	0.0000	0.0000	0.0000	2	
	7	0.0000	0.0000	0.0000	0.0000	0.0000	0.0000	0.0000	0.0000	0.0000	1	
	8	0.0000	0.0000	0.0000	0.0000	0.0000	0.0000	0.0000	0.0000	0.0000	0	8
9	0	0.9135	0.8337	0.7602	0.6925	0.6302	0.5730	0.5204	0.4722	0.4279	9	
	1	0.0830	0.1531	0.2116	0.2597	0.2985	0.3292	0.3525	0.3695	0.3809	8	
	2	0.0034	0.0125	0.0262	0.0433	0.0629	0.0840	0.1061	0.1285	0.1507	7	
	3	0.0001	0.0006	0.0019	0.0042	0.0077	0.0125	0.0186	0.0261	0.0348	6	
	4	0.0000	0.0000	0.0001	0.0003	0.0006	0.0012	0.0021	0.0034	0.0052	5	
	5	0.0000	0.0000	0.0000	0.0000	0.0000	0.0001	0.0002	0.0003	0.0005	4	
	6	0.0000	0.0000	0.0000	0.0000	0.0000	0.0000	0.0000	0.0000	0.0000	3	
	7	0.0000	0.0000	0.0000	0.0000	0.0000	0.0000	0.0000	0.0000	0.0000	2	
	8	0.0000	0.0000	0.0000	0.0000	0.0000	0.0000	0.0000	0.0000	0.0000	1	
	9	0.0000	0.0000	0.0000	0.0000	0.0000	0.0000	0.0000	0.0000	0.0000	0	9
		.99	.98	.97	.96	.95	.94	.93	.92	.91	x	n
							p						

TABLE C.5 Binomial probabilities (continued)   955

						p						
n	x	.01	.02	.03	.04	.05	.06	.07	.08	.09		
10	0	0.9044	0.8171	0.7374	0.6648	0.5987	0.5386	0.4840	0.4344	0.3894	10	
	1	0.0914	0.1667	0.2281	0.2770	0.3151	0.3438	0.3643	0.3777	0.3851	9	
	2	0.0042	0.0153	0.0317	0.0519	0.0746	0.0988	0.1234	0.1478	0.1714	8	
	3	0.0001	0.0008	0.0026	0.0058	0.0105	0.0168	0.0248	0.0343	0.0452	7	
	4	0.0000	0.0000	0.0001	0.0004	0.0010	0.0019	0.0033	0.0052	0.0078	6	
	5	0.0000	0.0000	0.0000	0.0000	0.0001	0.0001	0.0003	0.0005	0.0009	5	
	6	0.0000	0.0000	0.0000	0.0000	0.0000	0.0000	0.0000	0.0000	0.0001	4	
	7	0.0000	0.0000	0.0000	0.0000	0.0000	0.0000	0.0000	0.0000	0.0000	3	
	8	0.0000	0.0000	0.0000	0.0000	0.0000	0.0000	0.0000	0.0000	0.0000	2	
	9	0.0000	0.0000	0.0000	0.0000	0.0000	0.0000	0.0000	0.0000	0.0000	1	
	10	0.0000	0.0000	0.0000	0.0000	0.0000	0.0000	0.0000	0.0000	0.0000	0	10
12	0	0.8864	0.7847	0.6938	0.6127	0.5404	0.4759	0.4186	0.3677	0.3225	12	
	1	0.1074	0.1922	0.2575	0.3064	0.3413	0.3645	0.3781	0.3837	0.3827	11	
	2	0.0060	0.0216	0.0438	0.0702	0.0988	0.1280	0.1565	0.1835	0.2082	10	
	3	0.0002	0.0015	0.0045	0.0098	0.0173	0.0272	0.0393	0.0532	0.0686	9	
	4	0.0000	0.0001	0.0003	0.0009	0.0021	0.0039	0.0067	0.0104	0.0153	8	
	5	0.0000	0.0000	0.0000	0.0001	0.0002	0.0004	0.0008	0.0014	0.0024	7	
	6	0.0000	0.0000	0.0000	0.0000	0.0000	0.0000	0.0001	0.0001	0.0003	6	
	7	0.0000	0.0000	0.0000	0.0000	0.0000	0.0000	0.0000	0.0000	0.0000	5	
	8	0.0000	0.0000	0.0000	0.0000	0.0000	0.0000	0.0000	0.0000	0.0000	4	
	9	0.0000	0.0000	0.0000	0.0000	0.0000	0.0000	0.0000	0.0000	0.0000	3	
	10	0.0000	0.0000	0.0000	0.0000	0.0000	0.0000	0.0000	0.0000	0.0000	2	
	11	0.0000	0.0000	0.0000	0.0000	0.0000	0.0000	0.0000	0.0000	0.0000	1	
	12	0.0000	0.0000	0.0000	0.0000	0.0000	0.0000	0.0000	0.0000	0.0000	0	12
15	0	0.8601	0.7386	0.6333	0.5421	0.4633	0.3953	0.3367	0.2863	0.2430	15	
	1	0.1303	0.2261	0.2938	0.3388	0.3658	0.3785	0.3801	0.3734	0.3605	14	
	2	0.0092	0.0323	0.0636	0.0988	0.1348	0.1691	0.2003	0.2273	0.2496	13	
	3	0.0004	0.0029	0.0085	0.0178	0.0307	0.0468	0.0653	0.0857	0.1070	12	
	4	0.0000	0.0002	0.0008	0.0022	0.0049	0.0090	0.0148	0.0223	0.0317	11	
	5	0.0000	0.0000	0.0001	0.0002	0.0006	0.0013	0.0024	0.0043	0.0069	10	
	6	0.0000	0.0000	0.0000	0.0000	0.0000	0.0001	0.0003	0.0006	0.0011	9	
	7	0.0000	0.0000	0.0000	0.0000	0.0000	0.0000	0.0000	0.0001	0.0001	8	
	8	0.0000	0.0000	0.0000	0.0000	0.0000	0.0000	0.0000	0.0000	0.0000	7	
	9	0.0000	0.0000	0.0000	0.0000	0.0000	0.0000	0.0000	0.0000	0.0000	6	
	10	0.0000	0.0000	0.0000	0.0000	0.0000	0.0000	0.0000	0.0000	0.0000	5	
	11	0.0000	0.0000	0.0000	0.0000	0.0000	0.0000	0.0000	0.0000	0.0000	4	
	12	0.0000	0.0000	0.0000	0.0000	0.0000	0.0000	0.0000	0.0000	0.0000	3	
	13	0.0000	0.0000	0.0000	0.0000	0.0000	0.0000	0.0000	0.0000	0.0000	2	
	14	0.0000	0.0000	0.0000	0.0000	0.0000	0.0000	0.0000	0.0000	0.0000	1	
	15	0.0000	0.0000	0.0000	0.0000	0.0000	0.0000	0.0000	0.0000	0.0000	0	15
20	0	0.8179	0.6676	0.5438	0.4420	0.3585	0.2901	0.2342	0.1887	0.1516	20	
	1	0.1652	0.2725	0.3364	0.3683	0.3774	0.3703	0.3526	0.3282	0.3000	19	
	2	0.0159	0.0528	0.0988	0.1458	0.1887	0.2246	0.2521	0.2711	0.2818	18	
	3	0.0010	0.0065	0.0183	0.0364	0.0596	0.0860	0.1139	0.1414	0.1672	17	
	4	0.0000	0.0006	0.0024	0.0065	0.0133	0.0233	0.0364	0.0523	0.0703	16	
	5	0.0000	0.0000	0.0002	0.0009	0.0022	0.0048	0.0088	0.0145	0.0222	15	
	6	0.0000	0.0000	0.0000	0.0001	0.0003	0.0008	0.0017	0.0032	0.0055	14	
	7	0.0000	0.0000	0.0000	0.0000	0.0000	0.0001	0.0002	0.0005	0.0011	13	
	8	0.0000	0.0000	0.0000	0.0000	0.0000	0.0000	0.0000	0.0001	0.0002	12	
	9	0.0000	0.0000	0.0000	0.0000	0.0000	0.0000	0.0000	0.0000	0.0000	11	
	10	0.0000	0.0000	0.0000	0.0000	0.0000	0.0000	0.0000	0.0000	0.0000	10	
	11	0.0000	0.0000	0.0000	0.0000	0.0000	0.0000	0.0000	0.0000	0.0000	9	
	12	0.0000	0.0000	0.0000	0.0000	0.0000	0.0000	0.0000	0.0000	0.0000	8	
	13	0.0000	0.0000	0.0000	0.0000	0.0000	0.0000	0.0000	0.0000	0.0000	7	
	14	0.0000	0.0000	0.0000	0.0000	0.0000	0.0000	0.0000	0.0000	0.0000	6	
	15	0.0000	0.0000	0.0000	0.0000	0.0000	0.0000	0.0000	0.0000	0.0000	5	
	16	0.0000	0.0000	0.0000	0.0000	0.0000	0.0000	0.0000	0.0000	0.0000	4	
	17	0.0000	0.0000	0.0000	0.0000	0.0000	0.0000	0.0000	0.0000	0.0000	3	
	18	0.0000	0.0000	0.0000	0.0000	0.0000	0.0000	0.0000	0.0000	0.0000	2	
	19	0.0000	0.0000	0.0000	0.0000	0.0000	0.0000	0.0000	0.0000	0.0000	1	
	20	0.0000	0.0000	0.0000	0.0000	0.0000	0.0000	0.0000	0.0000	0.0000	0	20
		.99	.98	.97	.96	.95	.94	.93	.92	.91	x	n
						p						

						p						
n	x	.10	.15	.20	.25	.30	.35	.40	.45	.50		
2	0	0.8100	0.7225	0.6400	0.5625	0.4900	0.4225	0.3600	0.3025	0.2500	2	
	1	0.1800	0.2550	0.3200	0.3750	0.4200	0.4550	0.4800	0.4950	0.5000	1	
	2	0.0100	0.0225	0.0400	0.0625	0.0900	0.1225	0.1600	0.2025	0.2500	0	2
3	0	0.7290	0.6141	0.5120	0.4219	0.3430	0.2746	0.2160	0.1664	0.1250	3	
	1	0.2430	0.3251	0.3840	0.4219	0.4410	0.4436	0.4320	0.4084	0.3750	2	
	2	0.0270	0.0574	0.0960	0.1406	0.1890	0.2389	0.2880	0.3341	0.3750	1	
	3	0.0010	0.0034	0.0080	0.0156	0.0270	0.0429	0.0640	0.0911	0.1250	0	3
4	0	0.6561	0.5220	0.4096	0.3164	0.2401	0.1785	0.1296	0.0915	0.0625	4	
	1	0.2916	0.3685	0.4096	0.4219	0.4116	0.3845	0.3456	0.2995	0.2500	3	
	2	0.0486	0.0975	0.1536	0.2109	0.2646	0.3105	0.3456	0.3675	0.3750	2	
	3	0.0036	0.0115	0.0256	0.0469	0.0756	0.1115	0.1536	0.2005	0.2500	1	
	4	0.0001	0.0005	0.0016	0.0039	0.0081	0.0150	0.0256	0.0410	0.0625	0	4
5	0	0.5905	0.4437	0.3277	0.2373	0.1681	0.1160	0.0778	0.0503	0.0312	5	
	1	0.3280	0.3915	0.4096	0.3955	0.3601	0.3124	0.2592	0.2059	0.1562	4	
	2	0.0729	0.1382	0.2048	0.2637	0.3087	0.3364	0.3456	0.3369	0.3125	3	
	3	0.0081	0.0244	0.0512	0.0879	0.1323	0.1811	0.2304	0.2757	0.3125	2	
	4	0.0004	0.0022	0.0064	0.0146	0.0283	0.0488	0.0768	0.1128	0.1562	1	
	5	0.0000	0.0001	0.0003	0.0010	0.0024	0.0053	0.0102	0.0185	0.0312	0	5
6	0	0.5314	0.3771	0.2621	0.1780	0.1176	0.0754	0.0467	0.0277	0.0156	6	
	1	0.3543	0.3993	0.3932	0.3560	0.3025	0.2437	0.1866	0.1359	0.0938	5	
	2	0.0984	0.1762	0.2458	0.2966	0.3241	0.3280	0.3110	0.2780	0.2344	4	
	3	0.0146	0.0415	0.0819	0.1318	0.1852	0.2355	0.2765	0.3032	0.3125	3	
	4	0.0012	0.0055	0.0154	0.0330	0.0595	0.0951	0.1382	0.1861	0.2344	2	
	5	0.0001	0.0004	0.0015	0.0044	0.0102	0.0205	0.0369	0.0609	0.0938	1	
	6	0.0000	0.0000	0.0001	0.0002	0.0007	0.0018	0.0041	0.0083	0.0156	0	6
7	0	0.4783	0.3206	0.2097	0.1335	0.0824	0.0490	0.0280	0.0152	0.0078	7	
	1	0.3720	0.3960	0.3670	0.3115	0.2471	0.1848	0.1306	0.0872	0.0547	6	
	2	0.1240	0.2097	0.2753	0.3115	0.3177	0.2985	0.2613	0.2140	0.1641	5	
	3	0.0230	0.0617	0.1147	0.1730	0.2269	0.2679	0.2903	0.2918	0.2734	4	
	4	0.0026	0.0109	0.0287	0.0577	0.0972	0.1442	0.1935	0.2388	0.2734	3	
	5	0.0002	0.0012	0.0043	0.0115	0.0250	0.0466	0.0774	0.1172	0.1641	2	
	6	0.0000	0.0001	0.0004	0.0013	0.0036	0.0084	0.0172	0.0320	0.0547	1	
	7	0.0000	0.0000	0.0000	0.0001	0.0002	0.0006	0.0016	0.0037	0.0078	0	7
8	0	0.4305	0.2725	0.1678	0.1001	0.0576	0.0319	0.0168	0.0084	0.0039	8	
	1	0.3826	0.3847	0.3355	0.2670	0.1977	0.1373	0.0896	0.0548	0.0312	7	
	2	0.1488	0.2376	0.2936	0.3115	0.2965	0.2587	0.2090	0.1569	0.1094	6	
	3	0.0331	0.0839	0.1468	0.2076	0.2541	0.2786	0.2787	0.2568	0.2188	5	
	4	0.0046	0.0185	0.0459	0.0865	0.1361	0.1875	0.2322	0.2627	0.2734	4	
	5	0.0004	0.0026	0.0092	0.0231	0.0467	0.0808	0.1239	0.1719	0.2188	3	
	6	0.0000	0.0002	0.0011	0.0038	0.0100	0.0217	0.0413	0.0703	0.1094	2	
	7	0.0000	0.0000	0.0001	0.0004	0.0012	0.0033	0.0079	0.0164	0.0312	1	
	8	0.0000	0.0000	0.0000	0.0000	0.0001	0.0002	0.0007	0.0017	0.0039	0	8
9	0	0.3874	0.2316	0.1342	0.0751	0.0404	0.0207	0.0101	0.0046	0.0020	9	
	1	0.3874	0.3679	0.3020	0.2253	0.1556	0.1004	0.0605	0.0339	0.0176	8	
	2	0.1722	0.2597	0.3020	0.3003	0.2668	0.2162	0.1612	0.1110	0.0703	7	
	3	0.0446	0.1069	0.1762	0.2336	0.2668	0.2716	0.2508	0.2119	0.1641	6	
	4	0.0074	0.0283	0.0661	0.1168	0.1715	0.2194	0.2508	0.2600	0.2461	5	
	5	0.0008	0.0050	0.0165	0.0389	0.0735	0.1181	0.1672	0.2128	0.2461	4	
	6	0.0001	0.0006	0.0028	0.0087	0.0210	0.0424	0.0743	0.1160	0.1641	3	
	7	0.0000	0.0000	0.0003	0.0012	0.0039	0.0098	0.0212	0.0407	0.0703	2	
	8	0.0000	0.0000	0.0000	0.0001	0.0004	0.0013	0.0035	0.0083	0.0176	1	
	9	0.0000	0.0000	0.0000	0.0000	0.0000	0.0001	0.0003	0.0008	0.0020	0	9
		.90	.85	.80	.75	.70	.65	.60	.55	.50	x	n
						p						

TABLE C.5   Binomial probabilities (concluded)   957

						p						
n	x	.10	.15	.20	.25	.30	.35	.40	.45	.50		
10	0	0.3487	0.1969	0.1074	0.0563	0.0282	0.0135	0.0060	0.0025	0.0010	10	
	1	0.3874	0.3474	0.2684	0.1877	0.1211	0.0725	0.0403	0.0207	0.0098	9	
	2	0.1937	0.2759	0.3020	0.2816	0.2335	0.1757	0.1209	0.0763	0.0439	8	
	3	0.0574	0.1298	0.2013	0.2503	0.2668	0.2522	0.2150	0.1665	0.1172	7	
	4	0.0112	0.0401	0.0881	0.1460	0.2001	0.2377	0.2508	0.2384	0.2051	6	
	5	0.0015	0.0085	0.0264	0.0584	0.1029	0.1536	0.2007	0.2340	0.2461	5	
	6	0.0001	0.0012	0.0055	0.0162	0.0368	0.0689	0.1115	0.1596	0.2051	4	
	7	0.0000	0.0001	0.0008	0.0031	0.0090	0.0212	0.0425	0.0746	0.1172	3	
	8	0.0000	0.0000	0.0001	0.0004	0.0014	0.0043	0.0106	0.0229	0.0439	2	
	9	0.0000	0.0000	0.0000	0.0000	0.0001	0.0005	0.0016	0.0042	0.0098	1	
	10	0.0000	0.0000	0.0000	0.0000	0.0000	0.0000	0.0001	0.0003	0.0010	0	10
12	0	0.2824	0.1422	0.0687	0.0317	0.0138	0.0057	0.0022	0.0008	0.0002	12	
	1	0.3766	0.3012	0.2062	0.1267	0.0712	0.0368	0.0174	0.0075	0.0029	11	
	2	0.2301	0.2924	0.2835	0.2323	0.1678	0.1088	0.0639	0.0339	0.0161	10	
	3	0.0852	0.1720	0.2362	0.2581	0.2397	0.1954	0.1419	0.0923	0.0537	9	
	4	0.0213	0.0683	0.1329	0.1936	0.2311	0.2367	0.2128	0.1700	0.1208	8	
	5	0.0038	0.0193	0.0532	0.1032	0.1585	0.2039	0.2270	0.2225	0.1934	7	
	6	0.0005	0.0040	0.0155	0.0401	0.0792	0.1281	0.1766	0.2124	0.2256	6	
	7	0.0000	0.0006	0.0033	0.0115	0.0291	0.0591	0.1009	0.1489	0.1934	5	
	8	0.0000	0.0001	0.0005	0.0024	0.0078	0.0199	0.0420	0.0762	0.1208	4	
	9	0.0000	0.0000	0.0001	0.0004	0.0015	0.0048	0.0125	0.0277	0.0537	3	
	10	0.0000	0.0000	0.0000	0.0000	0.0002	0.0008	0.0025	0.0068	0.0161	2	
	11	0.0000	0.0000	0.0000	0.0000	0.0000	0.0001	0.0003	0.0010	0.0029	1	
	12	0.0000	0.0000	0.0000	0.0000	0.0000	0.0000	0.0000	0.0001	0.0002	0	12
15	0	0.2059	0.0874	0.0352	0.0134	0.0047	0.0016	0.0005	0.0001	0.0000	15	
	1	0.3432	0.2312	0.1319	0.0668	0.0305	0.0126	0.0047	0.0016	0.0005	14	
	2	0.2669	0.2856	0.2309	0.1559	0.0916	0.0476	0.0219	0.0090	0.0032	13	
	3	0.1285	0.2184	0.2501	0.2252	0.1700	0.1110	0.0634	0.0318	0.0139	12	
	4	0.0428	0.1156	0.1876	0.2252	0.2186	0.1792	0.1268	0.0780	0.0417	11	
	5	0.0105	0.0449	0.1032	0.1651	0.2061	0.2123	0.1859	0.1404	0.0916	10	
	6	0.0019	0.0132	0.0430	0.0917	0.1472	0.1906	0.2066	0.1914	0.1527	9	
	7	0.0003	0.0030	0.0138	0.0393	0.0811	0.1319	0.1771	0.2013	0.1964	8	
	8	0.0000	0.0005	0.0035	0.0131	0.0348	0.0710	0.1181	0.1647	0.1964	7	
	9	0.0000	0.0001	0.0007	0.0034	0.0116	0.0298	0.0612	0.1048	0.1527	6	
	10	0.0000	0.0000	0.0001	0.0007	0.0030	0.0096	0.0245	0.0515	0.0916	5	
	11	0.0000	0.0000	0.0000	0.0001	0.0006	0.0024	0.0074	0.0191	0.0417	4	
	12	0.0000	0.0000	0.0000	0.0000	0.0001	0.0004	0.0016	0.0052	0.0139	3	
	13	0.0000	0.0000	0.0000	0.0000	0.0000	0.0001	0.0003	0.0010	0.0032	2	
	14	0.0000	0.0000	0.0000	0.0000	0.0000	0.0000	0.0000	0.0001	0.0005	1	
	15	0.0000	0.0000	0.0000	0.0000	0.0000	0.0000	0.0000	0.0000	0.0000	0	15
20	0	0.1216	0.0388	0.0115	0.0032	0.0008	0.0002	0.0000	0.0000	0.0000	20	
	1	0.2702	0.1368	0.0576	0.0211	0.0068	0.0020	0.0005	0.0001	0.0000	19	
	2	0.2852	0.2293	0.1369	0.0669	0.0278	0.0100	0.0031	0.0008	0.0002	18	
	3	0.1901	0.2428	0.2054	0.1339	0.0716	0.0323	0.0123	0.0040	0.0011	17	
	4	0.0898	0.1821	0.2182	0.1897	0.1304	0.0738	0.0350	0.0139	0.0046	16	
	5	0.0319	0.1028	0.1746	0.2023	0.1789	0.1272	0.0746	0.0365	0.0148	15	
	6	0.0089	0.0454	0.1091	0.1686	0.1916	0.1712	0.1244	0.0746	0.0370	14	
	7	0.0020	0.0160	0.0545	0.1124	0.1643	0.1844	0.1659	0.1221	0.0739	13	
	8	0.0004	0.0046	0.0222	0.0609	0.1144	0.1614	0.1797	0.1623	0.1201	12	
	9	0.0001	0.0011	0.0074	0.0271	0.0654	0.1158	0.1597	0.1771	0.1602	11	
	10	0.0000	0.0002	0.0020	0.0099	0.0308	0.0686	0.1171	0.1593	0.1762	10	
	11	0.0000	0.0000	0.0005	0.0030	0.0120	0.0336	0.0710	0.1185	0.1602	9	
	12	0.0000	0.0000	0.0001	0.0008	0.0039	0.0136	0.0355	0.0727	0.1201	8	
	13	0.0000	0.0000	0.0000	0.0002	0.0010	0.0045	0.0146	0.0366	0.0739	7	
	14	0.0000	0.0000	0.0000	0.0000	0.0002	0.0012	0.0049	0.0150	0.0370	6	
	15	0.0000	0.0000	0.0000	0.0000	0.0000	0.0003	0.0013	0.0049	0.0148	5	
	16	0.0000	0.0000	0.0000	0.0000	0.0000	0.0000	0.0003	0.0013	0.0046	4	
	17	0.0000	0.0000	0.0000	0.0000	0.0000	0.0000	0.0000	0.0002	0.0011	3	
	18	0.0000	0.0000	0.0000	0.0000	0.0000	0.0000	0.0000	0.0000	0.0002	2	
	19	0.0000	0.0000	0.0000	0.0000	0.0000	0.0000	0.0000	0.0000	0.0000	1	
	20	0.0000	0.0000	0.0000	0.0000	0.0000	0.0000	0.0000	0.0000	0.0000	0	20
		.90	.85	.80	.75	.70	.65	.60	.55	.50	x	n

p

EXAMPLE: For $n = 12$, $p = 0.25$, and $x = 3$, $P(X = 3) = 0.2581$. For $n = 15$, $p = 0.55$, and $x = 10$, $P(X = 10) = 0.1404$.
TEXT REFERENCE: Use of this table is discussed on pp. 190–191.

TABLE C.6 Poisson probabilities

Entry is probability $P(X = x) = \dfrac{\lambda^x \exp(-\lambda)}{x!}$.

					λ				
x	.1	.2	.3	.4	.5	.6	.7	.8	.9
0	0.9048	0.8187	0.7408	0.6703	0.6065	0.5488	0.4966	0.4493	0.4066
1	0.0905	0.1637	0.2222	0.2681	0.3033	0.3293	0.3476	0.3595	0.3659
2	0.0045	0.0164	0.0333	0.0536	0.0758	0.0988	0.1217	0.1438	0.1647
3	0.0002	0.0011	0.0033	0.0072	0.0126	0.0198	0.0284	0.0383	0.0494
4	0.0000	0.0001	0.0003	0.0007	0.0016	0.0030	0.0050	0.0077	0.0111
5	0.0000	0.0000	0.0000	0.0001	0.0002	0.0004	0.0007	0.0012	0.0020
6	0.0000	0.0000	0.0000	0.0000	0.0000	0.0000	0.0001	0.0002	0.0003

					λ				
x	1.0	1.5	2.0	2.5	3.0	3.5	4.0	4.5	5.0
0	0.3679	0.2231	0.1353	0.0821	0.0498	0.0302	0.0183	0.0111	0.0067
1	0.3679	0.3347	0.2707	0.2052	0.1494	0.1057	0.0733	0.0500	0.0337
2	0.1839	0.2510	0.2707	0.2565	0.2240	0.1850	0.1465	0.1125	0.0842
3	0.0613	0.1255	0.1804	0.2138	0.2240	0.2158	0.1954	0.1687	0.1404
4	0.0153	0.0471	0.0902	0.1336	0.1680	0.1888	0.1954	0.1898	0.1755
5	0.0031	0.0141	0.0361	0.0668	0.1008	0.1322	0.1563	0.1708	0.1755
6	0.0005	0.0035	0.0120	0.0278	0.0504	0.0771	0.1042	0.1281	0.1462
7	0.0001	0.0008	0.0034	0.0099	0.0216	0.0385	0.0595	0.0824	0.1044
8	0.0000	0.0001	0.0009	0.0031	0.0081	0.0169	0.0298	0.0463	0.0653
9	0.0000	0.0000	0.0002	0.0009	0.0027	0.0066	0.0132	0.0232	0.0363
10	0.0000	0.0000	0.0000	0.0002	0.0008	0.0023	0.0053	0.0104	0.0181
11	0.0000	0.0000	0.0000	0.0000	0.0002	0.0007	0.0019	0.0043	0.0082
12	0.0000	0.0000	0.0000	0.0000	0.0001	0.0002	0.0006	0.0016	0.0034
13	0.0000	0.0000	0.0000	0.0000	0.0000	0.0001	0.0002	0.0006	0.0013
14	0.0000	0.0000	0.0000	0.0000	0.0000	0.0000	0.0001	0.0002	0.0005
15	0.0000	0.0000	0.0000	0.0000	0.0000	0.0000	0.0000	0.0001	0.0002

					λ				
x	5.5	6.0	6.5	7.0	7.5	8.0	9.0	10.0	11.0
0	0.0041	0.0025	0.0015	0.0009	0.0006	0.0003	0.0001	0.0000	0.0000
1	0.0225	0.0149	0.0098	0.0064	0.0041	0.0027	0.0011	0.0005	0.0002
2	0.0618	0.0446	0.0318	0.0223	0.0156	0.0107	0.0050	0.0023	0.0010
3	0.1133	0.0892	0.0688	0.0521	0.0389	0.0286	0.0150	0.0076	0.0037
4	0.1558	0.1339	0.1118	0.0912	0.0729	0.0573	0.0337	0.0189	0.0102
5	0.1714	0.1606	0.1454	0.1277	0.1094	0.0916	0.0607	0.0378	0.0224
6	0.1571	0.1606	0.1575	0.1490	0.1367	0.1221	0.0911	0.0631	0.0411
7	0.1234	0.1377	0.1462	0.1490	0.1465	0.1396	0.1171	0.0901	0.0646
8	0.0849	0.1033	0.1188	0.1304	0.1373	0.1396	0.1318	0.1126	0.0888
9	0.0519	0.0688	0.0858	0.1014	0.1144	0.1241	0.1318	0.1251	0.1085
10	0.0285	0.0413	0.0558	0.0710	0.0858	0.0993	0.1186	0.1251	0.1194
11	0.0143	0.0225	0.0330	0.0452	0.0585	0.0722	0.0970	0.1137	0.1194
12	0.0065	0.0113	0.0179	0.0263	0.0366	0.0481	0.0728	0.0948	0.1094
13	0.0028	0.0052	0.0089	0.0142	0.0211	0.0296	0.0504	0.0729	0.0926
14	0.0011	0.0022	0.0041	0.0071	0.0113	0.0169	0.0324	0.0521	0.0728
15	0.0004	0.0009	0.0018	0.0033	0.0057	0.0090	0.0194	0.0347	0.0534
16	0.0001	0.0003	0.0007	0.0014	0.0026	0.0045	0.0109	0.0217	0.0367
17	0.0000	0.0001	0.0003	0.0006	0.0012	0.0021	0.0058	0.0128	0.0237
18	0.0000	0.0000	0.0001	0.0002	0.0005	0.0009	0.0029	0.0071	0.0145
19	0.0000	0.0000	0.0000	0.0001	0.0002	0.0004	0.0014	0.0037	0.0084
20	0.0000	0.0000	0.0000	0.0000	0.0001	0.0002	0.0006	0.0019	0.0046
21	0.0000	0.0000	0.0000	0.0000	0.0000	0.0001	0.0003	0.0009	0.0024
22	0.0000	0.0000	0.0000	0.0000	0.0000	0.0000	0.0001	0.0004	0.0012
23	0.0000	0.0000	0.0000	0.0000	0.0000	0.0000	0.0000	0.0002	0.0006
24	0.0000	0.0000	0.0000	0.0000	0.0000	0.0000	0.0000	0.0001	0.0003
25	0.0000	0.0000	0.0000	0.0000	0.0000	0.0000	0.0000	0.0000	0.0001

TABLE C.6 **Poisson probabilities (concluded)**

					λ				
x	12	13	14	15	16	17	18	19	20
0	0.0000	0.0000	0.0000	0.0000	0.0000	0.0000	0.0000	0.0000	0.0000
1	0.0001	0.0000	0.0000	0.0000	0.0000	0.0000	0.0000	0.0000	0.0000
2	0.0004	0.0002	0.0001	0.0000	0.0000	0.0000	0.0000	0.0000	0.0000
3	0.0018	0.0008	0.0004	0.0002	0.0001	0.0000	0.0000	0.0000	0.0000
4	0.0053	0.0027	0.0013	0.0006	0.0003	0.0001	0.0001	0.0000	0.0000
5	0.0127	0.0070	0.0037	0.0019	0.0010	0.0005	0.0002	0.0001	0.0001
6	0.0255	0.0152	0.0087	0.0048	0.0026	0.0014	0.0007	0.0004	0.0002
7	0.0437	0.0281	0.0174	0.0104	0.0060	0.0034	0.0019	0.0010	0.0005
8	0.0655	0.0457	0.0304	0.0194	0.0120	0.0072	0.0042	0.0024	0.0013
9	0.0874	0.0661	0.0473	0.0324	0.0213	0.0135	0.0083	0.0050	0.0029
10	0.1048	0.0859	0.0663	0.0486	0.0341	0.0230	0.0150	0.0095	0.0058
11	0.1144	0.1015	0.0844	0.0663	0.0496	0.0355	0.0245	0.0164	0.0106
12	0.1144	0.1099	0.0984	0.0829	0.0661	0.0504	0.0368	0.0259	0.0176
13	0.1056	0.1099	0.1060	0.0956	0.0814	0.0658	0.0509	0.0378	0.0271
14	0.0905	0.1021	0.1060	0.1024	0.0930	0.0800	0.0655	0.0514	0.0387
15	0.0724	0.0885	0.0989	0.1024	0.0992	0.0906	0.0786	0.0650	0.0516
16	0.0543	0.0719	0.0866	0.0960	0.0992	0.0963	0.0884	0.0772	0.0646
17	0.0383	0.0550	0.0713	0.0847	0.0934	0.0963	0.0936	0.0863	0.0760
18	0.0255	0.0397	0.0554	0.0706	0.0830	0.0909	0.0936	0.0911	0.0844
19	0.0161	0.0272	0.0409	0.0557	0.0699	0.0814	0.0887	0.0911	0.0888
20	0.0097	0.0177	0.0286	0.0418	0.0559	0.0692	0.0798	0.0866	0.0888
21	0.0055	0.0109	0.0191	0.0299	0.0426	0.0560	0.0684	0.0783	0.0846
22	0.0030	0.0065	0.0121	0.0204	0.0310	0.0433	0.0560	0.0676	0.0769
23	0.0016	0.0037	0.0074	0.0133	0.0216	0.0320	0.0438	0.0559	0.0669
24	0.0008	0.0020	0.0043	0.0083	0.0144	0.0226	0.0328	0.0442	0.0557
25	0.0004	0.0010	0.0024	0.0050	0.0092	0.0154	0.0237	0.0336	0.0446
26	0.0002	0.0005	0.0013	0.0029	0.0057	0.0101	0.0164	0.0246	0.0343
27	0.0001	0.0002	0.0007	0.0016	0.0034	0.0063	0.0109	0.0173	0.0254
28	0.0000	0.0001	0.0003	0.0009	0.0019	0.0038	0.0070	0.0117	0.0181
29	0.0000	0.0001	0.0002	0.0004	0.0011	0.0023	0.0044	0.0077	0.0125
30	0.0000	0.0000	0.0001	0.0002	0.0006	0.0013	0.0026	0.0049	0.0083
31	0.0000	0.0000	0.0000	0.0001	0.0003	0.0007	0.0015	0.0030	0.0054
32	0.0000	0.0000	0.0000	0.0001	0.0001	0.0004	0.0009	0.0018	0.0034
33	0.0000	0.0000	0.0000	0.0000	0.0001	0.0002	0.0005	0.0010	0.0020
34	0.0000	0.0000	0.0000	0.0000	0.0000	0.0001	0.0002	0.0006	0.0012
35	0.0000	0.0000	0.0000	0.0000	0.0000	0.0000	0.0001	0.0003	0.0007
36	0.0000	0.0000	0.0000	0.0000	0.0000	0.0000	0.0001	0.0002	0.0004
37	0.0000	0.0000	0.0000	0.0000	0.0000	0.0000	0.0000	0.0001	0.0002
38	0.0000	0.0000	0.0000	0.0000	0.0000	0.0000	0.0000	0.0000	0.0001
39	0.0000	0.0000	0.0000	0.0000	0.0000	0.0000	0.0000	0.0000	0.0001

EXAMPLE: For $\lambda = 14$ and $x = 8$, $P(X = 8) = 0.0304$.
TEXT REFERENCE: Use of this table is discussed on pp. 194–195.

TABLE C.7 Cumulative probabilities of the exponential distribution

Entry is area a under the exponential curve from 0 to $x(a)$.

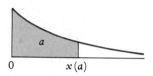

λx	.00	.01	.02	.03	.04	.05	.06	.07	.08	.09
0.0	0.0000	0.0100	0.0198	0.0296	0.0392	0.0488	0.0582	0.0676	0.0769	0.0861
0.1	0.0952	0.1042	0.1131	0.1219	0.1306	0.1393	0.1479	0.1563	0.1647	0.1730
0.2	0.1813	0.1894	0.1975	0.2055	0.2134	0.2212	0.2289	0.2366	0.2442	0.2517
0.3	0.2592	0.2666	0.2739	0.2811	0.2882	0.2953	0.3023	0.3093	0.3161	0.3229
0.4	0.3297	0.3363	0.3430	0.3495	0.3560	0.3624	0.3687	0.3750	0.3812	0.3874
0.5	0.3935	0.3995	0.4055	0.4114	0.4173	0.4231	0.4288	0.4345	0.4401	0.4457
0.6	0.4512	0.4566	0.4621	0.4674	0.4727	0.4780	0.4831	0.4883	0.4934	0.4984
0.7	0.5034	0.5084	0.5132	0.5181	0.5229	0.5276	0.5323	0.5370	0.5416	0.5462
0.8	0.5507	0.5551	0.5596	0.5640	0.5683	0.5726	0.5768	0.5810	0.5852	0.5893
0.9	0.5934	0.5975	0.6015	0.6054	0.6094	0.6133	0.6171	0.6209	0.6247	0.6284
1.0	0.6321	0.6358	0.6394	0.6430	0.6465	0.6501	0.6535	0.6570	0.6604	0.6638
1.1	0.6671	0.6704	0.6737	0.6770	0.6802	0.6834	0.6865	0.6896	0.6927	0.6958
1.2	0.6988	0.7018	0.7048	0.7077	0.7106	0.7135	0.7163	0.7192	0.7220	0.7247
1.3	0.7275	0.7302	0.7329	0.7355	0.7382	0.7408	0.7433	0.7459	0.7484	0.7509
1.4	0.7534	0.7559	0.7583	0.7607	0.7631	0.7654	0.7678	0.7701	0.7724	0.7746
1.5	0.7769	0.7791	0.7813	0.7835	0.7856	0.7878	0.7899	0.7920	0.7940	0.7961
1.6	0.7981	0.8001	0.8021	0.8041	0.8060	0.8080	0.8099	0.8118	0.8136	0.8155
1.7	0.8173	0.8191	0.8209	0.8227	0.8245	0.8262	0.8280	0.8297	0.8314	0.8330
1.8	0.8347	0.8363	0.8380	0.8396	0.8412	0.8428	0.8443	0.8459	0.8474	0.8489
1.9	0.8504	0.8519	0.8534	0.8549	0.8563	0.8577	0.8591	0.8605	0.8619	0.8633
2.0	0.8647	0.8660	0.8673	0.8687	0.8700	0.8713	0.8725	0.8738	0.8751	0.8763
2.1	0.8775	0.8788	0.8800	0.8812	0.8823	0.8835	0.8847	0.8858	0.8870	0.8881
2.2	0.8892	0.8903	0.8914	0.8925	0.8935	0.8946	0.8956	0.8967	0.8977	0.8987
2.3	0.8997	0.9007	0.9017	0.9027	0.9037	0.9046	0.9056	0.9065	0.9074	0.9084
2.4	0.9093	0.9102	0.9111	0.9120	0.9128	0.9137	0.9146	0.9154	0.9163	0.9171
2.5	0.9179	0.9187	0.9195	0.9203	0.9211	0.9219	0.9227	0.9235	0.9242	0.9250
2.6	0.9257	0.9265	0.9272	0.9279	0.9286	0.9293	0.9301	0.9307	0.9314	0.9321
2.7	0.9328	0.9335	0.9341	0.9348	0.9354	0.9361	0.9367	0.9373	0.9380	0.9386
2.8	0.9392	0.9398	0.9404	0.9410	0.9416	0.9422	0.9427	0.9433	0.9439	0.9444
2.9	0.9450	0.9455	0.9461	0.9466	0.9471	0.9477	0.9482	0.9487	0.9492	0.9497
3.0	0.9502	0.9507	0.9512	0.9517	0.9522	0.9526	0.9531	0.9536	0.9540	0.9545
3.1	0.9550	0.9554	0.9558	0.9563	0.9567	0.9571	0.9576	0.9580	0.9584	0.9588
3.2	0.9592	0.9596	0.9600	0.9604	0.9608	0.9612	0.9616	0.9620	0.9624	0.9627
3.3	0.9631	0.9635	0.9638	0.9642	0.9646	0.9649	0.9653	0.9656	0.9660	0.9663
3.4	0.9666	0.9670	0.9673	0.9676	0.9679	0.9683	0.9686	0.9689	0.9692	0.9695
3.5	0.9698	0.9701	0.9704	0.9707	0.9710	0.9713	0.9716	0.9718	0.9721	0.9724
3.6	0.9727	0.9729	0.9732	0.9735	0.9737	0.9740	0.9743	0.9745	0.9748	0.9750
3.7	0.9753	0.9755	0.9758	0.9760	0.9762	0.9765	0.9767	0.9769	0.9772	0.9774
3.8	0.9776	0.9779	0.9781	0.9783	0.9785	0.9787	0.9789	0.9791	0.9793	0.9796
3.9	0.9798	0.9800	0.9802	0.9804	0.9806	0.9807	0.9809	0.9811	0.9813	0.9815

λx	0.0	0.1	0.2	0.3	0.4	0.5	0.6	0.7	0.8	0.9
4.0	0.9817	0.9834	0.9850	0.9864	0.9877	0.9889	0.9899	0.9909	0.9918	0.9926
5.0	0.9933	0.9939	0.9945	0.9950	0.9955	0.9959	0.9963	0.9967	0.9970	0.9973
6.0	0.9975	0.9978	0.9980	0.9982	0.9983	0.9985	0.9986	0.9988	0.9989	0.9990
7.0	0.9991	0.9992	0.9993	0.9993	0.9994	0.9994	0.9995	0.9995	0.9996	0.9996
8.0	0.9997	0.9997	0.9997	0.9998	0.9998	0.9998	0.9998	0.9998	0.9998	0.9999
9.0	0.9999	0.9999	0.9999	0.9999	0.9999	0.9999	0.9999	0.9999	0.9999	0.9999

EXAMPLE: For $\lambda = 0.6$ and $x = 1.3$, $\lambda x = 0.6(1.3) = 0.78$ so $P(X \le 1.3) = 0.5416$.
TEXT REFERENCE: Use of this table is discussed on pp. 223–224.

TABLE C.8   Table of random digits

Row	1–5	6–10	11–15	16–20	21–25	26–30	31–35
				Column			
1	13284	16834	74151	92027	24670	36665	00770
2	21224	00370	30420	03883	94648	89428	41583
3	99052	47887	81085	64933	66279	80432	65793
4	00199	50993	98603	38452	87890	94624	69721
5	60578	06483	28733	37867	07936	98710	98539
6	91240	18312	17441	01929	18163	69201	31211
7	97458	14229	12063	59611	32249	90466	33216
8	35249	38646	34475	72417	60514	69257	12489
9	38980	46600	11759	11900	46743	27860	77940
10	10750	52745	38749	87365	58959	53731	89295
11	36247	27850	73958	20673	37800	63835	71051
12	70994	66986	99744	72438	01174	42159	11392
13	99638	94702	11463	18148	81386	80431	90628
14	72055	15774	43857	99805	10419	76939	25993
15	24038	65541	85788	55835	38835	59399	13790
16	74976	14631	35908	28221	39470	91548	12854
17	35553	71628	70189	26436	63407	91178	90348
18	35676	12797	51434	82976	42010	26344	92920
19	74815	67523	72985	23183	02446	63594	98924
20	45246	88048	65173	50989	91060	89894	36036
21	76509	47069	86378	41797	11910	49672	88575
22	19689	90332	04315	21358	97248	11188	39062
23	42751	35318	97513	61537	54955	08159	00337
24	11946	22681	45045	13964	57517	59419	58045
25	96518	48688	20996	11090	48396	57177	83867
26	35726	58643	76869	84622	39098	36083	72505
27	39737	42750	48968	70536	84864	64952	38404
28	97025	66492	56177	04049	80312	48028	26408
29	62814	08075	09788	56350	76787	51591	54509
30	25578	22950	15227	83291	41737	59599	96191
31	68763	69576	88991	49662	46704	63362	56625
32	17900	00813	64361	60725	88974	61005	99709
33	71944	60227	63551	71109	05624	43836	58254
34	54684	93691	85132	64399	29182	44324	14491
35	25946	27623	11258	65204	52832	50880	22273
36	01353	39318	44961	44972	91766	90262	56073
37	99083	88191	27662	99113	57174	35571	99884
38	52021	45406	37945	75234	24327	86978	22644
39	78755	47744	43776	83098	03225	14281	83637
40	25282	69106	59180	16257	22810	43609	12224
41	11959	94202	02743	86847	79725	51811	12998
42	11644	13792	98190	01424	30078	28197	55583
43	06307	97912	68110	59812	95448	43244	31262
44	76285	75714	89585	99296	52640	46518	55486
45	55322	07598	39600	60866	63007	20007	66819
46	78017	90928	90220	92503	83375	26986	74399
47	44768	43342	20696	26331	43140	69744	82928
48	25100	19336	14605	86603	51680	97678	24261
49	83612	46623	62876	85197	07824	91392	58317
50	41347	81666	82961	60413	71020	83658	02415

SOURCE: Excerpt from *Table of 105,000 Random Decimal Digits*. Interstate Commerce Commission, Bureau of Transport Economics and Statistics, May 1949.
TEXT REFERENCE: This table is discussed on pp. 243–244.

TABLE C.9 Table of random standard normal numbers

| Row | \multicolumn{10}{c}{Column} |
	1	2	3	4	5	6	7	8	9	10
1	0.464	0.137	2.455	−0.323	−0.068	0.296	−0.288	1.298	0.241	−0.957
2	0.060	−2.526	−0.531	−0.194	0.543	−1.558	0.187	−1.190	0.022	0.525
3	1.486	−0.354	−0.634	0.697	0.926	1.375	0.785	−0.963	−0.853	−1.865
4	1.022	−0.472	1.279	3.521	0.571	−1.851	0.194	1.192	−0.501	−0.273
5	1.394	−0.555	0.046	0.321	2.945	1.974	−0.258	0.412	0.439	−0.035
6	0.906	−0.513	−0.525	0.595	0.881	−0.934	1.579	0.161	−1.885	0.371
7	1.179	−1.055	0.007	0.769	0.971	0.712	1.090	−0.631	−0.255	−0.702
8	−1.501	−0.488	−0.162	−0.136	1.033	0.203	0.448	0.748	−0.423	−0.432
9	−0.690	0.756	−1.618	−0.345	−0.511	−2.051	−0.457	−0.218	0.857	−0.465
10	1.372	0.225	0.378	0.761	0.181	−0.736	0.960	−1.530	−0.260	0.120
11	−0.482	1.678	−0.057	−1.229	−0.486	0.856	−0.491	−1.983	−2.830	−0.238
12	−1.376	−0.150	1.356	−0.561	−0.256	−0.212	0.219	0.779	0.953	−0.869
13	−1.010	0.598	−0.918	1.598	0.065	0.415	−0.169	0.313	−0.973	−1.016
14	−0.005	−0.899	0.012	−0.725	1.147	−0.121	1.096	0.481	−1.691	0.417
15	1.393	−1.163	−0.911	1.231	−0.199	−0.246	1.239	−2.574	−0.558	0.056
16	−1.787	−0.261	1.237	1.046	−0.508	−1.630	−0.146	−0.392	−0.627	0.561
17	−0.105	−0.357	−1.384	0.360	−0.992	−0.116	−1.698	−2.832	−1.108	−2.357
18	−1.339	1.827	−0.959	0.424	0.969	−1.141	−1.041	0.362	−1.726	1.956
19	1.041	0.535	0.731	1.377	0.983	−1.330	1.620	−1.040	0.524	−0.281
20	0.279	−2.056	0.717	−0.873	−1.096	−1.396	1.047	0.089	−0.573	0.932
21	−1.805	−2.008	−1.633	0.542	0.250	−0.166	0.032	0.079	0.471	−1.029
22	−1.186	1.180	1.114	0.882	1.265	−0.202	0.151	−0.376	−0.310	0.479
23	0.658	−1.141	1.151	−1.210	−0.927	0.425	0.290	−0.902	0.610	2.709
24	−0.439	0.358	−1.939	0.891	−0.227	0.602	0.873	−0.437	−0.220	−0.057
25	−1.399	−0.230	0.385	−0.649	−0.577	0.237	−0.289	0.513	0.738	−0.300
26	0.199	0.208	−1.083	−0.219	−0.291	1.221	1.119	0.004	−2.015	−0.594
27	0.159	0.272	−0.313	0.084	−2.828	−0.439	−0.792	−1.275	−0.623	−1.047
28	2.273	0.606	0.606	−0.747	0.247	1.291	0.063	−1.793	−0.699	−1.347
29	0.041	−0.307	0.121	0.790	−0.584	0.541	0.484	−0.986	0.481	0.996
30	−1.132	−2.098	0.921	0.145	0.446	−1.661	1.045	−1.363	−0.586	−1.023
31	0.768	0.079	−1.473	0.034	−2.127	0.665	0.084	−0.880	−0.579	0.551
32	0.375	−1.658	−0.851	0.234	−0.656	0.340	−0.086	−0.158	−0.120	0.418
33	−0.513	−0.344	0.210	−0.735	1.041	0.008	0.427	−0.831	0.191	0.074
34	0.292	−0.521	1.266	−1.206	−0.899	0.110	−0.528	−0.813	0.071	0.524
35	1.026	2.990	−0.574	−0.491	−1.114	1.297	−1.433	−1.345	−3.001	0.479
36	−1.334	1.278	−0.568	−0.109	−0.515	−0.566	2.923	0.500	0.359	0.326
37	−0.287	−0.144	−0.254	0.574	−0.451	−1.181	−1.190	−0.318	−0.094	1.114
38	0.161	−0.886	−0.921	−0.509	1.410	−0.518	0.192	−0.432	1.501	1.068
39	−1.346	0.193	−1.202	0.394	−1.045	0.843	0.942	1.045	0.031	0.772
40	1.250	−0.199	−0.288	1.810	1.378	0.584	1.216	0.733	0.402	0.226
41	0.630	−0.537	0.782	0.060	0.499	−0.431	1.705	1.164	0.884	−0.298
42	0.375	−1.941	0.247	−0.491	−0.665	−0.135	−0.145	−0.498	0.457	1.064
43	−1.420	0.489	−1.711	−1.186	0.754	−0.732	−0.066	1.006	−0.798	0.162
44	−0.151	−0.243	−0.430	−0.762	0.298	1.049	1.810	2.885	−0.768	−0.129
45	−0.309	0.531	0.416	−1.541	1.456	2.040	−0.124	0.196	0.023	−1.204
46	0.424	−0.444	0.593	0.993	−0.106	0.116	0.484	−1.272	1.066	1.097
47	0.593	0.658	−1.127	−1.407	−1.579	−1.616	1.458	1.262	0.736	−0.916
48	0.862	−0.885	−0.142	−0.504	0.532	1.381	0.022	−0.281	−0.342	1.222
49	0.235	−0.628	−0.023	−0.463	−0.899	−0.394	−0.538	1.707	−0.188	−1.153
50	−0.853	0.402	0.777	0.833	0.410	−0.349	−1.094	0.580	1.395	1.298

SOURCE: Reprinted from W. H. Beyer, ed., *Handbook of Tables for Probability and Statistics,* 2nd ed., copyright The Chemical Rubber Company, 1968, p. 484, with the permission of CRC Press, Inc.

TABLE C.10   Confidence level associated with interval $L_r \leq \eta \leq U_r$

Entry is confidence level for given n and r.

n	1	2	3	4	5	6	7
2	.500						
3	.750						
4	.875	.375					
5	.938	.625					
6	.969	.781	.312				
7	.984	.875	.547				
8	.992	.930	.711	.273			
9	.996	.961	.820	.492			
10	.998	.979	.891	.656	.246		
11	.999	.988	.935	.773	.451		
12	1.000	.994	.961	.854	.612	.226	
13	1.000	.997	.978	.908	.733	.419	
14	1.000	.998	.987	.943	.820	.576	.209
15	1.000	.999	.993	.965	.882	.698	.393

EXAMPLE: For $n = 7$ and $r = 1$, confidence interval $L_1 \leq \eta \leq U_1$ has confidence coefficient 0.984.
TEXT REFERENCE: Use of this table is discussed on pp. 484–485.

TABLE C.11   Percentiles of the $D(n)$ distribution

Entry is $D(a;n)$ where $P[D(n) \leq D(a;n)] = a$.

n	.90	.95	.99	n	.90	.95	.99
5	.51	.56	.67	35	.20	.22	.27
10	.37	.41	.49	40	.19	.21	.25
15	.30	.34	.40	45	.18	.20	.24
20	.26	.29	.35	50	.17	.19	.23
25	.24	.26	.32	$n > 50$	$\frac{1.22}{\sqrt{n}}$	$\frac{1.36}{\sqrt{n}}$	$\frac{1.63}{\sqrt{n}}$
30	.22	.24	.29				

SOURCE: Tabulated values adapted by permission from Table 1 of L. H. Miller, "Table of Percentage Points of Kolmogorov Statistics," *Journal of the American Statistical Association*, Vol. 51 (1956), pp. 111–121.
EXAMPLE: $D(0.95;25) = 0.26$ so $P[D(25) \leq 0.26] = 0.95$.
TEXT REFERENCE: Use of this table is discussed on p. 537.

TABLE C.12   Durbin-Watson test bounds

$$\alpha = .05$$

	Number of Independent Variables ($p - 1$)									
	1		2		3		4		5	
n	d_L	d_U	d_L	d_U	d_L	d_U	d_L	d_U	d_L	d_U
15	1.08	1.36	0.95	1.54	0.82	1.75	0.69	1.97	0.56	2.21
16	1.10	1.37	0.98	1.54	0.86	1.73	0.74	1.93	0.62	2.15
17	1.13	1.38	1.02	1.54	0.90	1.71	0.78	1.90	0.67	2.10
18	1.16	1.39	1.05	1.53	0.93	1.69	0.82	1.87	0.71	2.06
19	1.18	1.40	1.08	1.53	0.97	1.68	0.86	1.85	0.75	2.02
20	1.20	1.41	1.10	1.54	1.00	1.68	0.90	1.83	0.79	1.99
21	1.22	1.42	1.13	1.54	1.03	1.67	0.93	1.81	0.83	1.96
22	1.24	1.43	1.15	1.54	1.05	1.66	0.96	1.80	0.86	1.94
23	1.26	1.44	1.17	1.54	1.08	1.66	0.99	1.79	0.90	1.92
24	1.27	1.45	1.19	1.55	1.10	1.66	1.01	1.78	0.93	1.90
25	1.29	1.45	1.21	1.55	1.12	1.66	1.04	1.77	0.95	1.89
26	1.30	1.46	1.22	1.55	1.14	1.65	1.06	1.76	0.98	1.88
27	1.32	1.47	1.24	1.56	1.16	1.65	1.08	1.76	1.01	1.86
28	1.33	1.48	1.26	1.56	1.18	1.65	1.10	1.75	1.03	1.85
29	1.34	1.48	1.27	1.56	1.20	1.65	1.12	1.74	1.05	1.84
30	1.35	1.49	1.28	1.57	1.21	1.65	1.14	1.74	1.07	1.83
31	1.36	1.50	1.30	1.57	1.23	1.65	1.16	1.74	1.09	1.83
32	1.37	1.50	1.31	1.57	1.24	1.65	1.18	1.73	1.11	1.82
33	1.38	1.51	1.32	1.58	1.26	1.65	1.19	1.73	1.13	1.81
34	1.39	1.51	1.33	1.58	1.27	1.65	1.21	1.73	1.15	1.81
35	1.40	1.52	1.34	1.58	1.28	1.65	1.22	1.73	1.16	1.80
36	1.41	1.52	1.35	1.59	1.29	1.65	1.24	1.73	1.18	1.80
37	1.42	1.53	1.36	1.59	1.31	1.66	1.25	1.72	1.19	1.80
38	1.43	1.54	1.37	1.59	1.32	1.66	1.26	1.72	1.21	1.79
39	1.43	1.54	1.38	1.60	1.33	1.66	1.27	1.72	1.22	1.79
40	1.44	1.54	1.39	1.60	1.34	1.66	1.29	1.72	1.23	1.79
45	1.48	1.57	1.43	1.62	1.38	1.67	1.34	1.72	1.29	1.78
50	1.50	1.59	1.46	1.63	1.42	1.67	1.38	1.72	1.34	1.77
55	1.53	1.60	1.49	1.64	1.45	1.68	1.41	1.72	1.38	1.77
60	1.55	1.62	1.51	1.65	1.48	1.69	1.44	1.73	1.41	1.77
65	1.57	1.63	1.54	1.66	1.50	1.70	1.47	1.73	1.44	1.77
70	1.58	1.64	1.55	1.67	1.52	1.70	1.49	1.74	1.46	1.77
75	1.60	1.65	1.57	1.68	1.54	1.71	1.51	1.74	1.49	1.77
80	1.61	1.66	1.59	1.69	1.56	1.72	1.53	1.74	1.51	1.77
85	1.62	1.67	1.60	1.70	1.57	1.72	1.55	1.75	1.52	1.77
90	1.63	1.68	1.61	1.70	1.59	1.73	1.57	1.75	1.54	1.78
95	1.64	1.69	1.62	1.71	1.60	1.73	1.58	1.75	1.56	1.78
100	1.65	1.69	1.63	1.72	1.61	1.74	1.59	1.76	1.57	1.78

TABLE C.12 Durbin-Watson test bounds (concluded)

$$\alpha = .01$$

	Number of Independent Variables ($p - 1$)									
	1		2		3		4		5	
	d_L	d_U	d_L	d_U	d_L	d_U	d_L	d_U	d_L	d_U
15	0.81	1.07	0.70	1.25	0.59	1.46	0.49	1.70	0.39	1.96
16	0.84	1.09	0.74	1.25	0.63	1.44	0.53	1.66	0.44	1.90
17	0.87	1.10	0.77	1.25	0.67	1.43	0.57	1.63	0.48	1.85
18	0.90	1.12	0.80	1.26	0.71	1.42	0.61	1.60	0.52	1.80
19	0.93	1.13	0.83	1.26	0.74	1.41	0.65	1.58	0.56	1.77
20	0.95	1.15	0.86	1.27	0.77	1.41	0.68	1.57	0.60	1.74
21	0.97	1.16	0.89	1.27	0.80	1.41	0.72	1.55	0.63	1.71
22	1.00	1.17	0.91	1.28	0.83	1.40	0.75	1.54	0.66	1.69
23	1.02	1.19	0.94	1.29	0.86	1.40	0.77	1.53	0.70	1.67
24	1.04	1.20	0.96	1.30	0.88	1.41	0.80	1.53	0.72	1.66
25	1.05	1.21	0.98	1.30	0.90	1.41	0.83	1.52	0.75	1.65
26	1.07	1.22	1.00	1.31	0.93	1.41	0.85	1.52	0.78	1.64
27	1.09	1.23	1.02	1.32	0.95	1.41	0.88	1.51	0.81	1.63
28	1.10	1.24	1.04	1.32	0.97	1.41	0.90	1.51	0.83	1.62
29	1.12	1.25	1.05	1.33	0.99	1.42	0.92	1.51	0.85	1.61
30	1.13	1.26	1.07	1.34	1.01	1.42	0.94	1.51	0.88	1.61
31	1.15	1.27	1.08	1.34	1.02	1.42	0.96	1.51	0.90	1.60
32	1.16	1.28	1.10	1.35	1.04	1.43	0.98	1.51	0.92	1.60
33	1.17	1.29	1.11	1.36	1.05	1.43	1.00	1.51	0.94	1.59
34	1.18	1.30	1.13	1.36	1.07	1.43	1.01	1.51	0.95	1.59
35	1.19	1.31	1.14	1.37	1.08	1.44	1.03	1.51	0.97	1.59
36	1.21	1.32	1.15	1.38	1.10	1.44	1.04	1.51	0.99	1.59
37	1.22	1.32	1.16	1.38	1.11	1.45	1.06	1.51	1.00	1.59
38	1.23	1.33	1.18	1.39	1.12	1.45	1.07	1.52	1.02	1.58
39	1.24	1.34	1.19	1.39	1.14	1.45	1.09	1.52	1.03	1.58
40	1.25	1.34	1.20	1.40	1.15	1.46	1.10	1.52	1.05	1.58
45	1.29	1.38	1.24	1.42	1.20	1.48	1.16	1.53	1.11	1.58
50	1.32	1.40	1.28	1.45	1.24	1.49	1.20	1.54	1.16	1.59
55	1.36	1.43	1.32	1.47	1.28	1.51	1.25	1.55	1.21	1.59
60	1.38	1.45	1.35	1.48	1.32	1.52	1.28	1.56	1.25	1.60
65	1.41	1.47	1.38	1.50	1.35	1.53	1.31	1.57	1.28	1.61
70	1.43	1.49	1.40	1.52	1.37	1.55	1.34	1.58	1.31	1.61
75	1.45	1.50	1.42	1.53	1.39	1.56	1.37	1.59	1.34	1.62
80	1.47	1.52	1.44	1.54	1.42	1.57	1.39	1.60	1.36	1.62
85	1.48	1.53	1.46	1.55	1.43	1.58	1.41	1.60	1.39	1.63
90	1.50	1.54	1.47	1.56	1.45	1.59	1.43	1.61	1.41	1.64
95	1.51	1.55	1.49	1.57	1.47	1.60	1.45	1.62	1.42	1.64
100	1.52	1.56	1.50	1.58	1.48	1.60	1.46	1.63	1.44	1.65

SOURCE: Reprinted, by permission of the Biometrika Trustees, from J. Durbin and G. S. Watson, "Testing for Serial Correlation in Least Squares Regression. II," *Biometrika*, Vol. 38 (1951), pp. 173 and 175.
EXAMPLE: For $n = 20$, $\alpha = 0.01$, and two independent variables, $d_L = 0.86$ and $d_U = 1.27$.
TEXT REFERENCE: Use of this table is discussed on p. 882.

APPENDIX

D DATA SETS

D.1 FINANCIAL CHARACTERISTICS

This data set contains information on net assets, net income, and net sales for 383 firms classified into eight industry groups.

Each line of the data set contains an identification number and information on seven other variables for a single firm. The eight variables are as follows:

Variable Number	Variable Name	Description
1	FIRM	Firm identification number (1–383)
2	INDUSTRY	Industry number coded as follows:
		1 Crude oil producers
		2 Textile products
		3 Textile apparel manufacturers
		4 Paper
		5 Electronic computer equipment
		6 Electronics
		7 Electronic components
		8 Auto parts and accessories
3	ASSETS (1)	Net assets in year 1 (in $ million)
4	ASSETS (2)	Net assets in year 2 (in $ million)
5	INCOME (1)	Net income in year 1 (in $ million)
6	INCOME (2)	Net income in year 2 (in $ million)
7	SALES (1)	Net sales in year 1 (in $ million)
8	SALES (2)	Net sales in year 2 (in $ million)

1	2	3	4	5	6	7	8
1	1	33.037	35.197	2.281	2.908	8.498	11.537
2	1	320.156	355.848	17.661	24.378	370.733	384.491
3	1	307.332	321.408	23.718	26.656	31.049	49.118
4	1	41.395	45.792	2.489	3.641	10.226	11.559
5	1	144.090	180.534	10.934	12.407	84.864	101.029
6	1	46.611	60.950	2.585	4.389	12.297	16.143
7	1	9.767	12.350	0.923	0.350	4.705	6.124
8	1	3.252	3.141	-0.174	-0.155	1.098	1.017
9	1	98.651	114.403	2.564	3.825	17.522	26.788
10	1	19.605	23.518	0.730	1.027	6.761	6.827
11	1	21.404	20.982	0.572	0.809	4.720	4.474
12	1	24.046	26.714	0.508	0.614	3.730	4.001
13	1	173.007	198.311	13.208	8.936	54.328	60.160
14	1	98.261	93.740	0.563	-0.792	13.442	13.269
15	1	6.958	5.708	-0.359	-0.558	2.479	1.521
16	1	23.682	23.996	-0.907	0.502	4.312	5.940
17	1	273.146	308.475	13.774	14.895	56.038	71.194
18	1	24.983	27.505	1.990	2.723	11.279	13.755
19	1	43.478	52.257	3.444	4.490	17.569	18.148
20	1	8.996	14.643	-0.176	0.270	1.107	1.953
21	1	307.403	322.025	11.521	20.836	68.764	67.442
22	1	365.377	363.423	7.070	10.577	40.995	48.238
23	1	17.987	27.188	1.064	1.631	5.173	6.310
24	1	446.506	480.794	31.259	37.461	128.664	146.655
25	1	67.860	94.029	4.824	6.101	17.067	20.074
26	1	321.744	547.854	80.528	85.031	182.449	202.052
27	1	255.357	312.386	13.991	17.638	127.806	149.302
28	1	157.964	237.078	5.631	7.459	41.197	45.764
29	1	18.203	23.921	0.099	0.190	4.679	5.594
30	1	141.333	255.155	17.173	20.555	118.166	123.335
31	1	40.345	42.460	2.464	2.740	5.326	5.509
32	1	20.308	23.382	2.028	2.966	4.097	4.814
33	1	3483.039	3458.764	-64.824	26.561	3557.532	3673.144
34	1	32.736	41.457	2.368	3.339	15.155	17.123
35	1	2085.829	2340.733	63.693	79.184	986.517	1093.102
36	1	20.862	22.203	1.355	1.957	4.783	11.781
37	1	31.069	33.268	1.420	1.808	2.716	3.532
38	1	89.658	90.445	1.351	0.394	11.331	11.260
39	1	27.925	48.831	2.133	2.581	9.669	11.885
40	1	730.017	772.251	5.802	6.900	182.627	194.079
41	1	69.692	87.668	1.696	2.742	17.690	24.122
42	1	14.672	23.909	1.679	0.883	5.006	4.906
43	1	32.733	24.743	-4.766	-0.419	7.270	5.346
44	1	53.121	63.963	2.079	2.708	9.596	10.996
45	1	3.517	3.318	0.020	-0.169	2.294	2.045
46	1	17.077	24.944	1.264	1.354	2.849	4.289
47	1	17.203	22.230	-0.026	1.580	4.929	5.334
48	1	52.296	62.005	-1.500	-2.318	5.165	6.365
49	1	59.748	60.552	1.933	3.020	14.828	16.871
50	1	14.660	14.579	0.408	0.687	2.529	3.282
51	1	36.516	38.493	2.921	1.322	9.442	9.608
52	1	316.345	374.205	-2.618	1.041	34.545	40.330
53	1	4.915	3.895	0.248	0.579	0.949	1.309
54	1	43.039	47.474	-1.887	-4.212	0.909	1.291
55	1	44.989	34.052	-4.253	-0.277	34.668	24.866
56	1	7.587	8.077	0.375	0.158	2.267	2.279
57	1	67.094	65.312	3.470	3.742	20.651	20.831
58	1	17.042	16.720	0.169	0.801	6.398	5.216
59	1	16.315	17.071	0.678	0.689	6.141	8.510
60	1	49.406	49.457	-5.277	0.655	22.226	21.437

1	2	3	4	5	6	7	8
61	2	98.291	105.288	7.814	9.729	23.778	26.808
62	2	120.361	129.735	7.722	7.054	194.511	206.686
63	2	86.612	80.549	3.178	2.732	97.496	95.772
64	2	78.705	88.090	3.819	3.939	128.598	125.086
65	2	1876.768	1945.958	54.190	66.969	2331.511	2451.761
66	2	103.526	111.185	3.730	5.577	170.813	234.989
67	2	242.555	278.809	23.159	21.080	393.291	441.342
68	2	271.921	283.952	9.486	11.119	430.817	447.119
69	2	2227.966	2630.037	118.818	188.623	2208.021	2601.829
70	2	177.614	153.874	6.144	6.087	81.555	84.332
71	2	378.621	372.959	-3.456	5.374	421.747	494.932
72	2	168.253	177.291	7.424	2.470	190.932	184.217
73	2	35.087	45.125	3.366	2.701	43.570	48.507
74	2	33.296	35.537	3.984	1.828	52.414	46.413
75	2	3.325	3.484	0.467	0.616	5.114	6.287
76	2	17.075	21.426	2.749	2.294	32.083	30.235
77	2	216.766	228.358	10.673	10.130	306.821	329.509
78	2	34.459	26.707	-0.941	0.576	53.361	44.423
79	2	8.806	12.289	1.569	1.176	10.966	15.003
80	2	37.476	50.316	1.667	2.095	29.588	40.524
81	2	119.556	125.491	5.018	6.896	182.525	218.244
82	2	73.948	81.329	5.983	7.090	61.690	69.895
83	2	40.027	40.460	2.716	0.274	68.498	55.874
84	2	25.240	18.237	0.212	-1.393	29.230	24.659
85	2	421.222	466.554	12.058	10.720	599.505	634.442
86	2	63.540	70.280	0.477	2.561	76.270	94.361
87	2	48.743	54.090	2.615	3.059	103.068	111.544
88	2	42.856	44.796	1.044	1.072	50.015	54.776
89	2	146.614	153.471	7.079	7.162	236.523	250.703
90	2	180.981	181.323	3.816	6.205	231.001	273.785
91	2	70.844	82.104	2.534	3.800	81.564	101.269
92	2	468.790	482.852	11.371	19.201	439.613	538.514
93	2	55.277	54.014	2.275	1.929	71.302	77.232
94	2	22.546	32.187	3.213	2.238	44.455	49.048
95	2	833.932	869.886	-0.867	20.994	1162.439	1279.287
96	2	171.237	201.282	15.247	9.659	221.647	241.229
97	2	1100.631	1173.978	22.140	20.669	996.824	1062.485
98	2	30.260	31.820	5.281	3.974	40.137	38.814
99	2	7.016	11.883	0.828	0.441	10.774	11.043
100	2	345.825	367.921	8.413	13.765	479.578	551.020
101	2	38.295	38.497	3.827	1.457	61.619	56.399
102	2	19.868	16.748	-0.421	0.134	20.733	22.599
103	2	145.220	153.824	-7.868	0.540	152.852	156.510
104	2	19.741	23.351	2.003	1.675	34.386	32.993
105	2	359.381	369.360	22.725	22.073	433.403	477.104
106	2	55.337	58.428	4.389	5.198	89.018	101.382
107	2	12.768	15.499	1.098	1.962	17.998	26.777
108	2	7.871	8.037	1.127	1.000	14.630	16.386
109	2	9.796	12.007	0.995	0.740	26.989	23.657
110	2	23.693	28.103	-0.497	1.181	39.135	50.421
111	2	23.205	21.246	-0.959	-0.312	20.879	23.698
112	2	24.434	25.793	2.514	2.797	41.562	41.965
113	3	52.288	67.090	2.218	0.680	73.175	84.460
114	3	48.435	57.810	1.196	1.528	48.429	59.755
115	3	74.096	104.981	9.403	2.502	88.318	78.569
116	3	27.379	24.694	-1.439	-1.898	29.088	27.760
117	3	55.835	66.343	4.922	5.515	87.476	106.024
118	3	242.812	303.727	17.905	20.268	396.723	465.007
119	3	148.376	155.930	4.791	4.385	236.810	268.119
120	3	415.396	427.906	15.732	18.221	674.024	738.672

1	2	3	4	5	6	7	8
121	3	21.657	26.793	1.188	1.544	39.635	57.129
122	3	18.498	21.041	1.903	2.076	46.088	49.336
123	3	65.791	81.270	0.668	0.510	102.888	116.853
124	3	28.219	33.504	0.775	1.717	35.752	44.686
125	3	149.359	136.670	8.139	-11.507	222.170	210.068
126	3	30.541	32.527	3.245	4.520	66.321	76.073
127	3	22.911	24.662	0.852	1.563	43.243	46.969
128	3	832.097	832.483	34.706	17.298	1764.240	1883.795
129	3	30.256	31.564	2.086	2.824	32.864	36.671
130	3	27.655	32.011	2.129	1.925	50.676	56.214
131	3	216.942	229.044	4.570	11.120	237.709	330.275
132	3	337.690	362.123	13.973	19.156	502.230	571.204
133	3	15.891	19.193	2.666	2.622	30.633	32.883
134	3	10.361	17.169	0.864	1.458	17.812	27.200
135	3	29.480	38.321	0.671	0.899	44.515	51.431
136	3	63.427	66.513	3.087	3.761	102.076	112.988
137	3	305.329	331.933	22.288	24.886	407.090	448.536
138	3	256.412	284.656	8.012	9.007	381.683	414.912
139	3	81.581	112.385	2.854	3.025	124.636	157.575
140	3	81.716	97.902	4.925	6.113	139.016	169.551
141	3	327.182	414.532	26.162	33.781	546.893	680.541
142	3	12.766	13.032	0.543	0.748	13.666	15.683
143	3	16.697	24.152	0.895	1.705	48.217	61.023
144	3	132.795	156.450	5.630	6.822	245.480	290.168
145	3	54.738	61.331	0.612	1.035	105.897	118.966
146	3	55.727	50.776	0.258	0.618	70.245	81.732
147	3	23.634	27.333	1.305	1.301	43.990	46.182
148	3	70.405	77.150	5.416	6.133	108.545	123.516
149	3	8.087	10.656	1.841	2.612	11.514	14.926
150	3	8.512	10.629	1.034	0.085	18.546	21.148
151	3	13.585	16.143	0.999	1.754	22.880	25.259
152	3	5.678	6.433	0.190	0.679	15.668	19.428
153	3	131.981	159.208	7.638	10.760	256.439	274.423
154	3	56.938	66.077	0.397	2.029	78.640	102.272
155	3	227.935	232.482	9.580	11.943	342.660	402.085
156	3	12.685	19.903	2.021	2.144	32.874	36.647
157	3	47.874	52.075	-1.069	1.156	38.075	49.028
158	3	95.637	116.532	6.440	7.490	179.816	192.254
159	3	82.075	85.915	9.783	10.965	154.724	172.012
160	3	88.359	100.690	4.057	5.634	154.888	196.680
161	3	28.130	38.817	2.763	3.841	30.108	40.127
162	3	60.762	58.273	0.771	1.477	85.118	87.549
163	3	12.898	13.940	2.438	2.300	26.583	24.775
164	3	13.769	14.909	1.040	0.425	21.777	21.395
165	3	108.794	139.760	5.114	5.939	85.822	84.229
166	3	19.011	25.038	1.143	1.457	25.318	29.369
167	3	183.176	219.225	19.873	23.540	372.702	401.209
168	3	29.655	39.998	3.310	4.444	38.295	50.903
169	3	212.932	233.429	9.343	12.458	352.894	381.388
170	3	56.958	54.895	2.520	2.263	81.279	79.415
171	3	23.844	26.703	2.036	2.331	42.379	48.497
172	3	10.981	10.976	1.426	1.094	21.230	21.998
173	3	9.692	13.096	0.535	1.230	26.622	31.720
174	3	9.468	13.334	2.792	2.979	30.912	38.223
175	3	17.057	22.557	1.546	1.886	42.827	55.133
176	3	25.695	25.526	2.485	2.137	69.533	57.854
177	3	6.090	6.808	0.244	0.348	15.926	19.719
178	3	33.182	38.151	1.696	2.128	55.619	63.864
179	3	3.479	4.374	0.733	0.822	8.636	10.179
180	3	11.818	11.987	0.922	0.626	34.672	33.195

1	2	3	4	5	6	7	8
181	3	12.220	16.128	0.689	1.066	27.842	36.828
182	3	14.957	15.548	0.377	0.306	31.217	30.129
183	3	10.427	12.913	0.738	1.345	19.093	25.784
184	3	8.240	10.938	1.894	2.580	26.476	31.406
185	3	41.907	42.941	3.740	5.017	55.481	69.854
186	3	14.525	17.563	2.531	2.954	19.825	22.887
187	3	23.706	30.623	1.462	3.676	51.307	62.905
188	4	43.060	47.135	2.734	3.687	36.353	43.569
189	4	994.853	1562.930	11.333	32.293	879.184	1887.666
190	4	1391.506	1471.317	42.559	59.544	1330.982	1501.588
191	4	688.492	697.367	14.175	23.582	697.108	757.052
192	4	106.703	108.282	1.665	2.905	60.862	64.626
193	4	595.544	622.023	19.236	24.566	479.868	553.250
194	4	483.549	486.714	4.340	6.282	499.886	532.888
195	4	2751.196	2802.900	93.590	138.694	2658.893	2825.996
196	4	1266.933	1300.139	42.779	75.060	1266.331	1364.198
197	4	1173.023	1169.280	31.466	35.165	1425.591	1523.867
198	4	443.204	482.408	14.236	22.390	477.943	509.535
199	4	1292.187	1387.853	30.853	55.782	1226.328	1367.681
200	4	1159.126	1171.518	35.612	52.115	1007.854	1098.618
201	4	30.312	31.779	−1.411	0.009	36.129	41.804
202	4	29.018	31.579	0.855	0.894	46.539	52.793
203	4	774.129	812.662	35.259	52.402	699.699	812.175
204	4	715.082	736.833	6.776	17.685	581.754	637.162
205	4	71.040	67.924	0.905	0.049	68.168	77.653
206	4	34.637	61.263	0.529	1.253	43.365	58.621
207	4	188.390	200.197	6.534	9.129	188.456	204.339
208	4	140.033	138.510	1.328	2.707	121.527	128.775
209	4	32.912	32.955	1.220	1.523	43.876	45.754
210	4	4.529	20.428	0.510	0.865	7.367	34.267
211	4	196.256	221.863	15.668	16.562	107.785	120.837
212	5	25.632	43.231	1.949	1.473	32.442	42.881
213	5	23.537	25.615	1.002	1.787	36.713	43.357
214	5	1.561	7.428	−1.052	1.161	0.556	9.076
215	5	32.403	46.818	2.252	5.261	20.462	40.937
216	5	100.240	89.427	0.583	1.971	68.700	80.715
217	5	202.692	259.761	14.310	20.655	198.246	253.197
218	5	67.605	73.362	2.660	1.558	101.272	103.194
219	5	2947.183	3025.063	88.722	103.402	2627.272	2869.352
220	5	429.474	427.247	−18.076	1.611	148.771	196.320
221	5	245.226	318.560	−0.635	−0.417	161.723	193.342
222	5	18.563	20.268	−1.916	−5.549	13.323	18.988
223	5	19.358	20.193	2.641	2.664	28.844	30.895
224	5	2232.655	2484.858	82.053	121.577	2462.315	3009.492
225	5	22.095	72.295	−5.515	3.164	4.946	35.520
226	5	29.426	28.844	−11.480	1.085	17.244	21.221
227	5	127.468	183.368	4.134	−18.051	98.294	90.477
228	5	4.948	12.141	−2.367	0.432	2.993	17.326
229	5	6.774	6.965	0.207	−0.128	8.756	7.960
230	5	15.032	18.720	0.166	0.458	5.643	7.942
231	5	4.103	6.225	−2.654	0.398	3.421	5.210
232	5	1.675	5.018	−0.170	0.425	2.916	6.581
233	5	6.719	8.278	−0.089	0.150	11.108	8.751
234	5	38.596	54.448	−9.167	−2.866	11.995	39.982
235	5	25.140	45.206	−6.001	−7.097	5.243	17.661
236	5	10.176	9.651	0.015	0.158	15.275	17.472
237	5	10.504	13.862	−5.063	−2.997	4.001	7.304
238	5	2.608	13.271	−2.577	−3.406	0.244	4.585
239	5	6.305	5.696	−0.721	0.363	6.704	6.029
240	5	21.044	29.722	−3.411	0.397	18.021	29.925

1	2	3	4	5	6	7	8
241	5	12.481	19.111	0.004	1.239	14.348	21.639
242	5	21.854	39.536	-6.966	0.649	5.434	29.502
243	5	24.700	46.787	1.396	2.331	10.855	20.521
244	5	2.627	4.724	-1.922	0.485	3.133	8.428
245	5	84.004	81.315	-0.594	0.551	52.551	58.084
246	5	22.095	72.295	-5.515	3.164	4.946	35.520
247	5	8.431	13.859	-1.910	1.554	11.278	21.128
248	5	11.413	17.487	0.952	1.200	25.596	29.834
249	6	53.546	51.957	1.040	-0.262	64.782	63.373
250	6	184.036	187.940	4.031	5.932	181.903	177.285
251	6	277.861	329.313	32.481	44.808	323.525	407.816
252	6	53.838	62.830	2.600	3.218	66.057	78.744
253	6	107.916	74.338	0.377	0.486	96.629	88.244
254	6	37.926	40.133	0.432	0.363	49.600	52.207
255	6	45.367	50.225	0.136	1.021	54.903	53.225
256	6	204.968	256.404	-10.585	10.430	260.669	302.260
257	6	401.301	444.267	7.347	12.485	372.515	430.150
258	6	75.383	78.563	0.124	1.412	114.122	124.160
259	6	59.889	70.428	-1.107	0.921	44.851	81.252
260	6	34.603	40.125	-0.269	0.405	24.867	37.557
261	6	28.712	30.841	0.251	0.824	43.508	46.930
262	6	42.916	72.028	2.754	5.021	80.719	133.688
263	6	826.922	852.024	47.500	55.584	1765.533	1977.791
264	6	134.881	145.097	-33.251	2.596	197.293	200.795
265	6	6.286	7.568	0.510	0.447	8.934	9.742
266	6	30.609	33.084	0.833	2.480	41.858	61.578
267	6	24.926	27.066	-1.374	1.567	42.449	48.981
268	6	2.946	17.348	0.131	1.007	6.263	15.119
269	6	5.177	7.451	0.358	0.394	6.352	8.192
270	7	491.453	359.448	-121.041	1.530	383.297	346.415
271	7	9.783	13.555	2.141	3.042	13.716	19.243
272	7	258.823	308.337	16.783	21.618	399.929	482.695
273	7	103.572	116.888	5.219	6.386	119.848	133.649
274	7	76.125	91.839	8.100	13.427	114.130	145.126
275	7	20.064	21.133	0.304	0.436	25.379	28.065
276	7	3.056	3.395	-0.524	0.135	4.917	6.094
277	7	5.797	5.563	0.250	0.263	4.926	5.584
278	7	35.080	41.402	1.527	1.955	33.776	37.660
279	7	9.651	11.212	-0.023	0.662	11.004	14.495
280	7	1.773	1.530	-0.381	-0.347	1.743	1.746
281	7	8.006	7.277	-0.151	-0.450	5.378	5.463
282	7	9.466	10.025	-0.576	0.259	11.754	16.598
283	7	39.955	44.133	-0.473	2.816	46.724	58.257
284	7	17.887	20.594	-1.012	0.119	19.080	24.358
285	7	9.280	10.831	0.576	0.883	14.646	18.355
286	7	47.550	47.781	-3.671	0.736	46.603	59.519
287	7	516.657	492.824	8.370	15.976	691.172	751.541
288	7	143.136	168.244	7.641	9.998	215.780	256.349
289	7	32.473	36.204	2.124	2.799	41.907	54.490
290	7	5.827	5.909	0.209	0.447	8.509	8.334
291	7	76.873	82.094	1.912	3.776	112.952	131.262
292	7	1066.853	1061.011	41.017	40.784	853.079	945.749
293	7	14.997	15.819	0.076	0.107	14.318	16.286
294	7	25.923	20.142	-0.298	-0.078	24.091	25.411
295	7	6.323	7.938	0.713	1.332	8.705	11.734
296	7	14.228	15.215	0.544	0.913	20.119	22.640
297	7	104.990	120.159	5.530	7.270	132.893	163.156
298	7	72.548	70.039	0.036	1.215	27.895	33.645
299	7	173.356	179.659	-10.831	0.900	159.120	197.973
300	7	7.470	8.586	0.278	-0.063	8.725	6.373

1	2	3	4	5	6	7	8
301	7	782.829	856.462	45.526	64.841	1031.749	1273.987
302	7	79.974	70.361	1.783	2.803	95.707	106.215
303	7	60.445	64.317	3.497	4.081	41.531	46.513
304	7	115.657	97.123	-2.241	-1.172	89.658	86.283
305	7	20.165	22.599	1.678	2.363	18.487	26.527
306	7	46.189	45.390	0.869	1.763	40.148	47.731
307	7	19.156	18.650	0.544	1.220	29.685	36.424
308	7	14.436	15.593	-0.649	0.398	15.104	21.765
309	7	29.106	28.966	0.463	-2.460	42.485	31.948
310	7	41.592	42.399	0.344	0.117	35.806	38.969
311	7	11.767	13.177	0.872	1.269	16.624	21.109
312	7	38.790	41.032	2.979	4.119	52.419	66.423
313	7	5.176	8.205	-4.242	-4.928	0.942	1.434
314	7	47.106	48.975	-1.933	1.654	63.322	74.446
315	7	8.145	12.899	0.633	0.998	10.328	12.196
316	7	19.962	29.624	-0.556	2.673	12.374	31.011
317	7	12.108	10.303	-1.935	-2.020	6.030	7.509
318	7	13.831	15.193	0.041	0.867	21.518	28.862
319	7	5.531	22.394	0.166	3.043	4.906	23.941
320	7	7.756	10.240	0.703	1.195	8.489	13.842
321	7	12.112	13.072	-0.138	0.173	12.853	13.025
322	7	14.881	16.331	0.616	0.790	13.302	16.332
323	7	3.147	8.936	-1.080	0.554	2.326	6.132
324	7	29.650	30.896	0.996	1.823	26.993	30.811
325	7	4.872	5.319	0.054	0.327	5.705	6.101
326	7	4.698	5.503	0.189	0.493	5.544	6.942
327	8	26.590	24.798	1.233	1.520	42.840	47.912
328	8	135.421	151.732	4.886	5.196	166.394	201.787
329	8	28.351	36.032	2.479	3.740	55.229	78.438
330	8	163.767	177.817	7.584	10.103	245.157	277.402
331	8	55.625	64.745	6.309	6.515	125.118	137.080
332	8	1619.015	1667.520	56.835	79.920	2159.730	2380.186
333	8	1300.048	1378.921	63.967	80.020	1550.054	1732.303
334	8	550.678	554.082	7.669	19.981	729.176	905.594
335	8	19.386	21.273	0.339	1.813	23.566	36.507
336	8	370.446	407.549	47.122	53.730	439.957	495.959
337	8	665.819	757.080	38.073	60.519	858.155	1109.792
338	8	1129.579	1279.626	77.004	95.029	1398.144	1651.144
339	8	45.738	59.208	1.922	3.275	52.758	68.441
340	8	271.663	318.553	17.948	19.668	363.968	391.314
341	8	13.335	20.886	1.434	1.843	37.071	49.414
342	8	74.743	99.174	5.457	7.077	82.798	93.439
343	8	408.271	540.327	19.320	26.291	461.924	636.240
344	8	20.415	21.497	0.936	1.314	30.617	32.713
345	8	76.976	90.557	3.370	8.109	109.582	154.571
346	8	195.265	213.925	11.321	17.285	271.581	334.040
347	8	68.306	76.343	3.506	5.053	98.653	118.558
348	8	586.732	613.967	66.840	71.024	716.704	802.459
349	8	179.264	218.990	6.973	15.256	325.448	365.481
350	8	71.284	85.826	5.912	5.816	146.322	149.899
351	8	29.886	25.449	0.198	0.467	31.795	32.767
352	8	132.485	166.323	24.014	27.626	153.326	172.155
353	8	32.521	31.236	0.586	0.899	36.569	36.162
354	8	134.451	160.434	12.932	15.551	240.858	295.538
355	8	293.471	329.299	15.574	19.229	386.584	452.600
356	8	136.326	147.780	3.669	6.360	204.375	205.760
357	8	88.918	131.481	5.212	7.707	145.113	210.109
358	8	5.366	7.156	-1.079	0.016	8.114	11.256
359	8	44.238	49.625	4.294	4.618	83.835	89.282
360	8	93.654	94.370	3.217	2.408	94.851	99.816

1	2	3	4	5	6	7	8
361	8	75.159	100.290	5.798	9.242	99.889	133.685
362	8	166.925	184.460	7.289	8.555	275.438	309.597
363	8	382.097	407.795	17.291	13.424	616.742	665.240
364	8	37.384	45.877	3.980	3.784	74.313	83.476
365	8	40.821	54.390	3.195	5.249	88.393	106.061
366	8	44.951	48.009	1.839	2.600	54.470	63.508
367	8	12.157	14.886	1.281	1.700	18.538	22.333
368	8	1506.522	1668.217	90.891	102.747	2088.407	2278.139
369	8	62.097	66.840	2.229	3.008	146.312	175.388
370	8	567.894	594.967	51.458	57.116	554.328	635.523
371	8	167.801	191.273	10.939	13.426	289.663	319.323
372	8	69.428	75.824	3.241	4.955	111.572	126.815
373	8	10.842	15.981	1.933	2.643	26.641	31.759
374	8	25.402	27.937	0.707	1.717	32.731	34.171
375	8	23.277	25.307	1.408	2.403	43.447	47.335
376	8	46.671	53.533	3.430	4.732	59.015	70.728
377	8	49.480	52.981	4.429	4.807	103.741	117.489
378	8	13.272	14.407	0.292	-0.028	32.507	35.084
379	8	24.986	25.325	0.324	0.088	30.752	31.196
380	8	64.899	73.919	4.386	7.440	108.813	129.973
381	8	83.894	108.963	9.303	14.074	90.550	96.532
382	8	28.883	36.202	3.166	4.031	59.360	72.837
383	8	8.668	8.514	0.205	0.108	18.535	21.288

 D.2   POWER CELLS

This data set contains the results of an experiment on 54 identical silver-zinc power cells. The experiment was conducted to measure two responses (number of discharge-charge cycles and failure mode) for each experimental cell. Four experimental control variables were employed (charge rate, discharge rate, depth of discharge, and ambient temperature), with particular levels of the four factors specified for each cell.

Each line of the data set contains an identification number and information on the two response variables and the levels of the four experimental factors for a single power cell. The seven variables are as follows:

Variable Number	Variable Name	Description
1	CELL	Cell identification number (1–54)
2	CYCLES	Number of discharge-charge cycles a cell survives before failure
3	MODE	Mode of failure (0 denotes shorting failure, 1 denotes low-voltage failure)
4	CHARGE	Charge rate (0.4, 1.0, 1.6 amperes)
5	DISCHRG	Discharge rate (1, 3, 5 amperes)
6	DEPTH	Depth of discharge (40, 80 percent of rated ampere hours)
7	TEMP	Ambient temperature (10°, 20°, 30°C)

SOURCE: The experimental setting and data are adapted from S. M. Sidik, H. F. Leibecki, and J. M. Bozek, *Cycles Till Failure of Silver-Zinc Cells with Competing Failure Modes—Preliminary Data Analysis,* NASA Technical Memorandum 81556, 1980.

1	2	3	4	5	6	7	1	2	3	4	5	6	7
1	138	1	0.4	1	40	10	28	16	1	1.0	3	80	10
2	325	1	0.4	1	40	20	29	96	1	1.0	3	80	20
3	246	0	0.4	1	40	30	30	141	0	1.0	3	80	30
4	101	0	0.4	1	80	10	31	3	1	1.0	5	40	10
5	281	1	0.4	1	80	20	32	383	0	1.0	5	40	20
6	150	0	0.4	1	80	30	33	469	1	1.0	5	40	30
7	142	1	0.4	3	40	10	34	2	1	1.0	5	80	10
8	179	1	0.4	3	40	20	35	308	1	1.0	5	80	20
9	398	0	0.4	3	40	30	36	385	0	1.0	5	80	30
10	44	1	0.4	3	80	10	37	55	1	1.6	1	40	10
11	170	1	0.4	3	80	20	38	73	1	1.6	1	40	20
12	314	1	0.4	3	80	30	39	216	1	1.6	1	40	30
13	12	1	0.4	5	40	10	40	3	1	1.6	1	80	10
14	103	1	0.4	5	40	20	41	69	1	1.6	1	80	20
15	503	0	0.4	5	40	30	42	62	1	1.6	1	80	30
16	10	1	0.4	5	80	10	43	35	1	1.6	3	40	10
17	145	1	0.4	5	80	20	44	15	1	1.6	3	40	20
18	159	1	0.4	5	80	30	45	305	0	1.6	3	40	30
19	78	1	1.0	1	40	10	46	2	1	1.6	3	80	10
20	122	1	1.0	1	40	20	47	13	1	1.6	3	80	20
21	283	1	1.0	1	40	30	48	159	0	1.6	3	80	30
22	8	1	1.0	1	80	10	49	3	1	1.6	5	40	10
23	143	1	1.0	1	80	20	50	6	1	1.6	5	40	20
24	130	0	1.0	1	80	30	51	233	0	1.6	5	40	30
25	33	1	1.0	3	40	10	52	2	1	1.6	5	80	10
26	80	1	1.0	3	40	20	53	7	1	1.6	5	80	20
27	217	0	1.0	3	40	30	54	2	1	1.6	5	80	30

 ## INVESTMENT FUND

This data set contains quarterly unit values of the equity and fixed-income components of an investment fund for an 11-year period. A quarterly unit value represents the value on the last day of the quarter of $1 invested at the inception of the fund. On January 1 of year 1, the unit values of the equity and fixed-income components were 1.0780 and 1.0549, respectively.

Each line of the data set contains an identification number, the year and quarter, and information on the values of the two fund components for a single quarter. The five variables are as follows:

Variable Number	Variable Name	Description
1	PERIOD	Period identification number (1–44)
2	YEAR	Year of series (1–11)
3	QUARTER	Quarter of year (1–4)
4	EQUITY	Unit value of equity component at last day of quarter
5	FIXED	Unit value of fixed-income component at last day of quarter

1	2	3	4	5		1	2	3	4	5
1	1	1	1.0620	1.0658		23	6	3	1.4399	1.6177
2	1	2	0.9699	1.0584		24	6	4	1.4668	1.6249
3	1	3	1.0322	1.0763		25	7	1	1.6055	1.6525
4	1	4	0.9702	1.0848		26	7	2	1.7342	1.7076
5	2	1	0.9944	1.0868		27	7	3	1.8593	1.7139
6	2	2	0.8842	1.0374		28	7	4	1.8624	1.6641
7	2	3	0.7435	1.0357		29	8	1	2.1086	1.6238
8	2	4	0.7794	1.0757		30	8	2	2.1378	1.7563
9	3	1	0.8943	1.1355		31	8	3	2.4167	1.7521
10	3	2	0.9251	1.1373		32	8	4	2.6049	1.7679
11	3	3	0.9438	1.1376		33	9	1	2.6610	1.8090
12	3	4	0.9425	1.1638		34	9	2	2.7680	1.7686
13	4	1	1.0418	1.2117		35	9	3	2.5949	1.7322
14	4	2	1.0502	1.2460		36	9	4	2.5433	1.9082
15	4	3	1.0394	1.2841		37	10	1	2.4322	1.9749
16	4	4	1.0038	1.3463		38	10	2	2.3985	2.0606
17	5	1	1.0418	1.4009		39	10	3	2.5211	2.1866
18	5	2	1.0656	1.4358		40	10	4	2.9551	2.4261
19	5	3	1.1246	1.4740		41	11	1	3.2605	2.5456
20	5	4	1.1550	1.5155		42	11	2	3.7334	2.6241
21	6	1	1.1852	1.5428		43	11	3	3.8861	2.5947
22	6	2	1.3033	1.5785		44	11	4	4.0496	2.6512

D.4   BEER SALES

This data set contains monthly time series for total domestic beer sales in a region and the average daily maximum and minimum temperatures for a five-year period. The temperatures are those recorded for a major city in the region.

Each line of the data set contains an identification number, the year and month, and information on sales and temperatures for a single month. The six variables are as follows:

Variable Number	Variable Name	Description
1	PERIOD	Period identification number (1–60)
2	YEAR	Year of series (1–5)
3	MONTH	Month of year (1–12)
4	SALES	Monthly total domestic beer sales in region (in thousands of hectoliters)
5	MAX	Monthly average of maximum daily temperatures (in degrees Celsius)
6	MIN	Monthly average of minimum daily temperatures (in degrees Celsius)

1	2	3	4	5	6
1	1	1	130.2	3.5	-3.2
2	1	2	137.0	6.9	0.8
3	1	3	171.4	11.5	2.9
4	1	4	178.5	13.0	4.7
5	1	5	211.9	16.9	8.6
6	1	6	227.4	19.9	10.3
7	1	7	253.8	22.7	13.0
8	1	8	254.6	22.1	13.2
9	1	9	152.5	19.4	11.5
10	1	10	177.2	13.9	7.1
11	1	11	161.8	8.8	0.8
12	1	12	179.3	8.6	3.1
13	2	1	156.8	3.7	-2.8
14	2	2	130.4	8.4	2.9
15	2	3	166.1	8.8	2.5
16	2	4	205.4	14.2	5.7
17	2	5	209.1	16.0	8.5
18	2	6	235.4	17.8	10.6
19	2	7	227.3	20.7	12.4
20	2	8	203.9	20.3	10.2
21	2	9	182.5	17.5	10.3
22	2	10	200.4	15.0	6.5
23	2	11	161.0	9.8	4.1
24	2	12	195.9	7.1	2.7
25	3	1	142.8	8.2	2.7
26	3	2	161.8	8.4	1.9
27	3	3	173.3	11.8	3.8
28	3	4	185.6	12.4	5.0
29	3	5	192.1	16.0	8.9
30	3	6	223.8	17.4	10.3
31	3	7	258.5	20.6	13.5
32	3	8	241.3	23.1	14.4
33	3	9	202.2	19.0	10.8
34	3	10	154.9	12.6	6.1
35	3	11	149.3	10.8	4.4
36	3	12	194.4	6.4	1.3
37	4	1	128.9	4.3	-0.5
38	4	2	150.7	7.0	1.4
39	4	3	173.4	9.3	1.6
40	4	4	187.0	11.8	3.5
41	4	5	214.0	16.4	7.8
42	4	6	252.0	21.2	12.2
43	4	7	239.2	20.8	13.2
44	4	8	231.0	20.7	12.9
45	4	9	158.5	18.7	10.5
46	4	10	161.9	13.7	6.7
47	4	11	155.2	7.7	0.7
48	4	12	175.2	6.9	1.3
49	5	1	139.2	9.0	3.5
50	5	2	141.8	9.5	3.2
51	5	3	195.9	11.8	4.5
52	5	4	196.4	13.4	5.0
53	5	5	192.6	17.9	9.4
54	5	6	220.9	19.1	11.4
55	5	7	200.8	20.5	12.7
56	5	8	242.7	21.8	13.6
57	5	9	165.2	17.5	9.4
58	5	10	174.4	13.0	5.9
59	5	11	159.5	9.9	5.2
60	5	12	167.9	3.8	-2.5

ANSWERS TO SELECTED PROBLEMS

CHAPTER 2

2.3 a.

Outcome	Number	Percent
A	38	63.3
I	7	11.7
C	5	8.3
S	3	5.0
G	2	3.3
O	5	8.3
Total	60	100.

c. A, 31.8

2.8 a., b.

1	5
2	55
3	0
4	
5	00000
6	05
7	055
8	00
9	00
10	00000
11	
12	5
13	
14	
15	0

c. (1) 15, 150 (2) 28 (3) 50 and 100

2.15 a.

Weekly Wages (dollars)	Number
360–under 380	8
380–under 400	16
400–under 420	21
420–under 440	10
440–under 460	5
Total	60

2.18 a. (1) 5.0 (2) 5.0 (3) no

b.

Response Time (minutes)	Percent of Alarms
2.5–under 7.5	10
7.5–under 12.5	36
12.5–under 17.5	30
17.5–under 22.5	18
22.5–under 27.5	6
Total	100
	(50)

2.22 a.

Less Than This Time (minutes)	Cumulative Percent of Alarms
2.5	0
7.5	10
12.5	46
17.5	76
22.5	94
27.5	100

b. 85

2.27 a. well, 2504   **b.** type of well, district

Type	Number of Wells	District	Number of Wells
New-field wildcat	521	I	1692
Other exploratory	265	II	143
Development	1718	III	602
Total	2504	IV	67
		Total	2504

2.30 a.

Phase of Flight	Cause of Error		
	M	O	Total
T	14	5	19
C	4	4	8
L	6	12	18
Total	24	21	45

b. (1) 25 (2) 67

c.

Cause of Error	Phase of Flight		
	T	C	L
M	74	50	33
O	26	50	67
Total	100%	100%	100%
	(19)	(8)	(18)

2.37 a. 7.5

b.

Packer:	1	2	3	4	5	6	7	8
Residual:	-1.5	-2.5	$+6.5$	-2.5	-2.5	-3.5	$+6.5$	-0.5

CHAPTER 3

3.1 731.75, 878.10

3.8 13.7

3.15 a. 746.00

3.22 a. 7.5–under 12.5

3.33 a. 9.583, 17.333   **b.** 21.389

3.40 7.750

3.47 28.378, 5.33

3.54 b. 155,315,596.2, 0.711

3.61 1/9

3.6 8.127

3.12 a. 687.67

3.19 13.167

3.29 985

3.36 a. 2063, 696

3.43 362,720.9, 602.26

3.51 a. 82.3

3.58 2063, 2.210

3.64

Number of cases (n)	6	3rd moment	29894.400
Mean	46.500	(Divisor $n - 1$)	
Std. deviation	38.960	4th moment	3657122.074
Minimum	8.	(Divisor $n - 1$)	
Maximum	108.	Std. 3rd moment	0.506
		Std. 4th moment	1.587

3.66 a.

Minimum	1st Quartile	Median	3rd Quartile	Maximum
0	275.5	746.0	971.5	2063

3.69 a.

	Minimum	1st Quartile	Median	3rd Quartile	Maximum
Group 1:	2	3	3.5	5	8
Group 2:	3	6	6	7	8

3.71 a. 1.06667, 1.15000, 1.11413   **b.** 1.1097

3.73 342.86

CHAPTER 4

4.4 a. $E_1 = \{o_2, o_3\}$, $E_1^* = \{o_1, o_4\}$
   b. $E_2 = \{o_1, o_2, o_3\}$

4.14 (1) $\{o_2, o_3, o_4, o_6, o_7, o_8\}$ (2) $\{o_5\}$ (3) $\{o_1, o_3, o_4, o_5\}$ (4) $\varnothing$

4.20 a. 6/7

4.25 a. univariate   **b.** 0.10   **c.** 0.20

4.27 a. (1) $P(A_1)$ (2) $P(A_1 \cap B_1)$ (3) $P(B_1 | A_1)$ (4) $P(A_1 | B_1)$
   c. a: (1) 0.2 (2) 0.2 (3) 1 (4) 0.67
      b: (1) 0 (2) 0.125 (3) 0.7

4.28 a.

B_i	$P(B_i)$
B_1	0.3
B_2	0.7
Total	1.0

b.

B_i	$P(B_i \mid A_2)$
B_1	0.125
B_2	0.875
Total	1.000

4.34 (1) 0.3 (2) 0.1 (3) 0.2

4.37 a. 0.01   **b.** 0.9596

4.45 a. 0.04, 0.008   **b.** 0.2

4.48 b.

	Conditional On	
	A_1	A_2
B_1	1	0.125
B_2	0	0.875
Total	1	1.000

4.51 a. 0.009

4.53 a. $P(A_1) = 0.2$, $P(A_2) = 0.8$, $P(B_1 \mid A_1) = 1$, $P(B_1 \mid A_2) = 0.125$, $P(A_1 \mid B_1) = 0.67$, $P(A_2 \mid B_1) = 0.33$

CHAPTER 5

5.3 b. (1) 0.60 (2) 0.80 (3) 0.35

5.7 b.

x:	0	1	2	3	4
$P(X \leq x)$:	0.60	0.80	0.90	0.95	1.00

$P(X \leq 3) = 0.95$

5.11 (1) 0.632 (2) 0.5

5.13 (1) 0.30 (2) 0.07 (3) 0.93 (4) 0.03 (5) 0.405

5.17 a.

x	$P(x)$
0	0.57
1	0.36
2	0.07
Total	1.00

b.

x	$P(X = x \mid Y = 0)$
0	0.541
1	0.405
2	0.054
Total	1.000

5.20 0.75

5.24 1.2875, 1.1347

5.28 a.

y:	−0.661	0.220	1.102	1.983	2.864
$P(y)$:	0.60	0.20	0.10	0.05	0.05

5.31 0.809

5.33 a. $Y = 9.50X$
   b. $E\{X\} = 3.5$, $\sigma\{X\} = 1.1619$, $E\{Y\} = 33.25$, $\sigma\{Y\} = 11.04$

c.

y:	9.50	19.00	28.50	38.00	47.50	57.00
$P(y)$:	0.05	0.15	0.25	0.40	0.10	0.05

5.37 a. 1.6, 1.140

b.

t:	0	1	2	3	4
$P(t)$:	0.20	0.27	0.32	0.15	0.06

5.38 a. 0.2, 1.140

b.

d:	−2	−1	0	1	2
$P(d)$:	0.08	0.18	0.35	0.24	0.15

$P(D = 0) = 0.35$, $P(D > 0) = 0.39$

5.42 a. 120,000, 360,000   **b.** 1581, 2739

5.45 a. (1) 0.917 (2) 0.083

5.49 a. 6.8, 35   **b.** −4.0   **c.** −2.0

5.52 a. −0.408

5.55 a. 1.10, 0, 1.62, 0.60   **b.** 1.10, 0, 1.22, 1.22

CHAPTER 6

6.5 a. $X = 0, 1, 2, 3, 4$ **b.** (1) 0.3164 (2) 0.2109 (3) 0.9492

6.8 (1) 0.3851 (2) 0.2355 (3) 0.2985 (4) 0.7442

6.11 b. 1.00, 0.750, 0.866

6.16 a. $T = 0, 1, 2, \ldots, 20$ **b.** (1) 0.1244 (2) 8 (3) 8.0

6.20 a. (1) 0.0302 (2) 0.2158   **b.** 3.5, 3.5

6.24 (1) 0.0771 (2) 0.8088 (3) 5.5

6.29 a. (1) 3.5 (2) 1.871 (3) 0.1850
6.31 a. (1) 0.0714 (2) 0.2143 (3) 0.3571   **b.** 6.5, 4.031
6.35 a. (1) 0.300 (2) 0.667   **b.** 1.8, 0.748

CHAPTER 7

7.1 b. 275, 208.33   **c.** (1) 0.4 (2) 0.5

7.4 a.

x:	0	3	6	9	12
$f(x)$:	0.0180	0.0807	0.1330	0.0807	0.0180

7.6 b. (1) $(X - 200)/10$ (2) $X/10$ (3) $X - 5$
7.8 a. (1) 0.5000 (2) 0.6628 (3) 0.8554 (4) 0.9938   **b.** (1) 0.5000 (2) 0.0668 (3) 0.1056
(4) 0.8276   **c.** (1) 0.00 (2) 1.32 (3) -1.32 **d.** (1) 2.326 (2) -1.645 (3) 1.960
7.11 a. (1) 0.8413 (2) 0.0918 (3) 0.8904   **b.** 2900.5   **c.** 2699.5 to 2900.5
7.15 a. $N(473, 169)$ **b.** 0.3121
7.17 a. 0.005   **b.** (1) 0.3935 (2) 0.7135 (3) 0.1353   **c.** 200

d.

x:	0	60	150	400	1000
$F(x)$:	0.0000	0.2592	0.5276	0.8647	0.9933

7.21 b. 0.333 hour   **c.** 0.5276

CHAPTER 8

8.3 37.00, 6.949

8.5 a.

x:	103	104	108	112
$P(x)$:	0.25	0.25	0.25	0.25

b. 106.75, 12.6875
8.7 b. 0.34, 0.6244
8.18 a. AB AC AD BC BD CD; six   **b.** 1/6

CHAPTER 9

9.1 0.234
9.6 (1) 1.74, 0.00949 (2) 1.74, 0.00424
9.16 0.003
9.18 a. (1) 0.9929 (2) 0.9858   **b.** 2.199 to 2.401, 2.229 to 2.371
9.24 a. 0.8164   **b.** 1.725 to 1.755   **c.** 1.706 to 1.774
9.27 a. 0.40, 0.663

b.

$\bar{X}$:	0	0.5	1	1.5	2
$P(\bar{X})$:	0.49	0.28	0.18	0.04	0.01

c. 0.40, 0.469

CHAPTER 10

10.9 a. 0.185   **b.** 4.84 and 5.58, approximately 95.4 percent   **c.** $4.906 \leq \mu \leq 5.514$
d. approximately 0.90 or 90 percent

10.16 $107.7 \leq \mu \leq 117.9$

10.22 \$9.81 million $\leq \tau \leq$ \$11.03 million

10.25 a. 152

10.29 a. $\mu \geq 4.906$

10.33 a. $-2.648 \leq X_{new} \leq 4.088$

10.37 a.

λ:	0	1	1.5	2	3
$L(\lambda)$:	0.0000	0.0226	0.0280	0.0244	0.0112

CHAPTER 11

11.10 a. $H_0: \mu \leq 60$, $H_1: \mu > 60$. If $z^* \leq 2.326$, conclude H_0; if $z^* > 2.326$, conclude H_1. $z^* = 2.160$. Conclude H_0.

11.12 a. $H_0: \mu \geq 0$, $H_1: \mu < 0$. If $z^* \geq -3.090$, conclude H_0; if $z^* < -3.090$, conclude H_1. $z^* = -13.350$. Conclude H_1.

11.14 a. $H_0: \mu = 8.50$, $H_1: \mu \neq 8.50$. If $|z^*| \leq 1.960$, conclude H_0; if $|z^*| > 1.960$, conclude H_1. $z^* = 2.538$. Conclude H_1.

11.20 a. 0.0154
b. If P-value ≥ 0.01, conclude H_0; if P-value < 0.01, conclude H_1. Conclude H_0.

11.22 a. 0.0000; one-sided
b. If P-value ≥ 0.001, conclude H_0; if P-value < 0.001, conclude H_1. Conclude H_1.

11.24 a. 0.0110; two-sided
b. If P-value ≥ 0.05, conclude H_0; if P-value < 0.05, conclude H_1. Conclude H_1.

11.27 a. $H_0: \mu = 110$, $H_1: \mu \neq 110$. If $|t^*| \leq 2.131$, conclude H_0; if $|t^*| > 2.131$, conclude H_1. $t^* = 1.167$. Conclude H_0.
b. $0.20 < P$-value < 0.50 (exact P-value $= 0.262$); two-sided

11.32 b.

μ:	2.15	2.45
$P(H_1; \mu)$:	0.239	0.879

d. H_0 is incorrect. $P(H_0; \mu = 2.45) = 0.121$. β risk.

11.33 a. If $\bar{X} \leq 2.372$, conclude H_0; if $\bar{X} > 2.372$, conclude H_1.   **b.** $P(H_1; \mu = 2.45) = 0.688$

11.38 $P(H_0; \mu = 70) \simeq 0.367$

11.40 $P(H_0; \mu = -6.0) \simeq 0.797$

11.42 $P(H_0; \mu = 10.0) \simeq 0.005$

11.45 **a.** 108

    b. If $z^* \le 1.645$, conclude H_0; if $z^* > 1.645$, conclude H_1. $z^* = 2.066$. Conclude H_1.

11.49 $1.268 \le \mu \le 1.732$. Conclude H_1. $\alpha = 0.05$.

CHAPTER 12

12.4 **a.**

x	$P(x)$	$\bar{p}$	$P(\bar{p})$
0	0.0000	0	0.0000
1	0.0000	1/9	0.0000
2	0.0000	2/9	0.0000
3	0.0001	3/9	0.0001
4	0.0008	4/9	0.0008
5	0.0074	5/9	0.0074
6	0.0446	6/9	0.0446
7	0.1722	7/9	0.1722
8	0.3874	8/9	0.3874
9	0.3874	1	0.3874

    c. (1) 0.3874 (2) 0.3874 (3) 0.9470

    d. $E\{X\} = 8.1$, $\sigma\{X\} = 0.9$, $E\{\bar{p}\} = 0.9$, $\sigma\{\bar{p}\} = 0.1$

12.7 **a.** no   **b.** yes   **c.** no   **d.** yes

12.9 **b.** $E\{X\} = 10.0$, $\sigma\{X\} = 2.8284$

    c. (1) 0.0262 (2) 0.0262 (3) 0.1428

    d. (1) 0.0262 (2) 0.0262 (3) 0.1428

12.11 **a.** $E\{\bar{p}\} = 0.40$, $\sigma\{\bar{p}\} = 0.030984$ **b.** (1) 0.0537 (2) 0.8926   **c.** 0.9988

12.14 (1) 0.2707 (2) 0.0526

12.16 **a.** $\bar{p} = 0.80$, $s\{\bar{p}\} = 0.026726$ **b.** $0.748 \le p \le 0.852$

12.21 **a.** 609   **b.** 714

12.27 **a.** $H_0: p \ge 0.70$, $H_1: p < 0.70$. If $z^* \ge -2.326$, conclude H_0; if $z^* < -2.326$, conclude H_1. $z^* = -1.091$. Conclude H_0.

    c. 0.1379; one-sided

12.31 **a.** If $\bar{p} \ge 0.593$, conclude H_0; if $\bar{p} < 0.593$, conclude H_1. $A = 0.593$.

    b. (1) 0.9686 (2) 0.4443 (3) 0.0099   **c.** $P(H_1; p = 0.60) = 0.4443$

12.35 **a.** 493

    b. $H_0: p \le 0.01$, $H_1: p > 0.01$. If $z^* \le 1.645$, conclude H_0; if $z^* > 1.645$, conclude H_1. $z^* = 4.017$. Conclude H_1.

12.39 $0.00067 \le p \le 0.368$

CHAPTER 13

13.3 **a.** 4290.50   **b.** $-14.04 \le \mu_2 - \mu_1 \le 54.04$

13.6 **a.** $926.92 \le \mu_2 - \mu_1 \le 2997.08$

13.9 **a.** H_0: $\mu_2 - \mu_1 = 0$, H_1: $\mu_2 - \mu_1 \neq 0$. If $|t^*| \leq 1.684$, conclude H_0; if $|t^*| > 1.684$, conclude H_1. $t^* = 0.989$. Conclude H_0.
b. $0.20 < P$-value < 0.50 (exact P-value $= 0.328$)

13.12 **a.** H_0: $\mu_2 - \mu_1 = 0$, H_1: $\mu_2 - \mu_1 \neq 0$. If $|z^*| \leq 1.645$, conclude H_0; if $|z^*| > 1.645$, conclude H_1. $z^* = 3.118$. Conclude H_1.
b. 0.0018

13.15 **a.** $1.821 \leq \mu_D \leq 2.119$

13.18 **a.** $-0.265 \leq \mu_D \leq -0.095$

13.21 **a.** H_0: $\mu_D \leq 0$, H_1: $\mu_D > 0$. If $t^* \leq 1.895$, conclude H_0; if $t^* > 1.895$, conclude H_1. $t^* = 25.08$. Conclude H_1.
b. P-value < 0.0005 (exact P-value $= 0.0000$ to four decimal places)

13.24 **a.** H_0: $\mu_D \geq 0$, H_1: $\mu_D < 0$. If $z^* \geq -2.326$, conclude H_0; if $z^* < -2.326$, conclude H_1. $z^* = -4.950$. Conclude H_1.
b. 0.0000 to four decimal places

13.28 **a.** $\bar{p}_2 - \bar{p}_1 = 0.011333$, $s\{\bar{p}_2 - \bar{p}_1\} = 0.021273$ **b.** $-0.0237 \leq p_2 - p_1 \leq 0.0463$

13.32 **a.** 0.44318
b. H_0: $p_2 - p_1 = 0$, H_1: $p_2 - p_1 \neq 0$. If $|z^*| \leq 1.645$, conclude H_0; if $|z^*| > 1.645$, conclude H_1. $z^* = 0.533$. Conclude H_0.
c. 0.5962

13.36 **a.** $0.801 \leq \sigma^2 \leq 1.945$ **b.** $0.895 \leq \sigma \leq 1.395$

13.40 If $X^2 \leq 59.34$, conclude H_0; if $X^2 > 59.34$, conclude H_1. $X^2 = 43.204$. Conclude H_0.

13.44 **a.** $0.560 \leq \sigma_2^2/\sigma_1^2 \leq 2.834$ **b.** $0.748 \leq \sigma_2/\sigma_1 \leq 1.684$

13.47 H_0: $\sigma_2^2 \leq \sigma_1^2$, H_1: $\sigma_2^2 > \sigma_1^2$. If $F^* \leq 2.20$, conclude H_0; if $F^* > 2.20$, conclude H_1. $F^* = 1.232$. Conclude H_0.

CHAPTER 14

14.1 **a.** Reject shipment. Accept shipment.
b. If $\bar{p} \leq 0.05$, conclude H_0 (accept shipment); if $\bar{p} > 0.05$, conclude H_1 (reject shipment).

14.4 **a.**

p:	0.02	0.10	0.20
$P(H_0;p)$:	0.9401	0.3918	0.0691

c. 0.0599, supplier's risk

14.7 **a.** (1) accept lot (2) reject lot (3) select second sample **b.** reject lot

14.10 **a.** $\mu_0 = 335$, LCL $= 314$, UCL $= 356$

14.15 **a.** $p_0 = 0.04$, LCL $= 0.00606$, UCL $= 0.07394$

b.

Month:	1	2	3	4	5	6
$\bar{p}$:	0.0333	0.0500	0.0333	0.0433	0.0433	0.0333
Month:	7	8	9	10	11	12
$\bar{p}$:	0.0367	0.0433	0.0467	0.0267	0.0300	0.0100

14.18 a. $\lambda_0 = 5.5$, UCL $= 12.54$ **c.** (1) 0.0044 (2) 0.7324

14.21 a.

k:	1	2	3	4	5	6	7	8	9	10
C_k:	5	−2	5	5	21	26	14	18	19	1
k:	11	12	13	14	15	16	17	18	19	20
C_k:	3	18	22	42	41	16	6	9	0	−7
k:	21	22	23	24	25	26	27	28		
C_k:	−2	8	3	0	16	25	18	16		

14.23 1.0153

14.25 $196.11 \leq \mu \leq 203.89$

14.28 a. 611   **b.** 1245

14.31 $0.455 \leq p \leq 0.645$

14.34 546

14.39 $452.05 \leq \mu \leq 486.52$

14.43 a. 300, 200, 100   **b.** 389, 162, 49

CHAPTER 15

15.1 a. 3   **b.** 0.998

15.3 a. 2.08   2.31   2.47   2.54   2.83   2.84   2.96   3.00   3.04, Md = 2.83
b. $2.31 \leq \eta \leq 3.00$

15.6 12

15.8 a. $1192 \leq \eta \leq 2040$

15.11 a. H_0: $\eta = 62.50$, H_1: $\eta \neq 62.50$; H_0: $p = 0.5$, H_1: $p \neq 0.5$ where $p = P(X_i > 62.50)$.
b. If $|z^*| \leq 1.645$, conclude H_0; if $|z^*| > 1.645$, conclude H_1. $z^* = 1.965$. Conclude H_1.
c. 0.0500

15.14 a. H_0: $\eta = 14$, H_1: $\eta \neq 14$; H_0: $p = 0.5$, H_1: $p \neq 0.5$ where $p = P(X_i > 14)$. If $2 \leq B \leq 8$, conclude H_0; if $B < 2$ or $B > 8$, conclude H_1. $B = 6$. Conclude H_0.
b. 0.0216

15.17 a. 78, 258, 249
b. H_0: $\delta \leq 0$, H_1: $\delta > 0$. If $z^* \leq 1.645$, conclude H_0; if $z^* > 1.645$, conclude H_1. $z^* = 3.952$. Conclude H_1.
c. 0.0000 to four decimal places

15.21 $0.4 \leq \eta_D \leq 1.7$

15.24 a. H_0: $\eta_D = 0$, H_1: $\eta_D \neq 0$. If $5 \leq B \leq 10$, conclude H_0; if $B < 5$ or $B > 10$, conclude H_1. $B = 12$. Conclude H_1.
b. 0.1186

15.27 a. −120, 120, 93
b. H_0: $\eta_D = 0$, H_1: $\eta_D \neq 0$. If $|z^*| \leq 1.645$, conclude H_0; if $|z^*| > 1.645$, conclude H_1. $z^* = 2.641$. Conclude H_1.

c. 0.0082

d.

Difference	Frequency
−1.0–under 0.0	3
0.0–under 1.0	5
1.0–under 2.0	4
2.0–under 3.0	3
Total	15

15.31 **a.** 18, 18.333

b. H_0: Sequence generated by a random process, H_1: Sequence generated by a process containing either persistence or frequent changes in direction. If $|z*| \leq 1.960$, conclude H_0; if $|z*| > 1.960$, conclude H_1. $z* = -0.154$. Conclude H_0.

15.34 **b.** $0.428 \leq \sigma \leq 5.378$

CHAPTER 16

16.1 **a.** (1) H_1: The distribution is not Poisson. (2) 7 (3) 18.48

b. (1) The distribution is not uniform with $a = 2$ and $b = 10$. (2) 5 (3) 9.24

c. (1) H_1: The distribution is not normal. (2) 9 (3) 16.92

16.4 H_0: The number of calls is Poisson distributed with $\lambda = 1.5$, H_1: The number of calls is not Poisson distributed with $\lambda = 1.5$. If $X^2 \leq 7.81$, conclude H_0; if $X^2 > 7.81$, conclude H_1. $X^2 = 23.92$. Conclude H_1.

16.10 H_0: The generated numbers are distributed as $N(0, 1)$, H_1: The generated numbers are not distributed as $N(0, 1)$. If $X^2 \leq 21.67$, conclude H_0; if $X^2 > 21.67$, conclude H_1. $X^2 = 8.00$. Conclude H_0.

16.16 **a., b.** $D(0.90; 10) = 0.37$. Selected values for the cumulative sample function and the confidence band are:

x:	−1	3	6	9	12	15	18	21	25	35
$S(x)$:	0.10	0.20	0.30	0.40	0.50	0.60	0.70	0.80	0.90	1.00
$L(x)$:	0	0	0	0.03	0.13	0.23	0.33	0.43	0.53	0.63
$U(x)$:	0.47	0.57	0.67	0.77	0.87	0.97	1.00	1.00	1.00	1.00

16.19 **a.** $D(0.90; 10) = 0.37$. Selected values for the confidence band are:

x:	0.4	2.6	4.4	4.9	10.6	11.3	11.8	12.6	23.0	40.8
$L(x)$:	0	0	0	0.03	0.13	0.23	0.33	0.43	0.53	0.63
$U(x)$:	0.47	0.57	0.67	0.77	0.87	0.97	1.00	1.00	1.00	1.00

b. H_0: $F(x) = 1 - \exp(-0.1x)$ for all x, H_1: $F(x) \neq 1 - \exp(-0.1x)$ for some x. Conclude H_0.

16.22 **a.** $F_0(9.3) = 0.548$

b. H_0: X is normally distributed with $\mu = 8.0$ and $\sigma = 10.5$, H_1: X is not normally distributed with $\mu = 8.0$ and $\sigma = 10.5$. If $D* \leq 0.23$, conclude H_0; if $D* > 0.23$, conclude H_1. $D* = 0.269$. Conclude H_1.

CHAPTER 17

17.1 **d.** 0.1372   **e.** 0.3087

17.4 **a.** $0.322 \leq p_1 \leq 0.512$
b. $H_0: p_2 = 0.20$, $H_1: p_2 \neq 0.20$. If $|z^*| \leq 2.576$, conclude H_0; if $|z^*| > 2.576$, conclude H_1. $z^* = 3.354$. Conclude H_1.

17.8 **a.** $H_0: p_1 = 0.35$, $p_2 = 0.45$, $p_3 = 0.20$; H_1: The p_i values are not those stated in H_0.
b. If $X^2 \leq 9.21$, conclude H_0; if $X^2 > 9.21$, conclude H_1. $X^2 = 17.54$. Conclude H_1.

c.

i:	1	2	3
$f_i - F_i$:	12.0	−27.0	15.0

17.12 **a.** H_0: Price movements in the two weeks are statistically independent, H_1: Price movements in the two weeks are not statistically independent. If $X^2 \leq 7.78$, conclude H_0; if $X^2 > 7.78$, conclude H_1. $X^2 = 77.35$. Conclude H_1.
b. Residuals $f_i - F_i$:

Movement in First Week	Movement in Second Week		
	Increase	No Change	Decrease
Increase	15.40	−9.40	−6.00
No change	−9.12	13.52	−4.40
Decrease	−6.28	−4.12	10.40

c.

Movement in Second Week	Conditional upon Movement in First Week		
	Increase	No Change	Decrease
Increase	0.80	0.14	0.09
No change	0.17	0.76	0.26
Decrease	0.03	0.10	0.65
Total	1.00	1.00	1.00

17.16 **a.** H_0: Probability distributions of grades in the two programs are identical, H_1: Probability distributions of grades in the two programs are not identical. If $X^2 \leq 9.49$, conclude H_0; if $X^2 > 9.49$, conclude H_1. $X^2 = 3.58$. Conclude H_0.

17.21 $0.979 \leq p_1/p_2 \leq 1.971$

CHAPTER 18

18.8 **a.** 0.7888   **b.** 17.5   **c.** (1) 1.8 (2) 0.5793

18.10 **a.** 1250   **b.** $\hat{Y} = 55.74 + 0.04540X$, 47.00

18.12 **b.** $\hat{Y} = 72.776 - 1.7726X$

18.16 **b.** 39.10

18.20 **a.**

i:	1	2	3	4	5	6	7	8
e_i:	−0.55	−3.23	−1.01	2.45	2.22	−0.87	−0.32	1.31

18.24 a.

Source	SS	df	MS
Regression	134.717	1	134.717
Error	25.283	6	4.214
Total	160.000	7	

b. (1) σ^2 (2) σ

18.28 b. 0.8420, −0.918

CHAPTER 19

19.1 $30.03 \leq E\{Y_h\} \leq 33.98$

19.8 a. $26.61 \leq Y_{h(\text{new})} \leq 37.40$

19.12 $-2.540 \leq \beta_1 \leq -1.005$

19.16 a. $H_0: \beta_1 = 0$, $H_1: \beta_1 \neq 0$. If $|t^*| \leq 2.447$, conclude H_0; if $|t^*| > 2.447$, conclude H_1. $t^* = -5.654$. Conclude H_1.
b. 0.0013

19.20 $H_0: \beta_0 = 0$, $H_1: \beta_0 \neq 0$. If $|t^*| \leq 3.707$, conclude H_0; if $|t^*| > 3.707$, conclude H_1. $t^* = 10.553$. Conclude H_1.

19.24 −2.150

19.29 a. $H_0: \beta_1 = 0$, $H_1: \beta_1 \neq 0$. If $F^* \leq 5.99$, conclude H_0; if $F^* > 5.99$, conclude H_1. $F^* = 31.97$. Conclude H_1.

19.33 a.

i:	1	2	3	4	5	6	7	8
e_i:	−0.55	−3.23	−1.01	2.45	2.22	−0.87	−0.32	1.31
$\hat{Y}_i$:	35.55	30.23	32.01	35.55	33.78	40.87	37.32	26.69

c.

i:	1	2	3	4	5	6	7	8
e_i':	−0.27	−1.58	−0.49	1.19	1.08	−0.42	−0.16	0.64

e_i'		f	F
$-\infty$ –under −0.718		1	2.0
−0.718–under 0		4	2.0
0 –under 0.718		1	2.0
0.718–under ∞		2	2.0
	Total	8	8.0

CHAPTER 20

20.1 a. $E\{Y\} = -5 + 0.15X_1 - 0.08X_2$ **c.** (1) 6.0 (2) −2.0 (3) 0.0668

20.4 a. X_1 **b.** (1) −0.5874 (2) 0.4977 (3) 0.7945

20.7 a. $\hat{Y} = 162.876 - 1.2103X_1 - 0.66591X_2 - 8.6130X_3$ **b.** 61.31

20.10 a.

Source	SS	df	MS
Regression	4133.633	3	1377.878
Error	2011.584	19	105.873
Total	6145.217	22	

b. 0.6727, 0.820   **c.** 0.6210, 0.788

20.13 b. 0.02539, −0.159

20.16 b.

i:	1	2	3	4	5	6	7	8
e_i':	−0.06	−1.16	0.23	0.16	0.41	−0.64	−1.34	−0.17

i:	9	10	11	12	13	14	15	16
e_i':	−0.75	0.06	1.35	1.18	−1.39	−1.65	1.28	−0.33

i:	17	18	19	20	21	22	23
e_i':	1.45	0.70	−1.19	0.71	0.36	0.58	0.22

e_i'	f	F
$-\infty$ –under −1.328	3	2.3
−1.328–under 0	7	9.2
0 –under 1.328	11	9.2
1.328–under ∞	2	2.3
Total	23	23.0

20.19 a. $H_0: \beta_1 = \beta_2 = \beta_3 = 0$, H_1: Not all $\beta_k = 0$, $k = 1, 2, 3$. If $F^* \le 5.01$, conclude H_0; if $F^* > 5.01$, conclude H_1. $F^* = 13.014$. Conclude H_1.

20.22 a. $-1.841 \le \beta_1 \le -0.579$
   b. $H_0: \beta_3 = 0$, $H_1: \beta_3 \ne 0$. If $|t^*| \le 2.093$, conclude H_0; if $|t^*| > 2.093$, conclude H_1. $t^* = -0.704$. Conclude H_0.

20.25 a. $50.99 \le E\{Y_h\} \le 71.63$ **b.** $37.43 \le Y_{h(\text{new})} \le 85.19$

20.29 b. $0.244 \le \beta_2 \le 0.432$

20.32 a.

i:	1	2	3	4	5	6	7	8	9	10	11	12
e_i:	−0.8	−0.7	−0.7	0.3	0.8	−0.3	1.7	0.2	0.3	−0.8	−1.0	1.0

b. $\hat{Y} = 4.084 + 5.8478X - 0.10736X^2$

i	X_i	X_i^2	Y_i
1	5	25	30
2	10	100	51
3	4	16	26

c. $H_0: \beta_2 = 0$, $H_1: \beta_2 \ne 0$. If $|t^*| \le 2.262$, conclude H_0; if $|t^*| > 2.262$, conclude H_1. $t^* = -2.803$. Conclude H_1.

20.35 b. $H_0: \beta_3 = 0$, $H_1: \beta_3 \ne 0$. If P-value ≥ 0.05, conclude H_0; if P-value < 0.05, conclude H_1. Conclude H_1.

20.42 b. Models containing variables (1) X_1, X_2, (2) X_1, X_3, (3) X_1, X_2, X_3

20.44 a. H_0: $\beta_2 = \beta_3 = 0$, H_1: Not both β_2 and β_3 equal zero. Full model: $E\{Y\} = \beta_0 + \beta_1 X_1 + \beta_2 X_2 + \beta_3 X_3$. Reduced model: $E\{Y\} = \beta_0 + \beta_1 X_1$. If $F^* \leq 3.52$, conclude H_0; if $F^* > 3.52$, conclude H_1. $F^* = 2.150$. Conclude H_0.

CHAPTER 21

21.5 a. (1) 3 (2) 24 (3) 27   **b.** (1) 3 (2) 76 (3) 79   **c.** (1) 4 (2) 11 (3) 15

21.7 a. (1) 15 (2) 5 (3) 88 (4) 88.0 (5) 81.0

b. videotapes

c.

Source	SS	df	MS
Treatments	930.0	2	465.0
Error	460.0	12	38.33
Total	1390.0	14	

21.11 a. H_0: $\mu_1 = \mu_2 = \mu_3$, H_1: Not all μ_j are equal. $F(2, 12)$.

b. If $F^* \leq 3.89$, conclude H_0; if $F^* > 3.89$, conclude H_1. $F^* = 12.13$. Conclude H_1.

21.15 a. $78.97 \leq \mu_1 \leq 91.03$ **b.** $-5.53 \leq \mu_2 - \mu_1 \leq 11.53$

21.19 $-7.50 \leq \mu_2 - \mu_1 \leq 13.50$, $7.50 \leq \mu_2 - \mu_3 \leq 28.50$, $4.50 \leq \mu_1 - \mu_3 \leq 25.50$

21.23 a. Residuals e_{ij}:

		j	
i	1	2	3
1	1	2	8
2	−3	−9	0
3	9	0	−5
4	−8	−1	4
5	1	8	−7

21.27 b.

i	X_{i1}	X_{i2}	Y_i
1	0	0	86
2	0	0	82
3	0	0	94
4	0	0	77
5	0	0	86
6	1	0	90
7	1	0	79
8	1	0	88
9	1	0	87
10	1	0	96
11	0	1	78
12	0	1	70
13	0	1	65
14	0	1	74
15	0	1	63

c. (1) $78.97 \leq \beta_0 \leq 91.03$ (2) $-5.53 \leq \beta_1 \leq 11.53$

CHAPTER 22

22.3

S_i	$P(S_i)$	A_j 0	1	2
0	0.4	0	-30	-60
1	0.5	-220	-180	-210
2	0.1	-440	-400	-360

22.9 b., c.

Act	T_1	T_2	T_3	Total Time	Maximum Time
1	W_1	W_2	W_3	108	43
2	W_1	W_3	W_2	109	43
3	W_2	W_1	W_3	96	35
4	W_2	W_3	W_1	95	37
5	W_3	W_1	W_2	100	36
6	W_3	W_2	W_1	98	36

Act 4 yields the smallest total. Act 3 yields the smallest maximum.

d. $10! = 3,628,800$

22.11 a. A_1, A_2, A_3, A_4 **b.** A_1, 200

c.

S_i	A_1	A_2	A_3	A_4
S_1	300	200	100	0
S_2	100	0	50	100
S_3	0	50	100	150

A_3 is the minimax regret act.

22.15 a. A_2, -142 **b.** 120

22.19 -126, 16

22.24 b. Optimal strategy is to undertake research and development and, if this alternative proves unsuccessful, to negotiate a merger arrangement. 80.6.

22.28 a. risk-seeking   **b.** (1) $EU = 0.2222$ (2) $EU = 0.0711$; prefers (1)

22.32 a. A_3: Half in each of V_1 and V_2, $EU(A_3) = 0.6668$

b.

Act:	A_1	A_2	A_3	A_4
Certainty Equivalent:	128.00	75.00	133.39	130.00

CHAPTER 23

23.1 a. A_1, 1300   **b.** 1450, 150

23.5 a.

$(6, K)$:	$(6, 0)$	$(6, 3)$	$(6, 6)$	$(6, -1)$
$EP(6, K)$:	1302	1325	1300	500

b. (6, 1), 1371   **c.** A_1

23.11 15

23.15 a.

n:	0	1	2	3	4
$BEP(n)$:	880.00	938.00	964.80	973.76	976.63

b. (1) $n = 0$. Optimal decision is to choose A_1. 880.00. (2) $n = 2$, (2, 0), 894.80

23.20 a.

S_i:	0	0.1	0.2	0.3	0.4
$P(S_i \mid X_0)$:	0.6261	0.2695	0.0525	0.0361	0.0158

b. A_1, 1872

CHAPTER 24

24.1 a.

Year:	1983	1984	1985
M.A.:	4.27	4.03	3.80

24.7 **a.** $T_t = 3.2900 + 0.051429X_t$, $X_t = 1$ at 1967, X in one-year units
c. 0.051429 or 51,429 metric tons
d. 4.6786, 5.380

24.11 a., b., c.

Year:	1967	1968	1969	1970	1971	1972	1973	1974	1975	1976
$C \cdot I$:	98.8	109.1	113.2	100.1	81.8	86.1	95.9	97.3	85.3	92.0
C:			100.6	98.1	95.4	92.2	89.3	91.3	94.8	99.2
I:			112.6	102.1	85.7	93.4	107.4	106.5	89.9	92.7

Year:	1977	1978	1979	1980	1981	1982	1983	1984	1985	1986
$C \cdot I$:	103.7	117.7	103.6	119.7	115.7	104.6	108.1	94.9	84.4	88.0
C:	100.5	107.4	112.1	112.3	110.3	108.6	101.5	·96.0		
I:	103.3	109.7	92.4	106.6	104.9	96.3	106.4	98.9		

24.16 a. $T'_t = 3.30527 + 0.015699X_t$, $X_t = 1$ at 1972, X in one-year units
c. 3.7 percent
d. 4809, 35,112

24.19

Year:	1972	1973	1974	1975	1976	1977	1978	1979
$C \cdot I$:	99.8	98.5	102.9	98.8	102.2	101.6	97.3	98.3

Year:	1980	1981	1982	1983	1984	1985	1986
$C \cdot I$:	98.3	100.2	99.7	103.8	99.6	100.5	98.9

24.22 a.

X_t:	2	3	9	10
T_t:	97.1	136.2	273.4	281.1

b. 40.3 percent, 2.8 percent
c. 300

24.27 1973, 1980. One cannot be certain if the end years of the cyclical component series (1969 and 1984) are turning points.

24.30 b.

Quarter	1981	1982	1983	1984	1985	1986
1		73.6	79.1	71.6	76.9	73.6
2		95.7	93.8	104.8	101.8	103.3
3	133.4	135.6	134.6	127.4	127.7	
4	96.2	91.1	95.6	94.1	93.3	

c.

Quarter:	1	2	3	4
Index:	73.1	101.1	132.4	93.4

24.39 a.

Quarter:	1	2	3	4
Deseasonalized Sales Revenue:	232.6	238.4	231.9	232.3

b. 929

24.43 a.

Quarter:	1	2	3	4
Forecast:	170	235	308	217

b. 229.8, 919.3

CHAPTER 25

25.1 a. 41.3   **b.** (1) 42.6 (2) 42.6 (3) 42.6

25.9 a. $A_{31} = 4.628$, $B_{31} = -0.0737$

Year:	32	33	34	35
Forecast:	4.55	4.48	4.41	4.33

25.15 a.

Quarter:	3	4
A_t:	235.5	235.3
B_t:	0.327	0.280
C_t:	1.301	0.936

b.

Year:	1987				1988			
Quarter:	1	2	3	4	1	2	3	4
Forecast:	176	236	307	221	177	237	309	222

c. 0.280, 0.922 or 92.2 percent

25.21 a. 305.6, -13.6 **b.** 131.29, 0.211   **c.** 87.44, -0.105 **d.** 232.14   **e.** 0.421

25.22 a. $215.6 \le Y_{h(new)} \le 248.7$

b. H_0: $\beta_4 = 0$, H_1: $\beta_4 \ne 0$. If $|t^*| \le 2.093$, conclude H_0; if $|t^*| > 2.093$, conclude H_1. $t^* = 9.10$. Conclude H_1.

25.29 b. H_0; $\rho \le 0$, H_1: $\rho > 0$. If $d > 1.78$, conclude H_0; if $d < 1.01$, conclude H_1; if $1.01 \le d \le 1.78$, the test is inconclusive. Conclude H_0.

25.33 b. $\hat{Y}'_t = 0.66145 X'_t$ **c.** 188.7

25.34 a. 18.6425, 10   **b.** $0.366 \le \beta_1 \le 0.957$

CHAPTER 26

26.1 a.

Year:	1982	1983	1984	1985	1986
(1) Monarch:	100.0	104.2	116.7	125.0	137.5
Emperor:	100.0	106.3	112.5	125.0	134.4
(2) Monarch:		104.2	112.0	107.1	110.0
Emperor:		106.3	105.9	111.1	107.5

b. 8.3, 7.1   **c.** Monarch, Monarch

26.4 a., b.

Year:	1984	1985	1986
Aggregate Cost:	12,233.80	11,915.50	11,705.00
Index:	100.0	97.4	95.7

26.7 a.

Year:	1984	1985	1986
Index:	100.0	97.4	95.7

26.13

Year:	1981	1982	1983	1984	1985	1986
Index:	92.8	95.6	96.4	97.5	100.0	101.4

26.15

Year:	1981	1982	1983	1984	1985	1986
Index:	90.9	100.9	100.0	109.5	119.0	136.9

26.21 a., b.

Year:	1979	1980	1981	1982	1983
Earnings (in 1979 dollars):	342.99	324.04	319.15	321.61	324.28
(1) Percent Relative:	100.0	94.5	93.0	93.8	94.5
(2) Link Relative:		94.5	98.5	100.8	100.8

c.

	Private Construction Workers	All Private Nonagricultural Workers
(1)	−5.5	−6.7
(2)	0.8	2.0

26.28 a.

Year:	1983	1984	1985	1986
Index:	100.0	93.9	111.0	123.3

b. 23.3

26.30

Year:	1983	1984	1985	1986
Index:	100.0	93.9	111.0	123.3

APPENDIX B

B.1   **a.** 4.61, 0.211   **b.** 2, 2   **c.** no
B.3   **a.** 15.51   **b.** 117.9
B.5   **a.** 2.015, −2.015   **b.** 0, 1.29
B.7   **a.** 1.341, −1.721   **b.** −2.326
B.11 **a.** 13.27, 0.0101   **b.** 1.67, 3.73   **c.** no
B.13 **a.** 3.23   **b.** 0.132

INDEX

TABLE C.2 Percentiles of the chi-square distribution

Entry is $\chi^2(a;\nu)$ where $P[\chi^2(\nu) \leq \chi^2(a;\nu)] = a$.

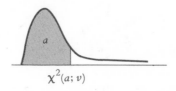

$$\chi^2(a;\nu)$$

df ν	.005	.010	.025	.050	.100	.900	.950	.975	.990	.995
1	0.0^4393	0.0^3157	0.0^3982	0.0^2393	0.0158	2.71	3.84	5.02	6.63	7.88
2	0.0100	0.0201	0.0506	0.103	0.211	4.61	5.99	7.38	9.21	10.60
3	0.072	0.115	0.216	0.352	0.584	6.25	7.81	9.35	11.34	12.84
4	0.207	0.297	0.484	0.711	1.064	7.78	9.49	11.14	13.28	14.86
5	0.412	0.554	0.831	1.145	1.61	9.24	11.07	12.83	15.09	16.75
6	0.676	0.872	1.24	1.64	2.20	10.64	12.59	14.45	16.81	18.55
7	0.989	1.24	1.69	2.17	2.83	12.02	14.07	16.01	18.48	20.28
8	1.34	1.65	2.18	2.73	3.49	13.36	15.51	17.53	20.09	21.96
9	1.73	2.09	2.70	3.33	4.17	14.68	16.92	19.02	21.67	23.59
10	2.16	2.56	3.25	3.94	4.87	15.99	18.31	20.48	23.21	25.19
11	2.60	3.05	3.82	4.57	5.58	17.28	19.68	21.92	24.73	26.76
12	3.07	3.57	4.40	5.23	6.30	18.55	21.03	23.34	26.22	28.30
13	3.57	4.11	5.01	5.89	7.04	19.81	22.36	24.74	27.69	29.82
14	4.07	4.66	5.63	6.57	7.79	21.06	23.68	26.12	29.14	31.32
15	4.60	5.23	6.26	7.26	8.55	22.31	25.00	27.49	30.58	32.80
16	5.14	5.81	6.91	7.96	9.31	23.54	26.30	28.85	32.00	34.27
17	5.70	6.41	7.56	8.67	10.09	24.77	27.59	30.19	33.41	35.72
18	6.26	7.01	8.23	9.39	10.86	25.99	28.87	31.53	34.81	37.16
19	6.84	7.63	8.91	10.12	11.65	27.20	30.14	32.85	36.19	38.58
20	7.43	8.26	9.59	10.85	12.44	28.41	31.41	34.17	37.57	40.00
21	8.03	8.90	10.28	11.59	13.24	29.62	32.67	35.48	38.93	41.40
22	8.64	9.54	10.98	12.34	14.04	30.81	33.92	36.78	40.29	42.80
23	9.26	10.20	11.69	13.09	14.85	32.01	35.17	38.08	41.64	44.18
24	9.89	10.86	12.40	13.85	15.66	33.20	36.42	39.36	42.98	45.56
25	10.52	11.52	13.12	14.61	16.47	34.38	37.65	40.65	44.31	46.93
26	11.16	12.20	13.84	15.38	17.29	35.56	38.89	41.92	45.64	48.29
27	11.81	12.88	14.57	16.15	18.11	36.74	40.11	43.19	46.96	49.64
28	12.46	13.56	15.31	16.93	18.94	37.92	41.34	44.46	48.28	50.99
29	13.12	14.26	16.05	17.71	19.77	39.09	42.56	45.72	49.59	52.34
30	13.79	14.95	16.79	18.49	20.60	40.26	43.77	46.98	50.89	53.67
40	20.71	22.16	24.43	26.51	29.05	51.81	55.76	59.34	63.69	66.77
50	27.99	29.71	32.36	34.76	37.69	63.17	67.50	71.42	76.15	79.49
60	35.53	37.48	40.48	43.19	46.46	74.40	79.08	83.30	88.38	91.95
70	43.28	45.44	48.76	51.74	55.33	85.53	90.53	95.02	100.4	104.2
80	51.17	53.54	57.15	60.39	64.28	96.58	101.9	106.6	112.3	116.3
90	59.20	61.75	65.65	69.13	73.29	107.6	113.1	118.1	124.1	128.3
100	67.33	70.06	74.22	77.93	82.36	118.5	124.3	129.6	135.8	140.2

SOURCE: Tabulated values adapted by permission from C. M. Thompson, "Table of Percentage Points of the Chi-Square Distribution," *Biometrika*, Vol. 32 (1941), pp. 188–189.

EXAMPLE: $\chi^2(0.900;4) = 7.78$ so $P[\chi^2(4) \leq 7.78] = 0.900$.

TEXT REFERENCE: Use of this table is discussed on p. 937.